CELLULAR AND MOLECULAR BIOLOGY OF HORMONE- AND NEUROTRANSMITTER-CONTAINING SECRETORY VESICLES

ANNALS OF THE NEW YORK ACADEMY OF SCIENCES
Volume 493

CELLULAR AND MOLECULAR BIOLOGY OF HORMONE- AND NEUROTRANSMITTER-CONTAINING SECRETORY VESICLES

Edited by Robert G. Johnson, Jr.

The New York Academy of Sciences
New York, New York
1987

Cover: The cover of the paperbound edition is a schematic drawing of an endocrine cell (left) and a neuron (right). Secretory vesicles are depicted in *red*. (From De Camilli and Navone, pp. 461-479.)

Library of Congress Cataloging-in-Publication Data

Cellular and molecular biology of hormone- and
neurotransmitter-containing secretory vesicles.

(Annals of the New York Academy of Sciences, ISSN 0077-8923 ; v. 493)
Proceedings of the Conference on the Cellular and Molecular Biology of Hormone and Neurotransmitter Containing Secretory Vesicles, held June 16-18, 1986, New York, N.Y.
Includes bibliographies and index.
1. Neuroendocrinology—Congresses. 2. Neurotransmitters—Congresses. I. Johnson, Robert G. (Robert Gahagen), 1952- . II. Conference on the Cellular and Molecular Biology of Hormone and Neurotransmitter Containing Secretory Vesicles (1986 : New York, N.Y.) III. Series.
Q11.N5 vol. 493 500 s 87-7706
[QP356.4] [599′.0188]
ISBN 0-89766-382-9
ISBN 0-89766-381-0 (pbk.)

PCP
Printed in the United States of America
ISBN 0-89766-381-0 (cloth)
ISBN 0-89766-382-9 (paper)
ISSN 0077-8923

ANNALS OF THE NEW YORK ACADEMY OF SCIENCES

Volume 493
April 24, 1987

CELLULAR AND MOLECULAR BIOLOGY OF HORMONE- AND NEUROTRANSMITTER-CONTAINING SECRETORY VESICLES[a]

Editor and Conference Chair
ROBERT G. JOHNSON, JR.

CONTENTS

[a] This volume is the result of the Conference on the Cellular and Molecular Biology of Hormone and Neurotransmitter Containing Secretory Vesicles, which was held by the New York Academy of Sciences on June 16-18, 1986, in New York, N. Y.

Part II. Structure and Composition of Secretory Vesicles

Poster Papers

Part III. Bioenergetics and Pharmacology of Neurotransmitter and Hormone Transport

Poster Papers

Part V. Vesicle Movement and Secretion

Poster Papers

Part VI. Exocytosis from the Perspective of the Secretory Vesicle

Poster Papers

Financial assistance was received from:

- BECKMAN INSTRUMENTS
- E. I. DU PONT DE NEMOURS & COMPANY, BIOMEDICAL PRODUCTS DEPARTMENT
- HOECHST-ROUSSEL PHARMACEUTICALS, INC.
- MERRELL DOW CORPORATION
- NATIONAL INSTITUTE OF MENTAL HEALTH, NATIONAL INSTITUTES OF HEALTH
- NATIONAL INSTITUTE OF NEUROLOGICAL AND COMMUNICATIVE DISORDERS AND STROKE, NATIONAL INSTITUTES OF HEALTH
- NATIONAL SCIENCE FOUNDATION
- SANDOZ, INC.
- SCHERING-PLOUGH
- STUART PHARMACEUTICALS/DIVISION OF ICI AMERICAS, INC.
- U. S. OFFICE OF NAVAL RESEARCH (GRANT NO. N00014–85–G–0077)

Preface

ROBERT G. JOHNSON, JR.

Howard Hughes Medical Institute
Departments of Medicine and Neurology
Massachusetts General Hospital
Harvard Medical School
Boston, Massachusetts 02114

Neuronal and endocrine tissues employ chemical messengers to communicate with other neurons or with target tissues. This chemical basis of the transmission of information between cells constitutes one of the fundamental governing principles in all of the biologically related disciplines, particularly neuroscience and endocrinology. A comprehensive understanding of the mechanisms involved in the regulation of synthesis, processing, packaging, release, and the effect of various intracellular and extracellular factors upon long- and short-term regulation of the chemical messengers is difficult to achieve because of the complexity and diversity of the neuroendocrine system. The difficulty in neuroendocrine research is thus to identify a solvable, important problem with possible universal applicability, to choose the appropriate technique, and to find the optimal model.

It has been known for several decades that the neurotransmitters and hormones which serve as the chemical messengers in the neuroendocrine system are not primarily identified as soluble components within the cytosol of neuroendocrine cells; rather they are localized to highly specialized secretory vesicles.

Neuroendocrine secretory vesicles, an essential component of neuroendocrine homeostasis, can therefore be studied in an isolated state, and findings can be readily extrapolated to the more complicated *in vivo* system.

This brings us to a question: Why hold a conference on neuroendocrine secretory vesicles? I believe the rationale for this conference is threefold. First, neuroendocrine secretory vesicles play an exceedingly important function in neuroendocrine homeostasis. Peptides that often function as hormones or neurotransmitters are packaged along with processing enzymes as the vesicles are formed. Once the proper gradients are generated, neurotransmitters and hormones accumulate into the vesicles (against apparently enormous concentration gradients), and are often metabolized to biologically active compounds. Within the granule, hormones and neurotransmitters are protected from degradative enzymes, stably stored, often undergo further posttranslational modification, and are quantally released via an exocytotic mechanism that is inexplicably linked to the movement of the vesicle and participation of the secretory vesicle membrane. In fact, most of the lifetime within the cell of the important neurotransmitters and hormones is localized within the secretory vesicle.

Second, subcellular organelles that contain a hormone or neurotransmitter have been treated for many years essentially as a "black box," a passive storage organelle where the hormone or neurotransmitter is packaged and stored prior to release. Indeed, many important neuroendocrine vesicles have not yet been isolated. In the past, interest has been concentrated on the synthesis of precursor molecules, effects of pharmacologic agents, and physiologic actions of the molecules after secretion. Countless meetings and symposia have dealt with secretion. The importance of vesicle movement and the role of the vesicle in the molecular response to the secretory stimulus, however, have for many years been unexplored. But neuroendocrine secretory vesicles are not passive storage organelles; they display dynamic functions. They originate in cells capable of plasticity, transport solutes against some of the largest gradients ever measured, actively process peptides and metabolize neurotransmitters or hormones within the intragranular space, move about the cell, and fuse with the plasma membrane.

Third, despite differing embryologic origins, the structure, composition, and function of neuroendocrine vesicles are remarkably similar and universal properties of packaging, metabolism, uptake, movement, and secretion are becoming discerned. The implication of this is that use of preparations of vesicles isolated from tissues rich in abundance can tell us something about less readily isolatable neurotransmitter- or hormone-containing vesicles.

The time has come to review neuroendocrine homeostasis from the viewpoint of the secretory vesicle. This volume, the proceedings of the first international meeting devoted solely to the secretory vesicle, outlines sequentially the life cycle of the secretory vesicle. In part I we review the biogenesis, plasticity, and cell biology of the secretory vesicles, in part II, their structure and composition. Part III deals with the transport function of the vesicles, part IV with the interesting array of peptides and processing enzymes. Part V takes up the question of how vesicles move within the cell. Finally, part VI examines exocytosis from the perspective of the vesicle.

A successful meeting of this type requires the talents of a number of individuals. Briefly, I would like to thank my advisory board for their help and suggestions, the scientific advisory committee of the New York Academy of Sciences, the conference department of the Academy, especially Ms. Ellen Marks and Ms. Renée Wilkerson, Mr. Bill Boland and his editorial staff at the Academy, and finally the governmental and private agencies who have generously contributed their support.

My most sincere thanks go to the speakers, discussants, and members of the audience. I believe this was a marvelous opportunity for investigators from a variety of disciplines to review and analyze the data presented, to debate their importance and physiologic implications, to come to agreement on universal properties of vesicles, and most importantly, to plan the next series of experiments.

Opening Remarks

BERTA SCHARRER

Department of Anatomy
Albert Einstein College of Medicine
Bronx, New York 10461

These opening remarks are intended to give some historical perspective to the topics dealt with at this conference. The justification for my having been selected for this retrospective task seems to be my long personal acquaintance with at least one class of secretory vesicles, those containing the active principles of peptidergic neurosecretory neurons. This is, however, not the only reason why I welcome this opportunity to speak to you. What I should like to underscore is the importance of structural considerations in our efforts to analyze the physiology and biochemistry of secretory systems, and I hope to exemplify this concept by highlighting some of the steps that, in the span of the past 50 years, have led to our current understanding of peptide-producing neurons.

It was a cytological discovery, the demonstration of "neurosecretory cells" in the hypothalamus by Ernst Scharrer in 1928, that set the stage for subsequent studies carried out with increasingly sophisticated methods. In these concentrated efforts focused on the analysis of the neuron's capacity for secretory activity, the search for structural correlates of functional states continues to bear fruit to this day. In fact, it is difficult to judge where our knowledge of neuropeptide function, and of secretory processes in general, would stand today without the contributions made by structural biology.

At the light microscopic level, neurosecretory peptides were first recognized in the form of cytoplasmic granules or droplets by their selective stainability, for example, with Gomori's aldehyde fuchsin technique. Subsequently, they could be identified with greater precision by means of immunocytochemistry, an approach that yielded significant insights into the evolutionary history and wide distribution of these neuroregulators.

A great step forward became possible after the advent of electron microscopy. Ultrastructurally, the stainable material turned out to consist of accumulations of cytoplasmic inclusions of smaller diameter which are referred to either as secretory granules, bounded by a membrane, or as secretory vesicles, with a specific biologically active content. In other words, we are dealing with two components, the membranous container and the material it encloses. As the topics to be covered during this conference demonstrate, the molecular configuration and functional significance of both entities are receiving equal attention.

Because of their electron density, as well as their "stainability," proteinaceous products represent convenient markers at both levels of magnification. The special structural configuration that distinguishes neurons from other cell types enables us to trace subsequent steps in the production of this material in different domains of the cell. The subcellular machinery required for the biosynthesis of the large precursor molecules of active neuropeptides has been localized in the perikaryon. It is here that the material arises at the level of the ribosome-studded endoplasmic reticulum, enters the intracisternal space, and is conveyed to the Golgi apparatus where it is packaged in the form of membrane-enclosed granules. Posttranslational processing, giving rise to active neuropeptides with the aid of specific enzymes, is considered to occur in the axon during the proximo-distal transport of the material to its site of release at the neuron terminal. What is of significance in the present context is that all of these steps leading to the exteriorization of the secretory product occur within membrane-enclosed intracytoplasmic compartments. The secretory vesicles, pinched off from Golgi cisternae, are part of this segregated cellular territory.

The release of neuropeptides by the process of exocytosis involving the opening and fusion of two membranes, the vesicle membrane and the axolemma, has been ascertained by the study of transmission electron micrographs in sequential order, and by freeze-fracture microscopy. The same approach has provided us with much valuable information on the dynamics of the reuptake of redundant membrane material following exocytosis, and on a related process, the interiorization of plasma membrane effecting the endocytotic uptake of neuroregulators such as enkephalins or catecholamines.

The use of electron microscopy also permits us to draw certain functional conclusions from the spatial relationships between sites of release and presumed sites of action of neuropeptides. Evidently, their activities are not restricted to neurohormonal interventions, as had been previously proposed. The discovery of synaptic (or synaptoid) peptidergic junctions with various types of effector cells, including neurons, calls attention to the specific roles neuropeptides may play as neurotransmitters, or "cotransmitters." The latter role has been suggested by the immunocytochemical demonstration, at the ultrastructural level, of the coexistence of neuropeptides with conventional nonpeptidergic neurotransmitters in one and the same axon terminal. In sum, a new chapter has been added to neurobiological research, in which neuropeptides now hold a central position.

By the same token, we have come to appreciate the importance of the intracellular membrane system in the production, packaging, transport, storage, and release of neurosecretory peptides, a system that is by no means static. Secretory vesicles can be considered to have a life cycle. They are fashioned from Golgi cisternae, are moved along axons, lose their structural identity during exocytosis, and presumably regain it by means of membrane recycling. The time has come to explore in molecular terms the specific structure of vesicular and other cytoplasmic membranes, and the mechanisms involved in their operation, including those based on proton pumps, protein phosphorylation, calcium channels, and calcium-binding proteins. As will become apparent, some of the insights gained from the analysis of peptidergic neurons also apply to nonproteinaceous secretory systems.

The Life Cycle of Catecholamine-storing Vesicles[a]

HANS WINKLER, MARTINA SIETZEN, AND MARIA SCHOBER

Department of Pharmacology
University of Innsbruck
A-6020 Innsbruck, Austria

INTRODUCTION

The catecholamine-storing vesicles can be classified into two types. One is represented by the chromaffin granules of the adrenal medulla and by the large dense-cored vesicles of sympathetic nerve. The small dense-cored vesicles of these nerves constitute a second type.

We will first discuss the adrenal chromaffin granules. These vesicles consist of a membrane and a soluble content destined for secretion (see refs. 1-3 for reviews). Two important functions of these membranes are (1) transport of the secretory products from the Golgi region to the plasma membrane where secretion by exocytosis takes place, and (2) accumulation of small molecules and involvement in the synthesis and modification of secretory products.

Before we discuss the biogenesis and the life cycle of chromaffin granules, we will describe the molecular properties of their secretory products.

SECRETORY CONSTITUENTS OF CHROMAFFIN GRANULES

The soluble content of these organelles, secreted by exocytosis, is made up of two groups of constituents[2,4]: (1) small molecules (catecholamines, nucleotides, ascorbic acid, and calcium), and (2) proteins and peptides. TABLE 1 gives the number of these molecules found in a single chromaffin granule. As we will point out below, the peptides are proteolytically processed within these organelles. It is therefore rather difficult to give the total number of molecules within a single granule. For the chromogranins only the number of the largest component still present in mature granules is given. As shown in FIGURE 1 this proprotein represents only part of the molecules originally present. For the enkephalins the total number of equivalents is given which are partly present within the granules as free molecules but to a greater extent[5] in

[a]This work was supported by the Dr. Legerlotz-Stiftung and by the Fonds zur Förderung der wissenschaftlichen Forschung (Austria).

precursors of varying sizes (see below). For neuropeptide Y no evidence for the presence of a precursor within mature granule has been obtained;[6] the figure therefore refers to the free peptide.

BIOSYNTHESIS OF PEPTIDES

As was just pointed out, proteins and peptides are major constituents of the soluble content of chromaffin granules. It is therefore not surprising that one major feature of the biogenesis of chromaffin granules is the biosynthesis and posttranslational processing of peptides, similar to that which has been established for several protein-secreting organs.[7] Thus these peptides are synthesized in the rough endoplasmic reticulum followed by transport to the Golgi region where they are packaged into newly formed granules.

BIOSYNTHESIS OF CHROMOGRANINS

The first results on the biosynthesis of chromogranins were obtained in 1971 in a pulse-labeling study[8,9] with perfused bovine adrenals (for an ultrastructural radioautography study, see ref. 10). At 1.5 min after ^{3}H-leucine labeling the newly synthesized chromogranins were present in the microsomal fractions, whereas at 45 min, they were found in a particle which after density gradient centrifugation equilibrated slightly above chromaffin granules. It was concluded that these lighter particles represented newly formed chromaffin granules which appeared less dense than the mature granules since they had not yet been filled with ATP and catecholamines (compare refs. 11 and 12).

More recently the biosynthesis of the chromogranins was studied in cell-free systems and with isolated chromaffin cells. Translation of bovine adrenal mRNA *in vitro* yielded two preproteins[13,14] for chromogranin A (but see ref. 15) that in the

TABLE 1. Secretory Constituents of a Single Chromaffin Granule (Bovine)

Constituent	Number of Molecules
catecholamines	3×10^6
nucleotides	930,000
ascorbic acid	120,000
calcium	90,000
chromogranin A	5,000
chromogranin B	80
chromogranin C	?
dopamine β-hydroxylase	140
total enkephalin equivalents	4,000
neuropeptide Y	428

The calculations on which these figures are based have been described previously.[4] Further minor constituents can be found in ref. 4.

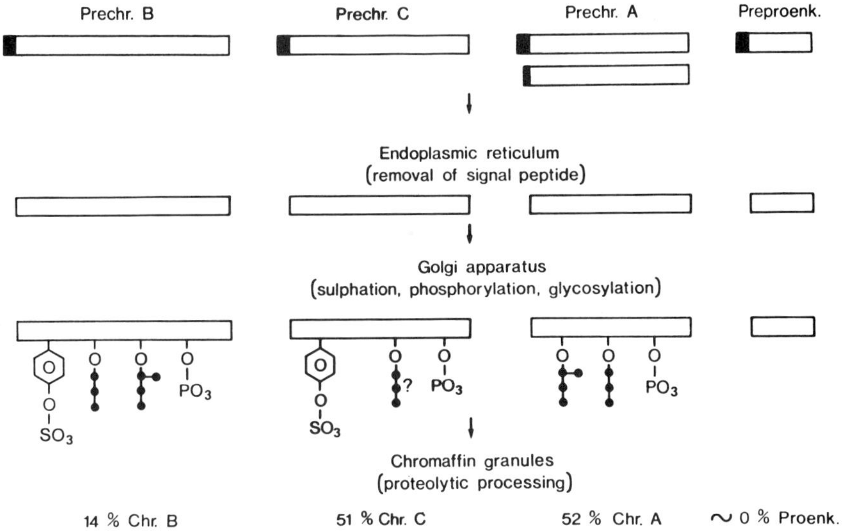

FIGURE 1. Biosynthesis of peptides in adrenal medulla. This figure compiles what is known about the biosynthesis of the chromogranins and enkephalins. It has not yet been established whether chromogranin C contains carbohydrate chains. The peptides are proteolytically processed within the granules. The values at the bottom of the figure indicate how much of the propeptide is still found in mature granules (for more information on chromogranins see ref. 29; for enkephalins see ref. 5)

presence of microsomes were converted into a single component.[14] For chromogranins B and C, single preproteins were synthesized.[14,16,17]

Early studies on the posttranslational modification of chromogranins employed ^{3}H-fucose and ^{35}S-sulfate. ^{3}H-fucose[18] was incorporated into newly formed chromaffin granules, however most of the label was recovered in a protein which had a high affinity for concanavalin (for an autoradiographic study see ref. 19). It was concluded that this protein represented dopamine β-hydroxylase, which in contrast to the chromogranins is rich in fucose.[20] ^{35}S-sulfate was found to be incorporated into proteoglycans[21] and into the chromogranins.[22]

In more recent studies we used isolated bovine chromaffin cells for biosynthesis studies.[14] After a pulse label with ^{3}H-leucine, prochromogranin A was apparently modified in the Golgi region, becoming slightly larger and more acidic. This could be explained by glycosylation in this cell compartment which is consistent with the presence of *O*-glycosidically linked carbohydrate chains in this molecule.[23] Chromogranin A is also phosphorylated, which results in five phosphate groups per molecule.[24] It is also sulfated,[22,25] probably on the carbohydrate chains.[17]

Chromogranin B is converted during cellular synthesis into a significantly more acidic protein (about 0.5 units change in pI).[14] This is most likely explained by a significant sulfation of this protein,[25,17] apparently at tyrosine residues,[17] and by phosphorylation at serine residues.[17] Chromogranin C becomes also sulfated at tyrosine residues and is also phosphorylated.[17] (Chromogranin C is called secretogranin II by these authors).

Another significant step in the posttranslational modification is the proteolysis occurring within chromaffin granules. In mature bovine chromaffin granules the originally synthesized chromogranins have become proteolytically processed to a varying degree (see FIG. 1). Chromogranin B is degraded to the largest extent with only 14%

of the proprotein remaining. This processing is apparently slow however. After a 2-hour chase period only a very limited degradation of newly synthesized chromogranins had occurred.[25] Since the half-life of chromaffin granules is about a week,[26] even a slow degradation process can lead to a significant processing.

FIGURE 1 compares the biosynthetic pathways of the chromogranins as just discussed with that of the enkephalins,[5] which are representative for the neuropeptides in these organelles. Apparently proenkephalin is not glycosylated, sulfated, or phosphorylated, however it shares with the chromogranins the prosttranslational processing by proteases. Mature granules contain at the most traces of the proenkephalin, whereas several large enkephalin-containing peptides are still present.[5,6,27,28] For neuropeptide Y no evidence of the presence of larger precursor molecules has been obtained.[6] Thus chromogranins and neuropeptides are processed to varying degrees, ranging from limited breakdown to a rather complete degradation. This raises the intriguing question of how the proteases involved in this process are regulated within the chromaffin granules.

BIOSYNTHESIS OF CATECHOLAMINES AND NUCLEOTIDES

Catecholamines

Amino acids are used in a twofold way to provide secretory constituents for chromaffin granules. As we have just discussed, these components are used for synthesizing peptides. In addition the amino acid tyrosine is transformed into the main hormones of the adrenal medulla, *i.e.,* the catecholamines (see FIG. 2). The pathways of this synthesis, first proposed by H. Blaschko, are well established.[30,31] One essential step, the conversion of dopamine to noradrenaline, occurs within chromaffin granules since the enzyme dopamine β-hydroxylase is found within these organelles.[31] This was confirmed by a study in which perfused bovine adrenals were pulse-labeled with ^{3}H-tyrosine.[9] Three minutes after the pulse most of the labeled catecholamines were already present in chromaffin granules and consisted of 15% dopamine, 81% noradrenaline, and 4% adrenaline, whereas 3 hours after the pulse only 0.4% ^{3}H-dopamine was left. Apparently ^{3}H-dopamine that enters the granules is efficiently transformed to noradrenaline (for further studies see ref. 26; for a more recent study see ref. 32). For the conversion to adrenaline, noradrenaline has to leak out of the granules in order to encounter the cytoplasmic enzyme phenylethanolamine-*N*-methyltransferase. Adrenaline formed in the cytoplasm must be pumped back into chromaffin granules. The half-life of this conversion process is about 30 hours, apparently depending on the slow leakage of noradrenaline from the granules into the cytoplasm.[32] A crucial step in these processes is the uptake of the catecholamines into the granules. As discussed in detail in several chapters of this volume, this occurs by an active transport involving a specific carrier driven by an electrochemical proton gradient.

Nucleotides

Several studies have established that the nucleotide pool of chromaffin granules can be radioactively labeled with precursors like ^{32}P-phosphate or ^{3}H-adenosine.[9,33–37]

This enabled us to study the origin of the granule ATP by a "pulse label" approach.[38] The time course of the labeling and the effect of cyanide, dinitrophenol, and atractyloside, which block mitochondrial ATP production, indicated that ATP was first

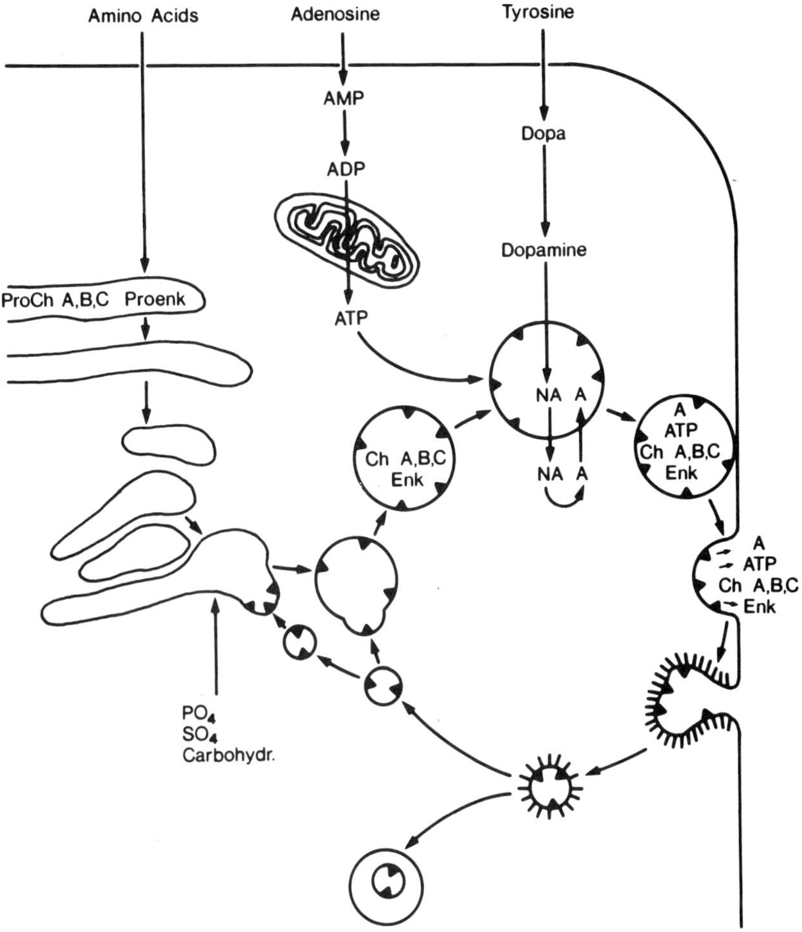

FIGURE 2. Life cycle of chromaffin granules. Amino acids are used by the chromaffin cells to synthesize chromogranins A, B, and C and neuropeptides including enkephalins. These products are modified in the Golgi region by incorporation of phosphate, sulfate, and carbohydrates and are packaged into new chromaffin granules. These organelles accumulate nucleotides and catecholamines in the cytoplasm and finally secrete their total content by exocytosis. The highly specialized membranes of chromaffin granules (indicated in the figure by *triangles*) are reused for several secretory cycles, thus providing a shuttle service from the Golgi region to the plasma membrane.

synthesized in mitochondria and was subsequently taken up by chromaffin granules. This led to the obvious conclusion that chromaffin granules must be able to accumulate this nucleotide and in fact in a series of studies[39-42] carrier-mediated transport of nucleotides into chromaffin granules was established. Some unresolved problems in connection with this uptake will be discussed in another paper by this author in this volume.

RELATIVE TIME COURSE OF THE BIOSYNTHESIS

After having pinched off from the Golgi stacks, the newly formed chromaffin granules already contain the chromogranins and neuropeptides (see FIG. 2). When do the small molecules like calcium, ATP, and catecholamines enter? Newly formed chromaffin granules are less dense than mature ones and can therefore be differentiated by density gradient centrifugation.[8,9] In pulse-labeling studies with ^{3}H-tyrosine or ^{3}H-adenosine there was no evidence that newly formed vesicles took up newly synthesized catecholamines or ATP preferentially.[9] In fact these components were taken up by all vesicles, indicating that even in mature vesicles there is a constant leakage followed by reuptake. In the absence of a preferential accumulation the filling of new vesicles with small molecules takes considerable time. In rat adrenal medulla newly synthesized vesicles marked by ^{35}S-sulfate-labeled proteoglycans required more than 24 hours in order to become as dense as mature ones.[22] A similar time course was found in experiments where the levels of catecholamines and ATP, first depleted by insulin hypoglycemia, were analyzed in the recovery period.[11] One can also calculate from the uptake rate for catecholamines (3.4 nmol/mg protein/min[43]) that it takes at a minimum 12 hours to fill a chromaffin granule (2.5 μmol catecholamine/mg protein[1]). Since leakage of catecholamines is known to occur, this theoretical figure will actually be too low and therefore is in reasonable agreement with the *in vivo* figure of about 24 hours.

During the filling period of new granules there seems to be a preferential accumulation of ATP versus the catecholamines, since the catecholamine: ATP ratios of new vesicles are lower than those in mature granules.[44] Unfortunately we have no direct experimental facts to tell us when calcium enters these granules. It has been speculated[26] that calcium is involved in the "condensation process" in new chromaffin granules by interacting with the acidic chromogranins. Recent data support this concept since it has been shown that these proteins bind calcium very efficiently.[45,46] Furthermore chromaffin granules can take up calcium by a carrier-mediated process[47] which depends on a sodium-calcium exchange.[48,49] Does this exchange occur early in the life cycle of a chromaffin granule? If so, then one should be able to detect calcium in new vesicles near the Golgi region by analyzing tissue sections, *e.g.,* by electron energy loss spectroscopy.

REGULATION OF BIOSYNTHESIS

Nervous stimulation of rat adrenal medulla by insulin or reserpine induces an increased synthesis of the enzyme dopamine β-hydroxylase,[50–52] which is a major constituent of chromaffin granules. On the other hand, hypophysectomy leads to a decline of dopamine β-hydroxylase levels in the adrenals,[53] probably by an increase in the degradation of the enzyme.[51] More recently it has been established that stimulation of isolated bovine chromaffin cells increases the amount of preproenkephalin mRNA in these cells, leading to increased enkephalin levels.[54,55] Enkephalins are also constituents of chromaffin granules, although rather minor ones (see ref. 4). Is only the synthesis of these components regulated, or can several or even most of the granule constituents be changed in their synthesis rate depending on various influences?

We have investigated this question by subjecting rats to various *in vivo* treatments (to be published). As was expected, dopamine β-hydroxylase levels were elevated after treatments (insulin and reserpine), leading to an increased activity of the splanchnic nerve. Enkephalin levels were also higher, which agrees with results on bovine chromaffin cells (see above) and with data on rats (reserpine,[56–58] insulin;[59] see also article by O. H. Viveros in this volume). On the other hand it has also been reported that denervation of the adrenals, and therefore reduced nervous activity, can elevate enkephalin levels in rat adrenals.[60,61]

In our study the level of another peptide (see TABLE 2), *i.e.,* neuropeptide Y, was also elevated, however much less than the enkephalins. Chromogranin A levels appeared to be uninfluenced by nervous activity. When compared to chromogranin A, however, chromogranin B was slightly increased after insulin and significantly elevated after reserpine pretreatment. Since chromogranin B is the major soluble protein of rat chromaffin granules,[28] these results indicate that nervous activity does not only induce a major membrane protein, *i.e.,* dopamine β-hydroxylase (see ref. 1) but also a main secretory component.

Hypophysectomy (see TABLE 2) led to the expected fall of dopamine β-hydroxylase activity and to a great decline of chromogranin A. We have not yet determined whether chromogranin A levels are lowered because of an increase in degradation (as it is the case for dopamine β-hydroxylase[51]) or a decrease in synthesis. Levels of chromogranin B do not decline significantly.

These complex results raise many questions. Let us discuss two points. First, chromogranins A and B, despite their physicochemical similarities, are apparently regulated separately. We cannot yet translate this into a functional meaning, however we think that this behavior is a strong indication of a specific function of each of these chromogranins. If these proteins are only involved in the formation and maintenance of the granules, a separate regulation of their levels would appear unnecessary. We therefore consider this result as an indication that each of the chromogranins has a specific function after exocytosis.

Second, what do the results, presented in TABLE 2, imply as far as the composition of a single chromaffin granule is concerned? Do the increased levels of chromogranin B, dopamine β-hydroxylase, and neuropeptides indicate that more vesicles are synthesized or do the same number of new vesicles have a different composition (see FIG. 3)? A final answer cannot yet be given, however the latter possibility seems more likely since it is supported by a morphometric study.[62] In these experiments adrenals were removed from rats which were treated with seven doses of reserpine and killed 36 hours after the last injection.

The chromaffin cells were analyzed morphometrically. There was a slight but not significant increase in the number of granules, but their size declined and thus the amount of granule membrane showed no change (see TABLE 3). A biochemical analysis revealed that the amount of dopamine β-hydroxylase per milligram of isolated granule membrane had significantly increased.[62]

These results support the view that after a depletion of vesicles within the adrenals caused by insulin or reserpine, new vesicles are formed that possess an increased amount of dopamine β-hydroxylase and probably also of neuropeptides and chromogranin B (see FIG. 3). This view is also supported by the determination of cytochrome b_{561} levels in our experiments (see TABLE 2). This major protein[3] of the granule membrane is not elevated after reserpine treatment, which again indicates that the amount of membrane has remained the same.

TABLE 2. Levels of Constituents of Chromaffin Granules in Rat Adrenals after Various Treatments[a]

	After		
	Insulin[b]	Reserpine[c]	Hypophysectomy[d]
Catecholamines	79	88	60
Chromogranin A	84	80	21
Chromogranin B	104	160	97
Dopamine β-hydroxylase	145	190	44
Cytochrome b_{561}	nd[e]	70	94
Total enkephalins	674	nd	95
Neuropeptide Y	163	nd	nd

[a] Results are expressed as a percentage of control levels. Chromogranins were determined with specific antisera using the immunodot procedure (see ref. 90). Figures are mean values of at least three animals (M. Sietzen, M. Schober, R. Fischer-Colbrie & H. Winkler, to be published).

[b] Insulin treatment: Rats were injected with 5 units/kg of insulin. After 2 h they were given glucose to stop hypoglycemia. They were killed 144 h later.

[c] Reserpine treatment: Rats were given 2.5 mg/kg reserpine three times on days 1, 3, and 5. They were killed 120 h after the first injection.

[d] Hypophysectomy: The rats were killed 14 days after hypophysectomy or after a control sham operation.

[e] nd = not determined.

What happens after hypophysectomy is more difficult to interpret. In a morphometric study[63] hypophysectomy led to a decline in the number of chromaffin granules. Two weeks after hypophysectomy, however, the number of adrenaline granules was only reduced by 30%. In our experiments catecholamine levels were also lowered by about 40%, whereas chromogranin B and cytochrome b_{561} showed no significant change. On the other hand both dopamine β-hydroxylase and chromogranin A levels were significantly lowered. Again it seems likely that the composition of vesicles has changed.

In conclusion, chromaffin granules have a complex composition and the adrenal medulla releases upon stimulation a rather complex mixture of secretory products. The composition of the granules and consequently of the secretory products can be modified differently by nervous activity and hypophysectomy. Apparently the "cocktail" of secretory products can be blended in different ways. Only when we understand more about the function of these various secretory products will we start to grasp the functional meanings of these exciting mechanisms.

TABLE 3. Morphometric Analysis of Chromaffin Cells after Reserpine

	Controls	Reserpine
Chromaffin granules		
Volume density relative to cytoplasm (%)	32.1	27.6
Mean cross-sectional area (μm^2)	0.062	0.047
Surface density of membranes relative to cytoplasm ($\mu m^2/\mu m^3$)	5.1	5.5
Number of granules per unit volume cytoplasm (μm^{-3})	16.8	20.3

Rats were injected with reserpine (4 mg/kg i.p.) 7 times at 48-hour intervals and killed 36 hours after the last injection (see ref. 62). Some of the morphometric results are shown.

MEMBRANES OF CHROMAFFIN GRANULES: RECYCLING CONTAINERS

Until now we have discussed the biosynthesis of the soluble constituents of chromaffin granules. All these components are secreted together by exocytosis (for review see ref. 64) which is brought about by a fusion of the membranes of chromaffin granules with the plasma membrane (see FIG. 2). In 1972[9] we observed in pulse-label experiments with perfused bovine glands that the synthesis rate of the membrane proteins was significantly lower than that of the secretory ones (for exocrine pancreas see ref. 65). Based on this finding we proposed that the membranes of chromaffin granules are reused for several secretory cycles, therefore only requiring a low resynthesis rate (see also ref. 26). This concept gained decisive support from studies with exogenous tracers in several endocrine tissues. During exocytosis compounds like cationized ferritin were applied from the outside, endocytosed, and finally ferritin-labeled membranes appeared in the Golgi region and newly formed vesicles.[66–68]

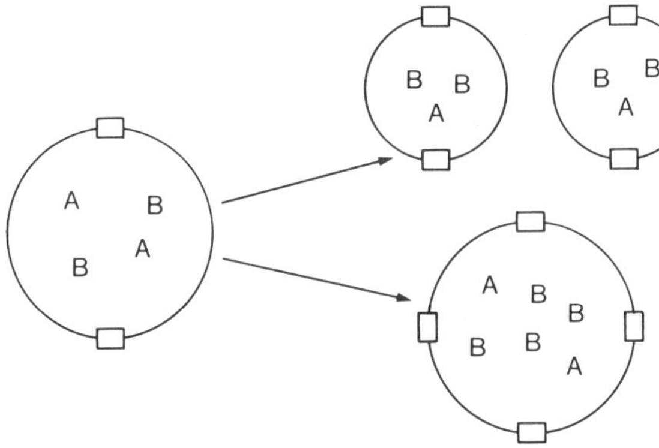

FIGURE 3. Regulation of the composition of chromaffin granules. As discussed in the text, an increase in the nervous activity to rat adrenal changes the levels of various granule components. This can be explained by two possible mechanisms: (i) more vesicles are produced with the same or reduced amount of some constituents, or (ii) the number of vesicles stays the same, but the amounts of some components have increased (see text). *A*, chromogranin A; *B*, chromogranin B. Dopamine β-hydroxylase is indicated in the membrane by *squares*.

These studies established the recycling of membranes, but did not yet prove that the membranes of secretory vesicles were specifically retrieved and recycled. In order to investigate this specific markers had to be used. Such markers were provided by antibodies against specific antigens of chromaffin granules, such as dopamine β-hydroxylase or glycoprotein III. In isolated bovine chromaffin cells these antigens become exposed on the cell surface during exocytotic activity and can be visualized by immunofluorescence.[69–73]

A quantitative evaluation by a fluorescence-activated cell sorter revealed a good correlation between the degree of secretory activity and the amount of granule antigens exposed on the cell surface.[73] A further incubation of the isolated cells after removal of the secretagogue led to a disappearance of the granule antigens from the cell surface.[71–73] Within 20 to 30 minutes antigens corresponding to about 15% of the granule pool were retrieved from the cell surface, apparently by uptake into the cell interior. Throughout this process the antigens were present on the cell surface in small patches. Apparently the granule antigens did not spread translationally in the plasma membrane, which represents a prerequisite for a specific and efficient retrieval of these membrane patches. The further fate of the retrieved membranes was studied at the ultrastructural level after the exposed membranes were gold-immunolabeled.[74] The immunolabeled membranes were retrieved via coated vesicles which subsequently shed their coat. Labeled vesicles were then found close to the Golgi region but not within the Golgi stacks. Finally, labeled membranes appeared in newly formed chromaffin granules.

We can conclude the following. During exocytosis (see FIG. 2) the membrane of chromaffin granules becomes part of the plasma membrane. The granule antigens do not spread within these membranes, but the total patch of the granule membrane is specifically and efficiently retrieved and some of these retrieved membranes are recycled into newly formed chromaffin granules. In this way the chromaffin cell uses the granule membrane as a container allowing the economic recycling of a membrane that is highly specialized for specific functions like the transport of protons, catecholamines, nucleotides, and calcium.

CATECHOLAMINE-STORING VESICLES IN SYMPATHETIC NERVES

In sympathetic nerves there are two types of vesicles, the large and the small dense-cored vesicles. The large vesicles resemble in most biochemical aspects adrenal chromaffin granules. Their secretory products (see FIG. 4) are noradrenaline, ATP, chromogranins A, B, and C, enkephalins, and neuropeptide Y.[29, 75–77] It seems likely that these vesicles take up catecholamines and nucleotides by the same mechanism as chromaffin granules do; they contain dopamine β-hydroxylase and therefore are able to synthesize noradrenaline from dopamine.[78]

Small vesicles apparently differ mainly in one aspect (see FIG. 4). They do not contain soluble proteins and peptides, *i.e.*, the chromogranins,[76] enkephalins[76,77] and neuropeptide Y.[77,79] They can, however, take up catecholamines and nucleotides[80] and have been shown to synthesize noradrenaline,[76] although the presence of dopamine β-hydroxylase necessary for this conversion step is controversial (see ref. 78). In any case, at present there is no convincing evidence to suggest that the membranes of these two types of vesicles differ. What can we conclude from these findings about the life cycle of these vesicles (see FIG. 5)? Large dense-cored vesicles, just like chromaffin granules, must arise from the Golgi region, which is responsible for sorting out proteins destined for export in all secretory cells. Subsequently these vesicles must be transported to the terminal by axonal flow.

Two possible origins have been suggested for the small vesicles. The first possibility is that small vesicles are formed by a pinch-off process from a membrane compartment within the axon or terminal (see FIG. 5, bottom part). Support for this concept was obtained in morphological studies.[81,82] Since small vesicles have a characteristic membrane composition, one has to postulate that this pinching off is not an unspecific process but allows the formation of vesicles with a specific membrane. Therefore this membrane compartment cannot simply be a smooth endoplasmic reticulum as defined by morphological criteria. In this connection it is interesting that the reticulum present in nerve axons has been considered to be related to Golgi cisternae at least by histochemical criteria[83] (see also refs. 84 and 85). Can specific vesicles bud off from such a Golgi complex extending into the axons? Questions like this one will probably become answerable if specific markers, *e.g.*, antibodies, are used to define the possible origin of small vesicles from such membrane compartments at the ultrastructural level.

The second possible origin of small vesicles is that they are formed from large ones during membrane retrieval after exocytosis (see FIG. 5, top part). Support for this concept, originally proposed by A. D. Smith in 1978,[86] can be seen in the fact that no significant differences have yet been found in the membrane composition of

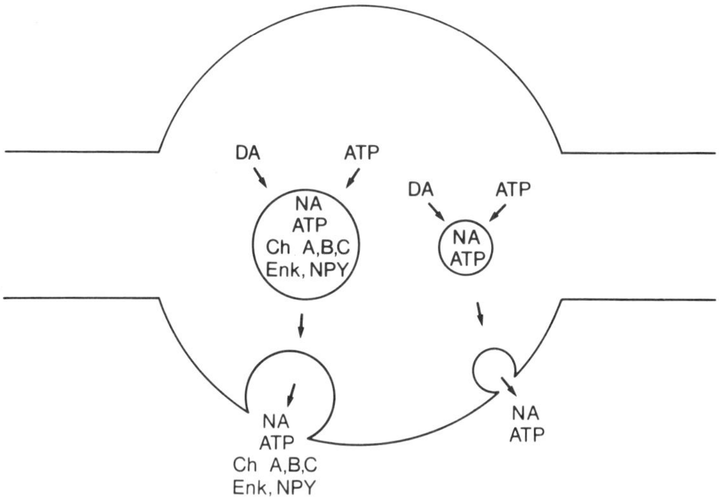

FIGURE 4. Storage vesicles in sympathetic nerve. This figure indicates that both small and large dense core vesicles in sympathetic nerve can take up nucleotides (*ATP*) and dopamine (*DA*), and both vesicles can convert dopamine to noradrenaline (*NA*). Only large dense core vesicles contain chromogranins, enkephalins, and neuropeptide Y however. Both vesicles secrete their content by exocytosis.

these two vesicle populations in sympathetic nerve. A common antigen for both types of vesicles has also been found in a rat brain preparation.[87] For the adrenal medulla specific retrieval of granule membranes after exocytosis has already been established (see above). Thus one would have to postulate that for the nerve, in contrast to the adrenal medulla, vesicles retrieved after exocytosis of the large dense core vesicles do not return to the Golgi region for refilling with peptides but fill up in the terminal with catecholamines and ATP. Since the terminal and the cell body are far apart in the neuron, such a mechanism seems a very economical variation of the recycling process in adrenal medulla, allowing the nerve to synthesize and secrete transmitter at a high rate in the terminal. It will be difficult to provide final proof for this elegant hypothesis however. In any case, a local reuse of the small vesicles for several secretory cycles has already been established. As shown by biochemical and morphological criteria, small vesicles can be depleted and refilled again.[88,89]

CONCLUSIONS

Adrenal chromaffin granules contain a complex mixture of secretory products including chromogranins A, B, and C, peptides, and small molecules. The relative concentration of the various constituents *in vivo* can be regulated by nervous activity

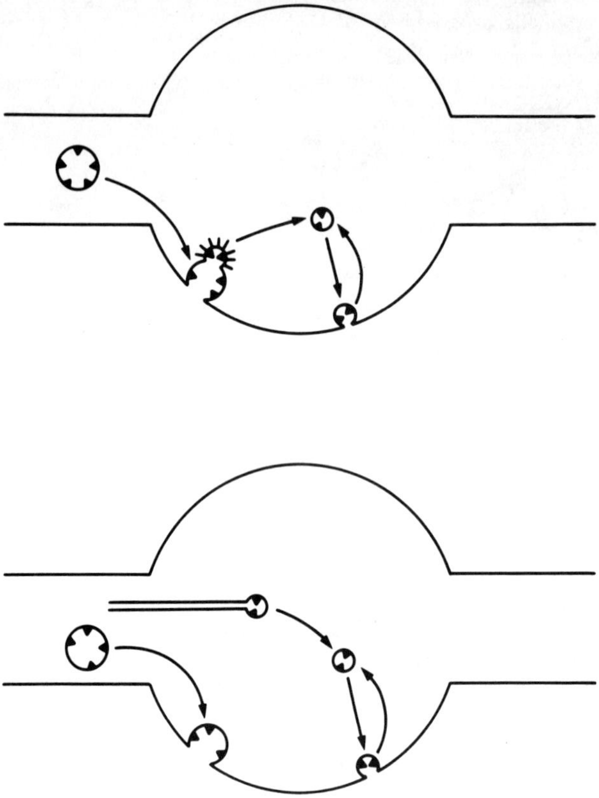

FIGURE 5. Two models of the origin of vesicles in sympathetic nerve. *Top:* Large dense core vesicles arrive in the terminal by axonal flow. After exocytosis the membranes of these vesicles are retrieved and used to form small dense core vesicles.[81] The small dense core vesicles can be reused for several cycles of secretion involving exocytosis, membrane retrieval, and filling with transmitter. The special composition of the membranes of the two vesicles is indicated by *triangles. Bottom:* Large dense core vesicles and small dense core vesicles have a separate origin. Large vesicles arrive by axonal flow. Small vesicles are formed from a membrane compartment (possibly smooth endoplasmic reticulum, or Golgi membranes; see text).

and corticosteroids. The membrane of chromaffin granules can be used for several secretory cycles, thus providing a shuttle service from the Golgi region to the plasma membranes. Therefore the chromaffin cell combines economy (use of the granule membrane as a recycling container) with great variability (blending of the secretory cocktail).

In sympathetic nerve the membranes of large dense core vesicles are probably used to form recycling small vesicles, which ensures a high rate of transmitter synthesis in the terminal.

REFERENCES

1. WINKLER, H. 1976. The composition of adrenal chromaffin granules: An assessment of controversial results. Neuroscience **1:** 65-80.
2. WINKLER, H. & E. WESTHEAD. 1980. The molecular organization of adrenal chromaffin granules. Neuroscience **5:** 1803-1823.
3. WINKLER, H. & S. W. CARMICHAEL. 1982. The chromaffin granule. *In* The Secretory Granule. A. M. Poisner & J. M. Trifaró, Eds.: 3-79. Elsevier Biomedical Press. Amsterdam.
4. WINKLER, H., D. K. APPS & R. FISCHER-COLBRIE. 1986. The molecular function of adrenal chromaffin granules: Established facts and unresolved topics. Neuroscience. **18:** 261-290.
5. UDENFRIEND, S. & D. L. KILPATRICK. 1983. Biochemistry of the enkephalins and enkephalin-containing peptides. Arch. Biochem. Biophys. **221:** 309-323.
6. FISCHER-COLBRIE, R., J. DIEZ-GUERRA, P. C. EMSON & H. WINKLER. 1986. Bovine chromaffin granules: Immunological studies with antisera against neuropeptide Y, metenkephalin and bombesin. Neuroscience. **18:** 167-174.
7. PALADE, G. E. 1975. Intracellular aspects of the process of protein synthesis. Science **189:** 347-358.
8. WINKLER, H., H. HÖRTNAGL, J. A. L. SCHÖPF, H. HÖRTNAGL & G. ZUR NEDDEN. 1971. Bovine adrenal medulla: Synthesis and secretion of radioactively labelled catecholamines and chromogranins. Naunyn-Schmiedebergs Arch. Exp. Pathol. Pharmakol. **271:** 193-203.
9. WINKLER, H., J. A. L. SCHÖPF, H. HÖRTNAGL & H. HÖRTNAGL. 1972. Bovine adrenal medulla: Subcellular distribution of newly synthesized catecholamines, nucleotides and chromogranins. Naunyn-Schmiedebergs Arch. Exp. Pathol. Pharmakol. **273:** 43-61.
10. BENCHIMOL, S. & M. CANTIN. 1982. Ultrastructural radioautography of synthesis and migration of proteins and catecholamines in the rat adrenal medulla. Cell Tissue Res. **225:** 293-314.
11. VIVEROS, O. H., L. ARQUEROS & N. KIRSHNER. 1971. Mechanism of secretion from the adrenal medulla. VII. Effect of insulin administration on the buoyant density, dopamine β-hydroxylase and catecholamine content of adrenal storage vesicles. Molec. Pharmacol. **7:** 444-454.
12. KIRSHNER, N. & O. H. VIVEROS. 1972. The secretory cycle of the adrenal medulla. Pharmacol. Rev. **24:** 385-398.
13. KILPATRICK, L., F. GAVINE, D. APPS & J. PHILLIPS. 1983. Biosynthetic relationship between the major matrix proteins of adrenal chromaffin granules. FEBS Lett. **164:** 383-388.
14. FALKENSAMMER, G., R. FISCHER-COLBRIE, K. RICHTER & H. WINKLER. 1985. Cell-free and cellular synthesis of chromogranin A and B of bovine adrenal medulla. Neuroscience **14:** 735-746.
15. SERCK-HANSSEN, G. & D. T. O'CONNOR. 1984. Immunological identification and characterization of chromogranins coded by poly(A) mRNA from bovine adrenal medulla and pituitary gland and human phaeochomocytoma. J. Biol. Chem. **259:** 11597-11600.
16. FISCHER-COLBRIE, R., C. HAGN, L. KILPATRICK & H. WINKLER. 1986. Chromogranin C: A third component of the acidic proteins in chromaffin granules. J. Neurochem. **47:** 318-321.
17. ROSA, P., A. HILLE, R. H. W. LEE, A. ZANINI, P. DECAMILLI & W. B. HUTTNER. 1985. Secretogranins I and II: Two tyrosine-sulfated secretory proteins common to a variety of cells secreting peptides by the regulated pathway. J. Cell Biol. **101:** 1999-2011.

18. GEISSLER, D., A. MARTINEK, R. U. MARGOLIS, J. A. SKRIVANEK, R. LEDEEN, P. KÖNIG & H. WINKLER. 1977. Composition and biogenesis of complex carbohydrates of ox adrenal chromaffin granules. Neuroscience **2:** 685-693.
19. BENCHIMOL, S. & M. CANTIN. 1978. Etude radioautographique de la synthèse et de la migration des glycoproteines dans les cellules de la médullo-surrenale du rat. Biol. Cell. **33:** 157-161.
20. FISCHER-COLBRIE, R., M. SCHACHINGER, R. ZANGERLE & H. WINKLER. 1982. Dopamine β-hydroxylase and other glycoproteins from the soluble content and the membranes of adrenal chromaffin granules: Isolation and carbohydrate analysis. J. Neurochem. **38:** 725-732.
21. MARGOLIS, R. U. & R. K. MARGOLIS. 1973. Isolation of chrondroitin sulfate and glycopeptides from chromaffin granules of adrenal medulla. Biochem. Pharmacol. **22:** 2195-2197.
22. BAUMGARTNER H., J. W. GIBB, H. HÖRTNAGL, H. SNIDER & H. WINKLER. 1974. Labelling of chromaffin granules in adrenal medulla with ^{35}S-sulfate. Mol. Pharmacol. **10:** 678-685.
23. KIANG, W. L., T. KRUSIUS, J. FINNE, R. U. MARGOLIS & R. K. MARGOLIS. 1982. Glycoproteins and proteoglycans of the chromaffin granule matrix. J. Biol. Chem. **257:** 1651-1659.
24. SETTLEMAN, J., R. FONSECA, J. NOLAN & R. H. HOGUE ANGELETTI. 1985. Relationship of multiple forms of chromogranin. J. Biol. Chem. **260:** 1645-1651.
25. FALKENSAMMER, G., R. FISCHER-COLBRIE & H. WINKLER. 1985. Biogenesis of chromaffin granules: Incorporation of sulfate into chromogranin B and into a proteoglycan. J. Neurochem. **45:** 1475-1480.
26. WINKLER, H. 1977. The biogenesis of adrenal chromaffin granules. Neuroscience **2:** 657-683.
27. PATEY, G., D. LISTON & J. ROSSIER. 1984. Characterization of new enkephalin-containing peptides in the adrenal medulla by immunblotting. FEBS Lett. **172:** 303-308.
28. FISCHER-COLBRIE, R. & I. FRISCHENSCHLAGER. 1985. Immunological characterization of secretory proteins of chromaffin granules: Chromogranins A, chromogranins B and enkephalin-containing peptides. J. Neurochem. **44:** 1854-1861.
29. HAGN, C., R. L. KLEIN, R. FISCHER-COLBRIE, B. H. DOUGLAS & H. WINKLER. 1986. An immunological characterization of five common antigens of chromaffin granules and of large dense-cored vesicles of sympathetic nerve. Neurosci. Lett. **67:** 295-300.
30. BLASCHKO, H. 1959. The development of current concepts of catecholamine formation. Ann. Rev. Pharmacol. **11:** 307-316.
31. KIRSHNER, N. 1975. Biosynthesis of the catecholamines. *In* The Adrenal Gland Handbook of Endocrinology, vol. 6. H. Blaschko, G. Sayers & A. D. Smith, Eds.: 341-355 American Physiological Society. Washington, D.C.
32. CORCORAN, J. J., S. P. WILSON & N. KIRSHNER. 1984. Flux of catecholamines through chromaffin vesicles in cultured bovine adrenal medullary cells. J. Biol. Chem. **259:** 6208-6214.
33. STJÄRNE, L., R. HEDQVIST & H. LAGERCRANTZ. 1970. Catecholamines and adenine nucleotide material in effluent from stimulated adrenal medulla and spleen: A study of exocytosis hypothesis for hormone secretion and neurostransmitter release. Biochem. Pharmacol. **19:** 1147-1158.
34. STEVENS, P., R. L. ROBINSON, K. VAN DYKE & R. STITZEL. 1972. Studies on the synthesis and release of adenosine triphosphate 8-^{3}H in the isolated perfused cat adrenal gland. J. Pharmacol. Exp. Ther. **181:** 463-471.
35. STEVENS, P., R. L. ROBINSON, K. VAN DYKE & R. STITZEL. 1975. Synthesis, storage and drug-induced release of ATP-8-^{3}H in the perfused bovine adrenal gland. Pharmacology **13:** 40-55.
36. TRIFARO, J. M. & J. DWORKIND. 1975. Phosphorylation of the membrane components of chromaffin granules: Synthesis of diphosphatidylinositol and presence of phosphatidylinositol kinase in granule membranes. Can. J. Physiol. Pharmacol. **53:** 479-492.
37. CORCORAN, J. J., S. P. WILSON & N. KIRSHNER. 1986. Turnover and storage of newly synthesized adenine nucleotides in bovine adrenal medullary cell cultures. J. Neurochem. **46:** 151-160.
38. PEER, L. J., H. WINKLER, S. R. SNIDER, J. W. GIBB & H. BAUMGARTNER. 1976. Synthesis

of nucleotides in adrenal medulla and their uptake into chromaffin granules. Biochem. Pharmacol. **25:** 311-315.
39. KOSTRON, H., H. WINKLER, L. J. PEER & P. KÖNIG. 1977. Uptake of adenosine triphosphate by isolated chromaffin granules: A carrier-mediated process. Neuroscience **2:** 159-166.
40. ABERER, W., H. KOSTRON, E. HUBER & H. WINKLER. 1978. A characterization of the nucleotide uptake by chromaffin granules of bovine adrenal medulla. Biochem. J. **172:** 353-360.
41. WEBER, A. & H. WINKLER. 1981. Specificity and mechanism of nucleotide transport in chromaffin granules. Neuroscience **6:** 2269-2276.
42. WEBER, A., E. W. WESTHEAD & H. WINKLER. 1983. Specificity and properties of the nucleotide carrier in chromaffin granules from bovine adrenal medulla. Biochem. J. **210:** 789-794.
43. TAUGNER, G. 1971. The effect of salts on catecholamine fluxes and adenosine triphosphatase activity in storage vesicles from the adrenal medulla. Biochem. J. **123:** 219-225.
44. SLOTKIN, T. A. & N. KIRSHNER. 1973. Recovery of rat adrenal amine stores after insulin administration. Mol. Pharmacol. **9:** 105-116.
45. BULENDA, D. & M. GRATZL. 1985. Matrix free Ca^{2+} in isolated chromaffin vesicles. Biochemistry **24:** 7760-7765.
46. REIFFEN, F. U. & M. GRATZL. 1986. Chromogranins, widespread in endocrine and nervous tissue bind CA^{2+}. FEBS Lett. **195:** 327-330.
47. KOSTRON, H., H. WINKLER, D. GEISSLER & P. KÖNIG. 1977. Uptake of calcium by chromaffin granules in vitro. J. Neurochem. **28:** 487-493.
48. KRIEGER-BAUER, H. I. & M. GRATZL. 1983. Effects of monovalent and divalent cations on Ca^{2+} fluxes across chromaffin secretory membrane vesicles. J. Neurochem. **41:** 1269-1276.
49. PHILLIPS, J. H. 1981. Transport of Ca^{2+} and Na^{+} across the chromaffin-granule membrane. Biochem. J. **200:** 99-107.
50. PATRICK, R. L. & N. KIRSHNER. 1971. Effect of stimulation on the levels of tyrosine hydroxylase, dopamine β-hydroxylase, and catecholamines in intact and denervated rat adrenal glands. Mol. Pharmacol. **7:** 87-96.
51. CIARANELLO, R. D., G. F. WOOTEN & J. AXELROD. 1975. Regulation of dopamine β-hydroxylase in rat adrenal glands. J. Biol. Chem. **250:** 3204-3211.
52. LIMA, L. & T. L. SOURKES. 1986. Reserpine and the monoaminergic regulation of adrenal dopamine β-hydroxylase activity. Neuroscience **17:** 235-245.
53. WEINSHILBOUM, R. & J. AXELROD. 1970. Dopamine β-hydroxylase activity in the rat after hypophysectomy. Endocrinology **87:** 894-899.
54. EIDEN, L. E., P. GIRAUD, H.-U. AFFOLTER, E. HERBERT & A. J. HOTCHKISS. 1984. Alternative modes of enkephalin biosynthesis regulation by reserpine and cyclic AMP in cultured chromaffin cells. Proc. Natl. Acad. Sci. **81:** 3949-3953.
55. WILSON, S. P., C. D. UNSWORTH & O. H. VIVEROS. 1984. Regulation of opioid peptide synthesis and processing in adrenal chromaffin cells by catecholamines and cyclic adenosine 3′:5′-monophosphate. J. Neurosci. **4:** 2993-3001.
56. SCHULTZBERG, M., J. M. LUNDBERG, T. HÖKFELT, L. TERENIUS, J. BRANDT, R. P. ELDE & M. GOLDSTEIN. 1978. Enkephalin-like immunoreactivity in gland cells and nerve terminals of the adrenal medulla. Neuroscience **3:** 1169-1186.
57. MOCCHETTI, I., A. GUIDOTTI, J. P. SCHWARTZ & E. COSTA. 1985. Reserpine changes the dynamic state of enkephalin stores in rat striatum and adrenal medulla by different mechanisms. J. Neurosci. **5:** 3379-3385.
58. BOHN, M. C., J. A. KESSLER, L. GOLIGHTLY & I. B. BLACK. 1983. Appearance of enkephalin-immunoreactivity in rat adrenal medulla following treatment with nicotinic antagonists or reserpine. Cell Tissue Res. **231:** 469-479.
59. UNSWORTH, C. D., S. P. WILSON & H. VIVEROS. 1984. Enkephalins in the adrenal medulla: Regulation of synthesis and processing by catecholamines and cAMP in bovine chromaffin cells in primary culture and by the splanchnic innervation in the adrenal of the rat. *In* Endocrinology. F. Labrie & L. Proul, Eds.: 993-998. Elsevier. New York.
60. LA GAMMA, E. F., J. E. ADLER & I. B. BLACK. 1984. Impulse activity differentially

regulates (Leu)enkephalin and catecholamine characters in the adrenal medulla. Science **224:** 1102-1104.
61. KILPATRICK, D. L., R. D. HOWELLS, G. FLEMINGER & S. UDENFRIEND. 1984. Denervation of rat adrenal glands markedly increases preproenkephalin mRNA. Proc. Natl. Acad. Sci. **81:** 7221-7223.
62. GAGNON, C., W. PFALLER, W. M. FISCHER, M. SCHWAB, H. WINKLER & H. THOENEN. 1976. Increased specific activity of membrane-bound dopamine β-hydroxylase in chromaffin granules after reserpine treatment. J. Neurochem. **28:** 853-856.
63. POHORECKY, L. & J. H. RUST. 1968. Studies on the cortical control of the adrenal medulla in the rat. J. Pharmacol. Exp. Ther. **162:** 227-238.
64. POLLARD, H. B., R. ORNBERG, M. LEVINE, K. KELNER, K. MORITA, R. LEVINE, E. FORSBERG, K. W. BROCKLEHURST, L. DUONG, P. I. LELKES, E. HELDMAN & M. YOUDIM. 1985. Hormone secretion by exocytosis with emphasis on information from the chromaffin cell system. Vitam. Horm. **42:** 109-196.
65. MELDOLESI, J., N. BORGESE, P. DE CAMILLI & B. CECCARELLI. 1978. Cytoplasmic membranes and the secretory process. Cell. Surf. Rev. **5:** 510-627.
66. HERZOG, V. & M. G. FARQUHAR. 1977. Luminal membrane retrieved after exocytosis reaches most Golgi cisternae in secretory cells. Proc. Natl. Acad. Sci. **74:** 5073-5077.
67. HERZOG, V. & H. REGGIO. 1980. Pathways of endocytosis from luminal plasmamembrane in rat exocrine pancreas. Eur. J. Cell Biol. **21:** 141-150.
68. SUCHARD, S. J., J. J. CORCORAN, B. C. PRESSMANN & R. W. RUBIN. 1981. Evidence for secretory granule membrane recycling in cultured adrenal chromaffin cells. Cell Biol. Int. Rep. **5:** 953-962.
69. WILDMANN, J., M. DEWAIR & M. MATTHAEI. 1981. Immunochemical evidence of exocytosis in isolated chromaffin cells after stimulation with depolarizing agents. J. Neuroimmunol. **1:** 353-364.
70. DOWD, D. J., C. EDWARDS, D. ENGLERT, J. E. MAZURKIEWICZ & H. Z. YE. 1983. Immunofluorescent evidence for exocytosis and internalization of secretory granule membrane in isolated chromaffin cells. Neuroscience **10:** 1025-1033.
71. PHILLIPS, J. H., K. BURRIDGE, S. P. WILSON & N. KIRSHNER. 1983. Visualization of the exocytosis/endocytosis secretory cycle in cultured adrenal chromaffin cells. J. Cell Biol. **97:** 1906-1917.
72. LINGG, G., F. FISCHER-COLBRIE, W. SCHMIDT & H. WINKLER. 1983. Exposure of an antigen of chromaffin granules on cell surface during exocytosis. Nature **301:** 610-611.
73. PATZAK, A., G. BÖCK, R. FISCHER-COLBRIE, K. SCHAUENSTEIN, W. SCHMIDT, G. LINGG & H. WINKLER. 1984. Exocytotic exposure and retrieval of membrane antigens of chromaffin granules: Quantitative evaluation of immunofluorescence on the surface of chromaffin cells. J. Cell Biol. **98:** 1817-1824.
74. PATZAK, A. & H. WINKLER. 1986. Exocytotic exposure and recycling of membrane antigens of chromaffin granules: Ultrastructural evaluation after immunolabeling. J. Cell Biol. **102:** 510-515.
75. KLEIN, R. L. 1982. Chemical composition of the large noradrenergic vesicles. *In* Neurotransmitter Vesicles. R. L. Klein, H. Lagercrantz & H. Zimmermann, Eds.: 133-173. Academic. London.
76. NEUMANN, B., C. J. WIEDERMANN, R. FISCHER-COLBRIE, M. SCHOBER, G. SPERK & H. WINKLER. 1984. Biochemical and functional properties of large and small-dense core vesicles in sympathetic nerves of rat and ox vas deferens. Neuroscience **13:** 921-931.
77. FRIED, G., L. TERENIUS, E. BRODIN, S. EFENDIC, G. DOCKRAY, J. FAHRENKRUG, M. GOLDSTEIN & T. HÖKFELT. 1986. Neuropeptide Y, enkephalin and noradrenaline coexist in sympathetic neurons innervating the bovine spleen. Cell Tissue Res. **243:** 495-508.
78. KLEIN, R. L. & A. K. THURESON-KLEIN. 1984. Noradrenergic vesicles. Molecular organization and function. Handb. Neurochem. **7:** 71-109.
79. FRIED, G., L. TERENIUS, T. HÖKFELT & M. GOLDSTEIN. 1985. Evidence for differential localization of noradrenaline and neuropeptide Y in neuronal storage vesicles isolated from rat vas deferens. J. Neurosci. **2:** 450-458.
80. ABERER, W., R. STITZEL, H. WINKLER & E. HUBER. 1979. Accumulation of (^{3}H)ATP in small dense-core vesicles of superfused vasa deferentia. J. Neurochem. **33:** 797-801.

81. DROZ, B., A. RAMBOURG & H. L. KOENIG. 1975. The smooth endoplasmic reticulum: Structure and role in the renewal of axonal membrane and synaptic vesicles by fast axonal transport. Brain Res. **93:** 1-13.
82. HOLTZMANN, E. 1977. The origin and fate of secretory packages, especially synaptic vesicles. J. Neuroscience **2:** 327-355.
83. QUATAKER, J. 1981. The axonal reticulum in the neurons of the superior cervical ganglion of the rat as a direct extension of the Golgi apparatus. Histochem. J. **13:** 109-124.
84. MATHEKE, M. L. & E. HOLTZMANN. 1984. The effects of monensin and or puromycin on transport of membrane components in the frog retinal photoreceptor J. Neurosci. **4:** 1093-1103.
85. BROADWELL, R. D. & B. J. BALIN. 1985. Endocytic and exocytic pathways of the neuronal secretory process and transsynaptic transfer of wheat germ agglutinin-horseradish peroxidase in vivo. J. Comp. Neurol. **242:** 632-650.
86. SMITH, A. D. 1978. Biochemical studies of the mechanism of release. *In* The Release of Catecholamines from Adrenergic Neurons. D. M. Paton, Ed.: 1-15. Pergamon Press. Oxford and New York.
87. FLOOR, E. & S. E. LEEMAN. 1985. Evidence that large synaptic vesicles containing substance P and small synaptic vesicles have a surface antigen in common in rat. Neurosci. Lett. **60:** 231-237.
88. WAKADE, A. R. 1979. Recycling of noradrenergic storage vesicles of isolated rat vas deferens. Nature **281:** 374-376.
89. WAKADE, A. R. & T. D. WAKADE. 1982. Biochemical evidence for reuse of noradrenergic storage vesicles in the guinea-pig heart. J. Physiol. **327:** 337-362.
90. JAHN, R., W. SCHIEBLER & P. GREENBARD. 1984. A quantitative dot-immunobinding assay for proteins using nitrocellulose membrane filters. Proc. Natl. Acad. Sci. **81:** 1684-1687.

DISCUSSION OF THE PAPER

M. J. GEISOW (*National Institute for Medical Research, London, U.K.*): I was pleased to see for the first time your elegant demonstration of recycling of the chromaffin granule membrane. Could you tell us please whether you feel that this recycling is selective, or is it the entire membrane that has been recycled?

H. WINKLER (*University of Innsbruck, Innsbruck, Austria*): We cannot address the issue of proteins in the membrane other than dopamine β-hydroxylase and glycoprotein III, which are recycled as shown.

W. VAN DER KLOOT (*SUNY, Stony Brook, N.Y.*): Do you have an indication from your immunological studies whether exocytosis can occur at a site of previous exocytosis, and if so, whether such release is preferential?

H. WINKLER: Since no antibodies are available to label a possible "exocytosis site" in the plasma membrane, we cannot answer this question. However, we have no evidence to suggest that granules fuse on the same site in a consecutive fashion. Such a membrane should give large patches of granule membranes in the plasma membrane during exocytosis, which we have not observed (see Patzak, A. & H. Winkler. 1986. J. Cell Biol. **102**(2): 510-515.)

The Molecular Basis of Phenotypic Choices in the Sympathoadrenal Lineage

PAUL H. PATTERSON

Division of Biology
California Institute of Technology
Pasadena, California 91125

As the papers in this volume make quite clear, secretory vesicles can contain peptides, proteins, purines, and possibly other molecules in addition to classical transmitters. It is also now a commonplace observation that a single neuron can synthesize, store, and release several of these molecules. In fact, individual neurons can change the particular complement of transmitter and peptide they produce as they move through different stages of development. This phenotypic plasticity was originally observed when the environment of neurons growing in culture was altered, and more recently changes in transmitter phenotype have also been found to represent stages of normal development *in vivo.* Such plasticity has been observed both in vertebrates and in invertebrates. This article will briefly describe some of the molecular signals that control changes in transmitter and vesicle phenotype during development.

THE SIF-CHROMAFFIN-NEURON CHOICE

One of the most thoroughly studied systems in this regard is the interconversion of the various members of the sympathoadrenal lineage. Adrenal chromaffin cells are small cells with very large vesicles and few or no processes, whereas sympathetic neurons have large cell bodies, small synaptic vesicles, and very long processes. Chromaffin cells secrete norepinephrine or epinephrine as their transmitter or hormone, and most sympathetic neurons secrete norepinephrine. A minority population of sympathetic neurons, which innervate particular targets such as the eccrine sweat glands, secrete acetylcholine as their transmitter. A third derivative of the lineage, the small intensely fluorescent (SIF) cells, forms a minor population of chromaffin-like cells in sympathetic ganglia that secrete primarily dopamine and norepinephrine and contain vesicles which are much larger than those in neurons, but which are significantly smaller than those in chromaffin cells.

These various derivatives can be interconverted by manipulation of the culture environment in which they are grown. The conversions that have been observed are illustrated in FIGURE 1. Doupe *et al.*[1] found that neonatal rat superior cervical ganglia contain a small population of apparently undifferentiated cells that can give rise to the full range of sympathoadrenal derivatives. These precursor cells do not display any of the characteristic ultrastructural features of SIF or chromaffin cells or neurons, but in the presence of 10^{-8} M corticosteroid hormone they develop into biochemically and morphologically normal SIF cells, whereas chromaffin cells are produced in 10^{-6} M corticosteroid.[1]

The SIF cells derived from the precursor cells can also be converted into neurons by removal of corticosteroid and addition of nerve growth factor (NGF).[1] These neurons are indistinguishable from normal noradrenergic sympathetic neurons, as judged by light and electron microscopy, and by biochemical and immunological criteria. Furthermore, chromaffin cells taken from the rat adrenal medulla can also grow processes and assume a neuronal morphology when NGF is added.[2] This chromaffin-neuron conversion can be as complete as the SIF-neuron conversion if an additional, as yet uncharacterized, factor in heart cell conditioned medium is added along with the NGF.[3] The chromaffin-neuron conversion is drawn as proceeding through a SIF cell intermediate stage (FIG. 1), because at early times after the addition of NGF cells are observed to contain vesicles of both the chromaffin and SIF cell type.[3] Cells containing both SIF cell and neuronal-size vesicles are also found. Both the chromaffin-neuron and SIF-neuron conversions are inhibited by corticosteroids.[2,3] Unlike the SIF-chromaffin cell interconversion, the SIF-neuron conversion does not

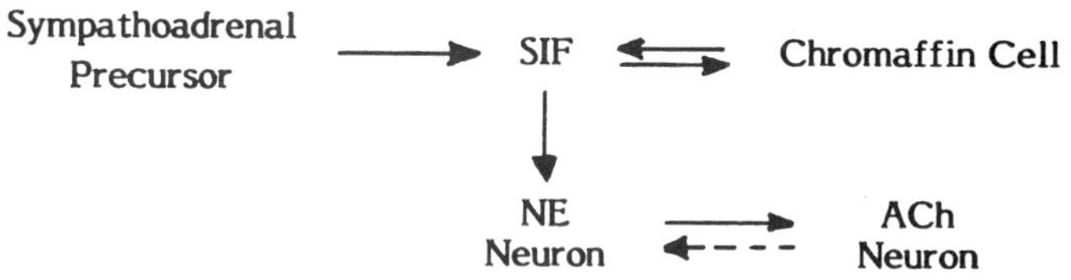

FIGURE 1

appear to be reversible; removal of NGF and addition of corticosteroid to sympathetic neurons results in neuronal death rather than a neuron-SIF conversion. Thus the decision to become a neuron appears to be irreversible, and once the SIF cells are exposed to NGF (in the absence of corticosteroid) they require it for survival.

These results show that corticosteroids and NGF can play both permissive and instructive roles in these phenotypic decisions. Corticosteroids are permissive in that they are required for the survival and differentiation of SIF and chromaffin cells. They are instructive in that they can influence the SIF-chromaffin cell and SIF-neuron decisions. Similarly, NGF is required for the survival of sympathetic neurons, and it can influence the SIF-neuron choice as well. It is interesting that survival factors are also used in differentiation choices.

Another feature of this lineage diagram (FIG. 1) is the central position of the SIF cells. These cells are derived from an undifferentiated precursor, and in turn can give rise to all of the other derivatives. Since the noradrenergic-cholinergic conversion was first observed in culture and subsequently was found also to occur as part of normal development *in vivo* (see below), we raised the possibility that the observed SIF cell interconversions may mean that these cells are the central intermediates of the lineage *in vivo.*[4]

THE CHOLINERGIC-NORADRENERGIC CHOICE

Noradrenergic neurons, obtained either from rat sympathetic ganglia or by conversion from SIF or chromaffin cells, can be converted to the cholinergic phenotype in culture.[1,3,5–7] This conversion has been followed in individual cells over time in microcultures, and intermediate, dual function neurons have been characterized.[8] Cholinergic properties can be induced by a glycoprotein secreted by various nonneuronal cells, including cultured heart cells.[9,10] This differentiation signal is of interest because it alters the type of synapse and transmitter made by the neurons without affecting their survival or growth.[9]

The protein has recently been purified to homogeneity; it has an apparent size of 45 kDa in both its native[11] and denatured[12] states. After deglycosylation, the core protein runs at 22 kDa on SDS gels.[12] Both the native and deglycosylated forms have the same 11 amino acid NH_2-terminal sequence (R. Aebersold, S. Kent, K. Fukada, L. Hood, and P. Patterson, unpublished), and antisera produced against a peptide of this sequence can specifically precipitate the native protein.[13] It will be of considerable interest to determine where and when the message for this protein is produced *in vivo,* and to use the antisera to determine the role of the protein in development. Since this molecule or a very similar one, can induce cholinergic properties in spinal cord and sensory neurons,[14,15] this differentiation signal could have widespread effects.

In addition to these classical transmitters, the peptide phenotype of cultured sympathetic neurons can also be experimentally influenced.[16] The neurons also synthesize, store, and release purines,[17,18] serotonin,[19] and an as yet uncharacterized transmitter.[20] Thus these neurons can utilize at least five different transmitters at synapses with cardiac myocytes, and it will be of interest to determine which of these are controlled by the "cholinergic" signal, and to identify the other differentiation cues involved.

The question is often raised as to whether this plasticity in choice of transmitter is simply a reflection of the developmental potential these neurons possess under artificial conditions. The answer appears to be negative because at least some of these phenotypic transitions are actually expressed during normal development. For instance, the subpopulation of sympathetic neurons that is cholinergic and innervates sweat glands in rat passes through a noradrenergic stage before becoming cholinergic.[21–24] In fact, the mature cholinergic neurons retain certain remnants of their developmental history; they still possess the high-affinity uptake system for catecholamines as well as low levels of tyrosine hydroxylase. These features of the converted cholinergic neurons were first observed in culture studies.[25–28] An *in vivo* counterpart for the presence of serotonergic transmission in cultured sympathetic neurons[19] has also recently been found; many postnatal sympathetic neurons transiently contain serotonin immunoreactivity.[29] Similar observations have been made in other areas of the nervous system. Substitution of one transmitter or peptide phenotype for another during normal development may also occur in mammalian enteric, sensory, and parasympathetic ganglia,[30–33] in neocortex[34] and retina[35] and in identified snail neurons.[36]

THE CHEMICAL-ELECTRICAL CHOICE

Another form of synaptic plasticity that can be manipulated in culture is the formation of low-resistance electrical junctions. When grown in normal, serum-containing medium, sympathetic neurons are rarely electronically coupled to one another. When grown in serum-free medium, however, they display a very high degree of coupling.[37] The formation of these junctions is not entirely due to the removal of serum, but rather coupling is promoted by the addition of the hormones contained in the serum-free medium.[38,39] Insulin is particularly effective in inducing coupling, even when it is added in the presence of serum. It is not clear if this type of electrotonic coupling represents a normal phase of sympathetic neuron development, as it does with many invertebrate neurons,[40] or if it is induced by unusual culture conditions. Detailed biochemical, electrophysiological, and ultrastructural examination of neurons grown in serum-free conditions has rarely been made.[41] Finally, it is of interest to point out that insulin is another example, like NGF and corticosteroids, of a survival factor that can also influence phenotypic choices. Thus it appears that there is no such thing as a "normal" neuronal phenotype in culture; phenotype is influenced by the presence or absence of a great many ingredients, some of which are required for survival and growth.

REFERENCES

1. Doupe, A. J., P. H. Patterson & S. C. Landis. 1985. Small intensely fluorescent cells in culture: Role of glucocorticoids and growth factors in their development and interconversions with other neural crest derivatives. J. Neurosci. **5:** 2143-2160.
2. Unsicker, K., B. Krisch, U. Otten & H. Thoenen. 1978. Nerve growth factor-induced fiber outgrowth from isolated rat adrenal chromaffin cells: Impairment by glucocorticoids. Proc. Natl. Acad. Sci. U.S.A. **75:** 3498-3502.
3. Doupe, A. J., S. C. Landis & P. H. Patterson. 1985. Environmental influences in the development of neural crest derivatives: Glucocorticoids, growth factors, and chromaffin cell plasticity. J. Neurosci. **5:** 2119-2142.
4. Landis, S. C. & P. H. Patterson. 1981. Neural crest cell lineages. Trends Neurosci. **4:** 172-175.
5. Ogawa, M., T. Ishikawa & A. Irimajiri. 1984. Adrenal chromaffin cells form functional cholinergic synapses in culture. Nature **307:** 66-68.
6. Bunge, R., M. Johnson & C. D. Ross. 1978. Nature and nurture in development of the autonomic neuron. Science **199:** 1409-1416.
7. Patterson, P. H. 1978. Environmental determination of autonomic neurotransmitter functions. Ann. Rev. Neurosci. **1:** 1-18.
8. Potter, D. D., S. C. Landis, S. G. Matsumoto & E. J. Furshpan. 1986. Synaptic functions in rat sympathetic neurons in microcultures. II. Adrenergic/cholinergic dual status and plasticity. J. Neurosci. **6:** 1080-1089.
9. Patterson, P. H., L. F. Reichardt & L. L. Y. Chun. 1975. Biochemical studies on the development of primary sympathetic neurons in cell culture. Cold Spring Harbor Symp. Quant. Biol. **40:** 389-397.
10. Patterson, P. H. & L. L. Y. Chun. 1977. Induction of acetylcholine synthesis in primary cultures of dissociated rat sympathetic neurons. I. Effects of conditioned medium. Dev. Biol. **56:** 263-280.
11. Weber, M. J. 1981. A diffusible factor responsible for the determination of cholinergic functions in cultured sympathetic neurons. J. Biol. Chem. **256:** 3447-3453.
12. Fukada, K. 1985. Purification and partial characterization of a cholinergic differentiation factor. Proc. Natl. Acad. Sci. U.S.A. **82:** 8795-8799.
13. Fukada, K. 1986. The production of antisera against a cholinergic differentiation factor. Soc. Neurosci. **12:** 106.5.
14. Giess, M. C. & M. J. Weber. 1984. Acetylcholine metabolism in rat spinal cord cultures: Regulation by a factor involved in the determination of the neurotransmitter phenotype of sympathetic neurons. J. Neurosci. **4:** 1442-1452.

15. MATHIEU, C., A. MOISAND & M. J. WEBER. 1984. Acetylcholine metabolism by cultured neurons from rat nodose ganglia: Regulation by a macromolecule from muscle-conditioned medium. Neuroscience **13:** 1373-1386.
16. KESSLER, J. A. 1985. Differential regulation of peptide and catecholamine characters in cultured sympathetic neurons. Neuroscience **15:** 827-839.
17. WOLINSKY, E. J. & P. H. PATTERSON. 1985. Potassium-stimulated purine release by cultured sympathetic neurons. J. Neurosci. **5:** 1680-1687.
18. FURSHPAN, E. J., D. D. POTTER & S. G. MATSUMOTO. 1986. Synaptic functions in rat sympathetic neurons in microcultures. III. A purinergic effect on cardiac myocytes. J. Neurosci. **6:** 1099-1107.
19. SAH, D. W. Y. & S. G. MATSUMOTO. Evidence for serotonin synthesis, uptake and release in some dissociated rat sympathetic neurons in culture. Submitted.
20. POTTER, D. D., S. G. MATSUMOTO, S. C. LANDIS, D. W. Y. SAH & E. J. FURSHPAN. 1986. Transmitter status in cultured sympathetic principal neurons: Plasticity, graded expression and diversity. Prog. Brain Res. **68:** 103-119.
21. LANDIS, S. C. & D. KEEFE. 1983. Evidence for neurotransmitter plasticity *in vivo:* Developmental changes in properties of cholinergic sympathetic neurons. Dev. Biol. **98:** 349-372.
22. YODLOWSKI, M. L., J. R. FREDIEU & S. C. LANDIS. 1984. Neonatal 6-hydroxydopamine treatment eliminates cholinergic sympathetic innervation and induces sensory sprouting in rat sweat glands. J. Neurosci. **4:** 1535-1548.
23. LANDIS, S. C., J. R. FREDIEU & M. YODLOWSKI. 1985. Neonatal treatment with nerve factor antiserum eliminates cholinergic sympathetic innervation of rat sweat glands. Dev. Biol. **112:** 222-229.
24. LEBLANC, G. & S. LANDIS. 1986. Development of choline acetyltransferase (CAT) in the sympathetic innervation of rat sweat glands. J. Neurosci. **6:** 260-265.
25. REICHARDT, L. F. & P. H. PATTERSON. 1977. Neurotransmitter synthesis and uptake by isolated sympathetic neurones in microcultures. Nature **270:** 147-151.
26. LANDIS, S. C. 1976. Rat sympathetic neurons and cardiac myocytes developing in microcultures: Correlation of the fine structure of endings with neurotransmitter function in single neurons. Proc. Natl. Acad. Sci. U.S.A. **73:** 4220-4224.
27. IACOVITTI, L., T. H. JOH, D. H. PARK & R. P. BUNGE. 1981. Dual expression of neurotransmitter synthesis in cultured autonomic neurons. J. Neurosci. **1:** 685-690.
28. WOLINSKY, E. J. & P. H. PATTERSON. 1983. Tyrosine hydroxylase activity decreases with induction of cholinergic properties in cultured sympathetic neurons. J. Neurosci. **3:** 1495-1500.
29. HAPPOLA, O., H. PAIVARINTA, S. SOINILA & H. STEINBUSCH. 1986. Pre- and postnatal development of 5-hydroxytryptamine-immunoreactive cells in the superior cervical ganglion of the rat. J. Aut. Nerv. Sys. **15:** 21-31.
30. TEITELMAN, G., T. H. JOH & D. J. REIS. 1978. Transient expression of a noradrenergic phenotype in cells of the rat embryonic gut. Brain Res. **158:** 229-234.
31. COCHARD, P., M. GOLDSTEIN & I. B. BLACK. 1978. Ontogenic appearance and disappearance of tyrosine hydroxylase and catecholamines in the rat embryo. Proc. Natl. Acad. Sci. U.S.A. **75:** 2986-2990.
32. PRICE, J. & A. W. MUDGE. 1983. A subpopulation of rat dorsal root ganglion neurons is catecholaminergic. Nature **301:** 241-243.
33. JONAKAIT, G. M., K. A. MARKEY, M. GOLDSTEIN & I. B. BLACK. 1984. Transient expression of selected catecholaminergic traits in cranial sensory and dorsal root ganglia of the embryonic rat. Devel. Biol. **101:** 51-60.
34. BERGER, B., C. VERNEY, P. GASPAR & A. FEBVRET. 1985. Transient expression of tyrosine hydroxylase immunoreactivity in some neurons of the rat neocortex during postnatal development. Dev. Brain Res. **23:** 141-144.
35. SCHNITZER, J. & A. C. RUSOFF. 1984. Horizontal cells of the mouse retinal contain glutamic acid decarboxylase-like immunoreactivity during early developmental stages. J. Neurosci. **4:** 2948-2955.
36. GESSER, B. P. & L. I. LARSSON. 1985. Changes from enkephalin-like to gastrin/cholecystokinin-like immunoreactivity in snail neurons. J. Neurosci. **5:** 1412-1417.

37. HIGGINS, D. & H. BURTON. 1982. Electrotonic synapses are formed by fetal rat sympathetic neurons maintained in a chemically defined medium. Neuroscience **7:** 2241-2253.
38. KESSLER, J. A., D. C. SPRAY, J. C. SAEZ & M. V. L. BENNETT. 1984. Determination of synaptic phenotype: Insulin and cAMP independently initiate development of electrotonic coupling between cultured sympathetic neurons. Proc. Natl. Acad. Sci. U.S.A. **81:** 6235-6239.
39. WOLINSKY, E. J., P. H. PATTERSON & A. L. WILLARD. 1985. Insulin promotes electrical coupling between cultured symapathetic neurons. J. Neurosci. **5:** 1675-1679.
40. GOODMAN, C. S. & N. C. SPITZER. 1979. Embryonic development of identified neurones: Differentiation from neuroblast to neurone. Nature **280:** 208-214.
41. WOLINSKY, E. J., S. C. LANDIS & P. H. PATTERSON. 1985. Expression of noradrenergic and cholinergic traits by sympathetic neurons cultured without serum. J. Neurosci. **5:** 1497-1508.

DISCUSSION OF THE PAPER

P. DE CAMILLI (*University of Milan, Milan, Italy*): I would like to make a comment about the possible physiologic function of the protein produced by the heart which induces the conversion of catecholaminergic neurons into cholinergic neurons. Might the function of this protein be synergistic with that of atrial naturatic factor? This factor produces vasodilation, an effect antagonistic to that produced by stimulation of the sympathetic system. The physiologic function of your heart factor could be to suppress (antagonize) the catecholaminergic system. Along this line, is there any homology between the known primary sequence of your factor and that of atrial naturatic factor (ANF)?

P. H. PATTERSON (*California Institute of Technology, Pasadena, Calif.*): Our NH_2-terminal sequence was compared to that of other known sequences and no homologies were found. The relationship of the cholinergic differentiation factor to ANF is unknown.

O. H. VIVEROS (*Wellcome Research Laboratories, Research Triangle Park, N.C.*): My first question is: Could you please identify the species and the age of the animals used for the culture of chromaffin cells under conditions where phenotype can be transformed by NGF in light of the fact that bovine adult chromaffin cells respond very poorly by extending processes when challenged by NGF?

My second question is the following: Chromaffin cells of different mammalian species synthesize, store and secrete a bewildering number of different neuropeptides. Though the adrenal medulla is expected to have the same or very similar function in the different species, the particular neuropeptides and their ratios to catecholamine are also widely different in different species. Would you care to comment on the functional significance and the signals that may be involved in determining this phenotypic variability in otherwise functionally, morphologically and ontogenically identical cells?

P. H. PATTERSON: The chromaffin cell to neuron conversion results were from two-week-old and adult rat adrenal medulla. It has been reported that adult bovine chromaffin cells are not converted to neurons. This and the many other striking species differences among members of this lineage are indeed puzzling. Some of the differences, such as the number of SIF cells in sympathetic ganglia, could be explained by the amount of phenotype influencing cues, such as glucocorticoid concentration. Other differences such as the bovine-rat problem just mentioned are unexplained.

J.-P. HENRY (*Institut de Biologie Physico-Chimique, Paris, France*): In the case of secretory cells with dual phenotypes, do you have secretory vesicles with hybrid morphology or functions?

P. H. PATTERSON: There is very good evidence for dual function neurons, some evidence for dual function varicosites (both dense core and clear vesicles present), and little evidence regarding dual function vesicles, except that ATP is found with classical transmitters. Perhaps Dr. Pietro De Camilli will have more to say about this issue later in the meeting.

L. M. ROMAN (*Howard Hughes Medical Institute, Dallas, Tex.*): In polarized epithelial cells characterized by two distinct plasma membrane domains, clear sorting events to the apical and basolateral domains exist. By use of viral membrane proteins and MDCK cells as a test system it also appeared that there seems to be a pH-dependent sorting event. Incubation of the cells with ammonium chloride or chloroquine also appears to disrupt sorting to the apical and basolateral surface-apical proteins.

Given the discussion of the plasticity of the neuroendocrine cells and the requirement of specific cell machinery for packaging into granules, has anyone looked by differential-subtractive hybridization into whether a battery of proteins is turned on with the change of phenotype; and if so, are there classes of "early" proteins which may represent receptors, or components of a granule packaging system?

P. H. PATTERSON: Dr. Kathy Sweadner found that the cholinergic and noradrenergic cultured synaptic neurons display different profiles of surface and secretory glycoproteins. Dr. Randy Pittman extended this comparison to several types of sensory neurons. Dr. Anne Zurn also demonstrated a difference in glycolipid composition between cholinergic and noradrenergic neurons. Dr. David Anderson recently isolated cDNA cones which are enriched in sympathetic neurons, when compared to adrenal chromaffin cells, and he has used these as lineage markers in developmental studies (*Cell,* Dec. 1986). In none of these cases is the function of the molecule of interest known.

M. LEVINE (*NIDDK, National Institutes of Health, Bethesda, Md.*): Agents such as glucocorticoids, nerve growth factor, and insulin seem to regulate phenotype of neuronal cells individually *in vivo. In vivo,* these cells presumably are exposed to several regulatory agents at once. How would regulation of development, therefore, proceed *in vivo*?

P. H. PATTERSON: There are several possibilities here. Dr. Story Landis and co-workers have obtained evidence that in the case of the cholinergic sympathetics, it is the sweat gland target tissue that induces the cholinergic phenotype. Thus, if there is a soluble signal, here it may be available only to those axons which innervate this target. In addition, Dr. Ed Hawrot finds a cholinergic signal on the surface of heart cells and Dr. Jack Kessler finds a cholinergic signal on the surface of neurons. Thus, the soluble molecule in cholinergic maturation may be a released form of a membrane bound protein. Finally, the action of the soluble signals is subject to activity. Dr. Pat Walicke and I showed that electrical stimulation of sympathetic nervous system makes it much less responsive to the cholinergic signal. Thus, there can be a one-to-one basis in their response to soluble proteins.

The Chromaffin Granule: A Model System for the Study of Hormones and Neurotransmitters[a]

JOHN H. PHILLIPS AND JAMES G. PRYDE

Department of Biochemistry
University of Edinburgh Medical School
Edinburgh, EH8 9XD, United Kingdom

During the past decade, there has been a dramatic increase in our understanding of processes involving membrane movements within cells. In particular, the use of enveloped viruses with mammalian cells has revolutionized the study of membrane glycoproteins, revealing the complex interrelationships of a series of intracellular compartments. These viruses ingeniously exploit endocytotic and exocytotic pathways, routes for membrane shuttling between points of loading and discharge that exist in virtually all cells. In spite of intensive study, however, many fundamental questions of mechanism remain to be solved, and in this commentary we review the use of the bovine adrenal medulla as a model tissue in which to study them.

The adrenal medulla is a true neuroendocrine tissue, derived from the neural crest; it contains cells that are packed with secretory granules, the sites of storage of the catecholamine hormones epinephrine and norepinephrine. Exploited by physiologists and pharmacologists for many years, its potential for biochemical approaches to the secretory response was strongly emphasized by Douglas,[1] who, in a series of classic studies, used the perfused adrenal gland in the formulation of the concept of "stimulus-secretion coupling," a process dependent on influx of extracellular Ca^{2+}, and showing analogies to stimulus-contraction coupling in skeletal muscle. This led to the recognition of the adrenal medulla as an ideal tissue for biochemical studies of the granules themselves. The early analytical studies of Hillarp[2] were followed by studies of granule transport properties[3] and purification[4] that formed the foundation for the mass of detailed biochemical information at our disposal today. However, the dynamic relationship between granules and other cell constituents could not be properly studied without one further tool, bringing the adrenal gland fully into the field of cell biology. This was the development of a method for establishing primary cultures of bovine chromaffin cells[5] and the concomitant establishment of the PC-12 cell line from a rat pheochromocytoma.[6]

Each cell in the adrenal medulla contains 10,000-30,000 chromaffin granules.[7-9] The granules are spherical, membrane-bounded organelles with an isotropic matrix containing high concentrations of protein, catecholamine, and nucleotides.[10,11] They are far from uniform in size.[12] Although the mean diameter of bovine granules is about 280 nm, a substantial proportion has a diameter greater than 400 nm; corrected diameters of mixed epinephrine and norepinephrine granules isolated from bovine

[a] This work was supported by grants from the Medical Research Council.

adrenal medulla are given by Coupland[12] as 280 ± 135 nm (n = 1383). Some rough calculations based on existing morphometric data are given in TABLE 1. These figures emphasize the high proportion of cellular membrane that is located in the secretory granules and partly explain why the adrenal medulla is such an attractive tissue for membrane biochemists. The chromaffin granules have high density and are easily purified on density gradients, after which hypotonic lysis frees the membranes from the soluble matrix components.

CHROMAFFIN GRANULE MEMBRANES AS SHUTTLES FROM THE GOLGI TO THE CELL SURFACE

A two-dimensional electrophoretic separation of the proteins of a preparation of purified bovine chromaffin cells in culture is dominated by Coomassie-blue-stained spots corresponding to the major matrix proteins of the secretory granule (FIG. 1). This is not surprising in view of the high volume of the cytoplasm that is occupied by granules (TABLE 1) and the concentration of protein (about 200 mg/ml) that is found within them. Labeling the cells with ^{35}S-methionine reveals the synthesis of a number of major polypeptides, in particular the chromogranins (FIG. 2). The precursor forms of these secreted proteins are seen to be labeled after 10 min,[13] but after 60 min of labeling, glycosylation and sulfation in the Golgi apparatus have converted these into characteristically heterogeneous species.

In contrast to this, there is little synthesis of components of the membranes of the granules. About 40% of granule membrane protein is found in two major constituents, dopamine β-hydroxylase (E.C. 1.14 17.1) and cytochrome b_{561}[11], neither of which can be seen in FIGURE 2, or in the products of translation of mRNA *in vitro.*[13,14] Although mRNA for dopamine β-hydroxylase is clearly present,[15,16] most studies on the biosynthesis of this enzyme have utilized the rapidly dividing rat PC-12 cells.[17,18] The low rate of synthesis of membrane proteins is consistent with the idea that membrane proteins are extensively reutilized after exocytosis,[19] and that we must think of the membranes as shuttles, returning from the cell surface to the perinuclear region of the cell, to be refilled there with secretory proteins that have undergone maturation and sorting in the Golgi apparatus (FIG. 3).

TABLE 1. Morphometric Analysis of Bovine Chromaffin Cells

The Adrenal Chromaffin Cell	
Diameter of a chromaffin cell	16 μm
Surface area of a spherical cell	800 μm²
The Adrenal Chromaffin Granule	
Diameter of a chromaffin granule	0.28 μm
Surface area of a granule	0.25 μm²
Number of granules in a cell	30,000
Volume of cell occupied by granules	16%
Total area of granule membrane in one cell	7,400 μm²

Data for bovine cells have been collected from several sources; assumptions made are discussed elsewhere.[7] A recent electron microscope study of the rat adrenal gland[9] yields the following values: epinephrine cells, 23,000 granules/cell, volume fraction 10%; norepinephrine cells, 33,000 granules/cell, volume fraction 14%.

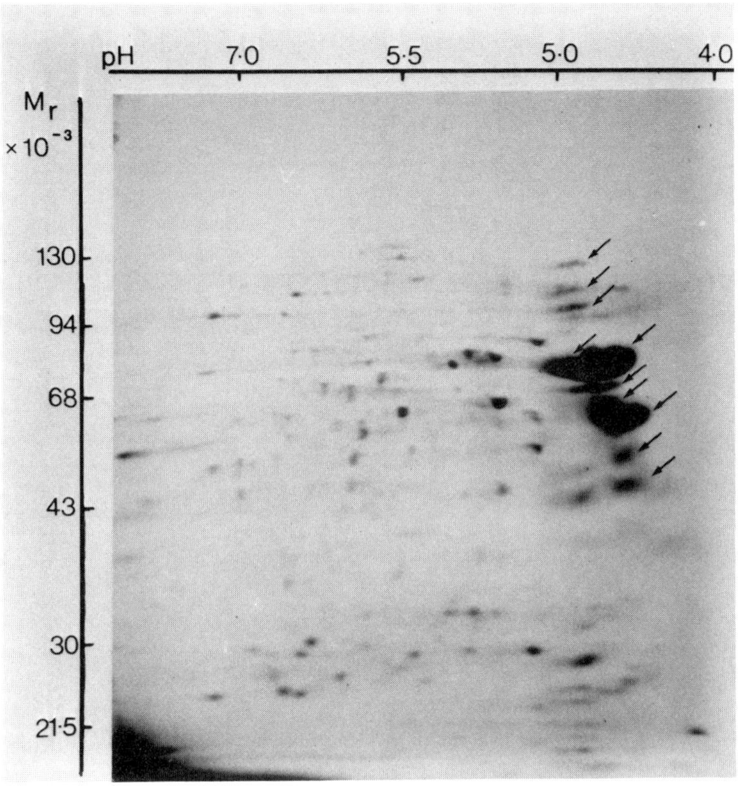

FIGURE 1. Two-dimensional gel electrophoresis of proteins of isolated bovine adrenal chromaffin cells. The gel was stained with Coomassie blue. Proteins of the chromaffin granule matrix (chromogranins) are *arrowed.*

The extent and rate of this shuttling have been studied in two ways. In studies in whole animals, adrenal glands can be stimulated to release massive amounts of catecholamine; this may be induced by injecting insulin into a fasted animal, when 80% of adrenal catecholamine is released within a few hours.[20,21] The recovery phase may be investigated by biochemical analysis or by electron microscopy.[22,23] The former is complicated by the fact that catecholamine biosynthesis appears to be rate limiting for the assembly of mature chromaffin granules,[8] but the morphological studies show the appearance of electron-translucent vesicles of a size not dissimilar to that of granules; these are distributed throughout the cytoplasm within a few hours of stimulation. The number of dense-cored chromaffin granules reaches a minimum about 12 hours after insulin injection, but then slowly increases until control values are reached again after 4 days,[23] the time by which cellular catecholamine content can be shown to be restored.[20]

The appearance of large translucent vesicles is preceded by formation of translucent microvesicles[23] which are probably themselves formed by a process of protein (clathrin?) coating of granule membranes after their insertion into the plasmalemma, followed by internalization as coated vesicles.[24,25] This process would appear to resemble

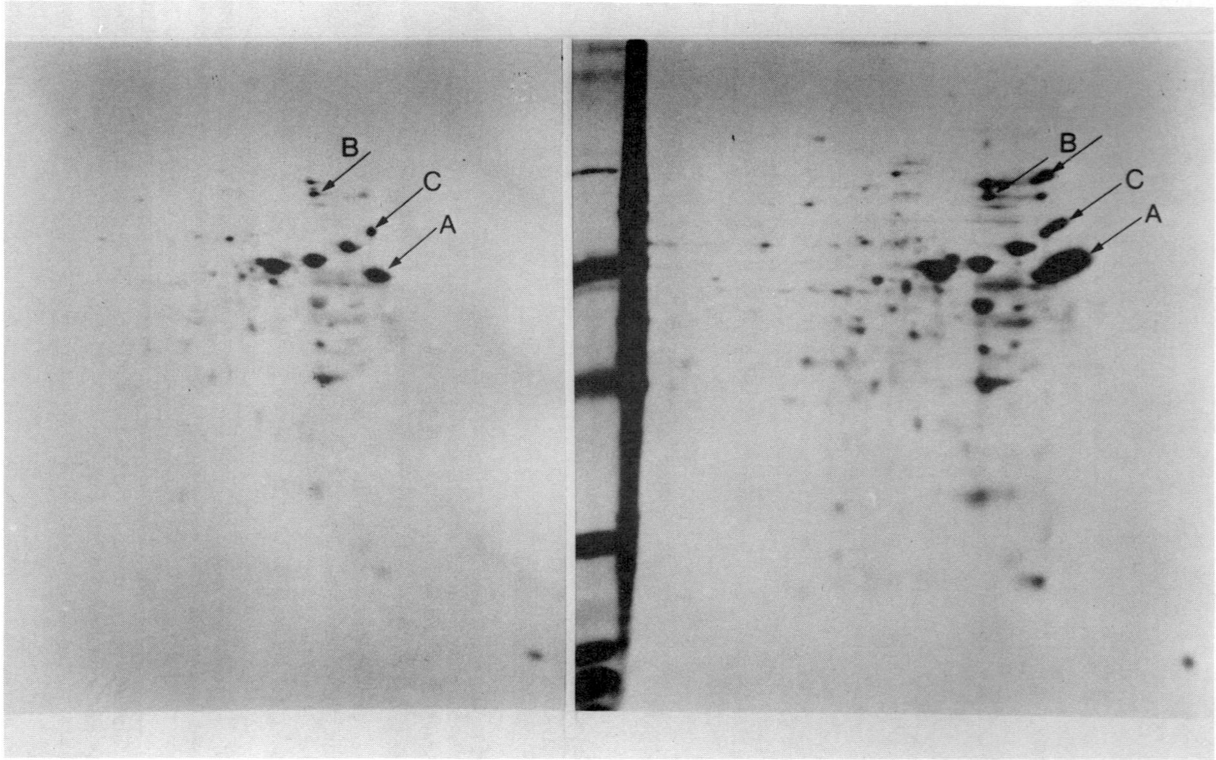

FIGURE 2. Autoradiograph of two-dimensional gel electrophoresis of proteins of isolated bovine adrenal chromaffin cells. Whole cells were incubated with ^{35}S-methionine for 10 min (*left*) or 60 min (*right*). Chromogranins A, B, and C and their putative precursors are labeled.

plasma membrane internalization during receptor-mediated endocytosis, and indeed, cell-surface concanavalin A receptors are internalized at a roughly similar rate.[26]

These studies are complemented by experiments with isolated cells, which release 10-40% of their content of hormone on stimulation with a variety of secretagogues. Below the surfaces of stimulated cells are seen empty vacuoles that often appear rather larger than typical chromaffin granules.[5,27,28] Antigenic sites on the inner surface of the granule membrane are exposed on the cell surface during exocytosis and can be labeled with antibodies. Surprisingly, the secretory granule membrane remains intact

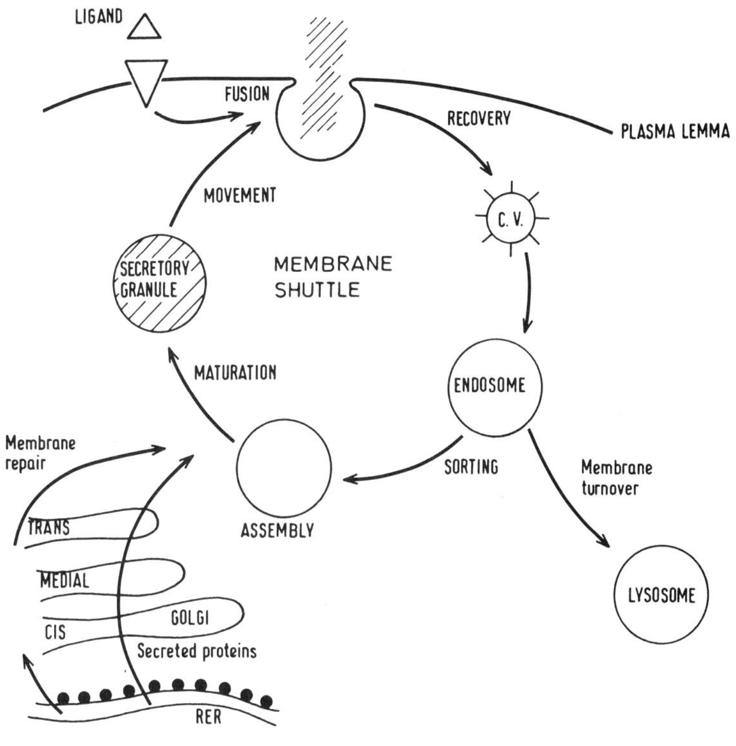

FIGURE 3. The exocytosis-endocytosis membrane shuttle; *c.v.*, coated vesicle.

on the cell surface while it is awaiting its turn for internalization. It can be revealed by fluorescent antibodies (FIG. 4), showing that its antigens do not diffuse at random. Following insertion of the membranes of 10-20% of the granules in a cell, recovery of membrane antigens back into the interior of the cell takes about 30-60 min (FIG. 4).[26,29]

These studies suggest that a chromaffin cell can internalize an area of granule membrane that is roughly equivalent in size to its own surface area within 30-60 min. This rate of removal of material from the surface is not dissimilar to the rate of turnover of plasma membrane estimated for macrophages by Z. A. Cohn and his colleagues,[30] a process calculated to be equivalent to the endocytotic removal of 3.1% of the plasma membrane per minute.

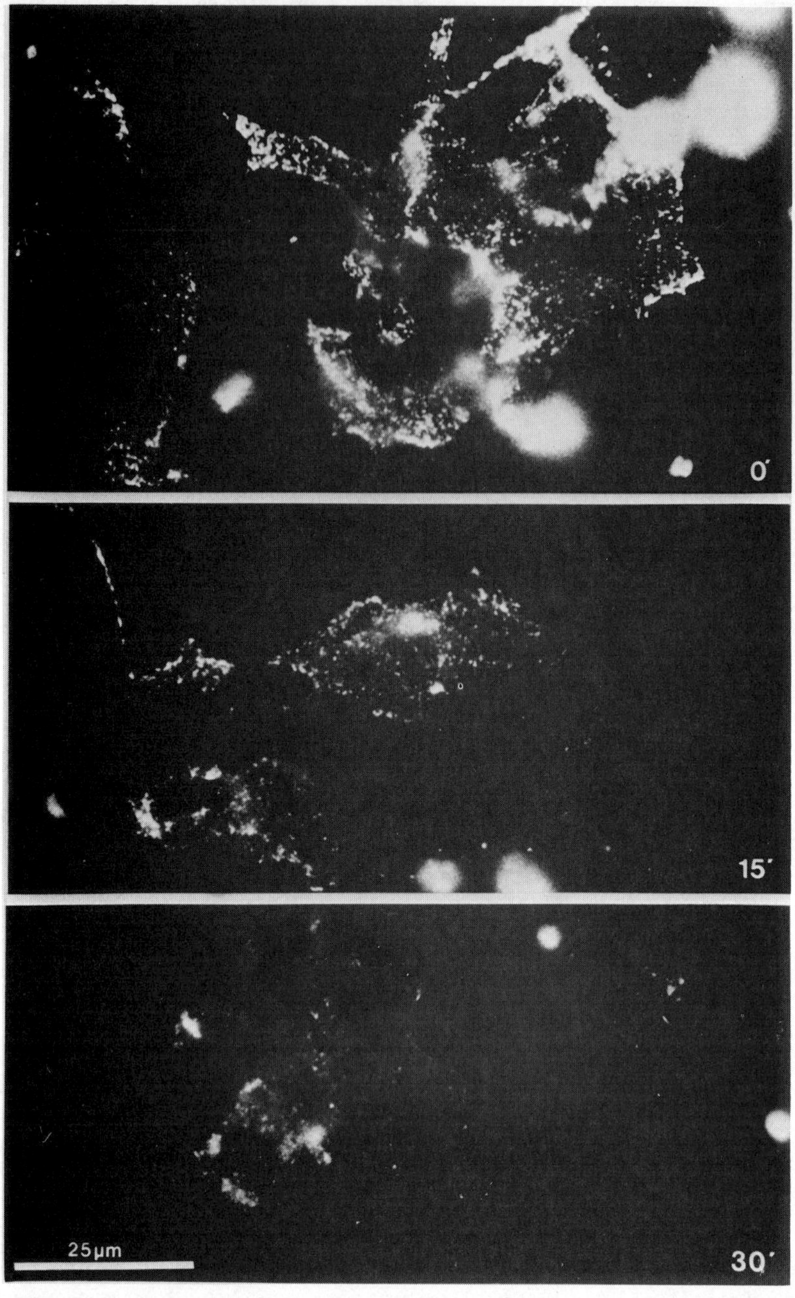

FIGURE 4. Internalization of chromaffin granule membranes. Bovine chromaffin cells were treated with 100 μM veratridine for 5 min at 37°C, then with medium containing 100 μM tetrodotoxin for the times shown. Anti-dopamine β-hydroxylase serum was then added for 5 min. This was revealed using fluorescein-labeled second antibody.

It has been argued that massive stimulation bears little relationship to the process of hormone release *in vivo,* and that the mechanisms involved may not in fact be the same. The studies with whole animals (in which granule recovery is complete within a few days of the insulin treatment) do show, however, that the cells have the potential *in vivo* for this extensive membrane recycling process.

There seems little reason to doubt that physiological release occurs from adrenal glands by exocytosis. The resting output of perfused intact glands has frequently been measured. If one takes typical values from the literature[31,32] it works out that about 0.05% of the adrenal medullary store of catecholamine is released from the gland per minute in the absence of stimulation. This is equivalent to exocytosis-endocytosis cycling of an area of membrane equivalent in size to the whole plasma membrane every 200 min. This release occurs primarily from one face of the cell, at which fairly extensive membrane sorting and recovery must therefore be occurring, even in the resting gland.

PROPERTIES OF THE CHROMAFFIN GRANULE MEMBRANE

How a secretory granule membrane receives its own specific secretory proteins from the Golgi apparatus, how it acquires its complement of small molecules by

TABLE 2. Catalytic and Structural Proteins of Adrenal Chromaffin Granule Membranes

Protein	Other Tissues in Which Protein Has Been Identified
Identified Polypeptides	
dopamine β-hydroxylase	
cytochrome b_{561}	posterior pituitary
H^+-ATPase	many neuroendocrine tissues
α-actinin	
Polypeptides Not Yet Identified	
catecholamine transporter	platelets, sympathetic neurons
nucleotide transporter	electroplax, sympathetic neurons
Ca^{2+}/Na^+ transporter	posterior pituitary
enkephalin convertase	
phosphatidylinositol kinase	pancreatic β-cells
ATPase II (unknown function)	posterior pituitary

specific transport processes, how it is translocated to the cell surface and then fuses in response to the appropriate biological signal, and how it is then specifically internalized, emptied of extracellular contaminants, and reutilized are fundamental problems. But it seems axiomatic that at least some of the solutions must be sought in a detailed understanding of the biochemistry of the granule membrane, particularly its protein components. To date, disappointingly few proteins of known function and composition have been identified (TABLE 2), but this merely reflects the fact that in general we know little about what sorts of enzyme to look for in trying to elucidate the above mechanisms.

A two-dimensional separation of chromaffin granule membrane proteins is shown in FIGURE 5. Many components stain poorly with Coomassie blue, but being glyco-

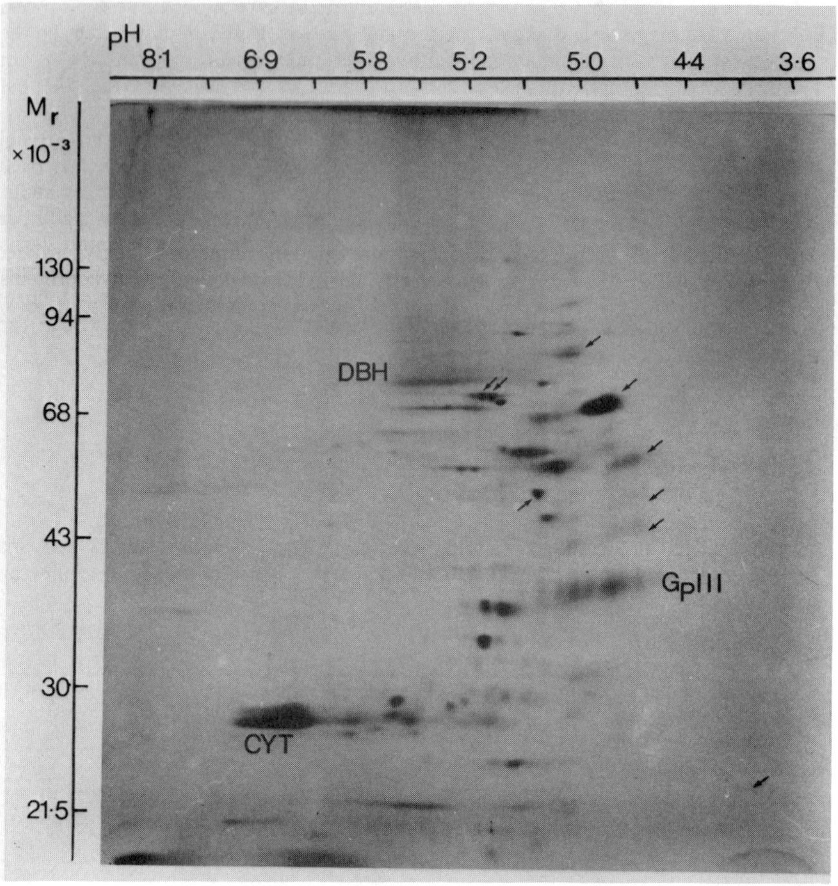

FIGURE 5. Two-dimensional gel electrophoresis of proteins of purified chromaffin granule membranes, stained with Coomassie blue. *Arrows* indicate adherent nonmembrane proteins. *DBH,* dopamine β-hydroxylase; *GpIII,* glycoprotein III[33]; *CYT,* cytochrome b_{561}.

sylated, can be revealed by concanavalin A (FIG. 6) or other lectins.[33] Neither technique gives one a quantitative estimate of the abundance of a constituent in the membrane. This type of experiment is complicated by the fact that many proteins (presumably from the cytosol and elsewhere in the tissue) adhere to chromaffin granules during their isolation. In fact, this is an important property that can be exploited in order to search for cytosolic proteins that have specific functions in the processes of the exocytosis-endocytosis cycle. In this way, several groups have identified families of cytosolic proteins that show Ca^{2+}-dependent binding to granule membranes.[34–36] We have used Na_2CO_3 washing[37,38] as a technique for removing adherent nonintrinsic proteins (FIG. 5), among which are traces of chromogranins, the major matrix components of the granules.

In order to simplify the mixture of proteins found in solubilized membranes we have used the Triton X-114 phase separation technique of Bordier.[39] Chromaffin granule membranes yield three phases after adding this detergent: a phospholipid-rich phase separates spontaneously, and a detergent-rich phase separates from the aqueous phase on warming to 30°C.[40] FIGURE 7, showing separations of the proteins and glycoproteins in these phases, demonstrates that most components have a characteristic distribution which can be exploited in subsequent purification steps.

The most hydrophobic phase is highly enriched in phospholipid and cholesterol and contains virtually all of ATPase I, the H^+-translocating ATPase (E.C.3.6.1.3) of the membrane,[41] as well as a few other polypeptides. In addition, it contains a characteristic group of glycoproteins, revealed in FIGURE 7b by decoration with lentil lectin (from *Lens culinaris*). The detergent-rich phase, which Bordier[39] suggested contained the intrinsic proteins of the membrane, contains numerous nonglycosylated polypeptides, including all the cytochrome (FIG. 7). It contains few glycoproteins, lentil lectin revealing only dopamine β-hydroxylase and an indistinct component above it known as glycoprotein II. The aqueous phase contains not only rather hydrophilic or amphiphilic peripheral proteins (such as adherent matrix components), but also many glycoproteins which, although intrinsic constituents, are maintained in solution by adherent detergent.[40] An example is glycoprotein III (M_r 38,000), the main component that is demonstrated by lentil lectin binding in FIGURE 7.

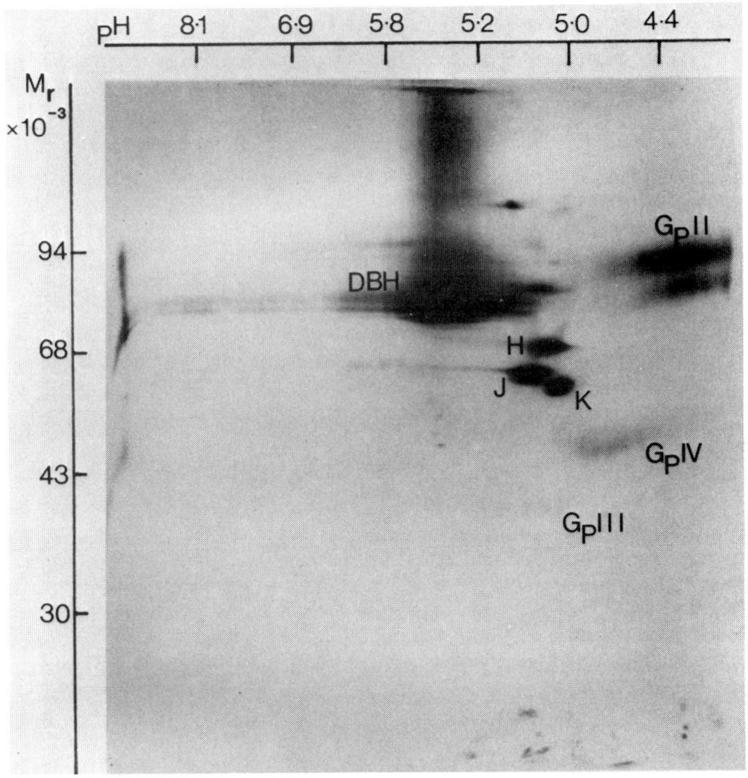

FIGURE 6. Autoradiograph of two-dimensional gel electrophoresis of proteins of purified chromaffin granule membranes decorated with ^{125}I-labeled concanavalin A. *DBH,* dopamine β-hydroxylase. Other glycoproteins are labeled according to Gavine *et al.*[33]

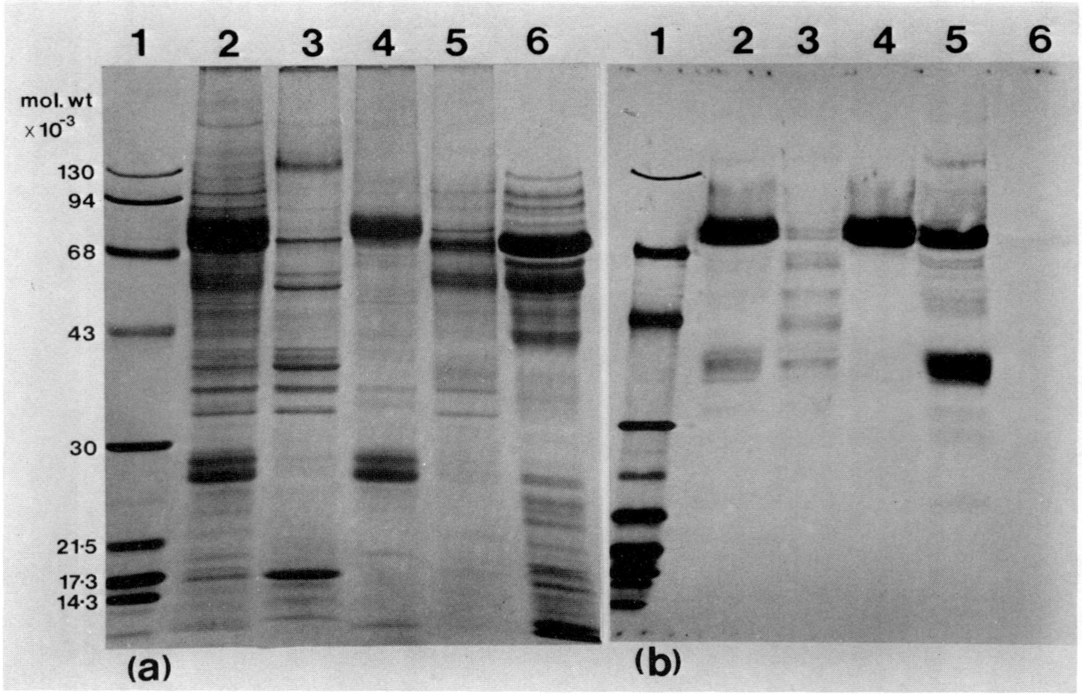

FIGURE 7. Fractionation of chromaffin granule membrane proteins using Triton X-114. (**a**) Gel stained with Coomassie blue. Tracks contain (1) standard proteins; (2) whole membranes; (3) phospholipid-rich phase; (4) detergent-rich phase; (5) aqueous phase; and (6) a lysate of chromaffin granules, containing the matrix proteins. (**b**) Autoradiograph of similar gel to **a**, but run with radioactive standard proteins, and decorated with ^{125}I-labeled *Lens culinaris* lectin.

Examination of two-dimensional gels suggests that chromaffin granule membranes probably contain about forty intrinsic polypeptides, about half of which are glycosylated. Gels are dominated by the glycoprotein dopamine β-hydroxylase (M_r 75,000 when reduced) and by the nonglycosylated cytochrome (M_r 26,000). These are considerably more abundant than other proteins and it is interesting to note that it is these two that have a specialized role in norepinephrine biosynthesis.[42,43] The remaining constituents are present in an abundance equivalent to perhaps 0.5-5% of the total membrane protein.

To put this into perspective, a simple calculation shows that fifty copies of a polypeptide of molecular weight 50,000 in one granule would represent 1% of the membrane protein. If a protein stains normally with Coomassie blue, this should be clearly visible on a gel. The membranes are therefore unlikely to be found to contain a bewildering variety of additional components, and it is thus a reasonable objective to try to purify or to raise antibodies against several of these constituents of unknown function in order to probe their role and distribution among intracellular compartments.[44]

Of course, not all polypeptides stain well with Coomassie blue or bind the most commonly used lectins. An important approach to identifying proteins that participate in general functions of the exocytosis-endocyosis cycle is to search for antigens, identifiable using monoclonal antibodies, that are found in secretory granule membranes from many tissues. Several have been reported recently: the antigens seem to be common to many neuroendocrine granules, but to be absent from exocrine granules. We have investigated three: SV-2, a transmembrane glycoprotein of M_r 105,000-110,000 originally identified in electroplax synaptic vesicles[45] (antibody kindly provided by Dr. R. G. Kelly); p65, an integral protein of M_r 65,000 from rat brain vesicles[46] (antibody kindly provided by Dr. J. L. Bixby); and synaptophysin, an integral glycoprotein of brain coated vesicles with M_r 38,000[47] (antibody kindly provided by Dr. B. Wiedenmann). All three are found in preparations of chromaffin granule membranes, although their presence in the granules *in vivo* has not yet been demonstrated unequivocally. All three turn out to be minor constituents, not identifiable with Coomassie-blue-stained bands or with any of the major lectin-binding proteins (except p65, which probably binds wheat germ agglutinin).

These, and proteins like them, would be attractive subjects for detailed biochemical investigation. That all are minor constituents suggests that, for further investigation and purification, a readily available source (such as chromaffin granules) is an absolute necessity.

SECRETED PROTEINS OF THE ADRENAL MEDULLA

It is among the secretory proteins that the most dramatic results have emerged recently. The genes for several chromaffin granule matrix polypeptides appear to be expressed in a wide variety of tissues, only the abundance of their products varying widely from one tissue to another. The most significant finding, however, is that all are confined to neural or endocrine tissues, and none have been detected in exocrine secretions (TABLE 3).

Chromogranin A, at least 50% of the protein of the bovine chromaffin granule matrix, has been shown to be identical to, or very closely similar to, parathyroid secretory protein I[48] and this finding has been followed by its detection by immunological methods in many other neuroendocrine tissues.[49–51] It has a distinctive distribution in the brain.[52] Chromogranin B, another acidic glycoprotein which has also been called secretogranin I,[53] is a minor component of the bovine granule, but a major component in other species, and is also widespread.[51,53] Both proteins exist in chromaffin granules as high molecular weight species together with smaller amounts of proteolytic degradation products, but in other tissues it seems likely that hydrolysis may be far more extensive,[54,55] unless this is an artifact of isolation. Chromogranin C, also called secretogranin II, is a third glycoprotein, not particularly abundant in chromaffin granules, but a major secreted protein in the bovine anterior pituitary.[53]

As yet the functions of the chromogranins are unknown. Can they possibly have specialized roles within such a variety of secretory granules, or is their function to be found outside the cell, exerted after secretion? It is possible that a solution to this outstanding problem may come from a knowledge of their sequences and proteolytic cleavage points.

The chromaffin granule has proved to be a gold mine for hunters of neuropeptides since the first discovery of enkephalins in the tissue a few years ago[56] and the demonstration that these are stored in chromaffin granules.[57] Proteolytic derivatives of proenkephalin are found in epinephrine-containing cells and derivatives of prodynorphin in norepinephrine cells.[58] Not only are these peptides abundant, the extent of precursor processing responds to metabolic conditions in the adrenal cell.[59,60] There are clearly several proteases within chromaffin granules, responsible for liberating active neuropeptides as well as for processing the chromogranins and secretogranins. Processing of these proteins is slow, occuring over several days[61]; indeed, the bulk of the enkaphalins are secreted from the cell as components of molecular weight between 3,000 and 22,000, rather than as penta-, hexa- or heptapeptides.[62] Chromogranin A is mainly secreted as the component with M_r 70,000 which is formed by cleavage of the molecule's signal sequence,[13] although some of it is degraded. The mechanism by which these intragranular proteases may be regulated[63] by cytosolic signals is clearly an important field for future development.

TABLE 3. Distribution of Matrix Proteins of Adrenal Chromaffin Granules in Other Tissues

Protein Family	Tissues in Addition to Adrenal Medulla
chromogranin A	CNS; anterior pituitary, sympathetic neurons, intestinal enterochromaffin cells, parathyroid, thyroid C cells, endocrine pancreas
chromogranin B (= secretogranin I)	CNS; anterior pituitary, sympathetic ganglia, intestinal enterochromaffin cells, endocrine pancreas
chromogranin C (= secretogranin II)	anterior pituitary
proenkephalin	CNS; sympathetic ganglia
prodynorphin	CNS; intestine

We probably do not yet know all the active components of the adrenal medullary secretion: somatostatin precursors, neuropeptide Y and vasoactive intestinal polypeptide have recently been added to the list.[64–66] Their physiological significance and the regulation of their synthesis have yet to be elucidated.

CONCLUSION

The accessibility of the adrenal medulla to a variety of experimental approaches makes it a most attractive model for investigations of neuroendocrine secretion. The bovine gland is firmly established as a source for purification of secretory granules and their membranes on a large scale, and the study of primary cultures of bovine cells permits the investigation of these components within their cellular environment. This interface is indeed the focus of modern study, which seeks to explain the biochemistry of membrane shuttling in the exocytosis-endocytosis cycle, and how the metabolic processes that occur within the lumen of the granules are regulated across this limiting membrane.

ACKNOWLEDGMENTS

We gratefully acknowledge help received from many colleagues, especially D. K. Apps, L. Kilpatrick, J. M. Percy, and S. L. Wood.

REFERENCES

1. Douglas, W. W. 1968. Br. J. Pharmacol. **34:** 451-474.
2. Hillarp, N.-A. 1959. Acta Physiol. Scand. **47:** 271-279.
3. Kirshner, N. 1962. J. Biol. Chem. **237:** 2311-2317.
4. Smith, A. D. & H. Winkler. 1967. Biochem. J. **103:** 480-482.
5. Fenwick, E. M., P. B. Fajdiga, N. B. S. Howe & B. G. Livett. 1978. J. Cell Biol. **76:** 12-30.
6. Greene, L. A. & A. S. Tischler. 1976. Proc. Natl. Acad. Sci. U.S.A. **73:** 2424-2428.
7. Phillips, J. H. 1982. Neuroscience **7:** 1595-1609.
8. Ungar, A. & J. H. Phillips. 1983. Physiol. Rev. **63:** 787-843.
9. Nordmann, J. J. 1984. J. Neurochem. **42:** 434-437.
10. Winkler, H. 1976. Neuroscience **1:** 65-80.
11. Winkler, H. & E. Westhead. 1980. Neuroscience **5:** 1803-1823.
12. Coupland, R. E. 1968. Nature. **217:** 384-388.
13. Falkensammer, G., R. Fischer-Colbrie, K. Richter & H. Winkler. 1985. Neuroscience **14:** 735-746.
14. Kilpatrick, L., D. K. Apps & J. H. Phillips. Unpublished.
15. Joh, T. H., E. E. Baetge, M. E. Ross & D. J. Reis. 1983. Cold Spring Harbor Symp. Quant. Biol. **48:** 327-336.
16. Sabban, E. L. & M. Goldstein. 1984. J. Neurochem. **43:** 1663-1668.
17. Sabban, E. L., L. A. Greene & M. Goldstein. 1983. J. Biol. Chem. **258:** 7812-7818.
18. McHugh, E. M., R. McGee & P. J. Fleming. 1985. J. Biol. Chem. **260:** 4409-4417.
19. Winkler, H. 1977. Neuroscience **2:** 657-683.
20. Patrick, R. L. & N. Kirshner. 1971. Mol. Pharmacol. **7:** 87-96.
21. Slotkin, T. A. & N. Kirshner. 1973. Mol. Pharmacol. **9:** 105-116.
22. Abrahams, S. J. & E. Holtzmann. 1973. J. Cell Biol. **56:** 540-558.
23. Koerker, R. L., W. E. Hahn & F. H. Schneider. 1974. Eur. J. Pharmacol. **28:** 350-359.
24. Nagasawa, J. & W. W. Douglas. 1972. Brain Res. **37:** 141-145.
25. Benedeczky, I. & A. D. Smith. 1972. Z. Zellforsch. **124:** 367-386.

26. PHILLIPS, J. H., K. BURRIDGE, S. P. WILSON & N. KIRSCHNER. 1983. J. Cell Biol. **97:** 1906-1917.
27. BURGOYNE, R. D., M. J. GEISOW & J. BARRON. 1982. Proc. R. Soc. Lond. B **216:** 111-115.
28. BAKER, P. F. & D. E. KNIGHT. 1981. Phil. Trans. R. Soc. Lond. B **296:** 83-103.
29. LINGG, G., R. FISCHER-COLBRIE, W. SCHMIDT & H. WINKLER. 1983. Nature **301:** 610-611.
30. STEINMAN, R. M., S. E. BRODIE & Z. A. COHN. 1976. J. Cell Biol. **68:** 665-687.
31. DOUGLAS, W. W. & R. P. RUBIN. 1961. J. Physiol. **159:** 40-57.
32. MARLEY, E. & W. D. M. PATON. 1961. J. Physiol. **155:** 1-27.
33. GAVINE, F. S., J. G. PRYDE, D. L. DEANE & D. K. APPS. 1984. J. Neurochem. **43:** 1243-1252.
34. CREUTZ, C. E., L. G. DOWLING, J. J. SANDS, C. VILLAR-PALASI, J. H. WHIPPLE & W. J. ZAKS. 1983. J. Biol. Chem. **258:** 14664-14674.
35. GEISOW, M. J. & R. D. BURGOYNE. 1982. J. Neurochem. **38:** 1735-1741.
36. BADER, M. F., T. HIKITA & J. M. TRIFARÓ. 1985. J. Neurochem. **44:** 526-539.
37. HOWELL, K. E. & G. E. PALADE. 1982. J. Cell Biol. **92:** 822-832.
38. HIGGINS, J. A. 1984. Biochem. J. **219:** 261-272.
39. BORDIER, C. 1981. J. Biol. Chem. **256:** 1604-1607.
40. PRYDE, J. G. & J. H. PHILLIPS. 1986. Biochem. J. **233:** 525-533.
41. PERCY, J. M., J. G. PRYDE & D. K. APPS. 1985. Biochem. J. **231:** 557-564.
42. SRIVASTAVA, M., L. T. DUONG & P. J. FLEMING. 1984. J. Biol. Chem. **259:** 8072-8075.
43. HARNADEK, G. J., E. A. REIS & D. NJUS. 1985. Biochemistry **24:** 2640-2644.
44. WOOD, S. L., D. K. APPS & J. H. PHILLIPS. 1985. Biochem. Soc. Trans. **13:** 710-711.
45. BUCKLEY, K. & R. G. KELLY. 1985. J. Cell Biol. **100:** 1284-1294.
46. MATTHEW, W. D., L. TSAVALER & L. F. REICHARDT. 1981. J. Cell Biol. **91:** 257-269.
47. WIEDENMANN, B. & W. W. FRANKE 1985. Cell **41:** 1017-1028.
48. COHN, D. V., R. ZANGERLE, R. FISCHER-COLBRIE, L. L. H. CHU, J. J. ELTING, J. W. HAMILTON & H. WINKLER. 1982. Proc. Natl. Acad. Sci. U.S.A. **79:** 6056-6059.
49. O'CONNOR, D. T., D. BURTON & L. J. DEFTOS. 1983. Life Sci. **33:** 1657-1663.
50. COHN, D. V., J. J. ELTING, M. FRICK & R. ELDE. 1984. Endocrinology **114:** 1963-1974.
51. FISCHER-COLBRIE, R., H. LASSMANN, C. HAGN & H. WINKLER. 1985. Neuroscience **16:** 547-555.
52. SOMOGYI, P., A. J. HODGSON, R. W. DE POTTER, R. FISCHER-COLBRIE, M. SCHOBER, H. WINKLER & I. W. CHUBB. 1984. Brain Res. Rev. **8:** 193-230.
53. ROSA, P., A. HILLE, R. W. H. LEE, A. ZANINI, P. DE CAMILLI & W. B. HUTTNER. 1985. J. Cell Biol. **101:** 1999-2011.
54. NOLAN, J. A., J. Q. TROJANOWSKI & R. HOGUE-ANGELETTI. 1985. J. Histochem. Cytochem. **33:** 791-798.
55. HUTTON, J. C., F. HANSEN & M. PESHAVARIA. 1985. FEBS Lett. **188:** 336-340.
56. SCHULTZBERG, M., J. M. LUNDBERG, T. HOKFELT, L. TERENIUS, J. BRANDT, R. P. ELDE & M. GOLDSTEIN. 1978. Neuroscience **3:** 1169-1186.
57. VIVEROS, O. H., E. J. DILIBERTO, E. HAZUM & K.-J. CHANG. 1979. Mol. Pharmacol. **16:** 1101-1108.
58. DUMONT, M., R. DAY & S. LEMAIRE. 1983. Life Sci. **32:** 287-294.
59. FLEMINGER, G., H.-W. LAHM & S. UDENFRIEND. 1984. Proc. Natl. Acad. Sci. U.S.A. **81:** 3587-3590.
60. HOOK, V. Y. H., L. E. EIDEN & R. M. PRUSS. 1985. J. Biol. Chem. **260:** 5991-5997.
61. FLEMINGER, G., E. EZRA, D. L. KILPATRICK & S. UDENFRIEND. 1983. Proc. Natl. Acad. Sci. U.S.A. **80:** 6418-6421.
62. LEWIS, R. V., A. S. STERN, S. KIMURA, J. ROSSIER, S. STEIN & S. UDENFRIEND. 1980. Science **208:** 1459-1461.
63. HOOK, V. Y. H. & L. E. EIDEN. 1985. Biochem. Biophys. Res. Commun. **128:** 563-570.
64. CORDER, R. & P. J. LOWRY. 1982. Biosci. Rep. **2:** 397-403.
65. CORDER, R., P. C. EMSON & P. J. LOWRY. 1984. Biochem. J. **219:** 699-706.
66. EIDEN, L. E., R. L. ESKAY, J. SCOTT, H. POLLARD & A. J. HOTCHKISS. 1983. Life Sci. **33:** 687-693.

DISCUSSION OF THE PAPER

J.-M. TRIFARÓ (*McGill University, Montreal, Canada*): The half-time for granule membrane internalization seems to be, according to your talk, about 40 minutes. However, experiments with the patch-clamp technique (intact cell conformation) in which changes in membrane capacity were measured seem to indicate much shorter retrieval times.

J. H. PHILLIPS (*University of Edinburgh, Edinburgh, U.K.*): I believe that this arises because of the extent of the challenge given to the cells. When approximately 10% of their granules discharge their contents at the cell surface the inserted membrane overwhelms the capacity of the cellular machinery for extremely rapid retrieval, so that many granule membranes have to await their turn. When few granules are discharged, recovery is likely to occur within seconds.

B. WIEDENMANN (*Institute of Cell and Tumor Biology, Heidelberg, Federal Republic of Germany*): What is the evidence that chromaffin membranes are associated with coated vesicle peptides during membrane retrieval as shown by one of the introductory diagrams, and secondly, have chromaffin-membrane-associated polypeptides been shown to be associated with coated pits?

J. H. PHILLIPS: Biochemical studies on the association of coated vesicle polypeptides with chromaffin granule membrane components are difficult to perform because they require large amounts of highly-purified coated vesicles. Furthermore, it is not easy to separate coated vesicles involved in different intracellular shuttling roles (*e.g.*, membrane assembly and membrane retrieval). Indeed, the only evidence that exists for coating of retrieved membranes is morphological (Geisow, M. J. *et al.* 1985. Eur. J. Cell Biol. **38:** 51-56; Nagasawa and Douglas[24]; Benedeczky and Smith[25]). Ideally, one should perform ultrastructural studies using immunoelectron microscopy with antibodies to chromaffin granule membranes and to clathrin.

O. H. VIVEROS (*Wellcome Research Laboratories, Research Triangle Park, N. C.*): Since this session is on biogenesis, sorting, and packaging of secretory vesicles, I would like to give a brief preview of information I will be presenting in the conference session on Peptides and Processing within Secretory Vesicles that we find very surprising. With Dr. Wilson and Dr. Kirshner we have recently published evidence that chromaffin cells in culture are unable to increase their dopamine β-hydroxylase (DBH) content (soluble and membrane bound) under conditions wherein opioid peptides are being increased 50-fold or more. Cell secretion is not enhanced (*i.e.*, using reserpine or tetrabenazine). Thus, it may be concluded that these cells increase soluble vesicular proteins and peptides with no formation of new vesicular membranes or formation of new vesicles. Therefore, the newly made proenkephalin-A-derived peptides have to be sorted into and packed into "mature vesicles" and would require a mechanism to insert these proteins into preexisting chromaffin vesicles (a vesicular shuttle system). These results may indicate that vesicles are in a dynamic state in the secretory cell, with several different mechanisms to change the content that eventually will be secreted: (1) formation of completely new vesicle (soluble and membrane components), (2) formation of content with reutilization of membranes previously emptied by a cycle of exocytosis, and (3) formation of soluble components (with relative ratios that may differ markedly depending on the recent history of the cell) that are packed into preexisting vesicles. A final consequence of these findings is that "mature granules" may be a useful conceptualization but do not truly exist in a metabolically active secretory cell.

J. H. PHILLIPS: Essentially nothing is known about the assembly stage of granule formation (see FIG. 3). However, we do not have to postulate that this is physically part of the Golgi apparatus of the cell. Instead, there may be a dynamic situation in which both newly synthesized secretory polypeptides and membrane components are shuttled from the Golgi to assembly points, which may themselves consist both of recycled membrane material and of "immature" granules. Thus, I prefer to think that each of your items 1-3 is an aspect of the same process. Such post-Golgi assembly has recently been suggested for insulin granules in the endocrine pancreas.

The concept of the "mature" granule may indeed be meaningless, as you suggest. The process of granule maturation may be highly responsive to cytosolic signals not only in terms of rates of synthesis of individual secretory products, but also of their processing within the granules.

Preproparathyroid Hormone

A Model for Analyzing the Secretory Pathway

KRISTINE M. WIREN, MASON W. FREEMAN,
JOHN T. POTTS, JR., AND HENRY M. KRONENBERG

Endocrine Unit
and
Department of Medicine
Harvard Medical School
Boston, Massachusetts 02114

Our laboratory has been involved for a number of years in investigations of the biology of parathyroid hormone (PTH). These investigations, ultimately aimed at understanding the physiological role of the hormone in calcium homeostasis in biochemical and molecular terms, have recently focused on the structure and function of the parathyroid hormone gene. These studies provide an opportunity to investigate the nature of the regulation of the transcription of the parathyroid hormone gene, the translation of the messenger RNA, and the posttranslational modification, packaging, storage, and secretion of the hormone by signals in the extracellular fluid. Study of the early events in biosynthesis of the hormone allows the definition of critical features of the early cotranslational events in the modification and directed transport of the parathyroid hormone precursor. The purpose of this review is to illustrate these studies with parathyroid hormone with the view that the findings have general relevance to an understanding of the common mechanisms used by all cells to secrete proteins. The report will focus on the area where our data are most extensive, the early events involved in the transport of parathyroid hormone from the polyribosomes into the cisternal space of the endoplasmic reticulum.

Parathyroid hormone, like other secreted polypeptides, is believed to traverse a specialized cellular pathway. This cellular pathway has been extensively defined at the ultrastructural level by Palade and his colleagues.[1] The mechanisms believed to explain the vectoral transport of proteins destined for secretion from the polyribosome to the cisternal space of the endoplasmic reticulum and thence the Golgi for packaging into secretory granules have been extensively studied and modeled in several laboratories over the last 5 to 10 years.[2,3] The amino-terminal extensions (called signal, pre-, or leader sequences) found on the precursors of virtually all secreted proteins share certain common structural features believed important for directing the proteins along the secretory pathway. As the signal sequence emerges from the ribosome, it binds to a signal recognition particle (SRP).[4] This particle, in turn, binds to an integral membrane protein of the endoplasmic reticulum called docking protein or signal recognition particle receptor.[5] Thus, the signal recognition particle provides the link that brings the polyribosome to the membrane of the rough endoplasmic reticulum. The binding of the signal recognition particle to the docking protein releases the polyribosome from the SRP-docking protein complex. By unknown mechanisms, the nascent protein precursor is then inserted into and across the membrane of the endoplasmic reticulum. The signal sequence is cleaved from the precursor by signal peptidase, an endopeptidase located on the inner surface of the membrane of the endoplasmic reticulum.

The signal sequence may play several roles in the complicated series of steps just described: the signal binds (reversibly) to SRP; the signal may be required for insertion of the precursor protein into the membrane, either through interaction with the lipid bilayer or perhaps with specific transport proteins; and the signal provides information that determines the signal peptidase cleavage site. These multiple functions may reside in either separate or overlapping portions of the signal. Although signal sequences do not share a common primary sequence, von Heinje[6] and others have noted certain regularities in the structure of signal peptides (see FIG. 1). Following a short, often positively charged hydrophilic sequence, all signals contain a hydrophobic stretch of 10-15 residues. The signal cleavage site comes 4-10 residues distal to the hydrophobic sequence. Small, neutral amino acids are invariably found at the −1 position just before the cleavage site (Ser, Ala, Gly, Cys, Thr, or Gln); small, neutral amino acids are usually located at −3 as well.

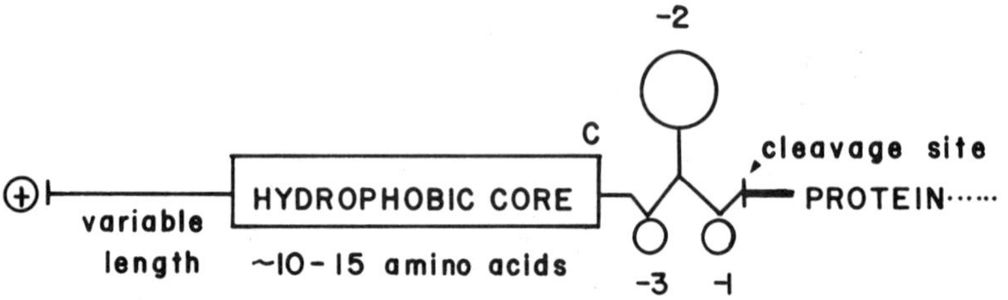

FIGURE 1. Typical features of signal sequences of secreted proteins. The *circled plus sign* indicates that the amino-terminal region of variable length usually has a net positive charge. The hydrophobic core region is *boxed. Small circles* represent the small amino acids found at positions −1 and −3 (with +1 being the first residue of the mature protein). The *large circle* at −2 indicates that the residue often has a bulky side chain.

Preproparathyroid hormone (preproPTH), the precursor of PTH, contains a typical amino-terminal signal sequence of 25 amino acids. Following the signal is the prohormone-specific (or pro) domain consisting of only six amino acids. The translocation and secretion of preproPTH follows a course similar to that of other secreted proteins. PreproPTH is virtually undetectable in intact cells, since the signal sequence is cleaved from the nascent chain cotranslationally as it traverses the endoplasmic reticulum. The pro sequence is removed intracellularly in the distal portion of the secretory pathway, *i.e.,* in the Golgi apparatus or secretory vacuole. ProPTH is not secreted from the normal parathyroid cell, nor does the prohormone-specific segment or any of its potential degradation products accumulate in the cell.[7,8] Thus, the function of this transitory biosynthetic intermediate is particularly enigmatic.

In order to characterize structure-function relationships between the precursor-specific regions of preproPTH and the secretory pathway, mutant preproPTH sequences have been constructed and expressed both in intact cells and *in vitro.* These deletion mutants can be categorized into three groups: (1) amino-terminal signal sequence mutants, (2) signal cleavage domain mutants, and (3) a mutant missing only sequences distal to the cleaved signal, *i.e.,* the pro region. These mutants were generated by one of two approaches, either using Bal 31 nuclease to create deletions of variable extent, or using synthetic oligonucleotide sequences to introduce specific deletions into the preproPTH sequence. The study of these deletion mutants helps to define the functional boundaries of the signal sequence and also to distinguish different or overlapping functions of specific regions in the precursor-specific sequences.

To create the first class of mutants, Bal 31 nuclease was used to remove progressively larger portions of the DNA encoding the first several amino acid residues of preproPTH (see FIG. 2). Since initiation of protein synthesis generally occurs at the most proximal ATG sequence, the removal of the normal initiator ATG was expected to lead to initiation of protein synthesis at the ATG normally encoding methionine +7 of preproPTH. Similar reasoning led to the construction of mutant genes encoding proteins beginning at methionine +11 and methionine +14 of preproPTH. The first deletion removes only the first few hydrophilic amino acids of the signal, the second removes an evolutionarily conserved lysine and shortens the hydrophobic core by one residue, and the third removes more of the hydrophobic core sequence.

The second set of deletion mutants, those in which the sequences around the cleavage site have been altered, also were created using Bal 31 nuclease digestion protocols and restriction enzyme digestion. Each of these mutants is missing the short pro sequence and a variable portion of the sequences proximal to the signal cleavage site. The most informative mutants, called m144 and Bov/Human fusion, are illustrated in FIGURE 2. The third type of deletion mutant is missing precisely the six-residue prohormone sequence. In this mutant, the signal sequence is intact and is linked directly to the mature PTH sequence; a synthetic 82 nucleotide fragment encoding only the leader sequence was used to replace the natural sequence encoding the signal plus prohormone-specific sequences (FIG. 2).

These mutant genes, as well as the normal sequence encoding preproPTH, have been introduced into eukaryotic secretory cells using retroviral vectors constructed by Richard Mulligan's group.[9,10] All cDNAs were inserted into a recombinant retroviral vector derived from a Moloney murine leukemia viral (MMLV) provirus. The MMLV virus has been modified by (1) removing essentially all protein-coding sequence from the provirus, but leaving intact the long terminal repeats (LTRs; for integration, promoter, and polyadenylation functions), the packaging sequence just downstream from the 5′ LTR, and the *env* gene splice donor and acceptor sequences; (2) introducing a selectable marker, the *E. coli* gene *gpt,* encoding xanthine-guanine phosphoribosyl transferase; and (3) introducing the preproPTH sequence into the region normally containing the *gag-pol* sequence. Recombinant retroviruses were packaged into infective particles without production of a helper virus using the Psi-2 cell lines previously described.[9] Media containing infectious particles that contain mutant and normal PTH DNA were then used to infect and transform the rat pituitary cell line GH_4C_1. Transformed clones expressing mutant PTH sequences and the *gpt* gene were obtained in selective media. These cell lines have been used to characterize the effect of the various mutations described above on the secretion of preproPTH.

When the normal preproPTH sequence was expressed in GH_4C_1 cells, the intracellular handling of the precursor mimicked that found in the parathyroid cell. FIGURE 3 shows the results of a pulse-chase analysis of such cells.[10] Cells were labeled for 15 minutes with ^{35}S-methionine, then cold methionine was added and cells were incubated for various periods of time. PTH-related peptides were isolated from cell extracts and media at each time point by immune precipitation. The proteins were detected by gel electrophoresis followed by fluorography. No preproPTH is seen at early times because the signal sequence is removed rapidly. In the cells, proPTH is the first PTH-related protein detected. With time, the proPTH is converted to PTH and the PTH is secreted from the cell. Thyrotropin-releasing hormone, which stimulates the release of prolactin form GH_4C_1 cells, also stimulates the release of PTH from the cells secreting PTH.

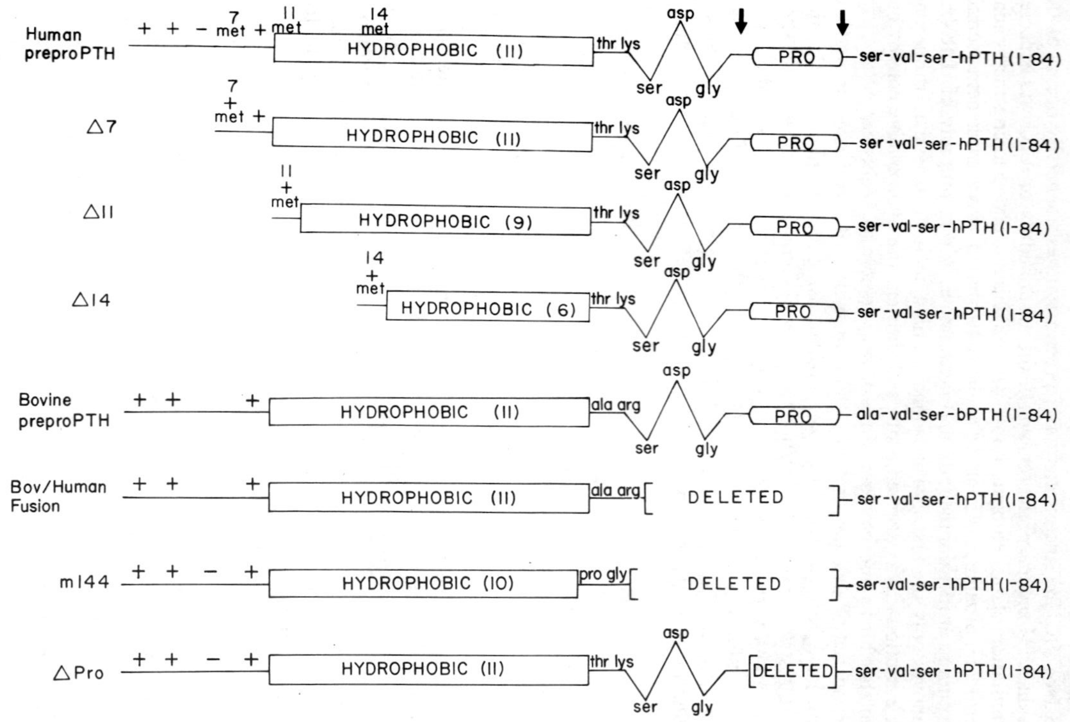

FIGURE 2. Mutant forms of preproparathyroid hormone. Symbols (+, −) and associated numbers indicate locations of amino-terminal charged residues; numbers in parentheses indicate number of residues in hydrophobic core. Symbols and abbreviations at far left indicate names given to specific mutants.

This physiologic response suggests that the PTH is secreted via a normal, regulated pathway in this cell line.

In order to analyze further the secretory phenotype of the various mutant PTH proteins on a more biochemical level, a linked transcription-translation assay in cell-free extracts has been employed. Plasmids that express the mutant PTH genes have been created by inserting mutant sequences into the pBR322-based expression vector pGL101,[11] which contains the prokaryotic *lac* promotor. Mutant PTH genes, which are inserted distal to the *lac* promoter, are then transcribed by *E. coli* polymerase *in vitro* and translated in a reticulocyte lysate with or without added microsomal preparations from canine pancreas. In addition, pGL101-based plasmids also transcribe the gene encoding penicillin resistance, pre-β-lactamase. Along with any PTH RNA, the pre-β-lactamase RNA is translated.[12] This protein precursor serves as an internal control for the amount of processing obtained in each reaction by measuring the conversion of pre-β-lactamase to β-lactamase in the same tube. PTH-related products are subjected to immune precipitation, followed by SDS-polyacrylamide gel electrophoresis and fluorography. This assay is especially well suited for describing and characterizing the very early events in the secretory pathway and complements the analysis from intact cells.

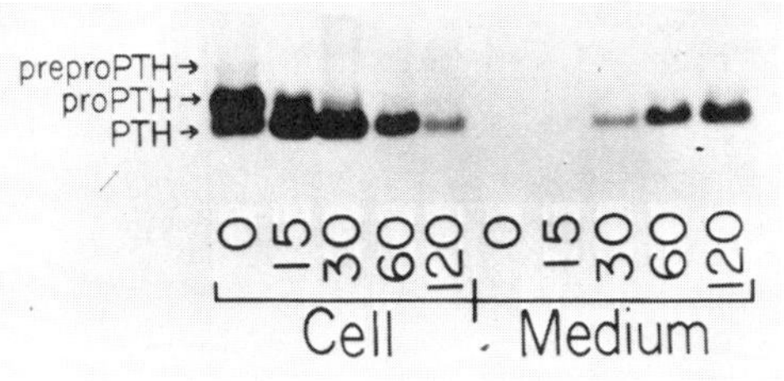

FIGURE 3. Pulse-chase analysis of GH_4C_1 cells expressing human preproPTH cDNA. See text for experimental protocol.

The PTH mutants shown in FIGURE 2 have been expressed using these assays, and the secretory phenotypes have been characterized. The three amino-terminal mutants help to define the functional boundary of the amino terminus of the signal sequence. In GH_4C_1 cell lines expressing these mutant PTH proteins, pulse-chase analysis has revealed that the mutant missing the first six residues has normal secretory behavior. The mutants missing more of the signal sequence are not processed normally however. Precursor protein encoded by these more defective mutants is detected intracellularly at early time points, but then turns over rapidly. In addition, no PTH-related products are detectable by immune precipitation from the media in pulse-chase experiments. Following a prolonged pulse, however, a small amount of PTH is secreted from the cells containing the mutant missing the first 10 residues of the signal.

The carboxy-terminal mutants have been expressed in the same assays. Both the Bov/Human fusion mutant and the m144 mutant are transported across the microsomal membrane inefficiently. In both cells and *in vitro,* cleavage of the Bov/Human fusion precursor is extremely inefficient; cleavage of the m144 precursor is more efficient. For both precursors, the cleavage occurs within the mature PTH sequence itself, following small, neutral amino acids. A portion of the inefficiently cleaved Bov/Human precursor protein is found trapped in the microsomal membrane and exhibits properties associated with integral membrane proteins.

The mutant missing the prohormone sequence but containing an intact signal sequence also does not behave normally. The protein is transported across the microsomal membrane with substantial efficiency, but the cleavage of the signal is inefficient. Consequently, the precursor protein is detected in the cell, and only small amounts of authentic PTH are secreted. A small amount of a protein slightly bigger than PTH is secreted as well.

These results suggest that the precursor regions of preproPTH have four domains of function. Much of the amino-terminal domain can be removed without loss of function (efficient processing of the shortened Δ7 mutant), though a possibly important role for an amino-terminal charged residue has not been evaluated in eukaryotic cells. The hydrophobic sequence is required for transport of the precursor across the membrane of the endoplasmic reticulum (inefficient processing of the Δ11 and Δ14 mutants). Sequences distal to the hydrophobic sequence, proximal to the signal cleavage site, are also required for efficient membrane transport and determine the efficiency of cleavage and location of the cleavage site used by signal peptidase (aberrant and inefficient processing of Bov/Human and m144 mutants). Finally, the sequences immediately distal to the signal also influence the efficiency and precise location of signal cleavage (aberrant and inefficient cleavage of Δpro mutant). The four domains of function just described can perhaps be generalized to all secreted proteins; proteins not containing prohormone-specific domains may have analogous sequences in the amino terminus of the mature molecule (see FIG. 1). Ultimately, studies of these and other mutants should permit more precise analysis of the biochemical basis of the directed transport of proteins across membranes.

REFERENCES

1. PALADE, G. 1975. Intracellular aspects of the process of protein synthesis. Science **189:** 347-358.
2. WICKNER, W. & H. F. LODISH. 1985. Multiple mechanisms of protein insertion into and across membranes. Science **230:** 400-407.
3. KELLEY, K. B. 1985. Pathways of protein secretion in eukaryotes. Science **230:** 25-32.
4. WALTER, P., I. IBRAHIMI & G. BLOBEL. 1981. Translocation of proteins across the endoplasmic reticulum. J. Cell Biol. **91:** 545-561.
5. MEYER, D. I., E. KRAUSE & B. DOBBERSTEIN. 1982. Secretory protein translocation across membranes: The role of the "docking protein." Nature **297:** 647-650.
6. VON HEIJNE, G. 1983. Patterns of amino acids near signal-sequence cleavage sites. Eur. J. Biochem. **133:** 17-21.
7. HABENER, J. F., T. D. STEVENS, G. W. TREGEAR & J. T. POTTS, JR. 1976. Radioimmunoassay of human proparathyroid hormone: Analysis of hormone content in tissue extracts and in plasma. J. Clin. Endocrinol. Metab. **42:** 520-530.
8. HABENER, J. F., H. T. CHANG & J. T. POTTS, JR. 1977. Enzymatic processing of proparathyroid hormone by cell-free extracts of parathyroid glands. Biochemistry **16:** 3910-3917.
9. MANN, R., R. C. MULLIGAN & D. BALTIMORE. 1983. Construction of a retrovirus packaging mutant and its use to produce helper-free defective retrovirus. Cell **33:** 153-159.
10. HELLERMAN, J. G., R. C. CONE & J. T. POTTS, JR., A. RICH, R. C. MULLIGAN & H. M. KRONENBERG. 1984. Secretion of human parathyroid hormone from rat pituitary cells infected with a recombinant retrovirus encoding preproparathyroid hormone. Proc. Natl. Acad. Sci. **81:** 5340-5344.
11. GUARENTE, I., G. LAUER, T. M. ROBERTS & M. PTASHNE. 1980. Improved methods for maximizing expression of a cloned gene: A bacterium that synthesizes rabbit β-globin. Cell **20:** 543-553.

12. KRONENBERG, H. M., B. J. FENNICK & T. J. VASICEK. 1983. Transport and cleavage of bacterial pre-β-lactamase by mammalian microsomes. J. Cell Biol. **96:** 1117-1119.

DISCUSSION OF THE PAPER

L. E. EIDEN (*NIMH, National Institutes of Health, Bethesda, Md.*): Are there problems with using deletion mutations in that failure to cleave may be due to spatial considerations?

J. T. POTTS, JR. (*Massachusetts General Hospital, Boston, Mass.*): You are correct, deletions versus substitution mutants may give different types of information. Our mutants m144 and Bov/human fusion may be an example of substitution more than deletion, we think. Introduction of proline near the junction point of the modified signal sequence and the hormone appears to improve cleavage efficiency. Substitution of hydrophobic for hydrophilic resides in the "hydrophilic core" region would be pertinent, for example; we have not performed such studies, however.

M. GRIMES (*University of Oregon, Eugene, Oreg.*): Are any of your signal sequence mutants secreted via the constitutive pathway of secretion?

J. T. POTTS: We have not yet been able to examine this issue although we plan to. GH_4 cells contain few secretory granules. Cells such as AtT-20 cells rich in such granules have been used by Drs. Moore, Kelly, and others for other proteins to study secretory pathways. We will need to use such cell lines to properly sort out secretory pathways meaningfully.

Factors Controlling Packaging of Peptide Hormones into Secretory Granules[a]

HSIAO-PING H. MOORE

Department of Physiology-Anatomy
University of California
Berkeley, California 94720

INTRODUCTION

An important feature of eukaryotic cells that distinguishes them from lower cells is their highly compartmentalized internal structures. To achieve such structural organization, cells must have mechanisms to segregate and sort their cellular components and to localize them in distinct membrane-bound compartments. Rules must therefore exist that govern whether, for example, a newly synthesized protein emerging from the ribosome should be transported to the nucleus or to the lysosomes. These rules are beginning to be unravelled in several systems, including proteins segregated to the mitochondria,[1] to the rough endoplasmic reticulum,[2] and to the lysosomes.[3] In this paper I will discuss recent progress in our understanding of how proteins are targeted to the secretory granules in peptide-hormone-secreting cells.

THE EXPERIMENTAL SYSTEM

Endocrine and neuronal cells package peptide hormones and neurotransmitters in specialized secretion granules. These granules are stored in the cytoplasm and fuse with the plasma membrane only when cells receive an external stimulus. Because secretion is regulated by physiological conditions, we call these granules "regulated secretory granules" to distinguish them from other types of secretory vesicles (see below). To study protein localization to this organelle, we looked for an endocrine cell line which maintained the ability to synthesize and package hormones in culture. The AtT-20 cells, derived from a mouse anterior pituitary tumor, retained many of the differentiated phenotypes of corticotrophs even after extensive passages *in vitro.* These cells synthesize proopiomelanocortin (POMC), the precursor to adrenocorti-

[a] This work has been supported by NIH research grant no. GM 35239-01 and NSF Presidential Young Investigator Award no. DCB 8451636.

cotrophic hormone (ACTH), process it to the mature peptides, package them in secretory granules, and release them in response to secretagogues.[4] The availability of such a cell line permits one to use biochemical, genetic, and molecular biological techniques for the study of protein localization to the regulated secretory granules.

TWO PATHWAYS OF SECRETION IN ENDOCRINE CELLS

Like all secretory proteins, newly synthesized peptide hormones are first sent to the rough endoplasmic reticulum (RER) by means of signal peptides and the signal recognition particle.[5] Two other classes of proteins, namely, lysosomal and plasma membrane proteins, also gain access to the lumen of the RER by similar mechanisms. Therefore, it is reasonable to assume that additional targeting information might be present on peptide hormones for subsequent sorting into the regulated secretory granules. There is little evidence, however, to support the existence of such sorting domains on any secretory proteins. The only class of proteins for which an active receptor-mediated mechanism has been identified is the lysosomal enzymes.[3] It is conceivable that plasma membrane proteins and secretory proteins are not actively diverted away by sorting receptors but are transported automatically to the cell surface by a bulk-flow process.

In some cell types, available data appear to be consistent with the above hypothesis. Immunoelectron microscopic studies of hepatocytes revealed that the vesicular stomatitis virus (VSV) membrane glycoprotein G and albumin en route to the cell surface are localized to the same regions of the Golgi stacks and in the same secretory vesicles.[6] Further, yeast mutants defective in the secretion of invertase are also defective in their ability to externalize a number of other membrane and secretory proteins.[7] These results suggest that, at least in these cell types, secretory proteins and plasma membrane proteins are not segregated from one another and their transport can be explained without invoking sorting domains.

Do these arguments also apply to the transport of peptide hormones in endocrine cells and neuronal cells? Are sorting domains required in directing hormones to the secretory granules? To answer these questions, we first determined if hormone-secreting cells sort their secretory and plasma membrane proteins. It is not unreasonable to speculate that different cell types may need to sort proteins differently. Endocrine cells, as discussed earlier, secrete hormones in a regulated fashion whereas the other cell types mentioned earlier secrete proteins constitutively. These differences in the mode of secretion may reflect differences in their mechanisms of protein transport as well. Indeed, two sets of experiments indicate that plasma membrane proteins and secretory proteins made by AtT-20 cells are segregated from one another during transit to the cell surface.

Gumbiner and Kelly[4] used antibodies against an endogenous viral membrane protein, gp70, to follow the secretory pathway of a plasma membrane protein and compared it with the pathway taken by the hormone adrenocorticotropin ACTH. They concluded that there are at least two distinct routes to the cell surface in this cell. While ACTH is packaged into regulated secretory granules and its secretion is accelerated by secretagogues, the viral protein is externalized by a different, so-called "constitutive" pathway. This protein is transported to the cell surface as soon as it is synthesized, *i.e.,* without first storage in the secretory granules. Furthermore, its externalization is not influenced by secretagogues. Some of the properties which distinguish the two pathways are summarized in TABLE 1. To test if these two pathways

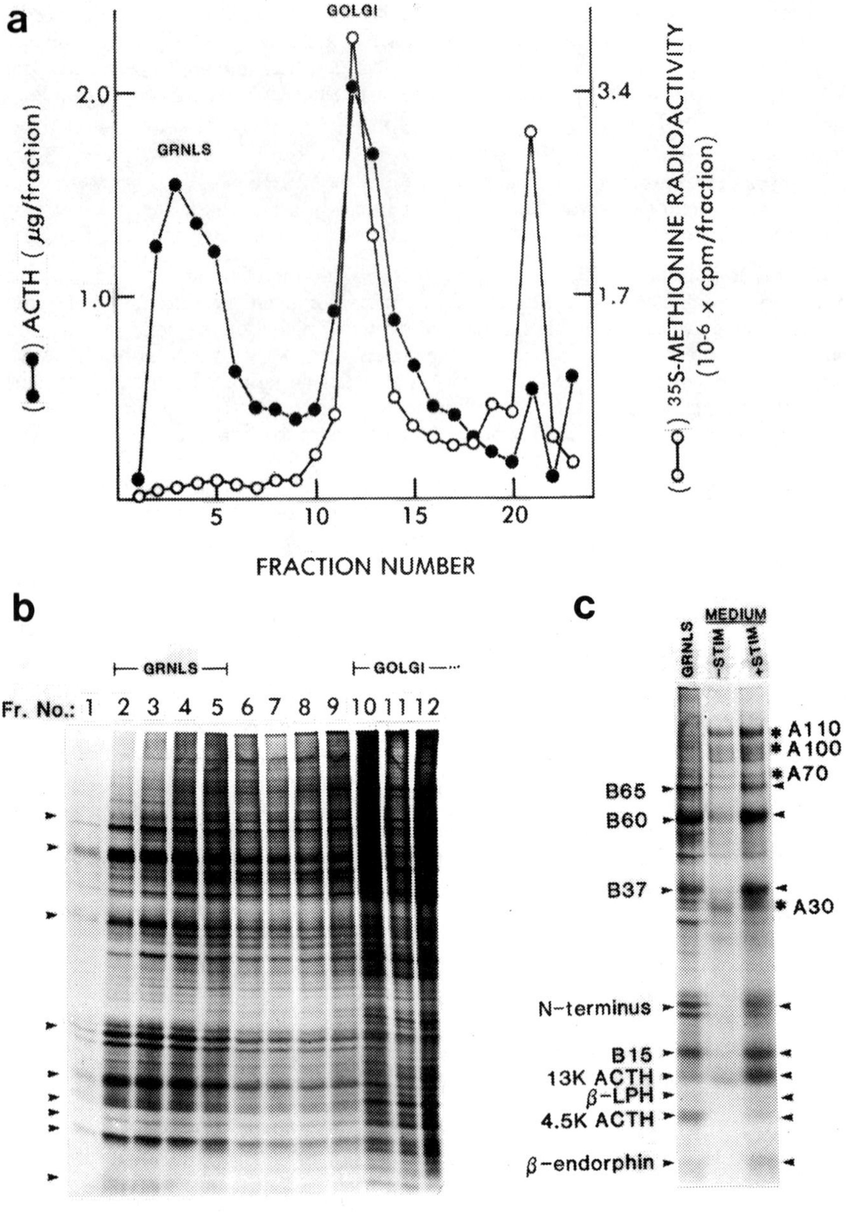

FIGURE 1. Secretion of ^{35}S-methionine-labeled endogenous proteins by AtT-20 cells. All secretory proteins fall into one of two classes. Class A proteins are not packaged into dense secretory granules, and are released continuously into the culture medium whether or not the cells are stimulated. In contrast, class B proteins are packaged in dense-core secretory granules and show enhanced secretion when cells are stimulated. **(a,b)** SDS polyacrylamide gel analysis of AtT-20 cell membranes fractionated on D20-Ficoll density gradients. *Arrows* indicate those

also provide general routes for the externalization of other proteins made by AtT-20 cells, we examined the pathways taken by all endogenous secretory products after labeling the cells with ^{35}S-methionine (FIG. 1). A number of radiolabeled molecules were found in the tissue culture medium and they all fall into one or the other of the two classes: those externalized by the constitutive pathway and those by the regulated pathway. Thus, in addition to gp70 and ACTH, other products in the secretory apparatus are also sorted into these two distinct pathways (FIG. 2). The presence of these two secretory pathways has also been found in other endocrine and neuronal cell lines. These include the growth-hormone-producing pituitary tumor GH3 cells,[8] the pheochromocytoma-derived PC-12 cells,[9] and the pancreatic insulin-producing HIT cell line (H.-P. H. Moore, unpublished results).

SORTING DOMAINS ON PEPTIDE HORMONES

The finding that two secretory pathways are present in one cell implies that address signals must be present on at least one class of the transported proteins to ensure

TABLE 1. Characteristics of the Two Secretory Pathways in AtT-20 Cells

Property	Constitutive Pathway	Regulated Pathway
Rate of externalization in absence of stimulation	rapid $t_{1/2}$ = 30-40 min	slow $t_{1/2}$ = 6-7 h
Increase in rate of secretion in response to secretagogues	no	yes
Secretory proteins packaged in dense secretory granules	no	yes
Requires external Ca^{2+}	no	yes
Inhibition by chloroquine[24]	no	yes

correct sorting. In order to identify such sorting signals, we utilized DNA transfection and *in vitro* mutagenesis techniques. The method involves altering the DNA sequences encoding secretory proteins *in vitro* by recombinant DNA techniques, and then expressing them in AtT-20 cells to test their effects on sorting. This approach relies on two factors: (1) AtT-20 cells should be capable of incorporating and expressing foreign DNA; and (2) the gene product to be assayed should be distinguishable from endogenous proteins. Both requirements have been fulfilled.

TABLE 2 shows the results of expression of a number of exogenous secretory proteins in AtT-20 cells by DNA transfection. Several conclusions can be drawn from these studies: (1) Heterologous hormones originating from another endocrine tissue

polypeptides that copurify with dense ACTH-containing secretory granules (GRNLS). (**c**) Comparison with polypeptides secreted from stimulated and unstimulated cells. The polypeptides enriched in the dense secretory granules are released in a secretagogue-dependent manner; their rate of secretion is greatly increased when cells are stimulated with 8-Br-cAMP. In addition to these proteins, the cells also secrete other proteins which are not enriched in the isolated secretory granule fractions (indicated by *). Secretion of this class of proteins is not affected by secretagogues.

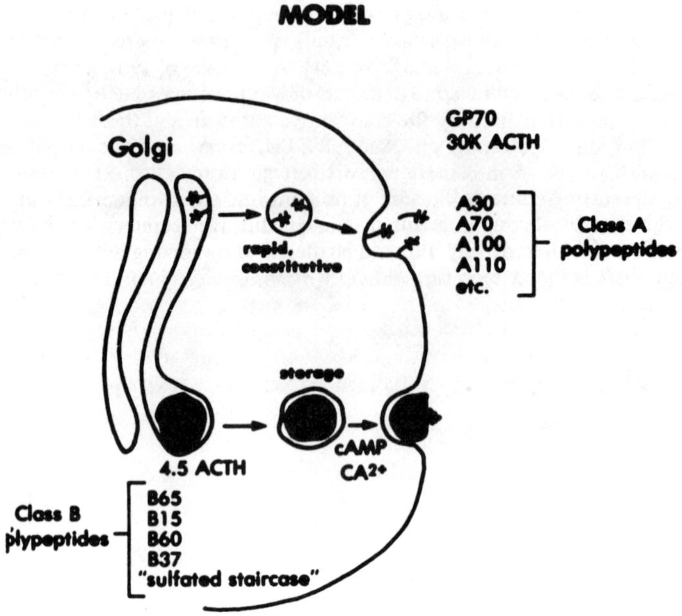

FIGURE 2. Segregation of proteins into two secretory pathways in AtT-20 cells. Peptide hormones and class B proteins (see FIG. 1) are packaged into dense secretory granules and stored intracellularly until cells are stimulated. Other proteins (class A polypeptides, see FIG. 1) are continuously sent to the cell surface by membrane vesicles that fuse constitutively with the plasma membrane. It should be noted that the site at which these proteins are sorted from one another has not been identified. The depiction of these two types of vesicles as originating from opposite ends of the same Golgi stack is therefore, at present, hypothetical.

can be sorted correctly by AtT-20 cells into ACTH-containing secretory granules. Thus, the sorting machinery appears to be conserved among different types of endocrine cells. (2) Since an exocrine secretory product, trypsinogen, is also secreted by the regulated pathway in this endocrine cell line, it is likely that exocrine cells share similar transport mechanisms. (3) Sorting by transfected cells is selective, because not all proteins transfected into AtT-20 cells are packaged into regulated secretory gran-

TABLE 2. Summary of Secretory Pathways Taken by Various Proteins Expressed in AtT-20 Cells

Protein	Origin	Pathway	Proteolytic Processing	References
Mouse ACTH	endogenous	regulated	yes	4, 20
MuLV, gp70	endogenous virus	constitutive	yes	4
Laminin	endogenous	constitutive	no	21
Human insulin	endocrine/pancreas	regulated	yes	22
Human growth hormone (hGH)	endocrine/pituitary	regulated	no	23
Rat trypsinogen	exocrine/pancreas	regulated	no	21
Truncated VSV G protein (TG)	exogenous virus	constitutive	no	23

ules. As discussed above, hormones and an exocrine protein expressed in AtT-20 cells by DNA transfection are packaged into regulated secretory granules. By contrast, a truncated fragment of a viral membrane protein, TG (structure shown in FIG. 3), prefers the constitutive route to the surface. (4) Proteolytic processing of the secretory proteins does not appear to correlate with its intracellular localization, because both secretory pathways can transport unprocessed and processed products.

With this ability of AtT-20 cells to sort foreign secretory proteins, we can now investigate the sorting domains. Sorting of proteins into the two pathways conceivably could be achieved by sorting domains on the constitutive secretory proteins (class A proteins in FIG. 2). Alternatively, it could be mediated by sorting domains on proteins destined to the regulated secretory granules (class B proteins). The other class of proteins would then be transported to the second pathway by a nonspecific, bulk-flow process. To address which class of secretory proteins is actively sorted, we made a fusion protein between a constitutively secreted protein and a regulated secreted protein.[10] If constitutive proteins contain sorting domains, then the presence of these domains on the fusion product should divert it to the constitutive pathway. The opposite result would be predicted if sorting domains reside on the regulated secreted proteins.

FIGURE 4 shows that the viral protein TG, normally secreted by the constitutive route, can be diverted to the regulated secretory pathway if it is attached to the human growth hormone sequence. These results have two important implications. First, they indicate that sorting domains are contained within peptide hormone sequences. Secondly, they suggest that proteins which exit by the constitutive route may lack any sorting signal. Without additional data we cannot rule out the alternative interpretation, namely, sorting domains also exist on constitutive secreted proteins but they are simply subordinate to the sorting domains of peptide hormones. At the present time, however, there is no need to invoke any sorting domain on constitutively secreted proteins in AtT-20 cells.

THE MECHANISM OF SORTING

Our current working hypothesis for sorting of secretory proteins in AtT-20 cells is diagramed in FIGURE 5. Peptide hormones are sorted from other secretory products by binding specifically to cellular carriers which recognize unique sorting domains on the hormones. These carriers then direct packaging of hormones into the regulated secretory granules. Proteins not sorted by the carriers are transported in bulk to the cell surface by the constitutive secretory route.

The precise nature of the sorting signals on peptide hormones has not yet been identified. The ACTH precursor is known to undergo several posttranslational modifications, including proteolytic processing,[11,12] sulfation,[13] phosphorylation,[14] and *N*-linked glycosylation.[15] One of these modifications could, in principle, serve as the sorting signal for packaging hormones into the regulated secretory granules. We now know that at least two of these modifications, *i.e.*, *N*-linked glycosylation and sulfation, are not essential for sorting of ACTH. Treatment of cells with tunicamycin, under conditions when 97% of the *N*-linked glycosylation is inhibited, does not have any detectable effect on the transport of ACTH.[16] It follows that sulfation, which occurs on *N*-linked oligosaccharides of ACTH (H.-P.H. Moore, unpublished results), is also

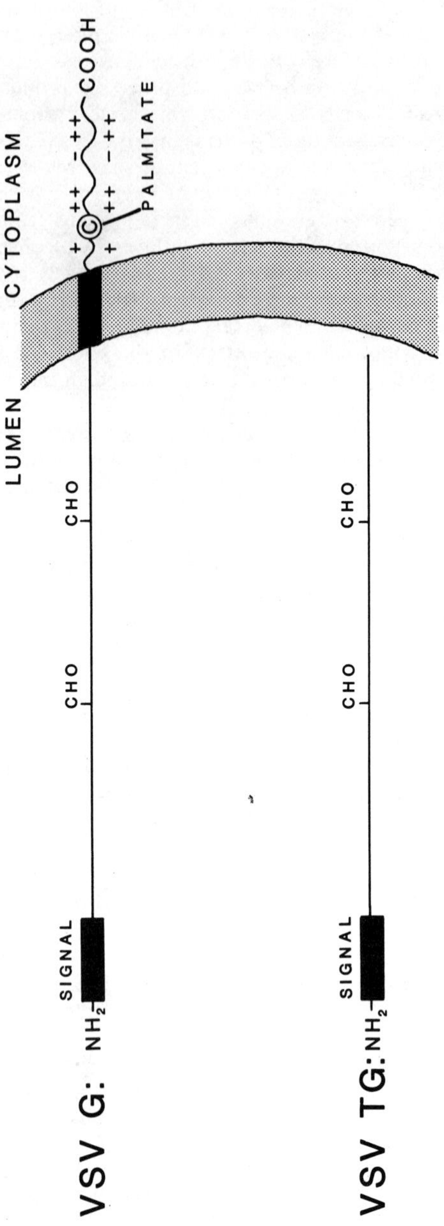

FIGURE 3. Structure of the membrane-bound and the truncated forms of the vesicular stomatitis virus (VSV) glycoprotein (G). The truncated G (TG) lacks the entire membrane anchor and the cytoplasmic domains and is secreted by cultured cells.[25]

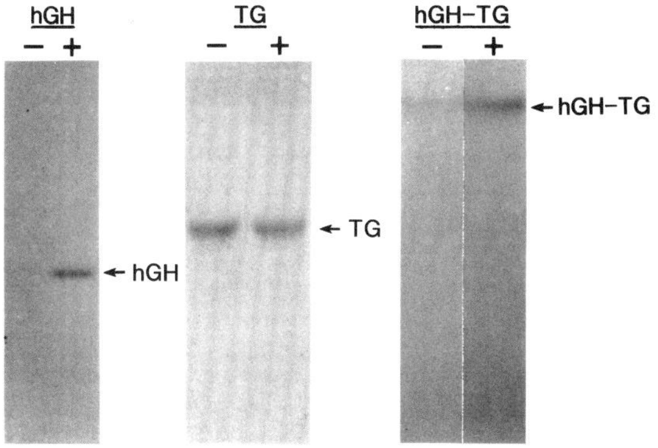

FIGURE 4. Pathways of secretion of hGH, TG, and their fusion polypeptide hGH-TG in transfected AtT-20 cells. The secretory pathway taken by each protein was determined by analyzing its rates of secretion from stimulated (+) and unstimulated (−) cells. The gels show that hGH is secreted by the regulated pathway and TG is secreted by the constitutive pathway. The fusion protein, hGH-TG, takes on the secretion pathway of hGH and is secreted by the regulated pathway.

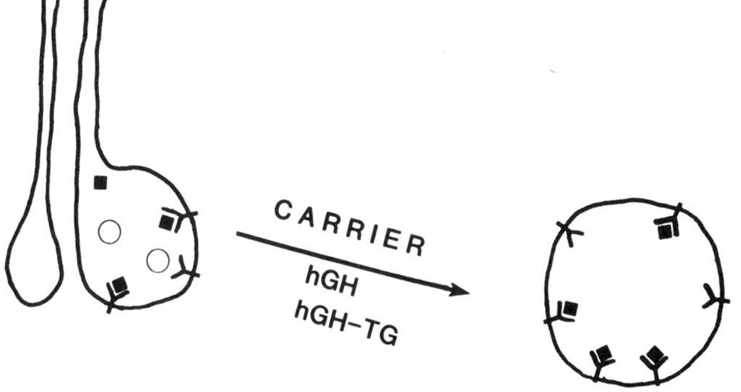

FIGURE 5. Hypothetical model for sorting of secretory proteins in AtT-20 cells. Sorting is envisaged to be mediated by cellular carriers that recognize sorting domains on proteins destined for the regulated secretory granules. Proteins that are not segregated away by this mechanism are transported in bulk to the constitutive pathway by a passive-flow process.

not involved in targeting. The role of the other posttranslational modifications in intracellular transport of hormones remains to be elucidated.

Virtually nothing is known about the cellular carriers that mediate the sorting of hormones. In the studies described in FIGURES 1 and 2, we noticed the presence of a granule-specific molecule whose characteristics are reminiscent of those of proteoglycans. Such molecules could, in principle, provide packaging and sorting functions by interacting with positively charged secretory proteins, as first suggested by investigators studying the zymogen granules from exocrine pancreas.[17,18] Burgess and Kelly[19] characterized this granule-specific proteoglycan in AtT-20 cells and showed, both by mutant and by inhibitor studies, that its biosynthesis is not crucial for sorting of ACTH. Other cellular carriers which could specifically interact with hormones must be sought.

CONCLUSIONS

Our studies showed that sorting domains on peptide hormones direct their packaging into the regulated secretory granules. Although these domains have not been precisely mapped out, the transfection experiments revealed a somewhat unexpected property. The observation that several peptide hormones are targeted to the same secretory granules in AtT-20 cells (TABLE 2) suggests that they may share similar sorting domains. Direct inspection of their primary amino acid sequences showed no obvious homology. This is not unprecedented, however, since signal sequences of all known secretory and membrane proteins lack apparent sequence homology, even though they appear to be recognized by the same SRP. Further, lysosomal enzymes preferentially phosphorylated by the *N*-acetylglucosaminylphosphotransferase also fail to show significant consensus sequence. Future deletional mapping of sorting domains on the individual hormone will be necessary to delineate the sorting signals.

SUMMARY

Endocrine, exocrine, and neuronal cells package only a subset of their secretory products into the electron-dense secretory granules. To investigate the factors controlling selective packaging of proteins into these granules, we utilized the mouse pituitary tumor cell line, AtT-20, which retained the capability to sort adrenocorticotropic hormone (ACTH) into secretory granules *in vitro.* Packaging of ACTH was blocked by treatment with weak bases, but was unaffected when *N*-linked glycosylation or sulfation was inhibited. To test whether the targeting information is specified by sorting domains present on peptide hormone sequences, we determined if a protein could be diverted to the dense secretory granules by attachment to a peptide hormone sequence. A plasmid DNA was constructed that encoded a hybrid protein in which a fragment of a viral membrane protein was fused to the carboxy terminus of human growth hormone. AtT-20 cells transfected with the hybrid were found to target it to dense secretory vesicles efficiently. These results support the hypothesis that sorting domains on peptide hormones direct their packaging into dense secretory vesicles.

ACKNOWLEDGMENT

The work described herein was performed in part with Dr. Regis B. Kelly.

REFERENCES

1. SCHATZ, G. & R. A. BUTOW. 1983. Cell **32:** 316-318.
2. WALTER, P., R. GILMORE & G. BLOBEL. 1984. Cell **38:** 5-8.
3. SLY, W. S. & H. D. FISCHER. 1982. J. Cell Biochem. **18:** 67-85.
4. GUMBINER, B. & R. B. KELLY. 1982. Cell **28:** 51-59.
5. WALTER, P. & G. BLOBEL. 1981. J. Cell Biol. **91:** 545-550.
6. STROUS, G. J. A. M., P. K. WILLEMSEN, J. W. SLOT, H. J. GEUZE & H. F. LODISH. 1983. J. Cell Biol. **97:** 1815-1822.
7. SCHEKMAN, R. 1982. Trends Biochem. Sci. **7:** 243-246.
8. GREEN, R. & D. SHIELDS. 1984. J. Cell Biol. **99:** 97-104.
9. SCHWEITZER, E. S. & R. B. KELLY. 1985. J. Cell Biol. **101:** 667-675.
10. MOORE, H.-P. H. & R. B. KELLY. 1986. Nature **321:** 443-445.
11. ROBERTS, J. L. & E. HERBERT. 1977. Proc. Natl. Acad. Sci. **74:** 5300-5304.
12. MAINS, R. E., B. A. EIPPER & N. LING. 1977. Proc. Natl. Acad. Sci. U.S.A. **74:** 3014-3018.
13. HOSHINA, H., G. HORTIN & I. BIOME. 1982. Science **217:** 63-64.
14. EIPPER, B. A. & R. E. MAINS. 1982. J. Biol. Chem. **257:** 4907-4915.
15. EIPPER, B. A., R. E. MAINS & D. GUENZI. 1976. J. Biol. Chem. **251:** 4121-4126.
16. KELLY, R. B., K. B. BUCKLEY, T. BURGESS, S. S. CARLSON, P. CARONI, J. E. HOOPER, A. KATZEN, H.-P. H. MOORE, S. R. PFEFFER & T. SCHROER. 1984. Cold Spring Harbor Symp. Quant. Biol. **48:** 697-707.
17. TARTAKOFF, A., L. J. GREENE & G. E. PALADE. 1974. J. Biol. Chem. **249:** 7420-7431.
18. REGGIO, H. A. & G. E. PALADE. 1978. J. Cell Biol. **77:** 288-314.
19. BURGESS, T. L. & R. B. KELLY. 1984. J. Cell Biol. **99:** 2223-2230.
20. MOORE, H.-P. H., B. GUMBINER & R. B. KELLY. 1983. J. Cell Biol. **97:** 810-817.
21. BURGESS, T. L., C. S. CRAIK & R. B. KELLY. 1985. J. Cell Biol. **101:** 639-645.
22. MOORE, H.-P. H., M. WALKER, F. LEE & R. B. KELLY. 1983. Cell **35:** 531-538.
23. MOORE, H.-P. & R. B. KELLY. 1985. J. Cell Biol. **101:** 1773-1781.
24. MOORE, H.-P. H., B. GUMBINER & R. B. KELLY. 1983. Nature **302:** 434-436.
25. ROSE, J. K. & J. E. BERGMANN. 1982. Cell **30:** 753-762.

DISCUSSION OF THE PAPER

O. H. VIVEROS (*Wellcome Research Labs, Research Triangle Park, N.C.*): I would like to call attention to the fact the secretory cells have not one, but may have several secretory compartments that can be differentially activated and have markedly different contents: chromaffin cells secrete from chromaffin vesicles, lysosomes, endoplasmic reticulum and cytosol; catecholamines-ATP-enkephalin-DBH, lysosomal enzymes, acetylcholinesterase, and ascorbate are all secreted by a Ca^{2+}-dependent, nicotine-receptor-mediated mechanism, in spite of originating in several different compartments. Thus, when considering sorting of proteins there may be signals, not only for a constitutive path and a regulated path, but there have to be signals of higher specificity to channel the corresponding proteins and peptides to particular subcellular secretory compartments.

H.-P. H. MOORE (*University of California, Berkeley, Calif.*): It is clear that there are subclasses of constitutive and regulated vesicles. MDCK cells, for example, have two types of constitutive vesicles, one destined for the apical surface and the other destined for the basolateral surface. By the same token, neuronal cells may have different regulated vesicles, some dense-core and some clear. I agree with you that in these cell types more complexed sorting must take place to direct the proper proteins into appropriate compartments.

I. GEFFEN (*Albert Einstein College of Medicine, Bronx, N.Y.*): This is in reference to a statement you have made regarding response of GH_3 cells to TRH in secreting prolactin or PTH that all secretory cells would respond to the same signals in secretion of any foreign-secreted protein introduced. Could you comment on further evidence supporting this statement?

H.-P. H. MOORE: Using immunoelectron microscopic techniques, Lelio Orci and I (in unpublished results) have evidence that foreign peptide hormones and the endogenous hormone ACTH are colocalized to the same secretory granules. This confirms the notion that the sorting machinery is conserved among different endocrine cells. Since the foreign hormone is packaged into the endogenous secretory granules, it is not too surprising that their secretion can now be regulated by secretagogues that normally affect secretion of endogenous hormones from these cells.

J. L. HO (*Tufts University, New England Medical Center, Boston, Mass.*): Recent evidence (published in *Nature* in the first week of June 1986) indicates that chloroquine and NH_4Cl affects not only granular pH but also amount of glycosylation of secretory proteins. Is there evidence in your system that chloroquine is not interfering with glycosylation?

H.-P. H. MOORE: Chloroquine has a slight effect on glycosylation of ACTH. However, since glycosylation does not appear to be essential for the sorting of ACTH, the effect of chloroquine on ACTH packaging cannot simply be attributed to its alteration of glycosylation.

M. J. GEISOW (*National Institute for Medical Research, London, U.K.*): I'd like to take up the suggestion of Hsiao-Ping that pH may be involved in sorting into the constitutive or triggered pathways—pH change is used generally to dissociate receptor-ligand complexes (acidic endosomes lead to generalized receptor-ligand dissociation). In the case of the transferrin receptor this is more resistant to dissociation from transferrin at low pH (the complex cycles intact). In the case of IgA transport in neonatal rats IgA associates with the receptor in the acidic gut, but dissociates in the neutral-alkaline environment of the blood stream.

M. LEVINE (*NIDDK, National Institutes of Health, Bethesda, Md.*): What evidence is there that what are termed "constitutively secreted substances" are actually within some vesicular compartment?

H.-P. H. MOORE: Immunoelectron microscopic evidences from several laboratories have shown that VSV G protein, albumin, and immunoglobulin molecules are localized within vesicular compartments before reaching the cell surface.

P. FLEMING (*Georgetown University, Washington, D.C.*): We have found the DBH is secreted from the PC12 cells by both constitutive and stimulus-coupled pathways, and that the constitutively secreted DBH is sulfated whereas the other DBH is not. Have you looked for differential sulfation of the proteins you have discussed today and could you comment on sulfation as a signal for constitutive secretion?

H.-P. H. MOORE: We found sulfated molecules are secreted by both constitutive and regulated pathways. This makes it unlikely that sulfation plays a role in sorting into the two pathways. Consistent with this notion is the observation that sulfation on ACTH can be inhibited by tunicamycin treatment without affecting sorting.

D. CASTLE (*Yale Medical School, New Haven, Conn.*): My first question is: Are there problems in interpreting sorting through use of chloroquine since elevation of pH_{in} may also inhibit processing of precursors to mature hormones? Secondly, how are Orci's results, published in *Cell* in 1985, interpreted where an antibody selectively recognizing proinsulin colocalizes to the same organelle, the forming granule, as mature insulin?

H.-P. H. MOORE: In response to your first question, if chloroquine simply inhibits processing, one should see precursor ACTH secreted by the regulated pathway by the drug-treated cells. This is not found. Chloroquine not only inhibited processing of ACTH, but also *diverted* the hormone from the regulated pathway to the constitutive pathway. To respond to your second question, according to our sorting model, proinsulin is sorted to a precursor compartment which matures into secretory granules. It is therefore expected that proinsulin and insulin should be localized to the same immature compartment, as observed by Orci.

Sidedness of the Major Membrane Proteins of Brain Synaptic Vesicles

ERIK FLOOR[a]

Department of Physiology
University of Massachusetts Medical School
Worcester, Massachusetts 01605

Synaptic vesicles purified from rat or cow brain contain a distinctive set of major proteins: protein Q, M_r = 74,000; R, ~ 61,000; S, 57,000; T ~ 40,000; U, 38,000; and V, 33,000.[1,2] Proteins R and T from rat brain are integral membrane proteins as shown by partitioning in Triton X-114 two-phase symptoms and by inextractability by alkali at low ionic strength.[1,2] The localization of the other major proteins is not known.

To identify externally facing proteins, synaptic vesicles were incubated with proteases. Proteins Q, R, S, and T were largely cleaved to smaller fragments by trypsin or chymotrypsin (FIG. 1). Protein U was partially cleaved by trypsin, and all of the proteins were degraded by proteinase K (FIG. 1).

These data provide information on protein sidedness if it can be shown that the proteases did not have access to the vesicle interior. Two kinds of indirect evidence of the integrity of the synaptic vesicle membrane during these experiments were gathered. First, by equilibrium density centrifugation the density of synaptic vesicles was found to be higher on a density gradient formed from a membrane permeant substance. Glycerol equilibrates with the intravesicular water space, which increases the vesicle density on a glycerol gradient, while large polysaccharides, *e.g.,* Ficoll, are excluded by intact membrane vesicles and thus do not affect vesicle density.[3] Synaptic vesicles from rat brain homogenates purified by chromatography on controlled-pore glass CPG-3000[1,2] had a density of ~1.06 g/cc after equilibrium density centrifugation (100,000 g, 17 h) on a 5-25% gradient of Ficoll (M_r ~ 400,000) in CK buffer and a density of ~1.12 g/cc after centrifugation on a 0-30% glycerol gradient containing 15% Nycodenz in the same buffer. Thus the synaptic vesicles seemed to be intact, *i.e.,* impermeant to Ficoll, before protease treatment. Second, treatment by trypsin or chymotrypsin under the conditions employed did not lyse synaptic vesicles in similar studies with vesicles containing known neurotransmitters.[4,5]

An interaction of proteins S and V at the outer surface of the synaptic vesicles was demonstrated directly by covalent cross-linking with the impermeant, thiol-cleavable reagent, 3,3′-dithiobis(sulfosuccinimidylpropionate) (DTSSP, Pierce). After cross-linking, vesicle proteins were analyzed by a two-dimentional (SDS) polyacrylamide gel electrophoretic procedure in which proteins are separated under nonreducing conditions, the cross-linker then cleaved with thiol reagent, and proteins separated

[a]Current address: Department of Anatomy, University of Wisconsin Medical School, 1300 University Avenue, Madison, Wis. 53706.

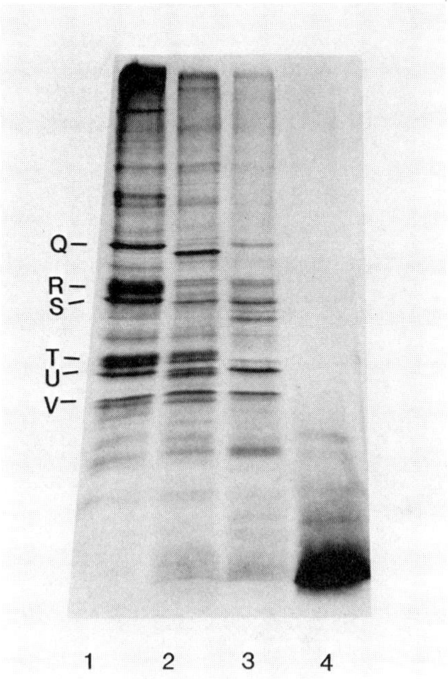

FIGURE 1. Sensitivity of synaptic vesicle proteins to proteases. Synaptic vesicles purified from rat brain homogenate and collected after chromatography on CPG-3000 in CK buffer[2] consisting of 0.16 M KCl, 5 mM $NaHPO_4$ (pH 6.6); 1 mM EGTA, 0.02% NaN_3, 1 mM dithiothreitol, and 1 μM 3,5-di-*tert*-butyl-4-hydroxybenzyl ether (BHBE) were pooled, divided into four equal portions of 90 μg total synaptic vesicle protein[7] each, and incubated 5 h at 4°C with proteases as indicated. The vesicles were then collected by centrifugation at 200,000 g for 30 min at 4°C and analyzed by electrophoresis on a 10% SDS polyacrylamide gel with Coomassie blue staining.[2] Lane 1, control (no additions); lane 2, trypsin (6 μg/ml, Sigma type XI) plus 2 mM $CaCl_2$; lane 3, α-chymotrypsin (10 μg/ml, Sigma type VII); and lane 4, proteinase K (3 μg/ml, Boehringer).

FIGURE 2. Cross-linking of synaptic vesicle proteins with DTSSP. Synaptic vesicles purified as described for FIGURE 1 were incubated in 50 μM DTSSP in 0.15 M KCl and 10 mM $KHPO_4$ (pH 6.6) for 10 min at room temperature. Then 25 mM Tris/HCl, pH 7.4, containing 20 mM *N*-ethylmaleimide was added and the vesicles pelleted and resuspended in SDS sample buffer (2% SDS, 63 mM Tris/HCl, pH 6.8, 5% 2-mercaptoethanol, and 10% glycerol) minus thiol reagent. After heating at 50°C for 3 min, the vesicle proteins were electrophoresed on a 10% polyacrylamide, 10 cm × 2.5 mm cylindrical tube gel. The gel was then removed, washed in SDS sample buffer, mounted in 1% agarose in SDS sample buffer and subjected to electrophoresis on a 1.5 mm, 12 × 14 cm 10% polyacrylamide slab gel, and stained with Coomassie blue. Rat brain synaptic vesicle proteins were electrophoresed in the second dimension as size references (*left lane*). The positions of cross-linked proteins S and V after two-dimensional separation are indicated by *arrows.* These spots fell on a vertical line which intersected the diagonal band on noncross-linked proteins (present also after control incubations without DTSSP) at a point that indicates a molecular weight of 89,000 for the cross-linked aggregate of proteins S, M_r = 57,000, and V, 33,000.

under reducing conditions.[6] Proteins S and V were cross-linked in a 1:1 complex by DTSSP and migrated as an 89,000 M_r aggregate in the first electrophoretic dimension (FIG. 2). This aggregate was not present after similar incubations without DTSSP. The level of cross-linking of proteins S and V was low, in part because of the need to limit the cross-linking reaction to prevent formation of very large, cross-linked protein aggregates.

In conclusion, these results suggest that the major synaptic vesicle proteins are exposed on the cytoplasmic vesicle surface and therefore imply that proteins Q, S, U, and V are peripheral proteins, bound to synaptic vesicles in macromolecular complexes with synaptic vesicle specific integral membrane proteins. Further cross-linking experiments may provide direct confirmation of this structural model.

REFERENCES

1. FLOOR, E., S. F. SCHAEFFER & S. E. LEEMAN. 1985. Soc. Neurosci. Abstr. **11:** 644.
2. FLOOR, E. & S. E. LEEMAN. 1986. Submitted.
3. GIOMPRES, P. E., S. J. MORRIS & V. P. WHITTAKER. 1981. Neuroscience **6:** 757-763.
4. WAGNER, J. A. & R. B. KELLY. 1979. Proc. Natl. Acad. Sci. U.S.A. **76:** 4126-4130.
5. FLOOR, E. & S. E. LEEMAN. 1985. Neurosci. Lett. **60:** 231-237.
6. WANG, K. & F. M. RICHARDS. 1975. J. Biol. Chem. **250:** 6622-6626.
7. SCHAFFNER, W. & C. WEISSMAN. 1973. Anal. Biochem. **56:** 501-514.

Hormonal Induction of a Heterogeneous Population of Storage Granules in GH_4C_1 Pituitary Tumor Cells

JONATHAN G. SCAMMELL,[a] THOMAS G. BURRAGE,[b] AND PRISCILLA S. DANNIES[b]

[a] *Department of Pharmacology*
University of South Alabama
College of Medicine
Mobile, Alabama 36688

[b] *Department of Pharmacology*
Yale University School of Medicine
New Haven, Connecticut 06510

The mechanism by which endocrine cells regulate the amount of newly synthesized hormone that is sorted into storage granules is unknown.[1] Model systems suitable for study have been lacking. We have developed a model system where the cellular content of a hormone and the number of storage granules can be regulated independently of hormone synthesis or degradation and have presented some of the details of this model here.

GH_4C_1 cells are a rat pituitary tumor cell strain that secretes prolactin and growth hormone, but stores very little of these hormones and contains almost no secretory granules. We plated these cells at low density (10 cells/mm^2) and maintained them in DMEM:Ham's F10 medium with 15% gelded horse serum. After three days, we treated the cells with 17β-estradiol (1 nM), insulin (300 nM), and epidermal growth factor (10 nM) for five days in the above medium and found that the cellular content of prolactin was increased by over 30-fold, but the accumulation of prolactin in the medium was increased only sixfold. Growth hormone storage and secretion were both decreased by the hormone treatment.

To determine if the increase in cellular prolactin was accompanied by an increase in secretory granules, we compared the numbers of granules in electron micrographs of ultrathin sections from untreated GH_4C_1 cells and from cells treated with the combined hormone regimen. Control cells contained 1.0 ± 0.2 granules per cell section ($n = 130$), while hormone-treated cells contained 45.6 ± 4.5 granules per cell section ($n = 130$), a nearly 50-fold increase in granule number (FIGS. 1A and 1B).

To determine if all of the granules contained prolactin, ultrathin sections of Lowicryl-embedded GH_4C_1 cells were stained for prolactin by the Protein-A-gold technique. The density of immunoreactive sites for prolactin was determined in 208 granules from five separate cells. Only 75% of the granules stained for prolactin; most of the granules that did not stain were the less electron-opaque granules (FIGS. 2A and 2B).

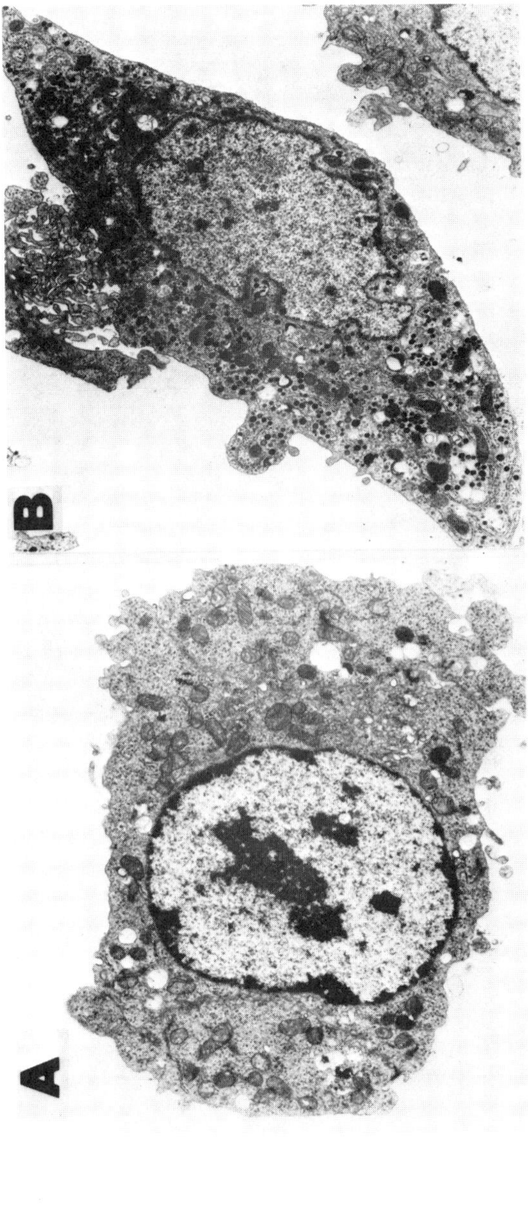

FIGURE 1. Electron micrographs of GH_4C_1 cells grown in (A) control medium or (B) medium supplemented with 17β-estradiol, insulin, and epidermal growth factor. Magnification: 8600×.

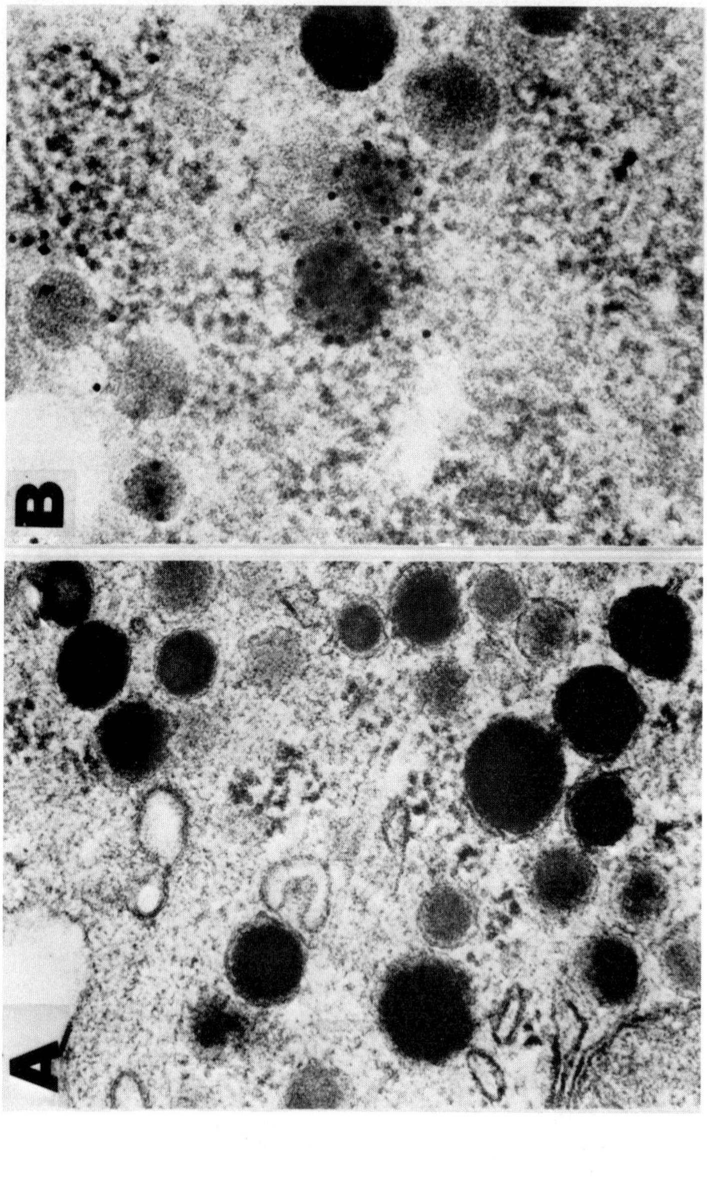

FIGURE 2. **(A)** Conventional electron microscopy of storage granules induced in GH_4C_1 cells by the administration of 17β-estradiol, insulin, and epidermal growth factor. **(B)** Immunoelectron microscopy of GH_4C_1 cells, treated as in **A**, using prolactin antiserum and Protein A gold. Magnification: 87,700×. (From Scammell *et al.*[4] Reprinted by permission from *Endocrinology*.)

The increase in secretory granules may occur in several ways. For example, hormone treatment may induce the formation of a factor that inhibits the release of storage granules, which then accumulate in the cytoplasm. Such a role has been suggested as a possible function of synapsin I in nervous tissue.[2] A second possibility is that hormone treatment may increase a signal or the recognition of a signal necessary to sort prolactin into storage granules. The evidence for these sorting mechanisms is discussed in the review by Kelly.[1] A third possibility is that the synthesis of granule components, such as the TSP 86/84 protein,[3] or the assembly of these components is rate limiting, and hormone treatment increases synthesis or facilitates assembly. These cells, with the ability to induce storage, offer us a model system to study the factors that govern the sorting of newly synthesized hormones into storage granules.

REFERENCES

1. KELLY, R. B. 1985. Pathways of protein secretion in eukaryotes. Science **230:** 35.
2. NESTLER, E. J. & P. GREENGARD. 1984. Substrate proteins. *In* Protein Phosphorylation in the Nervous System. Wiley. New York. p. 155.
3. ROSA, P., G. FUMAGALLI, A. ZANINI & W. B. HUTTNER. 1985. The major tyrosine-sulfated protein of the bovine anterior pituitary is a secretory protein present in gonadotrophs, thyrotrophs, mammotrophs, and corticotrophs. J. Cell Biol. **100:** 928-937.
4. SCAMMELL, J. G., T. G. BURRAGE & P. S. DANNIES. 1986. Hormonal induction of secretory granules in a pituitary tumor cell line. Endocrinology **119:** 1543-1548.

Somatostatin Is Targeted to Gonadotrophic Secretory Granules in Transgenic Mice Expressing a Metallothionein-Somatostatin Fusion Gene

PHILIP J. STORK,[a] MICHAEL J. WARHOL,[a]
MALCOLM J. LOW,[b] AND RICHARD H. GOODMAN[b]

[a] *Department of Pathology*
Brigham and Women's Hospital
Boston, Massachusetts 02115

[b] *Division of Endocrinology*
Tufts-New England Medical Center
Boston, Massachusetts 02111

INTRODUCTION

Peptide hormones are synthesized as large precursors, prehormones, that undergo a series of posttranslational modifications during intracellular packaging within secretory granules. The secretory granule contains the processing enzymes required for maturation of the prohormone and is designed to release its contents upon stimulation by the appropriate secretogogue.

Pituitary cell lines that correctly synthesize, process, and release endogenous hormone products through a regulated pathway are able to process and release the products of foreign pre-prohormone genes introduced by stable transfection. For example, AtT20 cells that normally synthesize and process ACTH precursors faithfully process proinsulin when the gene is introduced into the cells.

To assess the range of cell types able to process and package exogenously introduced pre-prohormones *in vivo,* transgenic mice were created by microinjection of a metallothionein-somatostatin fusion gene.[2] The most active site of somatostatin production and processing in this system is the anterior pituitary, a tissue that does not normally express somatostatin. Immunoreactive somatostatin within the anterior pituitary is found exclusively within gonadotrophs.[3]

The present study uses ultrastructural techniques to localize somatostatin to LH- and FSH-containing secretory granules with gonadotrophs of the anterior pituitary of transgenic mice. In addition, one of us (ML) has shown corelease of somatostatin and LH following stimulation of primary pituitary cultures from these mice by the gonadotrophic secretogogue LHRH.[3] These experiments demonstrate that a neurosecretory peptide encoded by a foreign gene can enter the regulated secretory pathway of pituitary cells from transgenic mice.

EXPERIMENTAL PROCEDURES

Immunocytochemistry

The pituitaries from transgenic mice were fixed *in situ* and embedded in Lowicryl resin. Serial thin sections (600 Å) were examined using the protein-A-colloidal gold technique.[4] Protein-A-gold complexes (14 nm) were used to visualize bound polyclonal antibodies directed to somatostatin, LH, FSH, ACTH, growth hormone, and prolactin (1:100 dilution). Reactions were carried out on the surface of nickel grids containing sequential thin sections of anterior pituitary.

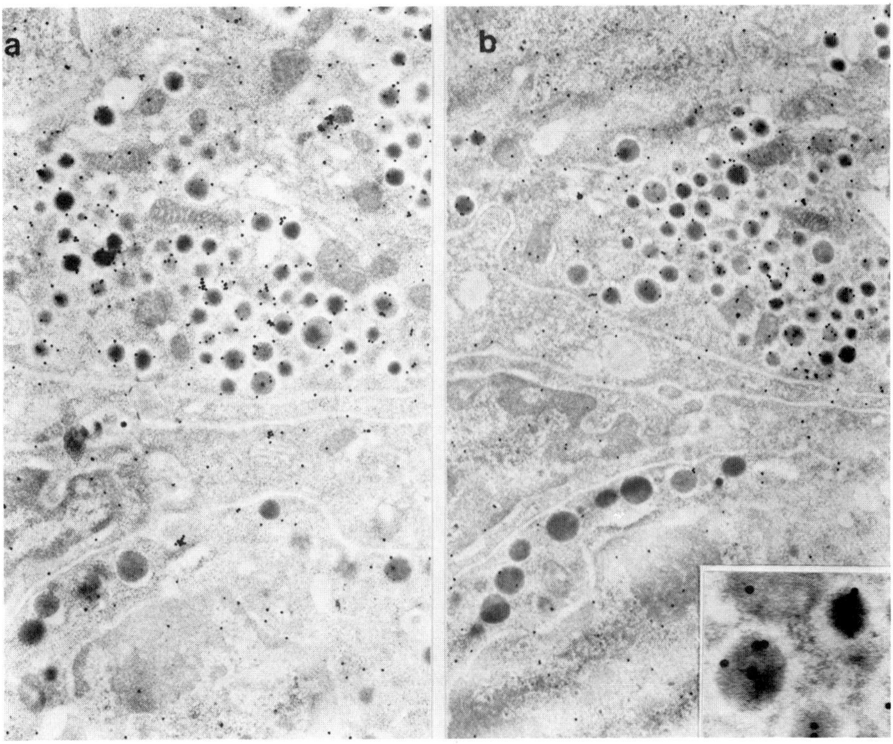

FIGURE 1. Serial sections of a gonadotrophic cell from a somatostatin-overproducing transgenic mouse. The gonadotrophin-producing cell has characteristic granules with central dark cores surrounded by a halo. Adjacent to the gonadotrophin cell is a growth-hormone-producing cell with larger, homogeneous granules. (**a**) Anti-FSH antiserum. The majority of granules in the cell are labeled with this antiserum. There is weak background staining over the nucleus and other organelles (original magnification: 8000×; reduced by 20%). (**b**) Antisomatostatin antiserum. The patterns of staining, both distribution and specificity, are identical to that seen with the anti-FSH. No staining is noted within the adjacent cell. (magnification: 8,000×). *Inset:* A higher power of the gonadotrophin granules demonstrating the peripheral halo. The dense granular matrix is labeled by gold particles (original magnification: 15,000×; reduced by 20%).

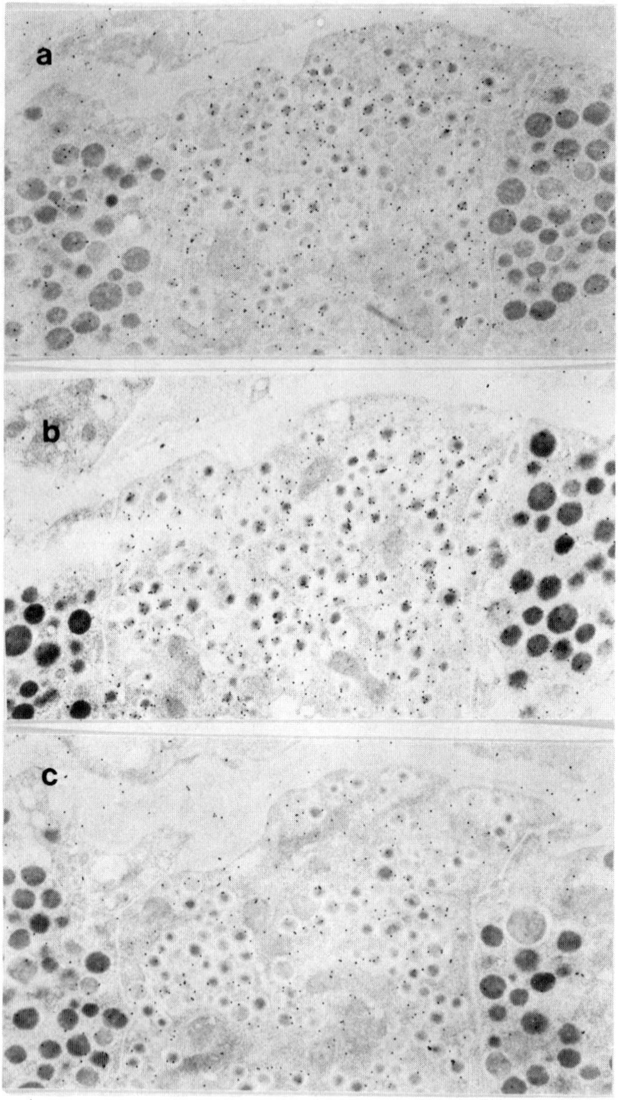

FIGURE 2. Serial sections of a single gonadotrophin-producing cell with two adjacent somatotrophic cells. **(a)** Antisomatostatin. The majority of the granules within this cell are labeled with this antiserum. The adjacent cells are unlabeled. There is weak background staining seen over the intercellular space (magnification: 6,000×). **(b)** Anti-LH antiserum. The majority of granules in the cell are also labeled with this antiserum in a pattern identical to that seen with the antisomatostatin (magnification: 6,000×). **(c)** Anti-FSH antiserum. A pattern of labeling identical to that seen with both the antisomatostatin and the anti-LH is observed (magnification: 6,000×).

RESULTS AND DISCUSSION

Ultrastructurally, the gold particles were clearly localized within secretory granules within somatostatin immunoreactive cells (FIG. 1, inset). Serial thin sections showed FSH and somatostatin (FIG. 1) and LH and somatostatin (FIG. 2) localized to the same cell and to the majority of secretory granules with a given section, suggesting that the same granule population contains somatostatin and the endogenous gonadotrophic hormones. Additional work has shown that LHRH causes a five-to-ten-fold increase in somatostatin secretion from primary pituitary cultures derived from these transgenic mice.

Taken together, these findings indicate that the products of such fusion genes in transgenic mice can enter the regulated secretory pathway of gonadotrophs. The expression of a related metallothionein growth hormone fusion gene in transgenic mice is also restricted to gonadotrophs,[3] suggesting that a fortuitous combination of DNA sequences common to both fusion genes accounts for this novel tissue-specific expression.

REFERENCES

1. MOORE, H.-P. H., M. D. WALKER, F. LEE & R. B. KELLY. 1983. Expressing a human proinsulin cDNA in a mouse ACTH-secreting cell. Intracellular storage, proteolytic processing, and secretion on stimulation. Cell **35:** 531-538.
2. LOW, M. J., R. E. HAMMER, R. H. GOODMAN, J. F. HABENER, R. D. PALMITER & R. L. BRINSTER. 1985. Tissue-specific posttranslational processing of pre-prosomatostatin encoded by a metallothionein-somatostatin fusion gene in transgenic mice. Cell **41:** 211-219.
3. LOW, M. J., P. J. STORK, R. E. HAMMER, R. L. BRINSTER, M. J. WARHOL, G. MANDEL & R. H. GOODMAN. 1986. Somatostatin is targeted to the regulated secretory pathway of gonadotrophs in transgenic mice expressing a metallothionein-somatostatin fusion gene. Submitted.
4. ROTH, J. *et al.* 1981. J. Histochem. Cytochem. **29:** 663-671.
5. LOW, M. J., R. M. LECHAN, R. E. HAMMER, R. L. BRINSTER, J. F. HABENER, G. MANDEL & R. H. GOODMAN. 1986. Gonadotroph-specific expression of metallothionein fusion genes in pituitaries of transgenic mice. Science **231:** 1002-1004.

Packaging of Pituitary Hormones

New Insights

A. ZANINI,[a] G. FUMAGALLI,[a] P. ROSA,[a,b]
M. G. COZZI,[a] AND W. B. HUTTNER[b]

[a] *CNR Center of Cytopharmacology*
University of Milan
Milan, Italy

[b] *European Molecular Biology Laboratory*
D-6900 Heidelberg
Federal Republic of Germany

The packaging of pituitary hormones into secretory granules employs mechanisms that have not yet been identified. In order to obtain information on this intracellular process, we have performed comparative studies on pituitary hormones (PRL, GH, LH, and FSH) and sulfated components, in particular two acidic tyrosine-sulfated secretory proteins, secretogranins I and II[1] (also called chromogranins B and C[2,3]).

By indirect immunofluorescence and double immunocytochemical labeling using protein-A-gold particles of different size, GH and PRL were found, in cow anterior pituitary, in specific somatotrophs and mammotrophs as well as in somatomammotrophs.[4,5] In these mixed cells the two hormones were packaged in different granules or in the same granules where they could be segregated in different portions of the granule content (FIG. 1).[5] In somatomammotrophs secretogranin II was present in the matrix of only a few secretory granules, usually less electron dense and smaller than the unlabeled ones; in somatotrophs it was completely absent.[4]

As far as LH and FSH are concerned, the two hormones were found by immunofluorescence in the same cells (as is well known from the literature). Secretogranin II was enriched in parallel with LH α and β subunits in subcellular fractions enriched in small granules, and was found by immunoelectron microscopy in the matrix of all secretory granules of gonadotrophs.[4] Cells containing secretogranin II also contained secretogranin I, and the distribution of the two tyrosine-sulfated proteins was not only overlapping but also complementary.[1] Biochemical studies (incubation *in vitro* of rat anterior pituitaries in the presence of [^{35}S]sulfate and analysis by 1D- and 2D-PAGE followed by fluorography of tissue homogenates) indicated that during development the secretogranins, such as sulfated LH α and β subunits, are present at highest levels in anterior pituitary glands of 14-day-old female rats, known for their high content in gonadotrophs. In these glands LHRH stimulated the *in vitro* release of both sulfated LH subunits and secretogranins (FIG. 2).[6]

From these observations it appears that the old concept "one cell, one hormone" is no longer likely; in fact not only LH and FSH can be packaged in the same cells, but also GH and PRL, as demonstrated by others in the rat as well.[7] The presence of the two hormones in different granules suggests the existence of mechanism(s) that sort them out from each other.[5] The presence in different cell types of acidic tyrosine-

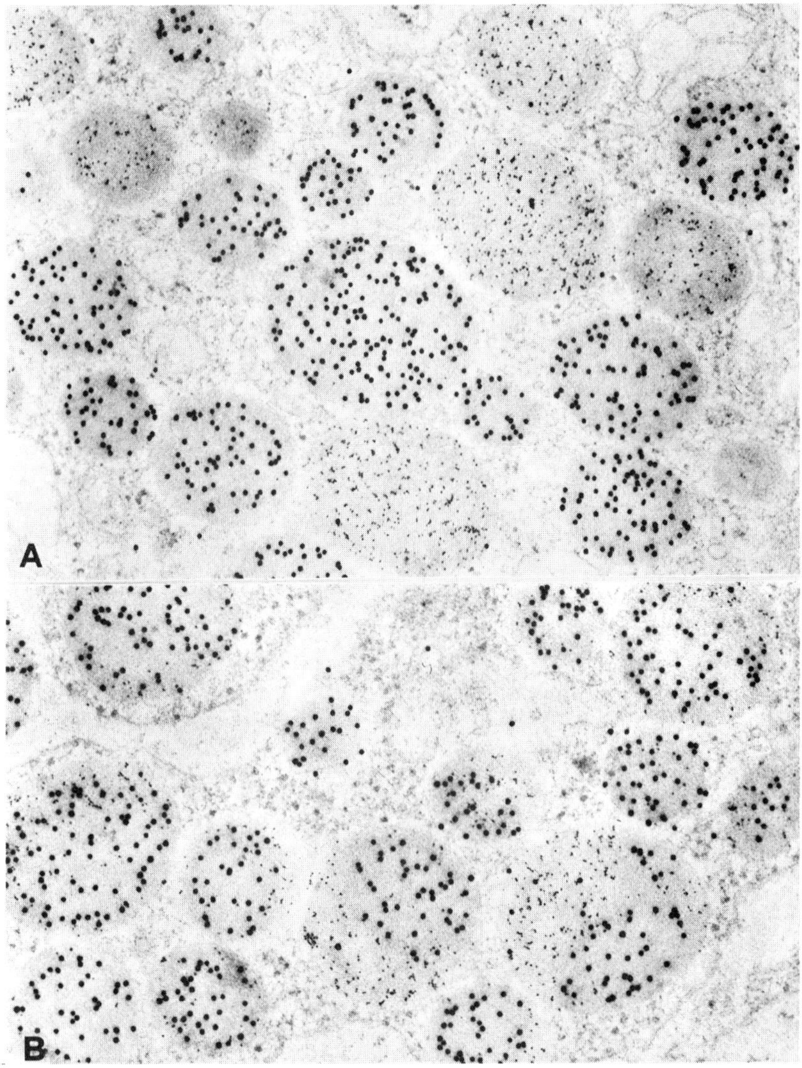

FIGURE 1. Ultrastructural localization, by double immunocytochemical labeling using protein-A-gold particles of different sizes, of GH (small gold particles) and PRL (large gold particles) in cow anterior pituitary. High magnification micrographs of somatomammotrophs. (**A**) Each granule is immunoreacting for one of the two hormones only (magnification: 38,500×). (**B**) Some granules have heterogeneous content with part composed of either GH or PRL (magnification: 41,500×). (From Fumagalli and Zanini.[5] Reprinted by copyright permission of the Rockefeller University Press.)

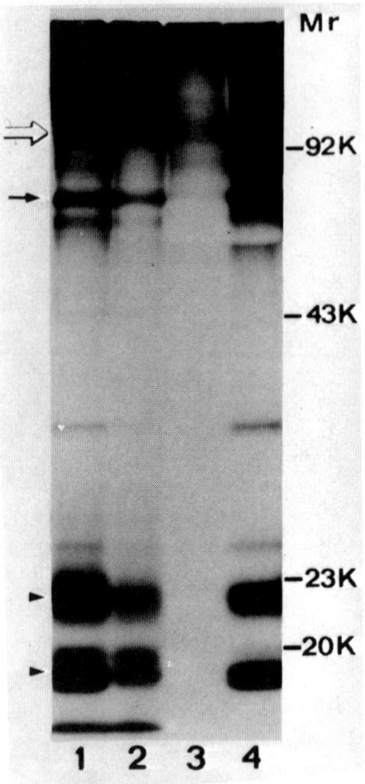

FIGURE 2. Fluorograms showing the effect of LHRH on the release of ^{35}S-labeled LH α and β subunits and secretogranins I and II from 14-day-old female rat adenohypophyses. Anterior pituitary slices, pulse-labeled for 90 min, were incubated for a total of 180 min in chase medium with or without LHRH (10 ng/ml). Lane 1: homogenate of pituitary slices incubated in the chase medium without LHRH; lane 2: homogenate of pituitary slices incubated in the chase medium with LHRH; lane 3: sulfated proteins released in the chase medium in the absence of LHRH; lane 4: sulfated proteins released in the chase medium in the presence of LHRH. *Open arrow:* secretogranin I; *arrow:* secretogranin II; *arrowheads:* sulfated LH α and β subunits. (From Cozzi and Zanini.[6] Reprinted by permission from Elsevier Science Publishers B.V.)

sulfated proteins, which are secreted together with the specific hormone via the regulated pathway, suggests that these "new" proteins might be involved in the sorting and packaging of different, though not all, pituitary hormones.

REFERENCES

1. Rosa, P., A. Hille, R. W. H. Lee, A. Zanini, P. De Camilli & W. B. Huttner. 1985. J. Cell Biol. **101:** 1999-2011.
2. Falkensammer, G., R. Fischer-Colbrie, K. Richter & H. Winkler. 1985. Neuroscience **14:** 735-746.
3. Fischer-Colbrie, R., C. Hagn, L. Kilpatrick & H. Winkler. 1986. J. Neurochem. **47:** 318-321.
4. Rosa, P., G. Fumagalli, A. Zanini & W. B. Huttner. 1985. J. Cell Biol. **100:** 928-937.
5. Fumagalli, G. & A. Zanini. 1985. J. Cell Biol. **100:** 2019-2024.
6. Cozzi, M. G. & A. Zanini. 1986. Mol. Cell. Endocrinol. **44:** 47-54.
7. Frawley, L. S., F. R. Boockfor & J. P. Hoeffler. 1985. Endocrinology **116:** 734-737.

Cholinergic Synaptic Vesicles from the Electromotor Nerve Terminals of *Torpedo*

Composition and Life Cycle

V. P. WHITTAKER

Abteilung Neurochemie
Max-Planck-Institut für biophysikalische Chemie
Göttingen, Federal Republic of Germany

THE ELECTROMOTOR SYSTEM OF THE ELECTRIC RAY

The electromotor systems of electric rays of genera *Torpedo* and *Narcine* are now well established models for the study of cholinergic transmission. The electrocytes of the electric organ are derived embryologically from muscle.[1,2] They are electrically inexcitable but respond to applied acetylcholine.[3,4] They receive a profuse cholinergic[3] innervation from the electric lobes, prominent paired nuclei on the dorsal surface of the brain stem just behind the cerebellum[1] containing the cell bodies of the electromotor neurones. The axons of these cells are heavily myelinated and travel in eight large nerve trunks (four on each side of the neural axis) between the gills into the electric organ. The electromotor cells themselves receive an axo-dendritic, contralateral input from the oval nuclei in the medulla.[5]

A single average-sized specimen of *Torpedo marmorata,* the most readily available species in Europe, provides about 400 g of electric organ containing 500-1000 times more synaptic material than muscle. The tissue is highly collagenous and difficult to homogenize in the conventional way, but freezing in liquid nitrogen renders it brittle and it may then be comminuted by crushing.[6] This process, besides breaking up the tissue, tears open the nerve terminals and by extracting the tissue fragments with iso-osmotic sucrose, saline, or sucrose-saline and removing coarse particles by centrifuging, a vesicle-rich cytoplasmic extract is obtained that is a suitable starting material for further purification by isopycnic continuous density-gradient centrifuging in a zonal rotor[6,7] or exclusion chromatography on porous glass beads[7,8] or Sephacryl.[9] In this way synaptic vesicles may be purified to a high and constant concentration of vesicle markers[7,10] Such vesicles are extremely rich in acetylcholine—over 6 nmol per milligram of protein, corresponding to an internal concentration of 0.9 M or 2×10^5 molecules per vesicle.[7] In addition vesicles contain a second small molecular mass constituent, ATP,[11] present in about 0.17 M concentration.

THE STRUCTURE OF THE UNPERTURBED SYNAPTIC VESICLE

The results of the work of my colleagues and myself on the structure of synaptic vesicles isolated from resting electromotor nerve terminals are summarized in FIGURE 1.[12,13] The vesicles are larger than those in ordinary motor nerve terminals or autonomic or central terminals: 90 versus 50 nm in diameter. All the lipid and almost all of the protein is assigned to the membrane.[7] The vesicle contains a vesicle-specific proteoglycan whose hydrophilic sulfonated carbohydrate residues are directed towards the core.[14] The membrane proteins are about five in number and are of molecular mass 160, 145, 45, 32, and 25 kDa.[10] At least four are functional membrane proteins and comprise a proton-translocating ATPase,[15] a Ca^{2+}, Mg^{2+}-stimulated ATPase,[16] an ADP-ATP carrier,[17,18] and an acetylcholine carrier.[19] The fifth, of molecular mass 42 kDa, is a nervous-system-specific form of actin.[20]

The internal pH of the vesicle is about 5.5[21] and the pH gradient probably plays an important part in the uptake of acetylcholine and the second small-molecular-mass component, ATP. Besides these, the vesicle contains appreciable amounts of Ca^{2+} and Mg^{2+}.[22] The nuclear magnetic resonance ^{31}P and ^{1}H spectra show that acetylcholine and ATP are essentially free in solution in the vesicle core[21,23] and osmotic pressure studies show that they exert an osmotic pressure.[24] Water-space measurements[25] show that there are three main water compartments, the osmotically active water (65% of total vesicle volume), the water bound to solutes (7%), and the membrane water (8%). The nonsolvated and hydrophobic components of the membrane account for 17% and the nonsolvated solutes for 3% of vesicle volume.

THE TRANSMITTER POOLS IN RESTING TERMINALS

The enzyme synthesizing acetylcholine, choline acetyltransferase, has a molecular mass of about 68 kDa and is present in the cytosol.[26] It utilizes choline and acetylcoenzyme A. The result of its activity is a small cytoplasmic pool of acetylcholine. This pool is continuously being replaced even in the resting terminal, since acetylcholine is being continuously lost from the terminal; extracellular transmitter is then rapidly hydrolyzed by acetylcholinesterase present in the synaptic cleft and the products of hydrolysis are taken up again by the terminal. As a result of this "futile recycling," the cytosolic pool of acetylcholine is easily labeled, using radioactive or deuterated choline or acetate. Interestingly, vesicles isolated from blocks of resting tissue exposed to labeled acetylcholine precursors do not themselves incorporate label, either in the form of the precursors or acetylcholine.[27,28] The proportion of tissue acetylcholine in the cytosolic pool can therefore be quite simply measured noninvasively[29] in experiments in which tissue blocks have been exposed to labeled choline by comparing the isotopic ratio r_{ACh} of tissue acetylcholine with that of its precursor choline r_{Ch}. The ratio of these ratios r_{ACh}/r_{Ch} is the proportion by which tissue acetylcholine synthesized from tissue choline has been diluted with nonexchangeable (vesicular) acetylcholine, *i.e.*, the proportion of acetylcholine in the cytosolic pool. This is 22 ± 3%.

THE EFFECT OF STIMULATION

Electrophysiologically, electromotor synapses closely resemble other motor synapses, *e.g.*, in frog muscle. Release of transmitter is quantized and stimulation synchronizes quantized release with the production of an excitatory postsynaptic potential (EPSP).[30] It is the summation in series and parallel of these normal-sized EPSPs occurring simultaneously in some 360,000 electrocytes stacked in some 500 columns on each side of the fish[1] that generates the electric organ discharge (about 40 V measured in air).

Stimulation mobilizes synaptic vesicles and causes them to release transmitter by

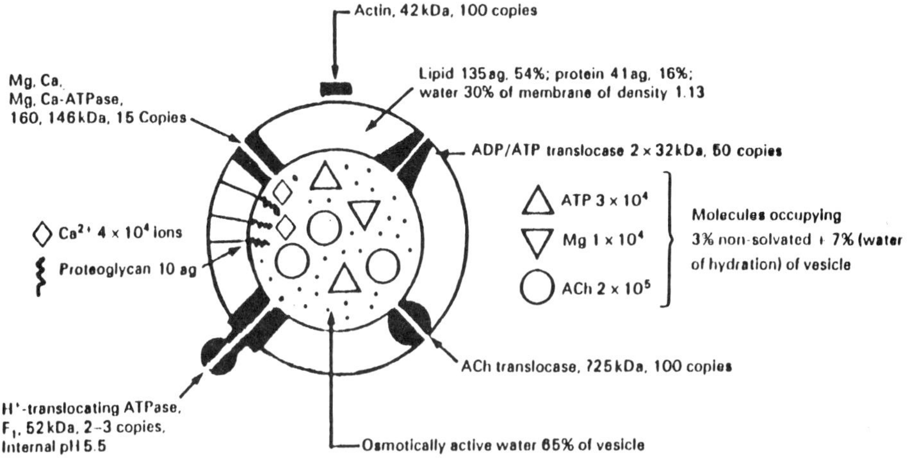

FIGURE 1. Structure of the synaptic vesicle. *ACh,* acetylcholine; ag$=10^{-18}$ g.

exocytosis. The vesicles reform and refill at the expense of cytoplasmic acetylcholine, which tends to fall during and after stimulation to about a third of its resting value. The reutilization of vesicles is shown by the fact that at low rates of stimulation (0.1 to 0.15 Hz), acetylcholine can be released for a long period without significant diminution of vesicle numbers. The recycling vesicles (identified by the uptake of dextran particles into their lumina), became smaller and denser than the reserve population however,[31,32] and so can be separated from them by density gradient centrifugation[31,33] or exclusion chromatography.[34] In experiments with labeled precursors, these recycling vesicles incorporate labeled acetylcholine from the cytosol, and on restimulation, the labeled acetylcholine is preferentially released.

Reversibility of the Effects of Stimulation

All the changes in the morphology and biophysical properties of vesicles just described are reversible.[32,33,35,36] FIGURE 2 shows the result of a recent experiment with perfused blocks of electric tissue. The electric organs were stimulated in anaesthetized fish *in vivo* via the electric lobe at 0.15 Hz for 3.33 h to generate a large pool of recycling vesicles and were then removed for perfusion with deuterated choline and allowed to recover. Cytoplasmic extracts of the blocks were prepared at suitable time intervals and submitted to continuous density gradient centrifuging in a zonal rotor. As can be seen in FIGURE 2, at 2 h endogenous acetylcholine was bimodally distributed in the gradient and newly synthesized deuterated acetylcholine had been taken up selectively by the denser, recycled vesicles. During the 16-h period of recovery, the depleted acetylcholine stores of the tissue were replenished and the recycled vesicles gradually recovered the biophysical properties of the reserve population as shown by the diminishing density of the labeled, recycled vesicle population (see insert). By 16 h, the recycled vesicles had regained the characteristics of the reserve population.

Basis for the Greater Density and Smaller Size of Recycled Vesicles

The increased density and smaller size of recycling or recently recycled vesicles can be precisely accounted for by a model in which partial reloading of vesicles emptied by exocytosis and reformed in a fully functional state by endocytosis is followed by osmotic dehydration in response to the reduced osmotic load. The lower osmotic load is apparent from the reduced osmotic fragility of recycled vesicles compared to that of reserve vesicles from stimulated or nonstimulated tissue[37] and their lower ratio of acetylcholine to a stable vesicle marker such as vesicular proteoglycan[36,38,14] (FIG. 3). The density changes induced in the two subpopulations of vesicles by changes in the osmotic pressure of the suspension medium are consistent with osmometer-like properties and a lower osmotic loading of the recycled subpopulation. Measurements of water space show a depleted water space in the recycled vesicles.

TABLE 1 lists the biophysical parameters of the two populations. Since those of the recycled vesicles vary with recovery time, two sets of values are given: those relating to zero and 2-hour recovery respectively.[24] There is no need to assume any change in the mass or composition of the vesicle membrane or in macromolecular constituents of the core which in any case probably consist only of glucosaminoglycan side chains of a proteoglycan firmly anchored in the membrane; indeed any such change would be inconsistent with the recovery, by the recycled vesicles, of the density and size of the reserve vesicles during recovery from stimulation.

FIGURE 2. The subpopulation of recycled vesicles present in blocks of stimulated electric organ and labeled with deuterated acetylcholine recover the biophysical properties of the reserve population after 16-h rest.[36] The physiological integrity of the blocks was preserved by perfusion. *Inset:* plot of gradient volume separating the peaks of reserve and recycling vesicles as a function of time.

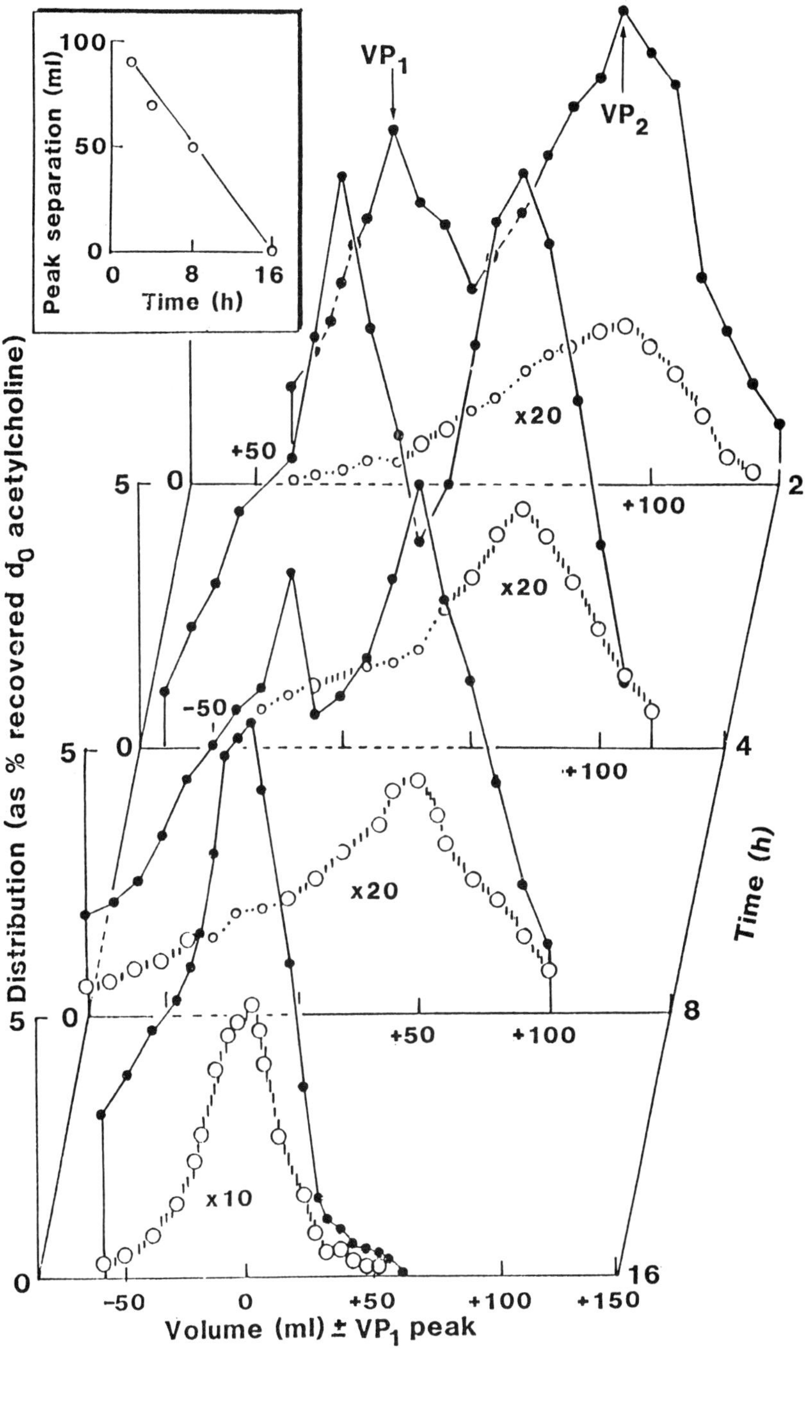
Peak separation (ml)
100
50
0
0
8
16
Time (h)
VP1
VP2
x20
x20
x20
x10
+50
+100
-50
+100
+50
+100
-50
0
+50
+100
+150
2
4
8
16
Time (h)
Distribution (as % recovered d0 acetylcholine)
5
0
Volume (ml) ± VP1 peak

TABLE 1. Parameters of Reserve (VP_1) and Recycled (VP_2) Vesicles at 800 mOsM

Symbol	Parameter	Units[a]	How Calculated	VP_1	VP_2 No Recovery	VP_2 2-h Recovery
	VESICLE					
d	diameter	nm	—	90[b]	71.1[c]	72.4[c]
V	volume	al	$\pi d^3/6$	0.382	0.188	0.199
ℓ	density	$fg \cdot al^{-1}$	—	1.056[d]	1.067[d]	1.066[d]
	MEMBRANE					
d_m	thickness	nm	$(d-d_c)/2$	4.1	8.7	8.7
V_m	volume	al	$V-V_c$	0.095	0.107	0.107
υ_m	volume fraction	—	$1-\upsilon_c$	0.25	0.57	0.54
ℓ_m	density	$fg \cdot al^{-1}$	—	1.13[e]	1.11[f]	1.11[f]
	CORE					
d_c	diameter	nm	$\sqrt[3]{6V_c/\pi}$	81.8	53.7	56.0
V_c	volume	al		0.287	0.081	0.092
υ_c	volume fraction	—		0.75	0.43	0.46
ℓ_c	density	$fg \cdot al^{-1}$	—	1.03	1.003	1.012
υ_G	glycerol space	—	—	0.65	0.42	0.44
υ_s	volume fraction of hydrated solutes[g]	—	—	0.10	0.01	0.02

[a] Note that 1 al (attoliter) = 10^{-18} l = 10^6 nm^3, 1 fg (femtogram) = 10^{-15} g, 1 fg/al = 1 g/ml.
[b] Ref. 11.
[c] Ref. 13.
[d] Isosmotic density gradient measurements of intact vesicle.
[e] Isosmotic density gradient measurements of osmotically collapsed ghost.[11]
[f] Assumes constant membrane mass and 12% increase in volume by imbibition of water.
[g] Determined by measuring the increase in the glycerol space on lysis, proportionally for VP_2 vesicles by the observed extent of reloading.

Labeling of the Transmitter Pools by Analogues

In theory, the preferentially released newly synthesized, labeled acetylcholine could not have released from recycling vesicles[39] but directly from the cytosolic compartment through some kind of gate, pore, or carrier.[40] Experiments with analogues of choline that are also substrates for choline acetyltransferase and are acetylated to acetylcholine analogues have, however, enabled the cytosolic pool and that in the recycling vesicles to be differentially labeled.[41] The possibility of doing this depends on the fact that the uptake of acetylcholine into vesicles is carrier mediated and that the carrier has its own distinctive specificity, different from that of choline acetyltransferase. Homocholine is one of several suitable choline analogues (FIG. 4). Both it and its acetylated product are taken up into recycling vesicles, but the ratio of homocholine to acetylhomocholine in the vesicular pool is much lower than that in the cytosol. At rest the analogues leak out of labeled tissue in the same ratio as they are found in the cytosol. On stimulation, however, the ratio in which the two transmitter analogues are released is *that in which they are present in the vesicles and not that of the cytosol.* The only way the proponents of the cytosolic release theory can accommodate this result would be to postulate that their hypothetical carrier changes its specificity on stimulation to that of the vesicular storage mechanism (or that new carriers are activated by stimulation). These assumptions seem far-fetched.

Recently the compound AH-5183, which in nanomolar concentrations blocks the uptake of acetylcholine into vesicles,[42] has been used to block vesicle recycling in perfused tissue. Under these circumstances, recycled vesicles are unable to take up labeled cytosolic acetylcholine (*i.e.*, recycling is effectively blocked) and the continued release of transmitter under these circumstances depends on the recruitment of the reserve pool.[43] The transmitter released is unlabeled; since it is inconceivable that any cytosolic gate could distinguish between labeled and unlabeled transmitter, the released transmitter must have been released directly from the reserve pool of vesicles via exocytosis.

ORIGIN OF CHOLINERGIC SYNAPTIC VESICLES

Except perhaps for a limited protein-synthetic activity by terminal mitochondria, it is generally accepted that the nerve terminal is dependent on the neuronal perikaryon for the synthesis of its proteins and organelles. In other systems there is evidence that synaptic vesicles are made in the cell body and conveyed to the terminal by axonal transport. The perikarya of the electromotor neurons, which are among the largest motor neurons found in vertebrates, are packed with rough and smooth endoplasmic reticulum, polysomes, and vesicles. Calculations show that each perikaryon has to maintain some 30 to 50 times its own volume of nerve terminal.

As mentioned earlier, synaptic vesicles contain a specific proteoglycan of the heparan sulfate type. This can be detected immunochemically, immunocytochemically, and by labeling *in vivo* with ^{35}S injected into the electric lobe. These techniques have been recently used to follow the synthesis of cholinergic synaptic vesicles in the cell bodies of the electromotor neurons and their export via the axon hillock and the nerve axons to the terminals in the electric organ. FIGURE 5a shows punctuate immunofluorescence staining of synaptic vesicles in the axon hillock area of an electromotor

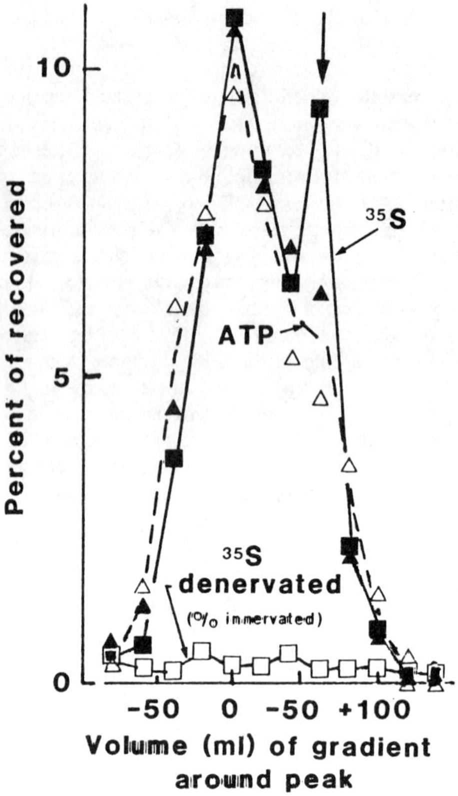

FIGURE 3. Isolation of synaptic vesicles on a continuous gradient in a zonal rotor after extraction from a frozen and crushed electric organ 48 h after *in vivo* injection of $^{35}SO_4^{2-}$ into the lobe (redrawn from ref. 14). Here ATP (▲,△) is used as a vesicle marker (nmol recovered, 3350 ± 670).[2] The distribution of ^{35}S in the gradient (total counts recovered, 6.6×10^5) is indicated by ■. *Open symbols* are the results with the contralateral organ, the nerves to which were cut before the injection of $^{35}SO_4^{2-}$. It will be noted that ^{35}S comigrates with the vesicle marker, indicating the presence of newly arrived vesicles in the vesicle pool, only when the afferent nerves are intact, though unlabeled vesicles are present (△). Although these vesicles were derived from an unstimulated organ, a small number of recycling vesicles are present, perhaps due to adventitious stimulation, as shown by the shoulder (*arrow*) on the dense side of the ATP peak. Note that the ATP: ^{35}S ratio is lower here than in the peak fractions, indicating that such vesicles are only partially reloaded with osmotically active small molecules.

perikaryon awaiting export down the axon[44]; FIGURE 5b shows the accumulation of stain above a ligature, FIGURE 5c its presence in the nerve terminal before ligation, and FIGURE 5d, its presence after ligation; the reduced staining reaction following ligation is apparent.[45] FIGURE 6a shows the distribution after separation on a zonal density gradient of covalently bound particulate ^{35}S extracted from nerve trunks while the wave of rapidly transported ^{35}S was passing down the axon following the injection of $^{35}SO_4^{2-}$ into the electric lobe.[46,47] FIGURES 6b, 6c, and 6d show the corresponding distribution of ^{35}S after extraction of vesicles from the electric organ, 18, 32, and 46 h after injection of ^{35}S into the lobe. It will be seen that initially the main peak of the highly sulfonated labeled vesicular proteoglycan does not coincide with that of the classical vesicle marker ATP; after 32 h it does. At the intermediate time points the labeled proteoglycan is bimodally distributed. It can be concluded from this experiment that vesicles arriving at the nerve terminal from the cell body have a low ATP (and acetylcholine) content, and perhaps because of this, a low density. After some hours

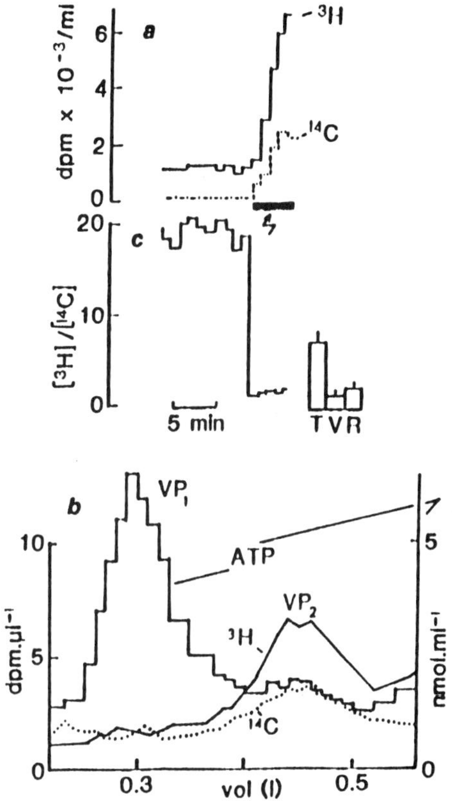

FIGURE 4. Experiments illustrating the incorporation of false transmitters into the fraction (VP_2) of recycling vesicles and their release from this fraction. (**a**) Blocks of electric organ were loaded with [^{3}H]homocholine and [^{14}C]choline after having been depleted of endogenous transmitter by stimulation through the nerve at 1 Hz for 30 min. On restimulation (10 Hz for 5 min) 8 h later in the presence of paraoxon (to stabilize released esters) and hemicholinium-3 (to inhibit reuptake of label), both radioactive labels were released, ^{3}H as a mixture of homocholine and acetylhomocholine, ^{14}C as acetylcholine. (**b**) The labels were incorporated exclusively in the fraction of recycling vesicles. (**c**) The ratio in which the true and false transmitters were released (block *R*) is the same as in the fraction of recycling vesicles (block *V*) and much lower (because of preferential vesicular uptake of acetylcholine relative to acetylhomocholine plus homocholine) than in whole tissue (block *T*). In contrast, the ratio in which the labels were released during the prestimulation resting period (*left*) is much higher than even the tissue ratio, reflecting a preponderance of homocholine and acetylhomocholine relative to acetylcholine in the cytoplasm. The tissue ratio observed is consistent with a 25% cytoplasmic acetylcholine pool and a 77% cytoplasmic homocholine plus acetylhomocholine pool. In other experiments, it was deduced that the proportion of acetylcholine that is cytoplasmic in resting tissue without any loading stimulus is 22%. Blocks are mean values of eight experiments; bars are SEMs.

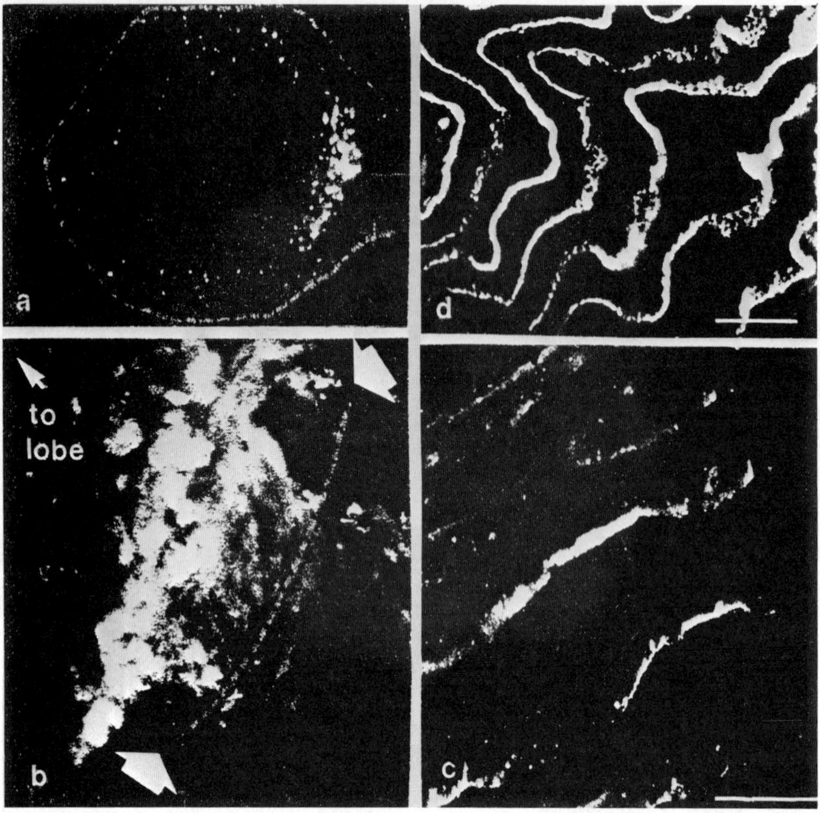

FIGURE 5. (**a**) An electromotor neuron cell body immunocytochemically stained with an antiserum raised against synaptic vesicle proteoglycan showing the distribution of synaptic vesicles in the perikaryon. The cell outline is indicated by the *dashed line.* Note the high concentration of vesicles in the axon hillock awaiting axonal transport. (**b**) Accumulation of immunoreactive material in electromotor axons above a ligature (between *broad arrows*). (**c,d**) Immunoreactive material is concentrated at the innervated faces of electrocytes but the amount present is reduced by ligation of the axon (**c**).

the vesicles become loaded with acetylcholine and acquire the density of reserve vesicles.[46,47]

CONCLUSIONS

Our conclusions regarding the transmitter pools and the recycling of synaptic vesicles are summarized in FIGURE 7. Vesicles are generated in the cell body and are conveyed by rapid transport to the terminal when they acquire acetylcholine and join

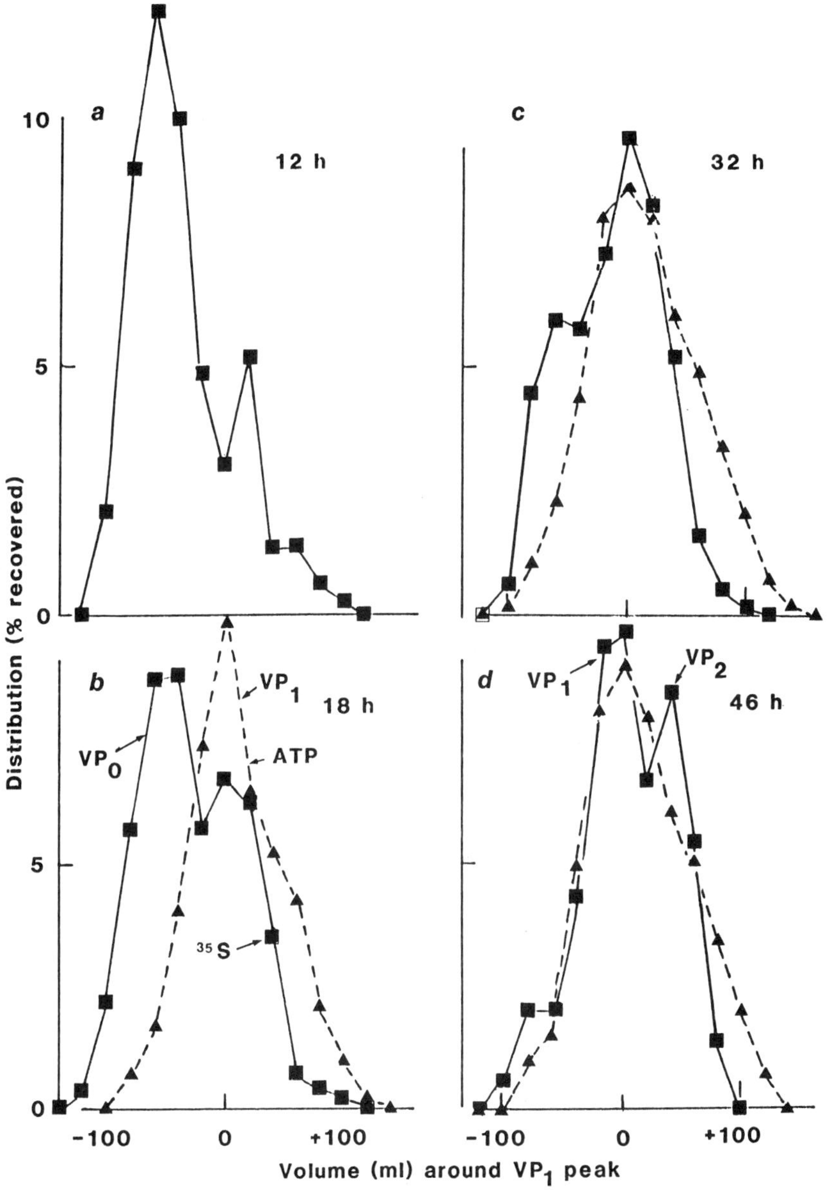

FIGURE 6. Separation of synaptic vesicles on a continuous density gradient in a zonal rotor (**a**) 12 h, (**b**) 18 h, (**c**) 32 h, and (**d**) 46 h after injecting $^{35}SO_4^{2-}$ into the lobe. Vesicles were extracted from (**a**) the frozen and crushed nerves or (**b,c,d**) electric organs. Note that at 18 h, the main (VP_0) peak of ^{35}S (■) in the gradient of terminal material has the same density as that of axonal material at 12 h but that by 32 h, the ^{35}S-labeled material mainly copurifies with the classical synaptic vesicle marker ATP (▲). For explanation see text.

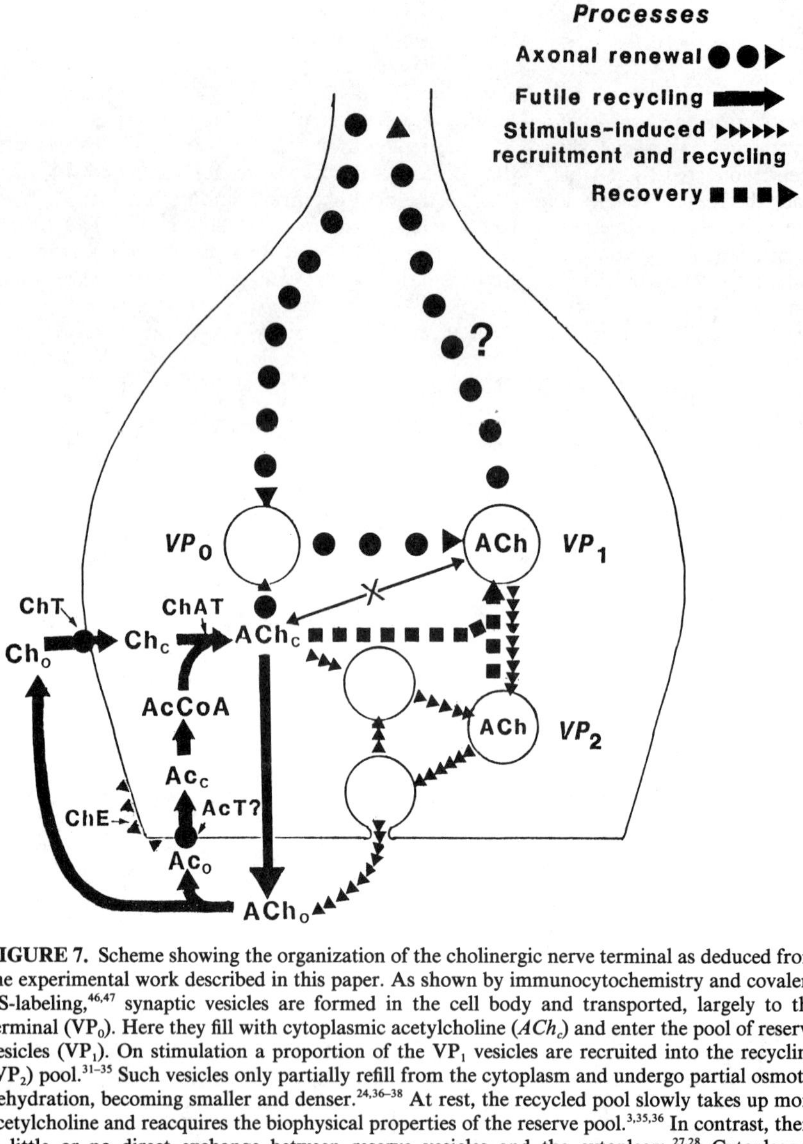

FIGURE 7. Scheme showing the organization of the cholinergic nerve terminal as deduced from the experimental work described in this paper. As shown by immunocytochemistry and covalent ^{35}S-labeling,[46,47] synaptic vesicles are formed in the cell body and transported, largely to the terminal (VP_0). Here they fill with cytoplasmic acetylcholine (ACh_c) and enter the pool of reserve vesicles (VP_1). On stimulation a proportion of the VP_1 vesicles are recruited into the recycling (VP_2) pool.[31–35] Such vesicles only partially refill from the cytoplasm and undergo partial osmotic dehydration, becoming smaller and denser.[24,36–38] At rest, the recycled pool slowly takes up more acetylcholine and reacquires the biophysical properties of the reserve pool.[3,35,36] In contrast, there is little or no direct exchange between reserve vesicles and the cytoplasm.[27,28] Cytoplasmic acetylcholine is subject to "futile recycling." Transmitter leaking out of the terminal (ACh_O) is rapidly hydrolyzed by acetylcholinsterase (*ChE*) in the cleft and the acetate (Ac_O) and choline (Ch_O) are salvaged; in both cases uptake is facilitated by carriers (*AcT, ChT*). Cytoplasmic choline (Ch_c) and acetate (Ac_c) are resynthesized to cytoplasmic acetylcholine by the soluble enzyme choline acetyltransferase (*ChAT*); acetate must however be first converted to acetylcoenzyme A (*AcCoA*). The cytoplasmic and VP_2 pools of transmitter can be specifically labeled by means of false transmitters and thus cytoplasmic release and that brought about by vesicle recycling can be readily distinguished (FIG. 4).[41] A similar cycle exists for synaptic vesicle ATP.[48] There is no evidence yet for antidromic transport of "worn out" vesicles, though by analogy with other systems this is likely to occur.

the pool of reserve vesicles. On stimulation, some of these are induced to undergo exocytosis, a process triggered by the ingress of free Ca^{2+} into the terminal. The molecular mechanisms involved in exocytosis are as yet poorly understood in any system. In the terminal it is a transient event and is followed by the retrieval of the vesicle in a functional state. The empty vesicle rapidly refills, perhaps by a process involving the exchange of inorganic cations acquired during exocytosis for cytosolic acetylcholine, but since this refilling process is partial and results in a reduced osmotic load, the refilled vesicles lose water to the cytosol, shrink, and become denser. They may recycle many times, but at rest they take up more acetylcholine and ATP (perhaps by utilizing the proton-gradient known to exist across their membranes[21]), reacquire water from the cytoplasm, and thus rejoin the population of reserve vesicles. Eventually synaptic vesicles must, like other organelles, "wear out." Just how such vesicles are disposed of, whether by leaving them unretrieved in the plasma membrane, which must then be adjusted in surface area by some other mechanism, such as membrane intake via coated vesicles, or by being passed back to the cell body by retrograde transport, remains obscure.

The model advanced in this article for vesicle recycling differs from earlier concepts in stressing the kinetic complexity of the system and the metabolic and functional heterogeneity of a class of organelles whose appearance in electron micrographs is deceptively uniform. Such a model is, however, more in line with what we know about cell biology in general and goes far to resolve paradoxes of transmitter synthesis, storage, and release such as the preferential release of newly formed transmitter once felt to create difficulties for a vesicular model.

REFERENCES

1. FRITSCH, G. 1890. Die elektrischen Fische nach neuen Untersuchungen anatomisch-zoologisch dargestellt. von Veit. Leipzig.
2. FOX, G. Q. & G. P. RICHARDSON. 1978. J. Comp. Neurol. **179:** 677-697.
3. FELDBERG, W. & A. FESSARD. 1942. J. Physiol. **101:** 200-216.
4. GRUNDFEST, H. 1957. Progr. Biophys. Biophs. Chem. **7:** 1-85.
5. SZABO, T. 1965. C. R. Soc. Biol. Paris **159:** 29-33.
6. WHITTAKER, V. P., W. B. ESSMAN & G. H. C. DOWE. 1972. Biochem. J. **128:** 833-846.
7. OHSAWA, K., G. H. C. DOWE, S. J. MORRIS & V. P. WHITTAKER. 1979. Brain Res. **161:** 447-457.
8. NAGY, A., R. R. BAKER, S. J. MORRIS & V. P. WHITTAKER. 1976. Brain Res. **109:** 285-309.
9. KRISTJANSSON, G. I., H. STADLER & U. WELSCHER. 1986. In preparation.
10. TASHIRO, T. & H. STADLER. 1978. Eur. J. Biochem. **90:** 479-487.
11. DOWDALL, M. J., A. BOYNE & V. P. WHITTAKER. 1974. Biochem. J. **140:** 1-12.
12. WHITTAKER, V. P. & H. STADLER. 1980. *In* Proteins of the Nervous System. 2d ed. R. A. Bradshaw & D. M. Schneider, Eds.: 231-255. Raven. New York.
13. WHITTAKER, V. P. 1986. Trends in Pharmacological Sciences. **7:** 312-315.
14. STADLER, H. & G. H. C. DOWE. 1982. EMBO J. **1:** 1381-1384.
15. HARLOS, P., D. A. LEE & H. STADLER. 1984. Eur. J. Biochem. **144:** 441-446.
16. BREER, H., S. J. MORRIS & V. P. WHITTAKER. 1977. Eur. J. Biochem. **80:** 313-318.
17. STADLER, H. & E. M. FENWICK. 1983. Eur. J. Biochem. **136:** 377-382.
18. LEE, D. A. & V. WITZEMANN. 1983. Biochemistry **22:** 6123-6230.
19. GIOMPRES, P. E. & Y. A. LUQMANI. 1980. Neuroscience **5:** 1041-1052.
20. ZECHEL, K. & H. STADLER. 1982. J. Neurochem. **39:** 788-795.
21. FÜLDNER, H. H. & H. STADLER. 1982. Eur. J. Biochem. **121:** 519-524.
22. SCHMIDT, R., H. ZIMMERMANN & V. P. WHITTAKER. 1980. Neuroscience **5:** 625-638.
23. STADLER, H. & H. H. FÜLDNER. 1980. Nature **286:** 293-294.

24. GIOMPRES, P. E. & V. P. WHITTAKER. 1986. Biochim. Biophys. Acta **822:** 398-409.
25. GIOMPRES, P. E., S. J. MORRIS & V. P. WHITTAKER. 1981. Neuroscience **6:** 757-763.
26. HEBB, C. O. 1972. Physiol. Rev. **52:** 918-957.
27. MARCHBANKS, R. M. & M. ISRAËL. 1971. J. Neurochem. **18:** 439-448.
28. KOSH, J. W. & V. P. WHITTAKER. 1985. J. Neurochem. **45:** 1148-1153.
29. WEILER, M., I. S. ROED & V. P. WHITTAKER. 1982. J. Neurochem. **38:** 1187-1191.
30. ERDÉLYI, L. & W. D. KRENZ. 1984. Comp. Biochem. Physiol. **79A:** 505-511.
31. ZIMMERMANN, H. & V. P. WHITTAKER. 1977. Nature **267:** 633-635.
32. ZIMMERMANN, H. & C. R. DENSTON. 1977. Neuroscience **2:** 715-730.
33. ZIMMERMANN, H. & C. R. DENSTON. 1977. Neuroscience **2:** 731-739.
34. GIOMPRES, P. E., H. ZIMMERMANN & V. P. WHITTAKER. 1981. Neuroscience **6:** 765-774.
35. GIOMPRES, P. E., H. ZIMMERMANN & V. P. WHITTAKER. 1981. Neuroscience **6:** 775-785.
36. ÁGOSTON, D. V., G. H. C. DOWE, W. FIEDLER, P. E. GIOMPRES, I. S. ROED, J. H. WALKER, V. P. WHITTAKER & T. YAMAGUCHI. 1986. J. Neurochem. **47:** 1584-1592.
37. GIOMPRES, P. E. & V. P. WHITTAKER. 1984. Biochim. Biophys. Acta **770:** 166-170.
38. WHITTAKER, V. P. 1984. Biochem. Soc. Trans. **12:** 561-576.
39. SUSZKIW, J. B., H. ZIMMERMANN & V. P. WHITTAKER. 1978. J. Neurochem. **30:** 1269-1280.
40. DUNANT, Y. & M. ISRAËL. 1985. Spektrum d. Wiss. **6:** 78-87.
41. LUQMANI, Y. A., G. SUDLOW & V. P. WHITTAKER. 1980. Neuroscience **5:** 153-160.
42. ANDERSON, D. C., S. C. KING & S. M. PARSONS. 1983. Mol. Pharmacol. **24:** 48-54.
43. SUSZKIW, J. B. & R. S. MANALIS. 1986. Abstract. Symposium on Cellular and Molecular Basis of Cholinergic Function, Buxton, U.K. May 11-16, 1986.
44. JONES, R. T., J. H. WALKER, H. STADLER & V. P. WHITTAKER. 1982. Cell Tiss. Res. **223:** 117-126.
45. AGOSTON, D. V. & J. M. CONLON. 1986. J. Neurochem. **47:** 445-453
46. STADLER, H., M. L. KIENE, P. HARLOS & U. WELSCHER. 1985. 36th Mosbach Colloquium: Neurochemistry. Springer. Heidelberg. pp. 55-65.
47. KIENE, M. L. 1986. Dissertation, Göttingen.
48. ZIMMERMANN, H., M. J. DOWDALL & D. A. LANE. 1979. Neuroscience **4:** 979-993.

DISCUSSION OF THE PAPER

N. MOREL (*CNRS, Gif sur Yvette, France*): Dr. M. Israël's group has shown preferential release of cytoplasmic acetylcholine in *Torpedo* electric organ and demonstrated the presence in the presynaptic plasma membrane of a protein able to translocate acetylcholine in reconstituted systems after activation by a calcium influx. So the mechanism of acetylcholine release appears to remain a questionable one.

V. P. WHITTAKER (*Max-Planck-Institut für biophysikalische Chemie, Göttingen, Federal Republic of Germany*): We believe the balance of evidence is in favor of vesicular release.

A. SCARPA (*Case Western Reserve University, Cleveland, Ohio*): What is the real definition of "light vesicles"? Are they different in size or content of water solutes?

V. P. WHITTAKER: The difference in the density between newly arrived (VP_0) and reserve (VP_1) vesicles may simply be accounted for by the difference in the density between a core containing inorganic ions (*e.g.,* KCl) and that containing organic ions (acetylcholine, cations, and ATP).

D. SULZER (*Columbia University, New York, N.Y.*): Would you please speculate on how the proteoglycan may be acting as a marker for vesicle recycling if the glycan moeity is on the external side of the plasma membrane?

V. P. WHITTAKER: The glycosaminoglycan moiety is on the inside of the vesicle and on the external (extracellular) side of the plasma membrane. Possibly these sugar side chains mutually interact (cross-linking by Ca^{2+} is a possibility), thereby keeping the patch of vesicle-derived membrane together and perhaps even providing an identifying signal for retrieval.

H. TAMIR (*Columbia University, New York, N. Y.*): How do you reconcile the short time span of release and relatively long time for recycling?

V. P. WHITTAKER: There is a misunderstanding here: We believe successive cycles of release and (partial) reloading (*i.e.*, vesicle recycling) occur very rapidly. By contrast, recovery (acquisition by recycling vesicles of the full complement of acetylcholine and ATP characteristic of reserve vesicles) is a slow process. The rate of this process is influenced by the general metabolic status of the tissue.

Lipid Composition and Orientation in Secretory Vesicles[a]

EDWARD W. WESTHEAD

Department of Biochemistry
University of Massachusetts
Amherst, Massachusetts 01003

The exocytotic release of hormones and neurotransmitters from subcellular storage vesicles begins with the elevation of cytosolic calcium concentration and ends with the storage vesicle open to the cell exterior through fusion of the vesicle membrane with the plasma membrane. No one step in the intervening process is understood at the molecular level. The function, if not the intention, of this report will be to focus attention on how little we know about the role of the membrane lipids in the fusion of the two membranes.

The fusion of two membranes requires an extreme reordering of the opposing lipid bilayers, very likely through intermediate formation on micelle-like structures. During reordering, lipids in the cytoplasmic monolayers which first make contact would form spherical inverted micelles with polar head groups in the centers of the spheres. Fatty acid chains would be in contact with neighboring spheres and with the fatty acid chains of the more distant monolayers of the two membranes. Evidence for the existence of such inverted micelles was developed through the study of artificial lipid bilayers. It was found that some phospholipids form only bilayers but that others are more likely to form the "hexagonal phase II" composed of inverted micellar structures as proposed for the membrane fusion intermediate. Phosphatidyl ethanolamine and phosphatidyl serine alone tend to form the hexagonal phase rather than bilayers, and cholesterol also favors reordering of the bilayer.[1] The process of fusion is also expected to be favored by a high degree of membrane fluidity, which is determined largely by the content of unsaturated fatty acids. The steric irregularity of the chain which is caused by double bonds lowers van der Waals contacts and reduces cohesion of the lipid phase. The presence in a membrane of lysolecithin (or other lysolipids) will cause dislocations in the lipid array and add a relatively hydrophilic element to the lipid phase. Addition of lysolecithin to a suspension of cells or lipid vesicles can promote fusion between them.[2] Free fatty acids too, if present in a membrane, should tend to induce rearrangement of membrane structure—after all, at physiological pH they are soap.

Glycolipids or gangliosides on the outer surfaces of cells serve as receptors or surface recognition markers.[3] Since the sialic acid portions of the carbohydrate have a high affinity for Ca^{2+} ion, gangliosides might play a role in exocytosis as membrane recognition sites or through their interaction with Ca^{2+}. The adrenal medulla has a ganglioside content almost as high as brain gray matter and much of that content is

[a]This work was supported by grants from the NSF (no. DMB-8309306) and the NIH (no. HL36704).

in the chromaffin granule membrane. The content of gangliosides in the granule is seven times higher than in mitochondria and three times higher than in the plasma membrane per milligram of protein, but the differences are significantly lower when expressed as the ratio to phospholipid.[4]

Not only the lipid content but the arrangement of lipids in the membrane should give us clues about the nature of the fusion process. Our expectation, necessarily naive at this time, is that lipids on the opposing faces, that is, the cytoplasmic faces of the plasma and vesicle membranes, will have special importance in the fusion process. Thus it becomes important to know the topography of lipid arrangement.

Before reviewing the lipid composition and arrangement of secretory vesicle membranes, we ought to recognize that the structure of the vesicle membrane is only half the picture. The plasma membrane is the other partner in the process of fusion, presumably at least an equal partner. The plasma membrane lipid is even more difficult to study, however, because it is more difficult to purify. A potentially more serious problem is that plasma membrane is topologically heterogeneous. This has been well documented for plasma membrane proteins and is probably true also for lipids. A comparative study of vesicle membranes is therefore a reasonable place to begin.

A high proportion of the information to be reviewed here has been obtained from study of the secretory vesicles from the adrenal medulla. These vesicles, which store and release catecholamines, are most commonly called chromaffin granules. They are easily prepared in high yield and purity. Twenty years ago Blaschko and co-workers[5] found that bovine chromaffin vesicle membranes had a remarkably high content of lysolecithin. Since lysolecithin has definite fusogenic properties and might play an important part in exocytosis, that finding has intrigued numerous investigators. Two important questions have been addressed: Is this high lysolecithin content an artifact caused by postmortem lipase action? Is the high lysolecithin content a general feature of secretory vesicle membranes? The answer to the first question seems to be no, at least with respect to chromaffin granules. Relatively recent experiments with vesicles from guinea pig[6] and rabbit glands,[7] which suggested that lysolecithin is produced postmortem, have been refuted on the basis that the vesicles from those species are highly impure.[8,9] Analysis of rat glands, frozen *in situ* in anesthetized animals, showed the same lysolecithin content as glands obtained and prepared in a conventional manner. Had the 20% lysolecithin content of the rat chromaffin granule membrane been produced during conventional gland preparation, the lysolecithin content of the gland would have increased 50%. Qualitative analysis for lysolecithin in quick-frozen glands from anesthetized animals of other species led to the same conclusion.[10]

To answer the second question we look first at TABLE 1 and see that lysolecithin makes up 7% to 17% of total phospholipid in the granule membranes of all animals from which reasonably pure chromaffin granules have been obtained, supporting the hypothesis that this lipid may be important in exocytotoic membrane fusion. The hypothesis is shaken however when we look at TABLE 2. These data from diverse secretory vesicles show that lysolecithin is high only in the membranes of isolated zymogen granules. These granules contain phospholipase, and in two cases[14,15] the authors presented evidence strongly suggesting that the lysolecithin is produced during isolation of the vesicles. In the other vesicles, no significant amount of lysolecithin was found. In the case of synaptic vesicles, the authors specifically looked for but did not detect lysolecithin. There is at present no evidence to suggest that the process of exocytosis is different in different secretory cells, and the virtual absence of lysolecithin in the membranes of some secretory vesicles is a serious impediment to a model of exocytosis that utilizes lysolecithin. It has been suggested that lysolecithin accumulates in vesicles during reutilization,[8] but there is now too little information on rates

TABLE 1. Lipid Classes of Chromaffin Granule Membranes as % Total Phospholipid

	Bovine					Horse	Pig	Rat	Human[c]
	ref. 5	ref. 11	ref. 12	ref. 13[a]	ref. 9[b]	(ref. 11)	(ref. 11)	(ref. 11)	(ref. 9)
PC	26	27	26	28	25	33	37	37	16
LPC[d]	17	17	17	11	16	7	11	15	23
PE	36	32	35	27	34	32	32	30	29
Sph	11	13	12	20	15	14	9	9	16
PS	9	10	8.9		7.9	11	9	11	8.5
PI			1.5		2.9				1.1
Cho/PL	0.53	0.53	0.40			0.48	0.48		

[a] PA given as 14%, *sic.*
[b] Bovine figures are averages from three types of isolation, all very similar.
[c] Pheochromocytoma, a tumor of the adrenal medulla, rich in chromaffin granules. Granules from pheochromocytoma contained 4.8% LPE.
[d] It is probable that in some cases any LPE present was included with LPC.

of reutilization in different cell types to test that hypothesis against the data in TABLE 2.

Free fatty acids also lower the stability of membranes and those have been found in isolated bovine chromaffin granule membranes, at 5% of the level of phospholipids, on a molar basis.[20] This was calculated from a mean value of about 2 μmol of phospholipid per milligram of protein.[21] It may be noted that the sum of LPC + PC in TABLE 1 is generally about 40% for all vesicles, suggesting that lysophosphatidylcholine is produced, at some point in vesicle synthesis, from phosphatidylcholine. This would produce free fatty acids, but those would be expected to escape rapidly from the membrane because of their high aqueous solubility. Since fatty acids make the granule leaky to protons and waste the proton gradient produced by ATP,[20] it seems likely that their concentration in the membrane *in vivo* is low.

If we look at TABLES 1 and 2 for other features of lipid composition common to secretory vesicles, we see that PE is also high and fairly constant, but PC and PE are the predominant lipids in most membranes. The content of other phospholipids is quite variable, but it is possible that part of this variability is in the separation techniques employed. In any case there is nothing in the phospholipid distribution in these tables which distinguishes the vesicle membranes as a class from mitochondria, microsomes, or plasma membrane fractions.

It is possible to recognize mitochondrial membranes by their content of cardiolipin, which is generally reported near 10 mole % of the phospholipid and would be twice that by weight or volume. This lipid, which is produced by mitochondrial enzymes, is low or absent in purified plasma membrane or membranes of other organelles. In fact it appears that even a few percent of cardiolipin is a good indicator of substantial mitochondrial contamination in a membrane fraction.

The cholesterol-to-phospholipid ratio in most highly purified synaptic membranes is over 0.5, at least 20% higher than usually reported for most other membrane fractions. The other fairly consistent feature of vesicle membranes is a higher lipid-to-protein ratio than most other membrane fractions.

TABLE 3 shows a comparison of the composition of synaptic vesicle membranes and plasma membranes. In the first column is the distribution of lipid classes. There is a striking similarity in the two membrane types, but it may be seen that the cholesterol-to-phospholipid ratio is substantially higher in the vesicle membranes. These data, and the data of TABLE 2, are in good general agreement with analyses of neuronal membranes collected earlier;[23] a similar comparison between the composition of the secretory vesicles and plasma membranes of bovine neurohypophysis led to the same conclusion.[24]

The fatty acid composition data of TABLE 3 relates to the question of lipid topography, or asymmetric distribution of lipid classes across the bilayer. The problem of determining lipid asymmetry is not a simple one and there is much conflict in published data.[25] It is, however, well established for the "universal model membrane," that of the human erythrocyte, that PS, PE, PI are on the cytoplasmic side of the membrane, and that most PC and sphingomyelin are on the outer surface. This puts the highly stable bilayer formers on the outside and the lipids most favoring the hexagonal phase on the cytoplasmic surface. If we now look at the distribution of saturated and highly unsaturated fatty acids among the lipid classes, we see a striking correlation. The 16:0, 18:0, and 18:1 fatty acids occupy about 90% of the positions in PC and Sph, with the 20:4 and 22:6 fatty acids rare on those head groups. Thus the outer leaflet is highly saturated if the red cell distribution of lipids also characterizes other plasma membranes. The highly unsaturated fatty acids occupy 35% to 50% of the positions available on PE, PS, and PI, indicating that we might expect the inner membrane leaflet to be much more fluid than the outer leaflet, in addition to having the hexagonal-phase-forming head groups.

TABLE 2. Lipid Classes of Other Secretory Vesicle Membranes, as % of Total Phospholipid

	pancreatic	*pancreatic*	*pancreatic*	*parotid*	*platelet*	*platelet*	*pituitary*	*synaptic*	*synaptic*
Reference	14[a,b]	15[b–d]	17[e]	16[f]	12[g]	9[h]	9[i]	17[j]	19[k,l]
PC	22	27	42	52	34	37	41	48	41
LPC	9	12	7	7	0	0.5	0	0	1
PE	32	15	35	21	32	23	26	31	25
Sph	24	9	9	9	17	25	22	4.5	12
PS	5	2	4.4	8.4	12	10	6.3	9	7.3
PI		15			2.5	4.3	3.2	2.6	3.7
Cho/PA	0.55		0.56					0.5	0.63

[a] Guinea pig pancreatic zymogen granules.
[b] Authors consider that lysolecithin is a postmortem artifact, see text.
[c] Pig pancreatic zymogen granules.
[d] PI value includes PA; authors also report 15% LPE.
[e] Ox pancreatic zymogen granules.
[f] Rabbit parotid zymogen granules.
[g] 5HT vesicles of rabbit platelets.
[h] 5HT granules of pig platelets.
[i] Bovine pituitary vesicles.
[j] Narcine synaptic vesicles.
[k] *Torpedo* synaptic vesicles.
[l] Authors also report 12% plasmenylethanolamine.

Reliable data on orientation of lipids in the plasma membrane of secretory cells are rare, but the better studies indicate that those membranes are indeed organized like the red cell. Fontaine and co-workers, for example, showed that the ethanolamine and serine are on the inner surface of the synaptic membrane with phosphatidylcholine on the outside.[26]

The information available on the lipid orientation of secretory vesicles is also very limited. As discussed above, lysolecithin has been considered a prime candidate for a role in the fusion process. Two studies of the orientation of lysolecithin in bovine chromaffin granule membranes agree that lysolecithin is located chiefly on the inner leaflet of the membrane. In that location it seems unlikely that it plays a role in the fusion process, at least not in the initial phase. The first study examined the lysolecithin

TABLE 3. Lipids of Rat Brain Synaptic Vesicles and Synaptosomal Membranes[a]

	%	16:0[b]	18:0	18:1	20:4	22:6
SYNAPTIC VESICLES						
PC	42	43	14	26	6	7
LPC	1.5	—	—	—	—	—
PE	36	7.4	26	8.6	17	31
Sph	5	5.7	88	—	—	—
PS	12	0.6	44	7.5	2.8	37
PI	3	8	40	5.6	33	2.6
Cho/PL[c]	0.58					
SYNAPTOSOMAL MEMBRANES						
	%	16:0[b]	18:0	18:1	20:4	22:6
PC	41	57	12	22	4.3	2.7
LPC	1.0	—	—	—	—	—
PE	34	6	29	7.8	7.3	32
Sph	5	5	88	—	—	—
PS	14	0.5	49	7.6	2.0	34
PI	4	10	38	8.5	37	5
Cho/PL[c]	0.44					

[a] Data are selected from those in original reference, Breckenridge *et al.*[21]
[b] The total of these selected five fatty acids will usually not equal 100%.
[c] Molar ratio.

orientation with a particulate enzyme, acetyl coenzyme A: lysolecithin acyltransferase, to add radioactive fatty acid to exposed lysolecithin.[27] Up to one third of the lysolecithin was reacylated within an hour at 4° when broken granules were used as substrate. When intact granules were used, however, only 3% as much lecithin was formed.

The second study took the opposite tack and digested accessible lysolecithin with lysophospholipase.[28] This soluble enzyme lysed about 70% of the total membrane lysolecithin in disrupted membranes during prolonged incubation at 30°. Under similar conditions only 10% of the lysolecithin of intact granules was hydrolyzed. The actual extent of reaction with these probes, and others, is not likely to be a quantitative measure of the orientation since other factors surely affect reactivity. An unknown amount of lipid is bound with unknown affinity to membrane proteins. There may also be other forms of organization or segregation not yet appreciated. This point is underscored by the fact that De Oliveira-Filqueiras[28] *et al.* had to heat granule membrane to 90° to make all of the lysolecithin accessible to the lipase.

The accessibility of the phospholipids of chromaffin granules has also been tested in other experiments with the chemical label trinitrobenzene sulfonate (TNBS) and with lipases.[29] Trinitrobenzene sulfonate reacted with only about 70% of the potentially reactive lipid of broken membranes, again suggesting some organization that sequesters the head groups stably and effectively. Sphingomyelinase, phospholipase A_2, and phospholipase C were also limited in their ability to digest lipids of either whole or broken granule membranes. In this study intact vesicles showed no strongly biphasic reaction rate that might give information on the transbilayer distribution of the reactive lipids, nor was there any difference in total reactivity between intact and broken membranes. Since these enzymes have been successfully used to examine lipid orientation in other membranes (*e.g.*, synaptic vesicles—see below), it may be that the vesicles had been damaged by a drop in osmolarity during preparation.

Orientation of glycolipids (gangliosides) has also been examined in chromaffin granule membranes. Although gangliosides constitute only 2 or 3 mole % of the membrane lipid, they could be an important element in the recognition or contact phase of the interaction with the plasma membrane. Westhead and Winkler[30] found that neuroaminidase was not able to remove measurable amounts of sialic acid from intact chromaffin granules, but was able to remove about 75% of the sialic acid from gangliosides as well as glycoproteins when the granules were lysed. One fourth of the gangliosides were resistant to neuraminidase whether the membranes were merely lysed and washed or they were repeatedly frozen and thawed to ensure thorough disruption. These data further support the concept that some significant fraction of the membrane lipids exists in a stable sequestered state.

Phosphorylation of phosphatidylinositol in chromaffin granule membranes has been studied by Phillips,[31] who concluded that this important phospholipid is located chiefly on the cytoplasmic face of the granule membrane.

The phospholipids of highly purified synaptic vesicles from electric organs have also been examined. The study of Deutsch and Kelly[18] directed chiefly to the lipid composition of *Torpedo* vesicles, showed that neither trinitrobenzene sulfonate nor isothionyl acetimidate (ITA) reacted completely with the membrane phospholipids. Roughly half of the PE and PS reacted with ITA, but even less reacted with the bulkier TNBS. Lysis of the vesicles increased the rate of reaction only 50% and the extent by less than 10%. The data did not provide much information about orientation.

In a later study, using narcine synaptic vesicles, Michaelson *et al.*[19] showed that TNBS made the vesicles leaky and could not be used for determination of orientation. Phospholipase C, on the other hand, was a satisfactory probe of orientation. In one minute, 45% of the achievable lipid hydrolysis took place with no vesicle lysis. The conclusions of this study were that on the external (cytoplasmic) surface is 100% of the PI, 77% of the PE, 45% of the plasmenylethanolamine, and 58% of the PC. Since PS is a poor substrate for phospholipase C, no conclusions could be drawn about its orientation.

The ganglioside and PI orientation of chromaffin granules and the orientations of PI and PE in synaptic vesicles support a model in which lipid orientation in the vesicle membrane is the same as that of the plasma membrane if one classifies them as oriented into or away from the cytoplasm. Within this framework one can then look again at the fatty acid composition of TABLE 3. From the combined data emerges a picture in which those lipids that form the most stable bilayers are in the membrane leaflet facing away from the cytoplasm. These same lipids have a great preponderance of saturated fatty acids. In both the vesicle and plasma membranes, the cytoplasmic surfaces, those that make initial contact with each other, have a high content of head groups that tend to form the hexagonal phase rather than the bilayer. PI, PS, and the sialic acid-containing gangliosides, which will all bind calcium ion, face the cytoplasm where calcium ion concentration fluctuates sharply during secretion. These lipids also have a high content of unsaturated fatty acids, lending them a fluidity which should help promote fusion. Certainly we have far too little data to feel secure with this picture, but it is an appealing one that may stimulate more work to substantiate or disprove it. A very interesting question is where those lipids are whose head groups are so thoroughly prevented from reacting with a variety of enzymes and chemical reagents.

ACKNOWLEDGMENTS

I am grateful to Ms. P. McQuilken for help with the literature search and to Dr. Richard Cameron, Yale University, who generously shared the results of his own literature search.

REFERENCES

1. Cullis, P. R., M. J. Hope, R. Nayar & C. P. S. Tilcock. 1985. In Phospholipids in the Nervous System, Vol. 2. Physiological Roles. L. A. Hurrocks, *et al.,* Eds.: 71-86. Raven Press, New York.
2. Poole, A. R., J. I. Howell & J. A. Lucy. 1970. Nature **227:** 810-813.
3. Ando, S. 1983. Neurochem. Internat. **5:** 507-537.
4. Sekine, M., T. Ariga, T. Miyatake, Y. Kuroda, A. Suzuki & T. Yamakawa. 1984. J. Biochem. **95:** 155-160.
5. Blaschko, H., H. Firemark, A. D. Smith & H. Winkler. 1967. Biochem. J. **104:** 545-549.
6. Sun, G. Y. 1979. Lipids **14:** 918-924.
7. Arthur, G. & A. Sheltawy. 1980. Biochem. J. **191:** 523-532.
8. Frischenschlager, I., W. Schmidt & H. Winkler. 1983. J. Neurochem. **41:** 1480-1483.
9. De Oliveira-Filqueiras, O. M., H. Van den Bosch, R. G. Johnson, S. E. Carty & A. Scarpa. 1981. FEBS Lett. **129:** 309-313.
10. Douglas, W. W., A. M. Poisner & J. M. Trifaro. 1966. Life Sci. **5:** 809-915.
11. Winkler, H., N. Strieder & E. Ziegler. 1967. Naunyn-Schmiedebergs Arch. Exp. Path. Pharmakol. **256:** 407-415.
12. DaPrada, M., A. Pletscher & J. P. Tranzer. 1972. Biochem. J. **127:** 681-683.
13. Balzer, H. & A. B. Khan. 1975. Naunyn Schmiedebergs Arch. Pharmakol. **291:** 319-333.
14. Meldolesi, J., J. D. Jamieson & G. E. Palade. 1971. J. Cell Biol. **49:** 130-149.
15. Rutten, W. J., J. J. H. M. De Pont, S. L. Bonting & F. J. M. Daemen. 1975. Eur. J. Bioch. **54:** 259-265.
16. Williams, M. A., M. K. Pratten, J. W. Turner & G. H. Cope. 1979. Histochem. J. **11:** 19-50.
17. White, D. A. & J. N. Hawthorne. 1970. Biochem. Jour. **120:** 533-538.
18. Deutsch, J. W. & R. B. Kelly. 1981. Biochemistry **20:** 378-385.
19. Michaelson, D. M., G. Barkai & Y. Barenholz. 1983. **211:** 155-162.
20. Husebye, E. S. & T. Flatmark. 1984. J. Biol. Chem. **259:** 15272-15276.
21. Breckenridge, W. C., I. G. Morgan, J. P. Zanetta & G. Vincendon. 1973. Biochim. Bioph. Acta **320:** 681-686.

22. WINKLER, H. & E. W. WESTHEAD. 1980. Neuroscience **5:** 1803-1823.
23. ROUSER, G., G. KRITCHEVSKY, A. YAMAMOTO & C. F. BAXTER. 1972. Adv. Lipid Res. **10:** 261-360.
24. VILHARDT, H. & G. HOMER. 1972. Acta Endocrinol. **71:** 635-648.
25. OP DEN KAMP, J. A. F. 1979. Ann. Rev. Biochem. **48:** 47-71.
26. FONTAINE, R. N., R. A. HARRIS & F. SCHROEDER. 1979. Life Sci. **24:** 395-399.
27. VOYTA, J. C., L. L. SLAKEY & E. W. WESTHEAD. 1978. Biochem. Biophys. Res. Comm. **80:** 413-417.
28. DE OLIVEIRA-FILQUEIRAS, O. M., A. M. H. P. VAN DEN BESSELAAR & H. VAN DEN GOSCH. 1979. Biochim. Biophys. Acta **558:** 73-84.
29. BUCKLAND, R. M., G. K. RADDA & C. D. SHENNAN. 1978. Biochim. Biophys. Acta **513:** 321-337.
30. WESTHEAD, E. W. & H. WINKLER. 1982. Neuroscience **7:** 1611-1614.
31. PHILLIPS. J. H. 1973. Biochem. J. **136:** 579-587.

DISCUSSION OF THE PAPER

L. M. ROMAN (*Howard Hughes Medical Institute, Dallas, Tex.*): This morning we heard about the stability of the synaptic vesicle membrane bound proteins existing as a patch at the plasma membrane following vesicle fusion. Given that gangliosides and glycolipids would be asymmetrically oriented in the synaptic vesicle (outer leaflet) which then would be localized to the outer leaflet in the plasma membrane, if such groups could interact with other externalized components such as glycosaminoglycans is there any indication that vesicle lipids may also be restricted in their mobility following fusion at the plasma membrane?

E. W. WESTHEAD (*University of Massachusetts, Amherst, Mass.*): Given the high concentration of Ca^{2+} outside the cell, and the strong tendency for that ion to be chelated by carboxylate groups, it would be surprising if glycolipids and glycoproteins were not cross-linked by Ca^{2+} where they are exposed to the cell exterior. Numerous experiments with chemical labels and enzymes show that some lipids of secretory vesicles are not reactive even where the vesicle has been lysed. This indicates that some kind of clustering or sequestration of lipids may exist in these membranes and that there is limited movement of lipids into and out of those protected regions. Together these data would suggest that substantial restriction of lipid mobility might occur after fusion, but much more experimentation is needed to more than suggest the possibility.

J.-P. HENRY (*Institut de Biologie Physico-Chimique, Paris, France*): In red blood cells, the asymmetry in phospholipid distribution between the two leaflets results from the effect of an ATP-dependent system. Does such a system exist in chromaffin granule membranes?

E. W. WESTHEAD: I have seen no experiments on the stability of lipid orientation in secretory vesicles. There have been very few reports of successful studies on lipid orientation in purified secretory vesicles.

P. J. FLEMING (*Georgetown University Medical Center, Washington, D.C.*): Since secretory vesicle membranes appear to be metabolically stable, is there a correlation with increased content of ether-containing phospholipids in secretory vesicle membranes?

E. W. WESTHEAD: Most lipid analyses of secretory vesicle membranes have been carried out with one-dimensional chromatography systems of limited ability to resolve different lipid classes and hence have not been very complete. Michaelson *et al.*[19] have reported 12 plasmenylthanolamines in narcine synaptic vesicles. In other analyses it is possible that ether-containing lipids were not resolved from other phospholipids.

Secretory Vesicle Cytochrome b_{561}: A Transmembrane Electron Transporter[a]

PATRICK J. FLEMING AND UTE M. KENT

Department of Biochemistry
Georgetown University Medical Center
Washington, D.C. 20007

INTRODUCTION

A *b*-type cytochrome was first recognized as a component of catecholamine storage granules in 1961.[1] This catecholamine-granule-associated cytochrome was differentiated functionally from other major cytochromes such as b_5 and P-450 in 1965.[2] During the 1970s Flatmark and colleagues characterized the cytochrome spectrally as b_{561}, showed that its midpoint potential was +140 mV, and that the cytochrome constituted as much as 15% of the chromaffin granule membrane.[3–5]

The first successful purification of cytochrome b_{561} in native form was reported by Apps *et al.*[6] in 1980 and the purified cytochrome was shown to have a pI different from most other chromaffin granule membrane proteins (pI = 6.2).[6] These authors also suggested that the cytochrome was a transmembrane protein and that it may mediate transmembrane electron transport for the reduction of dehydroascorbate inside the chromaffin granule. Regeneration of ascorbic acid inside the chromaffin granule is necessary because ascorbic acid is believed to be the physiological electron donor to monooxygenases found in the granule matrix. Njus *et al.* also suggested such a role for the cytochrome, and showed that chromaffin granules do have a mechanism for transmembrane electron transport, but that semidehydroascorbate is the acceptor for this membrane electron transfer system.[7–9] This latter fact is significant because semidehydroascorbate is the immediate product of the intravesicular dopamine β-hydroxylase reaction, and this radical must be rapidly reduced to regenerate ascorbic acid inside the chromaffin granule for continued norepinephrine biosynthesis.[10] Elsewhere in this volume, Njus *et al.* discuss the involvement of cytochrome b_{561} in the regeneration of ascorbic acid inside the chromaffin granule, and the fact that such a regeneration system is widespread in neuroendocrine secretory vesicles.

Our interest in cytochrome b_{561} has been the structure and function of the isolated protein, its wider role in nature, and its use as a model protein for investigating long-range biological electron transfer. In terms of its structure we know that the cytochrome is a single polypeptide of approximately 28-30 kDa with a blocked amino terminus[11]; that it is a transmembrane protein with the majority of lysyl and tyrosyl

[a] This work was supported by grant no. GM 27695 from the United States Public Health Service.

residues on the cytoplasmic side of the chromaffin granule membrane[12]; and that it has only a single heme according to spectral analyses.[5,11,13] In terms of its function, we know that the cytochrome does transfer electrons across a phospholipid bilayer.[14] Furthermore, the cytochrome is found associated with two intravesicular monooxygenases, dopamine β-hydroxylase and peptidyl α-amidating monooxygenase[15] and presumably is involved in supplying electrons to these enzymes. Questions we are currently pursuing are: What is the molecular structure of this redox protein that allows such long range, transmembrane electron transfer? Can the cytochrome donate electrons directly to putative enzyme acceptors such as dopamine β-hydroxylase and peptidyl α-amidating monooxygenase?

PURIFICATION OF MULTIPLE CYTOCHROME SPECIES

Since the function of cytochrome b_{561} is to catalyze transmembrane transfer, the assay for functionality must involve reconstitution of the cytochrome into membranes. This fact immediately limits the scope of purification schemes one may choose. Specifically, any detergents used during purification must be able to be removed for reconstitution of the protein. Although two other laboratories have published purification protocols for cytochrome b_{561}, both procedures use nondialyzable detergents.[6,16] Our laboratory set out to devise a purification method that would allow us to test functional activity of the purified cytochrome in phospholipid vesicles. Initially, we used hydrophobic chromatography in octyl glucoside,[11] and this method did result in pure cytochrome. Furthermore, we were able to use this cytochrome in reconstituted phospholipid vesicles to demonstrate that cytochrome b_{561} catalyzes transmembrane electron transport.[14] This result was important in establishing the function of the cytochrome, a function which had been hypothesized for a long time.

We have now devised a new purification protocol for cytochrome b_{561} that is fast, reproducible, and results in functionally active cytochrome. The major purification step in this new procedure is chromatofocusing, and alkyl glycoside detergents are used throughout so that the detergents may be removed for reconstitution of the cytochrome. Chromatofocusing separates proteins according to their respective pI values, and thus this procedure gives us a measure of the apparent pI of the native cytochrome. Initially, we obtained two species of cytochrome, one with pI of 7.1 and one with pI of 7.4. This result differs from the previously determined pI of the denatured cytochrome (pI = 6.2) and it was an exciting result because redox titration of the cytochrome has demonstrated heterogeneity.[3,13] Multiple forms of the cytochrome could account for such heterogeneity in redox titrations.

We subsequently discovered that the proportion of the two cytochrome species could be controlled by the concentration of dithiothreitol (DTT) present during the extraction and chromatography of the protein. TABLE 1 shows the change of cytochrome species obtained depending on the DTT concentration. These results are not due to the redox state of the heme because this is reduced to the same extent under all conditions shown. Rather, these results suggest that an intrachain disulfide, or other redox center besides the heme, becomes reduced at certain concentrations of DTT and this reduction changes the effective surface charge of the protein. These results are consistent with earlier data which also suggested that an intrachain disulfide exists in the cytochrome.[11] We are currently investigating whether or not these two charge species of cytochrome account for the heterogeneity previously observed in the redox titration of cytochrome b_{561}.

HOW MANY REDOX SITES IN CYTOCHROME b_{561}?

Cytochrome b_{561} is a single polypeptide with, apparently, only a single heme, yet it can transfer electrons across a phospholipid bilayer of 4-5 nm thickness. This fact and the above effects of DTT beg the question of whether or not the cytochrome contains other redox centers. We initiated a resolution of this question and the first investigation has been to screen for the presence of metal atoms in purified cytochrome using neutron activation analysis. This method allows for the detection of up to 15 elements, including the ones commonly found in proteins, with a single activation. This work was done in collaboration with Dr. Rolf Zeisler at the National Bureau of Standards, and the preliminary results show one molar equivalent of iron in the cytochrome and no other detectable elements. Thus, the previously determined stoichiometry of one heme per polypeptide was confirmed and no other metal redox centers are apparent. In view of the above results with DTT it may be possible that

TABLE 1. Effect of DTT on pI of Cytochrome b_{561}[a]

DTT[b]	pI
0	7.4
0.5	7.4
1.75	7.4 + 7.1
3.0	7.1

[a] Chromaffin granule membranes from the bovine adrenal medulla were extracted with 5% octyl glucoside in diethanolamine buffer, pH 9.4. After centrifugation and Millipore filtration, the extract was loaded onto a Pharmacia Mono-P chromatofocusing column. The pH gradient was developed by elution with a 1:10 dilution of Polybuffer 9-6 in 1% octyl glycoside. Various concentrations of DTT were present during the extraction. Column fractions were collected and assayed for cytochrome (OD 427 nm) and pH.

[b] Concentrations of DTT given in mM.

sulfhydryl-disulfide cycling is involved in an intraprotein electron transfer system, and we are currently investigating this possibility.

If heme is the single redox site in cytochrome b_{561}, one must explain how the cytochrome can transfer an electron over such a long distance as the thickness of a phospholipid bilayer. It may be that access to the heme is through a deep crevice or well and is limited to small molecules such as ascorbate or semidehydroascorbate. Alternately, one must not forget the possibility that cytochrome b_{561} uses electron tunneling to bridge the distance.

WHERE IS THE HEME?

One question that will need to be answered to understand how this cytochrome functions is the following: What is the location of the heme within the protein and within the membrane? No complete resolution of this question will be possible until we have a better picture of the polypeptide structure, but we do have one piece of information that relates to this point. We have found that electron transfer will occur rapidly to ferricyanide but not directly to cytochrome *c*. A small molecule redox mediator is required to obtain electron transfer from cytochrome b_{561} to cytochrome *c*. Although there are a number of possible interpretations for this difference in reactivity, it does suggest that the heme edge is not exposed in an unrestricted fashion on the surface of cytochrome b_{561}. Further studies with varying sizes of electron donors and acceptors should allow us to determine the size exclusion parameters for access to the redox center of this cytochrome.

DEMONSTRATING TRANSMEMBRANE ELECTRON FLOW TO DOPAMINE β-HYDROXYLASE

Although we have concentrated on cytochrome b_{561} as an electron transport protein *per se,* the physiologically relevant question is whether or not this cytochrome can supply electrons for the intravesicular enzymes dopamine β-hydroxylase and peptidyl α-amidating monooxygenase. Several studies have addressed this question indirectly with experiments using isolated chromaffin granules[17,18] or adrenal medullary cell preparations.[19,20] These studies have shown that ascorbate outside the chromaffin granule can stimulate dopamine β-hydroxylase activity inside the granule and that the external ascorbate does this by indirectly regenerating the internal supply of ascorbate for direct use by the hydroxylase. The transmembrane electron transport in isolated chromaffin granules demonstrated by Njus *et al.,*[8] which is associated with cytochrome b_{561} spectral changes,[9] could be responsible for this regeneration of internal ascorbate. To tie together these results it remains to demonstrate that one can recon-

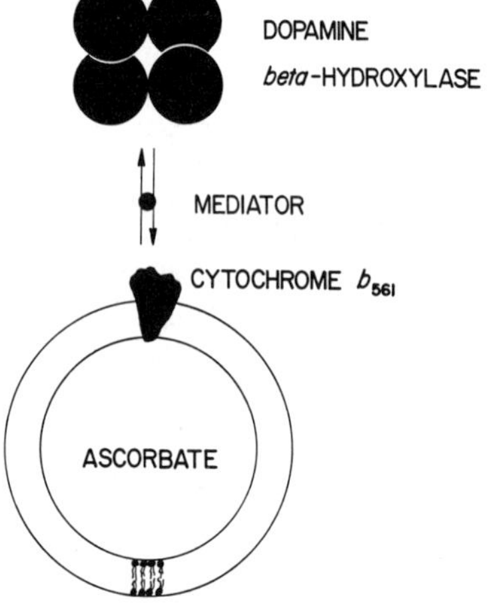

FIGURE 1. The experimental design to test transmembrane supply of electrons from ascorbate to dopamine β-hydroxylase.

TABLE 2. Reconstitution of Transmembrane Electron Transport to Dopamine β-Hydroxylase[a]

Assay Conditions	Octopamine Formed[b]
Vesicles + DBH	3.8
Vesicles + DBH + mediator	34.9
Trypsin treated vesicles + DBH + mediator	20.6
DBH + mediator	1.8

[a] Purified cytochrome b_{561} was reconstituted into ascorbate-loaded (150 mM) egg phosphatidyl choline vesicles by detergent dialysis in Hepes/NaCl buffer, pH 7.4. Ascorbate external to the vesicles was removed by gel filtration. The vesicles were mixed with ascorbate oxidase, tyramine, fumarate, catalase and dopamine β-hydroxylase in sodium acetate buffer, pH 5.5. In some cases 5 nmoles of potassium ferricyanide was added as an external redox mediator. The samples were incubated for 30 min at 25°C. Octopamine was assayed spectrophotometrically by oxidation to the hydroxybenzaldehyde with periodate. In one case the cytochrome reconstituted vesicles were treated with trypsin (30 min) and trypsin inhibitor (5 min) before gel filtration.

[b] The amount of octopamine is given in nanomoles.

stitute a transmembrane electron transfer system and stimulation of hydroxylase activity using purified, reconstituted cytochrome b_{561} and dopamine β-hydroxylase.

We are currently performing reconstitution experiments in an attempt to demonstrate transmembrane supply of electrons to dopamine β-hydroxylase using cytochrome b_{561} as the transmembrane electron transporter. A typical experiment is diagrammed in FIGURE 1. Purified cytochrome is reconstituted in phospholipid vesicles containing ascorbate at pH 7.4. The external pH is adjusted to 5.5 and purified soluble dopamine β-hydroxylase is added to the outside of the vesicles together with the necessary cofactors and tyramine as the catecholamine analogue substrate. Ascorbate oxidase is included outside the vesicles to oxidize any ascorbate that may leak from the vesicles. As shown in TABLE 2, we found that essentially no tyramine is hydroxylated by this system unless a redox mediator (ferricyanide) is included outside the vesicles to shuttle between cytochrome and hydroxylase.

The requirement of redox mediator to obtain hydroxylase activity fortuitously gives us a control against leakage of internal ascorbate. If any ascorbate were to leak out of the vesicles and escape the ascorbate oxidase, then hydroxylation of tyramine would be observed in the absence of ferricyanide. The fact that cytochrome b_{561} is necessary to obtain hydroxylation is demonstrated by two control experiments. Ascorbate-loaded vesicles without cytochrome b_{561} in the membrane do not support hydroxylase activity. And preincubation of the reconstituted cytochrome vesicles with trypsin for 30 minutes results in a 40% decrease in the amount of tyramine which is hydroxylated in the above experiment (TABLE 2). Experiments in which ascorbate is added as the external redox mediator also show similar results. In this latter case however, the blank values are very high due to the necessity of providing ascorbate in concentrations approaching the hydroxylase K_m of 1 mM.

We believe the results of these experiments conclusively show that cytochrome b_{561} is able to function in its putative role as a transmembrane electron transporter to supply electrons for dopamine β-hydroxylase activity. The results also support the need for regeneration of ascorbate inside the chromaffin granule because the cytochrome cannot donate electrons directly to the soluble form of the hydroxylase. We have not yet tested the potential ability of cytochrome b_{561} to donate electrons directly to the membrane-bound form of dopamine β-hydroxylase.

SUMMARY

The major function of cytochrome b_{561} is now clear. This transmembrane protein transports electrons across a secretory vesicle bilayer to supply electrons to monooxygenases inside the secretory vesicle.

Cytochrome b_{561} has been localized not only to adrenergic secretory vesicles, where it supplies electrons to dopamine β-hydroxylase, but also to peptidergic secretory vesicles that contain peptidyl α-amidating monooxygenase.[15,21] Thus, one would expect to find cytochrome b_{561} in the membranes of all neuroendocrine cells that contain amidated peptide secretory products. In addition, its wide occurrence as an integral membrane protein of secretory vesicles may make it useful for investigation of vesicle biogenesis and turnover.

One of the most important potential roles of cytochrome b_{561} is that it can be used as a model protein to investigate long-range biological electron transport. This cytochrome is a single polypeptide, which can be purified easily and reconstituted into a functional assembly. It also catalyzes an experimentally unambiguous transmembrane transport of electrons. A full molecular characterization of the structure and function of this cytochrome may provide insights into biological electron transfer which would otherwise be difficult or impossible to obtain.

REFERENCES

1. Spiro, M. J. & E. G. Ball 1961. J. Biol. Chem. **236:** 225-230.
2. Ichikawa, Y. & T. Yamano. 1965. Biochem. Biophys. Res. Comm. **20:** 263-268.
3. Flatmark, T. & O. Terland. 1971. Biochim. Biophys. Acta **253:** 487-491.
4. Terland, O., T. Silsan & T. Flatmark. 1974. Biochim. Biophys. Acta **359:** 253-256.
5. Terland, O. & T. Flatmark. 1980. Biochim. Biophys. Acta **597:** 318-330.
6. Apps, D. K., Pryde, J. G. & J. H. Phillips. 1980. Neuroscience **5:** 2278-2287.
7. Njus, D., M. Zallakian & J. Knoth. 1981. *In* Chemiosmotic Proton Circuits in Biological Membranes. V. P. Skulachev & Peter C. Hinkle, Eds. 365-374. Addison-Wesley. Reading, Mass.
8. Njus, D., J. Knoth, C. Cook & D. M. Kelley. 1983. J. Biol. Chem. **261:** 6429-6432.
9. Kelley, D. M. & D. Njus. 1986. J. Biol. Chem. **261:** 6429-6432.
10. Diliberto, E. J. & P. L. Allen. 1981. J. Biol. Chem. **256:** 3386-3393.
11. Duong, Le T. & P. J. Fleming. 1982. J. Biol. Chem. **257:** 8561-8564.
12. Duong, Le T. & P. J. Fleming 1984. Arch. Biochem. Biophys. **228:** 332-341.
13. Apps, D. K., M. P. Boisclair, F. S. Gavine & G. W. Pettigrew 1984. Biochim. Biophys. Acta **764:** 8-16.
14. Srivastava, M., Le T. Doung & P. J. Fleming. 1984. J. Biol. Chem. **259:** 8072-8075.
15. Duong, Le T., P. J. Fleming & J. T. Russell. 1984. J. Biol. Chem. **259:** 4885-4889.
16. Wakefield, L. M., A. E. G. Cass & G. K. Radda. 1984. J. Biochem. Biophys. Methods **9:** 331-341.
17. Levine, M., K. Morita, E. Heldman & H. B. Pollard. 1985. J. Biol. Chem. **260:** 15598-15603.
18. Beers, M. F., R. G. Johnson & A. Scarpa. 1986. J. Biol. Chem. **261:** 2529-2535.
19. Morita, K., M. Levine & H. B. Pollard. 1985. J. Neurochem. **46:** 939-945.
20. Levine, M., K. Morita & H. B. Pollard. 1985. J. Biol. Chem. **260:** 12942-12947.
21. Hortnagel, H., H. Winkler & H. Lochs. 1973. J. Neurochem. **20:** 977-985.

DISCUSSION OF THE PAPER

A. Scarpa (*Case Western Reserve University, Cleveland, Ohio*): Several years ago Dr. Johnson and I reported the presence of cytochrome b_{561} in isolated platelet granules. Such a finding will require a new additional explanation for the function that you have hypothesized in chromaffin and pituitary granules.

P. J. Fleming (*Georgetown University Medical Center, Washington, D.C*): Yes, Dr. Scarpa is referring to the fact that both polyclonal and monoclonal antibodies against chromaffin granule cytochrome b_{561} do not cross-react with the cytochrome in platelet vesicles. The platelet vesicles cytochrome is either a different cytochrome or is modified in such a way to cover the antigenic sites which our antibody recognizes. I have no idea, at the present time, what the function of the platelet cytochrome is.

E. L. Sabban (*New York Medical College, Valhalla, N.Y.*): What do you think might be the *in vivo* mediator of electron transport from the cytochrome to dopamine β-hydroxylase. Do you think that it is ascorbate?

P. J. Fleming: Yes, I believe that the internal ascorbic acid can be regenerated by the cytochrome after semidehydroascorbate is formed by dopamine β-hydroxylase. David Njus will present more data in the next talk relevant to this question.

R. M. Pruss (*NIMH, National Institutes of Health, Bethesda, Md.*): Considering that cytochrome b_{561} is an asymmetric molecule, can you discuss the directionality of electron transfer? Since the binding site for ascorbate is likely to be different on the inside and the outside of the granule, is electron transfer more efficient in one direction than in the other, (for example, outside to inside)?

P. J. Fleming: If the cytochrome accepts an electron from ascorbate (forming semidehydroascorbate) on one side of the membrane, and donates on electron to semidehydroascorbate (forming ascorbate) on the other side of the membrane, then the substrate and products are the same on both sides of the membrane. Both active sites of the cytochrome must bind both substrates and products. Therefore, to demonstrate transmembrane electron transport the asymmetric orientation of the cytochrome in the membrane is not important. Such asymmetric orientation will be important to obtain the kinetic parameters for the respective active sites on each of the cytochrome hydrophilic domains.

Mechanism of Ascorbic Acid Regeneration Mediated by Cytochrome b_{561}[a]

DAVID NJUS, PATRICK M. KELLEY, GORDON J. HARNADEK, AND YVONNE VIVAR PACQUING

Department of Biological Sciences
Wayne State University
Detroit, Michigan 48202

Although vitamin C (ascorbic acid) has been recognized as an essential nutrient for two centuries and has attracted considerable popular attention in recent years, its biological functions are still poorly understood. Ascorbic acid acts as a cofactor for the collagen-processing enzymes prolyl hydroxylase and lysyl hydroxylase. It may also act as a free-radical scavenger. A third important function, discussed here, is to serve as the electron donor for redox reactions occurring within secretory vesicles.

Secretory vesicles do not take up ascorbic acid at detectable rates.[1] Instead, they seem to have a mechanism for regenerating and reusing the preexisting internal ascorbic acid. Our hypothesis for the mechanism of ascorbic acid regeneration is illustrated in FIGURE 1. The ascorbate-requiring enzymes use ascorbic acid as a one-electron donor and produce the free radical, semidehydroascorbate. The latter will spontaneously disproportionate, two molecules reacting to yield one molecule of ascorbate and one of dehydroascorbate. To prevent this formation of irrecoverable dehydroascorbate, the vesicles must reduce semidehydroascorbate back to ascorbate before it disproportionates. This is the function of a membrane-bound protein, cytochrome b_{561}. Cytochrome b_{561} donates electrons to intravesicular semidehydroascorbate and, in turn, draws electrons from a cytosolic electron donor, probably ascorbic acid. Thus, cytochrome b_{561} functions as a transmembrane electron carrier to maintain redox equilibrium between cytosolic and intravesicular ascorbic acid.

ASCORBATE-REQUIRING REACTIONS IN SECRETORY VESICLES

If the above hypothesis is correct, then all secretory vesicles containing ascorbate-requiring enzymes should have cytochrome b_{561} in the membrane. Moreover, ascorbate

[a] This work was supported by NIH grant no. GM-30500. D. Njus is an Established Investigator of the American Heart Association.

should function as a one-electron donor and should be converted to semidehydroascorbate. These hypotheses may be tested in secretory vesicles containing either of the two known intravesicular ascorbate-requiring enzymes: dopamine β-hydroxylase and peptide amidating monooxygenase.

Dopamine β-hydroxylase, found in catecholamine-storing vesicles, uses ascorbic acid to hydroxylate dopamine to norepinephrine (**1**); Peptide amidating monooxygenase, found in peptide-storing vesicles, apparently uses ascorbic acid to amidate the carboxyl end of peptide hormones (**2**)[b]:

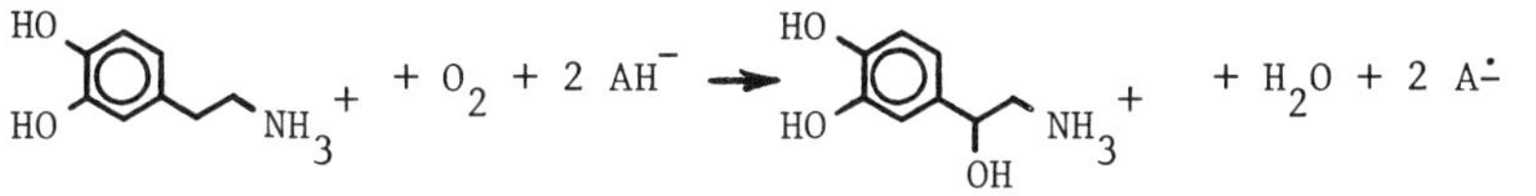

(1)

$$\text{R-NH-CH}_2\text{-COO}^- + O_2 + 2\ AH^- \rightarrow \text{R-NH}_2 + \text{OHC-COO}^- + H_2O + 2\ A^{\cdot -} \quad \textbf{(2)}$$

Chromaffin vesicles, which have dopamine β-hydroxylase, and neurohypophyseal secretory vesicles, which have peptide amidating monooxygenase, are examples of secretory vesicles harboring these two reactions. Both chromaffin vesicles and neurosecretory vesicles contain ascorbic acid,[2,3] and both have membrane-bound cytochrome b_{561}.[4,5] The membranes of both vesicles have the capacity to transfer electrons outward from internal ascorbate to an external electron acceptor.[3,6–8]

Since chromaffin vesicles are representative of organelles that store the catecholamines norepinephrine and epinephrine, cytochrome b_{561}-mediated ascorbate regeneration might also be expected in adrenergic synaptic vesicles in the central and peripheral nervous systems. Indeed, cytochrome b_{561} has been identified in synaptic vesicles from rat vas deferens[9] and bovine splenic nerve.[10] It seems reasonable to suppose that the neurohypophyseal secretory vesicles will be representative of organelles storing amidated peptide hormones. Amidated peptides include neuropeptide Y, α-MSH, substance P, calcitonin, gastrin, cholecystokinin, vasoactive intestinal polypeptide (VIP), pancreatic polypeptide, corticotropin-releasing factor, luteinizing-hormone-releasing hormone, growth-hormone-releasing factor and thyrotropin-releasing hormone as well as the neurohypophyseal hormones oxytocin and vasopressin.[11,12] In fact, amidation is so characteristic of biologically active peptides that it has been used to identify unknown peptides.[12] All secretory vesicles storing these amidated peptides and containing the peptide-amidating monooxygenase would be expected to have this ascorbate-regenerating system.

ALTERNATIVE FATES OF SEMIDEHYDROASCORBATE

To appreciate the functioning of the ascorbate-regenerating system, it is necessary to understand the properties of ascorbic acid. Although ascorbate can lose two elec-

[b] Abbreviations are: AH^-, ascorbate; $A^{\cdot -}$, semidehydroascorbate; A, dehydroascorbate; Cyt, cytochrome; FeCy, ferri/ferrocyanide; ox, oxidized; red, reduced.

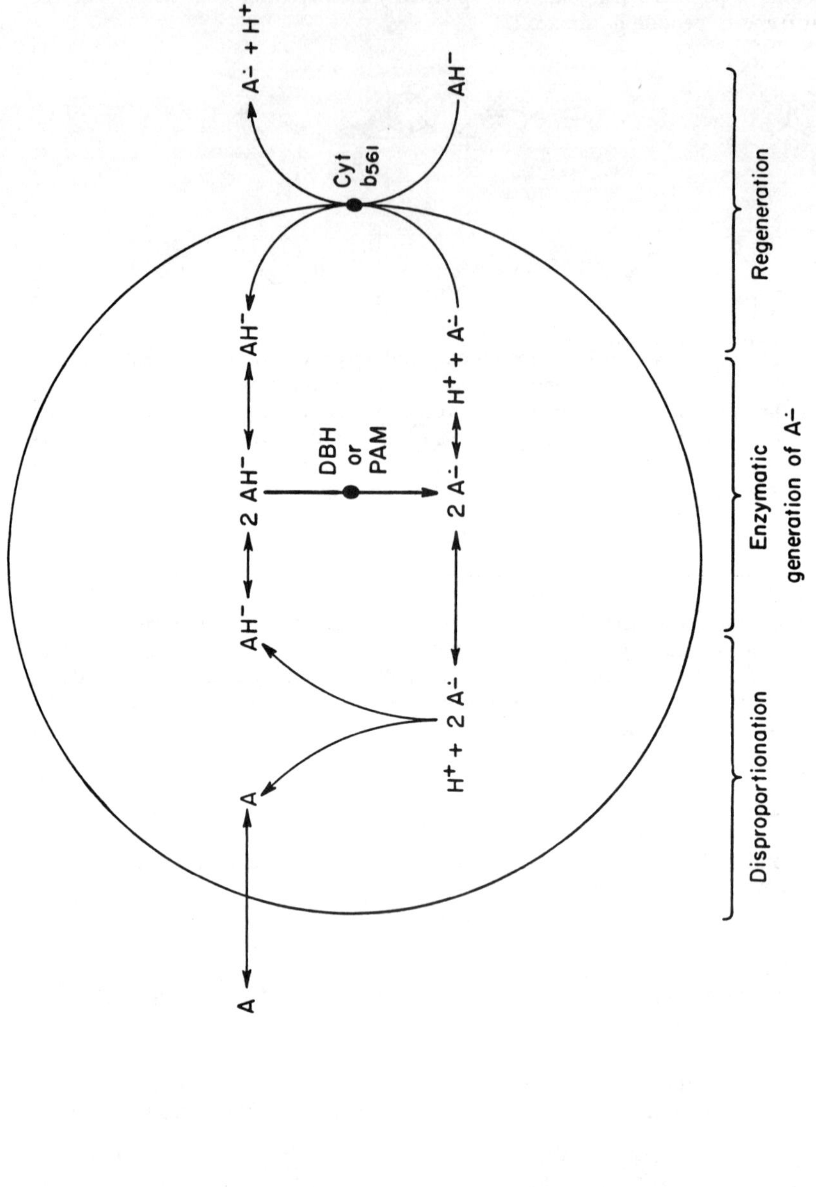

FIGURE 1. Pathways of ascorbate metabolism in secretory vesicles. *DBH,* dopamine β-hydroxylase; *PAM,* peptide amidating monooxygenase.

trons, it commonly functions as a one-electron donor. This was recognized many years ago in the cases of peroxidase and ascorbate oxidase,[13,14] but it was demonstrated only recently in the case of dopamine β-hydroxylase,[15,16] and has not yet been studied in the case of peptide amidating monooxygenase.[17,18] There is good reason why intravesicular enzymes, in particular, should use ascorbate as a one-electron donor. Whereas ascorbate and semidehydroascorbate are anionic and relatively impermeant, the fully oxidized species, dehydroascorbate, is a neutral compound and diffuses across chromaffin-vesicle membranes.[1] Therefore, if dehydroascorbate were produced in an organelle, it would escape and be lost. Moreover, dehydroascorbate can spontaneously hydrolyze to form 2,3-diketogulonic acid which can decay to several different products.[19] Since dehydroascorbate is both permeant and unstable, it would be wasteful to have it produced in an organelle. It would be advantageous to have intravesicular enzymes produce semidehydroascorbate instead and to have the latter quickly reduced back to ascorbate.

Semidehydroascorbate is an unusual free radical in that it is relatively unreactive. It tends to decay by disproportionation (FIG. 1) rather than by reacting with other compounds.[20] Since disproportionation produces dehydroascorbate, it is imperative that an ascorbate-regenerating system reduce semidehydroascorbate back to ascorbate faster than semidehydroascorbate can disproportionate. Disproportionation is a second-order process, so the rate depends strongly on the semidehydroascorbate concentration.[20,21] We will see that cytochrome b_{561} keeps intravesicular semidehydroascorbate levels low enough to make the rate of disproportionation small compared to the rate of regeneration.

The commonly cited standard reduction potential of ascorbic acid is that for the ascorbate-dehydroascorbate reaction. For biological systems, it is the one-electron oxidation that is of interest. The standard reduction potential for the ascorbate-semidehydroascorbate pair is +0.39 V at pH 6.0.[22] Since the reaction involves a proton, the reduction potential is pH-dependent: +0.42 V at pH 5.5 and +0.33 V at pH 7.0.

REVERSE ELECTRON TRANSFER IN CHROMAFFIN-VESICLE GHOSTS

As has often been the case in research on secretory vesicle membranes, the best subject for electron transfer studies has been the chromaffin vesicle, the catecholamine storage vesicle of the adrenal medulla. Because chromaffin vesicles contain reducing equivalents in the form of ascorbic acid and an extremely high concentration of catecholamines, we have thus far found it impossible to prepare membrane vesicles (ghosts) containing a trapped oxidant. It is relatively simple, however, to prepare ghosts containing ascorbic acid. Therefore, we chose to demonstrate electron transfer across the chromaffin-vesicle membrane by monitoring the oxidation of an external electron acceptor by internal ascorbate (FIG. 2).

The most direct method of assaying this reverse electron transfer is to follow the reduction of the external acceptor spectrophotometrically. Ascorbate-loaded chromaffin-vesicle ghosts will reduce either cytochrome *c* or ferricyanide.[8] We can be certain that this reduction is caused by internal ascorbate because we add ascorbate oxidase to eliminate any unsequestered ascorbate. At pH 7.0, ghosts loaded with 100 mM ascorbate reduce 60 μM cytochrome *c* at a rate of about 0.035 ± 0.010 μeq/

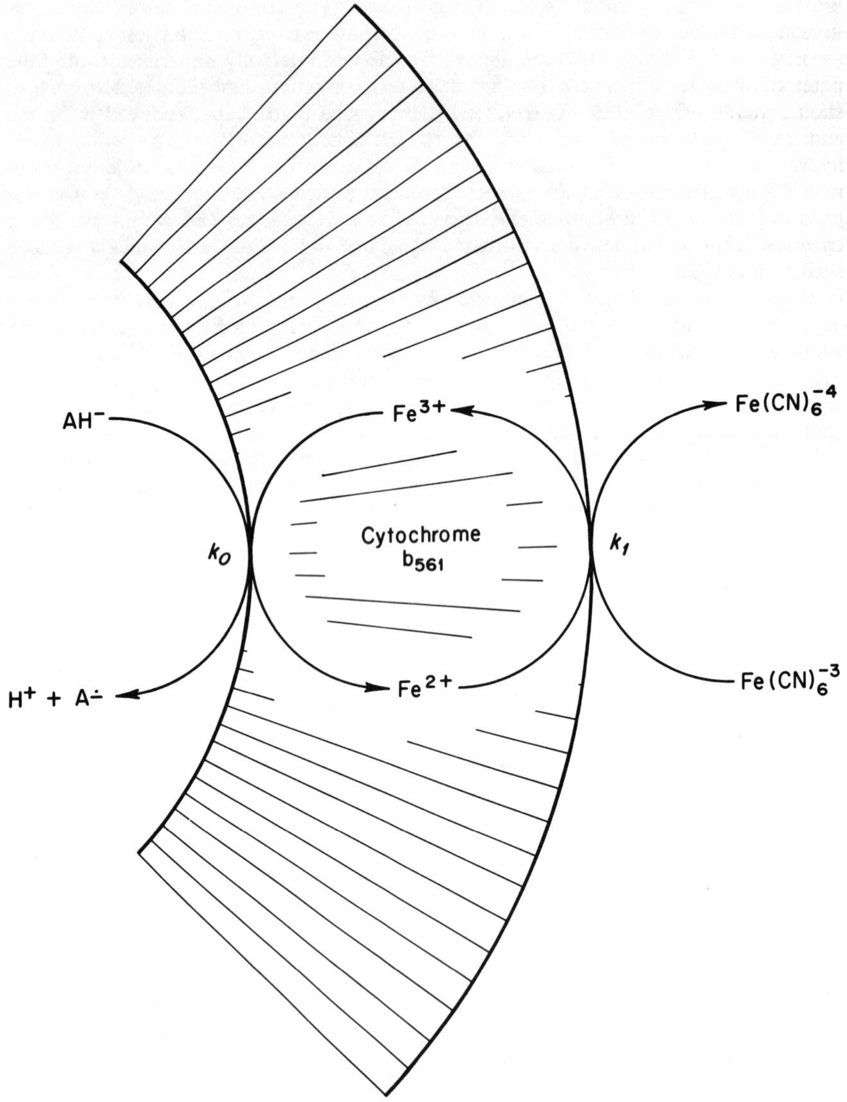

FIGURE 2. Reverse electron transfer.

min·mg protein and 200 μM ferricyanide at a rate of about 2.3 ± 0.3 μeq/min·mg protein. An extensive and continuing search has failed to yield any effective inhibitors of this electron transfer. Inhibitors tested include azide, cyanide, CO, NO, a variety of other inhibitors of oxidative and photophosphorylation, cytochrome P-450 inhibitors, and agents used for chemical modification of proteins.

Electron transfer can also be demonstrated by monitoring the change in membrane potential created by the charge transfer. The observation of a membrane potential should guarantee that electrons are being transferred across a well-sealed membrane and that neither the electron donor nor the acceptor is simply permeating. When ferricyanide or ferricytochrome *c* is added to a suspension of ascorbate-loaded chromaffin-vesicle ghosts, a membrane potential (interior positive) develops across the ghost membrane.[6] At 4°C, the membrane potential is stable and can be detected either by following the fluorescence of 1-anilinonaphthalene-8-sulfonate (ANS) or by measuring the distribution of the permeant anion SCN^-. This membrane potential cannot be elicited from ascorbate-free ghosts or by ferrocyanide added instead of ferricyanide.[7]

At physiological temperatures, a stable membrane potential can be produced by coupling the chromaffin-vesicle electron-transfer system to cytochrome oxidase using cytochrome *c*. The membrane potential is generated by transferring electrons from internal ascorbate to cytochrome *c*. Cytochrome *c* is then reoxidized by cytochrome oxidase. This membrane potential drives reserpine-sensitive norepinephrine transport, confirming the location of the electron transfer system in the chromaffin-vesicle membrane.[7]

CYTOCHROME b_{561} MEDIATES REVERSE ELECTRON TRANSFER

Cytochrome b_{561} is a high-potential *b*-type cytochrome which has been well characterized by the laboratories of Torgier Flatmark,[4,23] Patrick Fleming[24,25] and David Apps.[26,27] It is reduced by ascorbate and oxidized by semidehydroascorbate, ferricyanide and cytochrome *c*.[4,23,28] Cytochrome b_{561} is a likely candidate for a transmembrane electron carrier because it spans the chromaffin-vesicle membrane[24,26] and has a midpoint reduction potential (+100 to +140 mV) appropriate for this function.[4,27] Furthermore, there is a good correlation between the presence of cytochrome b_{561} and the occurrence of transmembrane electron transfer. Both chromaffin vesicle and neurohypophyseal secretory vesicle membranes contain cytochrome b_{561}[4,5] and both exhibit transmembrane electron transfer.[3,6–8] Chromaffin-vesicle ghosts reduce cytochrome *c* at a rate of 5 min^{-1} per cytochrome b_{561} while the value in neurohypophyseal secretory vesicles is 3.5 min^{-1}. Srivastava *et al.*[25] have reported transmembrane electron transfer catalyzed by purified cytochrome b_{561} reconstituted into phospholipid vesicles. The turnover number for cytochrome *c* reduction (570 min^{-1}) is anomalously high, however, so the experiment is unfortunately not as definitive as it could be.

The absorption spectrum of cytochrome b_{561} changes concomitantly with electron transfer across chromaffin-vesicle membranes.[28] In ascorbate-loaded chromaffin-vesicle ghosts, cytochrome b_{561} exhibits an absorption maximum at 561 nm, indicating that it is nearly completely reduced. Addition of ferricyanide causes the 561 nm absorption to diminish, consistent with oxidation of the cytochrome. The added ferricyanide becomes completely reduced if the amount added is less than the amount of intravesicular ascorbate. When this happens, cytochrome b_{561} returns to its reduced state. If an excess of ferricyanide is added, the intravesicular ascorbate becomes exhausted and cytochrome b_{561} remains oxidized. The spectrum of these absorbance changes correlates with the difference spectrum (reduced-oxidized) of cytochrome b_{561}. These results indicate that cytochrome b_{561} can accept electrons from internal ascorbate and pass them to external ferricyanide.

When ascorbate oxidase is added to a suspension of ascorbate-loaded ghosts, there is a brief drop in the absorbance at 561 nm, indicating that cytochrome b_{561} is transiently oxidized.[28] Since dehydroascorbate has no effect on the absorbance spectrum of cytochrome b_{561}, it is likely that this oxidation is caused by semidehydroascorbate generated by ascorbate oxidase acting on free ascorbate. Therefore, reduced cytochrome b_{561} can reduce semidehydroascorbate as it must if it is to function *in vivo* to regenerate internal ascorbate from semidehydroascorbate.

KINETICS OF REVERSE ELECTRON TRANSFER

By analyzing the kinetics of electron transfer from internal ascorbate to external ferricyanide, we may be able to deduce properties of the reverse reaction, inward electron flow. To analyze the electron transfer process, we divide it into two steps (FIG. 2): the reduction of cytochrome b_{561} by internal ascorbate **(3)** and the reduction of external ferricyanide by cytochrome b_{561} **(4)**. The respective rate constants for these two steps are k_0 and k_1:

$$\mathrm{AH^-} + \mathrm{Cyt}\ b_{561_{ox}} \xrightarrow{k_0} \mathrm{A}^{\dot{-}} + \mathrm{Cyt}\ b_{561_{red}} \quad \textbf{(3)}$$

$$\mathrm{Cyt}\ b_{561_{red}} + \mathrm{FeCy_{ox}} \xrightarrow{k_1} \mathrm{Cyt}\ b_{561_{ox}} + \mathrm{FeCy_{red}} \quad \textbf{(4)}$$

If we make the steady-state assumption ($d[\mathrm{Cyt}\ b_{561_{red}}]/dt = 0$), then the rate of ferricyanide reduction ($V = d[\mathrm{FeCy_{red}}]/dt$) is given by the following equation:

$$\frac{B}{V} = \frac{1}{k_0[\mathrm{AH^-}]} + \frac{1}{k_1[\mathrm{FeCy_{ox}}]},$$

where B is the total concentration of cytochrome b_{561} (oxidized plus reduced) expressed in units of membrane protein concentration (mg/ml). This analysis predicts that a maximum rate ($V_{max} = k_0[\mathrm{AH^-}]B$) will be achieved at high ferricyanide concentrations and that this rate is limited by the rate of electron transfer from internal ascorbate to cytochrome b_{561}. If this analysis is valid, then a plot of B/V against $1/[\mathrm{FeCy_{ox}}]$ should be linear. Moreover, the slope should be $1/k_1$ and the intercept should be $1/k_0[\mathrm{AH^-}]$. Plots of B/V vs. $1/[\mathrm{FeCy_{ox}}]$ are, in fact, linear. Values of k_1 and k_0 averaged from several different experiments are 6.1 ml/min·mg protein and 3.5×10^{-3} ml/

min·mg respectively. The value for k_0 should be considered an estimate because it assumes that the internal ascorbate concentration is that of the lysis medium. The true concentration is almost certainly less than this since some ascorbate will be lost through leakage and some will be oxidized. Because we are still uncertain about intravesicular ascorbate concentrations, we cannot be certain that the intercept of the double reciprocal plot is inversely proportional to $[AH^-]$. If the internal ascorbate concentration is less than assumed or if V_{max} is limited by something other than the rate of electron transfer from internal ascorbate to cytochrome b_{561}, then the actual value of k_0 will be larger. Our double-reciprocal analysis, therefore, should give us a minimum value for k_0 and is a promising approach to analyzing separately the kinetics of electron transfer at each surface of the membrane.

KINETICS OF ASCORBATE REGENERATION

Our objective is to understand the mechanism of intravesicular ascorbate regeneration. We can gain some insight into the kinetics of this process if we recognize that ascorbate regeneration is precisely the reverse of the intravesicular ascorbate oxidation (reverse electron transfer) analyzed above. In other words, if we know the equilibrium between ascorbate and cytochrome b_{561}, we can calculate k_{-0} (the rate constant for the reduction of internal semidehydroascorbate by cytochrome b_{561}) from k_0 (the rate constant for the reduction of cytochrome b_{561} by internal ascorbate). Approximate values for the necessary parameters can be taken from our preliminary studies and from the literature. This allows us to model the kinetics of the hypothesized ascorbate-regenerating system and to examine its kinetic feasibility.

The equilibrium constant (K_{eq}) for electron transfer between ascorbate and cytochrome b_{561} is related to the difference between the midpoint reduction potentials of the two redox pairs ($E^\circ_{Asc} - E^\circ_{Cyt}$) as given by the following equation:

$$K_{eq} = k_{-0}/k_0 = \exp [(E^\circ_{Asc} - E^\circ_{Cyt})(F/RT)].$$

Having values for k_0, E°_{Asc}, and E°_{Cyt} (TABLE 1) under the conditions found in chromaffin-vesicle ghosts (pH 6.0, 22°C), we may calculate k_{-0}. This rate constant for the reduction of internal semidehydroascorbate by cytochrome b_{561} is 125 ml/min·mg. For the sake of discussion, we will assume that this value is also an adequate estimate for k_{-0} in chromaffin vesicles *in vivo* although the conditions are somewhat different (pH 5.5, 37°C). Letting k_{-0} be 125 ml/min·mg, the cytochrome b_{561} concentration be 333 mg protein/ml internal volume, and the steady-state semidehydroascorbate concentration be 2×10^{-7} M,[15] we obtain an *in vivo* ascorbate regeneration rate of 1.4×10^{-4} M/s. Using the same semidehydroascorbate concentration and a disproportionation rate constant, k_{diss}, of 4×10^7 $M^{-1}sec^{-1}$, we obtain an *in vivo* disproportionation rate of 1.6×10^{-6} M/sec. Although these are clearly estimates, it appears that the rate of semidehydroascorbate regeneration is about two orders of magnitude faster than the rate of disproportionation. This is most significant because it verifies the kinetic feasibility of the ascorbate regeneration model.

THE CYTOSOLIC ELECTRON DONOR FOR ASCORBATE REGENERATION

The chromaffin cells of the adrenal medulla contain a high concentration of cytosolic ascorbate. Since ascorbate reduces cytochrome b_{561} and since NADH, NADPH and glutathione do not, it is reasonable to suppose that ascorbate might be the cytosolic electron donor for intravesicular ascorbate regeneration. Diliberto *et al.*[30] found that cytosolic ascorbate in chromaffin cells is maintained by an NADH-dependent semidehydroascorbate reductase activity located on the mitochondrial outer membrane. Wakefield *et al.*[31] suggested that this reductase, cytosolic ascorbate, cytochrome b_{561}, and internal ascorbate form an electron transfer chain linking cytosolic NADH to intravesicular dopamine hydroxylation. Recently, Beers *et al.*[32] have also published evidence that cytosolic ascorbate can serve as the electron donor for intravesicular ascorbate regeneration.

The reduction of intravesicular semidehydroascorbate by cytosolic ascorbate is

TABLE 1. Reaction Constants for Ascorbic Acid and Cytochrome b_{561}

Parameter	pH 5.5	pH 6.0	Ref.
k_{diss} (M^{-1} s^{-1})			
(0 ionic strength)	7×10^6	2×10^6	21
(0.045 M phosphate)	4×10^7	2×10^7	21
E° (ascorbate)	+0.42 V	+0.39 V	22
E° (cytochrome b_{561})	+0.14 V	+0.14 V	27
k_0 (ml/min·mg protein)	—	3.5×10^{-3}	this paper
k_{-0} (ml/min·mg protein)	—	125	this paper

thermodynamically favored because of the pH gradient and membrane potential existing across the chromaffin-vesicle membrane.[33] These energy gradients are created by a membrane-bound, inwardly directed, proton-translocating ATPase.[34] The membrane potential (inside positive) should draw electrons into the vesicles. The pH gradient should also contribute to the force driving electrons into the vesicles, because the ascorbate-semidehydroascorbate interconversion involves a proton. Lowering the pH shifts the equilibrium in favor of ascorbic acid and raises the midpoint reduction potential at pH 5.5 (the intravesicular pH) about 90 mV higher than at pH 7.0 (the cytosolic pH). Consequently, the pH gradient across the chromaffin-vesicle membrane should supply a force of about 90 mV driving electrons from cytosolic ascorbate to intravesicular semidehydroascorbate. It is noteworthy that neurohypophyseal secretory vesicles and many other secretory and synaptic vesicles have similar H^+-translocating ATPases and acidic interiors.[35,36] This electron transfer driven by a proton gradient is an interesting variation on the chemiosmotic concept of proton gradients generated by electron transfer.

SUMMARY

In summary, ascorbic acid serves as a one-electron donor for dopamine β-hydroxylase in chromaffin vesicles and probably for peptide amidating monooxygenase in neurohypophyseal secretory vesicles. It appears that the semidehydroascorbate that is produced is reduced by cytochrome b_{561} to regenerate intravesicular ascorbate. Cytochrome b_{561}, a transmembrane protein, is reduced in turn by an extravesicular electron donor, probably cytosolic ascorbic acid. It will be interesting to see whether other ascorbate-requiring enzymes in other organelles use a similar ascorbate-regenerating system to provide an intravesicular supply of reducing equivalents.

REFERENCES

1. TIRRELL, J. G. & E. W. WESTHEAD. 1979. Neuroscience **4:** 181-186.
2. INGEBRETSEN, O. C., O. TERLAND & T. FLATMARK. 1980. Biochim. Biophys. Acta **628:** 182-189.
3. RUSSELL, J. T., M. A. LEVINE & D. NJUS. 1985. J. Biol. Chem. **260:** 226-231.
4. FLATMARK, T. & O. TERLAND. 1971. Biochim. Biophys. Acta **253:** 487-491.
5. DUONG, L. T., P. J. FLEMING & J. T. RUSSELL. 1984. J. Biol. Chem. **259:** 4885-4889.
6. NJUS, D., J. KNOTH, C. COOK & P. M. KELLEY. 1983. J. Biol. Chem. **258:** 27-30.
7. HARNADEK, G. J., R. E. CALLAHAN, A. R. BARONE & D. NJUS. 1985. Biochemistry **24:** 384-389.
8. HARNADEK, G. J., E. A. RIES & D. NJUS. 1985. Biochemistry **24:** 2640-2644.
9. FRIED, G. 1978. Biochim. Biophys. Acta **507:** 175-177.
10. FLATMARK, T., H. LAGERCRANTZ, O. TERLAND, K. B. HELLE & L. STJARNE. 1971. Biochim. Biophys. Acta **245:** 249-252.
11. EIPPER, B. A., C. C. GLEMBOTSKI & R. E. MAINS. 1983. J. Biol. Chem. **258:** 7292-7298.
12. TATEMOTO, K., M. CARLQUIST & V. MUTT. 1982. Nature **296:** 659-660.
13. YAMAZAKI, I., H. S. MASON & L. PIETTE. 1960. J. Biol. Chem. **235:** 2444-2449.
14. YAMAZAKI, I. & L. H. PIETTE. 1961. Biochim. Biophys. Acta **50:** 62-69.
15. DILIBERTO, E. J., JR. & P. L. ALLEN. 1981. J. Biol. Chem. **256:** 3385-3393.
16. SKOTLAND, T. & T. LJONES. 1980. Biochim. Biophys. Acta **630:** 30-35.
17. BRADBURY, A. F., M. D. A. FINNIE & D. G. SMYTH. 1982. Nature **298:** 686-688.
18. EIPPER, B. A., R. E. MAINS & C. C. GLEMBOTSKI. 1983. Proc. Natl. Acad. Sci. U.S.A. **80:** 5144-5148.
19. TOLBERT, B. M. & J. B. WARD. 1982. Dehydroascorbic acid. *In* Ascorbic Acid: Chemistry, Metabolism, and Uses. P. A. Seib & B. M. Tolbert, Eds.: 101-123. American Chemical Society. Washington, D.C.
20. BIELSKI, B. H. J. 1982. Chemistry of Ascorbic Acid Radicals. *In* Ascorbic Acid: Chemistry, Metabolism, and Uses. P. A. Seib & B. M. Tolbert, Eds.: 81-100. American Chemical Society. Washington, D.C.
21. BIELSKI, B. H. J., A. O. ALLEN & H. A. SCHWARZ. 1981. J. Am. Chem. Soc. **103:** 3516-3518.
22. IYANAGI, T., I. YAMAZAKI & K. F. ANAN. 1985. Biochim. Biophys. Acta **806:** 255-261.
23. TERLAND, O. & T. FLATMARK. 1980. Biochim. Biophys. Acta **597:** 318-330.
24. DUONG, L. T. & P. J. FLEMING. 1984. Arch. Biochem. Biophys. **228:** 332-341.
25. SRIVASTAVA, M., L. T. DUONG & P. J. FLEMING. 1984. J. Biol. Chem. **259:** 8072-8075.
26. APPS, D. K., J. G. PRYDE & J. H. PHILLIPS. 1980. Neuroscience **5:** 2279-2287.
27. APPS, D. K., M. D. BOISCLAIR, F. S. GAVINE & G. W. PETTIGREW. 1984. Biochim. Biophys. Acta **764:** 8-16.
28. KELLEY, P. M. & D. NJUS. 1986. J. Biol. Chem. **261:** 6429-6432.

29. KNOTH, J., M. ZALLAKIAN & D. NJUS. 1981. Biochemistry **20:** 6625-6629.
30. DILIBERTO, E. J., JR., G. DEAN, C. CARTER & P. L. ALLEN. 1982. J. Neurochem. **39:** 563-568.
31. WAKEFIELD, L. M., A. E. CASS & G. K. RADDA. 1982. Fed. Proc. **41:** 893.
32. BEERS, M. F., R. G. JOHNSON & A. SCARPA. 1986. J. Biol. Chem. **261:** 2529-2535.
33. NJUS, D., M. ZALLAKIAN & J. KNOTH. 1981. The chromaffin granule: Proton-cycling in the slow lane. *In* Chemiosmotic Proton Circuits in Biological Membranes. V. P. Skulachev & P. C. Hinkle, Eds.: 365-374. Addison-Wesley. Reading, Mass.
34. NJUS, D., J. KNOTH & M. ZALLAKIAN. 1981. Proton-linked transport in Chromaffin Granules. *In* Current Topics in Bioenergetics, Vol. 11. D. R. Sanadi, Ed.: 107-147. Academic Press. New York.
35. RUSSELL, J. T. & R. W. HOLZ. 1981. J. Biol. Chem. **256:** 5950-5953.
36. NJUS, D. 1983. J. Autonom. Nerv. Syst. **7:** 35-40.

DISCUSSION OF THE PAPER

E. DILIBERTO (*Wellcome Research Laboratories, Research Triangle Park, N.C.*): Can the asymmetrical distribution of cytochrome b_{561} in the chromaffin vesicle membrane suggested by Dr. P. Fleming's data explain the difference in the rates of electron transfer obtained in the reconstituted system and chromaffin vesicle ghost experiments?

D. NJUS (*Wayne State University, Detroit, Mich.*): For some time now, we have been concerned about the apparent discrepancy between our rate of electron transfer measured in chromaffin-vesicle ghosts and Dr. Fleming's rate of electron transfer measured using reconstituted cytochrome b_{561}. Dr. Fleming has just informed me that they use ferricyanide as a mediator in measuring electron transfer to cytochrome *c*. We find that the rate of electron transfer in chromaffin vesicle ghosts is accelerated up to tenfold by ferricyanide. Taking this into account, the rate of electron transfer in the reconstituted system is probably not so different from the rate in chromaffin vesicle ghosts.

H. TAMIR (*Columbia University, New York, N.Y.*): Does the pH gradient of the vesicles control the activity of cytochrome b_{561}?

D. NJUS: In theory, the pH gradient across the vesicle membrane should control the rate of electron transfer between ascorbic acid on one side of the membrane and semidehydroascorbate on the other. Experimental evidence for this effect, however, has not yet been reported.

J.-P. HENRY (*Institut de Biologie Physico-Chimique, Paris, France*): Is electron transfer limiting the rate of dopamine hydroxylation?

D. NJUS: No, the chromaffin vesicle membrane's capacity for electron transfer is far greater than its dopamine β-hydroxylase activity. This is because the rate of electron transfer must be great enough not just to account for dopamine hydroxylation, but to reduce semidehydroascorbate before the latter can disproportionate. I imagine that cytochrome b_{561} *in vivo* is almost always in the reduced state and is poised to donate an electron to semidehydroascorbate immediately after the free radical is produced by dopamine β-hydroxylase.

M. LEVINE (*NIDDK, National Institutes of Health, Washington, D.C.*): In isolated chromaffin vesicle preparations only ascorbic acid enhances norepinephrine formation. Other reductants such as glutathione, homocystine, NADH, NADPH, and dithiothreitol are ineffective.[a] Likewise in intact cells only ascorbic acid enhances norepinephrine formation, as a measure of the ability to reduce cytochrome b_{561}.[b] These findings indicate that ascorbic acid *in vivo* is probably the cytosolic reducing substance for norepinephrine formation, presumably by way of cytochrome b_{561}.

D. NJUS: Thank you. Your data suggest that compounds unable to reduce cytochrome b_{561} are also incapable of supporting dopamine hydroxylation in chromaffin vesicles. This certainly strengthens the argument that electron transfer into the vesicles is mediated by cytochrome b_{561}.

R. M. PRUSS (*NIMH, National Institutes of Health, Bethesda, Md.*): Has anyone measured an effect of reserpine on either the pH gradient or the electrochemical proton gradient especially at high concentrations over and above that necessary to inhibit the carrier?

D. NJUS: Yes, at concentrations below 10 μM, reserpine has no effect on the pH gradient or membrane potential across chromaffin vesicle membranes. At reserpine concentrations above 10 μM, however, the membrane becomes more permeable and the pH gradient and membrane potential begin to dissipate. This is probably a detergent effect caused by the partitioning of the hydrophobic reserpine molecule into the lipid milieu of the membrane.

[a] Levine, M., K. Morita, E. Heldman & H. B. Pollard. J. Biol. Chem. **260:** 15598-15603 (1985).

[b] Levine, M. J. Biol. Chem. **261:** 7347-7356 (1986).

Chromogranins A, B, and C: Widespread Constituents of Secretory Vesicles[a]

R. FISCHER-COLBRIE, C. HAGN, AND M. SCHOBER

Department of Pharmacology
University of Innsbruck
A-6020 Innsbruck, Austria

The acidic soluble proteins of chromaffin granules have been collectively named chromogranins.[1] During the last 20 years these chromogranins have been analyzed and characterized in great detail (for review see refs. 2-5), their physicochemical properties and biosynthetic pathways established, and in 1986 the primary sequence of one chromogranin was obtained for the first time by Iacangelo *et al.*[6] Nonetheless, the physiological function of these characteristic proteins is still unknown. Originally it was thought that these proteins are involved in a storage complex with catecholamines. More recent studies demonstrating a widespread distribution of these proteins in several nonadrenergic endocrine and neuronal tissues have, however, called into question this finding. In this paper we summarize recent evidence demonstrating similarities of the chromogranins with neuropeptides and regulatory peptides which suggest that these proteins might have a function after secretion into circulation.

CHARACTERIZATION OF SOLUBLE PROTEINS OF CHROMAFFIN GRANULES

A powerful tool to separate these proteins is two-dimensional electrophoresis.[7] This method yields a complex pattern with about 40 polypeptides stained with Coomassie blue (see FIG. 1). Immunoblotting experiments with specific antisera (see FIG. 1), however, demonstrated that the majority of these proteins can be attributed to only three groups of acidic proteins (chromogranins A, B, and C) which make up about 87% of the total soluble proteins (see TABLE 6). All three groups are immunologically distinct; antisera raised against highly purified chromogranins do not cross-react in immunoblots with the other chromogranin groups (see FIG. 1). The presence of some common epitopes, however, cannot be excluded by this kind of experiment. In addition to chromogranins, two neuropeptides, *i.e.*, enkephalin-containing peptides derived from proenkephalin (1% of total soluble proteins; for review see ref. 8) and neuropeptide Y (0.2%; refs. 9-11), are present. Further constituents with a known function are dopamine β-hydroxylase (4%; see ref. 2), the enzyme catalyzing the

[a] This work was supported by the Dr. Legerlotz Foundation.

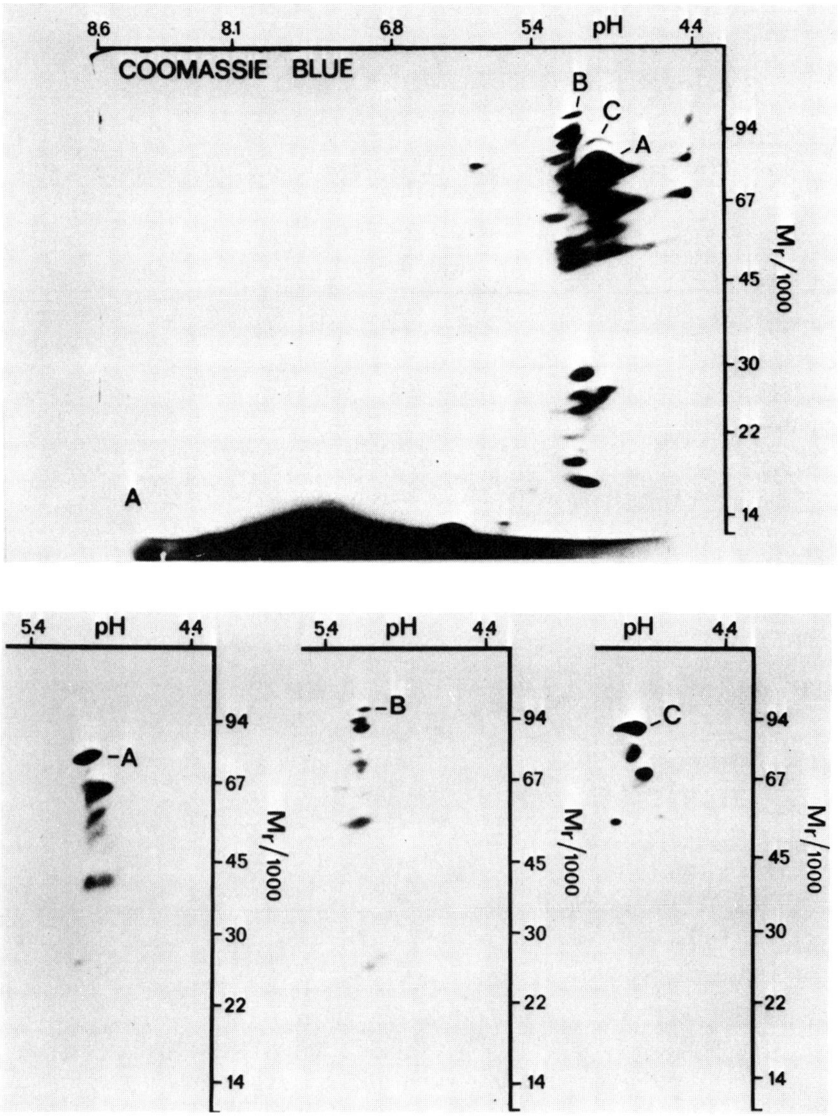

FIGURE 1. Immunological characterization of soluble proteins from bovine chromaffin granules. *Upper panel:* soluble proteins were subjected to two-dimensional electrophoresis and stained with Coomassie blue. *Lower panel:* immunoreplicas with antisera against chromogranin A, chromogranin B, and chromogranin C. *A,* chromogranin A; *B,* chromogranin B; *C,* chromogranin C.

biosynthesis of noradrenaline, and trypsin-like and carboxypeptidase-B-like proteases involved in processing of proenkephalin (for review see ref. 5).

PROPERTIES OF CHROMOGRANINS A, B, AND C

TABLE 1 summarizes some physicochemical properties and data on the biosynthetic pathway of these proteins (compare also the paper on life cycle by H. Winkler *et al.* in this volume). Chromogranin A originally isolated from chromaffin granules[12–14] is an acidic glycoprotein with a molecular weight of 70,000 to 75,000 as shown in analytical ultracentrifugation studies[13] and on SDS gels.[2] The sequence analysis performed by Iacangelo *et al.*,[6] however, revealed a true molecular weight of 48,200 for this protein. One reasonable explanation of this unusual behavior in SDS electrophoresis might be a decreased binding of SDS to this protein. Usually 1 g protein binds 1.4 g SDS. Reduced binding of SDS to chromogranin A due to the high amount of acidic amino acids present within this molecule (25% glutamic acid) would result in a significantly decreased migration in SDS gels, yielding higher apparent molecular weights. A similar phenomenon has been described for glycoproteins with highly branched carbohydrate chains.[15,16]

TABLE 1. Properties of Chromogranins and Proenkephalin

	Chromogranin A	Chromogranin B	Chromogranin C	Proenkephalin
Physicochemical Properties				
apparent M_r in SDS gels	75,000	100,000	86,000	29,000
molecular mass (kDa)	48,200[6]	nd[a]	nd	29,786[8]
number of amino acids	430[6]	nd	nd	263[8]
pI	5.0	5.2	5.0	nd
Biosynthesis				
number of preproteins	2[51]	1[51]	1[23]	1[8]
Posttranslational Modifications				
O-glycosylation (% carbohydrate)	5.4%[54,55]	5.7%[54]	nd	0%[56]
phosphorylation	yes[57,22]	yes[22]	yes[22]	nd
tyrosine sulfation	no[58,22]	yes[58,22]	yes[58,22]	nd
proteolytic processing within secretory granules	yes[18,35]	yes[18]	yes[23,22]	yes[18,8]
Widespread Distribution in Endocrine and Nervous Tissues	yes	yes	yes	yes

[a] nd = not determined.

The name chromogranin B was given recently[17,18] to a protein of apparent M_r 100,000 with a slightly more basic pI than chromogranin A.

The third chromogranin, chromogranin C, is a protein of apparent M_r 86,000 that was first detected in anterior pituitary by labeling with ^{35}S-sulfate.[19] Subsequently this protein was shown to be also present in adrenal medulla.[20] Recently the names TSP 86/84[21] and secretogranin II[22] were given to this protein. We prefer to call this protein chromogranin C,[5,23] since it is clearly a characteristic member of the chromogranin family (see below).

Recent work on chromogranins demonstrated that these proteins share several properties with neuropeptides, which we will discuss in more detail: (1) Chromogranins are present in chromaffin granules of all species so far analyzed. Parts of the sequence seem to be highly conserved, as antisera against bovine chromogranin cross-react quite well with proteins from other species including the frog. (2) Chromogranins are not only confined to the adrenergic system as originally thought for chromogranin A, but are widespread in several endocrine and neuronal tissues. (3) The biosynthesis of chromogranins is similar to neuropeptides. After synthesis in the endoplasmic reticulum the proproteins are modified significantly in the Golgi apparatus by glycosylation, phosphorylation, and sulfation (for discussion see article by H. Winkler *et al.* in this volume). Subsequently chromogranins are transported from the Golgi to the secretory vesicles where they are processed proteolytically. The degree of proteolytic processing and relative amounts of chromogranins in other tissues varies, as is known also for neuropeptides.

PRESENCE OF CHROMOGRANINS IN ADRENAL MEDULLA OF OTHER SPECIES

Two-dimensional electrophoresis followed by immunoblotting revealed the presence of chromogranins A, B, and C in the adrenal medulla of several species. Antisera against bovine chromogranins reacted quite well with chromogranins from most species tested, including rat and frog. Thus it seems reasonable that parts of the sequence of the chromogranins are highly conserved during evolution, as is known for neuropeptides. This is in agreement with sequence data available for the amino terminus which shows strong homology between bovine, human, and rat chromogranins.[24–26] Apparent molecular weights in other species differ slightly,[27] *e.g.,* human chromogranins have slightly higher molecular weights compared to bovine proteins.[28] In rat chromogranins A and C migrate in one-dimensional SDS gels into the same positions with an apparent M_r of 86,000.[22]

The relative distribution of chromogranins in adrenal medulla of other species, however, varies to a high degree. TABLE 2 summarizes the results. In chromaffin granules from ox, sheep, and horse, chromogranin A is the most prominent chromogranin group. In pig nearly as much chromogranin B as A was found, whereas in human, rat, and frog granules chromogranin B is the major chromogranin group. Thus the synthesis of chromogranins seems to be regulated differently in the adrenal medulla of several species. Similarly the expression of several neuropeptides known to be present in adrenal medulla is regulated. High amounts of enkephalins are found in chromaffin granules from cow and dog, whereas only traces of enkephalins are present in rat adrenal medulla.[29] High concentrations of neuropeptide Y have been discovered in mouse and cat adrenal medulla, less in man and guinea pig, and only

TABLE 2. Relative Amounts of Chromogranins in Various Species

bovine	A >> B >> C[18,22,23]
ovine	A >> B >> C[a]
equine	A > B[18]
porcine	A = B >> C[18]
rat	B >> A > C[18,22]
frog	B >> A[a]
human	B >> A > C[28]

Chromogranins A, B and C were identified with specific antisera in immunoblots. Relative amounts were deduced from Coomassie blue stains after a two-dimensional separation of these proteins.

[a] C. Hagn and R. Fischer-Colbrie, unpublished.

low concentrations were found in pig.[10] Neurotensin is found in high amounts only in granules from cat[30] and vasoactive intestinal polypeptide (VIP) only in those from frog.[31]

CHROMOGRANINS IN OTHER TISSUES

Chromogranin A was considered for many years to be an "adrenergic" protein, however in 1982 its presence in another endocrine tissue, the parathyroid gland, was established.[32] The occurrence of chromogranin A in several endocrine (pituitary, parathyroid gland, endocrine pancreas, gut) and neuronal[33–35] tissues is now well established (for a review see ref. 5). Recently chromogranins B and C were also shown to be widely distributed.[18,36,22]

Are the chromogranins always found together as in adrenal medulla? Most of these studies were performed by immunohistochemistry. Two pitfalls using this technique have to be considered: fixation of tissue sections seems critical for preservation of chromogranin-immunoreactive epitopes,[37,38] and if one uses polyclonal antisera great care has to be taken to obtain antisera specific for the individual chromogranins. Some controversial data (see below) are probably explained by these difficulties.

In the pituitary chromogranins occur in all three lobes (TABLE 3). Posterior pituitary contains only chromogranin C, in the intermediate lobe chromogranins A and C are present, and in the anterior lobe all three chromogranins are found.[36,39] More detailed studies of the anterior lobe revealed a colocalization of all three chromogranins with the glycoprotein-hormone-producing cells (gonadotrophs, thyrotrophs) in several species.[21,39–41] Only a few corticotrophs from ox and sheep but none from man or rat contain chromogranins. Mammotrophs and somatotrophs of several species and tumors derived from these cells[41] apparently do not contain chromogranins, although moderate staining of mammotrophs with anti-chromogranin C[21] and somatotrophs with anti-chromogranin A has been reported.[42]

In endocrine pancreas all three chromogranins are found (TABLE 4). Parathyroid gland contains only chromogranin A; chromogranins B and C are apparently absent, although immunohistochemically a weak reaction was reported for chromogranin B.[27,37] This weak reaction was considered to be an artifact,[36] since with immunoblotting experiments and antisera against chromogranin B adsorbed with chromogranin A no reaction was found.

In human thyroid gland chromogranin A was detected in the C cells and tumors derived from them. For bovine C cells controversial results were obtained (see TABLE 4). Thus it is not clear at the moment whether chromogranin A is present in ox C cells. A final answer might come from studies with isolated secretory granules from these cells.

In gut chromogranin A is found in most endocrine cells.[43,38,37] The enterochromaffin and gastrin cells contain in addition chromogranin B (see TABLE 4); no chromogranin

TABLE 3. Presence of Chromogranins in Pituitary[a]

	Chromogranin A	Chromogranin B	Chromogranin C
Anterior Lobe			
gonadotrophs			
bovine	+[42]		+[21]
ovine	+[39]	+[39]	+[39]
rat	+[40]		
human	+[41]		
thyreotrophs			
bovine			+[21]
ovine	+[39]	−[39]	±[39]
rat	+[40]		
human	+[41]		
corticotrophs			
bovine	−[42]		±[21]
ovine	±[39]	±[39]	±[39]
rat	−[40]		
human	−[41]		
mammotrophs			
bovine	−[42]		±[21]
ovine	−[39]	−[39]	−[39]
rat	−[40]		
human	−[41]		
somatotrophs			
bovine	±[42]		−[21]
ovine	−[39]	−[39]	−[39]
rat	−[40]		
human	−[41]		
Intermediate Lobe			
bovine	+[42,36]	−[36]	+[36]
ovine	+[39]	−[39]	+[39]
rat	−[40]		
human	−[41]		
Posterior Lobe			
bovine	−[42,36]	−[36]	+[36]
ovine	−[39]	−[39]	+[39]
rat	−[40]		
human	−[41]		

[a] + = positive; ± = a few cells are stained; − = no staining.

C is found in these cell types. Glicentin and pyloric D cell stain only for chromogranin C.[37]

In brain chromogranin A has a characteristic distribution that overlaps partially with various known neurotransmitters and neuropeptides but does not parallel any of them.[33] No detailed data are available for chromogranins B and C, however the

TABLE 4. Chromogranin Immunoreactivity in Several Endocrine Tissues

	Chromogranin A	Chromogranin B	Chromogranin C
Endocrine Pancreas			
glucagon A cells			
bovine	+[37]	+[37]	+[37]
human	+[37,38]	+[37]	+[37]
insulin B cells			
bovine	+[42]	+[37]	
human	−[38]		
insulin PP cells			
human	+[37], −[38]	−[37]	−[37]
insulin D cells			
human	−[38]		
Parathyroid Gland			
chief cells			
bovine	+[37,36,42]	+[37,27], −[36]	−[37,36]
human	+[37,59]		
Thyroid Gland			
C cells			
bovine	+[42], −[36,37]	−[36,37]	+[37], −[36]
human	+[37,59]		
Gut			
enterochromaffin cells			
bovine	+[37,36]	+[37,36]	−[37,36]
human	+[37,38,43]		
gastric EC-like cells			
bovine	+[37]	−[37]	−[37]
human	+[37,43]		
pyloric gastrin G			
bovine	+[37]	+[37]	−[37]
human	+[37,38]		
glicentin/PPY L cells			
bovine	−[37]	−[37]	+[37]
pyloric D cells			
bovine	−[37]	−[37]	+[37]

presence of B and C in the hippocampus, a region particularly rich in chromogranin A,[33] has been reported.[22]

Recently chromogranin-A-positive cells were also detected in spleen, lymph nodes, thymus, and fetal liver,[44] suggesting that this protein may provide a link between the neuroendocrine and immunological systems.

Thus we can conclude: Chromogranins are present in neuronal tissues and several but not all endocrine cells. Some of those endocrine cells contain all three chromogranins (chromaffin cells of the adrenal medulla, glycoprotein-hormone-producing cells of the anterior pituitary, A cells of the endocrine pancreas), some are positive for chromogranins A and B (enterochromaffin cells), whereas others contain only chromogranin A (chief cells of parathyroid gland) or chromogranin C (posterior pituitary, glicentin, and pyloric D cells of the gut).

Are the molecular properties of chromogranins in other tissues similar to adrenal medullary chromogranins? In order to test this we performed one- and two-dimensional immunoblots. No differences in apparent M_r and pI between chromogranins A, B, and C from adrenal medulla and anterior pituitary, endocrine pancreas, parathyroid gland, splenic nerve, and brain were found in either bovine or human tissues.[27,28,33,45] In contrast O'Connor and Frigon,[35] using gel-filtration studies, reported a somewhat smaller molecular size of bovine chromogranin A in pituitary. Others performing one-dimensional immunoblots found major chromogranin-A-immunoreactive proteins of M_r 21,000 in pancreas[26] and several other neuronal tissues.[34] One likely explanation for these latter results is proteolytic breakdown of chromogranin A. As discussed above, chromogranins are processed proteolytically by endogenous proteases. Proteases from other cell compartments (*e.g.*, lysosomes, zymogen granules for pancreas) also degrade chromogranins. Thus great care has to be taken to prevent proteolytic breakdown of chromogranins, *e.g.*, preparation of a soluble protein fraction from pituitary, hippocampus and pancreas resulted in pronounced proteolytic breakdown of chromogranins[27] (Fischer-Colbrie and Schober, unpublished data). The use of total particulate fractions or boiling of these fractions[22] in SDS sample buffer prior to electrophoresis as used in our studies seems to be more reliable for determination of the apparent molecular weight.

PROTEOLYTIC PROCESSING OF CHROMOGRANINS

Within chromaffin granules chromogranins are processed proteolytically by specific proteases present within this cell compartment yielding several proteins of smaller molecular weight cross-reacting immunologically (see FIG. 1). For proenkephalin a similar processing has been characterized in detail. Several proteases involved in processing of proenkephalin, mainly trypsin-like endopeptidases[46,47] and a carboxypeptidase-B-like exopeptidase,[48,49] have been isolated and characterized in detail. Much less is known about processing of chromogranins. Recent data available for chromogranin A, however, suggest a similar type of proteolytic processing for this protein.

Proteolytic processing of chromogranin A was first reported in 1974.[50] In this study an antiserum against chromogranin A labeled several proteins of smaller molecular weight in SDS gels. A more detailed analysis with a highly specific antiserum revealed the presence of eight major chromogranin-A-immunoreactive proteins in two-dimensional immunoblots with M_rs ranging from 71,000 to 17,400.[18] Tryptic digestion of these chromogranin-A-immunoreactive proteins yields polypeptides similar to those yielded by chromogranin A.[22,35] Recently the complete amino acid sequence of chromogranin A was reported for the first time[6] (see also article by Mark Grimes in this volume). Eight pairs of basic amino acids known as putative signals for trypsin-like endoproteases were found. Possible peptides arising after splitting of chromogranin A by such an enzyme are indicated in FIGURE 2. Each of these peptides could be correlated with one of the major chromogranin-A-immunoreactive proteins. Apparent

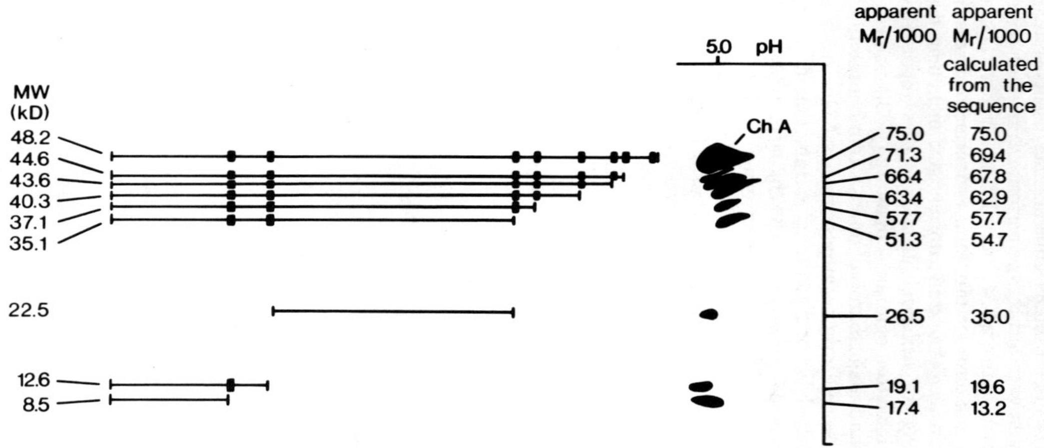

FIGURE 2. Proteolytic processing of chromogranin A. *Left:* schematical drawing of sites of basic pairs of amino acids within the sequence of chromogranin A.[6] *Right:* schematical drawing of an immunoblot with anti-chromogranin A.[18] Apparent M_rs of the major chromogranin-A-immunoreactive proteins as determined by SDS electrophoresis agree well with apparent M_rs of possible cleavage products as calculated from the sequence.

M_r calculated from the sequences agreed fairly well with apparent M_r obtained from SDS-gel electrophoresis. Thus proteolytic processing of chromogranin A seems to involve an overall mechanism similar to that described for proenkephalin, however further studies are necessary to establish whether or not similar proteases process both proteins.

Chromogranins B and C are also similarly processed. Thus there are four families of proteins (chromogranins A, B, and C and proenkephalin) present which are degraded by endogenous proteases present within chromaffin granules. The degree of proteolytic processing of each proprotein is different, however. Quantitative one-dimensional immunoblotting was performed to estimate the amount of proprotein present compared to proteins of smaller molecular weight. As shown in TABLE 5, about 49% of chromogranins A and C is degraded to proteins of smaller molecular weight and 51% remains unprocessed as proprotein. In contrast chromogranin B in bovine chromaffin granules is degraded to a greater degree: only 14% of chromogranin B is left undegraded. In splenic nerve and pituitary chromogranins A and C are degraded to a lesser degree, whereas the degree of proteolytic processing of chromogranin B is clearly different. Most of the proprotein (82.8% and 84.8% compared to 14.2%) remains undegraded. These differences in processing of chromogranin B compared to A and C suggest a specific mechanism of regulation for each of these proteins.

TABLE 5. Proteolytic Processing of Bovine Chromogranins

	% Proprotein Present (± SE; *n* = 8-30)		
	Chromogranin A	Chromogranin B	Chromogranin C
Chromaffin granules	51.0% ± 2.5	14.2% ± 1.2	50.7% ± 1.8
Splenic nerve	72.2% ± 2.7	82.8% ± 2.3	76.2% ± 1.9
Anterior pituitary	83.7% ± 2.5	84.8% ± 2.1	88.2% ± 1.2

The degree of proteolytic processing was estimated with quantitative immunoblotting. After one-dimensional electrophoresis proteins were transferred to nitrocellulose sheets and decorated with specific antisera against chromogranins followed by ^{125}I-protein A. Immunoreactive proteins were visualized by autoradiography, excised from the blot, and counted for radioactivity.

RELATIVE AMOUNTS OF CHROMOGRANINS IN OTHER TISSUES

In bovine chromaffin granules total chromogranin-immunoreactive proteins make up 87% of soluble proteins with chromogranins A composing 80%, chromogranins B 6% and chromogranins C 1% (TABLE 6). A similar analysis of large dense-cored vesicles of splenic nerve revealed a different pattern with chromogranins A making up 63%, chromogranins B 13%, and chromogranins C 10% of total soluble proteins. Thus splenic nerve contains 2.7 times more chromogranins B and 12.6 times more chromogranins C compared to bovine chromaffin granules. In another endocrine tissue (bovine anterior pituitary) chromogranin C is the most prominent chromogranin.[39,22]

In addition, as discussed above, the relative amounts of chromogranins in chromaffin granules from other species also vary (see TABLE 2). Thus different tissues and species contain different relative amounts of chromogranins. This variation of chromogranin levels in several secretory vesicles is consistent with a specific regulation of the synthesis of each chromogranin. (Several aspects of this topic are discussed in the article by H. Winkler *et al.* in this volume.)

POSSIBLE FUNCTIONS OF CHROMOGRANINS

The physiological function of the major secretory proteins of adrenal medulla, the chromogranins, is still unknown. Recent studies on these proteins have demonstrated

TABLE 6. Relative Amounts of Chromogranins in Chromaffin Granules and Splenic Nerve from Ox

	Chromogranins A	Chromogranins B	Chromogranins C	Total
Chromaffin granules	80%	6%	1%	87%
Splenic nerve	63%	13%	10%	86%

Relative amounts of chromogranins in chromaffin granules were calculated from data reported for the individual proprotein (chromogranin A: 40%[2]; chromogranin B: 0.9%[5]) and the degree of processing as shown in TABLE 5. Values for splenic nerve were obtained by comparative analysis of purified large dense-cored vesicles[45] and chromaffin granules with quantitative immunoblotting. It should be emphasized that this kind of calculation may only be considered to yield approximate values.

properties of chromogranins similar to those of precursors of neuropeptides, *e.g.*, proenkephalin: (1) Both chromogranins and neuropeptides are widespread in endocrine and nervous tissues. (2) The relative amounts of chromogranins in different tissues and species vary, as is also known for neuropeptides, probably due to a specific regulation of their synthesis. (3) The sequence of chromogranin A contains several pairs of basic amino acids known as putative signals for processing enzymes in neuropeptide precursors. Chromogranins are processed by these proteases to a certain degree, which differs in various tissues as also established for neuropeptides. Thus one could speculate that chromogranins are putative precursors of small bioactive peptides with, *e.g.*, regulatory, neurotrophic, or chemoattractive activities. In this context it is interesting to note that antisera directed against the amino terminus of the neuropeptide bombesin reacted specifically with chromogranin B.[11] An antiserum (Peninsula) against calcitonin gene-related peptide specifically labeled chromogranin C (unpublished observation). Thus peptides with similar sequence could be present within the chromogranins. Further studies are necessary to clarify this interesting observation.

ACKNOWLEDGMENT

We thank Drs. Lee Eiden and Mark Grimes for sharing the chromogranin A sequence with us prior to publication.

REFERENCES

1. BLASCHKO, H., R. S. COMLINE, F. H. SCHNEIDER, M. SILVER & A. D. SMITH. 1967. Secretion of a chromaffin granule protein, chromogranin from the adrenal gland after splanchnic stimulation. Nature **215:** 58-59.
2. WINKLER, H. 1976. The composition of adrenal chromaffin granules: An assessment of controversial results. Neuroscience **1:** 65-80.
3. WINKLER, H. & S. W. CARMICHAEL. 1982. The chromaffin granule. *In* The Secretory Granule. A. M. Poisner & J. M. Trifaró, Eds.: 3-79. Elsevier Biomedical. Amsterdam.
4. PHILLIPS, J. H. & D. K. APPS. 1979. Storage and secretion of catecholamines: The adrenal medulla. Int. Rev. Biochem. Physiol. Pharmacol. Biochem. **26:** 121-178.
5. WINKLER, H., D. K. APPS & R. FISCHER-COLBRIE. 1986. The molecular function of adrenal chromaffin granules: Established facts and unresolved topics. Neuroscience. **18:** 261-290.
6. IACANGELO, A., H. U. AFFOLTER, L. EIDEN, E. HERBERT & M. GRIMES. 1986. Bovine chromogranin A: Its sequence and the distribution of its messenger RNA in endocrine tissues. Nature. **323:** 82-86.
7. APPS, D. K., J. G. PRYDE & J. H. PHILLIPS. 1980. Cytochrome b_{561} is identical with chromomembrin B, a major polypeptide of chromaffin granule membranes. Neuroscience **5:** 2279-2287.
8. UDENFRIEND, S. & D. L. KILPATRICK. 1983. Biochemistry of the enkephalins and enkephalin-containing peptides. Arch. Biochem. Biophys. **221:** 309-323.
9. MAJANE, E. A., H. ALHO, Y. KATAOKA, C. H. LEE & H.-Y. T. YANG. 1985. Neuropeptide Y in bovine adrenal glands: Distribution and characterization. Endocrinology **117:** 1162-1168.
10. LUNDBERG, J. M., T. HÖKFELT, A. HEMSÉN, E. THEODORSSON-NORHEIM, J. PERNOW, B. HAMBERGER & M. GOLDSTEIN. 1986. Neuropeptide Y-like immunoreactivity in adrenaline cells of adrenal medulla and in tumors and plasma of pheochromocytoma patients. Reg. Peptides **13:** 169-182.
11. FISCHER-COLBRIE, R., J. DIEZ-GUERRA, P. C. EMSON & H. WINKLER. 1986. Bovine chromaffin granules: Immunological studies with antisera against neuropeptide Y, (met)enkephalin and bombesin. Neuroscience. **18:** 167-174.
12. HELLE, K. B. 1966. Some chemical and physical properties of the soluble protein fraction of bovine adrenal chromaffin granules. Molec. Pharmacol. **2:** 298-310.
13. SMITH, A. D. & H. WINKLER. 1967. Purification and properties of an acidic protein from chromaffin granules of bovine adrenal medulla. Biochem. J. **103:** 483-492.
14. SMITH, W. J. & N. KIRSHNER. 1967. A specific soluble protein from the catecholamine storage vesicles of bovine adrenal medulla. Purification and chemical characterization. Molec. Pharmacol. **3:** 52-62.
15. SEGREST, J. P., R. L. JACKSON, E. P. ANDREWS & V. T. MARCHESI. 1971. Human erythrocyte membrane glycoprotein: A re-evaluation of the molecular weight as determined by SDS polyacrylamide gel electrophoresis. Biochem. Biophys. Res. **44:** 390-395.
16. PODUSLO, J. F. 1981. Glycoprotein molecular-weight estimation using sodium dodecyl sulfate-pore gradient electrophoresis: Comparison of tris-glycine and tris-borate-EDTA buffer systems. Anal. Biochem. **114:** 131-139.
17. WINKLER, H., G. FALKENSAMMER, A. PATZAK, R. FISCHER-COLBRIE, M. SCHOBER & A. WEBER. 1984. Life cycle of the catecholaminergic vesicle: From biogenesis to secretion.

In Regulation of Transmitter Function: Basic and Clinical Aspects. E. S. Vizi & K. Magyar, Eds.: 65-73. Akadémiai Kiadó. Budapest.
18. FISCHER-COLBRIE, R. & I. FRISCHENSCHLAGER. 1985. Immunological characterization of secretory proteins of chromaffin granules: Chromogranins A, chromogranins B and enkephalin-containing peptides. J. Neurochem. **44:** 1854-1861.
19. ROSA, P. & A. ZANINI. 1981. Characterization of adenohypophysial polypeptides by two-dimensional gel electrophoresis. II. Sulfated and glycosylated polypeptides. Molec. Cell. Endocrinol. **24:** 181-193.
20. ROSA, P. & A. ZANINI. 1983. Purification of a sulfated secretory protein from the adenohypophysis. Immunochemical evidence that similar macromolecules are present in other glands. Eur. J. Cell Biol. **31:** 94-98.
21. ROSA, P., G. FUMAGALLI, A. ZANINI & W. B. HUTTNER. 1985. The major tyrosine-sulfated protein of the bovine anterior pituitary is a secretory protein present in gonadotrophs, thyrotrophs, mammotrophs and corticotrophs. J. Cell. Biol. **100:** 928-937.
22. ROSA, P., A. HILLE, R. H. W. LEE, A. ZANINI, P. DE CAMILLI & W. B. HUTTNER. 1985. Secretogranins I and II: Two tyrosine-sulfated secretory proteins common to a variety of cells secreting peptides by the regulated pathway. J. Cell. Biol. **101:** 1999-2011.
23. FISCHER-COLBRIE, R., C. HAGN, L. KILPATRICK & H. WINKLER. 1986. Chromogranin C.: A third component of the acidic proteins in chromaffin granules. J. Neurochem. **47:** 318-321.
24. HOGUE-ANGELETTI, R. A. 1977. Nonidentity of chromogranin A and dopamine β-monooxygenase. Arch. Biochem. Biophys. **184:** 364-372.
25. KRUGGEL, W., D. T. O'CONNOR & R. V. LEWIS. 1985. The amino terminal sequences of bovine and human chromogranin A and secretory protein I are identical. Biochem. Biophys. Res. Comm. **127:** 380-383.
26. HUTTON, J. C., F. HANSON & M. PESHAVARIA. 1985. β-granins: 21 kDa co-secreted peptides of the insulin granule closely related to adrenal medullary chromogranin. FEBS Lett. **188:** 336-340.
27. FISCHER-COLBRIE, R. H. LASSMANN, C. HAGN & H. WINKLER. 1985. Immunological studies on the distribution of chromogranin A and B in endocrine and nervous tissues. Neuroscience **16:** 547-555.
28. HAGN, C., K. W. SCHMID, R. FISCHER-COLBRIE & H. WINKLER. 1986. Chromogranin A, B and C in human adrenal medulla and endocrine tissues. Lab. Invest. **55:** 405-411.
29. HEXUM, T. D., H.-Y. T. YANG & E. COSTA. 1980. Biochemical characterization of enkephalin-like immunoreactive peptides of adrenal glands. Life Sci. **27:** 1211-1216.
30. GOEDERT, M., G. P. REYNOLDS & P. C. EMSON. 1983. Neurotensin in the adrenal medulla. Neurosci. Lett. **35:** 155-160.
31. LEBOULENGER, F., P. LEROUX, M.-C. TONON, D. H. COY, H. VAUDRY & G. PELLETIER. 1983. Coexistence of vasoactive intestinal peptide and enkephalins in the adrenal chromaffin granules of the frog. Neurosci. Lett. **37:** 221-225.
32. COHN, D. V., R. ZANGERLE, R. FISCHER-COLBRIE, L. L. H. CHU, J. J. ELTING, J. W. HAMILTON & H. WINKLER. 1982. Similarity of secretory protein I from parathyroid gland to chromogranin A from adrenal medulla. Proc. Natl. Acad. Sci. U.S.A. **79:** 6056-6059.
33. SOMOGYI, P., A. J. HODGSON, R. W. DE POTTER, R. FISCHER-COLBRIE, M. SCHOBER, H. WINKLER & I. W. CHUBB. 1984. Chromogranin immunoreactivity in the central nervous system. Immunochemical characterization, distribution and relationship to catecholamine and enkephalin pathways. Brain Res. Rev. **8:** 193-230.
34. NOLAN, J. A., J. Q. TROJANOWSKI & R. HOGUE-ANGELETTI. 1985. Neurons and neuroendocrine cells contain chromogranin: Detection of the molecule in normal bovine tissues by immunochemical and immunohistochemical methods. J. Histochem. Cytochem. **33:** 791-798.
35. O'CONNOR, D. T. & R. P. FRIGON. 1984. Chromogranin A, the major catecholamine storage vesicle soluble protein. J. Biol. Chem. **259:** 3237-3247.
36. LASSMANN, H., C. HAGN, R. FISCHER-COLBRIE & H. WINKLER. 1986. Presence of chromogranin A, B and C in bovine endocrine and nervous tissues: A comparative immunohistochemical study. Histochem. J. **18:** 380-386.

37. RINDI, G., R. BUFFA, F. SESSA, O. TORTORA & E. SOLCIA. 1986. Chromogranin A, B and C immunoreactivities of mammalian endocrine cells. Distribution, distinction from costored hormones/prohormones and relationship with the argyrophil component of secretory granules. Histochemistry. **85:** 19-28.
38. VARNDELL, I. M., R. V. LLOYD, B. S. WILSON & J. M. POLAK. 1985. Ultrastructural localization of chromogranin—A potent marker for the electron microscopical recognition of endocrine cell secretory granules. Histochem. J. **17:** 981-992.
39. RUNDLE, S., P. SOMOGYI, R. FISCHER-COLBRIE, C. HAGN, H. WINKLER & I. W. CHUBB. 1987. Chromogranin A, B and C: Immunohistochemical localization in ovine pituitary and the relationship with hormone-containing cells. Regulat. Pept. In press.
40. COHN, D. V., J. J. ELTING, M. FRICK & R. ELDE. 1984. Selective localization of the parathyroid secretory protein-I/adrenal medulla chromogranin A protein family in a wide variety of endocrine cells of the rat. Endocrinology **114:** 1963-1974.
41. LLOYD, R. V., B. WILSON, K. KOVACS & N. RYAN. 1985. Immunohistochemical localization of chromogranin in human hypophyses and pituitary adenomas. Arch. Path. Lab. Med. **109:** 515-517.
42. O'CONNOR, D. T., D. BURTON & L. J. DEFTOS. 1983. Chromogranin A: Immunohistology reveals its universal occurrence in normal polypeptide hormone producing endocrine glands. Life Sci. **33:** 1657-1663.
43. FACER, P., A. E. BISHOP, R. V. LLOYD, B. S. WILSON, R. J. HENESSY & J. M. POLAK. 1985. Chromogranin: A newly recognized marker for endocrine cells of the human gastrointestinal tract. Gastroenterology **89:** 1366-1373.
44. HOGUE ANGELETTI, R. & W. F. HICKEY. 1985. A neuroendocrine marker in tissues of the immune system. Science **230:** 89-90.
45. HAGN, C., R. L. KLEIN, R. FISCHER-COLBRIE, B. H. DOUGLAS, II & H. WINKLER. 1986. An immunological characterization of five common antigens of chromaffin granules and of large-dense cored vesicles of sympathetic nerve. Neurosci. Lett. **67:** 295-300.
46. LINDBERG, I., H.-Y. T. YANG & E. COSTA. 1982. An enkephalin-generating enzyme in bovine adrenal medulla. Biochem. Biophys. Res. **106:** 186-193.
47. MIZUNO, K., M. KOJIMA & H. MATSUO. 1985. A putative prohormone processing protease in bovine adrenal medulla specifically cleaving in between lys-arg sequences. Biochem. Biophys. Res. **128:** 884-891.
48. FRICKER, L. D. & S. H. SNYDER. 1982. Enkephalin convertase: Purification and characterization of a specific enkephalin-synthesizing carboxypeptidase localized to adrenal chromaffin granules. Proc. Natl. Acad. Sci. U.S.A. **79:** 3886-3890.
49. HOOK, V. Y. H., L. E. EIDEN & M. J. BROWNSTEIN. 1982. A carboxypeptidase processing enzyme for enkephalin precursors. Nature **295:** 341-342.
50. HÖRTNAGL, H., H. LOCHS & H. WINKLER. 1974. Immunological studies on the acidic chromogranins and on dopamine β-hydroxylase (E.C. 1.14.2.1) of bovine chromaffin granules. J. Neurochem. **22:** 197-199.
51. FALKENSAMMER, G., R. FISCHER-COLBRIE, K. RICHTER & H. WINKLER. 1985. Cell-free and cellular synthesis of chromogranin A and B of bovine adrenal medulla. Neuroscience **14:** 735-746.
52. KILPATRICK, L., F. GAVINE, D. APPS & J. PHILLIPS. 1983. Biosynthetic relationship between the major matrix proteins of adrenal chromaffin granules. FEBS Lett. **164:** 383-388.
53. MAJZOUB, J. A., H. M. KRONENBERG, J. T. POTTS, A. RICH & J. F. HABENER. 1979. Identification and cell-free translation of mRNA coding for a precursor of parathyroid secretory protein. J. Biol. Chem. **254:** 7449-7455.
54. FISCHER-COLBRIE, R., M. SCHACHINGER, R. ZANGERLE & H. WINKLER. 1982. Dopamine β-hydroxylase and other glycoproteins from the soluble content and the membranes of adrenal chromaffin granules: Isolation and carbohydrate analysis. J. Neurochem. **38:** 725-732.
55. KIANG, W. L., T. KRUSIUS, J. FINNE, R. U. MARGOLIS & R. K. MARGOLIS. 1982. Glycoproteins and proteoglycans of the chromaffin granule matrix. J. Biol. Chem. **257:** 1651-1659.
56. KILPATRICK, D. L., K. D. GIBSON & B. N. JONES. 1983. Is adrenal proenkephalin glycosylated? Arch. Biochem. Biophys. **224:** 402-404.

57. SETTLEMAN, J., R. FONSECA, J. NOLAN & R. HOGUE ANGELETTI. 1985. Relationship of multiple forms of chromogranin. J. Biol. Chem. **260:** 1645-1651.
58. FALKENSAMMER, G., R. FISCHER-COLBRIE & H. WINKLER. 1985. Biogenesis of chromaffin granules: Incorporation of sulfate into chromogranin B and into a proteoglycan. J. Neurochem. **45:** 1475-1480.
59. WILSON, B. S. & R. V. LLOYD. 1984. Detection of chromogranin in neuroendocrine cells with a monoclonal antibody. Am. J. Pathol. **115:** 458-468.

DISCUSSION OF THE PAPER

J. T. POTTS (*Massachusetts General Hospital, Boston, Mass.*): Could you speculate further about the biological role of chromogranins? Not all secretory vesicles contain them, hence there is a puzzle about postulating a universal intracellular mechanism in secretion. If the role is extracellular after secretion, the issue of specification is a problem. What are your thoughts on this matter?

R. FISCHER-COLBRIE (*University of Innsbruck, Innsbruck, Austria*): The regulation of the synthesis of chromogranins and the different degree of proteolytic processing seems to argue for a function after secretion (precursor of small active peptides). Their widespread distribution does not necessarily create a specificity problem. Some endocrine tissues (GH, PRL cells, thyroid cells) do not contain chromogranins, other tissues contain different relative amounts which in addition can be regulated by various stimuli, and proteolytic processing also differs, yielding different small peptides.

D. V. AGOSTON (*Max-Planck-Institut, Göttingen, Federal Republic of Germany*): Could you also detect the small biological active form of the bombesin using specific radioimmunoassay and HPLC? I think immunological cross-reactivity with a high molecular weight polypeptide using an antibody raised against an eight amino acid neuropeptide doesn't indicate any suggested function for chromogranins.

R. FISCHER-COLBRIE: This observation was not tested by ourselves but recently Dr. LeMaire detected bombesin and related peptides by RIA and reverse-phase HPLC in adrenal medulla. Sequence analysis of chromogranin B will establish if some of these peptides are found within the sequence of chromogranin B.

Presence of a Neuropeptide in a Model Cholinergic System[a]

DENES V. AGOSTON[b] AND J. MICHAEL CONLON[c]

[b] *Department of Neurochemistry*
[c] *Clinical Research Group for Gastrointestinal Endocrinology*
Max Planck Institute for Biophysical Chemistry
D-3400, Göttingen, Federal Republic of Germany

The presence of biologically active peptides in association with the classical low-molecular-mass transmitters has led to the concept of the coexistence of more than one type of transmitter at a nerve terminal.[1,2] One of the strongest pieces of evidence for the coexistence of neurosecretory substances is that of acetylcholine and the 28-amino-acid vasoactive intestinal polypeptide (VIP) belonging to the secretin-glucagon family of regulatory peptides.[2,3] The electromotor system of the electric ray *Torpedo marmorata* is the best characterized model cholinergic system,[4] and so we have examined this system for the occurrence of VIP or a VIP-like peptide.

Experiments were performed on adult *Torpedo marmorata.* Extraction and processing of the tissues as well as the performance of high-performance liquid chromatography (HPLC), radioimmunoassay (RIA), and immunohistochemistry were published elsewhere.[5–7]

We have detected VIP-like immunoreactivity (VIP-LI) in the cholinergic electromotor system of *Torpedo marmorata.* The immunohistochemistry revealed that the distribution of the VIP-LI in the electric lobes and electric organ (FIG. 1) is comparable to that of the stable cholinergic synaptic vesicle marker vesicle-specific proteoglycan.[5] Ligation of the electromotor nerves caused a marked accumulation of the VIP-like immunoreactivity in the lobes (+180%) and the proximal portions of the electromotor nerves (+130%) and a decrease in the electric organ (−50%) when measured by radioimmunoassay (RIA) using synthetic VIP (porcine sequence) as standard. VIP-like immunoreactivity in extracts of electric lobes, electromotor nerves, and electric organ was eluted from a semi-preparative reverse-phase HPLC column as a single peak with a similar retention time to porcine VIP (not shown). Rechromatography at higher resolution on an analytical column under near-isocratic conditions indicated diversity between the molecular forms of the VIP-like immunoreactivity extracted from electric lobe and electric organ (FIG. 2), suggesting the possibility of posttranslational processing. Dilution curves (not shown) provided evidence for a lack of homology with the synthetic peptide. Chemical characterization of an Elasmobranch VIP has recently been successful and has shown only NH_2-terminal decapeptide homology with the porcine VIP.[8] Purification and sequencing of the peptide is now being undertaken as is a study of its function in this model cholinergic system. In this connection, pharmacological[9] and electrophysiological[10] evidence has been published on the presynaptic modulatory effect of VIP in other cholinergic neurons.

[a] This work was supported by the Deutsche Forschungsgemeinschaft.

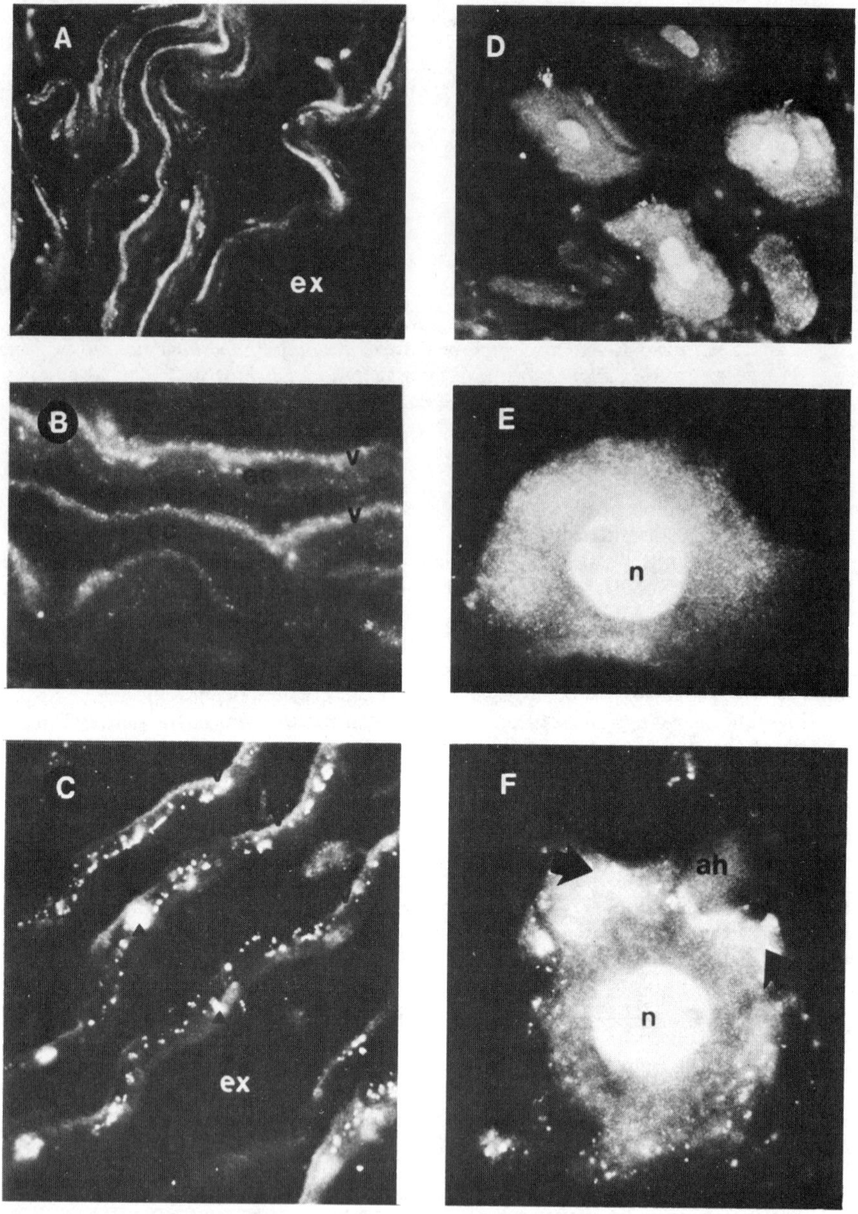

FIGURE 1. Immunofluorescence of antisera to VIP on transversely sectioned control *Torpedo* **(A,B,C)** electric organ and **(D,E,F)** electric lobe. Fluorescence labeling is found on the innervated, ventral surface (*v*) of **(A,B)** the electrocytes (*ec*) and in **(D,E)** the lobe cell bodies. **(C)** Transverse sections of *Torpedo* electric organ 10 days after ligation of the electromotor nerves. A remarkable decrease of immunostaining on the ventral surfaces (*v*) of the electrocytes is seen, but some immunofluorescence (*arrowheads*) remains. **(F)** Several bright spots (*arrows*) appeared in the cytoplasm of the electrocytes (*ec*). Accumulation of the immunoreactivity appears to be concentrated at the axon hillocks (*ah*) in the electric lobe (*arrows*) after constriction of the nerves. Strong nuclear staining can be seen; *n* = nucleus, *ex* = extracellular space.

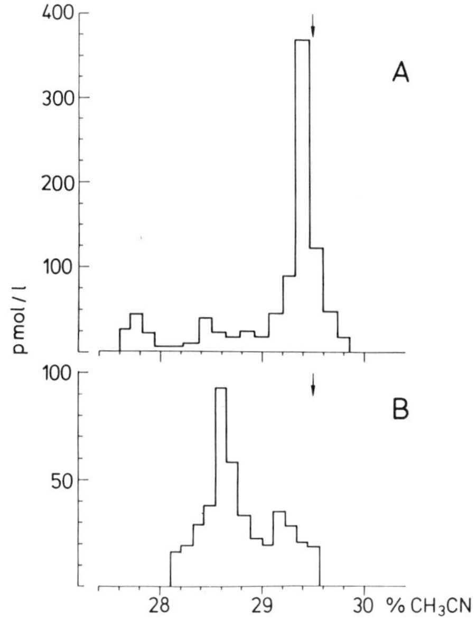

FIGURE 2. Rechromatography after preparative reverse-phase HPLC of the VIP-LI material extracted from (**A**) electric lobe and (**B**) electric organ on an analytical Supercosil-18-DB column under near-isocratic elution conditions. Fractions were assayed for VIP-LI using antibody G143. The *arrow* shows the retention time of synthetic VIP and the ordinates show the concentration of acetonitrile in the eluting solvent.

REFERENCES

1. LUNDBERG, J. M. & T. HOKFELT. 1983. Trends Neurosci. **6:** 325-333.
2. HOKFELT, T., O. JOHANSON & M. GOLDSTEIN. 1984. Science **225:** 1326-1334.
3. SAID, S. I. 1984. Peptides **5:** 143-150.
4. WHITTAKER, V. P. 1984. Biochem. Soc. Trans. **12:** 561-576.
5. AGOSTON, D. V., G. H. C. DOWE, W. FIEDLER, P. E. GIOMPREC, I. S. ROED, J. H. WALKER, T. YAMAGUCHI & V. P. WHITTAKER. 1985. J. Neurochem. In press.
6. CONLON, J. M., M. BALLMANN & R. LAMBERTS. 1985. Gen. Comp. Endocrinol. **58:** 150-158.
7. AGOSTON, D. V., M. BALLMANN, J. M. CONLON, G. H. C. DOWE & V. P. WHITTAKER. 1985. J. Neurochem. **45:** 398-406.
8. DIMALINE, R., M. C. THORNDYKE & J. YOUNG. 1986. Regul. Pept. **14:** 1-10.
9. BARTFAI, T. 1985. Trends Pharmacol. **6:** 331-334.
10. CHAN-PALAY, V. & S. L. PALAY, Eds. 1984. Coexistence of Neuroactive Substances in Neurones. Wiley. New York.

Antibodies to a Synthetic Peptide, Chromogranin$_{1\text{-}14}$

RUTH HOGUE ANGELETTI, MARY BILDERBACK, AND JIANG QIAN

Department of Neuropathology
University of Pennsylvania
Philadelphia, Pennsylvania 19104

Chromogranin has a characteristic amino-terminal sequence which is identical in both the bovine and human polypeptides ($CG_{1\text{-}14}$): Leu-Arg-Val-Asn-Ser-Pro-Met-Asn-Lys-Gly-Asp-Thr-Glu-Val-.[1,2] Polypeptide chains of similar amino-terminal sequences but differing molecular weights have been identified in the chromaffin vesicle.[3] It was thus proposed that they compose a family of related proteins.

A synthetic peptide corresponding to $CG_{1\text{-}14}$ was prepared and used as antigen for production of both polyclonal and monoclonal antibodies. Such probes would facilitate further study of this family of proteins, and provide antibodies that would react across species boundaries. Antibodies were successfully obtained by both methods. The polyclonal antiserum and monoclonal antibodies reacted in immunoblots of bovine, rat, and human tissues with multiple polypeptide chains. In bovine tissues (FIG. 1A), these correspond to the multiple chains identified by their parent M_r of 125,000, 100,000, 85,000, and 75,000. The M_r 65,000 form was, however, not detected. In human tissues (FIG. 1B), our previous polyclonal antisera against bovine chromogranin identify several proteins with M_r greater than 100,000, plus a lower M_r 70,000 band. The antipeptide antibodies also react with these very high molecular weight forms. The antipeptide antibodies reveal some additional polypeptides, however, with molecular weights ranging from 85,000 to 60,000.

Although the probes were designed to provide a straightforward analysis, they have instead reinforced the complexity of the chromogranin family. It should be emphasized that the immunoreactivity could be completed both by the isolated peptide and by purified chromogranin A. Thus, there are multiple related proteins in each species studied. The precise relationship among them can only be determined by detailed genetic analysis. Furthermore, the antibodies developed by these probes do function across species boundaries.

The monoclonal antibodies to $CG_{1\text{-}14}$ have also proved useful for biological studies. We have recently found that chromogranin is located in cells within immune tissues.[4] A putative neuroendocrine cell containing chromogranin has been isolated from spleen.[5] A second cell type in spleen, the large granular lymphocyte, also appears to contain chromogranin.[6] These cells are present in the pellet and buffy fractions, respectively, of Ficoll-separated spleen cells. FIGURE 2 shows immunohistochemical analysis of cytospin preparations of spleen buffy and pellet layers. The $CG_{1\text{-}14}$ monoclonal antibodies specifically detect a single cell type in each cell preparation.

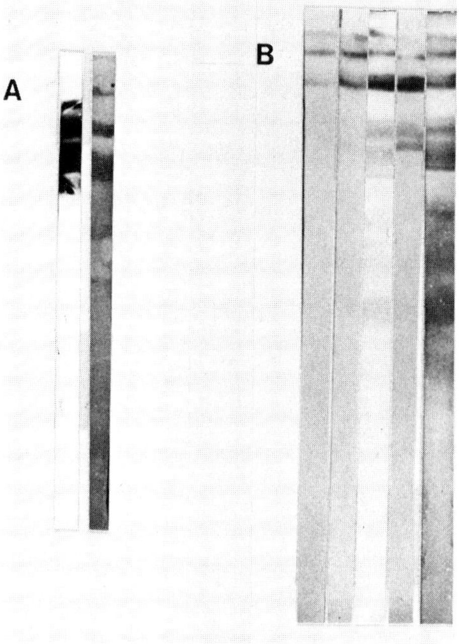

FIGURE 1. Immunoblots detected by anti-CG$_{1-14}$. **(A)** Bovine chromaffin granule lysate. *Left to right:* rabbit anti-bovine chromogranin, mouse monoclonal anti-CG$_{1-14}$ (5A12NT). **(B)** Human tissue. *Left to right:* rabbit anti-bovine chromogranin, monoclonal anti-human chromogranin, rabbit anti-CG$_{1-14}$ serum, monoclonal anti-CG$_{1-14}$ (7F5NT), monoclonal anti-CG$_{1-14}$ (5A12NT).

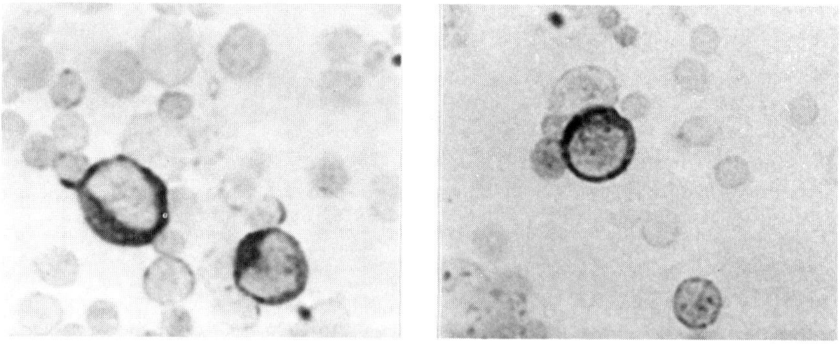

FIGURE 2. Immunocytochemistry of dissociated spleen cells with anti-CG$_{1-14}$ monoclonal antibodies. *Left,* buffy layer from a Ficoll-Hypaque gradient. *Right,* pellet layer.

REFERENCES

1. HOGUE ANGELETTI, R. 1977. Non-identity of dopamine beta-monoxygenase and chromogranin A. Arch. Biochem. Biophys. **184:** 364-372.
2. COHN, D. V., R. LANGERLE, R. FISCHER-COLBRIE, L. L. H. CHU, J. J. ELTING, J. W. HAMILTON & H. WINKLER. 1982. Proc. Natl. Acad. Sci. U.S.A. **79:** 6056-6059.
3. SETTLEMAN, J., R. FONSECA, J. NOLAN & R. HOGUE ANGELETTI. 1985. Relationship of multiple forms of chromogranin. J. Biol. Chem. **260:** 1645-1651.
4. HOGUE ANGELETTI, R. & W. F. HICKEY. 1985. A neuroendocrine marker in tissues of the immune system. Science **230:** 89-91.
5. HOGUE ANGELETTI, R. & W. F. HICKEY. Purification of a neuroendocrine cell from rat spleen. Cell. In press.
6. HENKART, P. & R. HOGUE ANGELETTI. Unpublished data.

Kinetics of Ferricyanide Reduction by Ascorbate-loaded Chromaffin-Vesicle Ghosts[a]

PATRICK M. KELLEY, YVONNE VIVAR PACQUING, AND DAVID NJUS

Department of Biological Sciences
Wayne State University
Detroit, Michigan 48202

Ascorbate-loaded chromaffin-vesicle ghosts will reduce external ferricyanide.[1,2] This ability to transfer electrons across the vesicle membrane is attributed to the membrane protein cytochrome b_{561}[2-4] and is thought to be used *in vivo* to take up electrons for the regeneration of intravesicular ascorbic acid (Njus *et al.*,[5] this volume). The reduction of ferricyanide by ascorbate-loaded ghosts is shown in FIGURE 1B.[2] Upon addition of ferricyanide, there is a rapid and nearly complete oxidation of cytochrome b_{561} (FIG. 1A). The cytochrome stays oxidized until almost no ferricyanide remains and then returns to its reduced state. When excess ferricyanide is added, the cytochrome becomes oxidized and remains so, since the internal supply of ascorbate is exhausted. This demonstrates that cytochrome b_{561} can accept electrons from ascorbic acid internally, transfer the electrons across the membrane, and reduce ferricyanide on the outside. This shows that cytochrome b_{561} can transfer electrons across the chromaffin-vesicle membrane, although it does not prove that the cytochrome is the only transmembrane electron carrier.

That cytochrome b_{561} is at least the principal transmembrane electron carrier is suggested by kinetic arguments. By recording rates of steady-state electron transfer from internal ascorbate to external ferricyanide, we can determine rate constants for the transfer of electrons from ascorbate to the membrane (k_0) and for the transfer of electrons from the membrane to ferricyanide (k_1) (see Njus et al.,[5] this volume). These rate constants are determined solely from measurements of the total rate of ferricyanide reduction and thus apply to total electron transfer across the chromaffin-vesicle membrane. If most or all of these electrons are being carried by cytochrome b_{561}, then the rate constants should also describe the kinetics of cytochrome b_{561} oxidation and reduction.

To test this, we have performed a computer simulation of the experiment shown in FIGURE 1. We assume that electron transfer occurs through four simple bimolecular reactions[b]:

[a]This work was supported by NIH grant no. GM-30500. D. Njus is an Established Investigator of the American Heart Association.

[b]Abbreviations are: AH^-, ascorbate; $A^{\dot{-}}$, semidehydroascorbate; A, dehydroascorbate; Cyt, cytochrome b_{561}; FeCy, ferri/ferrocyanide; ox, oxidized; red, reduced.

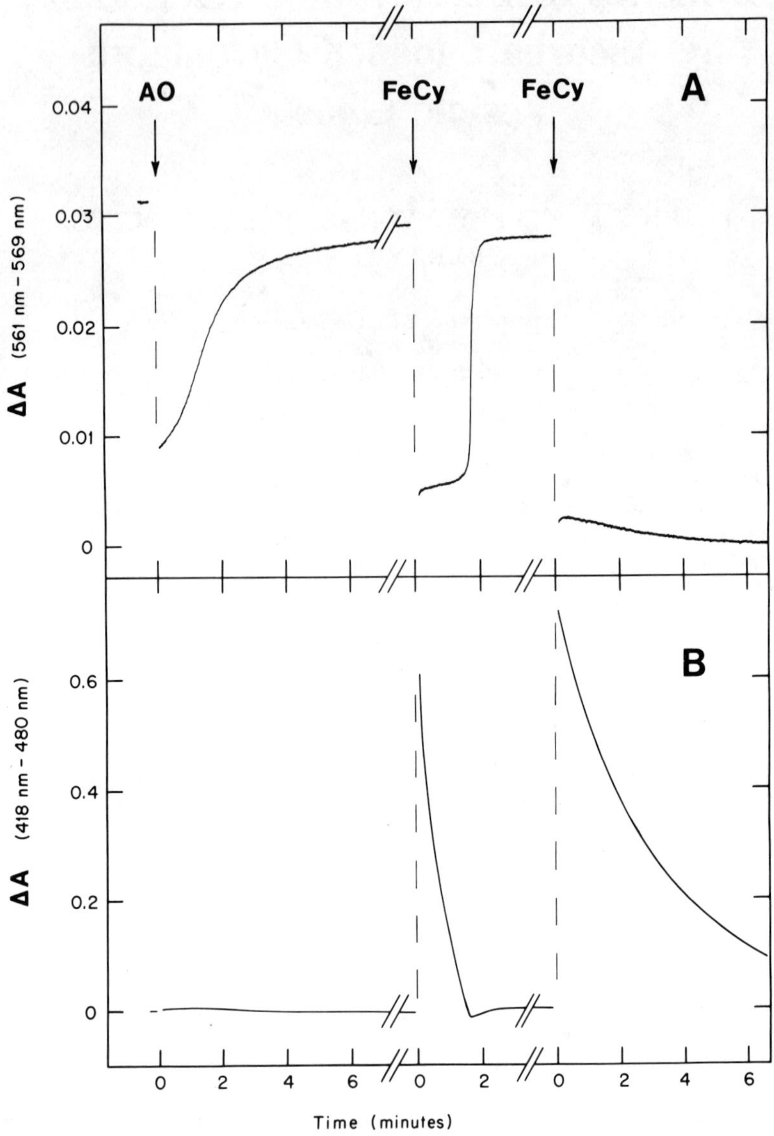

FIGURE 1. Time courses of cytochrome b_{561} oxidation and ferricyanide reduction. Chromaffin-vesicle ghosts (4.5 mg of protein) were suspended in 2.16 ml of 0.4 M sucrose, 10 mM Hepes, 250 μM KCN, pH 7.0. Ascorbate oxidase (8 units) was added at the time indicated *AO.* Potassium ferricyanide (1.78 μmol) was added at each of the two times indicated *FeCy.* **(A)** Absorbance monitored at 561-569 nm (cytochrome b_{561}). **(B)** Absorbance monitored at 418-480 nm (ferricyanide). (From Kelley and Njus.[2] Reprinted by permission from the *Journal of Biological Chemistry.*)

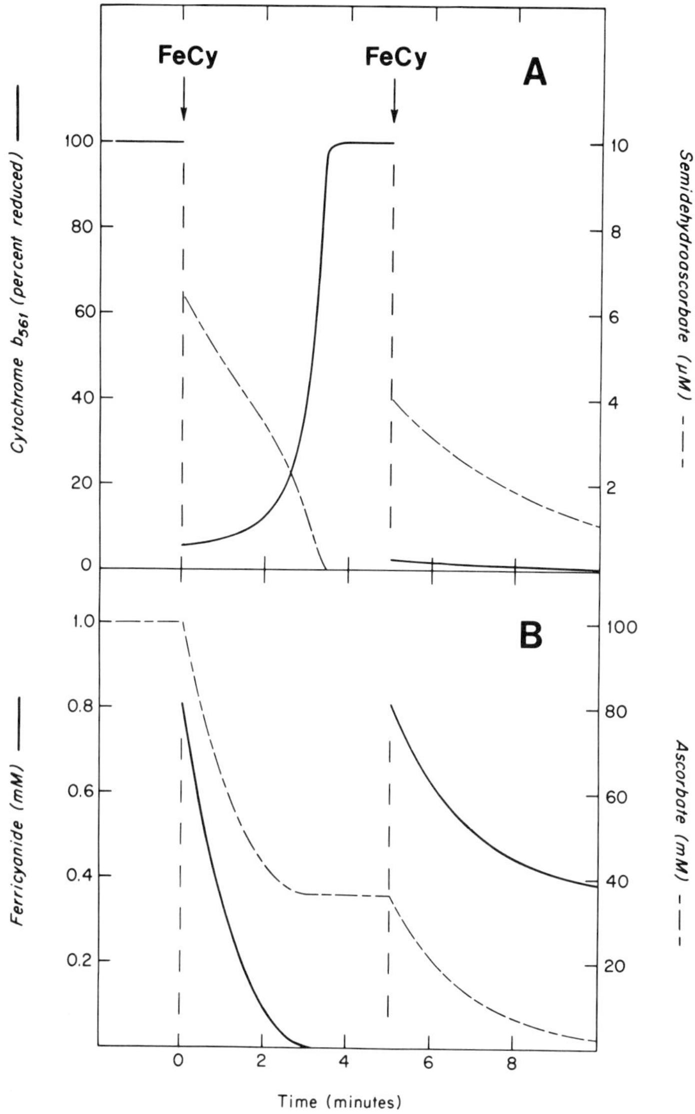

FIGURE 2. Computer simulation of the experiment shown in FIGURE 1, calculated as discussed in the text.

$$\text{(I)} \qquad FeCy_{ox} + Cyt_{red} \xrightarrow{k_1} FeCy_{red} + Cyt_{ox}$$

$$\text{(II)} \qquad AH^- + Cyt_{ox} \xrightarrow{k_0} A^{\dot{-}} + Cyt_{red} + H^+$$

$$\text{(III)} \qquad A^{\dot{-}} + Cyt_{red} + H^+ \xrightarrow{k_{-0}} AH^- + Cyt_{ox}$$

$$\text{(IV)} \qquad A^{\dot{-}} + A^{\dot{-}} + H^+ \xrightarrow{k_{diss}} AH^- + A$$

If we also assume that the concentrations of reduced cytochrome b_{561} and semidehydroascorbate remain at steady state, then the kinetics of electron transfer are described by the following equations:

$$(1) \qquad d[FeCy_{ox}]/dt = -k_1[FeCy_{ox}][Cyt_{red}]$$

$$(2) \qquad d[AH^-]/dt = (d[FeCy_{ox}]/dt)(V_o/2V_i)$$

$$(3) \qquad [Cyt_{red}] = (k_0[AH^-]P)/k_0[AH^-] + k_1[FeCy_{ox}] + k_{-0}[A^{\dot{-}}])$$

$$(4) \qquad [A^{\dot{-}}] = \sqrt{k_1[FeCy_{ox}][Cyt_{red}]/(2\,k_{diss}V_i)}$$

These equations can be numerically integrated to determine the time dependences of the concentrations of ferricyanide, ascorbate, and semidehydroascorbate and the redox state of cytochrome b_{561}; P, V_o, and V_i are the membrane protein concentration, the extravesicular volume, and the intravesicular volume respectively. Estimated values of the rate constants[5] are $k_0 = 3.5 \times 10^{-3}$ ml/min·mg protein, $k_1 = 6.1$ ml/min·mg, $k_{-0} = 125$ ml/min·mg, and $k_{diss} = 1.2 \times 10^6\ M^{-1}\ s^{-1}$.

We expect some error in estimating the intravesicular volume and ascorbate concentration as well as in determining the rate constants. Nevertheless, as shown in FIGURE 2, the theoretical simulation accurately reflects changes in both the ferricyanide concentration and the redox state of cytochrome b_{561}. The accurate prediction of ferricyanide reduction rates is not surprising since the kinetic parameters were deduced from measurements of ferricyanide reduction. The prediction of cytochrome b_{561} redox changes, however, indicates that the cytochrome is oxidized by ferricyanide and reduced by ascorbate at rates sufficient to account for all of the transmembrane electron transfer observed. This further substantiates the view that cytochrome b_{561} is responsible for transmembrane electron transfer for the purpose of regenerating intravesicular ascorbic acid.

REFERENCES

1. HARNADEK, G. J., E. A. RIES & D. NJUS. 1985. Biochemistry **24:** 2640-2644.
2. KELLEY, P. M. & D. NJUS. 1986. J. Biol. Chem. **261:** 6429-6432.
3. NJUS, D., J. KNOTH, C. COOK & P. M. KELLEY. 1983. J. Biol. Chem. **258:** 27-30.
4. SRIVASTAVA, M., L. T. DUONG & P. J. FLEMING. 1984. J. Biol. Chem. **259:** 8072-8075.
5. NJUS, D., P. M. KELLEY, G. J. HARNADEK & Y. V. PACQUING. 1986. Mechanism of ascorbic acid regeneration mediated by cytochrome b_{561}. This volume.

Identification of a Synaptic Vesicle Specific Protein in *Torpedo* and Rat[a]

GUNNAR INGI KRISTJANSSON, JOHN H. WALKER, AND HERBERT STADLER

Department of Neurochemistry
Max Planck Institute for Biophysical Chemistry
D-3400 Göttingen, Federal Republic of Germany

In recent years attempts have been made to identify the molecular components of the synaptic vesicle. The best understood are the molecular components of the synaptic vesicles isolated from the electric organ of the ray *Torpedo.* Their structure is thought to be relatively simple, consisting of six to eight major proteins and a proteoglycan core. The composition of mammalian synaptic vesicles is less well known. Because of their neurotransmitter heterogeneity, their composition is expected to be more complex. It can be postulated on the basis of their unique function, however, that they have molecular components in common. This should also apply to synaptic vesicles isolated from different species. In an attempt to identify unknown synaptic vesicle components conserved through evolution, we have raised antisera to synaptic vesicles isolated from *Torpedo* and used these antisera to identify a M_r 86,000 protein conserved between *Torpedo* and rat.

Synaptic vesicles from the electric organ of *Torpedo* were isolated by the method of Tashiro and Stadler[1] and used to raise an antiserum in guinea pigs. *Torpedo* synaptosomes isolated as described in Israël *et al.*[2] and separated on SDS-PAGE were found to contain a M_r 86,000 antigen. The antigen was also found in rat synaptosomes isolated using the method of Gray and Whittaker.[3] The antigen was not found in rat liver, heart, kidney, or adrenal glands.

To further localize the antigen, rat synaptosomes were subfractionated by detergent extraction and ultracentrifugation into synaptosomal membranes and synaptosomal cytosol-containing vesicles. The antigen was found in the cytosol-containing vesicle fraction. This fraction was further fractionated to yield a synaptic vesicle pellet and synaptosomal cytosol fraction. The antigen was found only in the synaptic vesicle pellet. The antigen was barely detected in highly purified synaptosomal membranes.

Immunohistochemical localization showed the antigen to be associated with synaptic terminals in diaphragm, cerebellum, hippocampus, and cerebral cortex. The antigen's evolutionary conservation and association with synaptic vesicles suggests that it is functionally important, possibly playing a role in synaptic transmission.

[a] This work was supported by grants from EMBO (to GIK), Deutsche Forschungsgemeinschaft (no. SFB 33, to JHW and HS) and the Max-Planck-Gesellschaft. GIK is now a Max-Planck-Gesellschaft Fellow.

REFERENCES

1. Tashiro, T. & H. Stadler. 1978. Eur. J. Biochem. **90:** 479-487.
2. Israël, M. *et al.* 1974. Biochem. J. **160:** 113-115.
3. Gray, E. G. & V. P. Whittaker. 1962. J. Anat. **96:** 79-87.

Ascorbic Acid Enhancement of Norepinephrine Biosynthesis in Chromaffin Cells and Chromaffin Vesicles

MARK LEVINE

Laboratory of Cell Biology and Genetics
NIDDK, NIH
Bethesda, Maryland 20892

Ascorbic acid is the best reducing substance for the isolated enzyme dopamine β-monooxygenase.[1] It has not been clear, however, whether ascorbic acid plays a similar role in the hydroxylation of dopamine *in situ.* Indeed, there have been several seemingly contradictory observations concerning dopamine β-monooxygenase and ascorbic acid in chromaffin cells and chromaffin vesicles, or chromaffin granules. Dopamine β-monooxygenase in the adrenal medulla is localized exclusively to the inner aspect of chromaffin vesicles.[2] By contrast, ascorbic acid has quite a different distribution, since nearly 85% of ascorbic acid in adrenal medulla is extragranular.[3] Furthermore, it has been suggested that ascorbic acid cannot be transported into isolated chromaffin granules.[4] Since the localization of the cofactor and the enzyme are divergent, and the enzyme appears to be inaccessible to extragranular ascorbate, I have addressed whether ascorbic acid can increase norepinephrine formation in chromaffin tissue, and how this could occur.

Cultured chromaffin cells were incubated with or without ascorbic acid for several hours, during which time the cells accumulated ascorbic acid approximately 10-fold greater than control.[5] The cells were then incubated with [^{14}C]tyrosine, and radiolabeled tyrosine, dopamine, and norepinephrine were measured (FIG. 1). Only norepinephrine biosynthesis was enhanced by ascorbic acid, with no changes in the other precursors.[5,6] During this same time virtually none of the radiolabel could be detected as epinephrine, regardless of ascorbate preincubation.[5]

Since ascorbic acid specifically increased only dopamine β-monooxygenase activity, the mechanism of ascorbic acid action was investigated using isolated chromaffin vesicles. It was first necessary to determine whether the action of ascorbic acid could be explained simply by entry of ascorbic acid into chromaffin granules. As shown in FIGURE 2A, [^{14}C]ascorbic acid was not transported into chromaffin vesicles. Also, ascorbic acid did not enter chromaffin granules in the presence and absence of dopamine and Mg-ATP.[5,7] Despite the inability of ascorbate to enter chromaffin granules, extragranular ascorbate enhanced intragranular [^{3}H]norepinephrine formation (FIG. 2B) without changing total [^{3}H]dopamine uptake (data not shown).

All of these data indicate that ascorbic acid indeed enhances dopamine β-monooxygenase activity in chromaffin cells. The mechanism of ascorbate action to form norepinephrine in chromaffin cells, however, is more complex than the action of

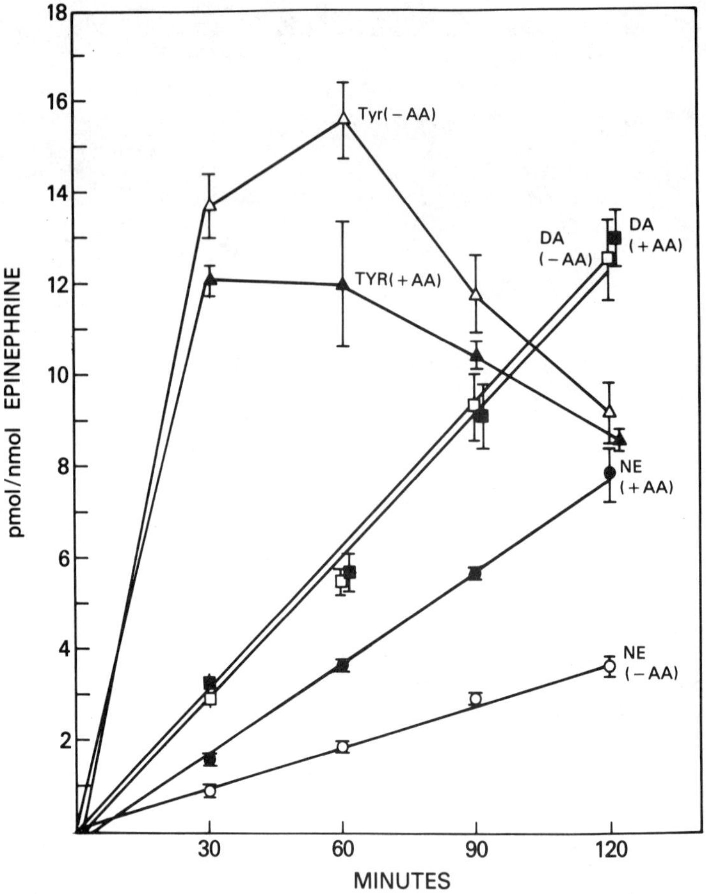

FIGURE 1. Ascorbic acid enhancement of norepinephrine formation in cultured chromaffin cells. Six-day-old chromaffin cells were preincubated with (*closed symbols*) or without (*open symbols*) 250 μM ascorbic acid for 3 hours. Ten μM [^{14}C]tyrosine was then added at time 0, and radiolabeled tyrosine (▲, △), dopamine (■, □), and norepinephrine (●, ○) were measured by HPLC and liquid scintillation spectrometry (from ref. 6).

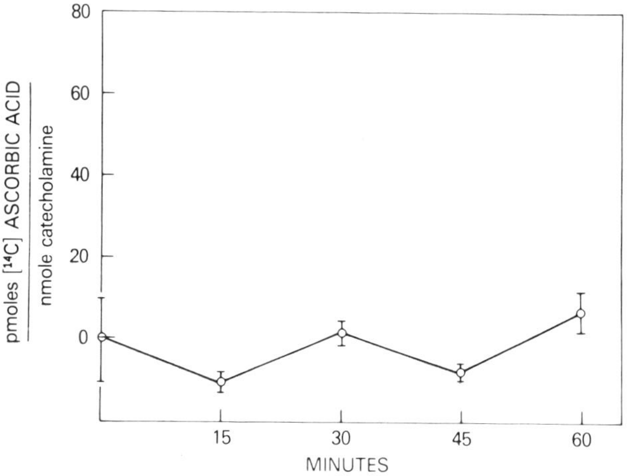

FIGURE 2A. Lack of ascorbic acid transport into chromaffin granules. Chromaffin granules isolated by differential centrifugation in sucrose were incubated with 2 mM ^{14}C-labeled ascorbic acid for varying times at 37°C. Ascorbic acid uptake was quantitated by liquid scintillation spectrometry. Mg-ATP and dopamine did not affect ^{14}C-labeled ascorbic acid uptake (data not shown). Data are from refs. 5 and 7.

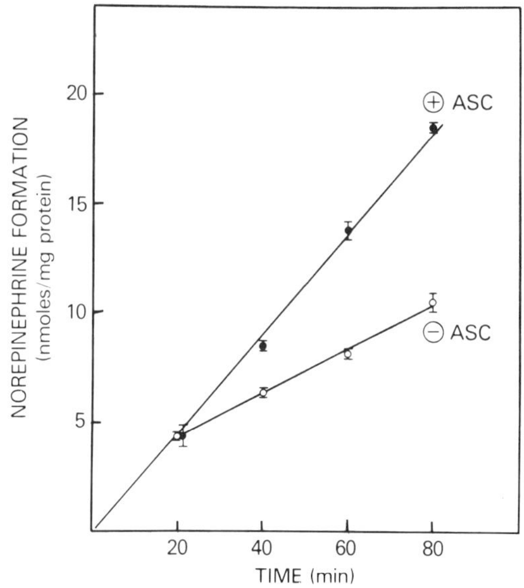

FIGURE 2B. Ascorbic acid enhancement of norepinephrine formation in isolated chromaffin granules. Isolated chromaffin granules were incubated with 50 μM [^{3}H]dopamine, 2.5 mM Mg-ATP, with (●) and without (○) 2 mM ascorbic acid. [^{3}H]Norepinephrine formation was measured at indicated times using differential elution from a cation exchange resin. Total [^{3}H] dopamine uptake was unaffected by ascorbic acid (data not shown). Data and details can be found in ref. 7.

ascorbic acid with the isolated enzyme. Although ascorbic acid is unable to enter chromaffin granules, extragranular ascorbate nonetheless enhances norepinephrine formation. Therefore, reducing equivalents from ascorbic acid must be transported across the chromaffin granule membrane and eventually to dopamine β-monooxygenase.

Our understanding of the mechanism of electron transfer from extravesicular ascorbic acid to intragranular dopamine β-monooxygenase is incomplete. One probable intermediate in electron transfer is cytochrome b_{561}, an integral chromaffin granule membrane protein.[8,9] It is not known, however, whether other membrane proteins also participate in electron transfer *in situ.* In addition, the intragranular electron acceptor is not known. One likely candidate is semidihydroascorbate, or ascorbate-free radical. It is not clear, however, whether this species mediates regeneration of intragranular ascorbate, and whether there is subsequent reduction of dopamine β-monooxygenase directly or indirectly. Mg-ATP might also be important in mediating electron transfer and subsequent norepinephrine formation, perhaps by maintaining $\Delta\Psi$, maintaining ΔpH, or by participating in the formation of another intragranular component. Experiments to test some of these possibilities are now in progress.

REFERENCES

1. FRIEDMAN, S. & S. KAUFMAN. 1965. J. Biol. Chem. **240:** 4763-4773.
2. LADURON, P. 1975. FEBS Lett. **52:** 132-134.
3. MORITA, K., M. LEVINE, E. HELDMAN & H. B. POLLARD 1985. J. Biol. Chem. **260:** 15112-15116.
4. TIRRELL, J. & E. WESTHEAD. 1979. Neuroscience **4:** 181-186.
5. LEVINE, M., K. MORITA & H. B. POLLARD. 1985. J. Biol. Chem. **260:** 12942-12947.
6. LEVINE, M. 1986. J. Biol. Chem. **261:** 7347-7356.
7. LEVINE, M., K. MORITA, E. HELDMAN & H. B. POLLARD. 1985. J. Biol. Chem. **260:** 15598-15603.
8. NJUS, D., J. KNOTH, D. COOK & P. M. KELLEY. 1983. J. Biol. Chem. **258:** 27-30.
9. SRIVASTAVA, M., L. T. DUONG & P. J. FLEMING. 1984. J. Biol. Chem. **259:** 8072-8074.

Characterization of a Presynaptic Membrane Protein Ensuring a Calcium-Dependent Acetylcholine Release

NICOLAS MOREL, MAURICE ISRAEL,
BERNARD LESBATS, SERGE BIRMAN, AND
ROBERT MANARANCHE

Department of Neurochemistry
CNRS, 91190 Gif sur Yvette, France

In *Torpedo* electric organ, acetylcholine (ACh) release was studied at four different levels: intact nerve electroplaque synapses, isolated nerve terminals (synaptosomes), resealed synaptosomal membrane sacs, and proteoliposomes into which presynaptic membrane proteins were incorporated. Electrical stimulation of nerve endings *in situ* in the whole tissue mobilized the cytoplasmic pool of neurotransmitter, which was released and renewed before any detectable change of the ACh pool contained in synaptic vesicles.[1,2] In response to a calcium influx triggered by various agents, isolated synaptosomes released most of their cytoplasmic ACh without any depletion of the vesicular ACh pool.[3] Synaptosomal ghosts refilled with ions and soluble ACh also exhibited a strictly calcium-dependent ACh release.[4] It was then possible to go one step further and to incorporate presynaptic membrane constituents into liposomal membranes made of synthetic lecithin, which have entrapped soluble ACh. The proteoliposomes thus obtained released ACh in response to a calcium influx.[5] A similar finding was observed using rat brain presynaptic membranes.[6] ACh release from proteoliposomes fulfilled several criteria expected for a physiologically relevant mechanism.[7] (1) It was calcium dependent; ACh was released only when calcium entered the proteoliposomes and magnesium could not replace calcium. (2) The kinetics of ACh release from proteoliposomes and synaptosomes were very similar. (3) A presynaptic membrane protein was involved that was not found in plasma membranes of the electroplaques. This integral membrane protein possessed an intracellular domain necessary for the ACh-translocating properties. (4) Finally, ACh was released from proteoliposomes in preference to choline or ATP when these substances were also present inside the liposomes with ACh.

Recently, we have used the reconstitution of ACh release in liposomes as a functional test to purify the membrane protein involved. This protein, which we propose to call mediatophore, was purified according to a three-step procedure: purification of the presynaptic plasma membrane according to a large-scale procedure,[8] extraction at alkaline pH, and solubilization in organic solvents. The purified protein was associated to lipids and still active (FIG. 1a). The calcium-dependent ACh release from proteoliposomes was proportional to the amount of mediatophore used for the re-

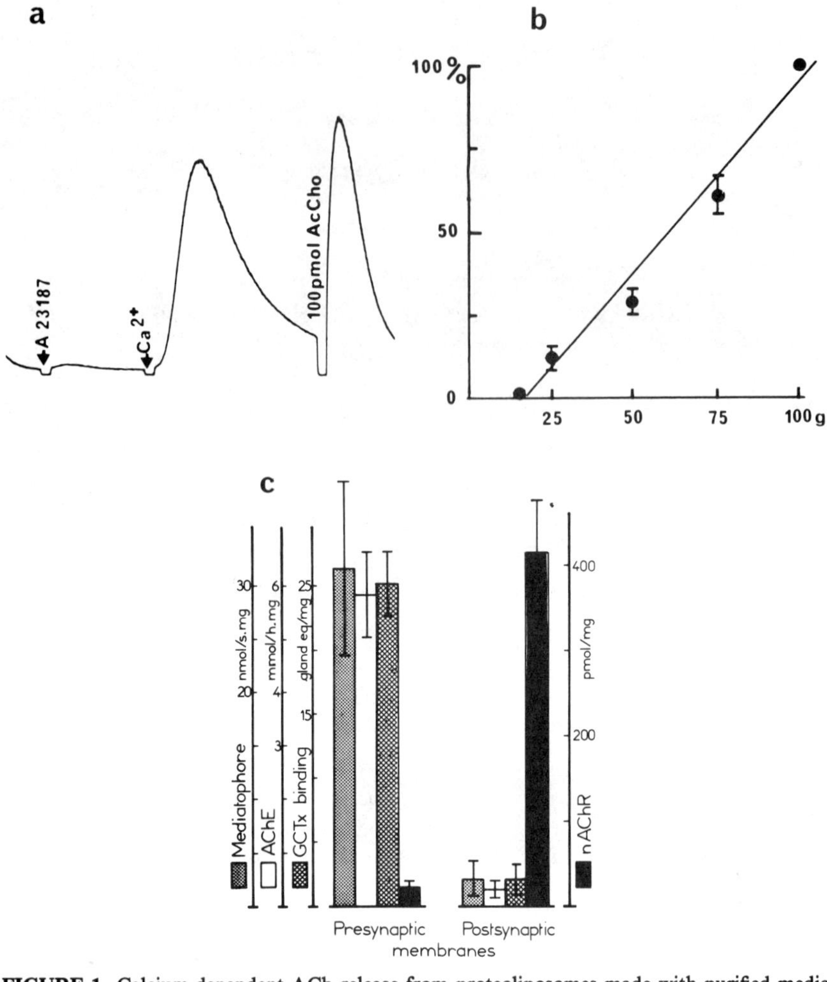

FIGURE 1. Calcium-dependent ACh release from proteoliposomes made with purified mediatophore. (**a**) Proteoliposomes (containing 1 nmol of occluded ACh) were added to the reaction mixture for the ACh chemiluminescent assay.[3] The release of ACh was elicited by the successive additions of calcium ionophore A23187 (7 μM) and calcium (10 mM) and recorded by monitoring the light emission. It was calibrated by injecting an internal ACh standard. (**b**) Calcium-dependent ACh release from proteoliposomes prepared with increasing amounts of purified mediatophore (derived from 20 to 100 g electric organ). The ACh release was estimated from the mean slope of the rising phase of the light emission. (**c**) The mediatophore purification procedure was applied to purified presynaptic or postsynaptic plasma membranes. Mediatophore activity was measured after reconstitution in proteoliposomes as in **b**. It was expressed per milligram of membrane protein for comparison with the other membrane markers, which were measured directly on the membrane preparations. Hydrophobic acetylcholinesterase activity and the binding of the presynaptic neurotoxin of Glycera convoluta[9] are presynaptic membrane markers, whereas nicotinic ACh receptor is specific for the postsynaptic innervated membrane.

constitution (FIG. 1b). It appeared specifically associated to the presynaptic plasma membrane and was not found in postsynaptic membranes (FIG. 1c). After lipid removal by ether precipitation, mediatophore activity was lost but the protein became water soluble as is the case for proteolipids. The lipid-free mediatophore was submitted to gel filtrations (5.2 nm Stokes radius) and to centrifugations in sucrose density gradients (9.8 ± 0.7 *S* sedimentation coefficient). This permitted the calculation of an apparent molecular weight of 210,000 ± 16,000. After solubilization in SDS and gel electrophoresis, a single band was observed (FIG. 2a), suggesting that the mediatophore is

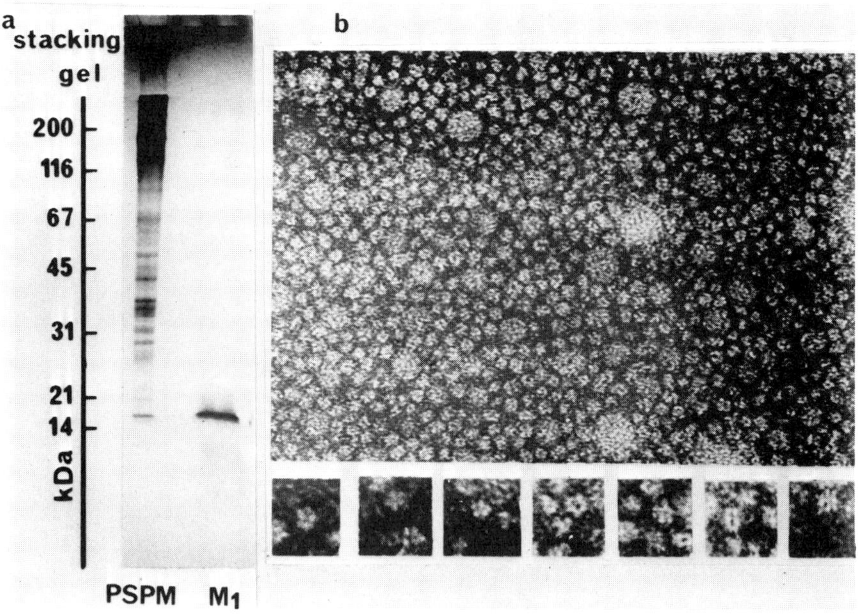

FIGURE 2. Identification of the mediatophore after delipidification. (**a**) Protein patterns of presynaptic plasma membranes (15 μg protein, PSPM) and of the purified mediatophore fraction (about 25 μg protein, M1). Electrophoresis was performed in a 7.5-20% acrylamide gradient gel in the presence of SDS without reducing agents; Coomassie blue staining. (**b**) Negative staining of the lipid free mediatophore in 2% neutral sodium phosphotungstate. *Insets* show high magnifications of face and profile views of the molecule taken from three independent experiments.

made of 17,000-Da subunits. When observed after negative staining (FIG. 2b), it looked like an 8-nm particle and seemed pentameric.

Therefore, the presynaptic plasma membrane contains a protein able to translocate ACh after activation by calcium. This protein, the mediatophore, has been purified. The possibility that ACh is released by this protein during synaptic activity has to be envisaged.

REFERENCES

1. ISRAEL, M., Y. DUNANT & R. MANARANCHE. 1979. Prog. Neurobiol. **13:** 237-275.
2. DUNANT, Y., G. J. JONES & F. LOCTIN. 1982. J. Physiol. (London) **325:** 441-460.
3. ISRAEL, M. & B. LESBATS. 1981. J. Neurochem. **37:** 1475-1483.
4. ISRAEL, M., B. LESBATS & R. MANARANCHE. 1981. Nature **294:** 474-475.
5. ISRAEL, M., B. LESBATS, N. MOREL, R. MANARANCHE, T. GULIK-KRZYWICKI & J. C. DEDIEU. 1984. Proc. Natl. Acad. Sci. U.S.A. **81:** 277-281.
6. MEYER, E. M. & J. R. COOPER. 1983. J. Neurosci. **3:** 987-994.
7. BIRMAN, S., M. ISRAEL, B. LESBATS & N. MOREL. 1986. J. Neurochem. **47:** 433-444.
8. MOREL, N., J. MARSAL, R. MANARANCHE, S. LAZEREG, J. C. MAZIE & M. ISRAEL. 1985. J. Cell Biol. **101:** 1757-1762.
9. MOREL, N., M. THIEFFRY & R. MANARANCHE. 1983. J. Cell Biol. **97:** 1737-1744.

Localization of Cytochrome b_{561} in Neuroendocrine Tissues That Contain Amidated Neuropeptides

REBECCA M. PRUSS[a]

Laboratory of Cell Biology
National Institute of Mental Health
Bethesda, Maryland 20892

Cytochrome b_{561} is one of the major proteins found in chromaffin granule membranes.[1] It is thought to transfer electrons to maintain reduced ascorbic acid within these granules.[2] Ascorbic acid is a cofactor for dopamine β-hydroxylase (DBH) and donates an electron during the conversion of dopamine to norepinephrine. Another ascorbic-acid-requiring enzyme that exists in secretory granules is the one described by Bradbury *et al.*[3] and Eipper *et al.*[4] that is involved in the COOH-terminal amidation of a large number of neuropeptides. Cytochrome b_{561} immunoreactivity has been found in the pituitary[5,6] as well as in adrenal medulla and sympathetic neurons.[5] I have examined the tissue distribution of cytochrome b_{561} by indirect immunocytochemistry in order to see whether this cytochrome is present in other neuroendocrine tissues and to try to correlate its presence to that of other known chromaffin granule markers.

Immunocytochemical analysis of cytochrome b_{561} immunoreactivity was performed using a monoclonal antibody that was found to be specific for this protein.[8] The antibody was originally screened for its specificity for chromaffin granules and thus named CG7. CG7 labels a 27-kDa chromaffin granule protein on immunoblots that is the same molecular weight as purified cytochrome b_{561} (FIG. 1a, lanes 1-3). Although a screen of other bovine tissues for cytochrome b_{561} by immunoblot analysis was negative (data not shown), CG7 did label a protein of the same molecular weight as cytochrome b_{561} in purified pituitary secretory granule membranes (FIG. 1a, lanes 4 and 5). The monoclonal antibody bound only to chromaffin-granule-containing subcellular fractions of bovine adrenal medulla and to fractions enriched for Golgi markers. Titration experiments indicated that cytochrome b_{561} is 10 to 100 times less abundant in pituitary vesicle membranes than in chromaffin granules. It would, therefore, be a relatively rare protein in other neuroendocrine tissues and thus be undetectable by immunoblot analysis. Immunocytochemistry can detect proteins that are concentrated in subcellular compartments. Since cytochrome b_{561} would likely be confined to secretory vesicle membranes, I examined its tissue distribution using 14-μm frozen sections of immersion-fixed bovine tissues. Sections were incubated with CG7-containing ascites fluid diluted 1:250 either alone or as a mixture with rabbit sera specific for other chromaffin cell markers, including chromogranin A, methionine enkephalin, DBH, phenylethanolamine *N*-methyltransferase, as well as a number of peptides, including vasoactive intestinal polypeptide (VIP), neuropeptide Y (NPY),

[a] Current address: Merrell Dow Research Institute, 2110 E. Galbraith Rd., Cincinnati, Ohio 45215.

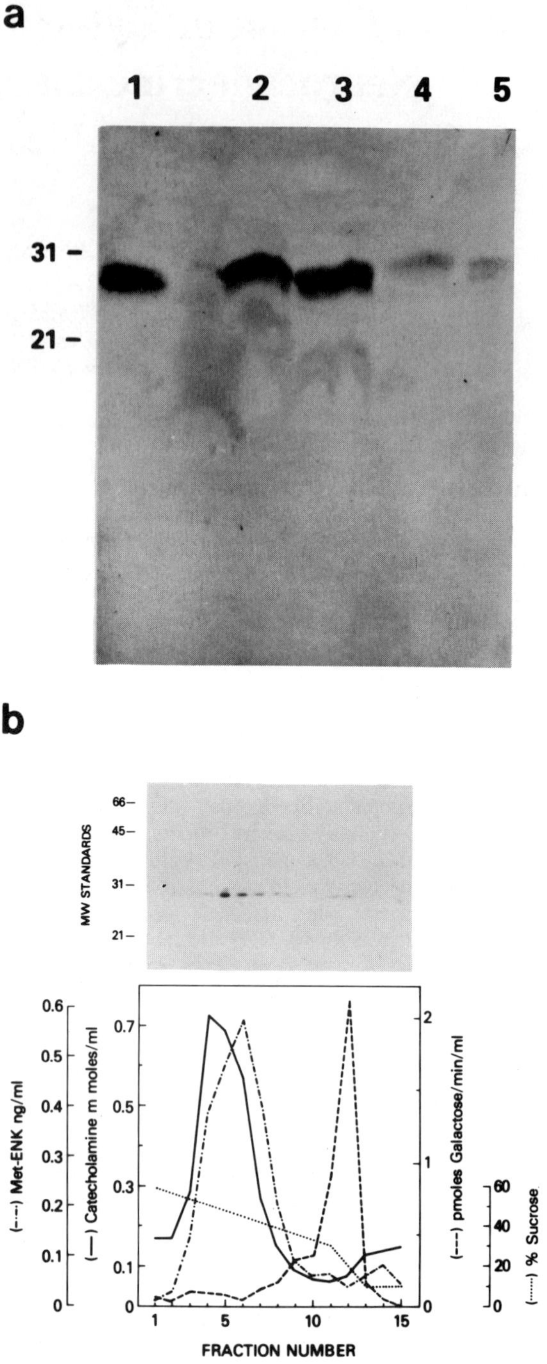

FIGURE 1. Identification of the protein antigen labeled by CG7. Immunoblots were prepared from 15% SDS polyacrylamide gels loaded with purified protein and membrane fractions.[8] Panel **a** contains: lane 1, purified cytochrome b_{561}; lane 2, chromaffin granule membranes; lane 3, whole chromaffin granules; lane 4, posterior pituitary neurosecretory vesicle (NSV) membranes; lane 5, salt-washed NSV membranes. Panel **b** is an immunoblot analysis of adrenal medulla membranes fractionated over sucrose gradients. Fractions were assayed for protein and a variety of subcellular markers prior to loading equivalent amounts of protein on consecutive lanes of the acrylamide gel.

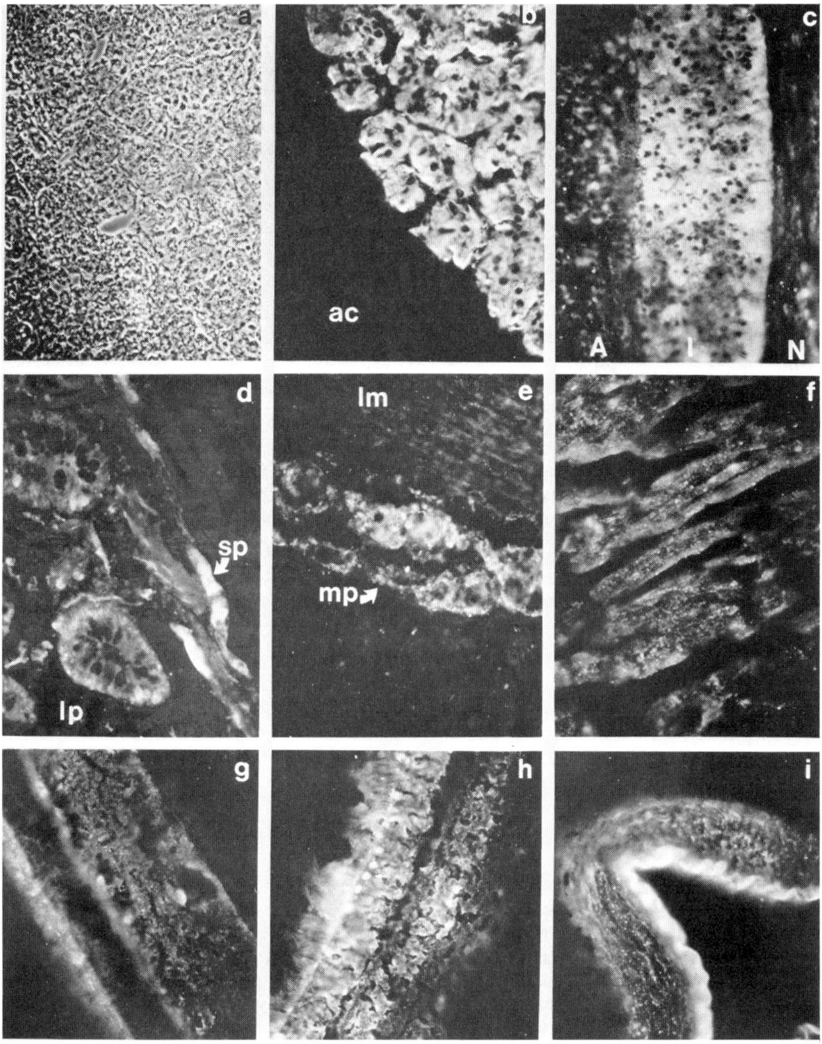

FIGURE 2. Immunocytochemical detection of cytochrome b_{561} in tissue sections. Bovine tissues were obtained from a local slaughterhouse, cut into 0.5-cm-thick pieces, and fixed by immersion overnight in 4% paraformaldehyde bufferred to pH 6.0 with 0.1 M potassium phosphate. The tissue was then saturated with 5% sucrose in 0.1 M potassium phosphate prior to cutting 14 μm frozen sections. Sections were mounted onto gelatin-coated slides, thawed, and incubated overnight at 4° in CG7 ascites fluid, diluted 1:250 in PBS plus 10% fetal calf serum. Sections were then washed with PBS and further incubated for 30 minutes at 37° in goat anti-mouse immunoglobulins coupled to fluorescein (GaMlg-Fl). When sections were double-labeled with a rabbit serum, those sections were incubated with a mixture of GaMlg-Fl and goat anti-rabbit immunoglobulins coupled to rhodamine. Sections were viewed with a 25× objective using a Zeiss Universal fluorescence microscope equipped with fluorescence excitation and barrier filters and phase contrast optics. Photographs were made from color slides using a Vivitar instant slide printer. Cytochrome-b_{561}-positive tissues included: (**a, b**) adrenal medulla (the phase contrast view in **a**), presented to demonstrate the absence of staining in the adrenal cortex (*ac* in **b**); (**c**) all three lobes of the pituitary—anterior (*A*), intermediate (*L*), and neural (*N*); (**d**) the submucus plexus (*sp* indicated with an *arrow*), as well as nerve fibers in the lamina propria (*lp*), and enteric crypt cells; (**e**) myenteric plexus (*mp*) and longitudial muscle (*lm*); (**f**) heart muscle fibers; (**g**) bovine retina (the inner nuclear layer is clearly labeled); (**h**) rhesus retina (showing labeling of the inner nuclear layer and cone cells indicated with *arrows*); and (**i**) vascular nerve fibers.

galanin, substance P, and insulin. The tissues examined included adrenal, liver, kidney, heart, pancreas, pituitary, parathyroid, spleen, retina, and gut.

Cytochrome b_{561} immunoreativity was found in all neuroendocrine tissues examined and in heart muscle fibers (FIG. 2). It was absent from adrenal cortex, liver, kidney, and spleen, as well as from the peptide-hormone-containing tissues of pancreas and parathyroid. In the gut and in vascular nerve fibers, cytochrome b_{561} immunoreactivity was found in association with many amidated neuropeptides, including VIP, NPY, galanin, and substance P. It was rarely found colocalized with chromogranin A, another major chromaffin granule protein. In fact, cytochrome b_{561} was striking in its absence from a number of chromogranin-A-positive tissues, including spleen, pancreatic islets, and parathyroid.[7] Although the pancreas and parathyroid contain peptide hormones, they are not amidated. In bovine retina, cytochrome b_{561} was found in the inner nuclear layer and in some ganglion and photoreceptor cells. Cone cells and amacrine cells were labeled by CG7 in sections of monkey retina. In pituitary, cytochrome b_{561} was localized in all cells of the intermediate lobe, in the majority, if not all, nerve fibers in the posterior lobe, and in a large percentage of anterior lobe cells. Cytochrome b_{561} was found in all cells of the bovine adrenal medulla.

All the tissues identified to contain cytochrome b_{561} by CG7 labeling contain either amidated neuropeptides or dopamine-derived catecholamines (or, in the case of the adrenal medulla, both these products). This distribution is correlated with the presence of ascorbic-acid-requiring enzymes within the secretory vesicles of these tissues. These data thus support the notion of a general role for cytochrome b_{561} as an electron carrier important for ascorbic acid regeneration in a wide range of neuroendocrine tissues.

ACKNOWLEDGMENTS

I would like to thank Drs. Ruth Hogue Angeletti, Dona Wong, Åke Rokaeus, Thomas O'Donohue, and Dominique Aunis for providing rabbit antisera used in these studies, Dr. Andrew Mariani for his gift of rhesus retina sections, Dr. Rhona Mirsky for advice and results concerning CG7 immunoreactivity in the gut, Dr. Patrick Fleming for a sample of purified cytochrome b_{561}, and Dr. James Russell for pituitary membrane fractions. Emily Shepard provided excellent technical assistance.

REFERENCES

1. WINKLER, H. & E. WESTHEAD. 1980. Neuroscience **5:** 1803-1823.
2. NJUS, D., J. KNOTH, C. COOK & P. M. KELLY. 1983. J. Biol. Chem. **258:** 27-30.
3. BRADBURY, A. F., M. D. A. FINNIE & C. G. SMYTH. 1982. Nature **298:** 686-688.
4. EIPPER, B. A., R. E. MAINS & C. C. GLEMBOTSKI. 1983. Proc. Natl. Acad. Sci. **80:** 5144-5148.
5. HORTNAGL, H., H. WINKLER & H. LOCHS. 1973. J. Neurochem. **20:** 977-985.
6. DUONG, L. T., P. J. FLEMING & J. T. RUSSELL. 1984. J. Biol. Chem. **259:** 4885-4889.
7. NOLAN, J. A., J. Q. TROJANOWSKI & R. HOGUE-ANGELETTI. 1985. J. Histochem. Cytochem. **33:** 791-798.
8. PRUSS, R. M. 1987. Neuroscience. In press.

Cholinergic Synaptic Vesicles Isolated from Motor Nerve Terminals from Electric Fishes to Rat

Molecular Composition and Functional Properties[a]

WALTER VOLKNANDT AND HERBERT ZIMMERMANN

AK Neurochemie
Zoologisches Institut der J. W. Goethe-Universität
6000 Frankfurt am Main 11, Federal Republic of Germany

The only tissue source from which synaptic vesicles containing solely acetylcholine (ACh) have previously been isolated is the electric organ of electric rays. Therefore, all results on the molecular characterization and functional properties of cholinergic vesicles are solely derived from an animal at the lowest step of vertebrate evolution (Elasmobranch fish). It is still an open question whether the general molecular properties also apply to cholinergic vesicles of higher vertebrates, especially mammals.

We report here the isolation of cholinergic synaptic vesicles from the electric organs of the evolutionarily further advanced bony fish *Electrophorus electricus* (electric eel) and *Malapterurus electricus* (electric catfish) as well as from the motor nerve terminals of the rat diaphragm. Vesicles enriched by sucrose density gradient centrifugation were further purified on columns of Sephacryl-1000. This was verified both by biochemical and electron microscopical criteria. Differences in diameter among synaptic vesicles from the various tissues were reflected by their elution pattern from the Sephacryl-1000 column. Synaptic vesicles from *Torpedo marmorata* with a diameter of about 90 nm are retarded less than synaptic vesicles from electric eel organ (about 60 nm) or vesicles from electric catfish electric organ and rat diaphragm (about 45 nm). Specific activities of ACh (nmol/mg protein) of chromatography-purified vesicle fractions were 36 (electric eel), 2 (electric catfish) and 1 (rat).[1]

Similar to *Torpedo* synaptic vesicles, vesicles from *Electrophorus, Malapterurus,* and rat were all found to store ATP in addition to ACh (molar ratios ACh/ATP between 9 and 12).

[a]This work was supported by the Deutsche Forschungsgemeinschaft (grant no. Zi 140/7-1).

By application of antibodies raised against purified synaptic vesicle proteoglycan[2] from either *Torpedo marmorata* or pig brain (gift of Dr. H. Stadler, Göttingen), the presence of proteoglycan in isolated cholinergic synaptic vesicles from electric eel and electric catfish electric organ and rat diaphragm could be demonstrated (FIG. 1). As revealed by immunocytochemistry the vesicular binding activity for the antiproteoglycan antibody is reflected by the specific association of the antigens with nerve terminals of the respective tissues.

Synaptic vesicles isolated from rat diaphragm contain binding activity for a monoclonal antibody (asv 48) raised against a 65-kDa protein of rat brain synaptic vesicles[3] (gift of Dr. L. F. Reichardt, FIG. 1). The antibody binding of isolated synaptic vesicles directly corresponds to results demonstrating specific staining of rat diaphragm nerve terminals after indirect immunofluorescence cytochemistry. Binding activity for asv 48 is absent from *Torpedo, Electrophorus,* and *Malapterurus* cholinergic synaptic vesicles.

As revealed by immunocytochemistry (FIG. 2) and immunotransferblotting, binding activity for an affinity-purified antibody against the vesicle-associated phosphoprotein synapsin I[4] (gift of Dr. T. Ueda) is present selectively in all cholinergic nerve endings of all tissues studied from *Torpedo* to rat. Nerve terminal varicosities are clearly depicted.

The results imply that both ATP and proteoglycan play a general role in cholinergic signal transmission from *Torpedo* to rat and also that synapsin I may be involved in the control of exocytotic release in electric organs.

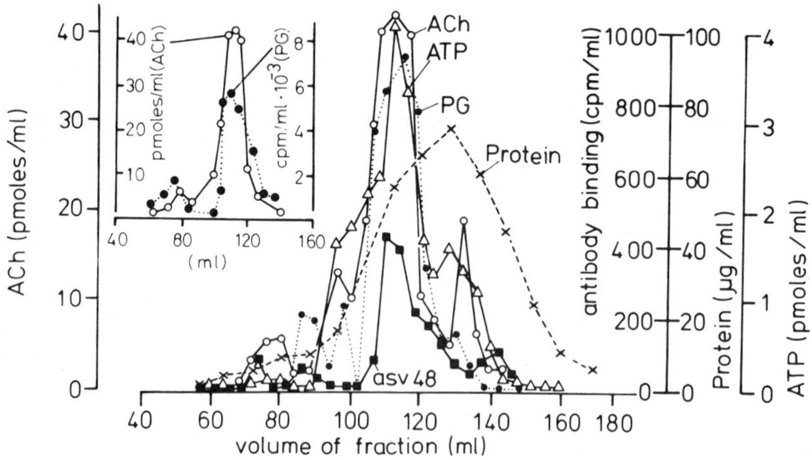

FIGURE 1. Separation of synaptic vesicles derived from rat diaphragm by chromatography on Sephacryl-1000. Note coincidence of elution of ACh, ATP, and binding activity for antiproteoglycan antiserum (PG) (*Torpedo* vesicle proteoglycan, or, *inset,* anti-pig brain vesicle proteoglycan) as well as for the monoclonal antibody asv 48. Binding activity is separated from void volume and main protein peak.

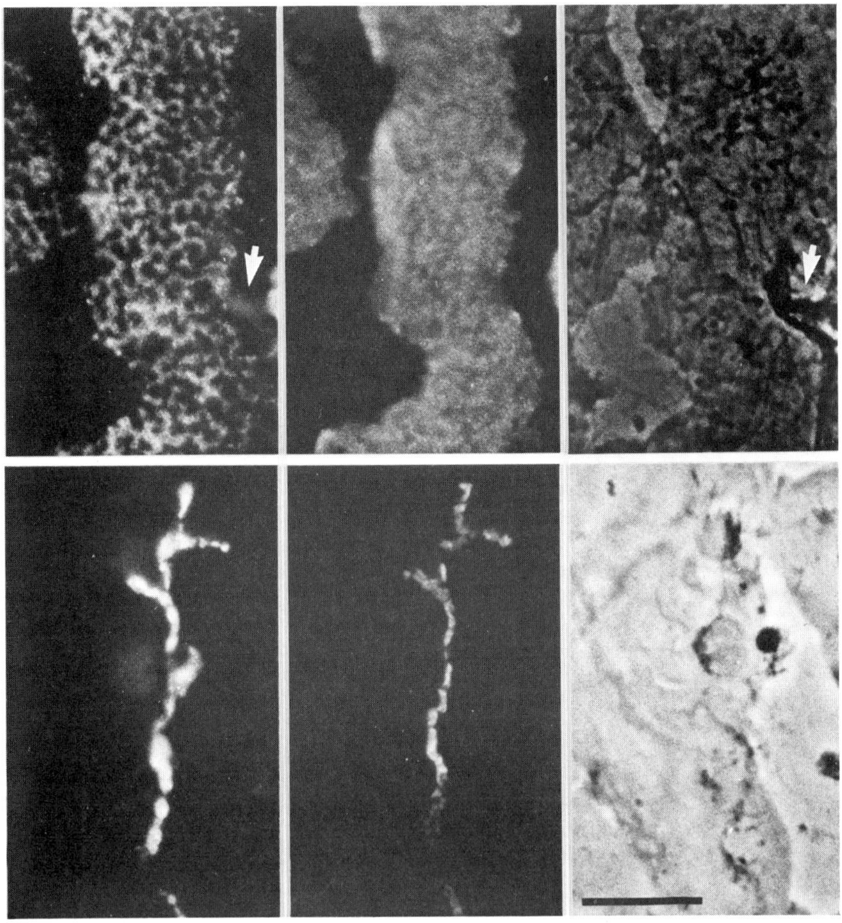

FIGURE 2. Binding of antisynapsin I antibody to nerve terminal varicosities (*left,* FITC conjugated second antibody), of rhodamine-conjugated α-bungarotoxin (*middle*), and corresponding phase contrast image (*right*). Electric organs are from *Torpedo marmorata* (*upper panel*) and *Electrophorus* (*lower panel*). *Arrows:* myelinated axon apparent with FITC staining and in phase contrast image but not with α-bungarotoxin staining. *Bar* = 20 μm.

REFERENCES

1. VOLKNANDT, W. & H. ZIMMERMANN. 1986. J. Neurochem. **47:** 1449-1462.
2. STADLER, H. & G. H. C. DOWE. 1982. EMBO J. **1:** 1381-1384.
3. MATTHEW, W. D., L. TSAVALER & L. F. REICHARDT. 1981. J. Cell Biol. **91:** 257-269.
4. UEDA, T. & S. NAITO. 1982. Prog. Brain Res. **56:** 87-103.

Proton Pumps and Chemiosmotic Coupling as a Generalized Mechanism for Neurotransmitter and Hormone Transport

ROBERT G. JOHNSON, JR.

Howard Hughes Medical Institute
Departments of Medicine and Neurology
Massachusetts General Hospital
Harvard Medical School
Boston, Massachusetts 02114

INTRODUCTION

One of the common themes that is emerging from the investigation of isolated neuroendocrine secretory vesicles is the almost universal existence of energy transducing membranes, electrochemical proton gradients, and chemiosmotic coupling mechanisms. The central component of this theme is the existence of an electrochemical proton gradient that is due to the operation of an anisotopically directed H^+-translocating ATPase within the membrane of the secretory vesicle. This proton pump catalyzes the vectorial movement of protons from the cytosol to the intragranular space resulting in the establishment of a transmembrane pH gradient (ΔpH, inside acidic) and transmembrane electrical gradient ($\Delta\Psi$, inside positive). The internal pH of these granules is maintained at pH 5.5, independent of the external pH, and *in vitro,* a membrane potential of 80 mV, inside positive, has been measured. Therefore, the magnitude of the electrochemical potential (defined as $\Delta\bar{\mu}_{H+} = \Delta\Psi + Z\Delta pH$, where $Z = RT/F$) can approach 180 mV.

Many of the important physiologic functions of endocrine secretory vesicles directly relate to the presence of transmembrane H^+ gradients. These include: (1) to maintain the intragranular chemical messengers in the unoxidized form; (2) to activate or inactivate processing enzymes; (3) to alter the physicochemical properties of the packaged neurotransmitter or hormone; and (4) to act as the driving force for accumulation of the neurotransmitter or hormone.

Investigation of the effect of the H^+ gradients upon amine transport has been influenced by documentation of the pivotal role of H^+-ATPases in establishing transmembrane H^+ gradients and their coupling to energy transduction and ion and metabolite transport in a variety of cells and subcellular organelles.[1–3]

The relationship between the $\Delta\bar{\mu}_{H+}$ and the equilibrium potential for the hormone or neurotransmitter ($\Delta\bar{\mu}_{A+}$) is predicted by the chemiosmotic hypothesis, which states that the energy for the substrate transport system is derived from the $\Delta\bar{\mu}_{H+}$ generated by the H^+-translocating ATPase.[4] A series of investigations from several laboratories have shown that coupling of transport to energy transduction proceeds via a specific transporter in equilibrium with the $\Delta\bar{\mu}_{H^+}$, which catalyzes neurotransmitter or hormone influx and H^+ efflux (antiport mechanism).[5-10] The magnitude of the equilibrium distribution gradients for the biogenic amines is one of the largest in any mammalian system: 135,000 to 1.

In this report, the salient features of chemiosmotic gradients as they relate to the best studied neuroendocrine secretory transport system, amine uptake in chromaffin granules, are reviewed. The following papers in this volume will more completely describe the properties of the proton pump and the amine transporter.

THE CHROMAFFIN GRANULE AS A MODEL FOR THE STUDY OF CHEMIOSMOTIC GRADIENTS

There are several practical reasons why the chromaffin granule has become the archetypal neuroendocrine secretory vesicle for investigation of the coupling between H^+ gradients and neurotransmitter or hormone uptake. First, almost limitless quantities of adrenal glands can be removed from slaughterhouse animals such as pigs, sheep, and cows, and the chromaffin cells within the adrenal of the medulla of these animals contain large numbers of chromaffin granules. Second, because of the extreme density of the chromaffin granule (1.18 g/ml), isolation is easily and rapidly achievable through differential and isotonic centrifugation density gradients. Third, primarily due to the density, the granules can be isolated in extremely high purity devoid of other subcellular organelles and fragmented membranes. Fourth, due to the large amounts of granules that can be isolated, it is feasible to prepare chromaffin ghosts formed by hyposmotic lysis of chromaffin granules, extensive washing, and reformation in isotonic media followed by extensive dialysis. This preparation offers many advantages over the chromaffin granule preparation in that there are no endogenous ion gradients, intragranular matrix binding components, or endogenous amines. Finally, the adrenal medulla arises embryologically from cells within the neuroectoderm that are pleuropotential, some differentiate to form sympathetic ganglia, others to form catecholamine-containing chromaffin cells. Underlying investigations of the chromaffin granule, therefore, is the implicit assumption that the principles governing physiologic transport in this system will be operant in less readily isolable adrenergic and cholinergic vesicles.

PRINCIPLES OF CHEMIOSMOTIC COUPLING MECHANISMS

One of the major scientific intellectual contributions over the past 25 years has been the resolution of the apparent paradox of how scalar energy forces could result in vectorial solute transport. This contribution, the chemiosmotic theory, has revolutionized both the understanding and experimental approach in the area of energy transduction in biological membranes.

The original chemiosmotic formulation is elegant in its simplicity.[4] The basic tenets are threefold (TABLE 1). First, a topologically closed insulating membrane exists with a low permeability to ions and solutes. Second, an anisotopically oriented H^+-ATPase

TABLE 1. Fundamentals of Chemiosmotic Coupling Mechanisms

- A topologically closed insulating biological membrane exists that maintains a low permeability to metabolites, solutes, and ions in general, and to hydrogen and hydroxyl ions, in particular.
- A vectorially oriented H^+-translocating ATPase within the biological membrane generates an electrochemical proton gradient.
- There are proton-linked antiport or symport systems within the membrane coupled to the electrochemical potential for ion or metabolite transport.

within the biologic membrane generates an electrochemical proton gradient. Third, proton-linked antiport or symport transporter systems within the membranes couple the electrochemical potential to ion or metabolic transport. The chromaffin granule satisfies each of these tenets.

Permeability and Buffering Capacity

The striking characteristic of the membrane of the chromaffin granule is its extremely low permeability to monovalent and divalent cations and small anions. In point of fact, the conductance to H^+ is at one order of magnitude lower than that of mitochondria, the subcellular organelle for which Mitchell formulated the chemiosmotic hypothesis.[4] In addition, the intragranular H^+ buffering capacity is very large, approaching 300 μmol H^+/pH unit/g dry weight. In contradistinction to the extremely low permeability to H^+, K^+, Na^+, Ca^{2+}, Mg^{2+}, and ATP, the chromaffin granule maintains a finite permeability to Cl^-, which may be important in the control of intragranular pH and transmembrane potential *in vivo.*

Representative experiments shown in FIGURE 1 indicate the variety of techniques that can be utilized to determine the permeability of isolated chromaffin granules. Behavior of the chromaffin granules as perfect osmometers in various media (NaCl, KCl, sucrose) indicated the impermeability of the membrane to one or both ions (FIG. 1A). The use of a permeable cation, ammonia, permitted evaluation of the distribution of the corresponding anion by monitoring the movement of the anion into the granule through sensitive spectrophotometric measurement of chromaffin granule swelling (FIG. 1B). The relative permeabilities can also be estimated by the use of ion-selective ionophores and spectrophotometric recording. The addition of carbonyl cyanide *p*-trifluoromethoxyphenylhydrazone (FCCP), an electrogenic proton ionophore, catalyzes the movement of protons down their electrochemical gradient, in this case generating an outward diffusion potential (negative inside potential, FIG. 1C). Thermodynamically, it would be favorable for the cation within the media, in this case K^+, to move down its electrochemical gradient with an inward flux. The lack of significant swelling of the granules, however, indicates the permeability of the membrane to K^+ is quite low. When valinomycin, an ionophore that mediates equilibration of potassium across biological membranes is added, a rapid swelling of the granules is measured, consistent with K^+ influx. The quantitation of H^+ and K^+ fluxes can be achieved with the use of sensitive potentiometric on-line recording with ion-sensitive and selective electrodes. The influx of K^+ into and efflux of H^+ from isolated chromaffin granules is shown in FIGURE 1D using a K^+ sensitive electrode and pH electrode, respectively. The net ion fluxes can be quantitated as shown.

These photometric, potentiometric, and radiochemical techniques indicate that even when one ion is rendered freely permeable to the chromaffin granules there is no corresponding movement of the counterion in spite of an exceedingly high gradient. The physiologic role of this extreme cation permeability is not completely understood. It is, however, reasonable that a subcellular organelle whose intragranular space contains these remarkably high concentrations would attempt to minimize any transmembrane ion fluxes that may affect the stability of the intragranular complex, resulting in efflux and redistribution of the biogenic amines.

The low proton permeability measured across the chromaffin granule membrane led to the idea that a pH gradient and perhaps transmembrane potential existed. Unfortunately, the intragranular space of the chromaffin granule is too small to accommodate a microelectrode. This necessitates indirect measurements of the pH gradient.

Measurement of ΔpH and ΔΨ

The difference between the intragranular pH and the pH of the medium in which isolated chromaffin granules are suspended has been measured by (1) lysis of the granules and calculation of the buffering capacity, (2) [^{14}C]methylamine distribution,

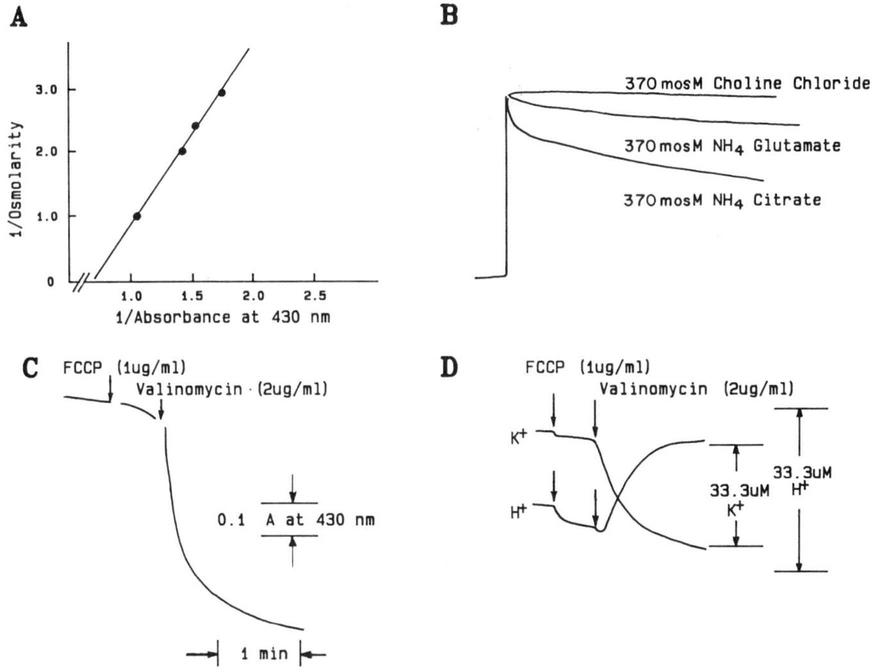

FIGURE 1. Summary of experiments investigating the permeability and regulation of volume in isolated chromaffin granules: **(A)** light scattering of chromaffin granules as a function of osmolarity; **(B)** light scattering of chromaffin granules in isotonic solutions of various inorganic salts; **(C)** swelling of chromaffin granules induced by various ionophores; and **(D)** simultaneous recordings of K^+ release and H^+ uptake induced by FCCP and valinomycin. (For further details see Johnson and Scarpa.[11])

(3) quenching of fluorescent amines, and (4) ^{31}P-NMR spectroscopy. The intragranular pH has been measured to be very acidic by all of these techniques.

If the detergent Triton X-100 is added to a stirred suspension of chromaffin granules suspended in a low concentration of buffer and the pH monitored with a standard pH electrode, a marked acidification is observed. This observation alone suggests a highly acidic matrix space; calculation of the buffering capacity coupled with a knowledge of the intragranular water space results in an approximate internal pH of 4.5 to 6.0 (FIG. 2A).

A more accurate quantitative method to measure the ΔpH is with a weak base such as methylamine. Methylamine is a weak base ($pK_a = 10.6$) that can only permeate biologic membranes in the deprotonated, uncharged form. Although at equilibrium the concentration of uncharged methylamine will be the same on both sides of a membrane, if one side of the membrane is more acidic, methylamine will become protonated on that side and will be effectively trapped in the acidic medium. When exposed to an organelle with an acidic interior relative to the external medium, the total equilibrium methylamine concentration will thus be much higher inside than outside. At near physiologic pH values, the ratio of internal to external methylamine closely approximates the ratio of internal to external H^+, *i.e.,* of the ΔpH. The ΔpH across the membrane of an organelle can thus be accurately and reproducibly quantitated by radioassay of the distribution of tracer amounts of [^{14}C]methylamine.[12] A small deviation from ideal methylamine distribution according to the ΔpH occurs because of methylamine binding to membrane and matrix components, but this deviation is confined to, at most, the equivalent of 0.2 pH units.[11]

When freshly isolated chromaffin granules are suspended in buffered sucrose, KCl, NaCl, Na-isethionate, or choline-Cl media at pH 7.0 and incubated with 10 μM [^{14}C]methylamine, a transmembrane ΔpH of 1.43 units is measured.[11] The measurements are consistent with an intragranular pH of 5.5, and additional experiments have shown that this acidic intragranular pH is independent of the external pH and of the ionic composition of the media. The acidic chromaffin granule interior is further evidence for an extremely low membrane conductance to protons and has been confirmed by numerous investigators to be pH 5.5-5.7.[13,14]

The endogenous ΔpH of isolated granules can be perturbed by protonophores and large concentrations of ammonia (and other primary amines; see FIG. 2B). FCCP catalyzes the electrogenic proton movement through the membrane down the electrochemical proton gradient for H^+. Because the chromaffin granule is impermeable to movement of anion efflux or cation influx, the efflux of H^+ induced by FCCP and the generation of a large diffusion potential limit further efflux. Ammonia, on the other hand, can permeate the chromaffin granule membrane in the uncharged form. If ammonia is present in the incubation medium in large amounts (>1 mM) reprotonation of the accumulated amine will occur, removing H^+ from the intravesicular solution. As the buffering capacity of the matrix space is exceeded, alkalinization of the intragranular space will occur stepwise and a decrease in the ΔpH will result.

Certain primary amines such as 9-aminoacridine and atebrin maintain a high native fluorescence when free in solution and can accumulate into chromaffin granules by the same mechanism as methylamine. Once trapped in the intragranular space, their fluorescence is quenched (FIG. 2C). By measuring the degree of quenching as well as the internal and external water spaces, the ΔpH using fluorescent dyes has produced values in good agreement with those obtained using [^{14}C]methylamine distribution.[15]

Because of the unusually high ATP content of the isolated chromaffin granule, ^{31}P-NMR can be used as another probe for measuring intragranular pH. Since the pK_a of the γ-P of ATP is within the physiologic range, the resonance of the γ-P is shifted according to the pH of the environment in which the ATP is located. Com-

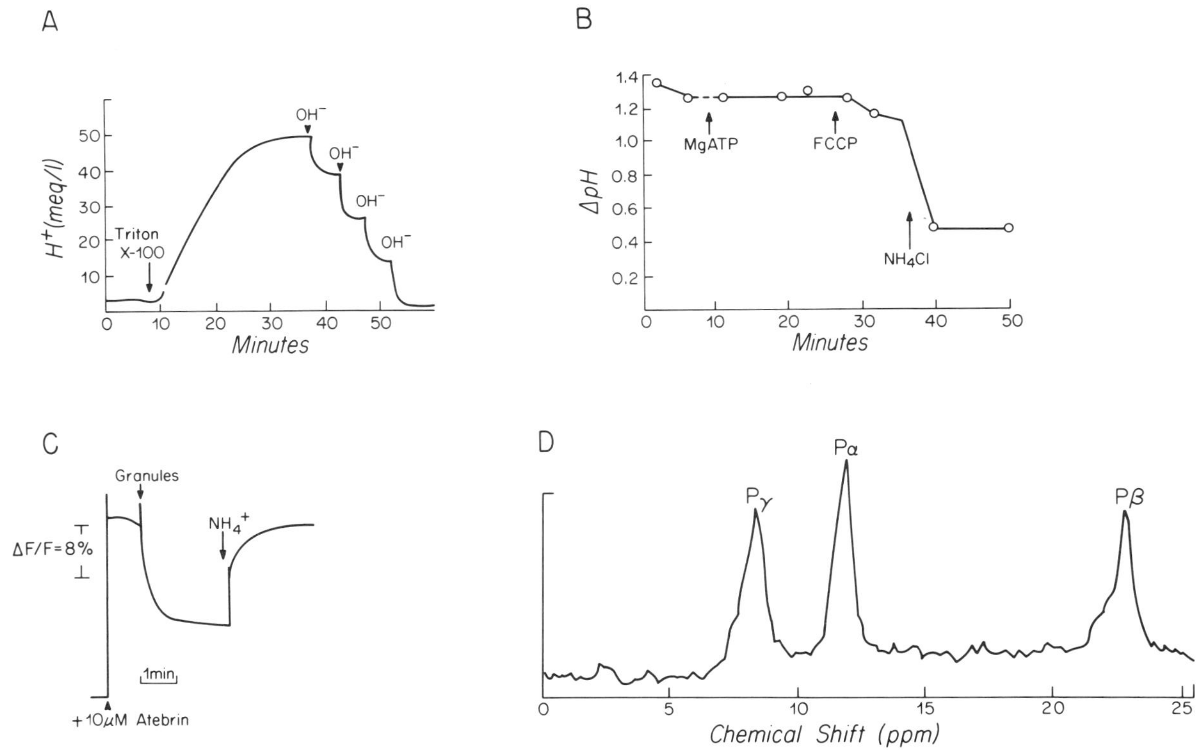

FIGURE 2. Summary of intragranular pH measurements in isolated chromaffin granules by **(A)** lysis of granules and calculation of the buffering capacity, **(B)** [^{14}C]methylamine distribution, **(C)** fluorescence dye distribution, and **(D)** [^{31}P]NMR. (**A** and **B** were adapted from Johnson and Scarpa,[19] **C** from Salama *et al.*,[15] and **D** from Pollard *et al.*[14])

parison of the shifts obtained in intact granules with those of model suspensions approximating them has made possible the determination of not only intragranular composition but also intragranular pH, with results comparable to those of the two other techniques (FIG. 2D).[16–18] Although ^{31}P-NMR has to date proved to be chiefly a confirmatory method, it is a potentially powerful tool in the study of ΔpH *in situ* or *in vivo.*

The transmembrane potential (ΔΨ) of chromaffin granules can be measured by radioisotopic tracers and spectrophotometric dyes. Thiocyanate (SCN^-) and tetramethylphenylphosphonium ($TPMP^+$) are lipophilic ions that can freely permeate membranes in the charged form to distribute according to the transmembrane Nernst potential. When incubated with a suspension of chromaffin granules or ghosts, the ratio of internal to external [^{14}C]SCN^- or [^{3}H]$TPMP^+$ will give an exact measure of the potential, positive or negative inside, respectively. These radioisotopic methods have been used with great success to quantitate reproducibly and sensitively the chromaffin granule ΔΨ, and have been independently verified in other organelles large enough to permit potential-recording electrodes (FIG. 3).[19–21]

The H^+-translocating ATPase

The existence of a ΔpH and ΔΨ across the membrane of the chromaffin granule led to the search for a proton pump. Initial experiments produced rapid conformation when it was found that: (1) addition of ATP to an isolated ghost preparation in which the internal and external pH values are initially identical results in the generation of a ΔpH, inside acidic,[23,24] (2) a fixed stoichiometry exists between ATP hydrolysis and H^+ translocation,[25,26] and (3) under appropriate conditions ATP can be synthesized at the expense of an imposed proton gradient.[27]

Originally, studies of the isolated ATPase were consistent with the H^+-ATPase being similar to the F_1-ATPase of mitochondria. However, elegant work by Cidon and Nelson showed that the granules were contaminated by mitochondria and the degree of contamination could account for the presence of mitochondrial-type ATPase.[28] Sodium bromide treatment proved to be a novel method for the separation of the mitochondrial ATPase from the chromaffin granule ATPase. The properties of the ATPase are discussed in the following paper in this volume (by Apps and Percy).

CHEMIOSMOTIC GRADIENTS AND AMINE TRANSPORT

Amine transport in intact chromaffin granules is complicated by the presence of existing gradients, large concentrations of endogenous catecholamines and ATP, and intragranular binding sites. For this reason, chromaffin ghosts formed by hyposmotic lysis of the chromaffin granules, extensive washing, and reformation under isotonic conditions has proved a superb model for the investigation of the effect of the ΔpH and ΔΨ on amine accumulation. By selection of the appropriate experimental conditions, *i.e.,* the presence of a permeable anion (chloride) or impermeable anion (isethionate or sucrose), the addition of ATP to the medium results in the generation of either a ΔpH alone or ΔΨ alone (FIG. 4). When the ghosts are formed in a medium containing an impermeant ion at one pH and resuspended at more basic pH, a membrane potential can be generated in addition to the existing ΔpH (FIG. 4).

Years of investigation into the empirical properties of amine transport have produced sufficient data to conclude comfortably that amine accumulation proceeds via a carrier-mediated process. The evidence includes: (1) structural and stereospecificity of uptake; (2) a Q_{10} of 4.6, (3) dependency upon the presence of ATP and Mg^{2+}, and (4) specific inhibition of transport by various compounds such as reserpine.

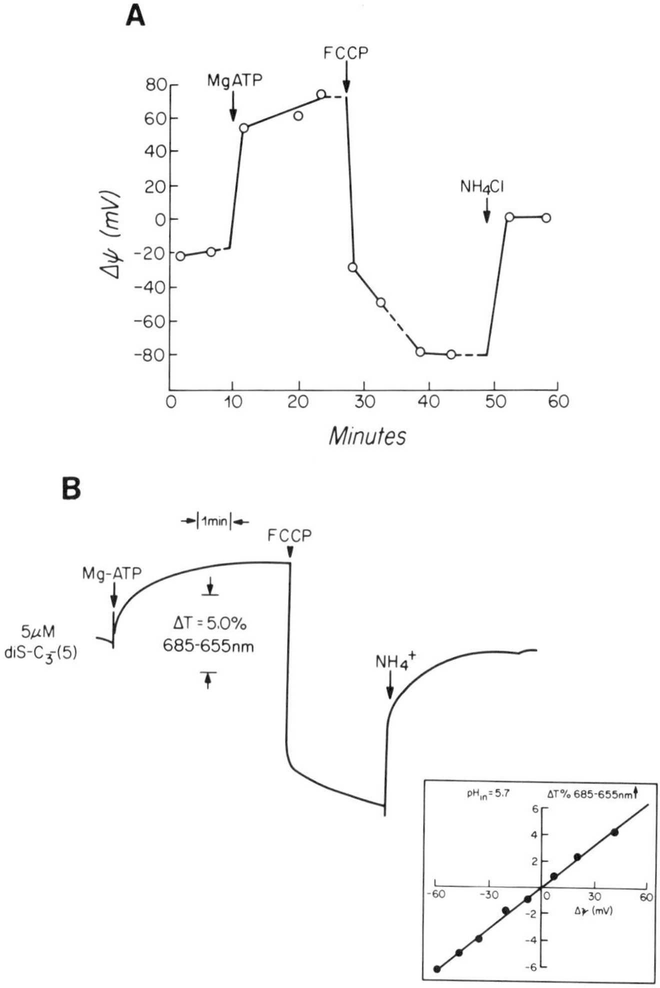

FIGURE 3. Summary of ΔΨ measurements in isolated chromaffin granules by (**A**) [^{14}C]SCN^- distribution and (**B**) Di(S)-C_3-(5) fluorescence dye distribution. (**A** was adapted from Johnson and Scarpa,[19] and **B** from Salama *et al.*[15])

In the example shown (FIG. 5), the time-resolved accumulation of dopamine into chromaffin ghosts is monitored on-line with a sensitive glassy carbon electrode. A rapid accumulation of dopamine into the chromaffin ghosts (observed as disappearance of dopamine from the medium) was measured. The accumulation reached a steady

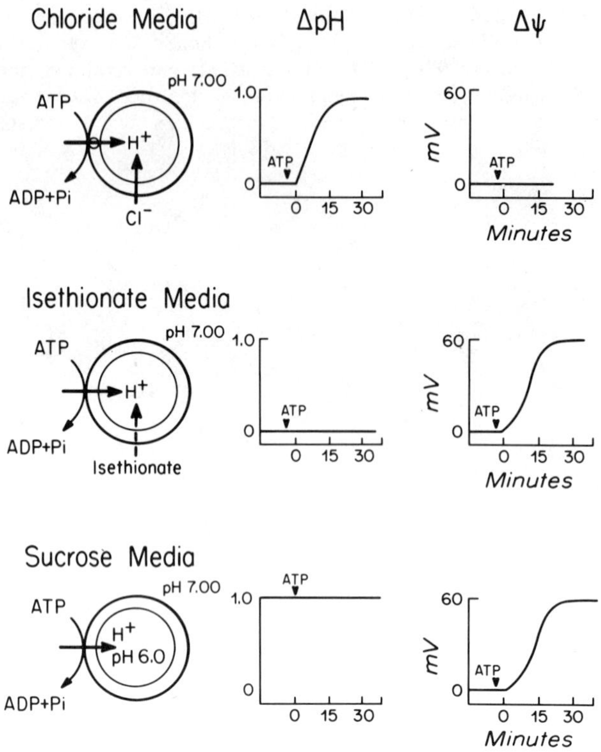

FIGURE 4. Schematic illustration of ATP-dependent generation of ΔpH, ΔΨ, or both in isolated chromaffin ghosts.

state after ten minutes whether the driving force was the ΔpH, ΔΨ, or both. Collapse of the respective gradients by NH_4^+, SCN^-, or both resulted in a prompt efflux of the accumulated amine to the baseline level. These results indicate that the proton concentration and/or electrical gradients are thermodynamically sufficient to drive amine accumulation. In addition, it is readily apparent that for a similar magnitude of the ΔpH and ΔΨ, a significantly higher steady-state accumulation is reached in the presence of the ΔpH.

These types of experiments can be used to provide the foundation for investigation of the precise relationship between the $\Delta\bar{\mu}_{H+}$ and the electrochemical gradient for amines ($\Delta\bar{\mu}_A$) on a quantitative level. The chemiosmotic hypothesis states that the energy for substrate transport systems is derived from the $\Delta\bar{\mu}_{H+}$ and generated from the proton pumping ATPase, and that coupling between the transported species (amine and proton, in this case) would approach equilibrium whenever

$$\Delta\bar{\mu}_A - n\Delta\bar{\mu}_{H+} = 0,$$

where n is the number of protons translocated in either the same direction as the amine (symport) or in the opposite direction (antiport). Thus, n is the stoichiometry of the reaction. Because of the ΔpH, which is acidic inside, and a ΔΨ, positive inside, generated by the H^+-translocating ATPase, the diffusion gradient for protons is in the outward direction in the chromaffin granule or ghost. Therefore, in the model for amine uptake based upon chemiosmotic processes, the influx of amines is coupled to H^+ efflux. More precisely, the H^+-amine antiport is defined by nonspontaneous obligatory coupling by the existence of a putative translocator. It is the value *n,* the stoichiometry of the H^+-amine antiport, which determines the relative contribution of the ΔpH and ΔΨ to the driving force.

In a series of experiments, the $\Delta\bar{\mu}_{H^+}$ and $\Delta\bar{\mu}_A$ were measured simultaneously under a wide range of conditions in an attempt to differentiate between various mathematically derived models for the driving force and possible H^+-amine stoichiometries. Amine uptake was measured in the presence of (1) ΔpH alone and (2) ΔΨ alone.

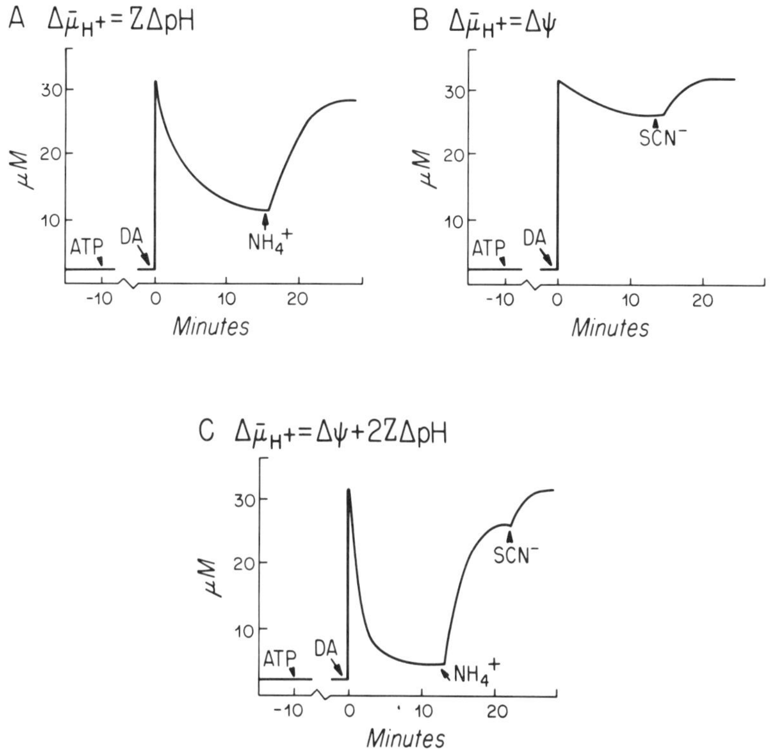

FIGURE 5. Kinetic measurement of dopamine accumulation into chromaffin ghosts using an amperometric electrode. Chromaffin ghosts for **A** and **B** were formed in a medium of 185 mM Na isethionate and 5 mM Hepes (pH 7.0) and resuspended in a reaction mixture consisting of (**A**) 185 mM KCl and 30 mM Hepes (pH 7.0) or (**B**) 185 mM isethionate and 30 mM Hepes (pH 7.0). Chromaffin ghosts for **C** were formed in a medium of 185 mM Na isethionate and 10 mM Mes (pH 5.5) and resuspended in a reaction medium of 185 mM Na isethionate and 30 mM Hepes (pH 7.0). The total volumes were 2.3 ml. After equilibration of the chromaffin ghosts with 3 mM ATP for ten minutes to ensure the electrochemical proton gradient had reached steady state, 30 μM dopamine was added. At the indicated times, 20 mM $(NH_4)_2$ SO_4 or 40 mM $KSCN^-$ was added to the incubation medium. Temperature was 37°C.

A reliable method for a titratable decrease in the membrane potential has been the addition of large concentrations of thiocyanate (1 to 60 mM) to the experimental medium.[20,21,28] In the first series of experiments (FIG. 6) ghosts were incubated in the presence of ATP and 60 mM thiocyanate in an attempt to dissipate completely any electrical gradients generated, thereby leaving only the ΔpH to drive accumulation.

Conversely, by incubating the ghosts with increasing concentrations of ammonia (1 to 50 mM), the magnitude of the ΔpH generated could be varied from 0 to 0.8 pH units; no ΔΨ was observed. ^{14}C-labeled 5-hydroxytryptamine (serotonin) was added to these suspensions of ghosts after establishment of the ΔpH or ΔΨ, and the distribution of the biogenic amine measured after apparent steady-state levels had been reached. In the representative experiment illustrated in FIGURE 6, curve *A*, the [^{14}C]serotonin accumulation under these conditions varied directly with the magnitude of the ΔpH, with a proportionality constant of 0.4. In each instance, the net [^{14}C]serotonin distribution after subsequent addition of 50 mM NH_4^+ was close to zero (data not shown). Data from 16 separate ghost preparations in which the ΔpH was the only driving force for biogenic amine accumulation, when plotted as amine distribution (mV) versus the ΔpH (mV), yielded a slope of 0.38 ± 0.09.

Conversely, when chromaffin ghosts were incubated in the presence of ATP and 60 mM ammonia, a ΔΨ of up to 50 mV and no ΔpH was measured. Thus, in the second series of experiments, only a ΔΨ was present to drive amine uptake. The addition of varied external concentrations of thiocyanate to the incubation medium resulted in a range of values for the gradient from 50 to 0 mV. In the representative experiment shown in FIGURE 6, curve *B*, [^{14}C]serotonin distribution measured under identical steady-state conditions, was proportional to the magnitude of the gradient with a constant of proportionality of 0.8. The observation that amine distribution slightly exceeds the magnitude of the gradient is probably secondary to limited internal binding or oxidation of the accumulated amine. Data from 10 separate ghost prepa-

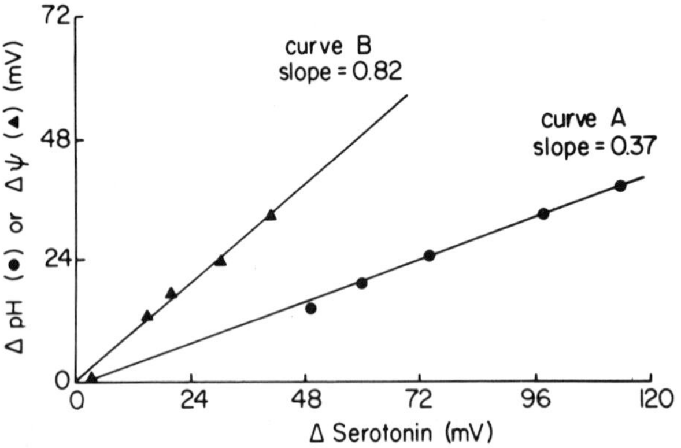

FIGURE 6. Relationship between ΔpH or ΔΨ and steady-state accumulation of [^{14}C]serotonin. *A:* ΔpH versus [^{14}C]serotonin accumulation. Chromaffin vesicles, formed in 185 mM sodium isethionate, 20 mM ascorbate plus 4 mM Tris/maleate at pH 7.0, were suspended in the incubation medium of similar composition with one half of the samples containing [^{14}C]methylamine and the other half [^{14}C]serotonin and various concentrations of NH_4^+ (0 to 30 mM). *B:* Experimental conditions were identical to **A,** except that [^{14}C]SCN^- was substituted for methylamine and sodium thiocyanate (0 to 60 mM) for NH_4^+. (For further details see Johnson *et al.*[7])

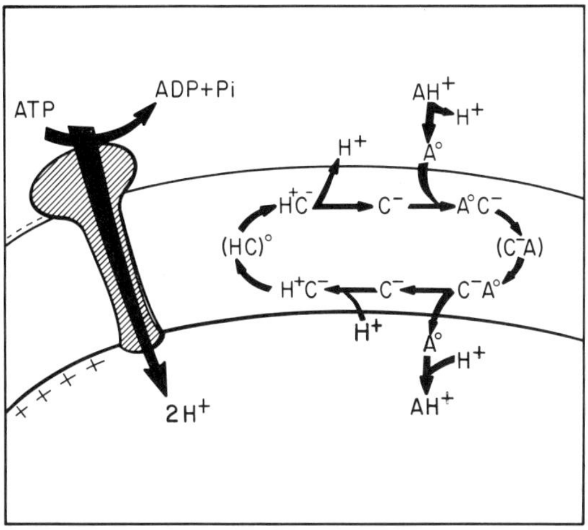

FIGURE 7. Schematic representation of catecholamine uptake into chromaffin granules coupled through energy transduction with chemiosmotic gradients. (For details see text.)

rations in which the gradient was the only driving force for amine accumulation yielded an average slope of 0.76 ± 0.06.

These results not only demonstrate that amine accumulation can be driven by either the ΔpH or ΔΨ, but in addition and of particular significance, indicate that over the same range of magnitudes of the driving force (0 to 50 mM), the accumulation of biogenic amines in the presence of a ΔpH alone is twice that observed in the presence of the ΔΨ of the same magnitude. These findings are consistent with an apparent driving force equal to $\Delta\Psi - 2Z\Delta\text{pH}$.

The physiological implications of this relationship can be explored (FIG. 7). If it is assumed that the internal pH of the chromaffin granule is 5.5 and the cytosolic pH 7.4, then a ΔpH of 1.9 pH units may exist *in vivo.* Isolated chromaffin granules can generate a membrane potential of up to 80 mV. Substituting values into the equation of $\Delta\bar{\mu}_A = \Delta\Psi - 2Z\Delta\text{pH}$, a transmembrane catecholamine gradient of 135,000 to 1 can be predicted. Generation of this enormous catecholamine gradient (by biologic standards) is based solely upon the existence of a $\Delta\bar{\mu}_{H+}$ and an amine transporter molecule able to cycle in response to the $\Delta\bar{\mu}_{H+}$. If the cytosolic concentration of catecholamines is 2 μM, as has been suggested by indirect measurements,[30] then the intragranular concentration would correspond to 270 mM, a value that approximates the concentration previously calculated from the measurements of catecholamine content.

The chemiosmotic coupling theory predicts only the thermodynamic relationships between the electrochemical proton gradient and the biogenic amine gradient. Investigation of the steady-state distribution of the biogenic amine as a function of driving force in the chromaffin granules does not provide information per se about the molecular mechanism of uptake, *i.e.,* the carrier charge, molecular species of amine transported, etc. The thermodynamic relationships, however, can be derived from modeling of the transported system. For example, it is thermodynamically favorable, because of the chemiosmotic gradients involved, for the inwardly directed complex (carrier plus substrate amine) to be neutral or negatively charged, and for the outwardly mobile complex to be neutral or positively charged and to bind protons. The initial step of the transport cycle would be hypothesized to be association of the substrate with the carrier at the external face of the membrane with the overall complex neutral or negatively charged. Equilibration of the complex between the inner membrane and outer membrane surfaces would be facilitated by a negatively charged complex because of the internally positive membrane potential, leading to dissociation of the biogenic amine from the carrier at the intragranular surface of the membrane. Association of the proton with the carrier and return of the complex to the outer membrane surface would therefore be facilitated by the membrane potential, if the carrier is neutral, or by the ΔpH, if the carrier is negatively charged. Four species of catecholamines exist: neutral, cationic, anionic, and zwitterionic. Models can be generated for each of these species (TABLE 2). For each amine species, the stoichiometry determines a unique driving force. Whereas several of the driving forces are the same for each amine species, the stoichiometry differs. Comparison of the equation produced from the empirical data with that from modeling indicates that for each amine charge there is only one carrier charge and stoichiometry that predict the driving force experimentally.

TABLE 2. Stoichiometry and Driving Force for H^+-Amine Antiport of Various Amine Species

Amine Charge	Driving Force	Carrier Charge	Stoichiometry H^+-Amine	Driving Force
A°	$n\Delta\Psi - (n+1)Z\Delta pH$	C°		$Z\Delta pH$
		C°	+1	$\Delta\Psi - 2Z\Delta pH$
		C°	+2	$2\Delta\Psi - 3Z\Delta pH$
		C^{-1}	+1	$\Delta\Psi - 2Z\Delta pH$
		C^{-1}	+2	$2\Delta\Psi - 3Z\Delta pH$
A^+	$(n-1)\Delta\Psi - nZ\Delta pH$	C^{-1}		$\Delta\Psi$
		C^{-1}	+1	$-Z\Delta pH$
		C^{-1}	+2	$\Delta\Psi - 2Z\Delta pH$
A^-	$(n+1)\Delta\Psi - (n+2)Z\Delta pH$	C°		$\Delta\Psi - 2Z\Delta pH$
		C°	+1	$2\Delta\Psi - 3Z\Delta pH$
		C°	+2	$3\Delta\Psi - 4Z\Delta pH$
$A^\pm$	$n\Delta\Psi - (n+1)Z\Delta pH$	C°		$Z\Delta pH$
		C°	+1	$\Delta\Psi - 2Z\Delta pH$
		C°	+2	$2\Delta\Psi - 3Z\Delta pH$

For example, if the anionic species were transported, the H^+-amine antiport stoichiometry would be zero, for the cation species, two, and for the neutral or zwitterionic species, one. Therefore, although chemiosmotic theory cannot distinguish the possible amine species transported, there is every indication that a unique stoichiometry is predicted by this. Thus, by actually measuring the number of protons effluxed per amine accumulated, the charge of the species transported may be determined. The possibility does exist, of course, that amine accumulation is coupled to other ion movements, but this possibility is limited by the low permeability of the membrane and the absence of significant gradients.

The effect of the internal and external pH upon the kinetics of amine accumulation has been measured in detail by several groups of investigators. The evidence supports the conclusion that it is the neutral form of the amine that is accumulated.

CHEMIOSMOTIC GRADIENTS IN OTHER NEUROENDOCRINE SECRETORY VESICLES

All neuroendocrine secretory vesicles that have been isolated and studied possess a H^+-translating ATPase within their respective membranes responsible for the generation and maintenance of chemiosmotic gradients. In the case of many of the biogenic amine-containing secretory vesicles, this electrochemical proton gradient serves as the driving force for accumulation of the biogenic amines which proceeds through a specialized membrane transporter in equilibrium with the electrochemical proton gradient. In other biogenic amine containing secretory vesicles, however, a transporter does not exist; accumulation occurs by diffusion of the uncharged amine through the lipid bilayer. At equilibrium, the concentration gradient of biogenic amines should equal the magnitude of the ΔpH. For example, if the intragranular pH of the secretory vesicle is 5.4 and the cytosolic pH is 7.4, the ΔpH would be 2.0 units and the predicted amine concentration gradient 100 to 1. Only agents or conditions which collapse or enhance the magnitude of the ΔpH affect amine distribution under these circumstances. In these particular secretory vesicles, the physiologic role of the biogenic amines is not well understood. The existence of an acidic intravesicular space may have important implications for the physicochemical state of the intravesicular peptides. For example, the solubility of the insulin-zinc complex at acidic pH values is quite low and therefore the complex precipitates, permitting an increase in the amount of insulin that can be packaged and stored (see TABLE 3). In cholinergic vesicles, a proton pump has also been demonstrated. The resulting proton electrochemical gradient is thought to drive acetylcholine accumulation. Since acetylcholine is a quaternary amine, however, the molecular mechanism of uptake must differ from that of the biogenic amines.

SUMMARY

Neuroendocrine secretory vesicles contain within their membranes a highly specialized H^+-translocating ATPase responsible for the generation and maintenance of an electrochemical proton gradient, ΔpH inside acidic, and ΔΨ inside positive. Coupled with a high internal buffering capacity and extremely low permeability of the membrane to protons, this proton pump can generate and maintain an intravesicular pH of 5.5, independent of the external pH, and transmembrane electrical potential of 60 mV. The chemiosmotic gradient has important implications for several functions of the secretory vesicles: (1) maintaining oxidizable substances (such as biogenic amines) in the unoxidized form; (2) stimulating (or inhibiting) peptide processing enzymes; (3) permitting precipitation of intravesicular protein complexes, thereby increasing the amount that can be stored within the vesicle; and (4) serving as the driving force for the uptake of certain hormones and neurotransmitters such as acetylcholine, biogenic amines, and ATP. By using the putative biogenic amine transporter as an example, it can be demonstrated that based purely upon the existence of a transporter in equilibrium with the electrochemical proton gradients, an amine concentration approaching 135,000 to 1 can be achieved. The bioenergetics of amine transport do not predict the molecular mechanism of amine translocation. By using kinetic analyses of amine accumulation under a variety of situations, however, initial information concerning the salient aspects of amine transport is being obtained.

TABLE 3. Neuroendocrine Secretory Vesicles with Chemiosmotic Gradients

	Amine	Peptide	Internal pH	Δ (mV)	H^+-ATPase	Reserpine-Sensitive Uptake	Ref.
BIOGENIC-AMINE-CONTAINING SUBCELLULAR ORGANELLE							
Chromaffin granule	NE, E	enkephalin	5.5	+60	yes	yes	11
Peripheral nerve granule	NE	enkephalin	5.5	+45	yes	yes	31
Platelet dense granule	5-HT	—	5.4	+35	yes	yes	6
Mast cell granule	histamine	heparin	5.8	—	yes	yes	32
Pancreatic β cell granule	5-HT	insulin	6.0	+60	yes	no	33
Anterior pituitary granule	DA/NE	GH/Prl	5.1	+60	yes	no	34
Posterior pituitary granule	—	VP/Oxy	5.5	+35	yes	no	35
ACETYLCHOLINE-CONTAINING SUBCELLULAR ORGANELLE							
Acetylcholine granules	Ach	enkephalin	5.0	+40	yes	no (AH5183)	36

REFERENCES

1. GREVILLE, G. D. 1969. Curr. Top. Bioenerg. **3:** 1-79.
2. KABACK, H. R. 1972. Biochim. Biophys. Acta **265:** 367-417.
3. NICHOLLS, D. 1982. Bioenergetics: An Introduction to the Chemiosmotic Theory. Academy Press. London.
4. MITCHELL, P. 1966. Chemiosmotic Coupling in Oxidative and Photosynthetic Phosphorylation. Glynn Research. Bodmin, Cornwall.
5. CARTY, S. E., R. G. JOHNSON & A. SCARPA. 1985. *In* The Enzymes of Biological Membranes, Vol. 3. A. Martonosi, Ed.: 449-495. Plenum. New York.
6. CARTY, S. E., R. G. JOHNSON & A. SCARPA. 1981. J. Biol. Chem. **256:** 11244-11250.
7. JOHNSON, R. G., S. E. CARTY & A. SCARPA. 1981. J. Biol. Chem. **256:** 5733-5780.
8. NJUS, D., J. KNOTH & M. ZALLAKIAN. 1981. Curr. Top. Bioenerg. **13:** 107-145.
9. NJUS, D. & G. K. RADDA. 1978. Biochim. Biophys. Acta. **463:** 219-244.
10. PHILLIPS, J. H. 1978. Biochem. J. **170:** 673-679.
11. JOHNSON, R. G. & A. SCARPA. 1976. J. Gen. Physiol. **68:** 601-631.
12. ROTTENBERG, H. 1975. J. Bioenerg. **7:** 61-64.
13. BASHFORD, C. L. *et al.* 1976. Neuroscience **1:** 399-412.
14. POLLARD, H. B. *et al.* 1976. J. Biol. Chem. **251:** 4544-4550.
15. SALAMA, G., R. G. JOHNSON & A. SCARPA. 1979. J. Gen. Physiol. **75:** 109-140.
16. CASEY, R. P. *et al.* 1977. Biochemistry **16:** 972-977.
17. NJUS, D. *et al.* 1978. Biochemistry **17:** 4337-4343.
18. POLLARD, H. B. *et al.* 1979. J. Biol. Chem. **254:** 1170-1177.
19. JOHNSON, R. G. & A. SCARPA. 1979. J. Biol. Chem. **254:** 3750-3760.
20. HOLZ, R. W. 1978. Proc. Natl. Acad. Sci. USA **75:** 5190-5194.
21. PHILLIPS, J. H. & Y. P. ALLISON. 1978. Biochem. J. **170:** 661-672.
22. JOHNSON, R. G., N. CARLSON & A. SCARPA. 1978. J. Biol. Chem. **253:** 15120-15121.
23. PHILLIPS, J. H. 1977. Biochem. J. **186:** 289-297.
24. INGEBRETSON, O. C. & T. FLATMARK. 1979. J. Biol. Chem. **254:** 3833-3839.
25. JOHNSON, R. G., M. F. BEERS & A. SCARPA. 1982. J. Biol. Chem. **257:** 10701-10707.
26. FLATMARK, T. & O. C. INGEBRETSON. 1977. FEBS Lett. **78:** 53-56.
27. ROISIN, M. P., D. SCHERMAN & J.-P. HENRY. 1980. FEBS Lett. **115:** 143-146.
28. CIDON, S. & N. NELSON. 1985. J. Biol. Chem. **258:** 2892-2898.
29. JOHNSON, R. G., S. CARTY & A. SCARPA. 1981. J. Biol. Chem. **257:** 10701-10707.
30. PERLMAN, R. L. & B. E. SHEARD. 1982. Biochim. Biophys. Acta **719:** 334-340.
31. MARON, R., B. I. KANNER & S. SCHULDINER. 1979. FEBS Lett. **98:** 237-240.
32. JOHNSON, R. G., S. E. CARTY, B. J. FINGERHOOD & A. SCARPA. 1980. FEBS Lett. **120:** 75-79.
33. HUTTON, J. C. 1982. Biochem. J. **204:** 171-178.
34. CARTY, S. E., R. G. JOHNSON & A. SCARPA. 1982. J. Biol. Chem. **257:** 7269-7273.
35. RUSSELL, J. T. & R. W. HOLZ. 1981. J. Biol. Chem. **256:** 5950-5953.
36. MICHAELSON, D. M. & I. ANGEL. 1980. Life Sci. **27:** 39-44.

The H^+-Translocating ATPase of Chromaffin Granule Membranes[a]

DAVID K. APPS AND JUDITH M. PERCY

Department of Biochemistry
University of Edinburgh Medical School
Edinburgh EH8 9XD, United Kingdom

1. INTRODUCTION

The ATP-dependent uptake of catecholamines by bovine adrenal chromaffin granules has been studied in detail, both in intact granules and in resealed chromaffin granule membrane "ghosts." Kinetic and thermodynamic studies have shown that the very large transmembrane concentration gradients are achieved by chemiosmotic coupling between an electrogenic H^+-translocating ATPase and a separate electrogenic catecholamine-H^+ antiporter.[1] In contrast to these bioenergetic studies, progress in the molecular characterization of the transport system has proved slow and controversial.[2] Treatment of chromaffin granule membranes with organic solvents was found to release an ATPase that was closely similar to the F_1-ATPase of mitochondria.[3] Subsequently[4] it was shown that this enzyme could be removed by NaBr treatment of the membranes, with retention of H^+-translocating activity. The F_1-like ATPase must therefore be a contaminant. Other separation methods have now been devised, in which solubilization of the H^+-translocator of chromaffin granule membranes is accompanied by its separation from contaminating mitochondrial ATPase; this has permitted preliminary studies of its structure, which has been found to be unlike those of mitochondrial or typical plasma membrane ATPases, but rather of a third type that may be characteristic of secretory vesicles and other acidic organelles.

2. CHROMAFFIN GRANULE MEMBRANES CONTAIN TWO ATPases

Solubilization of the ATPase activity of chromaffin granule membranes depends critically on the type and concentration of the detergent used; we have obtained good results with polyethyleneglycol dodecyl ethers ($C_{12}E_8$ and $C_{12}E_9$), using an initial detergent:protein ratio of 5:1. Exclusion chromatography of detergent-solubilized chromaffin granule membranes, on Sephacryl S-300 or Ultrogel AcA-22 columns in 0.1%

[a]This work was supported by research grants from the Medical Research Council of the United Kingdom.

$C_{12}E_8$, gives good resolution of the membrane proteins, and yields two peaks of ATPase activity, each containing a different set of polypeptides.[5] Inhibitor studies confirm that the two ATPase peaks contain different enzymes, and sensitivity to *N*-ethylmaleimide (NEM) and to vanadate (see section 4) have been used as diagnostic of the high- and low-molecular weight ATPases, respectively, hereafter termed ATPase I and ATPase II.

3. PREPARATIVE SEPARATION OF THE TWO ATPases

The nonionic detergent Triton X-114 efficiently solubilizes most membrane proteins; on warming of the solution to 30°, a phase separation occurs, which can be used as the basis of a fractionation procedure that separates intrinsic and extrinsic membrane proteins.[6] Chromaffin granule membrane ATPase activity is retained on solubilization with Triton X-114, and the solution yields three fractions[7,8]: (1) a precipitate that forms on standing, containing about 10 polypeptides and having NEM-sensitive ATPase activity; (2) a detergent-rich phase, containing the major membrane proteins dopamine β-hydroxylase and cytochrome b_{561}, and having vanadate-sensitive ATPase activity; (3) an aqueous phase, containing secretory proteins such as chromogranins and soluble dopamine β-hydroxylase, and having no ATPase activity. Triton X-114 fractionation thus provides a rapid and convenient method of completely separating ATPases I and II. An alternative method was discovered during purification of cytochrome b_{561}[9]: after solubilization with $C_{12}E_8$, addition of cholate and ammonium sulfate (0.8 M) results in the separation of a lipid-rich fraction containing ATPase I, while ATPase II remains in solution.[7] ATPase I purified by this method contains fewer polypeptides than the fraction, of similar specific activity, obtained by the Triton X-114 procedure (see section 6). The soluble fraction is used for the purification of ATPase II (L. F. McCallum, unpublished).

The distribution of mitochondrial F_1-ATPase in these fractions was investigated using an antiserum raised against beef heart F_1.[7] In each case, the ATPase I fraction contained no detectable F_1; the mitochondrial contaminant appeared in the same fraction as ATPase II.

4. PROPERTIES OF THE TWO ATPases

The effects of various inhibitors on the separated ATPases are shown in TABLE 1. ATPase I is sensitive to a number of inhibitors in common with mitochondrial H^+-translocating ATPase, such as *N,N'*-dicyclohexylcarbodiimide (DCCD), 4-chloro-7-nitro-benzofurazan (NbfCl), quercetin and tributyl tin, but not to specific inhibitors of F_1, such as efrapeptin, or F_0, such as oligomycin, and it is distinguishable by its sensitivity to NEM. ATPase II is most sensitive to vanadate, but is also inhibited to some extent by tributyl tin. The inhibitors NEM, quercetin, and vanadate have been used in experiments to determine the function of ATPase I (section 5).

The substrate specificity of the two enzymes was also investigated.[7] The ATPase activity of chromaffin granules is relatively nonspecific; a number of nucleoside triphosphates are hydrolyzed, ITP being the best substrate after ATP.[10] This is reflected in the properties of purified ATPase I, which also hydrolyzes ITP, GTP, and dGTP, this activity being inhibited by NEM. The partially purified ATPase II preparation also has ITPase activity, which is, however, insensitive to vanadate; it therefore appears not to be due to ATPase II itself.

5. ATPase I IS A PROTON PUMP

The H^+-translocating activity of chromaffin granule membrane ATPase has been demonstrated by measuring ATP-dependent acidification of liposomes containing the reconstituted enzyme.[11] The availability of inhibitors that discriminate between the two chromaffin granule ATPases allowed us to perform experiments with resealed chromaffin granule "ghosts," which contain the catecholamine translocator and therefore also carry out ATP-dependent catecholamine uptake.

TABLE 1. Effects of Various Inhibitors on the ATPase Activity of Chromaffin Granule Membranes, and on Partially Purified ATPases I and II

Inhibitor	Concentration (μM)	Inhibition (%) Membranes	ATPase I	ATPase II
Efrapeptin	2	9	0	7
Oligomycin	5	2	0	0
DCCD	25	55	80	5
NEM	20	40	95	0
NbfC1	25	84	84	4
Quercetin	30	36	78	15
Tributyl tin	2	72	98	29
Vanadate	2	22	9	60

FIGURE 1 shows the effect of quercetin on the ATPase activity of unfractionated chromaffin granule membranes, and on separated ATPase I and ATPase II. Quercetin selectively inhibits ATPase I, and ATP-driven uptake of 5-hydroxytryptamine (a good substrate for the catecholamine transporter) is also strongly inhibited (FIG. 1), in parallel with the inhibition of ATPase I. This suggests that ATPase I is the H^+-translocator that establishes the transmembrane ΔpH used in catecholamine uptake.

Measurement of the steady-state ΔpH by methylamine distribution (FIG. 2) shows that this is significantly affected by high concentrations of quercetin, when ATPase I is extensively inhibited, but that low concentrations have little effect. Catecholamine uptake is measured during the first 20 seconds after addition of substrates to the "ghosts," during the approach to the steady state, and is presumably inhibited because quercetin reduces the ΔpH during this period. Quercetin does not inhibit the amine transporter itself, as shown in experiments in which uptake of 5-hydroxytryptamine was driven by a ΔpH (about 2.1) imposed by a rapid pH jump, and in which quercetin up to 100 μM had no effect on the rate of amine uptake.

A more direct demonstration of H^+-translocation is given by the quenching of the fluorescence of the permeant weak base 9-amino-6-chloro-2-methoxyacridine (ACMA)

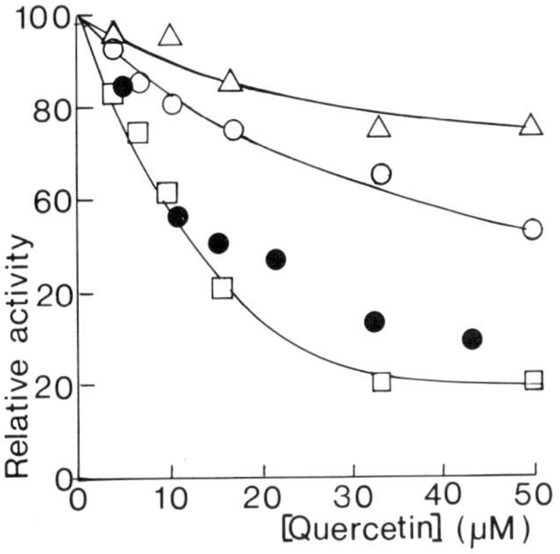

FIGURE 1. Effects of quercetin. Chromaffin granule membrane ATPase activity (○), ATPase I activity (□), and ATPase II activity (△) were measured by a coupled spectrophotometric assay, with 2 mM ATP, 10 mM Mg^{2+}, pH 7.0. Uptake of 5-hydroxytryptamine into resealed chromaffin granule "ghosts" (●) was measured by a membrane filtration assay with 58 μM 5-hydroxytryptamine, 6 mM ATP, 3 mM Mg^{2+}, pH 7.0, in 0.3 M sucrose, pH 7.0.

as it accumulates within acidic membrane vesicles. Quenching of ACMA fluorescence by chromaffin granule "ghosts" is dependent on ATP (or ITP) and can be completely blocked by NEM (FIG. 3); vanadate has no effect. These results confirm that H^+-translocation into the "ghosts" is brought about by ATPase I, and that ATPase II is not a proton pump.

6. SUBUNIT COMPOSITION OF ATPase I

The first (ATPase I) fraction from Triton X-114 fractionation of chromaffin granule membranes contains about 10 major polypeptides, as judged by SDS polyacrylamide gel electrophoresis (FIG. 4, lane 2), of which only six appear in the ammonium-sulfate insoluble fraction from the alternative $C_{12}E_8$ fractionation procedure (lane 5), which also contains ATPase I.

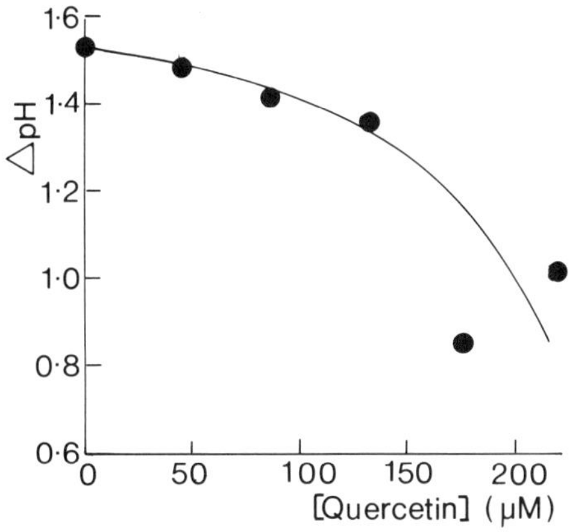

FIGURE 2. Effect of quercetin on the steady-state, ATP-dependent ΔpH in resealed chromaffin granule "ghosts," calculated from the distribution of [^{14}C]methylamine in presence of 6 mM ATP, 3 mM Mg^{2+}, 40 mM KI, pH 7.0.

The largest of these bands (about 150 kDa) is not present in unfractionated membranes (FIG. 4, lane 1) and disappears when the sample is heated to 100° before electrophoresis; it is therefore probably an aggregate of smaller components.

The other five bands (apparent molecular weights: 70,000, 57,000, 40,000, 33,000, and 16,000) are all fairly major components of chromaffin granule membranes, each composing 2-3% of the total protein. They copurify through other procedures such as exclusion chromatography, hydrophobic chromatography, and density gradient centrifugation (unpublished results), but at present only the largest (70 kDa) and the smallest (apparently 16 kDa, but probably 7 kDa: see section 9) have been shown, by the use of covalently binding inhibitors, to be definitely part of the H^+-ATPase complex (see section 7).

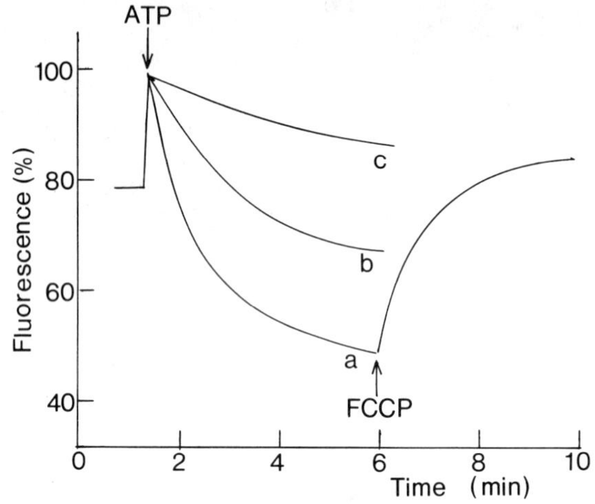

FIGURE 3. ATP-dependent quenching of ACMA fluorescence by resealed chromaffin granule "ghosts"; external pH 7.6. ATP (3 mM) and FCCP (5 μM) were added at the points shown. "Ghosts" were preincubated (5 min, 20°) with 0.5 μM ACMA and NEM at various concentrations: 0 (*a*); 10μM (*b*); 20 μM (*c*).

7. CHEMICAL LABELING OF ATPase I SUBUNITS

Among the various inhibitors of ATPase I (TABLE 1), both NEM and DCCD react covalently with polypeptides, and can therefore be used to identify the sites of inhibition, although both are potentially nonspecific in their reaction. We have used these inhibitors, in radiolabeled form, to examine the functions of the ATPase I subunits.

The ATPase activity of chromaffin granule membranes is maximally inhibited about 50% by NEM, with complete inhibition of H^+-translocation[12]; this is due to inactivation of ATPase I, the residual activity being due to ATPase II. Inactivation of purified ATPase I by NEM is shown in FIGURE 5; at this low NEM concentration (10 μM) complete inactivation is not achieved, although it does occur at higher

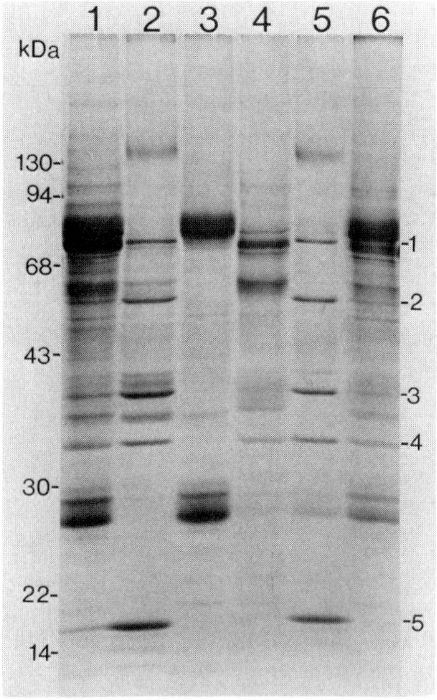

FIGURE 4. Fractionation of solubilized chromaffin granule membrane proteins: Coomassie-blue-stained SDS polyacrylamide gel of fractions from two different separation procedures. Lane 1, chromaffin granule membranes. Lanes 2-4, Triton X-114 fractionation (insoluble, detergent-rich and aqueous phases). Lanes 5 and 6, $C_{12}E_8$/cholate/ammonium sulfate fractionation (insoluble and soluble phases). Positions of molecular-weight standard proteins are shown on the left. Numbers on the right indicate the positions of the ATPase 1 subunits.

concentrations. Inclusion of ATP (2 mM) affords almost complete protection from inactivation. When chromaffin granule membranes are incubated with low concentrations (1-10 μM) of [^{3}H]NEM, a 70-kDa polypeptide becomes heavily labeled (FIG. 6, lane 1); fractionation of the labeled membranes with Triton X-114 confirms that this is the largest subunit of ATPase I. Densitometric scanning of the autoradiograph shows that about 75% of the incorporated label is in this one band, and that this labeling is reduced by about 50% by ATP, in short incubations. During more prolonged

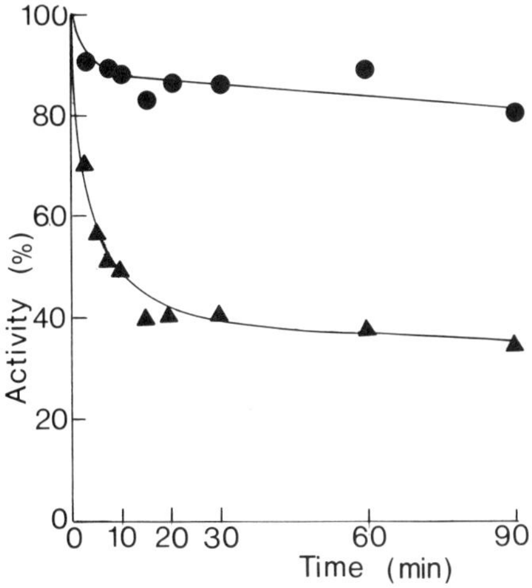

FIGURE 5. Time-course of inhibition of partially purified ATPase I by NEM (10 μM) in the absence of ATP (▲) or in the presence of 2 mM ATP (●).

exposure to NEM, however, the effect of ATP is less marked, suggesting that NEM may react rapidly with a group (probably an -SH group) in the ATP binding site, and more slowly with another group elsewhere in the polypeptide.

Treatment of the membranes with [^{14}C]DCCD labels principally a 16-kDa polypeptide (FIG. 6, lane 5); this also copurifies with ATPase I. Labeling of this band is enhanced by ATP.[13]

Despite their reactivity, NEM and DCCD have proved specific enough to label the susceptible subunits of ATPase I rather selectively when used in low concentrations with intact chromaffin granule membranes. They are much less specific, however, in their labeling of solubilized proteins. When purified ATPase I is treated with [^{3}H]NEM, the 70-kDa, 57-kDa, and 40-kDa bands become labeled, and [^{14}C]DCCD produces similar nonspecific labeling of the solubilized enzyme.

8. MEMBRANE TOPOGRAPHY OF ATPase I

The H^+-ATPase of chromaffin granules is inwardly directed, hydrolyzing extragranular ATP; the active site of ATPase I should therefore be exposed on the outer (cytoplasmic) face of the granule. The disposition of the ATPase I polypeptides has been tested by the use of impermeant protein labeling agents, and by proteolysis.

Lactoperoxidase-catalyzed radioiodination of intact chromaffin granules labels numerous membrane proteins, which are best resolved by two-dimensional gel electrophoresis (FIG. 7). They include cytochrome b_{561} and subunits 1 (70 kDa) and 2 (57 kDa) of ATPase I, this identification being confirmed by electrophoresis and autoradiography after Triton X-114 fractionation of the labeled membranes (FIG. 8). The exposure of these subunits on the cytoplasmic face of the granule was confirmed by

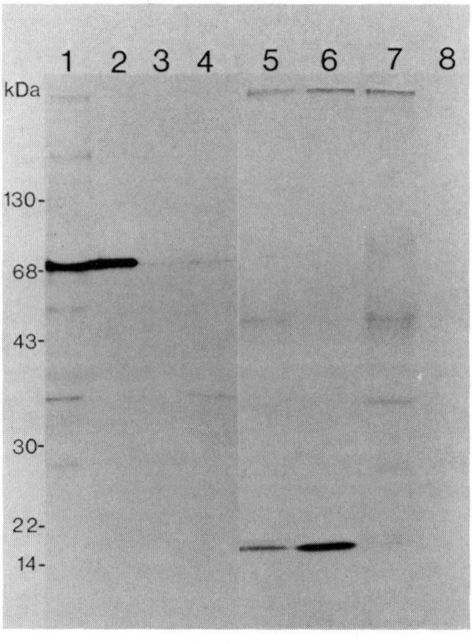

FIGURE 6. Chemical labeling of ATPase I subunits. Purified chromaffin granule membranes were treated with 10 μM [^{3}H]NEM (lanes 1-4) or with [^{14}C]DCCD (lanes 5-8), fractionated with Triton X-114, and the fractions subjected to SDS polyacrylamide gel electrophoresis and autoradiography. Lanes 1 and 5, unfractionated membranes; lanes 2 and 6, insoluble fraction; lanes 3 and 7, detergent-rich fraction; lanes 4 and 8, aqueous fraction.

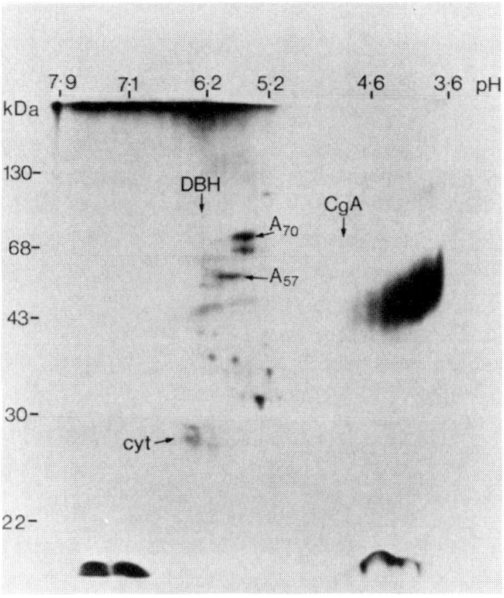

FIGURE 7. Radioiodination of proteins in chromaffin granules. Intact chromaffin granules (1 mg protein/ml) were labeled for 10 min at 20° with ^{125}I (0.1 mCi/ml) in presence of lactoperoxidase, glucose oxidase, and glucose. After further purification of the granules by centrifugation through 1.7 M sucrose, granule membranes were prepared and subjected to two-dimensional polyacrylamide gel electrophoresis, followed by autoradiography. The positions of ATPase 1, subunits 1 and 2, and of cytochrome b_{561} are shown. Note the absence of label in the membrane protein dopamine β-hydroxylase, and in the secretory protein chromogranin A, the positions of which are also shown.

their rapid degradation during trypsin treatment of intact granules (not shown). In addition, subunit 3 (40 kDa) is labeled by lactoperoxidase/$^{125}I^-$; this protein focuses poorly, and its identification in FIGURE 6 is therefore not possible. Its labeling is, however, clearly shown in one-dimensional electrophoresis (FIG. 8). Subunits 4 (33 kDa) and 5 (16 kDa) are not labeled, nor are they degraded by proteolytic treatment of chromaffin granules.

9. DISCUSSION

Apart from contaminating mitochondrial ATPase, chromaffin granule membrane preparations contain two ATPases that differ in their structure and inhibitor sensitivity. Several ways of separating them have been devised, the most rapid and convenient being phase separation after solubilization with Triton X-114.

The function of ATPase II is unknown. It is clearly not a proton pump, but may translocate some other ion. Until it is better characterized we cannot even be sure that it is a genuine granule constituent, but it is noteworthy that secretory vesicles of the posterior pituitary also contain two ATPases, similar in properties to those of chromaffin granules.[14]

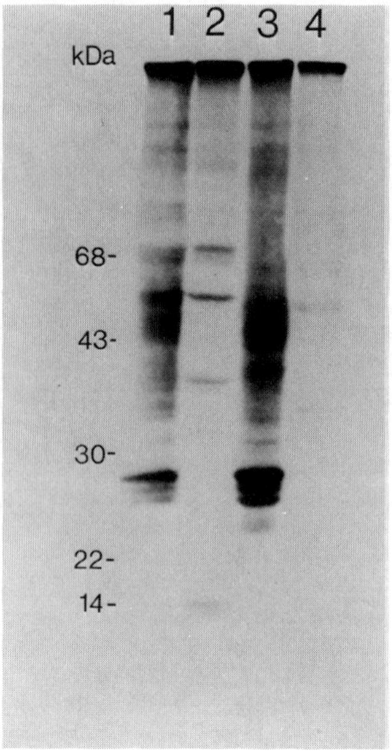

FIGURE 8. Triton X-114 fractionation of ^{125}I-labeled chromaffin granule membrane proteins. After labeling of intact granules (see FIG. 6), membranes were prepared and fractionated, and the fractions subjected to SDS polyacrylamide gel electrophoresis and autoradiography. Lanes are as in FIGURE 6; note the labeling of ATPase 1 subunits 1, 2, and 3 (lane 2) and of cytochrome b_{561} (lane 3).

The structure and function of ATPase I have been defined in more detail. ATP-dependent H^+-translocation and uptake of 5-hydroxytryptamine by resealed chromaffin granule "ghosts" are inhibited by NEM and quercetin, both relatively specific inhibitors of ATPase I. This confirms this enzyme's function as a proton pump, whereas the insensitivity of these processes to vanadate shows that ATPase II does not translocate protons.

ATPase I is structurally complex. It may contain as many as five types of subunit, since the five polypeptides discussed below copurify through a number of fractionation procedures. It is not yet clear, however, whether they all belong to the H^+-ATPase complex. Their properties are summarized in TABLE 2.

TABLE 2. Properties of ATPase I Subunits

Subunit	MW	Covalent Binding	[^{125}I]Lactoperoxidase Labeling	Trypsin Digestion
1	70,000	NEM	+	+
2	57,000		+	+
3	40,000		+	+
4	33,000		−	−
5	7,000	DCCD	−	−

Experiments with NEM and DCCD were performed with chromaffin granule membranes.

Subunit 1 (70 kDa) contains the ATP-hydrolyzing site, since it is specifically labeled by [^{3}H]NEM, whereas ATP inhibits this labeling and protects the ATPase from inactivation. Subunit 5 is the DCCD-reactive protein already characterized.[13] Its apparent molecular weight is 16,000, but since it has a higher electrophoretic mobility than its 8-kDa mitochondrial counterpart, its true molecular weight is probably about 7000, as suggested by amino-acid analysis. It is believed to be a constituent of the H^+-ATPase because DCCD inhibits ATPase activity and H^+-translocation, and ATP increases the rate of labeling of this protein by [^{14}C]DCCD. Like the DCCD-reactive subunit of mitochondrial ATPase, it may be part of a H^+-conducting channel, since low concentrations of DCCD enhance the ATP-dependent ΔpH in chromaffin granule "ghosts," suggesting an inhibition of passive H^+-conductance.[15]

The functions of the remaining three components (57 kDa, 40 kDa, and 33 kDa) are unknown, as is the subunit stoichiometry. In principle, chemical labeling might be used to determine the membrane concentrations of subunits 1 and 5, but there are uncertainties about the reaction stoichiometries of both NEM and DCCD.

ATPase I is clearly more complex than the simple H^+-ATPases of fungal plasma membranes.[16] In its size and structure, it bears a superficial resemblance to mitochondrial F_1F_0-ATPase, but it does not bind anti-F_1 serum, and there are also differences in inhibitor sensitivity (notably to NEM). A closer parallel is with the H^+-ATPases found in the vacuolar membranes of fungi and higher plants.[17–21] These have two large subunits (67-89 kDa and 57-64 kDa) and a smaller, DCCD-reactive subunit (14-18 kDa), and may therefore be similar in structure to chromaffin granule membrane ATPase I, although affinity labeling studies[18] suggest that there is an ATP-binding site in the 57-kDa subunit. Although not yet characterized structurally, the H^+-ATPases of Golgi membranes, coated vesicles, and lysosomes show similar properties of inhibitor sensitivity, and may also be related. Perhaps continued study of ATPase I of chromaffin granules will illuminate these less accessible ATPases.

REFERENCES

1. NJUS, D., J. KNOTH & M. ZALLAKIAN. 1981. Curr. Top. Bioenerg. **11:** 107-147.
2. WINKLER, H., D. K. APPS & R. FISCHER-COLBRIE. 1986. Neuroscience. **18:** 261-290.
3. APPS, D. K. & G. SCHATZ. 1979. Eur. J. Biochem. **100:** 411-419.
4. CIDON, S. & N. NELSON. 1983. J. Biol. Chem. **258:** 2892-2898.
5. APPS, D. K., J. G. PRYDE & R. SUTTON. 1983. Neuroscience **9:** 687-700.
6. BORDIER, C. 1981. J. Biol. Chem. **256:** 1604-1607.
7. PERCY, J. M., J. G. PRYDE & D. K. APPS. 1985. Biochem. J. **231:** 557-564.
8. PRYDE, J. G. & J. H. PHILLIPS. 1986. Biochem. J. **233:** 525-533.
9. APPS, D. K., J. G. PRYDE & J. H. PHILLIPS. 1980. Neuroscience **5:** 2279-2287.
10. KIRSHNER, N. 1972. J. Biol. Chem. **237:** 2311-2317.
11. CIDON, S., H. BEN-DAVID & N. NELSON. 1983. J. Biol. Chem. **258:** 11684-11688.
12. FLATMARK, T., M. GRONBERG, E. HUSEBYE & S. V. BERGE. 1982. FEBS Lett. **149:** 71-74.
13. SUTTON, R. & D. K. APPS. 1981. FEBS Lett. **130:** 103-106.
14. RUSSELL, J. T. 1984. J. Biol. Chem. **259:** 9496-9507.
15. APPS, D. K., J. G. PRYDE, R. SUTTON & J. H. PHILLIPS. 1980. Biochem. J. **190:** 273-282.
16. GOFFEAU, A. & C. W. SLAYMAN. 1981. Biochim. Biophys. Acta **639:** 197-223.
17. UCHIDA, E., Y. OHSUMI & Y. ANRAKU. 1985. J. Biol. Chem. **260:** 1090-1095.
18. MANOLSON, M. F., P. A. REA & R. J. POOLE. 1985. J. Biol. Chem. **260:** 12273-12279.
19. MANDALA, S. & L. TAIZ. 1985. Plant Physiol. **78:** 327-333.
20. LICHKO, L. P. & L. A. OKOROKOV. 1985. FEBS Lett. **187:** 349-353.
21. RANDALL, S. K. & H. SZE. 1986. J. Biol. Chem. **261:** 1364-1371.

DISCUSSION OF THE PAPER

S. CIDON (*Rockefeller University, New York, N.Y.*): First, while working with Dr. Nathan Nelson we have purified the proton ATPase from chromaffin granule membranes and in our hands the purified enzyme is composed of four major subunits, of 115, 72, 51, and 39 kDa. After purification, subunits 115 and 39 kDa can be labeled with [^{14}C]*N*-ethylmaleimide and subunit 115 binds [^{14}C]DCCD. Our current thinking is that subunit 115 kDa is the active site of the enzyme.

Second, we have not been able to find ATPase activity which was sensitive to vanadate in purified chromaffin granule membranes.

D. K. APPS (*University of Edinburgh Medical School, Edinburgh, U.K.*): In response to your first question, we do not find the 115-kDa polypeptide either in the purified ATPase or in the granule membranes. This may be because we do not routinely use protease inhibitors during the purification, but we have never found that inclusion of these inhibitors makes a difference. The fact that our ATPase is active without this polypeptide suggests that it may be unnecessary, or it may be the aggregate of other bands, which we have found under some conditions. One other difference is that we have labeled the membranes with [^{3}H]NEM or [^{14}C]DCCD, then fractionated them, whereas you used these reagents on the purified ATPase. This may account for the difference in labeling specificity.

In response to your second question: In chromaffin granule membranes, residual catecholamines reduce vanadate to vanadyl, which is not inhibitory. To see inhibition of the ATPase actively by vanadate, you must first solubilize and fractionate.

E. W. WESTHEAD (*University of Massachusetts, Amherst, Mass.*): Have you tried to compare quantitively the effect of ATP on inhibition of NEM labeling and NEM inhibition of the enzyme? I have tried different ATP concentrations in the hope of showing a good correspondence between the labeling and inhibition at less than maximal inhibitory concentration.

D. K. APPS: We have not found this. We have found that 2 mM ATP protects the activity to a greater extent than it inhibits labeling of subunit 1 by [^{3}H]NEM. This suggests that there may be more than one site of labeling within this subunit.

B. WIEDENMANN (*Institute of Cell and Tumor Biology, Heidelberg, Federal Republic of Germany*): How many copies of a proton pump do you expect per granule? I am asking since you consider one or several of the major integral membranes of a chromaffin granule membrane as candidate(s) for the subunits of the H^+-ATPase, using rather unspecific compounds like NEM and lactoperoxidase for the identification. Our work on the proton pump of coated vesicles, as well as work on other membranes in many laboratories, show clearly that the proton pump is only a minor membrane component, representing very few copies per granule membrane.

D. K. APPS: On the basis of Coomassie staining, subunit 1 is 1-2% of the membrane protein, equivalent to 30-60 copies per granule. There may, of course, be more than one copy per ATPase molecule. We are wary of using a potentially nonspecific label, such as NEM. Our justification is that (1) it labels essentially only one polypeptide in intact granule membranes; (2) ATPase activity is inhibited under the same conditions, and (3) ATP protects the activity and inhibits labeling. Lactoperoxidase-catalyzed iodination is, of course, not a specific label for ATPase or anything else. It was used to show which polypeptides are exposed on the cytoplasmic face of the granule.

The Amine Transporter from Bovine Chromaffin Granules: Photolabeling and Partial Purification[a]

S. SCHULDINER, R. GABIZON, Y. STERN, AND R. SUCHI

Department of Molecular Biology
Hebrew University
Hadassah Medical School
Jerusalem, Israel

PHOTOAFFINITY LABELS: AZIDO DERIVATIVES OF 5HT

Photoactive derivatives have been used to identify the transporter of bovine chromaffin granules,[1] pig platelet storage organelles, rat brain synaptic vesicles, and 5HT-containing organelles in rat basophilic leukemia cells (Gabizon *et al.*, in preparation and see also below).

The specificity of the first one, 4-azido, 3-nitrophenylazo-[5HT] (ANPA-5HT), is supported by the following criteria. (1) ANPA-5HT is a competitive inhibitor of amine transport in the dark. The apparent K_i is identical to the apparent K_m of transport of 5HT, its parent compound. The latter indicates that the transporter recognizes the modified substrate. (2) Upon illumination, ANPA-5HT photoinactivates the amine transport. Transport of unrelated neurotransmitters or the generation of a pH gradient across the membrane are not inhibited.[1] (3) The rate of photoinactivation is lower in the presence of other substrates of the transporter. The concentrations required to "protect" the transporter correlate well with the known affinities of the various substrates. From the dependence of the rate of photoinactivation on the concentration of ANPA-5HT an apparent kinetic constant can be calculated. This constant is similar to the apparent K_i. The agreement suggests that the inactivation of transport is a result of a sequence of events which starts when ANPA-5HT binds to the transporter and is thereafter activated by light. (4) The rate of labeling of a membrane suspension with [^{3}H]ANPA-5HT in the light is lower in the presence of other substrates of the transporter.

Recently, ANPA-5HT has been iodinated with ^{125}I. The resulting compound displays properties almost identical to those of ANPA-5HT and dramatically shortens the time required for visualization of the label. The pattern of labeling of the membrane

[a] The studies performed by our laboratories were supported by NIH grant no. NS16708.

polypeptides has been analyzed by separating them by SDS gel electrophoresis and subjecting the gel to fluorography or autoradiography. More than 80% of the label in the protein fraction is associated with a 48-kDa polypeptide and the labeling is inhibited by reserpine and by other substrates of the transporter. The concentration of reserpine required to inhibit labeling by 50% is similar to its apparent K_i of transport.[1,2]

Extraction of chromaffin granule membranes with 1% cholate renders a soluble fraction that can be reconstituted into liposomes.[3] The proteoliposomes thus obtained display the transporter activity. We have now found that in the presence of the detergent at concentrations above 0.5%, labeling of the 48-kDa polypeptide is inhibited. A second polypeptide (apparent M_r-56,000) is labeled under these conditions. The concentrations of the substrates required to prevent labeling are higher than those required in the intact membrane. It is worth mentioning that in at least two other detergents, octylglucoside and Triton X-100, labeling of the 48-kDa peptides is not inhibited even at concentrations that achieve full solubilization (Schuldiner *et al.*, unpublished observations).

An azido derivative of tetrabenazine, a specific inhibitor of transport, has been synthesized.[4] The compound, [^{3}H]TBA, binds reversibly to the granule membranes in the dark with a K_d of about 50 nM and a density of sites of 40-50 pmol/mg protein, consistent with reported densities of reserpine and dihydrotetrabenazine binding sites. Upon irradiation, TBA bound irreversibly to a polypeptide with an apparent M_r of 70,000. Since TBA and TBZ compete for the same binding site (R2 type), the authors suggest that the 70-kDa polypeptide is the R2 binding site.[4] This suggestion is supported by radiation inactivation studies in which it was found that the apparent M_r of the TBZ binding site is 65,000.

PURIFICATION OF THE TRANSPORTER

The putative amine transporter from bovine adrenal chromaffin granules has been partially purified in a single step utilizing affinity chromatography. A 5-hydroxytryptamine moiety has been coupled to a Sepharose-4B matrix in a position ortho to the hydroxy group. When membranes solubilized with sodium cholate are chromatographed on the above matrix, a polypeptide is highly enriched. The enrichment is dependent on the presence of the proper ligand on the matrix and is inhibited if the column is previously equilibrated with a soluble ligand. Enrichment of the above polypeptide is accompanied by an increase in the specific activity of the transporter as measured by its labeling by 4-azido, 3-nitrophenyl-azo-(5-hydroxytryptamine). The ability of reserpine, a competitive inhibitor of binding and transport, to inhibit labeling of the purified transporter correlates well with its known kinetic constants in the native membranes.

The polypeptide purified was first thought to be identical to the one previously identified as the putative transporter based on specific labeling by a photoaffinity label.[6] However, this conclusion has now been discarded on the basis of a more detailed analysis of the electrophoretic pattern and the realization that (1) in the presence of 1% sodium cholate, a 56-kDa peptide but not the 48-kDa is labeled with [^{125}I]ANPA-5HT, and (2) an antibody raised against the purified polypeptide cross-reacts both in the membrane and in cholate extracts with a protein that has a mobility of the 56-kDa and not the 48-kDa polypeptide.

AMINE STORAGE IN OTHER ORGANELLES

In addition to the catecholamine-rich chromaffin granules of adrenal medulla, other intracellular organelles from a wide variety of secretory cells accumulate biogenic amines. These include the adrenergic synaptic vesicles[7–9]; 5-hydroxytryptamine-containing secretory granules in platelets, enterochromaffin cells, and serotoninergic neurons[10–17]; dopamine-rich vesicles in dopaminergic neurons[13]; and histamine-containing granules in mast cells and basophils.[14] Where it has been studied, the mechanism of amine accumulation into isolated storage organelles or membrane vesicles derived therefrom always involves exchange of extravesicular amine with intravesicular H^+, catalyzed by a reserpine-sensitive transporter.[7,9,15–17] A H^+-pumping ATPase in the organelle membrane generates an electrochemical gradient of H^+ (acid and positive inside), which provides the driving force for amine-H^+ exchange. All the organelles studied display the same inhibitor sensitivity, irrespective of the amine normally stored within.[7] These findings have led to the proposal that a closely similar or identical protein catalyzes amine-H^+ exchange in all biogenic amine storage organelles.[15]

Binding of TBZOH and the apparent K_m for 5HT transport were measured in mice brain synaptic vesicles. The values obtained (3 nM and 0.8 μM) are very similar to the values measured in adrenal chromaffin granules.[18]

[^{3}H]Reserpine binding was measured in a synaptic vesicle preparation from bovine caudate nucleus.[15] Binding is of a high-affinity type (K_d = 1.2 nM; B_{max} = 3.3 pmol/mg protein) and is dependent on ATP and inhibited by protonophores. Substrates displace reserpine at concentrations similar to those required for inhibition of dopamine transport.

ANPA-5HT was used in an attempt to identify the transporter in various organelles. It inhibits ATP-driven reserpine-sensitive 5HT transport into membrane vesicles prepared from porcine platelet dense granules (PL), rat brain synaptic vesicles (SV), and histamine-containing granules of rat basophilic leukemia cells (RBL). In addition, it specifically labels a polypeptide in each of the above-mentioned preparations. The apparent molecular weight of the labeled protein band varies from approximately 52,000 to 34,000, depending on the source of vesicles. The largest variation occurs between SV and RBL vesicles, whereas in both PL and bovine chromaffin (CG) vesicles a 48-kDa polypeptide is labeled. This result indicates that the molecular weight variation is not due to species difference, since RBL and SV membranes are both derived from rat, whereas CG and PL vesicles are from cow and pig, respectively. A more likely explanation is that the differences result from either proteolysis or processing or both, since the 34-kDa band is occasionally observed in overexposed samples of labeled CG vesicles. The 52-kDa band may be similar to the 56-kDa band labeled in the presence of cholate in chromaffin granules. Still, the possibility that the differences reflect functionally similar but structurally different amine transporters cannot be ruled out at present. Analysis of the labeled polypeptides by peptide mapping is likely to help in resolving this issue.

TWO SITES: TWO PROTEINS?

Based on the pharmacological work described in the literature (see Henry *et al.*,[5] this volume), a consensus seems to be emerging that there are two distinct sites on the transporter: R1 and R2 sites. The various substrates of the transporter bind to each of the sites with different affinities. Some are of the R1 type, some of the R2 type. There are two substrates that bind to both sites with comparable affinities: meta-iodobenzylguanidine[19] and ANPA-5HT (unpublished results).

It is not clear yet whether each of the sites is located in different polypeptides, and if they are, whether they are both subunits of the transporter. The evidence that points to more than one subunit is based on the following: (1) derivatives of 5HT label a 48-kDa peptide in the intact membrane; (2) when the membrane is disrupted the same compounds label an additional peptide (56 kDa) and this peptide has been partially purified by affinity chromatography; (3) a derivative of TBZ labels a 70-kDa peptide; (4) radiation inactivation analysis of reserpine and TBZOH binding suggest different apparent M_rs for each of the binding proteins (40,000 and 65,000). The discrepancies between the two laboratories involved in this aspect of the project may be due to the preparations and/or techniques used, and at this stage it is still too early to conclude if they are real conflicts.

In any case, it is clear that the final assignment of activities should wait until each of the putative subunits is purified and reconstituted in an active form.

REFERENCES

1. GABIZON, R., T. YETINSON & S. SCHULDINER. 1982. Photoinactivation and identification of the biogenic amine transporter in chromaffin granules from bovine adrenal medulla, J. Biol. Chem. **257:** 15145.
2. SCHULDINER, S., R. GABIZON, R. MARON, R. SUCHI & Y. STERN. 1985. The amine transporter from bovine chromaffin granules. Ann. N. Y. Acad. Sci. **456:** 268.
3. MARON, R., H. FISHKES, B. I. KANNER & S. SCHULDINER. 1979. Solubilization and reconstitution of the catecholamine transporter from chromaffin granules. Biochemistry **18:** 4781.
4. ISAMBERT, M. F. & J. P. HENRY. 1985. Photoaffinity labeling of the tetrabenazine binding sites of bovine chromaffin granule membranes. Biochemistry **24:** 3660.
5. HENRY, J. P., B. GASNIER, M.-P. ROISIN, M.-F. ISAMBERT & DANIEL SCHERMAN. 1986. Molecular pharmacology of the monoamine transporter of the chromaffin granule membrane. This volume.
6. GABIZON, R. & S. SCHULDINER. 1985. The amine transporter from bovine chromaffin granules. J. Biol. Chem. **260:** 3001.
7. IVERSEN, L. L. 1975. Uptake processes for biogenic amines. *In* Handbook of Psychopharmacology, Vol. 3. L. L. Iversen, S. D. Iversen & S. H. Snyder. Eds.: 381-442. Plenum. New York.
8. KIRSHNER, N. 1974. Molecular organization of the chromaffin vesicles of the adrenal medulla. Adv. Cytopharmacol. **2:** 265.
9. TOLL, L. & B. C. HOWARD. 1978. Role of Mg^{2+}-ATPase and a pH gradient in the storage of catecholamines in synaptic vesicles. Biochemistry **17:** 2517.
10. TRANZER, J. P., M. DAPRADA & A. PLETSCHER. 1966. Ultrastructural localization of 5-hydroxytryptamine in blood platelets. Nature **212:** 1574.
11. DAPRADA, M. & A. PLETSCHER. 1969. Differential uptake of biogenic amines by isolated 5-hydroxytryptamine organelles of blood platelets. Life Sci. **8:** 65.
12. BEAVEN, M. A., A. H. SOLL & K. J. LEWIN. 1982. Histamine synthesis by intact mast cells from canine fundic mucosa liver. Gastroenterology **82:** 254.
13. WHITTAKER, V. P. 1971. Subcellular localization of neurotransmitters. Adv. Cytopharmacol. **1:** 319.
14. MOTA, I., W. T. BERALDO, A. G. FERRI & L. V. JUNQUEIRA. 1954. Intracellular distribution of histamine. Nature **174:** 698.

15. RUDNICK, G., H. FISHKES, P. J. NELSON & S. SCHULDINER. 1980. Evidence for two distinct serotonin transport systems in platelets. J. Biol. Chem. **255:** 3638.
16. MARON, R., B. I. KANNER & S. SCHULDINER. 1979. The role of a transmembrane pH gradient in 5-hydroxytryptamine uptake by synaptic vesicles from rat brain. FEBS Lett **98:** 237.
17. FISHKES, H. & G. RUDNICK. 1982. Bioenergetics of serotonin transport by membrane vesicles derived from platelet dense granules. J. Biol. Chem. **257:** 5671.
18. SCHERMAN, D. & J. P. HENRY. In press.
19. DESPLANCHES, G. & J. P. HENRY. 1986. Uptake of metaiodobenzylguanidine by bovine chromaffin granule membranes. Mol. Pharmacol. **29:** 275.

DISCUSSION OF THE PAPER

L. EIDEN (*National Institute of Mental Health, Washington, D.C.*): Dr. Schuldiner, have you detected the 56-kDa protein immunologically in any of the other vesicle preparations that contain the 45-kDa proteins?

S. SCHULDINER (*Hebrew University, Jerusalem, Israel*): The antibody against the 56-kDa protein shows some cross-reactivity with a protein (52 kDa) from rat brain synaptic vesicles. It does not cross-react with vesicles from either platelets or rat basophilic leukemia cells.

J. H. PHILLIPS (*University of Edinburgh, Edinburgh, U.K.*): Have you tried treating intact granules with proteases to investigate whether binding of the inhibitor is protease-sensitive, or whether the 45-kDa protein is accessible on the granule surface?

S. SCHULDINER: In some preliminary experiments we have treated membrane vesicles with trypsin. Inhibition of transport and labeling followed in a similar pattern.

Molecular Pharmacology of the Monoamine Transporter of the Chromaffin Granule Membrane[a]

JEAN-PIERRE HENRY, BRUNO GASNIER, MARIE-PAULE ROISIN, MARIE-FRANÇOISE ISAMBERT, AND DANIEL SCHERMAN

Institut de Biologie Physico-Chimique
75005 Paris, France

INTRODUCTION

In the adrenal medulla, the hormones adrenaline and noradrenaline are stored in subcellular structures, the chromaffin granules, from which they are released into the bloodstream. The size of the cytoplasmic pool of the hormones is much smaller than that of the chromaffin granules, and moreover, the hormone concentration in the granules is four orders of magnitude larger than that of the cytoplasm.[1] This catecholamine concentration gradient originates in an active hormone transport across the chromaffin granule membrane which utilizes the cytoplasmic ATP as a source of energy.[2,3] During the last few years, progress has been made in our knowledge of the biophysics of the uptake system. Two different entities are involved: an inward proton pump driven by the cytoplasmic ATP, which acidifies and polarizes the granule matrix and thus generates a proton electrochemical gradient ($\Delta\mu H^+$),[4,6] and a monoamine transporter that utilizes the proton electrochemical gradient generated by the H^+ pump to accumulate catecholamines.[7–11] This transporter is not very specific since it translocates not only catecholamines such as dopamine, noradrenaline, or adrenaline, but also indoleamines such as 5-hydroxytryptamine (5-HT).[12] The true substrate of the transporter is the neutral form of these amines, which is of minor abundance at physiological pH, and the transporter catalyzes the electrogenic antiport of a proton against a neutral amine.[13–16]

The transporter is difficult to study by biochemical techniques because it catalyzes vectorial transport between two compartments of unequal size. Its purification requires its solubilization and its reconstitution into liposomes suitable for flux studies.[17,18] In our group, we have followed an alternate approach based on pharmacological techniques. In these techniques, which are generally used to study receptors, specific ligands are synthesized and the receptors are defined as specific ligand binding sites. Two classical inhibitors of the monoamine uptake system, tetrabenazine (TBZ) and reserpine

[a]This work was supported by the Centre National de la Recherche Scientifique (grant no. UA 1112) and by contracts from the Institut National de la Santé et de la Recherche Médicale (contract no. 83.6014) and the Ministère de la Recherche et de la Technologie (contract no. 83.C.0915)

(RES),[12] are analogues, respectively, of dopamine and 5-HT (FIG. 1). We have shown that they act primarily at the level of the monoamine transporter.[19] We have studied the binding of tritiated derivatives of these compounds to bovine chromaffin granule membranes. We took advantage of the fact that tetrabenazine and reserpine binding occurred specifically on the transporter to get information on its structure and on the mechanism of amine translocation.

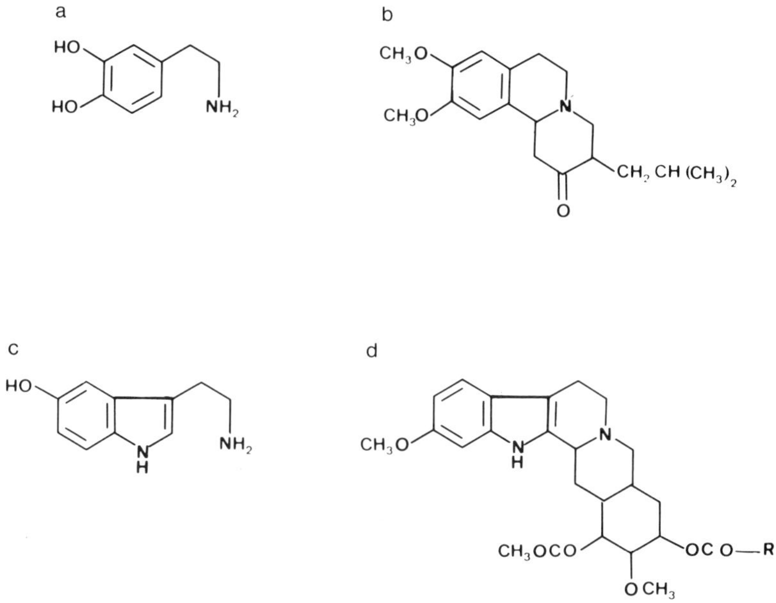

FIGURE 1. Structures of (**a**) dopamine, (**b**) tetrabenazine, (**c**) serotonin and (**d**) reserpine. R = 3,4,5-trimethoxybenzoyl.

DIHYDROTETRABENAZINE AND RESERPINE BINDING TO BOVINE CHROMAFFIN GRANULE MEMBRANES

Tetrabenazine (TBZ) has been known for a long time as an amine-depleting drug in the central nervous system.[20] It is very efficient, but its effect is of short duration. The depleting effect of TBZ has been attributed to an inhibitory effect of the drug on the vesicular monoamine uptake system.[21] Structurally, TBZ can be considered as an analogue of dopamine, though its pK_a is 6.0, which is about three units lower than that of catecholamines.[22] At physiological pH, TBZ is more soluble in lipid bilayers than noradrenaline, partly because it is in its neutral form. By reduction of the keto group of TBZ, two diastereoisomers of dihydrotetrabenazine (TBZOH) are obtained in different proportion. The most abundant isomer has a pK_a of 7.5.[22] TBZ and TBZOH are potent inhibitors of the monoamine transporter, characterized by an IC_{50} (concentration giving a 50% inhibition of ATP-dependent monoamine uptake) of about

3 nM (TABLE 1 and ref. 23). The isomers have been labeled by reducing TBZ with tritiated borohydride. The binding of [^{3}H]TBZOH to the chromaffin granule membrane has been studied using the more active isomer of [^{3}H]TBZOH.[23] [^{3}H]TBZOH binds to one class of sites characterized by a dissociation constant K_d of 3 nM and a density (B_{max}) of 60 pmol/mg of protein (TABLE 1). Subcellular fractionation experiments have shown that these sites are located only on chromaffin granule membranes. The excellent correlation existing between the occupancy of these sites and the inhibition of monamine uptake indicates that they are associated with the monoamine transporter. [^{3}H]TBZOH is displaced from its binding sites by various substrates and inhibitors of uptake. For inhibitors, the EC_{50} (concentration displacing 50% of the ligand) for [^{3}H]TBZOH displacement is identical to the IC_{50} for noradrenaline uptake. It should be stressed, however, that substrates displace very poorly [^{3}H]TBZOH. In a Michaelian uptake system with one substrate binding site, a substrate displaces theoretically 50% of a bound inhibitor at a concentration equivalent to its K_m. Such a result is not observed with [^{3}H]TBZOH (TABLE 1). This point will be discussed later.

Several conclusions have been drawn from the study of [^{3}H]TBZOH binding. The turnover number of the transporter, calculated from the maximal velocity of the transport and the density of [^{3}H]TBZOH binding sites, is 35 molecules per minute, a figure obtained with purified membranes. This value is underestimated since membrane purification leads to some loss of vectorial transport without concomitant loss of [^{3}H]TBZOH binding sites, as discussed later. A more exact figure is 140 molecules per minute, obtained on intact granules. With some assumptions, it is also possible to estimate the number of [^{3}H]TBZOH binding sites, and thus the number of monoamine transporters per granule. Our figure of about 20 molecules per granule is of the same order of magnitude as that proposed for the H^+ pump, though data on the pump are still uncertain.[24] Assuming a molecular weight of 70,000 for the polypeptide bearing [^{3}H]TBZOH binding sites, we have calculated that this polypeptide represents only 0.5% of the membrane protein. Thus, the transporter is a minor component of the membrane.

The fact that [^{3}H]TBZOH binds to the vesicular monoamine transporter makes it a valuable tool in neurobiology. For instance, it has been shown that in the central nervous system, brain homogenates contain only one class of binding sites for [^{3}H]TBZOH, with a K_d of 3 nM. Subcellular fractionation studies indicate that these sites are borne by synaptic vesicles.[26] Thus [^{3}H]TBZOH can be considered as a specific presynaptic marker. Moreover, a survey of the regionalization of [^{3}H]TBZOH binding sites pointed out that these sites were only associated with monoaminergic neurons (dopaminergic, noradrenergic, and serotoninergic). In fact, in eight different brain areas, it was calculated that the ratio of monoamine to [^{3}H]TBZOH binding sites was independent of the nature of the monoamine and of the brain area considered. This result was taken as an indication that all vesicles are filled with monoamines at their maximal capacity and that the system of monoamine uptake is not a rate-limiting step in vesicle homeostasis.[25]

Reserpine (RES) is another inhibitor of monoamine uptake, which acts at the level of the monoamine transporter.[12] In the central nervous system, it induces a depletion of monoamines, and this effect lasts longer than that of TBZ.[20] A tritiated reserpine of high specific activity has recently been commercialized, which allows binding studies. Results from our group[27] and from Deupree and Weaver[28] have indicated differences between the characteristics of [^{3}H]TBZOH and [^{3}H]RES binding. Saturation isotherms show that [^{3}H]RES binds to two different classes of sites. The high-affinity site R_1 has a K_d of 0.5 nM and a density of 10 pmol/mg of protein, whereas the low-affinity site R_2 has a K_d of 20 nM and a density of 60 pmol/mg of protein (TABLE 1 and ref. 27). Another difference between [^{3}H]TBZOH and [^{3}H]RES binding is that the kinetics of [^{3}H]RES binding are dependent upon $\Delta\mu H^+$ generation. [^{3}H]RES binding is about two orders of magnitude slower in the absence of ATP than in its presence, though plateau values are apparently similar (TABLE 1). The identity of the inhibition constant K_i of RES for ATP-dependent noradrenaline uptake and of the K_d of site R_1 indicates that the high affinity site R_1 is involved in monoamine translocation. From experiments on displacement of [^{3}H]RES by TBZ, it has been concluded that the low-affinity site R_2 is identical to the [^{3}H]TBZOH binding site. [^{3}H]RES binding thus introduces a new class of sites, site R_1, which differs from the [^{3}H]TBZOH binding site (site T). [^{3}H]RES is displaced from R_1 site by noradrenaline or serotonin at concentrations equivalent to their respective K_m. In purified membrane preparations, R_1 site is thus defined as a TBZ-resistant[27] or as a noradrenaline-sensitive [^{3}H]RES binding site.[28] This information is summarized in TABLE 1 and in FIGURE 2.

TABLE 1. TBZ and RES Binding and Inhibition of Noradrenaline Uptake[a]

		RES	
	TBZOH	−ATP	+ATP
BINDING EXPERIMENTS			
Kinetics of association ($t_{1/2}$, min, at 8 nM)	2	500	5
Binding at equilibrium			
Sites T : K_d (nM)	3	25	18
B_{max} (pmol/mg prot)	58	59	49
Sites R_1 : K_d (nM)	—	0.7	0.3
B_{max} (pmol/mg prot)	—	7	13
Displacement by noradrenaline (EC_{50}, μM)	1200		20
UPTAKE EXPERIMENTS			
Inhibition of ATP dependent noradrenaline uptake (K_i, nM)	3	—	0.5

[a] Data are taken from refs. 23 and 27.

SITES T AND R_1 ARE BOTH INVOLVED IN MONOAMINE TRANSLOCATION

The presence of two pharmacologically distinct sites on transporters has already been reported. The mitochondrial adenine nucleotide translocase has a site that binds the inhibitor bongkrekic acid and another one that binds the inhibitor atractyloside.[29] Nevertheless, one-site transporters also exist, such as the *E. coli* lactose transporter.[30] In the case of the platelet 5-HT uptake system, different drugs bind to the transporter, and it has been proposed that two of them, imipramine and paroxetin, bind to different sites.[31,32] Another interpretation is that the two drugs bind to the same site, and that this site is a regulatory site, somehow different from the translocation site.[33] This interpretation is based on binding data,[34] and also on physiological data,[35] dissociating imipramine binding and 5-HT uptake. Similarly, it has been proposed that desipramine, an inhibitor of neuronal noradrenaline uptake, binds to a regulatory site with a low affinity for noradrenaline.[36] Is such an interpretation possible in the case of the monoamine transporter of chromaffin granules? In this interpretation, the low affinity of

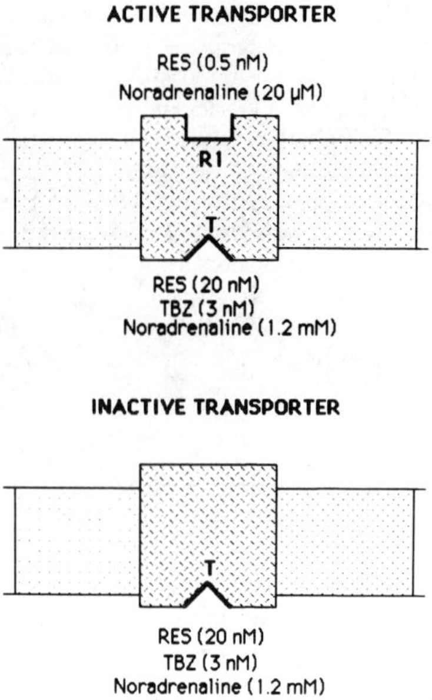

FIGURE 2. Intact chromaffin granules contain mostly the active form of the transporter, whereas the inactive form may represent 80-90% of the transporter in purified membranes.

site T for the substrates (EC_{50} of [^{3}H]TBZOH displacement by noradrenaline is 1.2 mM) might indicate that this site is an inhibitory site modulated by an endogenous TBZ-like molecule, which would control the catecholamine content of chromaffin granules. On the other hand, site R_1 (from which [^{3}H]RES is displaced with an EC_{50} of 20 μM by noradrenaline) would be the translocation site. However, results obtained with a new substrate of the monoamine transporter, *meta*-iodobenzylguanidine (MIBG), make this hypothesis unlikely. MIBG has been used as an adrenal medulla imaging agent in scintigraphic studies.[37] Though this compound is neither an indoleamine nor a catecholamine, the specificity and the stability of the labeling suggested that it was stored in chromaffin granules. We have studied the transport of [^{131}I] by chromaffin granule membranes and this compound was shown to be a substrate of the ATP-dependent system of monoamine uptake.[38] At pH 7.5, the K_m for MIBG and noradrenaline are 2 and 5 μM, respectively. In addition, MIBG and noradrenaline are translocated by the same transporter since noradrenaline is a competitive inhibitor of MIBG uptake.

We have compared the efficiency with which noradrenaline and MIBG displace [^{3}H]TBZOH and [^{3}H]RES from their binding sites (TABLE 2). The results obtained with MIBG are opposite those obtained with noradrenaline: MIBG displaces [^{3}H]TBZOH from its sites with a good efficiency (EC_{50} of the order of the K_m), but this compound is very inefficient in displacing [^{3}H]RES. Thus, with this new substrate, site T has normal behavior, making the hypothesis of a regulatory site untenable. It may also be noted that the results obtained in brain on the regionalization of [^{3}H]TBZOH binding[25] argue against the existence of a regulation of the monoamine transporter since all monoaminergic vesicles are filled to their maximal capacity, which is probably determined by the H^+ electrochemical gradient.

If sites T and R are both translocation sites, they should have a defined stoichiometry, the more likely hypothesis being an equal distribution of the two sites. In intact granule preparations, binding and titration experiments indicated the same density for sites T and R_1.[27] Membrane vesicles gave a different result, sites T being more abundant than sites R_1 (TABLE 1). The data obtained with membrane preparations are interpreted as resulting from the inactivation of an important fraction of [^{3}H]RES binding sites without concomitant inactivation of sites T (FIG. 2). This inactivation occurred during the osmotic lysis of chromaffin granules and it was accompanied by a loss of the ATP-dependent uptake activity (unpublished results). It might originate in a topological problem, since both ATP-dependent uptake and [^{3}H]RES binding require $\Delta\mu H^+$ generation and right-side-out vesicles.

CHARACTERIZATION OF SITES T AND R_1

[^{3}H]TBZOH and [^{3}H]RES are displaced from their respective sites by molecules structurally very different from one another. We found for some of them, however, a striking correlation between their potency to displace [^{3}H]TBZOH and a physicochemical parameter, their lipophilicity (FIG. 3). The lipophilicity was estimated from the octanol-water partition coefficient. This linear relationship was observed for six TBZ derivatives with an octanol-water partition coefficient extending over three orders of magnitude. It was also observed for five substrates (adrenaline, noradrenaline, tyramine, 5-HT, and MIBG), the partition coefficient of which differed from that of TBZ by five orders of magnitude from the hydrophobic TBZ to the hydrophilic adrenaline. This relationship indicates that TBZ derivatives and substrates have a similar affinity for site T when their concentration in the membrane phase is taken into account. The fact that membrane concentrations govern the occupancy of site T demonstrates that this site is hydrophobic. By assuming a specific volume of the membrane, it is possible to calculate an intrinsic dissociation constant of 0.4 μM for TBZ. It should be noted that RES, haloperidol, and chlorpromazine give experimental points located above the regression line, indicating a low intrinsic affinity of these drugs for site T.

The hydrophobicity of site T might be important considering the fact that this site binds the neutral form of TBZ and noradrenaline.[13–16] This result is somewhat paradoxical since, for the physiological substrate, this form is of minor abundance at neutral pH.[14] Nevertheless, the neutral form of the amines is more hydrophobic than the protonated one and its local concentration at the vicinity of site T is thus higher than that of the cationic species.

TABLE 2. Comparison of MIBG and Noradrenaline Properties[a]

Compound	K_m (μM)	EC_{50} Site R_1 (μM)	EC_{50} Site T (μM)
Noradrenaline	10	30	2000
MIBG	5	300	25

[a] Measurements were performed at pH 7.5, in the presence of 2.5 mM ATP and 1.25 mM $MgSO_4$. Data are taken from ref. 38.

At the level of site R_1, there is no relationship between the potency of drugs and substrates to displace [^{3}H]RES and their hydrophobicity. Thus, these displacement experiments confirmed the existence of two different binding sites on the monoamine transporter and showed that these two sites have different environments. D. Njus and his collaborators have reported that reserpic acid, an anionic impermeant derivative obtained by alkaline hydrolysis of RES, inhibited monoamine uptake only when added outside of the vesicle.[39] This result suggests an easy access to site R_1 from the external aqueous phase. Moreover, it indicates that the monoamine transporter is a protein with a fixed orientation in the membrane, and not a mobile carrier. It also suggests that the two translocation sites have a defined topology with respect to the membrane.

Differences between sites R_1 and T are also illustrated by chemical modification of chromaffin granule membranes by dicyclohexylcarbodiimide (DCCD). DCCD inhibits ATP-dependent noradrenaline uptake.[40–43] This reagent has some specificity for carboxylic acids and it inhibits H^+ translocation by the chromaffin granule H^+ pump as it does with the mitochondrial ATPase complex. It also inhibits the monoamine transporter as shown by experiments of ΔpH-driven monoamine uptake where the H^+ pump is bypassed.[42,44,45] DCCD inhibits not only H^+ pumps but also vectorial systems involved in H^+ translocation such as cyt *c* oxidase,[46] NAD transhydrogenase,[47] and the cyt b-c_1 complex.[48] Its effect on the monoamine transporter might be of this type since the transporter catalyzes a H^+ neutral amine antiport. In DCCD-treated membranes, binding of both [^{3}H]RES and [^{3}H]TBZOH is abolished.[45] Nevertheless, the dose dependency of the two inhibitions is different, indicating that the two sites are modified independently. This is confirmed by the fact that preincubation of the membranes with TBZ protected site T but not site R_1 against DCCD modification. The inhibition of site T was quite specific since the modification of one residue resulted in the loss of [^{3}H]TBZOH binding.

IDENTIFICATION OF THE POLYPEPTIDE CHAINS BEARING SITES T AND R_1

The polypeptidic chain bearing site T has been identified by photoaffinity.[49] The probe used, TBA, is an arylazido derivative of TBZ (FIG. 4). This compound has been tritiated either on the TBZ moiety or on the GABA used to link TBZ to the arylazido group. It is a competitive inhibitor of [^{3}H]TBZOH binding in the dark, characterized by a K_d of 50 nM. Its binding to the chromaffin granule transporter was difficult to study because of high nonspecific binding. To overcome this difficulty, bound and free [^{3}H]TBA were separated by centrifugation on short SP-Sephadex columns.[50] The illumination at 435 nm of [^{3}H]TBA complexes previously isolated by rapid centrifugation on SP-Sephadex columns results in the formation of a covalent tritiated complex with a yield of 25%. Analysis by electrophoresis on SDS polyacrylamide gels shows this complex to be associated with a 65-kDa polypeptide chain. This chain is a minor component of the membrane protein.

A similar approach was followed by S. Schuldiner and collaborators.[51] Using a probe derived from 5-HT, they labeled a polypeptide chain with a molecular weight of 45,000. Since the difference in molecular weight between the two labeled polypeptides cannot be accounted for by differences in experimental conditions, we have proposed that the monoamine transporter is an oligomer.[49] This hypothesis was tested by the technique of inactivation by ionizing radiation, in collaboration with C. Ellory at the

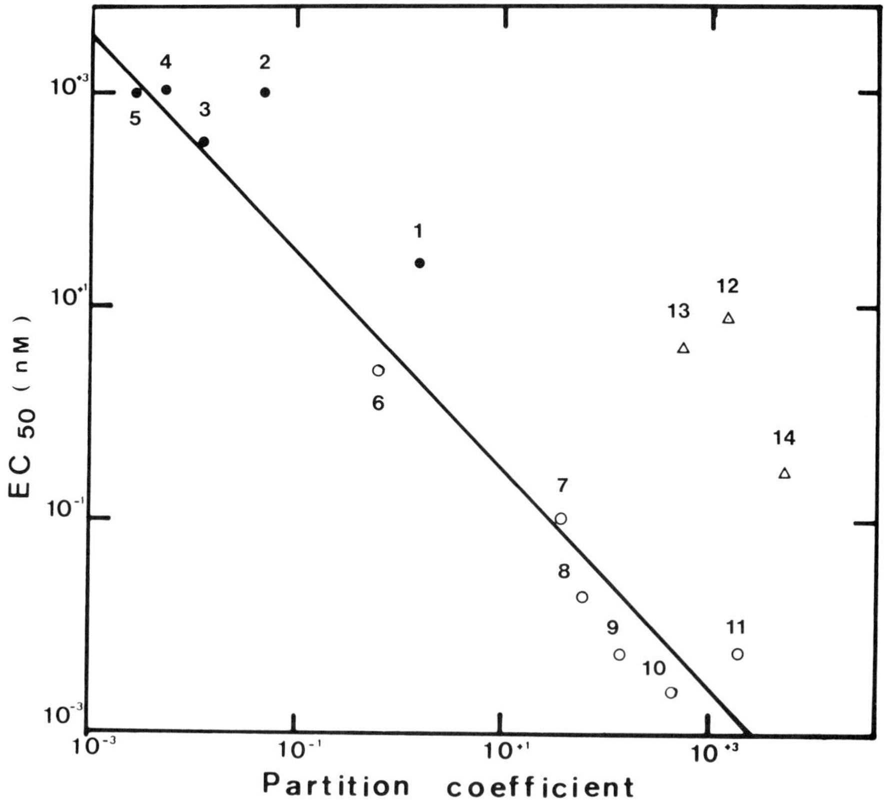

FIGURE 3. Correlation between the partition coefficient in octanol-water media of various substrates and inhibitors and their potency to displace [^{3}H]TBZOH from its binding site. ●, substrates: *1,* MIBG; *2,* tyramine; *3,* 5-HT; *4,* noradrenaline; *5,* adrenaline; ○, TBZ derivatives: *6,* succinyl TBZ; *7,* 2-amino TBZ; *8,* acetyl TBZ; *9* and *10,* isomers of TBZOH; *11,* TBZ; Δ, other inhibitors: *12,* haloperidol; *13,* chlorpromazine; *14,* RES.

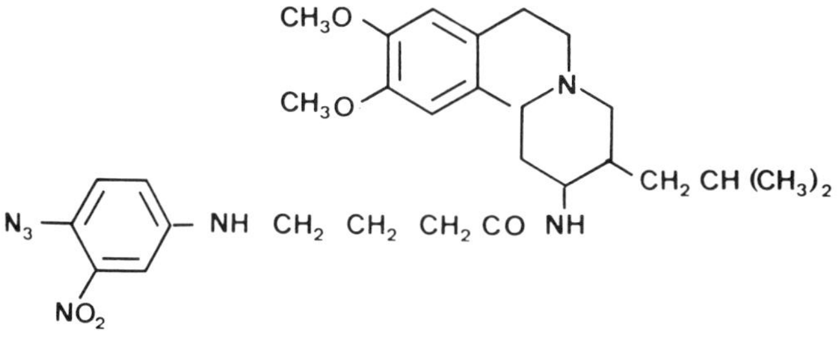

FIGURE 4. Structure of TBA.

University of Cambridge. In this technique, the sample is irradiated by high energy electrons generated by a linear accelerator and it is assumed that the target molecules are completely inactivated by a single hit of incident electrons.[52] Under these conditions, there is a linear relationship between the log of the remaining activity and the dose of radiation. The functional molecular weight of the target can be derived from the slope of the inactivation curve. The inactivation of [^{3}H]TBZOH binding activity followed a single exponential, from which a molecular weight of 68,000 was derived (TABLE 3). The inactivation of [^{3}H]RES binding was more complex, but under optimal conditions for assay of R_1 site, the inactivation was also exponential, corresponding to a molecular weight of 40,000. These experiments support the contention that the monoamine transporter has an oligomeric structure, with two types of subunits of molecular weights 68,000 and 40,000, bearing sites T and R_1, respectively (FIG. 5).

In our interpretation, the probe used by Gabizon *et al.*[51] would have labeled sites R_1, since this probe is derived from 5-HT, which, as noradrenaline, has a higher affinity for site R_1 than for site T (TABLE 1). Gel filtration of the solubilized transporter[53] suggests a high molecular weight since [^{3}H]TBZOH binding sites are eluted at about the same position as dopamine β-hydroxylase (MW 300,000). The transporter might thus be an oligomer of the $(\alpha\beta)_n$ type.

The hypothesis of an oligomeric structure for a transporter is not unique. The *E. coli* lactose transporter and the adenine nucleotide translocator are homodimers.[54,55] It has been hypothesized that an oligomeric structure gives some functional advantages to transport proteins.[56] An intersubunit channel might be used as a translocation path for the substrate. Moreover, the conformation changes required for translocation by a fixed transporter might result from a change in subunit association. Evidence for this hypothesis has recently been obtained by radiation inactivation of the lactose transporter.[54] It should also be noted that the platelet 5-HT transporter is an oligomer, according to radiation inactivation experiments with paroxetin and imipramine.[31] The two ligand binding sites are borne by different polypeptidic chains with molecular weights of 68,000 and 86,000, respectively. A structure of the $(\alpha\beta)_2$ type has been proposed.[31]

MECHANISM OF MONOAMINE TRANSLOCATION

The central question is how amine translocation is coupled to the proton electrochemical gradient. Active transports are generally accounted for by alternating access or gated pore mechanisms.[57–59] The substrate binds first to a site accessible only from

TABLE 3. Radiation Inactivation of Chromaffin Granule Membranes

Activity Tested	D_{37}[a] (Mrad ± SD)	Molecular Weight[b] (rad ± SD)
[^{3}H]TBZOH binding	9.6 ± 1.2	67.5 ± 8.8
[3H]RES binding	17.7 ± 2.0	36.7 ± 3.9
Cytochrome b_{561}	25.9 ± 3.7	25.2 ± 3.6
Dopamine β-hydroxylase	5.5 ± 0.5	117.5 ± 0.5

[a] D_{37} is the radiation dose responsible for reducing the measured activity to 37% of that found in the unexposed controls.

[b] Molecular weights calculated from D_{37} according to ref. 52.

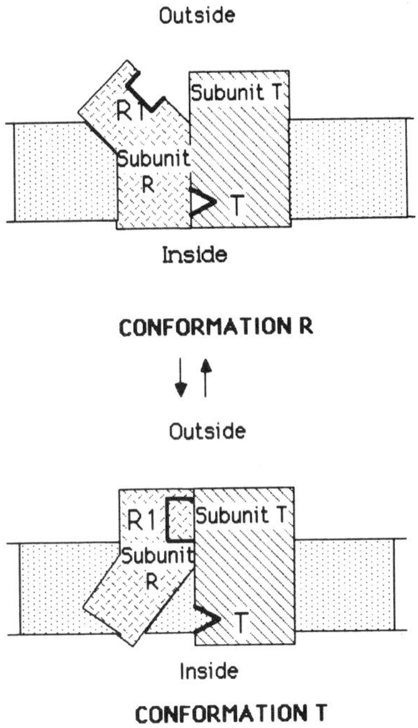

FIGURE 5. Model of the monoamine transporter. Conformation R is stabilized in the presence of RES and ATP; conformation T is stabilized in the presence of TBZ. Site T is considered to be more hydrophobic than site R_1.

one side of the membrane, the charge compartment. This site is a high-affinity site (charge site). A conformation change then occurs, which is coupled to movements of the driving species and which results in the translocation of the transported species. In the new conformation, the substrate is bound to a low-affinity site facing the discharge compartment from which it will dissociate (discharge site). Experimental evidence supports a mechanism of this type in the case of the ATP-ADP transporter.[60,61] In the case of the monoamine transporter, some data favor such a model (FIG. 5). Two different binding sites are involved in the translocation process. These two sites have a different affinity for the substrate.[27] Site R_1, which is the high-affinity site, is accessible from the cytoplasm (charge compartment).[39] Site T is hydrophobic and therefore buried in the membrane. Experiments are now in progress to determine the accessibility of site T. The fact that the kinetics of [^{3}H]RES binding is $\Delta\mu H^+$ dependent[27,28] suggests a $\Delta\mu H^+$-dependent conformational change of site R_1. This site would exist in a binding and a nonbinding conformation (FIG. 2). In the absence of an ATP-induced $\Delta\mu H^+$, site R_1 would be in the nonbinding conformation and [^{3}H]RES binding would only occur through a slow rate-limiting transconformational change. The $\Delta\mu H^+$ generated in the presence of ATP would accelerate this reaction, resulting in the opening of R_1 sites and fast [^{3}H]RES binding.

For site T to be a discharge site according to an alternate gated pore model would imply an open and a closed state. The closed state would be associated with the charge conformation, *i.e.,* that observed in the presence of ATP and RES. This behavior is not observed with purified membrane preparations, where [^{3}H]TBZOH binding is not suppressed in the presence of ATP and RES. The results of classical *in vivo* experiments, however, lead us to reinvestigate this point. As pointed out before, RES has a monoamine-depleting effect in brain that lasts longer (several days) than that of TBZ (24 h).[20] This effect results from a slower dissociation rate of RES.[62] If TBZ is administered to rats just before RES, the monoamine depletion has the time course of a TBZ-induced depletion.[63] From this experiment, it was concluded that TBZ and RES competed for the same site.[64] Recent experiments suggest that those data can be reconciled with our binding data. These recent experiments have been performed on intact granule preparations because purified membranes have a large fraction of sites R_1 in an inactive state (FIG. 2), and thus uncoupled from sites T. In the absence of a $\Delta\mu H^+$, RES displaces [^{3}H]TBZOH inefficiently. This displacement is likely to occur at the level of site T itself, since this site has a low affinity for RES (TABLE 1). The presence of a $\Delta\mu H^+$ shifted the displacement curve toward lower RES concentrations by about one order of magnitude. Since the site R_1 is blocked in the binding conformation in the presence of a $\Delta\mu H^+$ and of RES, this result is interpreted as indicating that when the transporter is in this conformation, site T is in a nonbinding state. Therefore in the active transporter, the opening of sites T and R_1 would be exclusive processes. The transporter would exist in two different conformations, conformation R with sites R_1 in an open state and site T closed, and conformation T with site T open and site R_1 closed (FIG. 5). When the experiment was repeated with purified membranes, a biphasic displacement of [^{3}H]TBZOH by RES was observed in the presence of ATP: RES displaced efficiently [^{3}H]TBZOH from a small fraction (10-20%) of its binding sites corresponding to the active transporter molecules. Thus, sites T and R_1 appear to be independent in purified membranes only because a majority of sites T have been uncoupled during the preparation.

To summarize our model, we propose that the monoamine transporter of chromaffin granule membranes has two different sites borne by two different polypeptide chains. We propose that the two chains are coupled to one another and that the transporter exists in two different conformations according to the gated pore mechanism.

ACKNOWLEDGMENTS

We thank B. Girard for help in the preparation of the manuscript. We are indebted to the Service Vétérinaire des Abattoirs de Mantes for collecting bovine adrenals.

REFERENCES

1. PHILLIPS, J. H. 1982. Neuroscience **7:** 1595-1609.
2. KIRSHNER, N. 1962. J. Biol. Chem. **237:** 2311-2317.
3. CARLSSON, A., N. Å HILLARP & B. WALDECK. 1962. Acta Physiol. Scand. **59** (Suppl. 215): 1-38.
4. CASEY, R. P., D. NJUS, G. K. RADDA & P. A. SEHR. 1977. Biochemistry **16:** 972-977.
5. PHILLIPS, J. H. & Y. P. ALLISON. 1978. Biochem. J. **170:** 661-672.
6. SCHERMAN, D. & J.-P. HENRY. 1980. Biochim. Biophys. Acta **599:** 150-166.
7. JOHNSON, R. G. & A. SCARPA. 1979. J. Biol. Chem. **254:** 3750-3760.
8. SCHERMAN, D. & J.-P. HENRY. 1980. Biochim. Biophys. Acta **601:** 664-677.
9. APPS, D. K., J. G. PRYDE & J. H. PHILLIPS. 1980. FEBS Lett. **111:** 386-390.
10. KANNER, B. I., I. SHARON, R. MARON & S. SCHULDINER. 1980. FEBS Lett. **111:** 83-86.
11. KNOTH, J., K. HANDLOSER & D. NJUS. 1980. Biochemistry **19:** 2938-2942.

12. PLETSCHER, A. 1976. Bull. Schweiz Akad. Med. Wiss. **32:** 181-190.
13. SCHERMAN, D. & J.-P. HENRY. 1981. Eur. J. Biochem. **116:** 535-539.
14. SCHERMAN, D. & J.-P. HENRY. 1983. Mol. Pharmacol. **23:** 431-436.
15. RAMU, A., M. LEVINE & H. POLLARD. 1983. Proc. Natl. Acad. Sci. U.S.A. **80:** 2107-2111.
16. JOHNSON, R. G., S. E. CARTY & A. SCARPA. 1981. J. Biol. Chem. **256:** 5773-5780.
17. MARON, R., H. FISHKES, B. I. KANNER & S. SCHULDINER. 1979. Biochemistry **18:** 4781-4785.
18. ISAMBERT, M.-F. & J.-P. HENRY. 1981. Biochimie **63:** 211-219.
19. SCHERMAN, D. & J.-P. HENRY. 1980. Biochem. Pharmacol. **29:** 1883-1890.
20. PLETSCHER, A., A. BROSSI & K. F. GEY. 1962. Int. Rev. Neurobiology **4:** 275-305.
21. PLETSCHER, A. 1977. Br. J. Pharmacol. **59:** 419-424.
22. SCHERMAN, D. & J.-P. HENRY. Biochimie **64:** 915-921.
23. SCHERMAN, D., P. JAUDON & J.-P. HENRY. 1983. Proc. Natl. Acad. Sci. U.S.A. **80:** 584-588.
24. PERCY, J. M., J. G. PRYDE & D. K. APPS. 1985. Biochem. J. **231:** 557-564.
25. SCHERMAN, D., G. BOSCHI, R. RIPS & J.-P. HENRY. 1986. Brain Res. **370:** 176-181.
26. SCHERMAN, D. 1986. J. Neurochem. **47:** 331-339.
27. SCHERMAN, D. & J.-P. HENRY. 1984. Mol. Pharmacol. **25:** 113-122.
28. DEUPREE, J. D. & J. A. WEAVER. 1984. J. Biol. Chem. **259:** 10907-10912.
29. VIGNAIS, P. V. 1976. Biochim. Biophys. Acta. **456:** 1-38.
30. KABACK, H. R. 1983. J. Membr. Biol. **76:** 95-112.
31. MELLERUP, T. E., P. PLENGE & M. NIELSEN. 1985. Eur. J. Pharmacol. **106:** 411-414.
32. PLENGE, P. & T. E. MELLERUP. 1984. Biochim. Biophys. Acta **770:** 22-28.
33. COSTA, E., D. M. CHUANG, M. L. BARBACCIA & O. GANDOLFI. 1983. Experientia **39:** 855-858.
34. SEGONZAC, A., R. RAISMAN, T. TATEISHI, H. SCHOEMAKER, P. E. HICKS & S. Z. LANGER. 1986. J. Neurochem. **44:** 349-355.
35. AHTEE, L., M. S. BRILEY, R. RAISMAN, D. LEBREC & S. Z. LANGER. 1981. Life Sci. **29:** 2323-2329.
36. LANGER, S. Z., L. TAHRAOUI, R. RAISMAN, S. ARBILLA, M. NAJAR & J. DEDEK. 1984. *In* Neuronal and Extraneuronal Events in Autonomic Pharmacology. W. W. Fleming *et al.,* Eds.: 37-49. Raven. New York.
37. WIELAND, D. M., J. WU, L. E. BROWN, T. J. MANGNER, D. P. SWANSON & W. H. BEIERWALTES. 1980. J. Nucl. Med. **21:** 349-353.
38. GASNIER, B., M.-P. ROISIN, D. SCHERMAN, S. COORNAERT, G. DESPLANCHES & J.-P. HENRY. 1986. Mol. Pharmacol. **29:** 275-280.
39. CHAPLIN, L., A. H. COHEN, P. HUETTL, M. KENNEDY, D. NJUS & S. J. TEMPERLEY. 1985. J. Biol. Chem. **260:** 10981-10985.
40. BASHFORD, C. L., R. P. CASEY, G. K. RADDA & G. A. RITCHIE. 1976. Neuroscience **1:** 399-412.
41. GIRAUDAT, J., M.-P. ROISIN & J.-P. HENRY. Biochemistry **19:** 4499-4505.
42. APPS, D. K., J. G. PRYDE, R. SUTTON & J. H. PHILLIPS. 1980. Biochem J. **190:** 273-282.
43. CIDON, S. & N. NELSON. 1983. J. Biol. Chem. **258:** 2892-2898.
44. SCHULDINER, S., H. FISHKES & B. I. KANNER. 1978. Proc. Natl. Acad. Sci. U.S.A. **75:** 3713-3716.
45. GASNIER, B., D. SCHERMAN & J.-P. HENRY. 1985. Biochemistry **24:** 1239-1244.
46. CASEY, R. P., M. THELEN & A. AZZI. 1979. Biochem. Biophys. Res. Comm. **87:** 1044-1051.
47. PENNINGTON, R. H. & R. R. FISHER. 1981. J. Biol. Chem. **256:** 8963-8969.
48. DEGLI ESPOSITI, M., E. M. M. MEIER, J. TIMONEDA & G. LENAZ. 1983. Biochim. Biophys. Acta **725:** 349-360.
49. ISAMBERT, M.-F. & J.-P. HENRY. 1985. Biochemistry **24:** 3660-3667.
50. PENEFSKY, H. S. 1977. J. Biol. Chem. **252:** 2891-2899.
51. GABIZON, R., T. YETIZON & S. SCHULDINER. 1982. J. Biol. Chem. **257:** 15145-15150.
52. KEMPNER, E. S. & W. SCHLEGEL. 1979. Anal. Biochem. **92:** 2-10.
53. SCHERMAN, D. & J.-P. HENRY. 1983. Biochemistry **22:** 2805-2810.
54. GOLDKORN, T., G. RIMON, E. KEMPNER & R. H. KABACK. 1984. Proc. Natl. Acad. Sci. U.S.A. **81:** 1021-1025.

55. BLOCK, M. R., G. ZACCAI, G. J. M. LAUQUIN & P. V. VIGNAIS. 1982. Biochem. Biophys. Res. Comm. **109:** 471-477.
56. KLINGENBERG, M. 1981. Nature. **290:** 449-454.
57. SINGER, S. J. 1974. Ann. Rev. Biochem. **43:** 805-833.
58. KLINGENBERG, M. 1976. *In* The Enzymes of Biological Membranes: Membrane Transport, Vol. 3. A. N. Martonosi, Ed.: 383-438, Plenum. New York.
59. TANFORD, C. 1983. Proc. Natl. Acad. Sci. U.S.A. **80:** 3701-3707.
60. AQUILA, H. & M. KLINGENBERG. 1982. Eur. J. Biochem. **122:** 141-145.
61. BLOCK, M. R., G. J. M. LAUQUIN & P. V. VIGNAIS. 1983. Biochemistry **22:** 2202-2208.
62. GIACHETTI, A., R. A. HOLLENBECK & P. A. SHORE. 1974. Naunyn-Schmiedeberg's Arch. Pharmacol. **283:** 263-275.
63. CARLSSON, A. & M. LINDQVIST. 1966. Acta Pharmacol. Toxicol. **24:** 112-120.
64. STITZEL, R. E. 1977. Pharmacol. Rev. **28:** 179-205.

DISCUSSION OF THE PAPER

G. RUDNICK (*Yale University School of Medicine, New Haven, Conn.*): We've studied the plasma membrane serotonin transporter extensively, and from all of our data, antidepressants like imipramine are strictly competitive inhibitors of serotonin transport, and serotonin competes with their binding. We've concluded (and others are coming around to this opinion) that serotonin and imipramine bind to either the same site, or mutually exclusive sites.

J.-P. HENRY (*Institut de Biologie Physico-Chimique, Paris, France*): Thank you for your comment. I was referring to the work of S. Langer and of E. Costa.

L. EIDEN (*National Institute of Mental Health, Washington, D.C.*): Dr. Henry, can you measure the size of the whole transporter with radiation inactivation? Do radiation inactivation experiments for ligand binding and enzymatic activity give similar results for model proteins?

J.-P. HENRY: No, the transport assay requires intact vesicles and cannot be performed on lyophilized irradiated vesicles. An instance of a transport protein, the *E. coli* lactose transporter, has been described in which the transporter was assayed after reconstitution (see ref. 54). But there are several instances in which binding activities were used for molecular weight determinations: β-adrenergic and muscarinic receptors and Na^+ and Ca^{2+} ion channels.

A. SCARPA (*Case Western Reserve University, Cleveland, Ohio*): Inactivation analysis is a powerful technique which has also "taken to the cleaners" several investigators studying monomer-dimer of functional ATPases. In your case, wouldn't it be more proper to use activity of the transporter rather than the binding of ligand to calculate oligomer vs. dimer? After irradiation you could have a peptide which can bind one but not another ligand and learn little on the *in situ* functional molecular weight of the transporter.

J.-P. HENRY: Unfortunately, irradiated vesicles are not suitable to assay vectorial activities such as monoamine uptake. To perform the type of experiments that you suggest, it would be necessary to reconstitute the transporter in lipid vesicles.

H. TAMIR (*Columbia University, New York, N.Y.*): Was nonhydrolyzable ATP used with reserpine to test its effect on uptake?

J.-P. HENRY: No, we used the H^+ ionophore CCCP to block the effect of the H^+ pump by collapsing the H^+ gradient.

Chromaffin Vesicle Function in Intact Cells[a]

NORMAN KIRSHNER, JAMES J. CORCORAN, BYRON CAUGHEY, AND MIRA KORNER

Department of Pharmacology
Duke University Medical Center
Durham, North Carolina 27710

The physiological role of chromaffin vesicles in adrenal medullary cells has been elucidated through studies of the isolated chromaffin vesicles and through studies of evoked secretion employing intact animals, isolated perfused glands, and adrenal medullary cell cultures.[1-4] The chromaffin vesicles contain high concentrations of catecholamines and ATP and they contain the enzyme dopamine β-hydroxylase, which catalyzes the oxidation of dopamine to norepinephrine. They are directly involved in the exocytotic secretion of epinephrine and norepinephrine, and the vesicle membranes are equipped with an ATP-dependent transport system for translocating catecholamines into the vesicles against a high concentration gradient. Although the individual reactions in the formation of norepinephrine and epinephrine have been well characterized, there was only little information on the integrated function of this pathway at the time the studies reported here were undertaken. Isolated chromaffin vesicles can take up ATP, but little was known about the accumulation of ATP or its turnover in chromaffin vesicles of intact cells. Studies in which we have examined the flux of catecholamines through chromaffin vesicles[5] and ATP metabolism in cultures of intact bovine adrenal medullary cells[6,7] are reported here.

CATECHOLAMINE SYNTHESIS AND TURNOVER

The pathway for the formation of norepinephrine and epinephrine in adrenal medullary cells is shown in FIGURE 1. Dopamine is formed in the cytosol from tyrosine through the sequential action of tyrosine hydroxylase and aromatic L-amino-acid decarboxylase. Dopamine is transported by the ATP-dependent catecholamine transporter into chromaffin vesicles, where it is oxidized to form norepinephrine. Norepinephrine then exits from the chromaffin vesicle into the cytosol, where it is *N*-methylated to form epinephrine, which is then taken back into the chromaffin vesicle via the amine transporter and stored.

To obtain kinetic information on that portion of the pathway involving chromaffin vesicles, cells were pulse-labeled with [^{3}H]dopamine or [^{3}H]norepinephrine and the amounts of [^{3}H]dopamine, [^{3}H]norepinephrine, and [^{3}H]epinephrine present in the

[a] The work reported here was supported by NIH grant nos. AM05427 and AM33037.

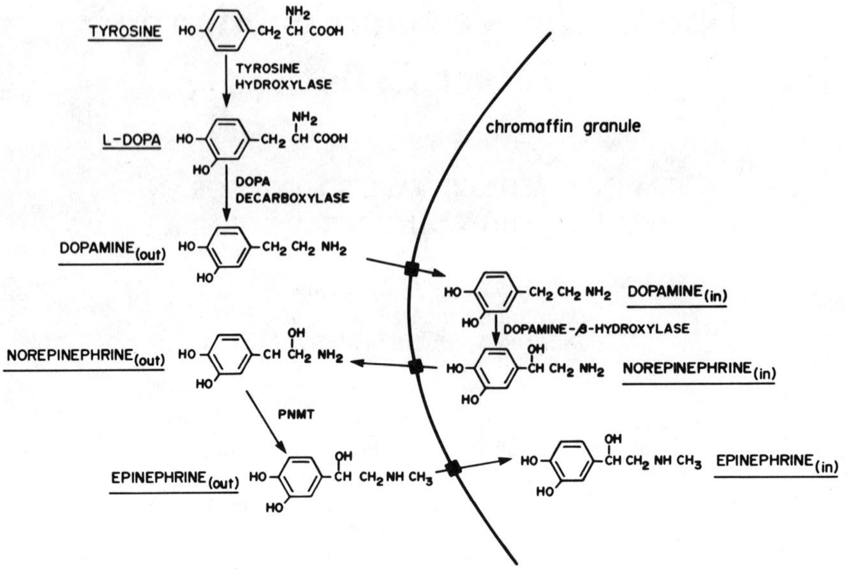

FIGURE 1. Flow of catecholamines through chromaffin vesicles.

cells were determined immediately after labeling and for periods of time up to 10 days.[5] The results of one such experiment in which cells were pulse-labeled for 2 hours with [^{3}H]dopamine is shown in FIGURE 2. [^{3}H]dopamine was readily taken up by cells and converted to [^{3}H]norepinephrine. Catecholamines were separated by high-performance liquid chromatography and fractions were collected at 12-s intervals for determination of radioactivity. At the end of the labeling period, the radioactivity present in the cell consisted of 3% [^{3}H]epinephrine, 41% [^{3}H]norepinephrine, and 56% [^{3}H]dopamine. [^{3}H]dopamine declined to very low levels at 24 hours after labeling while [^{3}H]norepinephrine reached a peak at this time and thereafter declined in a biphasic manner. Graphic analysis of the data (FIG. 2B) showed that the fast phase declined with a $t_{1/2}$ of 1.3 days and represented the conversion of norepinephrine to epinephrine, while the slow phase with a $t_{1/2}$ of 4.8 days represented the turnover of norepinephrine in norepinephrine cells. Once formed, [^{3}H]epinephrine declined with a $t_{1/2}$ of 17.3 days. More detailed kinetic studies of the conversion of [^{3}H]dopamine to [^{3}H]norepinephrine showed a biphasic disappearance of [^{3}H]dopamine and a biphasic appearance of [^{3}H]norepinephrine. About 15% of the [^{3}H]dopamine in the cells at the end of the labeling period was converted to [^{3}H]norepinephrine with a $t_{1/2}$ of 30 min and the remainder was converted to norepinephrine with a $t_{1/2}$ of 5.5 to 6.5 hours. When cells were labeled with [^{3}H]norepinephrine, kinetic constants for the disappearance of [^{3}H]norepinephrine and appearance and turnover of [^{3}H]epinephrine were similar to those observed in cells labeled with [^{3}H]dopamine. In these studies 20-38% of the radioactivity in the cells was present as [^{3}H]epinephrine immediately after labeling, while the remainder of the [^{3}H]norepinephrine was converted to [^{3}H]epinephrine with a $t_{1/2}$ of 0.8 to 1.2 days. The percent of total radioactivity in the cells present as [^{3}H]epinephrine immediately after labeling was independent of the pulse period but did vary with the cell preparation. In one set of experiments, after a 15-min pulse of [^{3}H]norepinephrine 18% of the radioactivity was present as [^{3}H]epinephrine, and after a 120-min pulse 20% of the radioactivity was present as [^{3}H]epinephrine, although the total amount of [^{3}H]norepinephrine taken up by the cells had increased more than fivefold. These results indicate that [^{3}H]norepinephrine was rapidly converted to [^{3}H]epinephrine in transit through the cytosol before being transported into the chromaffin vesicles.

We could not directly assess the rate of transport of catecholamine into chromaffin vesicles in the intact cells. Subcellular fractionation studies of cells immediately after labeling with [^{3}H]dopamine or [^{3}H]norepinephrine, however, showed that the distribution of [^{3}H]catecholamines was very similar to the distribution of endogenous amines; 58-65% of the [^{3}H]catecholamines and 63-67% of the endogenous catecholamines were present in the P_2 fraction that contained the chromaffin vesicles. These data suggest that transport into the vesicles is relatively rapid.

The rate of efflux of catecholamines from chromaffin vesicles was determined by treating cell cultures with reserpine immediately after labeling with [^{3}H]nor-

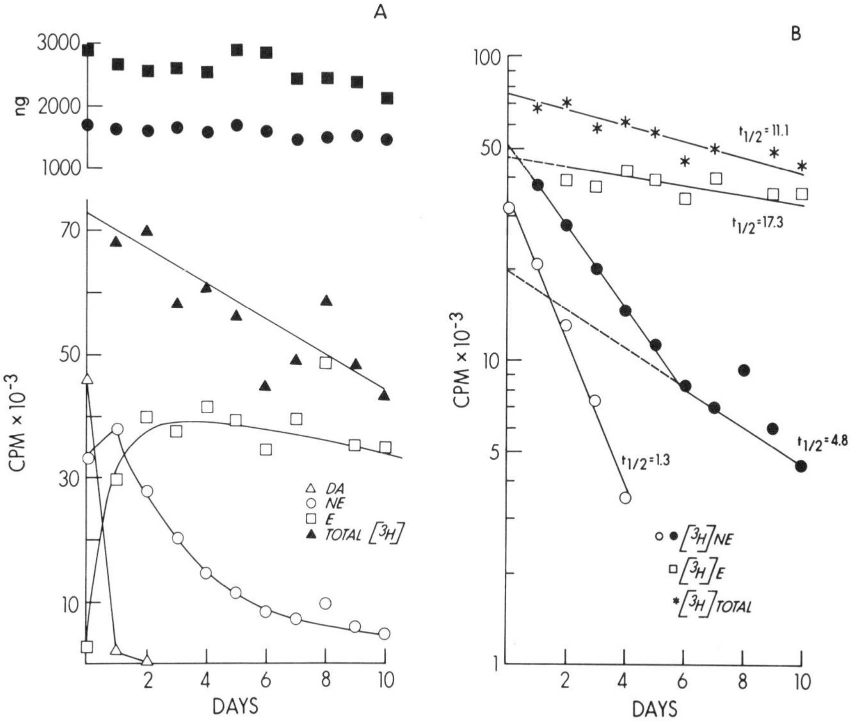

FIGURE 2. Conversion of [^{3}H]dopamine to [^{3}H]norepinephrine and [^{3}H]epinephrine. Cells were labeled for 2 h with [^{3}H]dopamine and examined immediately after the wash cycle and at the times indicated. **(A)** Linear presentation of the data. In the upper portion of the figure the endogenous levels of norepinephrine and epinephrine are shown. **(B)** Semilogarithmic presentation of the data. The points indicated by *open circles* represent the conversion of norepinephrine to epinephrine and were obtained by subtraction of the extrapolated values of the slow component from the experimental data points.

epinephrine. Reserpine blocks the transport of catecholamines into chromaffin vesicles and the resultant decline in endogenous and labeled catecholamines is a measure of the unidirectional leakage from the vesicles. In four experiments, the average $t_{1/2}$ values for the decline of endogenous and [^{3}H]norepinephrine were 27 ± 3 hours and 24 ± 2 hours respectively, and the average $t_{1/2}$ values for the decline of endogenous and [^{3}H]epinephrine were 33 ± 4 hours and 35 ± 5 hours respectively. Thus, the rate-limiting step in the conversion of norepinephrine to epinephrine is the rate of leakage of norepinephrine from chromaffin vesicles and not the *N*-methylation reaction catalyzed by phenylethanolamine-*N*-methyl transferase. In cells treated with reserpine no additional formation of [^{3}H]epinephrine was observed after the pulse period, whereas parallel experiments carried out with labeled cells not treated with reserpine showed no decline in endogenous catecholamines and the usual pattern of conversion of [^{3}H]norepinephrine to [^{3}H]epinephrine during the 48-hour experimental period.

The kinetic parameters determined in these studies are summarized in TABLE 1. In resting cell cultures epinephrine turns over very slowly with a half-life of 15 ± 2

TABLE 1. Kinetic Parameters for Chromaffin Vesicle Reactions in Cultured Bovine Adrenal Medullary Cells

Process	$t_{1/2}$ (Hours)
Vesicular uptake	fast
DA NE	5.6-6.5
Vesicular leakage	
NE	24±2
E	35±5
N-methylation	fast
NE turnover	
E cells (NE E)	29±2
NE cells	115±7
E turnover	353±46

days. The turnover of norepinephrine was biphasic. The first phase with a $t_{1/2}$ of 1.2 ± 0.1 days was due to the conversion of norepinephrine to epinephrine; the second phase with a $t_{1/2}$ of 4.8 ± 0.3 days represents the turnover of norepinephrine in norepinephrine cell types. The more rapid loss of norepinephrine from norepinephrine type cells than of epinephrine from epinephrine type cells may be due to the greater leakage rate of norepinephrine from chromaffin vesicles. Within the chromaffin vesicles dopamine is converted to norepinephrine with a $t_{1/2}$ of 5.5 to 6.5 hours. Rates of uptake into chromaffin vesicles and *N*-methylation of norepinephrine were not determined directly, but the data indicate that these are rapid processes compared to the β-hydroxylation reaction and leakage of catecholamines from chromaffin cells. These results provide an explanation of reports that after depletion of the catecholamine content of cat and rabbit adrenal glands, recovery of the epinephrine content lags behind the recovery of norepinephrine by several days.[8,9]

The bovine adrenal medulla contains two types of chromaffin cells, those storing primarily epinephrine and those storing primarily norepinephrine. Assuming steady-state kinetics, an estimate of the pool sizes of dopamine and norepinephrine utilized for epinephrine synthesis and the pool sizes of dopamine and norepinephrine in norepinephrine cells can be made. In the steady state the pool size of the substrate is directly proportional to the $t_{1/2}$ of the reactions for the appearance and disappearance of the product. In our cell preparations the total catecholamine content consisted of 67% epinephrine, 31.5% norepinephrine, and 1.5% dopamine. Assuming that epinephrine is in a single pool and a total catecholamine content of 100 nmols/10^6 cells, and using $t_{1/2}$ values of 360 h for the loss of epinephrine, 29 h for the overall conversion of norepinephrine to epinephrine, and 6 h for the oxidation of dopamine to norepinephrine, we obtain pool sizes of 1.1 nmol and 5.4 nmol respectively for dopamine and norepinephrine in epinephrine cells. Thus, 18% of the norepinephrine and 73% of the dopamine is present in epinephrine cells. Using a $t_{1/2}$ of 115 hours for the loss of norepinephrine from norepinephrine cells and pool sizes of 26 nmol and 0.4 nmol respectively for norepinephrine and dopamine in norepinephrine cells, we estimate the $t_{1/2}$ for the β-hydroxylation reaction in this cell type to be 1.8 h, considerably less than the overall $t_{1/2}$ of 6 h.

ATP METABOLISM IN CHROMAFFIN CELLS AND CHROMAFFIN VESICLES

Since chromaffin vesicles contain large amounts of ATP, it was of interest to examine the turnover of ATP in this organelle. Previous short-term studies have shown that perfused adrenal glands or cultured adrenal medullary cells rapidly incorporate [^{3}H]adenosine or $^{32}P_i$ into ATP that can be found in chromaffin vesicles.[10,11] We have used cultured adrenal medullary cells doubly labeled with [^{3}H]adenosine and $^{32}P_i$ to examine the metabolism of ATP by these cells and the stability of ATP in chromaffin vesicles.[6]

In general, cells were pulse-labeled with [^{3}H]adenosine and $^{32}P_i$ for 1 hour. At various times up to 10 days after labeling, perchloric acid extracts of the cells were prepared and their endogenous and labeled adenine nucleotide content were determined after separation by high-performance liquid chromatography or by thin-layer chromotography.[6,7] FIGURE 3 shows the results of one study in which the labeled ATP content was followed over a 10-day period. The declines in [^{32}P]ATP and [^{3}H]ATP were essentially identical. There was an initial fast falloff with a $t_{1/2}$ of 3.5 to 4.5 h during the first 10 hours followed by a much slower phase with a $t_{1/2}$ of 11.7 days for [^{32}P]ATP and 12.3 days for [^{3}H]ATP. Extrapolation of the slow phase to zero time indicated that 35-40% of the labeled ATP initially present accumulated in the slow compartment. In a second similar experiment using a different cell preparation, the $t_{1/2}$ values of the slow phase for [^{32}P]ATP and [^{3}H]ATP were 6.7 and 7.4 days, respectively.

As shown in FIGURE 3, the decline in total ^{32}P-labeled compounds and total ^{3}H-labeled compounds closely paralleled the decline of the respectively labeled ATP. It was shown that after labeling cells with [^{3}H]dopamine or [^{3}H]norepinephrine, the slow decline in total [^{3}H]catecholamine closely paralleled the decline in [^{3}H]epinephrine.[5] Thus, measuring the decline in total ^{32}P-labeled compounds and ^{3}H-labeled compounds provides a rapid method for closely approximating the decline in correspondingly labeled ATP or epinephrine. In studies in which cells were doubly labeled with [^{3}H]norepinephrine and $^{32}P_i$, the decline in total [^{3}H]catecholamines closely paralleled the decline in total ^{32}P-labeled compounds, indicating the turnover of ATP in the slow compartment was the same as the turnover of catecholamines.

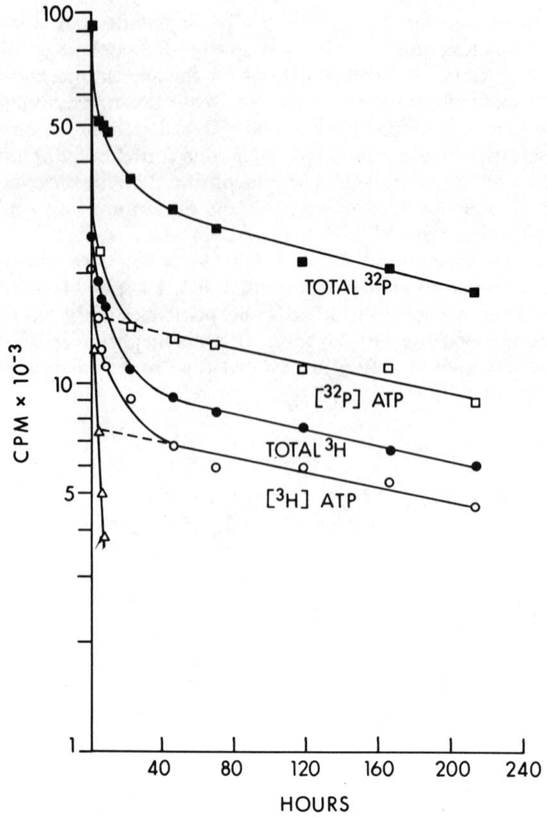

FIGURE 3. Turnover of labeled ATP in bovine adrenal medullary cells. Cells were labeled for 1 hour with $^{32}P_i$ and [^{3}H]adenosine and examined at the times indicated. Total ^{32}P and total ^{3}H refer to the total amounts of perchloric-acid-soluble radioactivity.

Subcellular fractionation studies were carried out to determine the localization of the labeled nucleotides immediately after and 3 days after a 1-hour pulse of [^{3}H]adenosine and $^{32}P_i$. The results are shown in FIGURE 4. The distribution of catecholamines and endogenous ATP was the same at both time points. Approximately 68% of the catecholamines and 50% of the ATP was present in the P_2 fraction. Immediately after labeling, the P_2 fraction contained only 4.4% of the total [^{32}P]ATP and 2.9% of the [^{3}H]ATP while the S_2 fraction contained 92% of the [^{32}P]ATP and 96% of the [^{3}H]ATP. Seventy-two hours after labeling, the distribution of [^{32}P]ATP and endogenous ATP was the same; approximately 50% of the compounds were present in the P_2 fraction and 30% in the S_2 fraction. The distribution of endogenous ATP and [^{3}H]ATP were, however, markedly different; only 25% of the [^{3}H]ATP was present in the P_2 fraction and 65% was present in the S_2 fraction. In the 0 to 24 hour interval following the labeling (data not shown), there was a threefold increase in the absolute amount of [^{32}P]ATP and a twofold increase in the amount of [^{3}H]ATP in the P_2 fraction while the total amounts of [^{32}P]ATP and [^{3}H]ATP had declined to 31% and 25% respectively of their values determined immediately after labeling.

The P_2 fraction contains mitochondria as well as chromaffin vesicles. To further localize the endogenous and labeled ATP, the P_2 fraction prepared from cells immediately and 72 hours after labeling were centrifuged through a Renograffin density gradient. This gradient clearly separates chromaffin vesicles (fractions 1-10) from mitochondria (fractions 14-18).[6] At both time points, 85-95% of the endogenous and labeled ATP and 85% of the catecholamines were present in fractions 1 to 10, while only trace amounts were detected in the mitochondrial fractions. Thus, practically all of the labeled and endogenous ATP in the P_2 fraction is present in chromaffin vesicles.

From the catecholamine and ATP content of the P_2 fraction we can estimate the percent of the total cellular ATP contained within the chromaffin vesicles. Practically all of the cellular catecholamines are localized in the chromaffin vesicles and practically all of the ATP in the P_2 fraction is present in chromaffin vesicles. Normalizing the recovery of catecholamines in the P_2 fraction to 100% and applying the same normalization factor to the ATP content, we estimate 75% of the total cellular ATP is present in chromaffin vesicles. Using NMR spectroscopy, Bevington *et al.*[12] estimated that the chromaffin vesicles of isolated bovine adrenal medullary cells contain 75% of the total ATP content. Applying similar consideration to the distribution of labeled

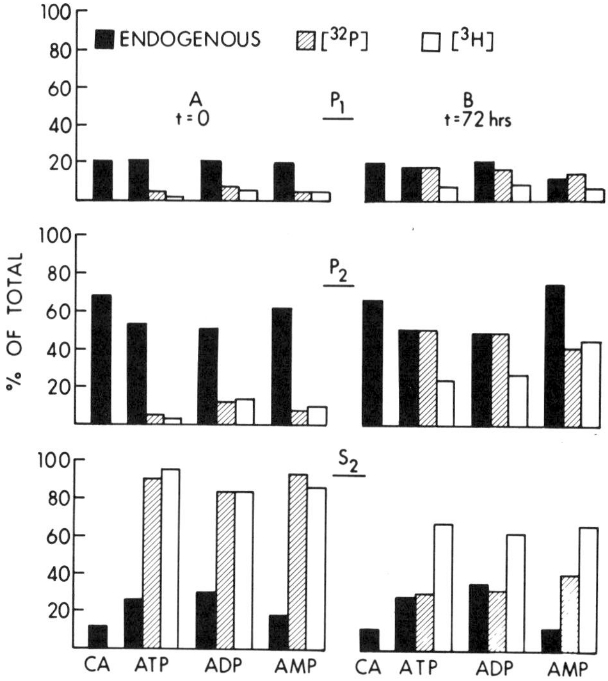

FIGURE 4. Distribution of catecholamines and endogenous and labeled adenine nucleotides in subcellular fractions. Cells were labeled for 1 hour with $^{32}P_i$ and [^{3}H]adenosine and examined immediately after labeling and 72 hours after labeling. The distribution was calculated as the percentage of the total of each component present in the fraction. P_1 is the 1000-g pellet, P_2 is the 20,000-g pellet, and S_2 is the 20,000-g supernatant.

ATP, we estimate that 3 days after labeling, 75% of the [^{32}P]ATP and 36% of the [^{3}H]ATP is present in the chromaffin vesicle pool and 25% of the [^{32}P]ATP and 64% of the [^{3}H]ATP is present in an extravesicular pool.

The presence of a persistent pool of extravesicular-labeled ATP with a turnover time similar to that of the chromaffin vesicle pool was confirmed by additional studies. Akkerman *et al.*[13] have shown that addition of a combination of metabolic inhibitors (NaCN, 2-deoxyglucose, and D-glucono-1,5-lactone) to platelets results in immediate cessation of energy production and a rapid decline in ATP, which reflects the velocity of energy consumption immediately prior to addition of the inhibitors. Addition of the metabolic inhibitors to adrenal medullary cells immediately after labeling resulted in a rapid loss of labeled ATP (FIG. 5). A more detailed kinetic analysis (FIG. 6) at early time points after addition of the inhibitors gave a $t_{1/2}$ of 7 min for the decline in labeled ATP compared to a $t_{1/2}$ of 3.5 to 4.5 hours in untreated cells. Within 30 minutes the labeled pools of both [^{32}P]ATP and [^{3}H]ATP were depleted by more than 90%, while the endogenous pool of ATP was depleted by only 25%. Addition of the metabolic inhibitors to cells 48 hours after labeling also caused an initial rapid depletion of the labeled ATP with a $t_{1/2}$ of 5 to 8 minutes, but resulted in only a 70% depletion of the [^{3}H]ATP even after a 90-minute exposure and only a 25% depletion of the [^{32}P]ATP and endogenous ATP. These values are similar to those found for the distribution of endogenous and labeled ATP into vesicular and extravesicular pools

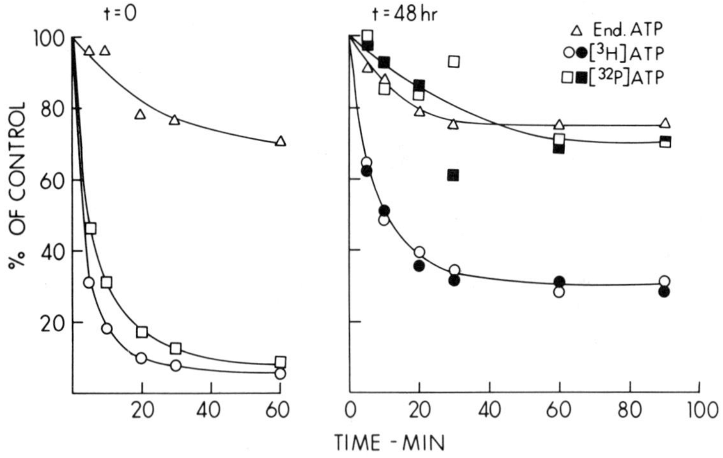

FIGURE 5. Effect of metabolic inhibitors on ATP content of cultured bovine adrenal medullary cells. Cells were labeled for 1 hour at 37° with [^{3}H]adenosine and $^{32}P_i$. At the end of the labeling period, cells were rinsed. Two plates of cells (24 well cluster plates) were reincubated for 48 hours in normal culture medium. A third plate of cells (t = 0) was incubated with Locke solution at room temperature immediately after labeling. Twelve of the wells received 1 mM NaCN, 30 mM DG, and 10 mM GL; the remaining 12 wells served as uninhibited controls. At the indicated times the medium was withdrawn from duplicate well and PCA extracts of the cells prepared. The PCA extracts were examined for endogenous ATP (△) by HPLC and for [^{3}H]ATP (○,●) and [^{32}P]ATP (□,■) by thin-layer chromatography. The two plates of cells which had been incubated for 48 hours after labeling were treated similarly. One plate (12 wells) received CN, DG, and GL as above (*open symbols*) and the other plates received 1 μM antimycin, 30 mM DG, and 10 mM GL (*closed symbols*). The remaining 12 wells from each plate served as uninhibited controls.

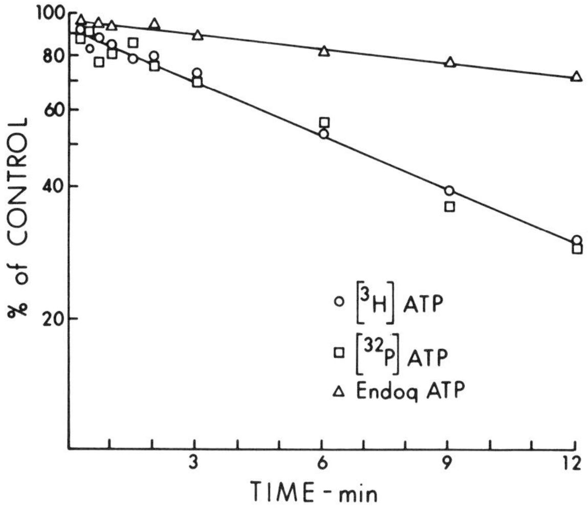

FIGURE 6. Effect of metabolic inhibitors on ATP turnover. Cells were labeled for 1 hour with $^{32}P_i$ and [^{3}H]adenosine. After washing, antimycin (4 μg/ml) and 2-deoxyglucose (30 mM) were added and the cells examined at the times indicated. The data are expressed as percentage of uninhibited controls. These studies were carried out at room temperature (23°C).

by subcellular fractionation. Thus, the metabolic inhibitors deplete the extravesicular pool of ATP but have no effect on the vesicular pool. Use of the metabolic inhibitors is a convenient and reliable tool for differentiating metabolic and vesicular or nonmetabolic pools of ATP.

A second method for differentiating vesicular from cytosolic pools of ATP is by treatment of the cells with digitonin. Digitonin in Ca^{2+}-free medium permeabilizes the plasma membrane, allowing the cytosolic content, including large protein molecules, to diffuse out of the cell.[14,15] With appropriate concentrations of digitonin, 10-20 μM, the chromaffin vesicles remain intact. Treatment of cells with digitonin immediately after labeling with [^{3}H]adenosine and $^{32}P_i$ resulted in a 98% loss of the acid-soluble ^{32}P-labeled compounds and an 88% loss of the acid-soluble ^{3}H-labeled compounds within 18 minutes. Exposure of the cells to digitonin 48 hours after labeling resulted in only a 55% loss of acid-soluble ^{32}P-labeled compounds and a 76% loss of acid-soluble ^{3}H-labeled compounds.

Examination of the cell content at the end of the incubation period showed a 95% loss of [^{3}H]ATP and a 90% loss of [^{32}P]ATP in cells exposed to digitonin immediately after labeling, similar to that observed after treatment with metabolic inhibitors. Forty-eight hours after labeling, exposure to digitonin resulted in a 79% loss of [^{3}H]ATP but only a 32% loss of [^{32}P]ATP. At both times there was only a 20-23% loss of endogenous ATP. Most of the [^{3}H]ATP which was lost from the cells was recovered in the medium as a combination of [^{3}H]ATP and [^{3}H]ADP, but only about 50% of the [^{32}P]ATP was recovered as a combination of [^{32}P]ADP and [^{32}P]ATP. Only 1-2% of the catecholamine content was released by the digitonin treatment.

The persistent labeled pool of extravesicular [^{3}H]ATP is most likely due to recycling of the [^{3}H]adenosine moiety of ATP and is similar in bovine adrenal medullary cells and bovine adrenal cortical cells (FIG. 7). In this study cortical cells were prepared in the same manner as that used for medullary cells and maintained under the same conditions. After removal of the medulla, which was used for preparing medullary cell cultures, the cortex was scraped clean of medullary particles, minced, and further digested with collagenase. Both cortical and medullary cells were plated at a density

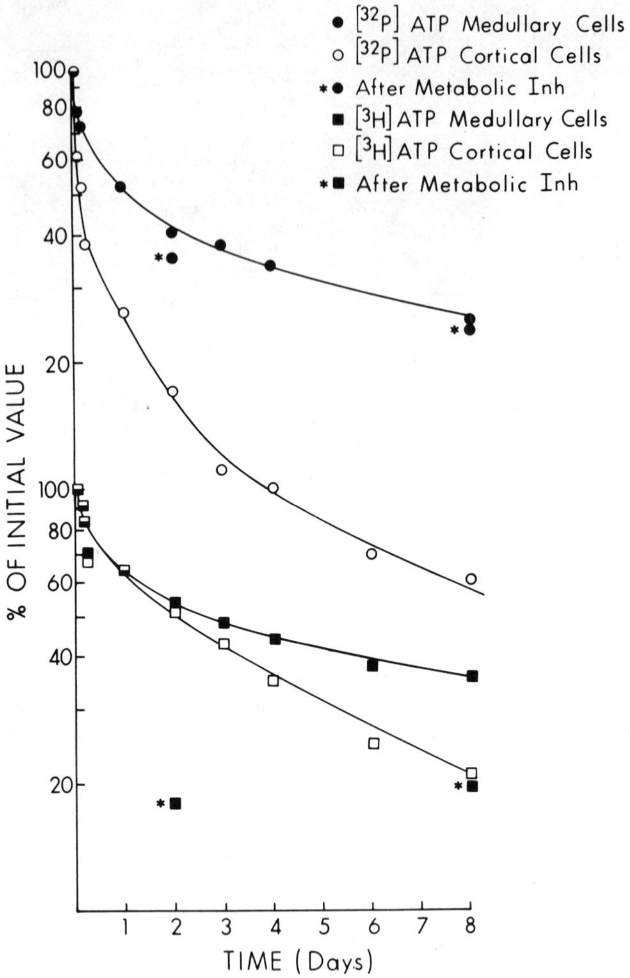

FIGURE 7. Turnover of ATP in bovine adrenal medullary and adrenal cortical cells. Cells were labeled with $^{32}P_i$ and [^{3}H]adenosine for 1 hour. After washing, cells were examined at the indicated times. At 2 days and 8 days duplicate wells were treated with antimycin (4 μg/ml) and 2-deoxyglucose (30 mM) for 30 min at 37° (*) before perchloric acid extracts were made.

of 5 $\times$ 10^5 cells per 2 cm^2. Cells were labeled with [^{3}H]adenosine and [^{32}P]P_i and examined at different times for up to 8 days. To determine the amounts of extravesicular ATP, cells were treated with antimycin and 2-deoxyglucose for 30 min prior to extraction with perchloric acid. Treatment of the cells with inhibitors 2 days after labeling resulted in a 95% depletion of [^{32}P]ATP in the cortical cells but only a 13%

depletion of [^{32}P]ATP in the medullary cells. Treatment of the cells with inhibitors 8 days after labeling resulted in almost complete depletion of the residual [^{32}P]ATP in cortical cells but no depletion in the medullary cells. The endogenous pool of ATP in cortical cells was depleted by 90% but only by 17-24% in medullary cells. Thus, the observed differences in the decline of [^{32}P]ATP is due to the sequestration of ATP in the chromaffin vesicles.

Treatment of cortical cells with the inhibitors also resulted in a 95% loss of [^{3}H]ATP at both 2 days and 8 days after labeling, while treatment of medullary cells resulted in a 66% loss of [^{3}H]ATP 2 days after labeling and a 45% loss 8 days after labeling.

An evaluation of the salvage pathway of purine metabolism was also obtained by determining the amount of [^{3}H]GTP present. Between 2 and 8 days the amount of [^{3}H]GTP expressed as a percent of [^{3}H]ATP remained constant in both cortical and medullary cells and contained 3.6 and 2.9%, respectively, of the radioactivity present in [^{3}H]ATP.

The accumulation of labeled ATP within chromaffin vesicles presents some puzzling features. Twenty-four and 72 hours after labeling, the specific activity of [^{32}P]ATP within the chromaffin vesicles was constant and the same as that in the extravesicular pool.[6] At these same times, the specific activity of [^{3}H]ATP in the vesicular pool was also constant but the specific activity of [^{3}H]ATP in the extravesicular pool was ninefold and fivefold higher respectively at 24 and 72 hours. From the data shown in FIGURE 7, even after 8 days the specific activity of the extravesicular pool of [^{3}H]ATP was fourfold higher than that of the vesicular pool. Thus the vesicular and extravesicular pools of ATP do not seem to exchange. The mechanism by which chromaffin vesicles acquire ATP is not clear. Evidence has been presented that ATP can be transported into vesicles by a carrier-mediated process; however, this carrier is not specific for adenine nucleotides but applies also to guanine, cytidine, and uridine nucleotides, and shows little preference for the monophosphate, diphosphate or triphosphate derivatives.[11,16,17] ATP within chromaffin vesicles can also become labeled by transphosphorylation and ATP-^{32}P exchange reactions.[18–20] These latter reactions would not lead to an accumulation of ATP within vesicles but they may explain the disparity between the accumulation of [^{32}P]ATP and [^{3}H]ATP in chromaffin vesicles.

REFERENCES

1. WINKLER, H. 1977. The biogenesis of adrenal chromaffin granules. Neuroscience **2:** 657-683.
2. WINKLER, H. & E. WESTHEAD. 1980. The molecular organization of adrenal chromaffin granules. Neuroscience **5:** 1803-1823.
3. PHILLIPS, J. H. 1982. Dynamic aspects of chromaffin granule structure. Neuroscience **7:** 1595-1609.
4. UNGAR, A. & J. H. PHILLIPS. 1983. Regulation of the adrenal medulla. Physiol. Rev. **63:** 787-843.
5. CORCORAN, J. J., S. P. WILSON & N. KIRSHNER. 1984. Flux of catecholamines through chromaffin vesicles in cultured bovine adrenal medullary cells. J. Biol. Chem. **259:** 6208-6214.
6. CORCORAN, J. J., S. P. WILSON & N. KIRSHNER. 1986. Turnover and storage of newly synthesized adenine nucleotides in bovine adrenal medullary cell cultures. J. Neurochem. **46:** 151-160.
7. CORCORAN, J. J., M. KORNER, B. CAUGHEY & N. KIRSHNER. 1986. Metabolic pools of ATP in cultured bovine adrenal medullary chromaffin cells. J. Neurochem. In press.

8. BUTTERWORTH, K. R. & M. MANN. 1957. The adrenaline and noradrenaline content of the adrenal gland of the cat following depletion by acetylcholine. Br. J. Pharmacol. **12:** 415-421.
9. BERTLER, A., N.-A. HILLARP & E. ROSENGREN. 1960. Some observations on the synthesis and storage of catecholamines in the adrenaline cells of the supradrenal medulla. Acta Physiol. Scand. **50:** 124-131.
10. STEVENS, P., R. L. ROBINSON, K. VAN DYKE & R. STITZEL. 1972. Studies on the synthesis and release of adenosine triphosphate-8-^{3}H in isolated perfused cat adrenal gland. J. Pharmacol. Exp. Therap. **181:** 463-471.
11. PEER, L. J., H. WINKLER, S. R. SNIDER & J. W. GIBB. 1976. Synthesis of nucleotides in adrenal medulla and their uptake into chromaffin granules. Biochem. Pharmacol. **25:** 311-315.
12. BEVINGTON, A., R. W. BRIGGS, G. K. RADDA & K. R. THULBORN. 1984. Phosphorus-31 nuclear magnetic resonance studies on pig adrenal glands. Neuroscience **11:** 281-286.
13. AKKERMAN, J.-W.N., G. GORTER, L. SCHRAMA & H. HOLMSEN. 1983. A novel technique for rapid determination of energy consumption in platelets. Biochem. J. **210:** 145-155.
14. WILSON, S. P. & N. KIRSHNER. 1983. Calcium-evoked secretion from digitonin-permeabilized adrenal medullary chromaffin cells. J. Biol. Chem. **258:** 4995-5000.
15. DUNN, L. A. & R. W. HOLZ. 1983. Catecholamine secretion from digitonin-treated adrenal medullary chromaffin cells. J. Biol. Chem. **258:** 4989-4993.
16. WEBER, A. & H. WINKLER. 1981. Specificity and mechanism of nucleotide uptake by adrenal chromaffin vesicles. Neuroscience **6:** 2269-2276.
17. WEBER, A., E. WESTHEAD & H. WINKLER. 1983. Specificity and properties of the nucleotide carrier in chromaffin granules from bovine adrenal medulla. Biochem. J. **210:** 789-794.
18. TAUGNER, G. 1974. Enzymatic phosphoryl-group transfer by a fraction of the water-soluble proteins of catecholamine storage vesicles. Naunyn-Schmiedeberg's Arch. Pharmacol. **285:** R82.
19. TAUGNER, G. & I. WUNDERLICH. 1979. Partial characterization of a phosphoryl group transferring enzyme in the membrane of catecholamine storage vesicles. Naunyn-Schmiedeberg's Arch. Pharmacol. **309:** 45-58.
20. TAUGNER, G., I. WUNDERLICH & D. JUNKER. 1980. Reversibility of ATP hydrolysis in catecholamine storage vesicles from bovine adrenal medulla. Naunyn-Schmiedeberg's Arch. Pharmacol. **315:** 129-138.

DISCUSSION OF THE PAPER

G. RUDNICK (*Yale University School of Medicine, New Haven, Conn.*): Is the incorporation of ATP into chromaffin granules affected by agents which stimulate or inhibit granule secretion?

N. KIRSHNER (*Duke University Medical Center, Durham, N.C.*): There is no data relevant to this question.

R. HOLZ (*University of Michigan Medical School, Ann Arbor, Mich.*): Does tetrabenazine decrease the number of secretory granules as measured by soluble and particulate dopamine β-hydroxylase?

N. KIRSHNER: There is no change in soluble or particulate dopamine β-hydroxylase.

M. GRIMES (*University of Oregon, Eugene, Oreg.*): Is there any evidence that the drugs tetrabenazine or reserpine inhibit reuptake of other small molecules in the secretory vesicle, like ascorbate?

N. KIRSHNER: I don't know of any studies that addressed this question.

H. B. POLLARD (*NIDDK, National Institutes of Health, Bethesda, Md.*): In recent studies by myself and Dr. Eduardo Rojas, we incidently measured cytosolic and granular ATP using a luciferin-luciferase detection system. We discriminated cytosolic ATP from the total by permeabilizing the plasma membrane using digitonin or high-voltage dielectric breakdown. The granule pool was then made available by adding Triton-X100. In no case did we get more than 5-10 of the total ATP as cytosolic. By contrast, you claim that 25 of the total is cytosolic. Is this perhaps due to culture conditions? We use chromaffin cells in suspension culture to eliminate endothelial cell concentration. Endothelial cells do contain substantial amounts of ATP. Could this be raising your apparent cytosolic ATP compartment?

N. KIRSHNER: I don't think culture conditions or cell contamination can account for the differences between your data and ours. Bevington *et al.*,[12] using NMR spectroscopy, also found that about 25% of the ATP control of freshly prepared adrenal medullary cells were present in the extravesicular compartment.

Acetylcholine Transport: Fundamental Properties and Effects of Pharmacologic Agents[a]

STANLEY M. PARSONS, BEN A. BAHR, LAWRENCE M. GRACZ, ROSE KAUFMAN, WAYNE D. KORNREICH, LENA NILSSON, AND GARY A. ROGERS

Department of Chemistry
University of California, Santa Barbara
Santa Barbara, California 93106

INTRODUCTION

Acetylcholine (ACh) is found in association with synaptic vesicle fractions obtained from mammalian brain, diaphragm and myenteric plexus, squid head ganglion, PC12 cells, and electric eel, catfish, and ray electric organs. Much well-known evidence supports the hypothesis that evoked release of ACh from nerve terminals occurs from the synaptic vesicle stores by means of calcium-dependent exocytosis. Thus, storage of ACh by the vesicles is an important step in cholinergic neurotransmission, and it is of fundamental importance that we understand this process. In particular, we would like to identify all of the enzymes, proteins, and small molecules which are involved, understand the electrical and osmotic transport relationships among the small molecules, and determine whether physiological regulation operates at this level of the cholinergic synapse. Progress on these molecular aspects of cholinergic vesicle function has been difficult until relatively recently. Biochemical studies are now possible, due in large measure to the efforts of Whittaker,[1] Kelly,[2] and co-workers, who demonstrated that one can isolate milligram quantities of highly purified homogeneously cholinergic synaptic vesicles from the electric organ of marine rays such as *Torpedo.*

[a] This work was supported by a grant from the Muscular Dystrophy Association and by grant no. NS15047 from the National Institute of Neurological and Communicative Disorders and Stroke.

THE CHOLINERGIC VESICLE ATPase

Torpedo vesicle membrane contains a cytoplasmically oriented ATPase which has kinetic and inhibitor properties similar to those of the mitochondrial ATPase.[3–9] For example, Ca^{2+} will support the ATP hydrolysis activity almost as well as Mg^{2+}; neither Na^+, K^+, nor Cl^- are specifically required; the hydrolysis rate as a function of the ATP concentration exhibits biphasicity; bicarbonate and certain other complex anions allosterically stimulate the activity; and dicyclohexylcarbodiimide and 4-chloro-7-nitrobenzo-oxadiazole inhibit the activity. The vesicle ATPase is not identical to the mitochondrial ATPase, however, since it is not inhibited by oligomycin[5] and it is glycosylated.[10] Because this is the only ATPase in *Torpedo* vesicles that has been described, and because the ATPase is stimulated about 20% by exogenous ACh,[3] it has been proposed that a mitochondrial-like ATPase is responsible for active transport of ACh.[3]

THE ACh TRANSPORTER

The *Torpedo* synaptic vesicle membrane exhibits a high permeability coefficient of 2×10^{-9} cm/s for ACh,[11] which suggests that a carrier for ACh is present in the membrane. The properties of the putative carrier, however, depend strongly on whether the ATPase is activated. In the absence of MgATP, ACh is taken up electrophoretically by interior negatively charged vesicles, but the uptake process exhibits little specificity with respect to choline and other organic cations.[11,12] We refer to this process as passive transport. In the presence of MgATP, ACh is taken up with specificity and to a much greater extent (FIG. 1) into vesicles which are hypothesized to be interior positively charged (see model below). We refer to the MgATP-dependent uptake as active transport. It can achieve concentration gradients for [^{3}H]ACh of as much as 50 to 1. The rather drastic apparent changes in the electrical and substrate specificity characteristics of passive compared to active transport suggest substantial reorganization of the ACh transporter upon activation of the vesicle ATPase.

The active transport process is saturable with respect to the ACh concentration[13] and is inhibited by low temperatures and osmotic shock.[14] Chemical reagents which inhibit the vesicle ATPase inhibit the active transport of ACh,[15] and bicarbonate ion stimulates active transport of ACh.[16] Active transport also can be sustained by Ca^{2+} in the absence of Mg^{2+},[14] and neither Na^+, K^+, nor Cl^- are specifically required.[16] These parallels in the properties of ACh active transport and the ATPase suggest that the ATPase drives active transport of ACh through a saturable transporter which is not linked to common specific ions such as Na^+, K^+, and Cl^-.

FUNDAMENTAL MODEL OF ACh TRANSPORT

A sulfhydryl-reactive mercurial blocks active transport of ACh without affecting the ATPase activity,[4,15] and mitochondrial uncouplers stimulate the ATPase in a manner coincident with inhibition of ACh active transport.[17,18] These and other results demonstrate that the ATPase and ACh transport activities can be functionally dis-

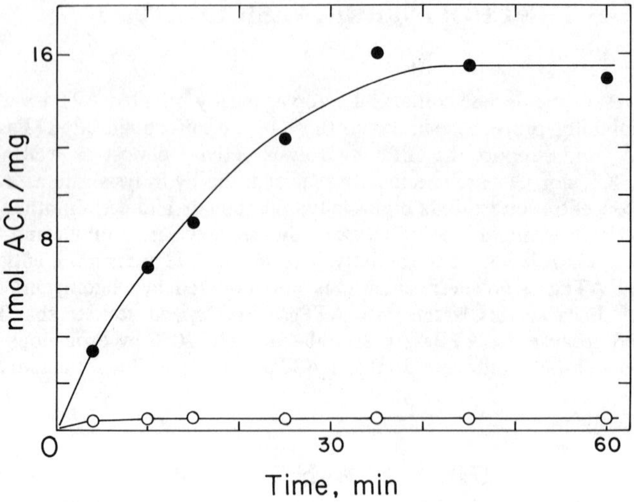

FIGURE 1. Time dependence for transport of ACh by cholinergic synaptic vesicles. VP_1 *Torpedo californica* electric organ synaptic vesicles were isolated as described.[18] Briefly stated, this involves differential sedimentation velocity pelleting, equilibrium buoyant density banding, and controlled-pore glass bead filtration of vesicles in isosmotic glycine-sucrose solutions. Pure VP_1 synaptic vesicles (0.675 mg vesicle protein/ml) treated 1 h with 0.15 mM diethyl-*p*-nitrophenylphosphate in 0.60 M glycine, 0.20 M *N*-2-hydroxyethylpiperazine-*N'*-2-ethanesulfonic acid, 1 mM EDTA, 1 mM EGTA, and 0.02% KN_3 titrated to pH 7.4 with 0.80 M KOH were incubated with 62.5 μM [^{3}H]ACh in the absence (○) or presence (●) of 5 mM MgATP, 20 mM $KHCO_3$, 2 mM $MgCl_2$, and an ATP regenerating system composed of 5 mM potassium phosphoenolpyruvate and pyruvate kinase for the indicated times, after which 45 μl was applied to a Millipore filter, which was washed by vacuum filtration. (From Anderson *et al.*[35] Reprinted by permission from the *Journal of Neurochemistry.*)

sociated from each other. Thus, separate proteins probably are involved. We have formulated a preliminary working model of the ACh active transport system, which is shown in FIGURE 2. The ATPase pumps protons into the vesicle. Evidence for the required internal acidification has been reported.[9,19–20] If this is the primary transport event, the vesicles presumably become internally positively charged as well as acidified. A separate ACh transporter is hypothesized to exist. It would draw upon the electrochemical proton gradient to drive ACh uptake, perhaps via exchange of protons for ACh. Howard,[21] Michaelson,[19] and co-workers also have presented evidence for a proton-coupled ACh storage system.

Although this model is consistent with the available evidence, it is nevertheless certain that it is an incomplete picture even at the basic level of ion movements. When beginning with a recently formed or recovered empty vesicle, a total of about 40,000 molecules of ACh, equivalent to an internal concentration of about 0.6 M, must be transported into each vesicle before it is fully loaded.[22] This can occur only if coupled ion movements maintain approximate electroneutrality and isosmolarity. Thus, a more complete model of ACh storage will include additional ions. The identities of these other ions and the transport systems responsible for their movements are not known, although it is possible that ATP[22] and an ATP translocase[23] might fulfill these functions in part.

ACh STORAGE IS PHARMACOLOGICALLY UNIQUE

The vesicle ATPase can be studied readily since it is an enzyme and it can be monitored even after solubilization of the membrane by detergent.[7] The hypothesized ACh transporter is more difficult to study since active transport of ACh depends not only on the transporter per se but also on the ATPase and the integrity of the vesicle membrane. Initial ambiguity in the site of action of reagents that affect ACh active transport is a problem, and of course the characteristic function of the transporter is lost in a detergent-solubilized state. An ideal solution to the practical difficulties encountered during study of the ACh transporter might be found by exploiting a storage-specific cholinergic pharmacology or toxicology. None are known for certain, but progress in the development of a new pharmacology is being made (see below).

The first step was to determine whether the pharmacological profile of the ACh active transport system is unique. Known cholinergic drugs were screened for effects.[24] A number of inhibitors of active transport were found, but the rank orders were in no case similar to those for action of the drugs on their known primary targets, such as nicotinic or muscarinic receptors (TABLE 1). Also, many noncholinergic drugs were screened for effects, but nothing significant was found.[24,25] Thus the ACh active transport system is pharmacologically unique.

THE AH5183 DRUG FAMILY

None of the drugs in TABLE 1 are particularly potent against ACh transport, and there was no reason to suppose that any of them would be particularly useful to us. However, the compound DL-*trans*-2-(4-phenylpiperidino) cyclohexanol (AH5183) was shown in the late 1960s and early 1970s to block neuromuscular transmission with unusual characteristics of action, and Marshall[26] hypothesized that the drug acts to block ACh storage by synaptic vesicles. More recently AH5183 has been shown to block evoked release of newly synthesized ACh from rat[27,28] and mouse[29] brain preparations, cat superior cervical ganglion,[30,31] PC12 cells,[32] frog neuromuscular junction,[33] and *Torpedo* electric organ synaptosomes[34] without significantly affecting high-affinity choline uptake, ACh synthesis, or Ca^{2+} influx. AH5183 was tested for its effect on

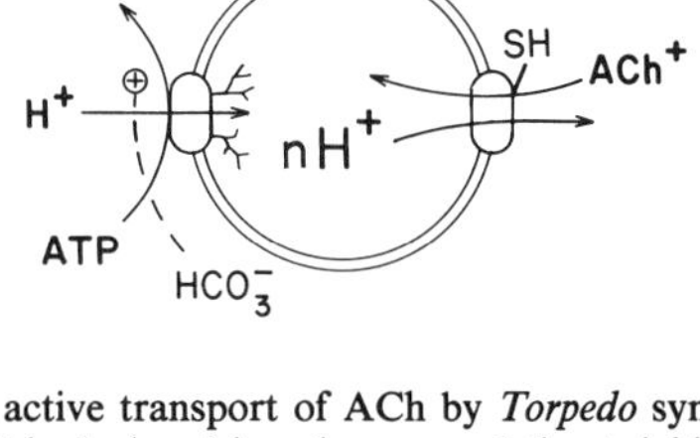

FIGURE 2. Model for the active transport of ACh by *Torpedo* synaptic vesicles. The number of protons pumped per ATP hydrolyzed is unknown and the stoichiometry of the hypothesized proton-ACh exchange is unknown. Glycosidic trees on the ATPase are indicated. ACh uptake is sensitive to a mercurial which does not affect the ATPase activity, suggesting the possible presence of a critical sulfhydryl group in the hypothesized ACh transporter.

TABLE 1. Inhibition of ACh Transport by Characterized Drugs[a]

Drug	ACh Transport IC_{50} (μM)[b]	Primary Target ED_{50} (μM)[c]
Nicotinic effectors		
Quinacrine	0.4	1.5
d-Tubocurarine	10	0.1-0.2
Phencyclidine	30	0.1-3
Procaine	100	1,000
Decamethonium	200	0.7-2
Nicotine	300	80
Hexamethonium	> 1,000	2-60
Gallamine	> 10,000	0.2-8
Carbamylcholine	> 10,000	0.4-45
Succinylcholine	> 10,000	
Muscarinic effectors		
Atropine	10	0.002
Oxotremorine	10	0.5
Pilocarpine	10	7
Bethanechol	> 10,000	200
Choline acetyltransferase inhibitors and substrates		
N-Hydroxyethyl-4-(1-napthylvinyl)pyridinium	5	0.6
N-Methyl-4-(1-napthylvinyl)pyridinium	30	0.5
Acetylcholine	200	
S-Acetylcoenzyme A	> 1,000	47
Coenzyme A	> 1,000	75
Choline	20,000	1,900
High-affinity choline uptake inhibitors		
Hemicholinium-15	100	7.7
Hemicholinium-3	300	0.06
Acetylcholinesterase inhibitors		
Physostigmine	300	0.06
Tetraethylpyophosphate	> 10,000	0.01

[a] Pure VP_1 vesicles were allowed to actively transport [^{3}H]ACh (50 μM) for 30 min in the presence of variable concentrations of drug.

[b] The concentration of drug that resulted in 50% inhibition of [^{3}H]ACh uptake.

[c] The concentration of drug that half-saturates the primary target for the drug, which is listed above it. The ED_{50} values were obtained from the literature.[24]

active transport of ACh by purified *Torpedo* synaptic vesicles with a striking result (FIG. 3). The drug was the most potent inhibitor found among the 80 which initially were screened; its IC_{50} value is about 40 nM[24] (see TABLE 2 for the structure). The drug has no effect on the ATPase activity. Thus, AH5183 is a candidate for study of the ACh transporter or other proteins important to ACh storage, assuming that it interacts with a specific binding site. In early work this was an important reservation since the drug is a tertiary amine. It was conceivable that the small amount of unprotonated drug that would be present could diffuse into the vesicle and neutralize the internal protons, thus destroying the driving force for ACh active transport.

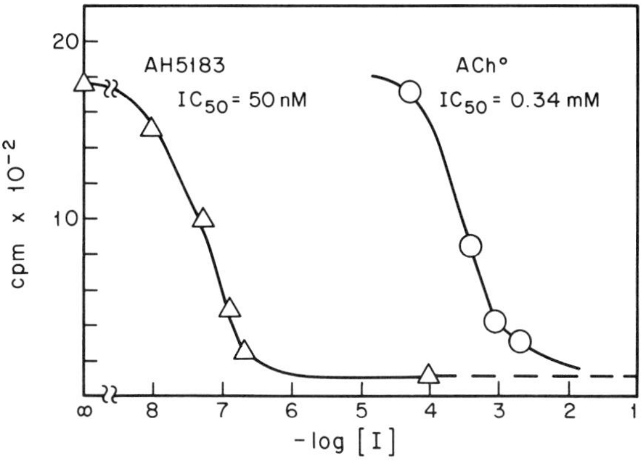

FIGURE 3. Inhibition of [^{3}H]ACh active transport. Active transport was carried out by pure VP_1 vesicles (0.3 mg/ml) in the presence of the indicated concentrations of either nonradioactive ACh (○) or nonradioactive AH5183 (△) for 30 min at 23°C and pH 7.8. The amount of transported [^{3}H]ACh was determined by Millipore filter vacuum filtration. Hyperbolic titration curves were fit to the data by nonlinear regression analysis and the best fits are plotted here on the negative logarithm of inhibitor concentration scale. The IC_{50} value for nonradioactive ACh was 0.34 mM and the IC_{50} value for AH5183 was 50 nM.

Inferential evidence that a receptor for AH5183 exists has been obtained from structure-activity studies. Over 70 new compounds that are analogues of AH5183 have been synthesized and screened for their effects on active transport of ACh by *Torpedo* vesicles. Some analogues that define which alterations do not impair pharmacological potency or that lead to useful derivatives are listed in TABLE 2. In the following discussion we refer to rings A, B, C, and D that correspond to the phenyl, piperidinyl, cyclohexanol, and, in some compounds, benzo rings, respectively, when reading left to right as shown in TABLE 2.

Comparisons of the IC_{50} value for AH5183 with those for compounds 2 and 7 show that a hydrocarbon moiety in the ring A region is necessary for drug potency and with those for compounds 3, 4, 5, and 6, that polar groups in the para position are unfavorable unless offset by additional hydrophobic groups. Data for compounds 10, 11, and 12, which vary in ring B structure, point up two pertinent facts. First, an attempt to have AH5183 more closely resemble the structure of choline by making the nitrogen quaternary lowers potency by 25-fold. Secondly, relatively small changes which would be expected to alter the dihedral angle between rings A and B also lower potency. Thus, by exchanging carbon for nitrogen, as in the pipirazine ring, or by introducing a double bond in conjugation with ring A, a geometry is favored in which the phenyl ring is pseudocoplanar with ring B. This relationship appears to be unfavorable.

The absolute necessity of ring C is demonstrated by the very low potency exhibited by compound 13. Furthermore, removal of the hydroxyl group or replacement with a sulfhydryl SH (compounds 14 and 15) lowers potency by a factor of 50. By examining structures of compounds 8 and 9, we see that the putative receptor exhibits at least 50-fold enantioselectivity, as expected of an optically active binding site. Compounds 16, 18, 19, and 20 all bear on the question of the conformational structure of ring C with regard to the amine and the alcohol. Compound 20, our most potent drug to

TABLE 2. Analogues of AH5183

		IC_{50} (nM)			IC_{50} (nM)
AH5183	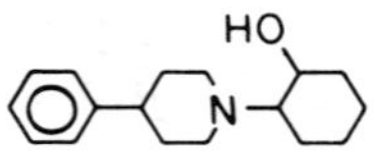	40	15.	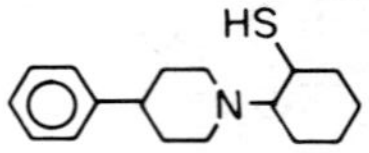	2,000
2.	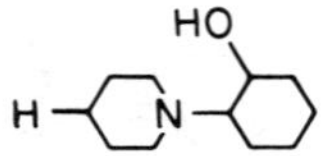	70,000	16.	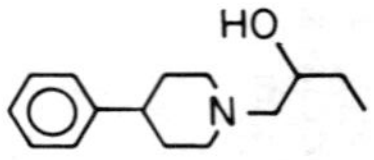	300
3.	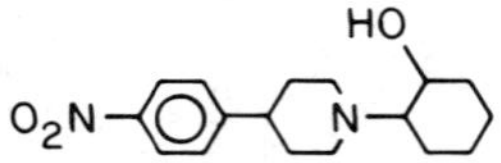	200	17.	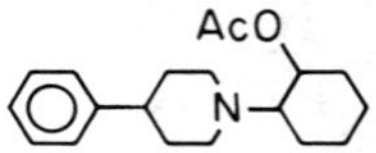	2,000
4.	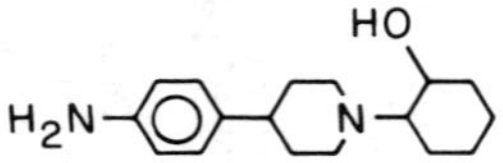	2,000	18.	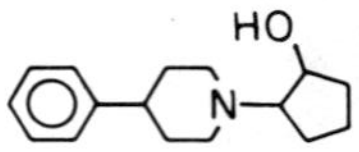	2,000
5.	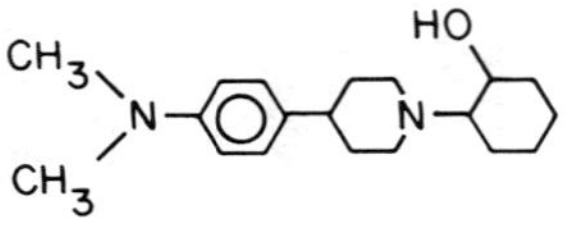	300	19.	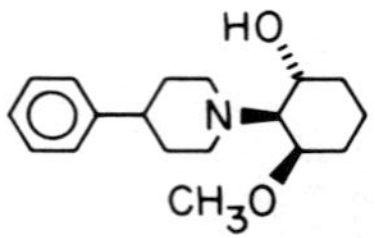	2,000
6.	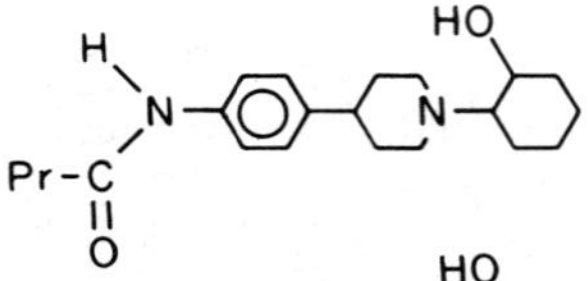	3,000	20.	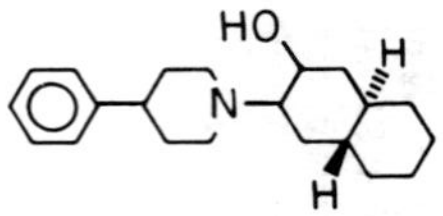	10
7.	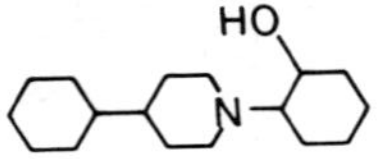	40	21.	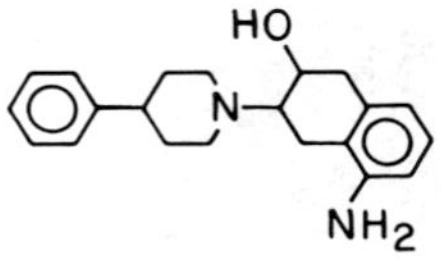	100

8.	R, R	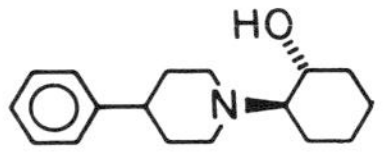	1,000	22.	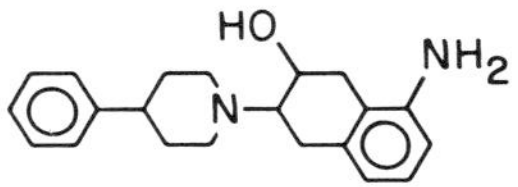	300
9.	S, S	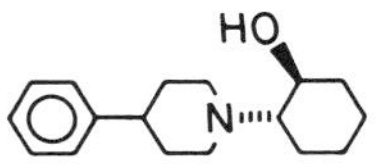	20	23.	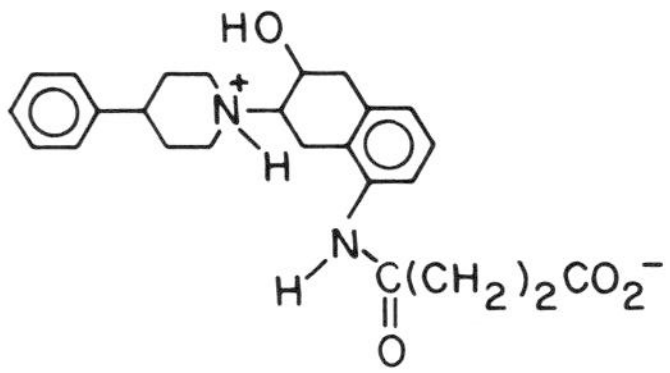	50
10.		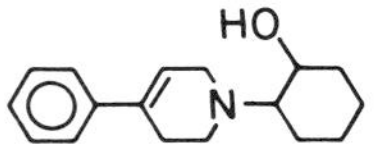	300	24.	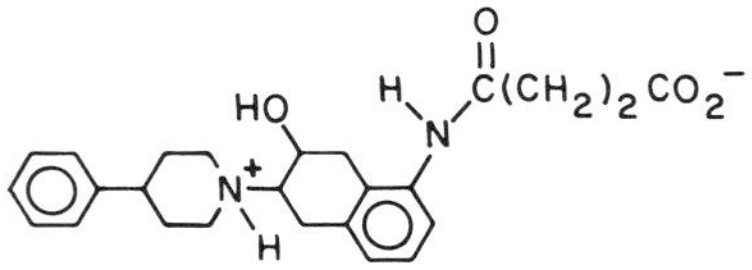	500
11.		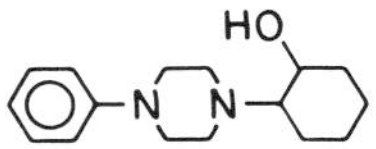	200	25.	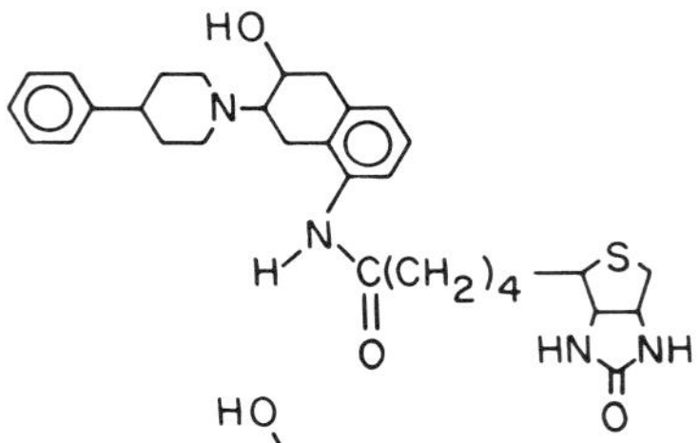	140
12.		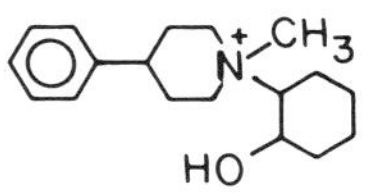	1,000	26.	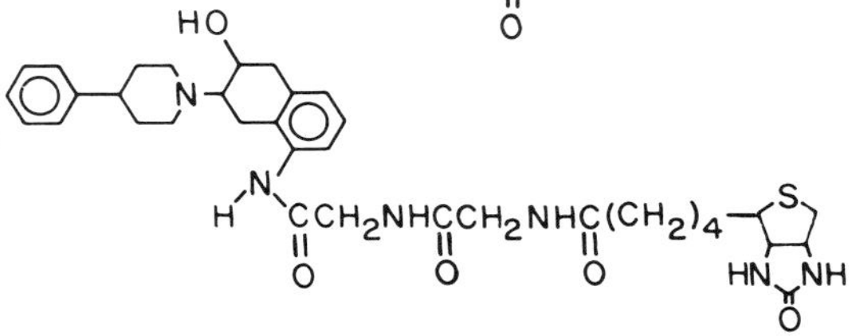	250
13.			90,000			
14.		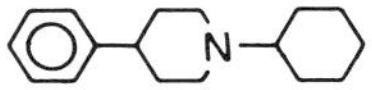	2,000			

date, locks the amine and alcohol groups into the diaxial conformation. Other substitutents or changes in ring C which would less favor a diaxial conformation have an adverse effect on drug activity. We conclude, therefore, that the drug binding site must have a moderately rigid spatial requirement for the amino-alcohol pair. In addition, an attempt to mimic ACh by acetylation of the alcohol (compound 17) lowers potency by 50-fold. Overall, the results strongly suggest the existence of a fastidious stereospecific receptor for AH5183.

Based on the observation that an additional ring D fused to ring C is spatially allowed (that is, compound 20), we have explored the effects of various aromatic ring systems at this site since from the chemical perspective aromatic rings are easier to modify than saturated rings. Comparison of data from isomers 21 and 22 and several like derivatives of each isomer has convinced us that the anilino group of 22 and anything attached to it experiences adverse steric interactions with the receptor. By contrast there appears to be little limit to the size of the group that can be attached to the anilino nitrogen of compound 21. These observations led us to synthesize compound 23. It is shown in the table as the zwitterion structure, which will be the overwhelmingly predominant species at the pH of the assay medium. Since the vesicle membrane should not be permeable to the zwitterion, we conclude that the AH5183 receptor resides on the external surface of the synaptic vesicle. Seeking further to explore potentially useful modifications of compound 21, we attached the biotinyl moiety (compounds 25 and 26) in order to capitalize on the biotin-avidin couple. While the bulk of the biotinyl group is tolerated, as evidenced by the respectable IC_{50}s of 140 and 250 nM, it appears that binding to the drugs by both the receptor and avidin is mutually exclusive (data not shown). This behavior possibly can be exploited to study reactivation of ACh active transport after rapid "destruction" of active drug by complexation with avidin.

THE MECHANISM OF AH5183 BLOCKADE

In steady-state initial velocity active transport measurements using purified *Torpedo* vesicles, AH5183 is a noncompetitive inhibitor of uptake with respect to ACh.[13] This result means that the drug can not inhibit uptake by binding to the ACh binding site on the outside of the transporter, since that would give competitive inhibition. Also, if active transport of ACh is allowed to proceed for a while before AH5183 is added, at least 10 molecules of the newly transported ACh are released from the vesicles for each AH5183 receptor that could be occupied.[35] This amplification effect suggests strongly that the drug acts indirectly to block ACh active transport.

THE AH5183 RECEPTOR

The structure-function studies provided the information required to produce the potent *l*-optical isomer of AH5183 in highly tritiated form.[36] With the availability of *l*-[^{3}H]AH5138 it has become possible to study the receptor directly. In a typical preparation of *Torpedo* vesicles there are about 3.7 ± 0.3 enantioselective saturable receptors for AH5138 per vesicle that exhibit an equilibrium dissociation constant of

34 nM and a first-order dissociation rate constant of 0.23 min^{-1} at 23°C.[36] ACh has no significant effect on the binding of *l*-[^{3}H]AH5183 to the receptor (FIG. 4). Thus, ACh does not bind to the AH5183 site, and ACh binding to the transporter active site is not linked conformationally to the AH5183 site.

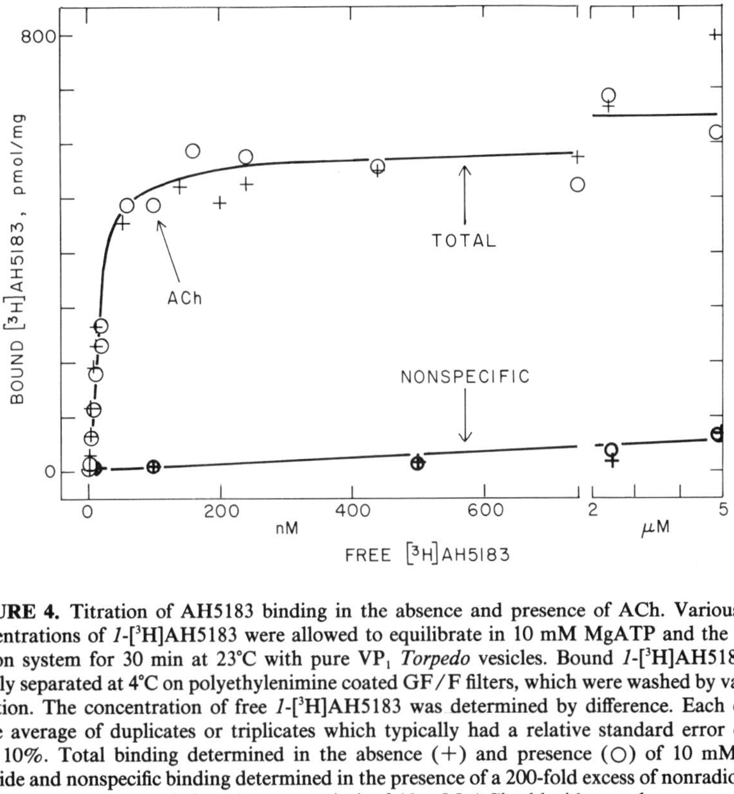

FIGURE 4. Titration of AH5183 binding in the absence and presence of ACh. Various total concentrations of *l*-[^{3}H]AH5183 were allowed to equilibrate in 10 mM MgATP and the regeneration system for 30 min at 23°C with pure VP_1 *Torpedo* vesicles. Bound *l*-[^{3}H]AH5183 was rapidly separated at 4°C on polyethylenimine coated GF/F filters, which were washed by vacuum filtration. The concentration of free *l*-[^{3}H]AH5183 was determined by difference. Each datum is the average of duplicates or triplicates which typically had a relative standard error of less than 10%. Total binding determined in the absence (+) and presence (○) of 10 mM ACh chloride and nonspecific binding determined in the presence of a 200-fold excess of nonradioactive AH5183 in the absence (+) and presence (○) of 10 mM ACh chloride are shown.

CHEMICAL STRUCTURE OF THE AH5183 RECEPTOR

Although the above results demonstrate the presence of a receptor in cholinergic vesicles for AH5183, they provide no direct information on its chemical nature. For example, we wish to determine whether it is a protein. Treatment of *Torpedo* vesicles with protein modification reagents provided the answer, as shown in TABLE 3. Binding of *l*-[^{3}H]AH5183 to its receptor was blocked completely by 1 mM concentrations of the lipid-insoluble sulfhydryl-group modifier *p*-chloromercuribenzene sulfonate and the lipid-soluble carboxyl-group modifier dicyclohexylcarbodiimide, but not completely by even 10 mM concentrations of reagents which react with other types of amino acid side chains. Also, the water-soluble carboxyl-group modifier 1-ethyl-3-(3-dimeth-

TABLE 3. Percent Inhibition of AH5183 Binding by Protein Modification Reagents[a]

Reagent	Concentration 1 mM	10 mM
4-(Chloromercuri)benzene-sulfonate	100	100
Trinitrobenzenesulfonate	−5	−35
N-Ethyl-5-phenylisoxazolium-3′-sulfonate	−20	−8
Iodoacetate	−21	11
D,L-Dithiothreitol	14	26
Dicyclohexylcarbodiimide	99	99
1-Ethyl-3-(3-dimethyl-aminopropyl)carbodiimide	44	56
Phenylglyoxal	37	31
4-Chloro-7-nitro-benzooxadiazole	69	94
Glutaraldehyde	32	56
4-Bromomethylbenzoate	32	42
	Time 2 h	24 h
Proteinase K	0	58
	Time 4 h	24 h
Proteinase K, cholate	90	100

[a] Pure VP_1 synaptic vesicles (0.11 mg/ml) were incubated in the presence of the indicated concentrations of each chemical reagent for 1 hour at 23°. The vesicle suspension then was equilibrated for 60 min with 450 nM total concentration of *l*-[^{3}H]AH5183 and the percentage of bound drug determined relative to untreated control. Negative percent inhibition means that drug binding was stimulated. Synaptic vesicles (0.11 mg/ml) were incubated in the presence of 0.15 mg proteinase K/ml or vesicles (0.07 mg/ml) solubilized in 0.5% sodium cholate were incubated in the presence of 0.1 mg proteinase K/ml, both at 23°, for the indicated times before the amount of bound drug was determined.

ylaminopropyl)-carbodiimide was a poor inhibitor of the AH5183 receptor. Proteinase K, a nonspecific protease, had little effect on *l*-[^{3}H]AH5183 binding in native vesicles, even after 24 hours of proteolysis, but in sodium-cholate-solubilized vesicles it rapidly destroyed the binding (TABLE 3). This behavior indicates that the AH5183 receptor is a protein that contains a critical sulfhydryl group accessible from the outside and a critical carboxyl group accessible from the lipid phase.

VP_1 AND VP_2 SYNAPTIC VESICLES

Cholinergic synaptic vesicles are known to occur in two forms, namely the resting fully loaded VP_1 vesicles studied above and the recycling empty VP_2 vesicles.[37] It is of interest to know if the AH5183 receptor also is present in VP_2 vesicles. This was examined at the density gradient step in the vesicle purification scheme where both types of vesicles still are present. FIGURE 5 demonstrates that a major peak of ACh transport and AH5183 binding activities banded about fraction 6, which had a density of 1.055 g/cm^3. This peak represents the classical VP_1 vesicles. In addition to that peak, another broad peak of ACh transport and AH5183 binding activities banded about fraction 17, which had a density of 1.08 g/cm^3. Apparent ACh transport in all the fractions was blocked by either AH5183 (10 μm) or the protonophoric uncoupler nigericin (1 μM). Thus, all of the observed ACh transport exhibited properties characteristic of authentic vesicular transport. Furthermore, the denser species were shown to be slightly smaller than VP_1 vesicles by means of size-exclusion chromatography of [^{3}H]ACh-labeled VP_1 vesicles mixed with [^{14}C]ACh-labeled dense species (not shown). The ACh transport and size properties of the denser species are those expected of VP_2 synaptic vesicles.[37]

Thus, VP_2 vesicles both transport ACh and bind AH5183. Close examination of FIGURE 5, however, reveals that the ratio of ACh transport to AH5183 binding observed for VP_2 vesicles is about 3.5-fold greater than for VP_1 vesicles. Although we can not say in this partially pure preparation of vesicles which marker has been altered, it appears that the properties of either ACh transport (including the ATPase) or the AH5183 receptor differ in VP_1 and VP_2 vesicles.

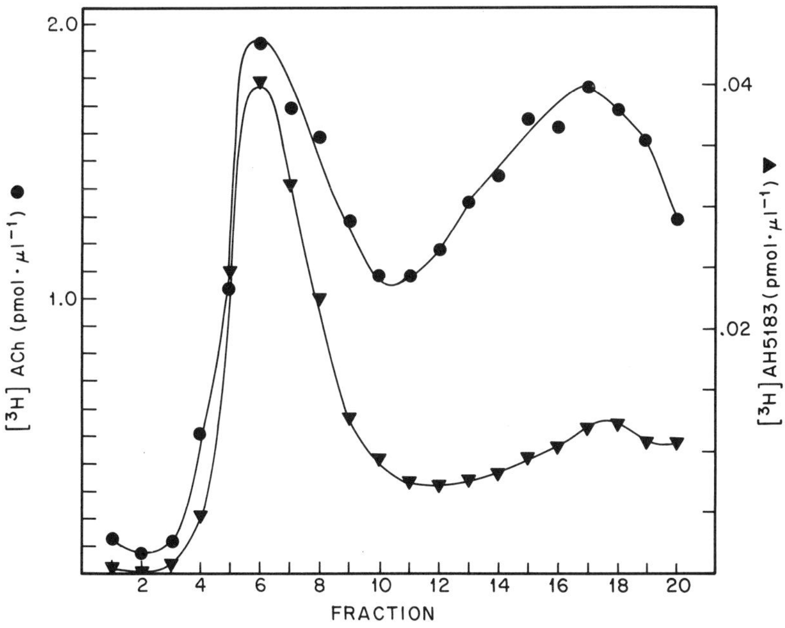

FIGURE 5. Specific AH5183 binding and transport of ACh in a density gradient containing crude synaptic vesicles. A crude synaptic vesicle pellet obtained by differential sedimentation velocity pelleting of homogenized electric organ was resuspended in isosmotic (0.8 M) buffer, centrifuged at 4°C at 196,000 $\times$g for 3.5 h in an isosmotic density gradient composed of glycine, sucrose, and Metrizamide, and fractionated. The bottom of the tube is to the right. The density of each fraction was determined from the refractive index, and *l*-[^{3}H]AH5183 binding at 500 nM total concentration and [^{3}H]ACh (50 μM) transport activity in 30 min at 23° was measured for each fraction.

CONCLUSION

Three components of the cholinergic vesicle ACh storage system have been described. These are the proton pumping ATPase, the ACh transporter, and the AH5183 receptor. All of the components appear to be cytoplasmically exposed proteins. It now is clear that the drug receptor binding site does not overlap the ACh binding site in the transporter, but whether the separate ligand sites lie in the same protein (the transporter) or in different proteins is not known. Clearly, what is required now is direct identification of the proteins corresponding to the three known components of ACh storage so we can interpret the drug-induced phenomena that are observed.

Regardless of its physical relationship to the ACh transporter, however, the AH5183 receptor may represent a unique intracellular mechanism by which cholinergic activity can be modulated. This includes not only quantal evoked release activity but also nonquantal leakage of ACh from terminals. Edwards *et al.*[38] have presented good evidence that the AH5183 receptor is linked to nonquantal leakage of ACh from motoneurons. A similar receptor-regulated tonic leakage of ACh from central neurons, if it occurs, could be an important determinant of neuronal excitability due to the summation of excitatory and inhibitory inputs characteristic of the CNS. The observation that some aspect of the ACh storage system of VP_2 synaptic vesicles is different from that of VP_1 vesicles suggests that this system is physiologically regulated in a long-lived manner. Thus, although we have no details yet, it seems that the ACh storage system exhibits the type of complexity expected if the AH5183 receptor does play an *in vivo* regulatory role.

REFERENCES

1. NAGY, A., R. R. BAKER, S. J. MORRIS & V. P. WHITTAKER. 1976. Brain Res. **109:** 285-309.
2. CARLSON, S. S., J. A. WAGNER & R. B. KELLY. 1978. Biochemistry **17:** 1188-1199.
3. BREER, H., S. J. MORRIS & V. P. WHITTAKER. 1977. Eur. J. Biochem. **80:** 313-318.
4. MICHAELSON, D. M. & I. OPHIR. 1980. J. Neurochem. **34:** 1483-1490.
5. ROTHLEIN, J. E. & S. M. PARSONS. 1979. Biochem. Biophys. Res. Commun. **88:** 1069-1076.
6. ROTHLEIN, J. E. & S. M. PARSONS. 1980. Biochem. Biophys. Res. Commun. **95:** 1869-1874.
7. ROTHLEIN, J. E. & S. M. PARSONS. 1982. J. Neurochem. **39:** 1660-1668.
8. DIEBLER, M. F. & S. LAZEREG. 1985. J. Neurochem. **44:** 1633-1641.
9. HARLOS, P., D. A. LEE & H. STADLER. 1984. Eur. J. Biochem. **144:** 441-446.
10. BATTEIGER, D. L. & S. M. PARSONS. 1986. Neurochem. Int. **8:** 249-253.
11. CARPENTER, R. S., R. KOENIGSBERGER & S. M. PARSONS. 1980. Biochemistry **19:** 4373-4379.
12. CARPENTER, R. S. & S. M. PARSONS. 1978. J. Biol. Chem. **253:** 326-329.
13. BAHR, B. A. & S. M. PARSONS. 1986. J. Neurochem. **46:** 1214-1218.
14. PARSONS, S. M., R. S. CARPENTER, R. KOENIGSBERGER & J. E. ROTHLEIN. 1982. Fed. Proc. **41:** 2765-2768.
15. PARSONS, S. M. & R. KOENIGSBERGER. 1980. Proc. Natl. Acad. Sci. U.S.A. **77:** 6234-6238.
16. KOENIGSBERGER, R. & S. M. PARSONS. 1980. Biochem. Biophys. Res. Comm. **94:** 305-312.
17. ANDERSON, D. C., S. C. KING & S. M. PARSONS. 1982. Biochem. Biophys. Res. Comm. **103:** 422-428.
18. ANDERSON, D. C., S. C. KING & S. M. PARSONS. 1982. Biochemistry **21:** 3037-3043.
19. MICHAELSON, D. M. & I. ANGEL. 1980. Life Sci. **27:** 39-44.
20. FULDNER, H. H. & H. STADLER. 1982. Eur. J. Biochem. **121:** 519-524.
21. TOLL, L. & B. D. HOWARD. 1980. J. Biol. Chem. **255:** 1787-1789.
22. MORRIS, S. J. 1980. Neuroscience **5:** 1509-1516.

23. LEE, D. A. & V. WITZEMANN. 1983. Biochemistry **22:** 6123-6130.
24. ANDERSON, D. C., S. C. KING & S. M. PARSONS. 1983. Mol. Pharmacol. **24:** 48-54.
25. ANDERSON, D. C., S. C. KING & S. M. PARSONS. 1983. Mol. Pharmacol. **24:** 55-59.
26. MARSHALL, I. G. 1970. Br. J. Pharmac. **38:** 503-516.
27. JOPE, R. S. & G. V. W. JOHNSON. 1986. Mol. Pharmac. **29:** 45-51.
28. OTERO, D. H., F. WILBEPIN & E. M. MEYER. 1985. Brain Res. **359:** 208-214.
29. CARROLL, P. T. 1985. Brain Res. **358:** 200-209.
30. COLLIER, B. & S. A. WELNER. 1985. J. Neurochem. **45:** 210-218.
31. COLLIER, B., S. A. WELNER, J. RICNY & D. M. ARAUJO. 1986. J. Neurochem. **46:** 822-830.
32. MELEGA, W. P. & B. D. HOWARD. 1984. Proc. Natl. Acad. Sci. U.S.A. **81:** 6536-6538.
33. VAN DER KLOOT, W. 1986. Pfluegers Arch. **406:** 83-85.
34. MICHAELSON, D. M., M. BURSTEIN & R. LICHT. 1986. J. Biol. Chem. **261:** 6831-6835.
35. ANDERSON, D. C., B. A. BAHR & S. M. PARSONS. 1986. J. Neurochem. **46:** 1207-1213.
36. BAHR, B. A. & S. M. PARSONS. 1986. Proc. Natl. Acad. Sci. U.S.A. **83:** 2267-2270.
37. ZIMMERMANN, H. & D. R. DENSTON. 1977. Neuroscience **2:** 715-730.
38. EDWARDS, C., V. DOLEZAL, S. TUCEK, F. VYSKOCIL & H. ZEMKOVA. 1985. Proc. Natl. Acad. Sci. U.S.A. **82:** 3514-3518.

DISCUSSION OF THE PAPER

W. VAN DER KLOOT (*State University of New York at Stony Brook, Stony Brook, N.Y.*): You can incorporate about 260 acetylcholine molecules into a vesicle that normally encloses tens of thousands of acetylcholine molecules. What are your thoughts about the difference in numbers?

S. M. PARSONS (*University of California, Santa Barbara, Calif.*): There are both practical factors which we know to limit uptake and philosophical comforts. The vesicles hydrolyze and deplete the ATP supply in these test tube experiments in about an hour so that acetylcholine pumping ceases. The *Torpedo* in contrast requires about 3 days to replenish its acetylcholine stores after exhaustive stimulation. Also, our typical assay is subsaturating in [^{3}H]acetylcholine and so optimal conditions are not obtained. Although our acetylcholine transport model contains some part of the truth, we know that it is incomplete since there must be coupled ion movements which maintain electroneutrality and isosmolarity. We don't know the identity of these ions, and it is possible that our uptake is limited by a missing ion.

Philosophically, biochemists must exhibit some humility in the early stages of subcellular investigation since we do terrible insult to the tissue in obtaining our materials. It is possible that damage has been done or that important peripheral protein factors are partially removed. Thus, I think it's important that we demonstrate qualitatively relevant behavior for the acetylcholine storage system, and work toward eventual quantitative reconstitution of the system. Clearly what we are studying reflects the *in vivo* system since AH5183 and many analogues of variable potency exhibit consistent behavior in a wide range of intact preparations.

J.-P. HENRY (*Institut de Biologie Physico-Chimique, Paris, France*): I have two questions. First, what is the effect of AH5183 on the ATPase activity and the electrochemical proton gradient? Second, have you looked at the subcellular distribution of drug binding activity?

S. M. PARSONS: In answer to your first question, AH5183 has no effect on the ATPase or on the $\Delta\mu H^+$. And in answer to your second question, some AH5183 binding activity appears to be present in a membrane fragment fraction of cytoplasmic membrane origin.

Enkephalin Uptake into Cholinergic Synaptic Vesicles and Nerve Terminals[a]

DANIEL M. MICHAELSON AND
DANIELA WIEN-NAOR

Department of Biochemistry
Tel Aviv University
Ramat Aviv 69978, Israel

INTRODUCTION

The opioid peptides, Leu-enkephalin and Met-enkephalin, are stored in a variety of peripheral and central neurons, where they are thought to act as neurotransmitters and/or neuromodulators (for review see ref. 1). The action of enkephalins, like that of other neurotransmitters, is terminated by specific degradative enzymes (*e.g.*, enkephalinases and aminopeptidases). However, whereas classical neurotransmitters, (*e.g.*, acetylcholine [ACh] and catecholamines), are recycled presynaptically following their release, it has generally been assumed that the released enkephalins are replenished via axoplasmic transport of newly formed peptides from the perikarya.[1–3] Previous studies show that some neurons contain enkephalins together with either ACh or catecholamines and that the classical neurotransmitters and the opioid peptides are coreleased following presynaptic stimulation.[4–7] It is not yet known, however, if these neurons replenish their opioid peptides only by axoplasmic transport or whether released enkephalins can be recycled.

Recent findings show that the cholinergic neurons of the *Torpedo* electric organ contain an enkephalin-like peptide[8] as well as regulatory presynaptic opiate receptors,[9] thus suggesting that the *Torpedo* electric organ is a useful model for the study of the life cycle of enkephalin-like peptides within the cholinergic nerve terminal. In the present study we utilized this model to investigate the possibility that enkephalins can be recycled following their release. This was done by examining the possibility that externally added enkephalins can be taken up by isolated *Torpedo* nerve terminals (synaptosomes) and that they can be subsequently transported into synaptic vesicles.

[a] This work was supported in part by grants from the U.S. National Institutes of Health (MH 40294) and from the U.S.-Israel Binational Science Foundation (85/0097).

EXPERIMENTAL PROCEDURES

Uptake of [^{3}H]Leu-Enkephalin and [^{3}H][D-Ala^2, D-Leu^5]Enkephalin into **Torpedo** *Synaptosomes*

Cholinergic synaptosomes were isolated from homogenates of freshly excised *Torpedo* electric organs by differential and sucrose gradient centrifugation as previously described.[10] The purified synaptosomes were centrifuged (175,000 g $\times$ 45 min) and resuspended (20-25 mg protein/ml) in modified *Torpedo* buffer (modified TB: 250 mM NaCl, 4.8 mM KCl, 2.4 mM $MgCl_2$, 10 mM glucose, 260 mM sucrose, 1.2 mM Na phosphate buffer, pH 7.2), after which they were kept at 4°C for up to 3 h prior to use. Uptake of the radiolabeled enkephalins was initiated by a 1:1 dilution of the synaptosomes with modified TB, which contained the specified concentrations of either [^{3}H][D-Ala^2-D-Leu^5]enkephalin ([^{3}H]DADL-enkephalin) or [^{3}H]Leu-enkephalin (2 μCi/ml). Unless otherwise specified, the reaction mixture also contained captopril (10 μM) and puromycin (50 μM), which inhibit endogenous peptidase activities[11] and which were added 10 min prior to the initiation of uptake. The amount of radioactivity taken up by the synaptosomes at each time point was determined by placing aliquots (20 μl in triplicate) of the reaction mixture on 0.45 μ Millipore filters (HAWP) and rapidly washing (4 $\times$ 2.5 ml) with ice-cold buffer. The amount of radioactivity retained on the filter was then measured by scintillation spectrometry. When the effects of opioids and opiates were examined they were added to the synaptosomes 3 min prior to the initiation of uptake. When the effects of salts were examined the synaptosomes were treated as outlined above except that they were resuspended and kept in an isotonic glucose buffer (0.8 M glucose, 1.2 mM Na phosphate buffer, pH 7.2). The indicated salts were added to the synaptosomes from isotonic stock solutions at $t = 0$. Results presented are means $\pm$ SEM of the indicated number of experiments.

Uptake of [^{3}H][D-Ala^2, D-Leu^5]Enkephalin into Synaptic Vesicles

Synaptic vesicles were purified from frozen *Torpedo* electric organs by differential and density gradient centrifugation and by permeation chromatography on a controlled pore glass beads column as previously described.[12] Prior to the uptake experiment the vesicles were incubated for 16 h at 25°C in a glycine buffer (0.8 M glycine, 50 mM Hepes, pH 7.4). This resulted in depletion of their endogenous neurotransmitter store. The vesicles were then washed by centrifugation (250,000 g $\times$ 1 h) and resuspended (2-3 mg protein/ml) in the same buffer. Uptake was initiated by a 1:1 dilution of the vesicles with glycine buffer which contained DADL-enkephalin at the indicated concentrations and [^{3}H]DADL-enkephalin (2 μCi/ml). The reaction mixture also contained captopril (10 μM) and puromycin (50 μM). The amount of radioactivity taken up by the vesicles was determined by filtration on 0.45 μ Millipore filters as described above. When the effects of NaCl and ATP were examined they were added to the vesicles isotonically at $t = 0$, whereas morphine was added 2 min prior to the initiation of uptake. Results presented are the means $\pm$ SEM of the indicated number of experiments.

Thin-layer chromatography (TLC) of the opioid peptides was performed as described in Altstein *et al.*[11] Protein was determined according to Lowry *et al.*[13]

RESULTS

Uptake of Enkephalins into Cholinergic Nerve Terminals

Addition of [^{3}H]Leu-enkephalin (2 μM) to *Torpedo* synaptosomes suspended at 25°C in modified TB containing puromycin (50 μM) and captopril (10 μM) resulted in accumulation of radiolabel by the nerve terminals (FIG. 1). Uptake was linear for 5 min (0.14 ± 0.01 pmol/mg protein/min; $n = 4$) and then leveled off to a plateau (5.3 ± 0.2 pmol/mg protein/min; $n = 4$) with a $t_{1/2}$ of ~20 min. Examination of the temperature dependence of the uptake revealed that it was about tenfold lower at 4°C than at 25°C (FIG. 1).

The *Torpedo* electric organ contains endogenous aminopeptidase and enkephalinase activities;[11,14] it was therefore pertinent to determine the extent to which [^{3}H]enkephalins are hydrolyzed by the synaptosomes. This was first attempted by

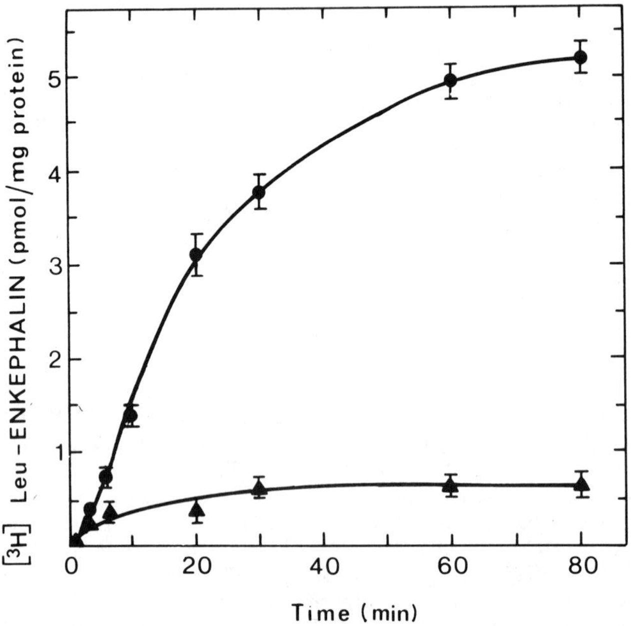

FIGURE 1. Time course of [^{3}H]enkephalin (2 μM) uptake by cholinergic synaptosomes purified from the *Torpedo* electric organ. Uptake was measured at 25°C (●) and 4°C (▲) in modified TB containing puromycin (50 μM) and captopril (10 μM), as described in EXPERIMENTAL PROCEDURES. Results presented are the means ± range of two experiments.

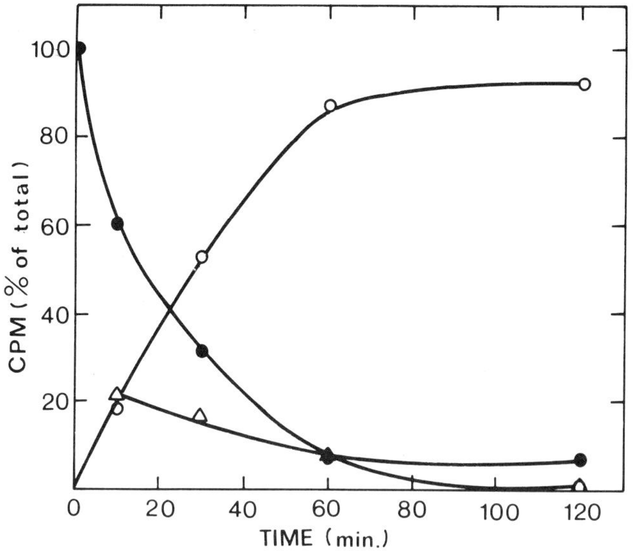

FIGURE 2. Time course of the hydrolysis of [^{3}H]Leu-enkephalin by *Torpedo* synaptosomes. Purified *Torpedo* electric organ synaptosomes were incubated at 25°C with [^{3}H]Leu-enkephalin (2 μM) as described in EXPERIMENTAL PROCEDURES but without the addition of puromycin and captopril to the buffer. At the indicated times the reaction was stopped by 1:100 dilution with ethyl acetate:isopropanol:acetic acid:water (40:40:1:19), and [^{3}H]Leu-enkephalin and its radiolabeled hydrolysis products were separated by TLC and identified as in ref. 11. [^{3}H]Leu-enkephalin, [^{3}H]Tyr-Gly-Gly and [^{3}H]-Tyr are represented respectively by ●, △, and ○. The amount of radioactivity applied to the TLC plate per lane was 100% = 10^5 dpm.

measurements in the absence of inhibitors of the synaptosomal aminopeptidase and enkephalinase activities. Analysis of the media following incubation of [^{3}H]Leu-enkephalin (2 μM) with the synaptosomes under the same conditions as those employed for the uptake experiment (FIG. 1), but in the absence of captopril and puromycin, revealed that the [^{3}H]Leu-enkephalin was rapidly hydrolyzed to [^{3}H]tryosine and [^{3}H]Tyr-Gly-Gly (FIG. 2). Half of the [^{3}H]Leu-enkephalin was hydrolyzed at $t = 15$ min; [^{3}H]tyrosine was formed with $t_{1/2} = 25$ min, and by $t = 60$ min accounted for virtually all of the added radioactivity (> 90%); the level of [^{3}H]Tyr-Gly-Gly peaked (20%) at $t = 10$ min and decayed with $t_{1/2} = 50$ min. These findings show that [^{3}H]Leu-enkephalin is hydrolyzed primarily by the aminopeptidase and to a smaller extent by the synaptosomal enkephalinase. The aminopeptidase activity was only partially blocked by puromycin (50 μM), which increased the $t_{1/2}$ for [^{3}H]Leu-enkephalin hydrolysis from 15 to 30 min (not shown). Thus, under these conditions the hydrolysis of [^{3}H]Leu-enkephalin by the residual synaptosomal aminopeptidase activity and the uptake of radioactive material by the nerve terminals both occurred in the same time scale. In order to circumvent the difficulties which these findings present, we decided to pursue a different approach and to study the uptake of enkephalins into cholinergic nerve terminals by investigating the translocation of the stable enkephalin analogue [^{3}H][D-Ala^2,D-Leu^5]enkephalin ([^{3}H]DADL-enkephalin) into *Torpedo* synaptosomes.

The rate and extent of accumulation of radiolabel by *Torpedo* synaptosomes following incubation with [^{3}H]DADL-enkephalin are shown in FIGURE 3. Uptake was linear for 5 min (0.073 ± 0.005 pmol/mg protein/min at 25°C, n = 9) and then leveled off to a plateau (1.2 ± 0.2 pmol/mg protein at t = 30 min; n = 10) with a $t_{1/2}$ of ≈10 min. Virtually all the radioactivity taken up by the synaptosomes (95 ± 3% of the total, n = 2) was lost within 1 min following hypotonic dilution, whereas only a small fraction (11 ± 6% of the total; n = 2) was released following an equivalent dilution with isotonic buffer (FIG. 4). Addition of Triton X-100 (0.5%) to synaptosomes preloaded with [^{3}H]DADL-enkephalin also caused a rapid loss of virtually all the synaptosomal radioactivity (FIG. 4). These findings suggest that the accumulated radiolabel is indeed transported into the nerve terminal. Analysis by TLC[11] revealed that virtually all the accumulated radioactivity as well as that present in the external media (> 90%) co-chromatographed with a [^{3}H]DADL-enkephalin standard (not shown).

Examination of the dependence of the initial rate of [^{3}H]DADL-enkephalin uptake on its concentration (1-100 μM) yielded a biphasic curve. At low [^{3}H]DADL-enkephalin levels (< 30 μM) the initial rates of uptake increased linearly, whereas at higher peptide concentrations it increased less steeply and approached saturation (FIG. 5). Measurements of the extent of [^{3}H]DADL-enkephalin uptake (at t = 30 min) showed that it increased linearly with the external peptide concentration (1-100 μM). Examination of the temperature dependence of the initial rate of [^{3}H]DADL-enkephalin (2 μM) influx revealed that it was much slower (sevenfold) at 4°C than at 25°C (FIG.

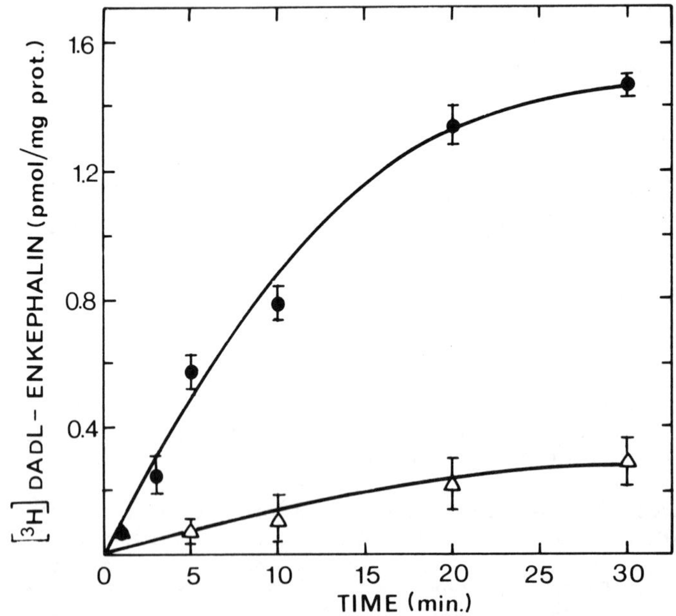

FIGURE 3. Time course of [^{3}H]DADL-enkephalin (2 μM) uptake by synaptosomes purified from the *Torpedo* electric organ. Uptake was measured at 25°C (●) and 4°C (Δ) in modified TB as described in EXPERIMENTAL PROCEDURES. Results presented are the means ± SEM of three experiments.

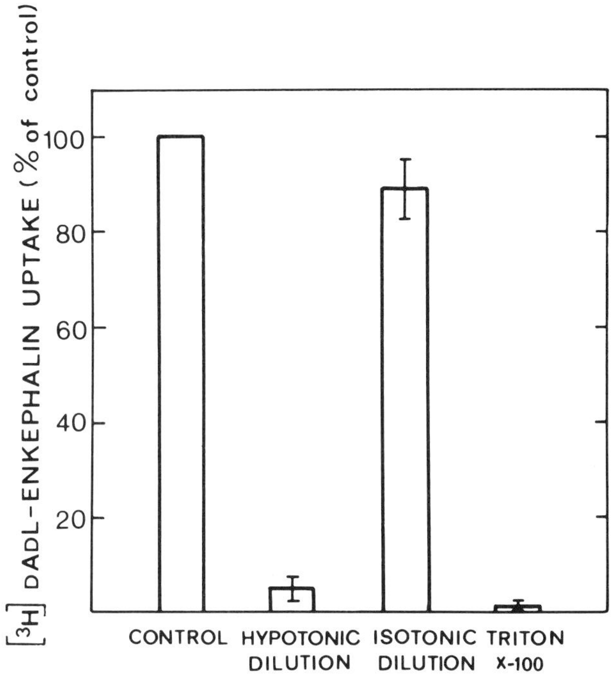

FIGURE 4. Effect of hypotonic and isotonic dilutions and of Triton X-100 on the accumulation of [^{3}H]DADL-enkephalin (2 μM) within purified *Torpedo* synaptosomes. Uptake was measured at 25°C in modified TB until a plateau was reached (t = 30 min), after which the nerve terminals were diluted 1:100 with water or isotonic buffer (0.8 M glucose, 1.2 mM Na phosphate, pH 7.4) or isotonic buffer containing Triton X-100 (0.5% w/v). The radioactivity retained within the synaptosomes 1 min after dilution was determined by filtration, as described in EXPERIMENTAL PROCEDURES. Results presented are the means ± range of two experiments. Control (100% = 1.42 ± 0.03 pmol/mg protein, n = 2) corresponds to the amount of [^{3}H]DADL-enkephalin taken up by the synaptosomes prior to the dilution.

3). These findings suggest that [^{3}H]DADL-enkephalin is translocated into the synaptosomes by a transport system.

The synaptosomal [^{3}H]DADL-enkephalin transport system was further characterized by examining the effects of specific cations and anions upon it. Comparison of the initial rates of [^{3}H]DADL-enkephalin (2 μM) uptake into synaptosomes suspended in modified TB (FIG. 3) and in a low ionic strength high glucose buffer (0.8 M glucose, 1.2. mM Na phosphate, pH 7.2) revealed that it was slower by 50 ± 20% (n = 5) in the latter. The anion and cation specificity of this effect were investigated by examining the effects on the uptake of [^{3}H]DADL-enkephalin into synaptosomes suspended in the glucose buffer of the isosmotic addition of specific salts. As shown in TABLE 1, the addition of NaCl (200 mM) and KCl (200 mM) accelerated the initial rate of [^{3}H]DADL-enkephalin uptake by respectively 52 ± 7% and 230 ± 56% (n = 3), whereas the addition of the membrane-impermeable anion 2-[*N*-morpholino]ethanesulfonate (MES) in the form of Na MES (200 mM) or K MES (200 mM) enhanced the initial rate of uptake by respectively 21 ± 18% and 112 ± 5% of control (n = 3). These findings show that the rate of [^{3}H]DADL-enkephalin uptake is stimulated by monovalent cations (K^+ > Na^+) and that monovalent anions also accelerate the uptake (Cl^- > MES) but to a smaller extent (TABLE 1). $CaCl_2$ and $MgCl_2$ (5 mM) accelerated [^{3}H]DADL-enkephalin uptake into synaptosomes suspended in the low ionic strength glucose buffer by about 50% but had no effect when added to synaptosomes suspended in a high ionic strength buffer (*e.g.,* 200 mM KCl, not shown). This suggests that bivalent cations also stimulate [^{3}H]DADL-enkephalin uptake, that they do so at least as potently as Na^+, and that the effect of bivalent cations can be masked by the presence of saturating levels of monovalent salts.

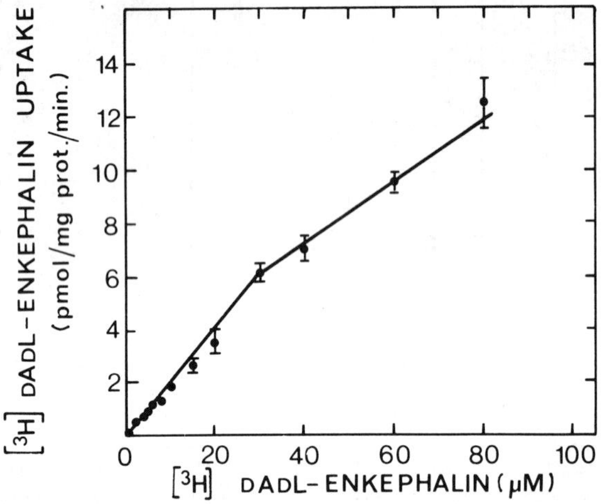

FIGURE 5. Dependence of the initial rate of [^{3}H]DADL-enkephalin uptake by purified *Torpedo* synaptosomes on ligand concentration. Uptake was measured at 25°C in modified TB and the initial rates were calculated from the slopes of the linear phase of uptake (0-5 min), as described in EXPERIMENTAL PROCEDURES. Results presented are the means ± SEM of at least three experiments.

Preliminary examination of the effects of salts on the extent of [^{3}H]DADL-enkephalin uptake revealed that KCl (200 mM) increased the extent of uptake by 92 ± 30% (n = 3), whereas NaCl (200 mM) had no effect (16 ± 20%, n = 3, TABLE 1). High levels of K^+ depolarize the synaptosomes. It is therefore tempting to suggest that the marked effects of KCl on [^{3}H]DADL-enkephalin uptake are due to K^+-induced membrane depolarization. However, since the concentration dependence of the effects of KCl (0-300 mM) on the rates and extents of [^{3}H]DADL-enkephalin uptake did not yield a linear relationship between uptake and log [KCl] (FIG. 6), the stimulatory effects of KCl on [^{3}H]DADL-enkephalin uptake cannot be attributed solely to depolarization of the presynaptic plasma membrane by K^+ but also to additional as yet uncharacterized mechanisms.

TABLE 1. Effect of Monovalent Cations and Anions on [^{3}H]DADL-Enkephalin Uptake into Purified *Torpedo* Synaptosomes[a]

Addition (200 mM)	Rate of Uptake (% Increase)	Extent of Uptake (% Increase)
NaCl	52 ± 7	16 ± 12
Na MES[b]	21 ± 18	nd
KCl	230 ± 56	92 ± 13
K MES	112 ± 5	nd[c]

[a] Uptake was measured at 25°C using [^{3}H]DADL-enkephalin (2 μM) and purified *Torpedo* synaptosomes suspended in 0.8 M glucose, 1.2 mM Na phosphate, pH 7.2, to which each of the indicated salts (200 mM) was added isotonically at t=0. The initial rates and extents of uptake were determined as outlined in FIGURE 6. Results presented (means ± SEM of three experiments) are percent increase in the initial rate (100% = 0.034 ± 0.01 pmol/mg protein/min, n=5) and in the extent (100% = 0.6±0.04 pmol/mg protein at t=30 min, n=5) of [^{3}H]DADL-enkephalin uptake into control synaptosomes without added salts.

[b] MES = 2-[*N*-morpholino]ethanesulfonate.

[c] nd = not determined.

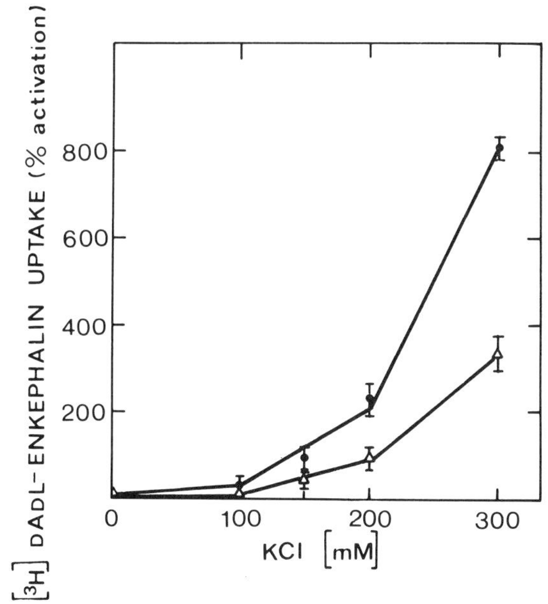

FIGURE 6. Concentration dependence of the effects of KCl on the initial rate (●) and the extent (△) of [^{3}H]DADL-enkephalin (2 μM) uptake into purified *Torpedo* synaptosomes. Uptake was measured at 25°C with synaptosomes suspended in 0.8 M glucose, 1.2 mM Na phosphate, pH 7.2, to which KCl (0-300 mM) was added isotonically at t = 0. The initial rates and the extents of uptake at the plateau (t = 30 min) were determined as described in EXPERIMENTAL PROCEDURES. Results presented (means ± SEM of three experiments) are percent increase in the initial rate (100% = 0.034 ± 0.01 pmol/mg protein/min, n = 5) and in the extent (100% = 0.6 ± 0.04 pmol/mg protein at t = 30, n = 5) of [^{3}H]DADL-enkephalin uptake into control synaptosomes without added KCl.

Pharmacological Characterization of the Synaptosomal [^{3}H]DADL-Enkephalin Transport System

This was pursued by determining the effects of specific opiate receptor agonists and antagonists and of the enkephalin fragments produced by the synaptosomal aminopeptidase and enkephalinase activities on [^{3}H]DADL-enkephalin uptake.

Incubation of *Torpedo* synaptosomes with the opiate agonist morphine (0-40 μM) resulted in a dose-dependent acceleration of the initial rate of [^{3}H]DADL-enkephalin uptake into the nerve terminals. Maximal activation (70 ± 5%; $n = 3$) was obtained at morphine concentrations upward of 20 μM with a half-maximal effect at ~10 μM (FIG. 7). The stimulatory effect of morphine (20 μM) was abolished by the opiate receptor antagonist naloxone (20 μM), which by itself did not affect uptake (FIG. 8). These findings suggest that the activation of [^{3}H]DADL-enkephalin uptake by morphine is mediated by an opiate receptor.[8] Examination of the effects of other opiate receptor ligands on [^{3}H]DADL-enkephalin uptake showed that it was not affected by 20 μM levels of Leu-enkephalin, DAGO, dynorphin, or β-endorphin; that it was slightly inhibited by DADL-enkephalin; and that it was accelerated by the opiate ethylketocyclozocine (EKC, TABLE 2). The inhibitory effect of DADL-enkephalin (20 μM) on the uptake of [^{3}H]DADL-enkephalin (2 μM) is probably due to the saturating

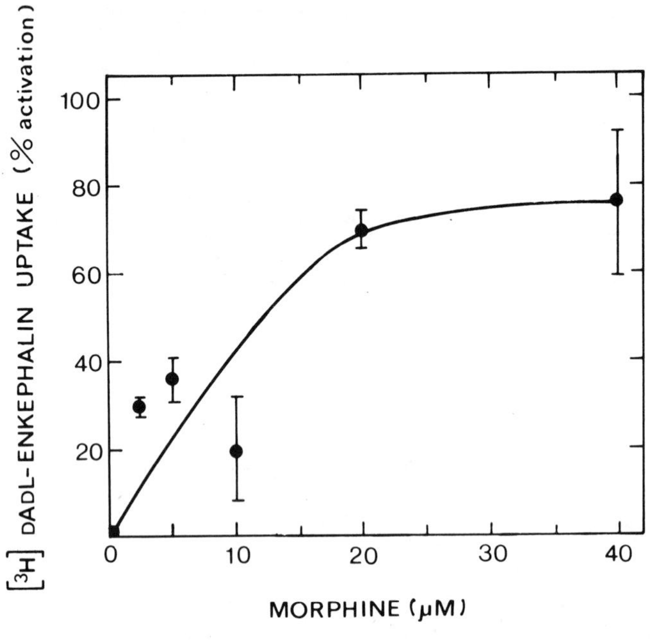

FIGURE 7. Concentration dependence of the effects of morphine on the initial rate of [^{3}H]DADL-enkephalin (2 μM) uptake into purified *Torpedo* synaptosomes. Initial rates of uptake were measured at 25°C in modified TB, which contained the indicated concentrations of morphine as described in EXPERIMENTAL PROCEDURES. Results presented (means ± SEM of three experiments) are percent activation relative to synaptosomes without added morphine (100% = 0.08 ± 0.013 pmol/mg protein/min, $n = 10$).

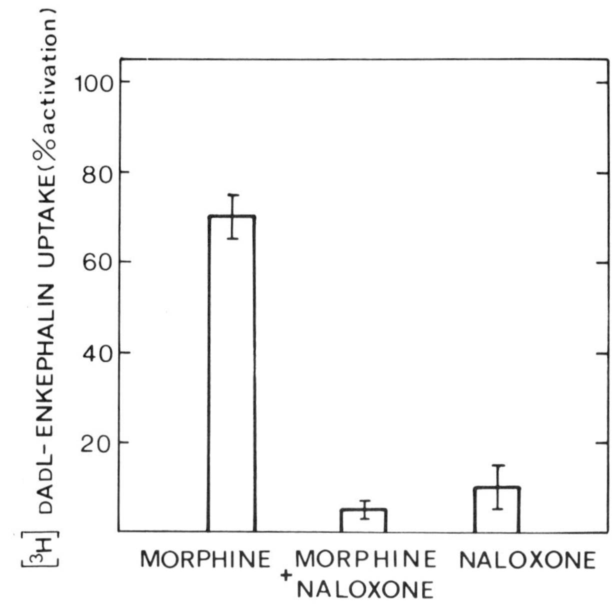

FIGURE 8. Effects of morphine, naloxone, and morphine plus naloxone on the initial rate of [^{3}H]DADL-enkephalin uptake into purified *Torpedo* synaptosomes. The initial rates of [^{3}H]DADL-enkephalin (2 μM) uptake were measured at 25°C in modified TB in the presence and absence of morphine (20 μM), morphine (20 μM) plus naloxone (20 μM), and naloxone (20 μM) as described in EXPERIMENTAL PROCEDURES. Results presented (means ± SEM of three experiments) are percent activation relative to synaptosomes without added morphine or naloxone (100% = 0.07 ± 0.15 pmol/mg protein/min, n = 5).

kinetics of the synaptosomal uptake system (see FIG. 5). The activation of uptake by EKC became saturated at ≈40 μM (106 ± 13%) and was half maximal at EKC ~15 μM.

The possibility that enkephalin breakdown products affect the translocation of [^{3}H]DADL-enkephalin into the nerve terminals was examined by investigating the effects on the uptake of the opioid peptide of [des-Tyr]enkephalin and tyrosine (which are produced from Leu-enkephalin by the synaptosomal aminopeptidase) and of Tyr-Gly-Gly and Phe-Leu (produced by the synaptosomal enkephalinase). Addition of [des-Tyr]enkephalin (0-120 μM) to the synaptosomes resulted in a dose-dependent acceleration of the initial rate of [^{3}H]DADL-enkephalin uptake. Maximal activation (112 ± 17%, n = 2) was obtained at [des-Tyr]enkephalin concentrations upward of 40 μM, while a half maximal effect was obtained at 20 μM (FIG. 9). This activation was not blocked by naloxone (not shown). In contrast to the marked stimulatory effect of [des-Tyr]enkephalin on [^{3}H]DADL-enkephalin uptake, tyrosine (20 μM), Tyr-Gly-Gly (20 μM) and Phe-Leu (20 μM) had virtually no such effects (activation by respectively 32 ± 30%, 9 ± 5%, and 35 ± 15%; n = 4). Thus the opioid peptide transport system seems to be activated by a hydrolytic product of the synaptosomal aminopeptidase but not by products of the enkephalinase.

TABLE 2. Effects of Opiates and Opioid Peptides on the Rate of [^{3}H]DADL-Enkephalin Uptake into *Torpedo* Synaptosomes

Ligand (20 μM)	Initial Rate of [^{3}H][D-Ala^2,D-Leu^5]Enkephalin Uptake (% of Control)	
Leu-enkephalin	116 ± 16	(3)
DADL-enkephalin	75 ± 10	(3)
DAGO	95 ± 14	(3)
Dynorphin$_{1-13}$	107 ± 17	(3)
β-endorphin	139 ± 20	(4)
EKC	173 ± 13	(5)
Morphine	170 ± 13	(5)

The initial rates of [^{3}H]DADL-enkephalin (2 μM) uptake were measured at 25°C in modified TB in the presence and absence of the indicated ligands (20 μM), as described in EXPERIMENTAL PROCEDURES. Results presented (means ± SEM of the indicated number of experiments) are percentage of the initial rate of uptake of control synaptosomes without added ligands (100% = 0.073 ± 0.005 pmol/mg protein/min; n=9).

Uptake of [^{3}H]DADL-Enkephalin into Torpedo *Synaptic Vesicles*

Incubation of purified *Torpedo* synaptic vesicles with [^{3}H]DADL-enkephalin (2 μM) at 25°C resulted in the accumulation of radiolabel by the vesicles (FIG. 10). Uptake was linear for up to 15 min (0.14 ± 0.01 pmol/mg protein/min; n = 11) and then leveled off to a plateau (2.9 ± 0.2 pmol/mg protein/min, n = 8). When vesicles preloaded with [^{3}H]DADL-enkephalin were diluted with water (1:100), about half of the vesicle-associated radioactivity was lost within 1 min of the dilution (51 ± 5% of the total, n = 3) while only a small fraction (11 ± 4% of the total, n = 3) was released upon equivalent dilution with isotonic buffer. Addition of Triton X-100 (0.1%) to vesicles preloaded with [^{3}H]DADL-enkephalin resulted in a rapid loss of virtually all the vesicle-associated radioactivity (> 95%). Analysis by TLC[11] revealed that virtually all the vesicle-bound radioactivity (> 90%) co-chromatographed with a [^{3}H]DADL-enkephalin standard. These findings show that the accumulated radiolabel is [^{3}H]DADL-enkephalin and support the contention that it is transported into the vesicles. It should be noted that, as in the case of synaptosomes, incubation of *Torpedo* vesicles with [^{3}H]Leu-enkephalin resulted in the hydrolysis of the opioid peptide by endogenous aminopeptidase and enkephalinase activities (not shown). We therefore continued the present study of the uptake of enkephalins into *Torpedo* synaptic vesicles utilizing [^{3}H]DADL-enkephalin.

Examination of the dependence of the initial rate of [^{3}H]DADL-enkephalin uptake (determined from the slope of the uptake curve during the first 15 min) on the ligand concentration revealed that it reached saturation (FIG. 11) and that within the concentration range investigated (2-30 μM) the results were best fitted by a single transport system with K_T = 12 ± 1 μM and V_{max} = 0.85 ± 0.15 pmol/mg protein/min (n = 3). Examination of the temperature dependence of the initial rate of [^{3}H]DADL-enkephalin (2 μM) influx revealed that it was much slower (sixfold) at 4°C than at 25°C (FIG. 10). These findings suggest that [^{3}H]DADL-enkephalin is taken up into the vesicles by facilitated diffusion via a saturable transport system.

Addition of MgATP, which drives the active transport of ACh into the vesicles,[15] resulted in a small but significant acceleration of the initial rate of [^{3}H]DADL-enkephalin (2 μM) uptake (25 ± 3% and 42 ± 5% at respectively 10 mM and 20 mM MgATP, n = 4). Similar results were obtained when the effects of CaATP alone or in conjunction with MgATP were examined (not shown). Morphine (20 μM) and NaCl (200 mM), which stimulate the uptake of [^{3}H]DADL-enkephalin into the synaptosomes (FIGS. 6 and 7), had no effect on its rate of uptake into the synaptic vesicles (activation by respectively −4 ± 9% and 7 ± 6%, n = 4).

DISCUSSION

The present findings show that isolated *Torpedo* electric organ synaptosomes and synaptic vesicles can take up externally added [^{3}H]DADL-enkephalin and suggest that the translocation of [^{3}H]DADL-enkephalin into the nerve terminals and its uptake into synaptic vesicles are both mediated by temperature-dependent saturable transport systems. These observations are qualitatively similar to findings recently obtained with mammalian preparations.[16–18] The apparent K_m of the *Torpedo* vesicular transport

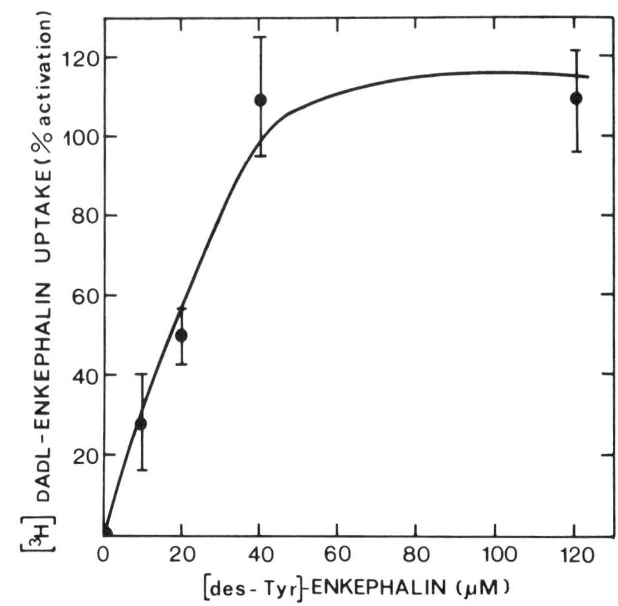

FIGURE 9. Concentration dependence of the effects of [des-Tyr]enkephalin on the initial rate of [^{3}H]DADL-enkephalin (2 μM) uptake into purified *Torpedo* synaptosomes. Initial rates of uptake were measured at 25°C in modified TB, which contained the indicated levels of [des-Tyr]enkephalin as described in EXPERIMENTAL PROCEDURES. Results presented (means ± SEM of three experiments) are percent activation relative to synaptosomes in the absence of [des-Tyr]enkephalin (100% = 0.06 ± 0.01 pmol/mg protein/min, n = 4).

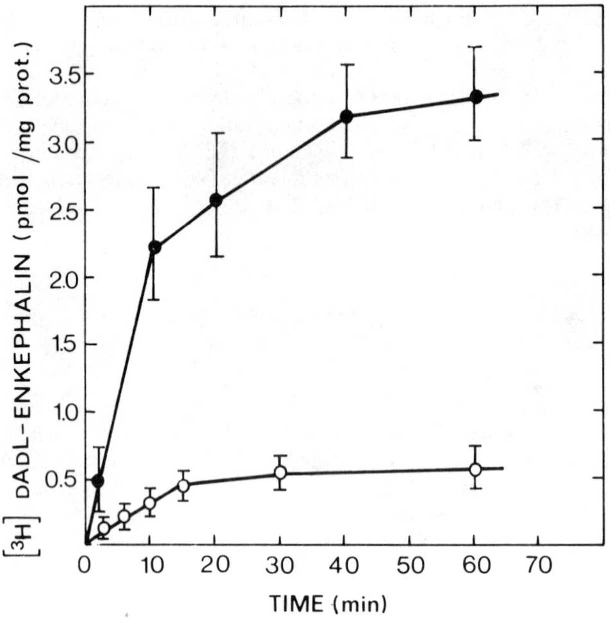

FIGURE 10. Time course of [^{3}H]DADL-enkephalin uptake by cholinergic synaptic vesicles purified from the *Torpedo* electric organ. Uptake was measured at 2 μM by filtration at 25°C (●) and at 4°C (○), as described in EXPERIMENTAL PROCEDURES. Results presented are the means ± SEM of three experiments.

system (12 ± 1 μM), however, is respectively five- and fiftyfold larger than those of the transport systems that translocate Met-enkephalin into bovine chromaffin granules[17] and Leu-enkephalin into brain synaptic vesicles,[16] while the rate of translocation of [^{3}H]DADL-enkephalin into *Torpedo* vesicles (V_{max} = 0.85 ± 0.15 pmol/mg protein/min) is respectively five- and eightfold faster than the rate of uptake of Met- and of Leu-enkephalin into bovine chromaffin granules and synaptic vesicles. These differences may be due either to pharmacological and biochemical differences between *Torpedo* and mammalian vesicular enkephalin transport systems, or alternatively, to the different radiolabled enkephalin derivatives employed.

The possibility that [^{3}H]DADL-enkephalin is taken up by membranous structures other than *Torpedo* synaptic vesicles and nerve terminals cannot yet be ruled out. However, the homogeneity of the *Torpedo* electric organ and the high degree of purity of the vesicular and the synaptosomal preparations[10,19] suggest that [^{3}H]DADL-enkephalin is indeed taken up by cholinergic nerve terminals and synaptic vesicles. This assertion is also supported by the finding that the rates of [^{3}H]DADL-enkephalin uptake into the vesicles, like those of ACh uptake,[20] are highest in the purified synaptic vesicles (*e.g.,* the specific rate of [^{3}H]DADL-enkephalin (2 μM) uptake by the crude vesicular fraction P_3[19] is sixfold lower than by the purified vesicles) and that the rates of [^{3}H]DADL-enkephalin uptake into purified synaptosomes are higher than their rates of uptake into the crude synaptosomal fraction P_2 (not shown). The similarity of the specific rates of uptake of [^{3}H]DADL-enkephalin into the nerve terminals and into the synaptic vesicles, as well as the differences in their responses to monovalent cations and morphine (TABLE 3, FIG. 1), suggest that the synaptosomal and vesicular transport systems are different and that the accumulation of [^{3}H]DADL-enkephalin by the synaptosomal fraction is not due to a contamination of free synaptic vesicles.

The stimulation by ATP of [^{3}H]DADL-enkephalin uptake into *Torpedo* synaptic vesicles ($\approx$ 40%) is much weaker than that observed with bovine brain synaptic vesicles[16] and is detectable only at high ATP levels (> 10 mM). These differences may be attributable to experimental conditions (*e.g., Torpedo* vesicles are purified in an EGTA buffer and therefore lose their membrane-associated calmodulin[19,21]), or alternatively to inherent differences between *Torpedo* and mammalian vesicles. Furthermore, it is not yet known whether the effects of ATP on *Torpedo* vesicles are mediated by the vesicular Ca^{2+}/Mg^{2+} ATPase. Future experiments utilizing ATP analogues and inhibitors of the vesicular Ca^{2+}/Mg^{2+} ATPase[22] will further our understanding of the biochemical mechanisms underlying the effect of ATP on the uptake of enkephalins into *Torpedo* synaptic vesicles.

The pharmacological experiments, in which the effects of opiates on the *Torpedo* synaptosomal and vesicular transport systems were examined, yielded the following findings: (1) morphine activates the uptake of [^{3}H]DADL-enkephalin into synaptosomes in a dose-dependent, naloxone-reversible fashion but does not affect the uptake of opioid peptides into the synaptic vesicles (FIGS. 7, 8 and TABLE 3); and (2) the synaptosomal transport system is activated by the nonopioid peptide [des-

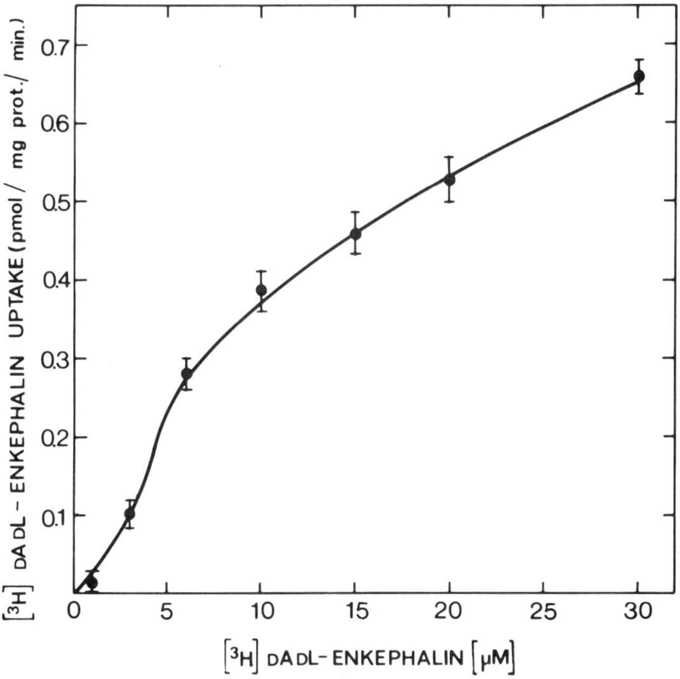

FIGURE 11. Dependence of the initial rate of [^{3}H]DADL-enkephalin uptake by *Torpedo* synaptic vesicles on ligand concentration. Initial rates were calculated from the slopes of the linear phase (0-15 min) of uptake as described in EXPERIMENTAL PROCEDURES. Results presented are the means ± SEM of three experiments.

Tyr]enkephalin. (The effects of [des-Tyr]enkephalin on the vesicular transport system have not yet been investigated.) Previous studies from our laboratory have shown that *Torpedo* nerve terminals contain presynaptic opiate receptors whose activation by morphine inhibits ACh release in a naloxone reversible fashion.[8,9] It is therefore tempting to suggest that the effects of morphine on the synaptosomal [^{3}H]DADL-enkephalin uptake system are mediated by these presynaptic opiate receptors. The finding that morphine has no effect on the vesicular transport system is consistent with this interpretation since the receptors are situated in the presynaptic membrane and therefore do not copurify with the synaptic vesicles. The findings that the effects of [des-Tyr]enkephalin on the synaptosomal uptake are not reversed by naloxone and that the concurrent addition of saturating levels of this peptide and of morphine results in an additive effect (results to be published) suggest that the effects of [des-Tyr]enkephalin on the nerve terminal, unlike those of morphine, are not mediated by an opiate receptor. This is in accordance with previous observations which indicate that removal of the amino terminal tyrosine results in a naloxone-insensitive nonopioid

TABLE 3. Comparison of the Characteristics of [^{3}H]DADL-Enkephalin (2 μM) Uptake into *Torpedo* Synaptosomes and Synaptic Vesicles

	Synaptosomal [^{3}H]DADL-Enkephalin Uptake	Vesicular [^{3}H]DADL-Enkephalin Uptake
Initial rate (V_{in}) (pmol/mg protein/min)	0.073 ± 0.005	0.14 ± 0.01
Extent of uptake (pmol/mg protein)	1.2 ± 0.2	2.9 ± 0.2
Effects on V_{in} of:		
NaCl (200 mM)	activation	none
Morphine (10 μM)	activation	none
ATP (20 mM)	—	activation

This table was compiled from Figures 1, 7, 10, TABLE 1, and data detailed in the text.

peptide.[23] It should be noted that in view of the evident complexity of the synaptosomal [^{3}H]DADL-enkephalin uptake system, it is possible that the observation that the opioid peptides tested do not affect uptake (TABLE 2) could result from a balance between an activating effect, similar to that of either morphine or [des-Tyr]enkephalin, and an inhibition due to competition between the opioid peptides and [^{3}H]DADL-enkephalin for the opiate peptide binding site(s) on the synaptosomal transport system. It may also be due to hydrolysis of the externally added peptides by endogenous peptidases.

On the basis of the present findings, it is tempting to suggest that *Torpedo* nerve terminals transport enkephalin and enkephalin-like peptides *in vivo* across the presynaptic membrane and that these peptides can then be translocated from the presynaptic cytosol into the synaptic vesicles. Thus the finding that the rate of uptake of opioid peptides into the nerve terminal can be accelerated by activating the presynaptic opiate receptors and by the addition of [des-Tyr]enkephalin suggests that enkephalin uptake may be regulated *in vivo* by two mechanisms, one of which is activated by the intact peptide and the other by its major hydrolytic product. Future studies aimed at investigating the uptake as well as the release of enkephalins and enkephalin-like peptides from *Torpedo* nerve terminals and synaptic vesicles will help in unraveling the biochemical mechanisms underlying these transport systems and their roles in the recycling and the release of opioid peptides from cholinergic neurons.

REFERENCES

1. WHITE, J. D., K. D. STEWART, J. E. KRAUSE & J. F. MCKELVY. 1985. Biochemistry of peptide secreting neurons. Physiol. Rev. **65:** 553-606.
2. JONES, B. N, J. E. SHIVELY, D. L. KILPATRICK, K. KOGIMA & S. UDENFRIEND. 1982. Enkephalin biosynthetic pathway: A 5300 dalton adrenal protein polypeptide that terminates at its COOH and with the sequence Met-enkephalin-Arg-Gly-COOH. Proc. Natl. Acad. Sci. U.S.A. **79:** 1313-1315.
3. MAINS, R. E., B. A. EIPPER, C. C. GLENBOTSKI & R. M. DOVES. 1983. Strategies for the biosynthesis of bioactive peptides. Trends Neurosci. **6:** 229-235.
4. SCHULTZBERG, M., T. HOKFELT & J. M. LUNDBERG. 1982. Coexistence of classical transmitters and peptides in the central and peripheral nervous systems. Br. Med. Bull. **38:** 309-313.
5. CHAMINADE, M., A. S. FOUTZ & J. ROSSIER. 1983. Corelease of enkephalins and precursors with catecholamines by the prefused cat adrenal in situ. Life Sci. **33:** 19-23.
6. PALMER, M. R., A. SEIGER, B. J. HOFFER & L. OLSON. 1983. Modulating interactions between enkephalin and catecholamines: Anatomical and physiological substrates. Fed. Proc. **42:** 2934-2945.
7. VIVEROS, O. H., E. J. DILIBERTO & A. J. DANIELS. 1983. Biochemical and functional evidence for the cosecretion of multiple messengers from single and multiple compartments. Fed. Proc. **42:** 2923-2928.
8. MICHAELSON, D. M., G. MCDOWALL & Y. SARNE. 1984. The *Torpedo* electric organ is a model for opiate regulation of acetylcholine release. Brain Res. **305:** 173-176.
9. MICHAELSON, D. M., G. MCDOWALL & Y. SARNE. 1984. Opiates inhibit acetylcholine release from *Torpedo* nerve terminals by blocking Ca^{2+} influx. J. Neurochem. **43:** 614-618.
10. MICHAELSON, D. M. & M. SOKOLOVSKY. 1978. Induced acetylcholine release from active purely cholinergic *Torpedo* synaptosomes. J. Neurochem. **30:** 217-230.
11. ALTSTEIN, M., Y. DUDAI & Z. VOGEL. 1984. Enkephalin degrading enzymes are present in the electric organ of *Torpedo californica*. FEBS Lett. **166:** 183-188.
12. DAY, N. C., D. WIEN & D. M. MICHAELSON. 1985. Saturable [D-Ala^2, D-Leu^5]enkephalin transport into cholinergic synaptic vesicles. FEBS Lett. **183:** 25-28.
13. LOWRY, O. H., N. J. ROSENBROUGH, A. L. FARR & R. J. RANDALL. 1951. Protein measurements with the folin phenol reagent. J. Biol. Chem. **113:** 265-275.
14. TURNER, A. J. & M. J. DOWDALL. 1984. The metabolism of neuropeptides: Both phosphoramidon-sensitive and captopril-sensitive metallopeptidases are present in the electric organ of *Torpedo marmorata*. Biochem. J. **222:** 255-259.
15. PARSONS, S. M. & R. KOENIGSBERGER. 1980. Specific stimulated uptake of acetylcholine by *Torpedo* electric organ synaptic vesicles. Proc. Natl. Acad. Sci. U.S.A. **77:** 6234-6238.
16. TAKEDA, M., F. TAKEDA, F. MATSUMOTO, R. TANAKA & K. KONNO. 1982. Divalent cation, ATP-dependent [^{3}H]Leu-enkephalin uptake by synaptic vesicle fraction isolated from bovine caudate nucleus. Brain Res. **234:** 319-326.
17. TAKEDA, F., M. TAKEDA, A. SHIMADA & K. KONNO. 1985. ATP-dependent [^{3}H]Met-enkephalin uptake by bovine adrenal chromaffin granule membranes. Brain Res. **344:** 220-226.
18. ROY, B. P., I. JAMAL & J. GO. 1982. Synaptic mechanism of methionine-enkephalin uptake. Life Sci. **31:** 2307-2310.

19. MICHAELSON, D. M. & I. OPHIR. 1980. Purification and characterization of synaptic vesicles from the electric organ of *Torpedo ocellata. In* Monographs in Neural Sciences, Vol. 7. E. Heldman, A. Levy, Y. Gutman & Z. Vogel, Eds.: 19-29. Karger. Basel.
20. MICHAELSON, D. M. & I. ANGEL. 1980. Saturable acetylcholine transport into purified cholinergic synaptic vesicles. Proc. Natl. Acad. Sci. U.S.A. **78:** 2048-2052.
21. REPHAELI, A. & S. PARSONS. 1982. Calmodulin stimulation of $^{45}Ca^{2+}$ transport and protein phosphorylation in cholinergic synaptic vesicles. Proc. Natl. Acad. Sci. U.S.A. **79:** 5783-5787.
22. MICHAELSON, D. M. & I. OPHIR. 1980. Sidedness of (calcium, magnesium) adenosine triphosphatase of purified *Torpedo* synaptic vesicles. J. Neurochem. **34:** 1483-1490.
23. MORELY, J. S. 1983. Chemistry of opioid peptides. Br. Med. Bull. **39:** 5-10.

DISCUSSION OF THE PAPER

O. H. VIVEROS (*Wellcome Research Laboratories, Research Triangle Park, N.C.*): In the course of our early experiments on the secretion of enkephalins, we found that the perfused dog adrenal gland takes up [^{3}H]Leu-enkephalin and [^{14}C]inulin and that when the gland is challenged with acetylcholine it cosecretes endogenous catecholamines and opioid peptides as well as the newly taken-up [^{14}C]inulin and [^{3}H]Leu-enkephalin (intact pentapeptide as confirmed by HPLC).

When similar experiments were repeated with primary cultures of chromaffin cells, we failed to demonstrate uptake and release of [^{3}H]enkephalin and [^{14}C]inulin into these cells. Thus the "bulk" uptake and release of extracelluar markers seen in perfused adrenals seem to be localized to extra chromaffin cell glandular elements (vascular wall elements? cortical cells? connective tissue cells?). While it is perfectly possible that *Torpedo* synaptosomes and vesicles may have a specific mechanism for neuropeptide uptake, I just want to stress the fact that finding uptake and receptor-induced release of a putative transmitter or neurohormone into or from neural or endocrine tissue is not a definite criterion to conclude that uptake and release is specific, or into the neural or endocrine cell present in a complex tissue.

D. M. MICHAELSON (*Tel Aviv University, Ramat Aviv, Israel*): I agree that the demonstration of uptake of a putative neurotransmitter is not by itself a definite criteria for specificity. However, preliminary findings from our laboratory, in which the uptake of [^{3}H]DADL-enkephalin (2 μm) and [^{3}H]inulin (2 μm) into *Torpedo* synaptosomes were compared, revealed that the rate and extent of the accumulation of [^{3}H]inulin are respectively 120-fold smaller than those of [^{3}H]DADL-enkephalin, and that the accumulation of [^{3}H]inulin, unlike that of the opioid peptide, is not affected by morphine. These findings suggest that the opioid peptides examined are indeed translocated into *Torpedo* nerve terminals by a specific mechanism.

Regarding the localization of the accumulated enkephalin, it seems that the adrenal gland and the electric organ differ in that in the electric organ the peptides are taken up by the neurons, whereas in the adrenal gland they accumulate in extrachromaffin-cell elements.

D. V. AGOSTON (*Max Planck Institute, Göttingen, Federal Republic of Germany*): What is the indication that your opioid or opioid-like peptide originates from the neuronal part of the electric organ? People tend to forget that 90% of the total volume of the electric organ is electrocytes. So far there is no experimental evidence that your peptide is present either in the electric lobe, which synthesizes all the functional cholinergic components, or in the electromotor nerves, which transport the synthesized components. There are, however, some indications that electrocytes are able to produce some "factors" with peptides. I agree that the question of the function of the uptake systems as well as of the presence of the peptide is still open.

D. M. MICHAELSON: Drs. Unsworth and Johnson have reported in this meeting that met-enkephalin is present in the electric lobe. This finding which we have recently confirmed and the observation that *Torpedo* synaptosomes contain an enkephalin-like peptide show that met-enkephalin, and possibly other opioid peptides, are indeed localized within the cholinergic neurons of the electric organ. These findings, as you correctly stated, do not exclude the possibility that the electroplaques also contain opioid peptides.

Composition and Transport Function of Membranes of Chromaffin Granules

Established Facts and Unresolved Topics

H. WINKLER

Department of Pharmacology
University of Innsbruck
A-6020 Innsbruck, Austria

In this short comment I would like to point out which problems connected with the function of the membranes of chromaffin granules are apparently resolved and which ones deserve further study. An understanding of these functions in molecular terms depends on a thorough knowledge of the composition and topology of the membranes involved, and I will therefore discuss this topic first.

COMPOSITION AND TOPOLOGY OF MEMBRANES OF CHROMAFFIN GRANULES

Two dimensional electrophoresis has been established[1] as an essential tool for characterizing the membrane proteins of chromaffin granules. Identified components with a known function include dopamine β-hydroxylase, cytochrome b_{561}, subunits of the H^+-ATPase (see below), actin, and α-actinin. Some functional proteins which have not yet been identified on 2D gels are the transporters for catecholamines, nucleotides, and calcium, and the carboxypeptidase H (enkephalin convertase).[2]

Knowledge of the topology of the membranes is limited. Two facts seem established: (1) the carbohydrate chains of the membrane glycoproteins face the interior of the granule[3,4]; and (2) a major part of the cytochrome b_{561} molecule is present on the outer face of chromaffin granules.[4–6] Evidence has been presented that most of the other proteins (including the ATPase and the phosphatidylinositol kinase) are also exposed on the outside of these organelles.[4] It is not clear, however, to what extent the various molecules are present in this position. In fact, Zaremba and Hogue-Angeletti[7] claim that the majority of proteins predominantly face the granule interior and have only some portion of their polypeptide chain facing the cytosol. Thus we cannot yet decide whether the major part of the proteins covers the outer face of the granules, is present within the membranes, or faces the interior of the organelles. This lack of precise knowledge also applies to the major membrane protein, *i.e.,* the dopamine β-hydroxylase. There is agreement that the carbohydrate chains and the active center face the inside,[8] but no such agreement exists on whether or not a hydrophobic part of the enzyme spans the membrane (thus reaching the outer surface).[4,7,9]

Future studies will have to identify the function of further membrane proteins as seen in 2D electrophoresis and to establish a more exact topology of these proteins within the membrane.

For the lipids the basic topological arrangement is the bimolecular leaflet.[8] A characteristic constituent of the phospholipids is lysolecithin. A claim[11] that the presence of lysolecithin in chromaffin granules is a postmortem artifact has been disproved.[12] This phospholipid is present in the outer part of the bimolecular leaflet.[13] This topology is shared by the gangliosides.[14] The biogenesis of lysolecithin in chromaffin granules is not yet elucidated. Chromaffin granules apparently contain only low amounts, if any, of phospholipase A_2.[15,16] Frischenschlager[16] found an enzyme in highly purified chromaffin granules that corresponded in its properties to the phospholipase A_2 described in synaptic vesicles.[17] The activity was low, however, and it proved difficult to exclude the possibility that its presence in granule preparations was due to contamination with plasma membranes also containing phospholipase A_2 activity. On the other hand, chromaffin granules have been reported to contain a relatively high activity of lysophospholipase,[18] which should reduce the level of lysolecithin in chromaffin granules. Thus, the formation and breakdown of lysolecithin in chromaffin granules remains mysterious although the high concentration of this characteristic phospholipid was discovered twenty years ago.[19]

CATECHOLAMINE TRANSPORT

The large ratios (up to 10^4) between the catecholamine concentration within the granules and in the cytoplasm are achieved by chemiosmotic coupling between a H^+-translocating ATPase and a separate catecholamine translocase (see refs. 20 and 21). The carrier transports not only dopamine, noradrenaline, and adrenaline, but also 5-hydroxytryptamine.[22] The amine content of chromaffin granules is therefore not determined by a carrier of high specificity, but by a limited supply of components in the cytoplasm (see FIG. 1). The catecholamines are the major amines in the cytoplasm of chromaffin cells. They are formed there by the enzymes necessary for their synthesis; 5-hydroxytryptamine is not synthesized in these cells and only traces apparently picked up by the chromaffin cells from the outside are found within chromaffin granules.[23]

As far as the molecular properties of the amine carrier are concerned, some confusion has arisen from the studies of two groups. Different approaches indicated quite different molecular weights (45,000 versus 70,000[24,25]). It seems, however, that this problem has now been resolved (see paper in this volume by J.-P. Henry). Apparently the carrier consists of two subunits. Future studies will establish the exact molecular arrangement of these two subunits and their interaction during amine translocation.

[H^+]ATPase

Cidon and Nelson[26] established that the proton-pumping ATPase present in chromaffin granules differed from the F_1/F_0 ATPase found in mitochondria. Purification of this enzyme was attempted by two groups.[27,28] Three bands of 70-72, 57, and 39-41

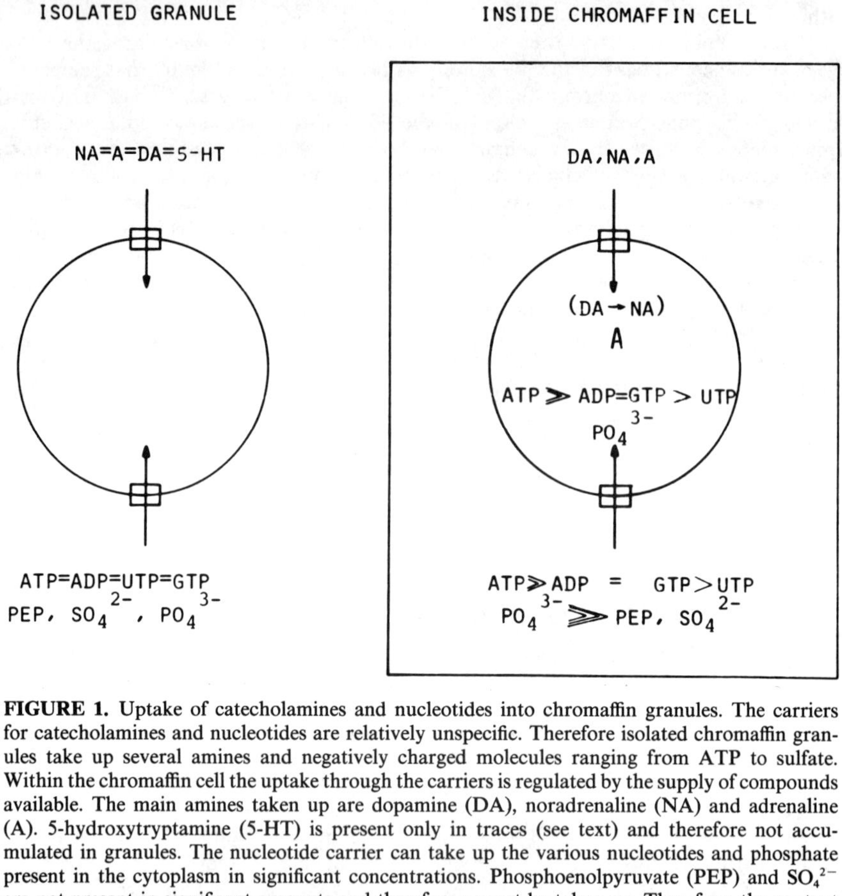

FIGURE 1. Uptake of catecholamines and nucleotides into chromaffin granules. The carriers for catecholamines and nucleotides are relatively unspecific. Therefore isolated chromaffin granules take up several amines and negatively charged molecules ranging from ATP to sulfate. Within the chromaffin cell the uptake through the carriers is regulated by the supply of compounds available. The main amines taken up are dopamine (DA), noradrenaline (NA) and adrenaline (A). 5-hydroxytryptamine (5-HT) is present only in traces (see text) and therefore not accumulated in granules. The nucleotide carrier can take up the various nucleotides and phosphate present in the cytoplasm in significant concentrations. Phosphoenolpyruvate (PEP) and SO_4^{2-} are not present in significant amounts and therefore cannot be taken up. Therefore, the content of chromaffin granules in amines and nucleotides mirrors the concentration of these compounds in the cytoplasm.

kDa were found to be common by both groups. An additional 115-kDa unit which bound dicyclohexylcarbodiimide was found by Cidon and Nelson[28] and considered to represent the active site of the enzyme, whereas a 16-kDa subunit possibly representing a hydrophobic membrane-spanning part was considered to be part of the enzyme by the other group.[27] Apparently both groups have isolated the same enzyme and future studies should now establish the exact subunit structure and its relationship to an apparently analogous ATPase found in coated vesicles and fungal vacuolar membranes (see ref. 2).

One additional problem remains to be solved. Schmidt *et al.*[29] observed the occurrence of stalked particles on the surface of chromaffin granules. In negative staining they appeared very similar to the stalked particles that represent the F_1/F_0 ATPase in mitochondria. Since at that time the granule ATPase was considered to be similar to the mitochondrial enzyme, the stalked particles on chromaffin granules were interpreted as the morphological substrate of the granule ATPase. As discussed above, however, the granule enzyme has no apparent biochemical relationship to the mitochondrial enzyme. What is then the function of these stalked particles on chromaffin granules? Do they represent the H^+-ATPase, which differs immunologically and biochemically from the mitochondrial enzyme, but still might be similar in its structure? In this case the granule enzyme should consist of an extramembrane ATP-splitting part and a hydrophobic membrane-spanning domain. Do these stalked particles represent a second ATPase apparently present in these organelles (see ref. 2), or do they have no relationship to any ATPase? This is an interesting question which still has to be resolved.

NUCLEOTIDE TRANSPORT

Chromaffin granules possess a saturable uptake system for nucleotides.[30–33] The carrier is relatively nonspecific: in addition to ATP, other nucleotides, phosphoenolpyruvate, and even sulfate and phosphate are transported.[33] Thus the presence of the various nucleotides and phosphate in the granule matrix reflects the cytoplasmic content of these anions and their transport by an unspecific carrier (see FIG. 1). The nucleotide carrier of chromaffin granules is inhibited by the amino acid reagent phenylglyoxal, indicating that arginine groups are present in a strategic position.

These experiments were performed with intact granules, which have a high internal concentration of ATP. Therefore it could not be established whether this uptake worked against a concentration gradient or simply exchanged nucleotides. Unfortunately experiments on granule "ghosts" did not resolve this question. Evidence for the presence of a nucleotide carrier could be obtained, however no significant ATP gradient could be established.[34] There is no doubt that *in vivo* chromaffin granules have to accumulate ATP, leading to a concentration gradient of at least 15-30. Why do ghosts fail to accumulate ATP in a similar fashion? Does lysis of granules for producing ghosts disturb the arrangement of the carrier within the membrane? Is a factor within the granules necessary for the transport, *e.g.*, calcium, which may enter these organelles early in biogenesis? Is it necessary to exchange ATP with another ion? Since granules have a high positive membrane potential, one would expect that chloride ions enter these granules under *in vivo* conditions. Is it a mechanism to extrude these chloride ions by exchanging them with ATP via the carrier? In this case experiments with ghosts not containing such an exchange ion in the interior could not accumulate ATP. Obviously we do not yet understand one basic feature of the nucleotide transport system.

CALCIUM TRANSPORT

Chromaffin granules contain significant concentrations of calcium.[35] The mechanism of entrance, however, remains controversial. Several groups maintain that calcium enters granules by exchange with sodium without any ATP requirement.[36–39] The fact that the proteins of the granule matrix, the chromogranins, bind calcium very effectively[40] could contribute to the accumulation of a relatively high total calcium amount in the granules. On the other hand, Burger *et al.*[41] (see also ref. 42) claim that chromaffin granules accumulate calcium by an ATP-dependent mechanism, although it has been suggested[2] that this can be explained by mitochondrial and microsomal contamination.

ELECTRON TRANSPORT

It seems well established that cytochrome b_{561} transports electrons from the cytosol into the granule interior in order to reduce semidehydroascorbate and to regenerate intravesicular ascorbate[43,44] (see also ref. 2). Ascorbate is thought to serve as an electron donor for dopamine β-hydroxylase and for the peptidylglycine α-amidating monooxygenase, which also occurs in chromaffin granules.[45]

CONCLUSIONS

It seems that the intensive research carried out during recent years has established many facts as far as the molecular function of the membrane of chromaffin granules is concerned. Several topics that remained unresolved for some time are just being solved or are on the brink of being solved.

REFERENCES

1. APPS, D. K., J. G. PRYDE & J. H. PHILLIPS. 1980. Cytochrome b_{561} is identical with chromomembrin B, a major polypeptide. Neuroscience **5:** 2279-2287.
2. WINKLER, H., D. K. APPS & R. FISCHER-COLBRIE. 1986. The molecular function of adrenal chromaffin granules: Established facts and unresolved topics. Neuroscience. **18:** 261-290.
3. HUBER, E., P. KÖNIG, G. SCHULER, W. ABERER, H. PLATTNER & H. WINKLER. 1979. Characterization and topography of the glycoproteins of adrenal chromaffin granules. J. Neurochem. **32:** 35-47.
4. ABBS, M. T. & J. H. PHILLIPS. 1980. Organization of the proteins of the chromaffin granule membrane. Biochim. Biophys. Acta **595:** 200-221.
5. KÖNIG, P., H. HÖRTNAGL, H. KOSTRON, H. SAPINSKY & H. WINKLER. 1976. The arrangement of dopamine β-hydroxylase (EC 1.14.2.1) and chromomembrin B in the membrane of chromaffin granules. J. Neurochem. **27:** 1539-1541.

6. DUONG, L. T. & P. J. FLEMING. 1984. The asymmetric orientation of cytochrome b_{561} in bovine chromaffin granule membranes. Arch. Biochem. Biophys. **228:** 332-341.
7. ZAREMBA, S. & R. HOGUE-ANGELETTI. 1985. A reliable method for assessing topographical arrangement of proteins in the chromaffin granule membrane. Neurochem. Res. **10:** 19-32.
8. WINKLER, H. & E. WESTHEAD. 1980. The molecular organization of adrenal chromaffin granules. Neuroscience **5:** 1803-1823.
9. SLATER, E. P., S. ZAREMBA & R. HOGUE-ANGELETTI. 1981. Purification of membrane-bound dopamine β-monooxygenase from chromaffin granules: Relation to soluble dopamine β-monooxygenase. Arch. Biochem. Biophys. **211:** 288-296.
10. WINKLER, H. & S. W. CARMICHAEL. 1982. The chromaffin granule. *In* The Secretory Granule. A. M. Poisner & J. M. Trifaro, Eds.: 3-79. Elsevier Biomedical Press. Amsterdam.
11. ARTHUR, G. & A. SHELTAWY. 1980. The presence of lysophosphatidylcholine in chromaffin granules. Biochem. J. **191:** 523-532.
12. FRISCHENSCHLAGER, I., W. SCHMIDT & H. WINKLER. 1983. Is lysolecithin an in vivo constituent of chromaffin granules? J. Neurochem. **41:** 1480-1483.
13. VOYTA, J. C., L. L. SLAKEY & E. W. WESTHEAD. 1978. Accessibility of lysolecithin in catecholamine-secretory vesicles to acyl CoA: lysolecithin acyl transferase. Biochem. Biophys. Res. Commun. **80:** 413-417.
14. WESTHEAD, E. W. & H. WINKLER. 1982. The topography of gangliosides in the membrane of the chromaffin granule of bovine adrenal medulla. Neuroscience **7:** 1611-1614.
15. EYSTEIN, S. H. & T. FLATMARK. 1986. Phospholipase activities in chromaffin granule ghosts of the bovine adrenal medulla (abstract). Third International Symposium Chromaffin Cell Biology, Coolfont.
16. FRISCHENSCHLAGER, I. 1985. Lysolecithin und Phospholipase A_2 in den Chromaffinen Granula des Nebennierenmarkes und Stimulus-Secretion-Coupling. pp. 1-193. Ph.D. thesis, Innsbruck, Austria.
17. MOSKOWITZ, N., S. PUSZKIN & W. SCHOOK. 1983. Characterization of brain synaptic vesicle phospholipase A_2 activity and its modulation by calmodulin, prostaglandin E_2, prostaglandin F_2, cyclic AMP, and ATP. J. Neurochem. **41:** 1576-1586.
18. FRANSON, R. C. & H. VAN DEN BOSCH. 1982. Lysophospholipase activity of bovine adrenal medulla. Biochim. Biophys. Acta. **711:** 75-82.
19. BLASCHKO, H., H. FIREMARK, A. D. SMITH & H. WINKLER. 1967. Lipids of the adrenal medulla: Lysolecithin, a characteristic constituent of chromaffin granules. Biochem. J. **104:** 545-549.
20. NJUS, D., J. KNOTH & M. ZALLAKIAN. 1981. Proton-linked transport in chromaffin granules. Curr. Top. Bioenerg. **11:** 107-147.
21. PHILLIPS, J. H. 1982. Dynamic aspects of chromaffin granule structure. Neuroscience **7:** 1595-1609.
22. PHILLIPS, J. H. 1974. Steady-state kinetics of catecholamine transport by chromaffin granule ghosts. Biochem. J. **144:** 319-325.
23. HOLZWARTH, M. A. & M. S. BROWNFIELD. 1985. Serotonin coexists with epinephrine in rat adrenal medullary cells. Neuroendocrinology **41:** 230-236.
24. GABIZON, R., T. YETINSON & S. SCHULDINER. 1982. Photoinactivation and identification of the biogenic amine transporter in chromaffin granules from bovine adrenal medulla. J. Biol. Chem. **257:** 15145-15150.
25. ISAMBERT, M.-F. & J.-P. HENRY. 1985. Photoaffinity labelling of the tetrabenazine binding sites of bovine chromaffin granule membranes. Biochemistry **24:** 3660-3667.
26. CIDON, S. & N. NELSON. 1982. Properties of a novel ATPase enzyme in chromaffin granules. J. Bioenerg. Biomembr. **14:** 499-512.
27. PERCY, J. M., J. G. PRYDE & D. K. APPS. 1985. Isolation of ATPase I, the proton translocator of chromaffin granule membranes. Biochem. J. **231:** 557-564.
28. CIDON, S. & N. NELSON. 1986. Purification of N-ethylmaleimide sensitive ATPase from chromaffin granule membranes. J. Biol. Chem. **261:** 9222-9227.
29. SCHMIDT, W., H. WINKLER & H. PLATTNER. 1982. Adrenal chromaffin granules: Evidence for an ultrastructural equivalent of the proton-pumping ATPase. Eur. J. Cell Biol. **27:** 96-104.

30. KOSTRON, H., H. WINKLER, L. J. PEER & P. KÖNIG. 1977. Uptake of adenosine triphosphate by isolated adrenal chromaffin granules: A carrier-mediated transport. Neuroscience **2:** 159-166.
31. ABERER, W., H. KOSTRON, E. HUBER & H. WINKLER. 1978. A characterization of the nucleotide uptake by chromaffin granules of bovine adrenal medulla. Biochem. J. **172:** 353-360.
32. WEBER, A. & H. WINKLER. 1981. Specificity and mechanism of nucleotide transport in chromaffin granules. Neuroscience **6:** 2269-2276.
33. WEBER, A., E. A. WESTHEAD & H. WINKLER. 1983. Specificity and properties of the nucleotide carrier in chromaffin granules from bovine adrenal medulla. Biochem. J. **210:** 789-794.
34. GRÜNINGER, H. A., D. K. APPS & J. H. PHILLIPS. 1983. Adenine nucleotide and phosphoenolpyruvate transport by bovine chromaffin granule "ghosts." Neuroscience **9:** 917-924.
35. WINKLER, H. 1976. The composition of adrenal chromaffin granules: An assessment of controversial results. Neuroscience **1:** 65-80.
36. KOSTRON H., H. WINKLER, D. GEISSLER & P. KÖNIG. 1977. Uptake of calcium by chromaffin granules in vitro. J. Neurochem. **28:** 487-493.
37. KRIEGER-BAUER, H. & M. GRATZL. 1982. Uptake of Ca^{2+} by isolated vesicles from adrenal medulla. Biochim. Biophys. Acta. **691:** 61-70.
38. PHILLIPS, J. H. 1981. Transport of Ca^{2+} and Na^{+} across the chromaffin-granule membrane. Biochem. J. **200:** 99-107.
39. YOON, P. S. & R. R. SHARP. 1985. Ca^{2+} and proton transport in chromaffin granule membranes: A proton NMR study. Biochemistry **24:** 7269-7273.
40. BULENDA, D. & M. GRATZL. 1985. Matrix free Ca^{2+} in isolated chromaffin vesicles. Biochemistry **24:** 7760-7765.
41. BURGER, A., W. NIEDERMAIER, R. LANGER & U. BODE. 1984. Further characteristics of the ATP-stimulated uptake of calcium into chromaffin granules. J. Neurochem. **43:** 806-815.
42. GRAFENSTEIN, H. R. K. V. & E. NEUMANN. 1983. ATP-stimulated accumulation of calcium by chromaffin granules and mitochondria from the adrenal medulla. Biochem. Biophys. Res. Commun. **117:** 245-251.
43. NJUS, G., J. KNOTH, C. COOK & P. M. KELLEY. 1983. Electron transfer across the chromaffin granule membrane. J. Biol. Chem. **258:** 27-30.
44. SRIVASTAVA, M., L. T. DUONG & P. J. FLEMING. 1984. Cytochrome b-561 catalyzes transmembrane electron transfer. J. Biol. Chem. **259:** 8072-8075.
45. DILIBERTO, E. J., F. S. MENNITI, J. KNOTH, O. H. VIVEROS & S. KIZER. 1986. Regeneration of ascorbate for dopamine β-hydroxylation and peptidyl-glycine α-amidation (abstract). Third International Symposium Chromaffin Cell Biology. Coolfont.

The Vacuolar ATPase Is Responsible for Acidifying Secretory Organelles

GARY RUDNICK

Department of Pharmacology
Yale University School of Medicine
New Haven, Connecticut 06510

Many of us are familiar with the way that cells make use of ion gradients across the plasma membrane for signaling purposes, as in the propagation of the action potential in excitable cells, and for doing osmotic work, as in Na^+-dependent transport systems in epithelial cells. This discussion will describe a process used in intracellular organelles for signaling, for regulation, and for driving transport processes. This is the ATP-driven pumping of hydrogen ions, or protons, from the cytoplasm to the lumen of a variety of intracellular organelles.

To begin, it is important to distinguish between this H^+-pumping ATPase and the H^+-driven ATP synthetase of mitochondrial inner membrane. Under physiological conditions, these enzymes work in opposite directions. The mitochondrial ATPase uses a H^+ potential to make ATP, whereas the ATPase in other organelles uses ATP to acidify the organelle interior. There are many differences between these two enzymes, but at this point it is useful to note the difference in their function.

ROLE OF THE ATPase IN CELL PHYSIOLOGY

We probably do not know all the cellular functions served by this pump, but we know some of them, and these alone would make the enzyme a key component in cell physiology. One can get an idea of its importance by where it is found. There are a series of membrane-bound intracellular organelles that constitute the vacuolar system. All of these organelles are involved, in one way or another, with the movement of macromolecules to or from the cell exterior, and most, if not all, of them contain this ATP-driven H^+ pump. For this reason, the enzyme is referred to as the vacuolar ATPase.

Receptor-mediated Endocytosis

The uptake of many macromolecules, including nutrients and hormones, involves binding to cell-surface receptors, internalization into coated vesicles and then endosomes, and delivery to the lysosome. In many cases, the ligand and receptor dissociate to allow reutilization of receptor, which is recycled back to the plasma membrane. The majority of ligands that bind to cell surface receptors are dissociated from those receptors at the low endosomal pH generated by H^+-pumping into the endosome.

Lysosomal Hydrolysis

Most of the lysosomal hydrolytic enzymes have low pH optima and work best when the lysosome is acidified. For example, the rate of lysosomal protein degradation is faster at lower lysosomal pH. The H^+-pumping ATPase is responsible for acidifying the lysosomal interior.

Protein Glycosylation

There is no direct evidence that the ATPase is required for protein glycosylation, but there are some tempting observations that can be interpreted as suggesting a role for transmembrane ion gradients in glycosylation and processing in the Golgi apparatus. These include the inhibition of glycosylation reactions by ionophores and other agents that dissipate intracellular pH differences, and the observation of glycosylation defects in mutant cell lines defective in endosomal acidification.

Biogenic Amine Accumulation

Synaptic vesicles and secretory granules participate in the recycling of biogenic amines and other neurotransmitters in neurosecretory cells. It is from these organelles that the transmitters are released by fusion with the plasma membrane. Released transmitter can be taken up at the plasma membrane or resynthesized and then reaccumulated in secretory granules. Biogenic amine secretion granules like adrenal chromaffin granules are acidified by the H^+-pumping ATPase. In their membrane, they also contain an amine transporter that exchanges internal H^+ ions for cytoplasmic amines. This transport system is responsible for the high concentration of internal amine (up to 0.5 M).

Neuropeptide Processing

Many enzymes required for neuropeptide processing in secretory granules have low pH optima. Some of the granules studied, but not all, are acidified by the vacuolar ATPase. Carboxy-terminal amidation of peptides also uses the H^+ potential generated by the vacuolar ATPase. The low internal pH and electrically positive interior facilitate reduction of internal ascorbate through cytochrome b_{561}.

Although platelet and chromaffin granules and a few other secretory granules contain the vacuolar ATPase, at least two exocrine secretory granules do not. These are the amylase-secreting granules of parotid and pancreatic acinar cells. In collaboration with David Castle and Peter Arvan, we tried quite hard to demonstrate the presence of a vacuolar ATPase in parotid granules and concluded that if the ATPase is present in these organelles, it is at levels of only a few percent that of chromaffin granules. The resting pH of these granules is quite high, about 6.8, and is not changed by adding ATP. We used accumulation of the lipophilic cation SCN^- as a measure of electrical potential, and saw little potential generation relative to chromaffin granules, which generate large, stable potentials under the same conditions. Finally, one does not see any Nbd-Cl sensitive ATP hydrolysis under conditions where over 75% of the chromaffin granule ATP hydrolysis is blocked.

Epithelial H^+ Transport

In the kidney, trans-epithelial H^+ flux responsible for urinary acidification is thought to be driven by the same vacuolar ATPase that is reversibly inserted into the plasma membrane by a process very similar to secretion.

BIOENERGETICS OF THE ATPase

Basically, this enzyme hydrolyzes cytoplasmic ATP and uses the energy liberated in that reaction to move H^+ ions from the cytoplasm to the vesicle lumen. In bioenergetic terms this has two consequences: the vesicle lumen becomes acidic and also positively charged with respect to the cytoplasm. For net H^+-pumping to occur, counterion movement (anion influx or cation efflux) must dissipate the transmembrane potential. The possibility exists that incorporation or activation of counterion permeabilities may regulate organelle acidification *in vivo.*

NEW TYPE OF ATP-DRIVEN ION PUMP

Prior to the discovery of the vacuolar ATPase, it was possible to classify all ATP-driven ion pumps as belonging to one of two classes, the E_1E_2 type and the F_1F_o type.

E_1E_2 Type Pumps

The E_1E_2 pumps are typified by the Na^+, K^+-ATPase, but members of this class also pump H^+ and Ca^{2+} in addition to Na^+ and K^+. All of these enzymes have, as an integral part of their catalytic cycle, a phosphorylated enzyme intermediate. They are probably closely related evolutionarily, since the primary sequence around the phosphorylation site is highly conserved. All of these enzymes are inhibited strongly by vanadate ion and share a common structure for the catalytic subunit, which is a transmembrane protein of approximately 100,000 Da.

F_oF_1 Type Pumps

In contrast, the F_1F_o-ATPases all pump H^+ (although they work in the opposite direction physiologically). They do not involve obligatory phosphoenzyme formation

TABLE 1. Characteristics of Ion-transporting ATPases

	F_1F_o	E_1E_2	Vacuolar
Ion pumped	H^+	Na^+, K^+, Ca^{2+}, H^+	H^+
E~P	no	yes	?
M_r	380,000 (F_1)	100,000	134,000
Location	extrinsic (F_1)	intrinsic	intrinsic
Inhibitors			
VO_3^-	no	yes	no
N_3^-	yes	no	no
N-ethylmaleimide	no	no	yes

and apparently transfer the terminal phosphate of ATP directly to water. All of these enzymes are inhibited by azide ion and share a highly conserved and interchangable catalytic subunit that is part of a larger complex of 380,000 to 400,000 Da. This F_1 complex binds reversibly to the F_o portion, which resides in the membrane and which, in the absence of F_1, acts as a H^+ conductance.

INHIBITOR SPECIFICITY

TABLE 1 summarizes the characteristics of the vacuolar ATPase that differentiate it from other ATP-driven ion pumps. The vacuolar ATPase is insensitive to inhibitors of F_1F_o and E_1E_2 ATPases, in particular azide and vanadate. It is much more sensitive than these enzymes to sulfhydryl reagents such as *N*-ethylmaleimide, and recently we have discovered that omeprazole, the H^+,K^+-ATPase inhibitor, also inactivates the vacuolar ATPase. The molecular mass of detergent-solubilized vacuolar ATPase is about 134,000 Da and it is an intrinsic membrane protein. This is distinct from F_1F_o-ATPase. The vacuolar ATPase pumps only protons, so far as we know. The mechanism of ATP hydrolysis and its coupling to proton pumping are still unknown. In fact, once we understand its mechanism, we may classify it as a subclass of one of the other ATPase types.

In membrane vesicles derived from chromaffin granules, we can examine the inhibitor sensitivity of H^+-pumping by measuring acridine orange fluorescence quenching. This fluorescence quenching is initiated by addition of Mg-ATP and terminated by FCCP, which allows the protons that were pumped in to leak out again. Again, there is no inhibition relative to the appropriate control for azide, efrapeptin, oligomycin, ouabain, or vanadate. NEM and Nbd-Cl, another sulfhydryl reagent, completely inactivate, and DCCD, which inhibits ATP-driven H^+ pumps of all types, also inactivates, although only at concentrations higher than those required to inhibit F_1F_o-ATPase.

We can also measure ATP hydrolysis by the same chromaffin granule membrane vesicles. Mitochondrial inhibitors such as azide, efrapeptin, and oligomycin inhibit 10-15% of the activity, vanadate and ouabain inhibit 6-7%, DCCD inhibits almost half, and most of the activity is inhibited by either NEM or Nbd-Cl.

The correspondence between ATPase inhibition and transport inhibition is more than coincidental. If we look at the time course of NEM inhibition and at each time point measure both ATP hydrolysis and ATP-driven serotonin transport, we see that both activities are lost in parallel until all of the transport activity is inactivated but about 20% of the ATPase remains. This 20% is probably accounted for by contaminating mitochondrial and Na^+,K^+-ATPase activities.

Results presented elsewhere in this volume suggest that the chromaffin granule ATPase is an intrinsic membrane protein of 134,000 Da, and is likely to be transmembrane in its orientation. This is very different from F_1, which is a soluble protein of 320-400 kDa that binds to the F_o membrane portion.

The vacuolar ATPase is, therefore, a new class of enzymes, structurally and functionally distinct from previous classes of ion-transporting ATPase, which is responsible for pumping H^+ into a wide variety of intracellular organelles, including secretory vesicles.

The Proton Pump of Synaptic Vesicles

HERBERT STADLER

Department of Neurochemistry
Max Planck Institute for Biophysical Chemistry
Göttingen, Federal Republic of Germany

Synaptic vesicles are the secretory organelles of the nerve terminal that store and release neurotransmitters. In the case of the acetylcholine-containing synaptic vesicles from electric organs, the best-characterized vesicle preparation at present, it was found that high concentrations of acetylcholine are stored in a fluid phase[1] at an acidic pH around 5.5.[2] These findings suggested that a proton ATPase might be present in the vesicle membrane, creating an electrochemical potential necessary for accumulation of transmitter.

In agreement with these findings, experiments using the technique of [^{14}C]methylamine uptake on isolated highly purified vesicles of *Torpedo* electric organ showed that ATP-dependent acidification of the vesicle interior occurred, indicating the presence of an ATP-driven proton pump. Further experiments showed in addition that the vesicle membrane contains a ouabain- and oligomycin-insensitive Mg-ATPase that might be part of the proton pump.[3]

Similar experiments were carried out using a synaptic vesicle preparation from guinea pig brain freed from membrane contaminants by chromatography on Sephacryl S 1000. As in the case with *Torpedo* electric organ vesicles, these vesicles showed ATP-dependent acidification of the interior and association with a ouabain- and oligomycin-insensitive Mg-ATPase, indicating as well the presence of an ATP-driven proton pump. Furthermore, morphological examination of the preparation using a quick freeze, deep-etch, rotary-shadowing technique showed that protrusions were present on the surface of the vesicle, suggesting that they might be part of the vesicular proton pump analogous to proton pumps in other systems.[4]

Altogether these findings provide evidence that synaptic vesicles are equipped with one or more copies of ATP-driven proton pumps that create an electrochemical gradient over the vesicle membrane necessary for uptake and storage of transmitter. Isolation and detailed characterization of the composition of the proton pump have not been carried out yet.

REFERENCES

1. Stadler, H. & H. H. Füldner. 1980. Nature **286:** 293-294.
2. Füldner, H. H. & H. Stadler. 1982. Eur. J. Biochem. **121:** 519-524.
3. Harlos, P., D. A. Lee & H. Stadler. 1984. Eur. J. Biochem. **144:** 441-446.
4. Stadler, H. & S. Tsukita. 1984. EMBO J. **3:** 3333-3337.

Electrochemical Proton Gradients in Amine- and Peptide-containing Subcellular Organelles

S. E. CARTY,[a] R. G. JOHNSON, JR.,[b] AND A. SCARPA[c]

[a] *Department of Surgery*
Milton S. Hershey Medical Center
Hershey, Pennsylvania 17033

[b] *Howard Hughes Medical Institute*
Massachusetts General Hospital
Harvard Medical School
Boston, Massachusetts 02114

[c] *Department of Physiology and Biophysics*
Case Western Reserve University
Cleveland, Ohio 44106

The data presented here were an outgrowth of our recent work with the adrenal chromaffin granule, work which has helped to elucidate the mechanism for its accumulation and storage of the catecholamines epinephrine, norepinephrine, and dopamine. In brief, catecholamines were found to be transported into the chromaffin granule by a carrier-mediated process linked to the presence of a transmembrane electrochemical proton gradient. This gradient, composed of both a transmembrane pH differential (acidic inside) and a transmembrane potential (positive inside), was found to be generated by the action of a membrane proton-translocating ATPase similar to the F_1 ATP-ase of mitochondria.[1]

Techniques perfected for investigation of amine transport in the chromaffin granule were then applied to a variety of less easily isolated organelles.[2–5] The isolated chromaffin granules of a human pheochromocytoma can be seen on electron microscopy to bear a marked resemblance to dense granules isolated from porcine platelets and bovine anterior pituitary glands, and also to granules seen *in situ* in the tissue of a human vasoactive-intestinal-peptide-secreting tumor (see FIG. 1). The organelles are all electron dense, membrane-limited structures possessing homogenous matrices.

The electromicrographic similarities of the chromaffin granule to other organelles invited comparison of their membrane permeability and electrical properties. Dense granules were recovered from pheochromocytoma, platelets, anterior pituitary, and VIP-secreting tumor and were purified from other subcellular organelles using isotonic density gradient methods. As measured by distribution of [^{14}C]methylamine, the isolated granules maintained internal pHs of 5.1, 5.4, 5.1, and 5.0, respectively, when suspended in isotonic media of pH 5.0-8.0 (TABLE 1). Their transmembrane pH gradients could in each case be completely collapsed by addition of NH_4^+. In the presence of MgATP, each type of organelle was found to generate a transmembrane potential of 90 mV, 40 mV, 70 mV, and 20 mV respectively, as measured by

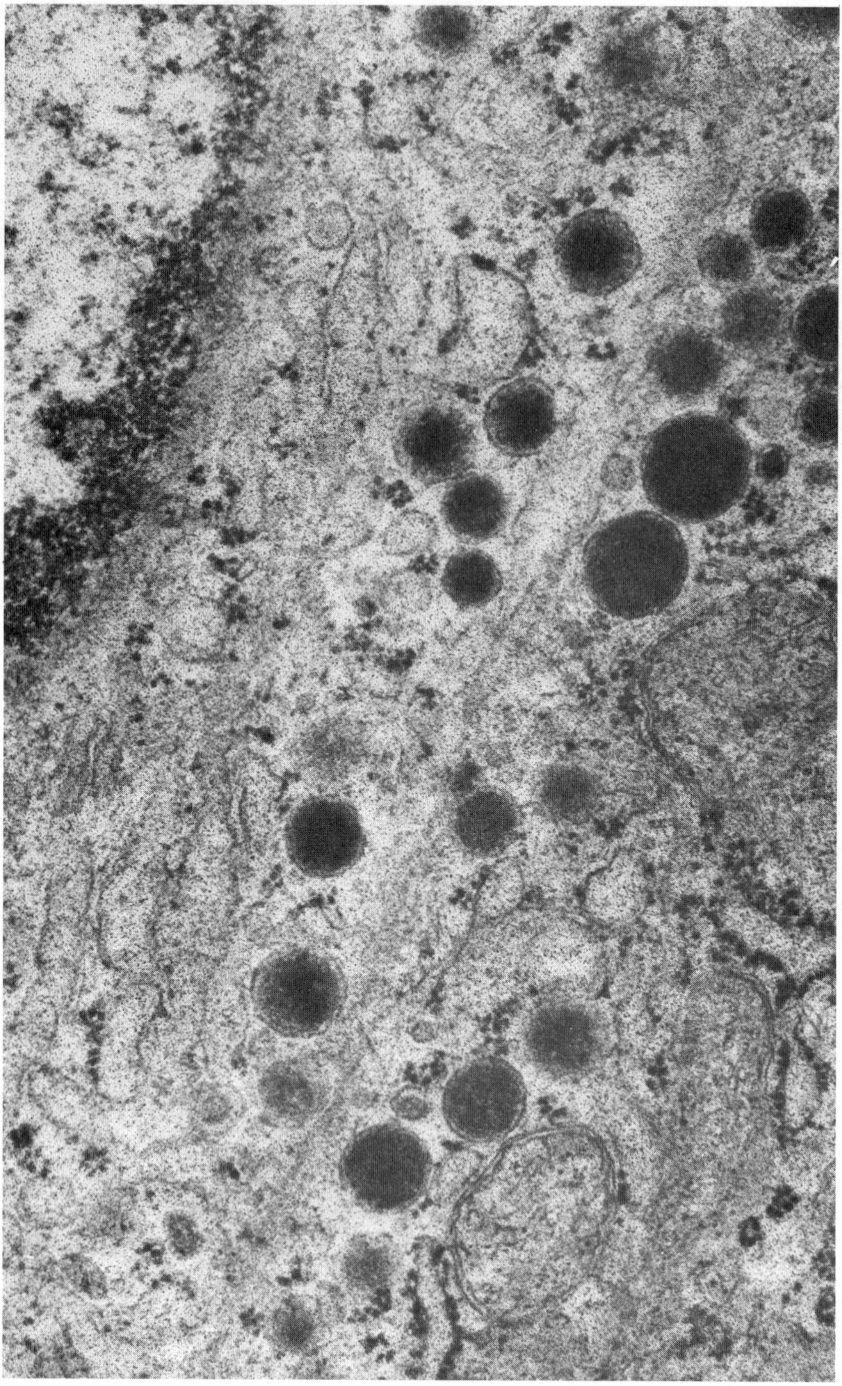

FIGURE 1. Electron micrograph of fresh tissue from a human pancreatic vasoactive intestinal peptide tumor. Magnification: 60,000 ×.

TABLE 1. Internal pH and Transmembrane Potential of Subcellular Storage Organelles

Organelle	Internal pH	Amine	Transmembrane Potential	Peptide
Chromaffin granule	5.1	norepinephrine epinephrine dopamine	90 mV	enkephalin
Platelet granule	5.4	serotonin	40 mV	?
Anterior pituitary granule	5.1	dopamine	70 mV	GH/PRL TSH/FSH/LH
Mast cell granule	6.1	histamine	—	multiple
VIP-oma granule	5.0	serotonin	20 mV	VIP

[^{14}C]thiocyanate distribution methods. In each case, the generation of a transmembrane potential was completely inhibited by trimethyl tin, an F_1-proton ATPase inhibitor. In experiments investigating the properties of mast cell granules, intact mast cells were isolated from rats and the intragranular pH was determined using both [^{14}C]methylamine and fluorescent dye distribution techniques. The internal pH of mast cell granules was estimated by these methods by be 6.1; transmembrane potential was not measured.

These and other experiments have suggested that each granule type described here possesses a membrane highly impermeable to protons and a proton-translocating ATPase capable of generating an electrochemical proton gradient. In chromaffin granules, we have shown this gradient to drive specific uptake of catecholamines, as discussed above, and in other work we have demonstrated that the electrochemical proton gradient drives specific transport of serotonin into platelet granules by a nearly identical mechanism.[2] The significance of the electrochemical proton gradient in anterior pituitary and VIP-oma granules, however, is still enigmatic since these organelles contain predominatedly peptides, with relatively few amines. It may be hypothesized that their gradients provide a driving force for the storage of peptides; their acidic intragranular pH may additionally serve to provide a stable storage milieu for peptides, amines, and enzymes prior to release. Despite their widely differing compositions, functions, and embryologic origins, these subcellular organelles appear to possess an identical mechanism for energy-linked transport, which suggests a universal physiology for the accumulation and storage of neurotransmitter and hormonal amines and peptides.

REFERENCES

1. CARTY, S. E., R. G. JOHNSON & A. SCARPA. 1985. The Enzymes of Biological Membranes, Vol. 3. A. Martinosi, Ed.: 449-495. Plenum. New York.
2. CARTY, S. E., R. G. JOHNSON & A. SCARPA. 1981. J. Biol. Chem. **256:** 11244-11250.
3. JOHNSON, R. G., S. E. CARTY & A. SCARPA. 1982. Biochem. Biophys. Acta **716:** 366-376.
4. CARTY, S. E., R. G. JOHNSON & A. SCARPA. 1982. J. Biol. Chem. **257:** 7269-7273.
5. JOHNSON, R. G., S. E. CARTY, B. J. FINGERHOOD & A. SCARPA. 1980. FEBS Lett. **120**(1): 75-79.

Molecular Weight and Hydrodynamic Properties of the Chromaffin Granule ATPase

GARY E. DEAN, PAMLEA J. NELSON, WILLIAM S. AGNEW, AND GARY RUDNICK

Yale University School of Medicine
New Haven, Connecticut 06510

Extraction of chromaffin granule membranes with the nonionic detergent $C_{12}E_9$ in the presence of phosphatidyl serine (PS) solubilizes ATPase activity, which in the intact membrane is coupled to H^+ pumping. Equilibrium sedimentation experiments performed in a table-top, air-driven ultracentrifuge indicate that the solubilized ATPase exists in a particle with roughly equal amounts (by weight) of protein and detergent. The particle partial specific volume, $\bar{v}(o)$, and total mass, $m(o)$, were determined by sedimentation in media where various amounts of H_2O were replaced with D_2O. Similar determinations were made for $C_{12}E_9$-PS mixed micelles. Gel filtration of the solubilized ATPase gave a Stokes radius of 43 ± 4 Å and an axial ratio of ~1 was calculated.

From the combined results of these sedimentation equilibrium and gel permeation chromatography studies, all of the major physical parameters of the $C_{12}E_9$-solubilized chromaffin granule ATPase may be deduced. These are presented in TABLE 1. We have made the conventional assumption that the partial specific volume of the protein element is 0.73 cm^3/g.

The large amount of detergent and lipid associated with the enzyme strongly suggests that it is closely associated with the lipid bilayer by hydrophobic interactions.

TABLE 1. Hydrodynamic Properties of the Chromaffin Granule ATPase[a]

Particle partial specific volume	$\bar{v}(o)$	0.829 cm^3/g
Particle mass	$m(o)$	264,000 ± 13,000 Da
Detergent partial specific volume	$\bar{v}(d)$	0.93 ± 0.02 cm^3/g
Lipid-detergent mixed micelle mass	$m(d)$	83,000 ± 30,000 Da
Protein mass in particle	$m(\rho)$	134,000 ± 7,000 Da
Detergent-lipid mass in particle	$m(o)-m(\rho)$	130,000 ± 20,000 Da
Detergent-lipid/protein mass ratio	δ	0.97
Stokes radius	R_s	43 ± 4 Å
Frictional coefficient	f/f_o	0.98
Axial ratio	a/b	1.0
Sedimentation coefficient	$S_{20,w}$	9.2 S
Diffusion coefficient	$D_{20,w}$	5.0 F

[a] Measurements and calculations are described in the text; the value of $\bar{v}(\rho)$ = 0.73 cm^3/g has been assumed.

This contrasts with the F_1F_0-ATPases, whose hydrolytic F_1 components are relatively water-soluble proteins that associate with the membrane-bound F_0 component by protein-protein interactions. The chromaffin granule ATPase is also much smaller than the F_1F_0-ATPase, with the hydrolytic F_1 portion alone a multisubunit complex of ~400,000 Da. In contrast, our best estimate of the chromaffin granule ATPase's molecular mass, 134,000 Da, is closer to the molecular masses of various E_1E_2-ATPases, which are also intrinsic membrane proteins. Despite these apparent structural similarities, the chromaffin granule ATPase cannot be classified as an E_1E_2 enzyme, since, like all other vacuolar ATPases studied, it is vanadate-insensitive.

The preparation of $C_{12}E_9$-solubilized chromaffin granule ATPase used here is not active in reconstituting H^+ pumping in liposomes. This observation raises the possibility that the 134,000-Da ATPase may not represent the entire ATP-driven H^+ pump and that subunits which dissociate in $C_{12}E_9$ are required for H^+ translocation. Given an axial ratio of 1.0, it is likely that the enzyme is deeply embedded in the lipid-detergent micelle, as expected for a transmembrane protein. Furthermore, the ATPase mass is sufficient, by analogy to the E_1E_2 ATPases, to account for an ATP-driven ion pump. It is premature, however, to conclude that the $C_{12}E_9$-solubilized ATPase represents the holoenzyme.

Increased cAMP or Ca^{2+} Second Messengers Reproduce Effects of Depolarization on Adrenal Enkephalin Pathways

E. F. LA GAMMA,[a] J. D. WHITE,[b] J. G. McKELVY,[b]
AND I. B. BLACK[a]

[a] *Division of Developmental Neurology*
Cornell University Medical College
New York, New York 10021

[b] *Department of Neurobiology and Behavior*
State University of New York at Stony Brook
Stony Brook, New York 11794

Trans-synaptic activity and membrane depolarization differentially regulate catecholamines and colocalized opiate peptides in the rat adrenal medulla.[1–6] Depolarization prevents the 50-fold rise in adrenal Leu-enkephalin and preproenkephalin mRNA in culture.[1,2] In order to better characterize signal transduction processes that regulate adrenal transmitters, we examined the role of Ca^{2+} and cyclic nucleotide second-messenger systems on enkephalin regulation.

ROLE OF CALCIUM IONS IN ENKEPHALIN REGULATION

Explanted medullae were depolarized in the presence of EGTA or the calcium ion channel blockers verapamil or its methoxy-derivative, D600. Inhibition of Ca^{2+} influx prevented the effects of KCl-induced depolarization on the rise in Leu-enkephalin immunoreactivity (Leu-enk) and on preproenkephalin mRNA (prepro-EK). Increasing intracellular Ca^{2+} with the ionophore A23187, in the absence of depolarizing agents, reproduced the effects of depolarization. In contrast, medullae grown in the presence of A23187, but in Ca^{2+}-free medium, showed similar increases in prepro-EK mRNA and Leu-enk, indicating an absolute requirement for Ca^{2+} (FIG. 1). In addition, KCl effects could be partially inhibited by the calmodulin and protein kinase C antagonist, trifluoperazine. However, KCl effects were not antagonized by the preferential calmodulin inhibitors W7, W13, or calmidizolium, even at doses 10-fold higher than required to prevent calmodulin-dependent effects.

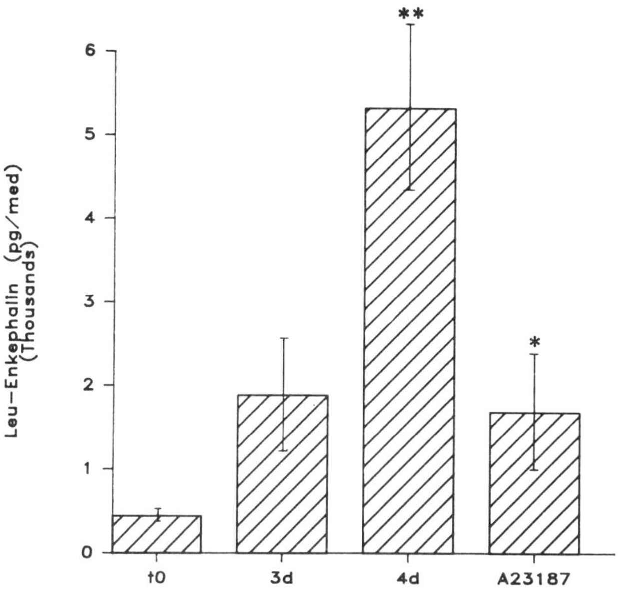

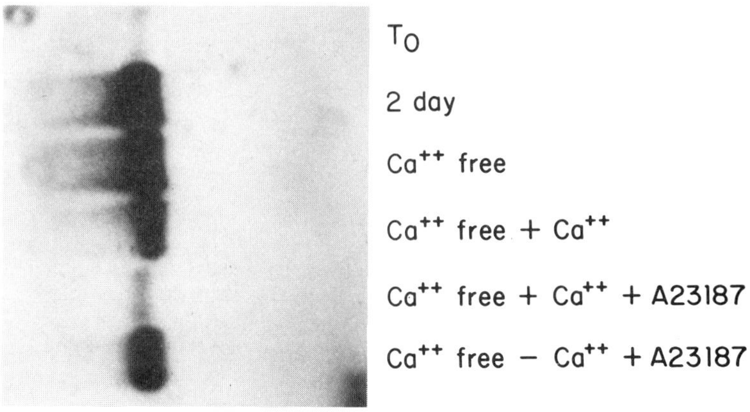

FIGURE 1. Medullae were grown for 2 days in the presence of the calcium ionophore A23187, which prevented the rise in enkephalin peptide (*top*) and preproenkephalin mRNA (*bottom*) in a Ca^{2+}-dependent fashion; T_O, noncultured medullae. Note: Top of gel is to the right with separation to the left.

EFFECTS OF INCREASED CYCLIC NUCLEOTIDES

Increased cAMP levels (forskolin, 25 μM, cholera toxin, 25 μg/ml, 8-Br-cAMP, 1 mM) reproduced the effects of depolarization, preventing the rise of Leu-enk in culture above zero time values. In contrast, increased cGMP (8-Br-cGMP, db-cGMP, 5 mM each) had no effect compared to 4-day control explants.

In summary, these data suggest that inhibitory effects of trans-synaptic activity and depolarization on enkephalin pathways occur through Ca^{2+} and cAMP second-messenger mechanisms. This is of particular interest since these treatments are known to increase tyrosine hydroxylase phosphorylation and activation, resulting in augmented catecholamine biosynthesis. Therefore, it appears that the same or similar molecular mechanisms may result in differential regulation of these colocalized transmitter systems.

REFERENCES

1. La Gamma, E. F., J. E. Adler & I. B. Black. 1984. Science **224:** 1102-1104.
2. La Gamma, E. F., J. D. White, J. E. Adler, J. E. Krause, J. F. McKelvy & I. B. Black. 1985. Proc. Natl. Acad. Sci. U.S.A. **82:** 8252-8255.
3. Bohn, M. C., J. A. Kessler, L. Golightly & I. B. Black. 1983. Cell Tissue Res. **231:** 469-479.
4. Fleminger, G., R. D. Howells, D. L. Kilpatrick & S. Udenfriend. 1984. Proc. Natl. Acad. Sci. U.S.A. **81:** 7985-7988.
5. Fleminger, G., H-W. Lahm & S. Udenfriend. 1984. Proc. Natl. Acad. Sci. **81:** 3587-3590.
6. Kilpatrick, D. L., R. D. Howells, G. Fleminger & S. Udenfriend. 1984. Proc. Natl. Acad. Sci. **81:** 7221-7223.

Neuronotrophic Activity in Adrenal Glands: Effect of Denervation

JOSE R. NARANJO, BRADLEY C. WISE, ITALO MOCCHETTI, AND ERMINIO COSTA

Fidia-Georgetown Institute for the Neurosciences
Washington, D.C. 20007

In the seventies, neuronotrophic activities were described in different peripheral organs, including the adrenal glands.[1-3] In the adrenal gland this neuronotrophic activity was initially reported as nerve growth factor (NGF),[3] and recently, the receptor for NGF and NGF mRNA were demonstrated in this organ.[4,5] However, an adrenal neuronotrophic activity, which is located in the chromaffin granules, fails to be antagonized by anti-NGF antibodies.[6]

Since trophic activity of different neuronotrophic factors includes the stimulation of various transmitter biosyntheses and the induction of transmitter-synthesizing enzymes,[7-9] we have investigated the adrenal neuronotrophic activity in terms of its ability to alter the expression of the proenkephalin system in cultured bovine chromaffin cells, and the activity of the enzyme choline acetyltransferase in cultured rat fetal septal cells.

An increase of the rat adrenal proenkephalin system in the absence of catecholamine[10] or neuropeptide Y[11] changes was reported from various laboratories[10,12] after section of the splanchnic nerve. This prompted us to study whether or not denervation of the adrenal gland changes its basal neuronotrophic activity.

Treatment of bovine chromaffin cells plated at low density with the adrenal PBS extract (see figure legend) from control innervated glands induced an increase in the content of Met[5]-enkephalin-immunoreactivity (ME-IR). This effect was significant 24 hours after the addition (FIG. 1A), and peaked at 48 hours of treatment (FIG. 1B). The increase in ME-IR was not due to an enhanced survival of the cells, and was preceded by an increase in the accumulation of proenkephalin mRNA (data not shown). Adrenal extracts from denervated glands produced the same effect on ME-IR content with a similar time course (FIG. 1). However, extracts from denervated glands were 3 to 5 times more potent than those from control glands (data not shown).

Treatment of rat fetal septal cells in culture with the adrenal PBS extracts resulted in an increase of the choline acetyltransferase activity in these cells (TABLE 1). Adrenal extracts from denervated glands, also in this system, was significantly more potent than adrenal extract from control innervated glands. In summary, a neuronotrophic activity present in adrenals regulates the dynamic state of the proenkephalin system in the gland, and has anabolic effects in a population of cholinergic neurons from the CNS. This activity, which increases after splanchnic denervation, may result from an induction of a preexisting factor(s) or from the disinhibition of a previously repressed gene(s).

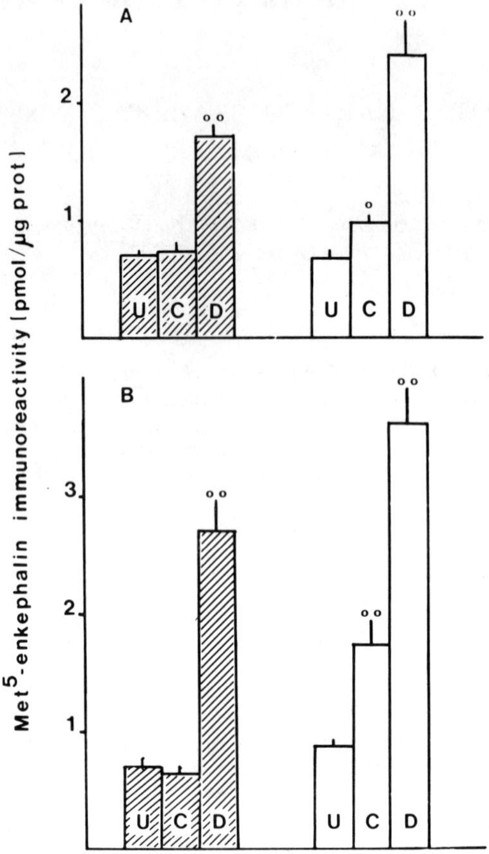

FIGURE 1. Effect of rat adrenal PBS extracts on the content of Met[5]-enkephalin immunoreactivity in cultured bovine chromaffin cells. Unilateral section of the splanchnic nerve was performed in 200-250 g Sprague-Dawley rats. Forty-eight hours after surgery, adrenal glands were removed and homogenized in one volume of 1 × phosphate-buffered saline (PBS). Homogenates were centrifuged for 15 minutes at 4°C in a microcentrifuge. Supernatants (adrenal PBS extracts) were collected, aliquoted, and frozen. Protein content (BioRad microassay) served as the quantitative parameter for the amount of extract used in the treatments. Bovine chromaffin cells were plated at a high (125,000 cells/cm^2, *hatched bars*) or at a low (62,500 cells/cm^2, *open bars*) density. After 2 days the culture medium was changed and the cells were treated with adrenal PBS extracts (10 μg of protein) from control (*C*) or denervated (*D*) glands. PBS was added to untreated (*U*) cells. Panel **A** shows the effect after 24 hours and panel **B** after 48 hours of treatment; (*o*, $p < 0.05$; *oo*, $p < 0.01$.)

TABLE 1. Effect of Adrenal PBS Extracts on ChAT Activity in Cultured Fetal Septal Cells

Treatment	n	Choline Acetyltransferase Activity (pmol/90 min/5 $\times$ 10^4 Cells)
Untreated	4	36.2 ± 1.1
Adrenal extract		
Control	4	54.6 ± 0.9[a]
Denervated	4	62.0 ± 2.1[a,b]

Septal cells were prepared from 17-day-old rat fetuses and cultured in serum-free defined media (Wise *et al.,* in preparation) in 96-multiwell plates at a plating density of 50,000 cells/well/50 μl. The treatment (4 μg of protein) was applied on the second and repeated on the fourth day of culture. Choline acetyltransferase activity was assayed two days after the last administration.

[a] $p < 0.001$ vs. untreated.

[b] $p < 0.02$ vs. control extract.

REFERENCES

1. JOHNSON, D. G., S. D. SILBERSTEIN, I. HANBAUER & I. J. KOPIN. 1972. J. Neurochem. **19:** 2025-2029.
2. CHAMLEY, J. H., I. GOLLER & G. BURNSTOCK. 1973. Dev. Biol. **31:** 362-379.
3. HARPER, G. P., F. L. PEARCE & C. A. VERNON. 1976. Nature **261:** 251-253.
4. NAUJOKS, K. W., S. KORSCHING, H. ROHRER & H. THOENEN. 1982. Dev. Biol. **92:** 365-379.
5. SHELTON, D. L. & L. F. REICHARDT. 1984. Proc. Natl. Acad. Sci. U.S.A. **81:** 7951-7955.
6. UNSICKER, K. 1986. Third Int. Symp. on Chromaffin Cell Biology.
7. ALOE, L. & R. LEVI-MONTALCINI. 1979. Proc. Natl. Acad. Sci. U.S.A. **76:** 1246-1250.
8. SCHWARTZ, J. P. & E. COSTA. 1979. Brain Res. **170:** 198-202.
9. ACHESON, A. L., K. W. NAUJOKS & H. THOENEN. 1984. J. Neurosci. **4:** 1771-1780.
10. FLEMINGER, G., H-W. LAHM & S. UDENFRIEND. 1984. Proc. Acad. Sci. U.S.A. **81:** 3587-3590.
11. HIGUCHI, H. & H-Y.T. YANG. 1986. J. Neurochem. **46:** 1658-1660.
12. SCHULTZBERG, M., J. M. LUNDBERG, T. HOKFELT, J. BRANDT, R. P. ELDA & M. GOLDSTEIN. 1978. Neuroscience **3:** 1169-1189.

Increases in the Size of Acetylcholine Quanta Are Blocked by Treatment with AH5183

WILLIAM VAN DER KLOOT

Departments of Physiology & Biophysics and Pharmacological Sciences
State University of New York at Stony Brook
Stony Brook, New York 11794

A number of treatments have been found to increase the size of MEPPs and MEPCs at the frog neuromuscular junction. For example, when a preparation is soaked for two hours in a hypertonic Ringer (200 mM NaCl in place of 120 mM), after return to normal Ringer the quantal size is roughly doubled.[1] If the hypertonic solution is made with gluconate in place of chloride, quantal size is increased fivefold.

Quantal size is also roughly doubled after 1 hour in solutions containing epinephrine (1-50 μM), norepinephrine (10-50 μM), 8-(4-chloro-phenylthio)-adenosine-3′,5′-monophosphate (50 μM), insulin (10-100 mU/ml) or ACTH (1.5 μM).[2]

All of these increases in size are reversible; quantal size returns to normal within 4 hours after the end of the treatment.

None of these treatments appears to affect markedly the ACh-gated channel in the end-plate membrane (TABLE 1).

AH5183, 2-(4-phenylpiperdino)cyclohexanol, blocks neuromuscular transmission by inhibiting active ACh uptake into synaptic vesicles.[3,4] It also blocks the nonquantal ACh release at the neuromuscular junction.[5] Soaking muscles in AH5183 does not change quantal size for several hours, so long as an anti-ChE is present.[6]

AH5183 (1-10 μM) partially blocks the increase in quantal size caused by the hormones and by pretreatment with hypertonic solutions. For example, a typical experiment 1 h pretreatment with 10 μM epinephrine increased quantal size 2.0-fold; adding 10 μM AH5183 to the epinephrine solution, the increase was only 1.3-fold. In 10 mU/ml insulin the increase was 2.4-fold. With insulin + AH4183 the increase was only 1.4-fold. Similar results were obtained, in a total of 22 experiments, with

TABLE 1. Effects of Pretreatment in 100 mU/ml Bovine Insulin on the ACh-gated Channel of the Frog End Plate[2]

	Control	Insulin-pretreated
Open time (ms)	3.1	3.0
Conductance (pS)	23	22
Reversal potential (mV)	+1.8	−1.6
Bungarotoxin bound (μM/g)	2.6	2.3

1.5 μM ACTH, 50 μM 8-(4-chloro-phenylthio)-adenosine-3′,5′-monophosphate, and with 2 h pretreatment in Ringer made hypertonic with 200 mM NaCl.

Since the treatments appear to have little effect on the ACh-gated channels in the end-plate membrane, and since the increases are antagonized by AH5183, an inhibitor of ACh accumulation into vesicles, it is concluded that the treatments act to increase the quantity of ACh/quantum.[7] It seems likely that the quantity of ACh incorporated into a secretory vesicle can be regulated.

REFERENCES

1. VAN DER KLOOT, W. & T. E. VAN DER KLOOT. 1985. Experentia **41:** 47-48.
2. VAN DER KLOOT, W. 1986. Brain Res. **376:** 378-381.
3. MARSHALL, I. G. 1970. Brit. J. Pharm. **40:** 68-77.
4. ANDERSON, D. H., S. C. KING & S. M. PARSONS. 1983. Molec. Pharm. **24:** 55-59.
5. EDWARDS, C., V. DOLEŽAL, S. TUČEK, H. ZEMKOVÁ & F. VYSKOČIL. 1985. Proc. Natl. Acad. Sci. U.S.A. **82:** 3514-3518.
6. VAN DER KLOOT, W. 1986. Pflügers Arch. **406:** 83-85.
7. VAN DER KLOOT, W. 1986. J. Physiol. **373:** 66P.

The Role of Secretory Granules in Peptide Biosynthesis[a]

RICHARD E. MAINS, EDWARD I. CULLEN, VICTOR MAY, AND BETTY A. EIPPER

Department of Neuroscience
The Johns Hopkins University School of Medicine
Baltimore, Maryland 21205

Secretory granules are the major site of synthesis and/or storage of many small, bioactive substances such as catecholamines, acetylcholine, histamine, and serotonin. Secretory granules are also the final site in the subcellular route taken by secreted peptides. As detailed in many of the other articles in this volume, secretory granules are complex structures, containing biosynthetic enzymes (*e.g.*, dopamine β-hydroxylase, or DBH; peptidyl-glycine α-amidating monooxygenase, or PAM; carboxypeptidase H, or CPH),[1-4] cofactors for enzymes (*e.g.*, ascorbic acid),[5,6] and carrier proteins (*e.g.*, chromogranins),[7,8] in addition to the bioactive substances to be secreted.

In this article, we first wish to review briefly a few of the constraints under which peptide-producing cells operate, and then to discuss our present studies on a purified biosynthetic processing enzyme (PAM) and the biosynthetic process catalyzed by this enzyme (peptide α-amidation) as it occurs *in vivo* and *in vitro.* Some key questions to be addressed include: Which biosynthetic steps occur in secretory granules, which occur upstream of the granule in the endoplasmic reticulum or Golgi, and which can occur in several locations? How are cofactors known to be in granules and known to be essential for the function of biosynthetic enzymes transported into the granules? What is the intragranular environment like during active biosynthetic processing and subsequent peptide storage? To what extent can a cell regulate the functions of its granules in response to various signals?

CELL BIOLOGY OF BIOACTIVE PEPTIDE PRODUCTION

A currently accepted model of the major cellular events involved in the synthesis, storage, and secretion of bioactive peptides is shown in FIGURE 1. Preproproteins containing the amino acid sequence of the bioactive neuroendocrine peptides are synthesized on membrane-bound ribosomes. The nascent polypeptide chains are inserted into the lumen of the rough endoplasmic reticulum (RER). This initial translocation step represents an important difference between the production of bioactive peptides destined for storage in secretory granules and the biosynthesis of conventional

[a]This work was supported by U.S. Public Health Service grant nos. AM-32948, AM-32949, DA-00266, AM-07269, DA-00097, and DA-00098.

small molecule transmitters and hormones such as acetylcholine and the catecholamines. The latter bioactive molecules or their immediate precursors can be synthesized and transported directly into mature secretory granules. For most soluble secreted polypeptides the 20-40 amino acid hydrophobic signal peptide at the NH_2 terminus of the preproprotein must interact with a signal recognition particle and its receptor in order to initiate the process of vectorial synthesis into the lumen of the RER and the subsequent proteolytic cleavage of the signal peptide.[20,21] The attachment of asparagine-linked oligosaccharide chains to the precursor occurs either during or very shortly after the vectorial synthesis of the polypeptide into the RER lumen.[9,10] Protein folding and disulfide bond formation occur rapidly and may be essential for subsequent processing steps.

Precursors are translocated to the *cis* face of the Golgi. A significant amount of maturation occurs as the precursors are transported across the Golgi cisternae to the *trans* face for eventual packaging into secretory granules. During their transit through the Golgi, precursors can be subjected to many of the enzymatic modifications depicted in FIGURE 1: oligosaccharide maturation, *O*-glycosylation, phosphorylation, sulfation of tyrosyl residues or oligosaccharides, hydroxylation of lysine residues, and probably initial endoproteolytic processing.[10–12,22,23] Many of the modifications that occur in the Golgi cisternae are not uniquely associated with the production of bioactive peptides.

In addition to these enzymatic modifications, one or a series of very important sorting steps must occur in the Golgi. It is clear that soluble molecules destined for lysosomes or for maturation and storage in secretory granules arrive at their final destinations with a remarkably high degree of fidelity.[10,12,13,24] How this sorting is accomplished and exactly where it occurs (leaving the ER, somewhere in the Golgi, or just at the exit side [*trans*] of the Golgi) are matters of current debate. Another current question concerns the site of the initial endoproteolytic cleavage of the proproteins. Earlier studies indicated that endoproteolytic cleavages began in the RER or Golgi, despite the increased requirements this early occurrence of proteolytic cleavage would impose on the sorting process.[16–18,25]

EVENTS THAT OCCUR IN SECRETORY GRANULES IN PEPTIDE-PRODUCING CELLS

Several biosynthetic events that clearly occur in secretory granules are also noted in FIGURE 1. These biosynthetic maturation steps include endoproteolytic cleavage of large precursors and biosynthetic intermediates into smaller peptides, exoproteolytic trimming of one or both ends of the resultant peptides, and final modifications of the NH_2- or COOH-terminal amino acids, including α-*N*-acetylation, *O*-acetylation, formation of pyroglutamic acid residues, and α-amidation.[4,19,26,27] In addition, both electron microscopic and biochemical data strongly argue that condensation of intragranular material and potentially important changes in the physical state of the peptides occur during storage in granules.[12,14,15,28] In the few cases where it is possible to estimate the amount of time required to turn over the peptide stores in a tissue, it is clear that the largest amount of time by far is spent in secretory granules. In primary cultures of anterior or intermediate pituitary cells, corticotropes and melanotropes secrete hormone at a basal rate of less than 0.5% of cell hormone content per hour, indicating that it would take a week to replace the peptide stored in the cells.[26,29,30] Younger granules may well differ significantly from older granules.

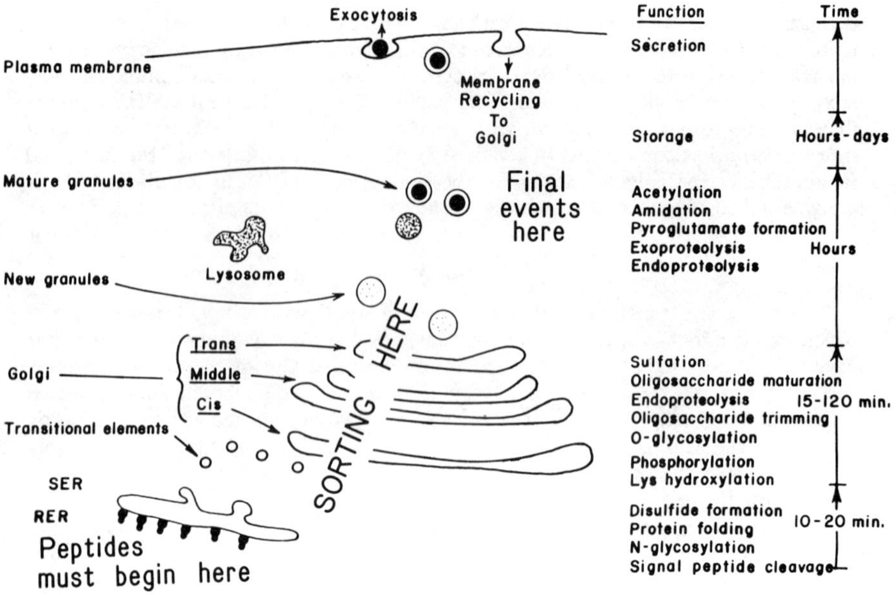

FIGURE 1. Commonly accepted events in the intracellular processing of secreted peptides. (Adapted from refs. 9-19.)

Before considering specific data on one biosynthetic enzyme and the corresponding cellular events, it is important to recall some of the simple observations diagrammed in FIGURE 2. In this example, a secretory granule of 100-nm diameter is considered. A sphere of this diameter has a volume of about 5×10^{-19} liters. If the spherical granule contained one free proton (1 H^+) at any given moment in time, its content of free H^+ would be about 1.6×10^{-24} moles; thus the concentration of free H^+ in the granule is about 3 μM, corresponding to a pH of about 5.5. The internal pH of secretory granules is frequently reported to be in the range of 5.5 to 6.0, although some granules have a substantially higher internal pH.[19,31,32] For example, parotid granules have an internal pH of 6.8, which corresponds to a couple of free H^+ for the size of these large parotid granules.[33] In considering the effects of intragranular pH, the rather substantial buffering capacity of the constituents of various granules must be taken into account.[34,35]

Continuing with the example in FIGURE 2, if the granule contains only a single copy of a processing enzyme, the enzyme concentration will be about 3 μM. Chromaffin granules are much larger than the model granule in FIGURE 2, and contain about 24 copies of DBH per granule.[36] The concentration of DBH in the chromaffin granules is roughly 3 μM. Although one copy of enzyme per granule may seem low, very few studies of purified enzymes are ever performed at an enzyme concentration as high as 3 μM (for an enzyme of 50 kDa, that would be 150 μg/ml of pure enzyme).

If the granule contains 3000 copies of a 15-kDa peptide precursor (roughly the median size for peptide preproproteins) or its product peptides, then the concentration of precursor-derived material would be nearly 10 mM or 150 mg/ml protein. Since secretory granules are not believed to contain more than 150 mg/ml protein,[14,15,28] 3000 copies of precursor-derived material per granule is an upper limit for a 100-nm diameter granule. It is important to note that this concentration of precursor-derived material (10 mM) is far above the level of peptide substrate tested in the vast majority of enzyme assays and far above the Michaelis constant (K_m) of most putative processing enzymes for their peptide substrates; thus the single copy of enzyme would be expected to operate at its maximal velocity (V_{max}). In addition, even a single copy of a relatively slow enzyme with a turnover rate of only 1 s^{-1} could complete the biosynthetic task required in about an hour. The processing events that are known to occur in secretory granules take place over a time course of a few to several hours.[25,29]

SUMMARY OF PROPERTIES OF A PURIFIED POSTTRANSLATIONAL PROCESSING ENZYME

A large percentage of all known bioactive peptides are α-amidated; conversely, all α-amidated peptides identified to date are related to bioactive neuroendocrine peptides.[26,37–40] In general, the COOH-terminal α-amide group is essential for full bioactivity of the peptide. Interestingly, the ability to α-amidate peptides is rather uniquely associated with secretory granules.[1–3,39] Hence, α-amidation is one of the few posttranslational processing steps thought to be unique to neuroendocrine peptides.

PAM is believed to catalyze the α-amidation of peptides by the following general scheme. The peptide substrate must have a COOH-terminal glycine residue, but a variety of amino acids can be located in the penultimate position (*e.g.*, Trp, Val, Pro, Glu). The enzyme has been purified about 20,000-fold from extracts of frozen bovine neurointermediate pituitary.[38] The purified enzyme is a collection of closely related proteins with apparent molecular weights in the range of 38,000 to 50,000. The enzyme requires the presence of molecular oxygen and is presumed to consume oxygen. Enzyme activity is stimulated by added copper ions and inhibited by diethyldithiocarbamate.[1,38] PAM is presumed to function in a manner analogous to DBH,[41] with enzyme-bound

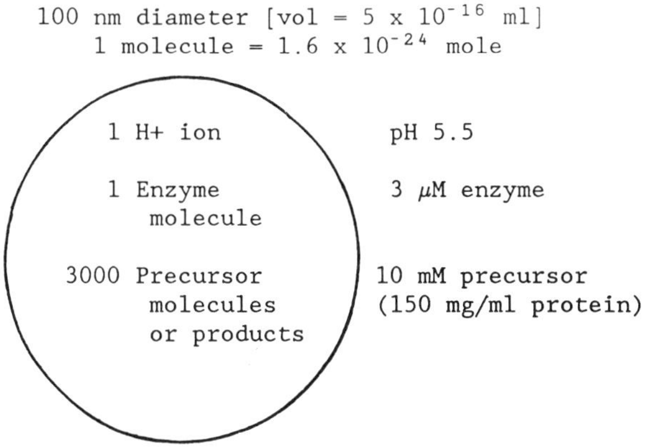

FIGURE 2. A model secretory granule. A spherical secretory granule with a diameter of 100 nm is depicted. The granule is arbitrarily depicted as containing 1 free proton, one molecule of a posttranslational processing enzyme, and 3000 molecules of a polypeptide precursor of molecular mass 15 kDa.

copper undergoing repeated reduction by ascorbate and oxidation during the activation of molecular oxygen and oxidation of the substrate to produce the α-amide group and glyoxylate. Quantitative studies have demonstrated that the enzyme does consume nearly one mole of ascorbate for every mole of amidated peptide product formed. Similarly, glyoxylate is produced during the reaction.

PAM is a soluble, secretory granule-associated enzyme and is secreted along with peptide by AtT-20 mouse pituitary tumor cells. The ratio of enzyme activity to peptide in spent medium is very similar to the ratio in isolated secretory granules.[2] PAM activity, as assayed in a test tube under optimized conditions, is regulated in parallel with pro-ACTH/endorphin production by glucocorticoids in the anterior lobe of the pituitary and by dopamine agonists and antagonists in the intermediate lobe of the pituitary.[2,40]

Pro-ACTH/endorphin can give rise to two α-amidated product peptides, αMSH and joining peptide. Joining peptide terminates in a Glu-NH_2 and is a major product in ACTH/endorphin cells in the anterior pituitary, intermediate pituitary, and arcuate nucleus[29,42]; α-melanotropin terminates in a Val-NH_2, and is a major product in ACTH/endorphin cells in the intermediate pituitary and arcuate nucleus but not in the anterior pituitary.[29,42] As model substrates to investigate the ability of PAM to produce αMSH and joining peptide, we have utilized the tripeptides D-Tyr-Val-Gly and D-Tyr-Glu-Gly. As shown in FIGURE 3, purified PAM will produce the appropriate amidated peptides from both precursor peptides. Interestingly, the optimal pH for conversion of the Glu-containing substrate is pH 5.5, whereas the optimal pH for the Val-containing substrate is pH 9. Since electron microscopic immunocytochemical localization studies have consistently demonstrated the colocalization of all of the various product peptides in the same secretory granules, this points to an intriguing problem concerning the actual environment in which PAM functions in secretory granules. Similar questions have arisen before concerning how PAM and CPH can function in the same secretory granules when their pH optima in test tube assays differ so greatly.[2,4] To try to resolve this question, we decided to examine the functioning of PAM in intact cells, paying particular attention to its ability to α-amidate the two corresponding natural substrates.

STUDIES ON THE EFFECT OF ASCORBATE ON PEPTIDE AMIDATION IN CULTURED INTERMEDIATE PITUITARY MELANOTROPES

When intermediate pituitary cells are placed in culture, they rapidly lose their normal content of ascorbic acid and their ability to produce α-amidated peptides.[43–45] The cellular supply of ascorbate is dependent on uptake of ascorbate from the medium (or from plasma *in vivo*), and *in vivo* the pituitary and adrenal contain the highest concentration of ascorbate in the body. The uptake of ascorbate by anterior pituitary, intermediate pituitary, and AtT-20 cells is a saturable process that takes about a third to a half of a day to reach completion with extracellular ascorbate levels comparable to plasma values. The cells can concentrate ascorbate roughly 20-40 times over the concentration of ascorbate in the medium. An example of an uptake study using [^{14}C]ascorbate with AtT-20 mouse pituitary tumor cells is shown in FIGURE 4. The uptake process is sodium-dependent and is not blocked by excess levels of extracellular glucose. The chromaffin cells of the adrenal medulla have a very similar ascorbate uptake system.[5,6] It is not at all clear how many tissues that produce α-amidated peptides possess this type of ascorbate uptake system. The uptake of labeled ascorbate into AtT-20 cells is effectively competed by the presence of unlabeled reduced ascorbate in the medium (FIGURE 5). Addition of millimolar amounts of dehydroascorbate to the medium has little or no effect on the accumulation of labeled ascorbate by the cells.

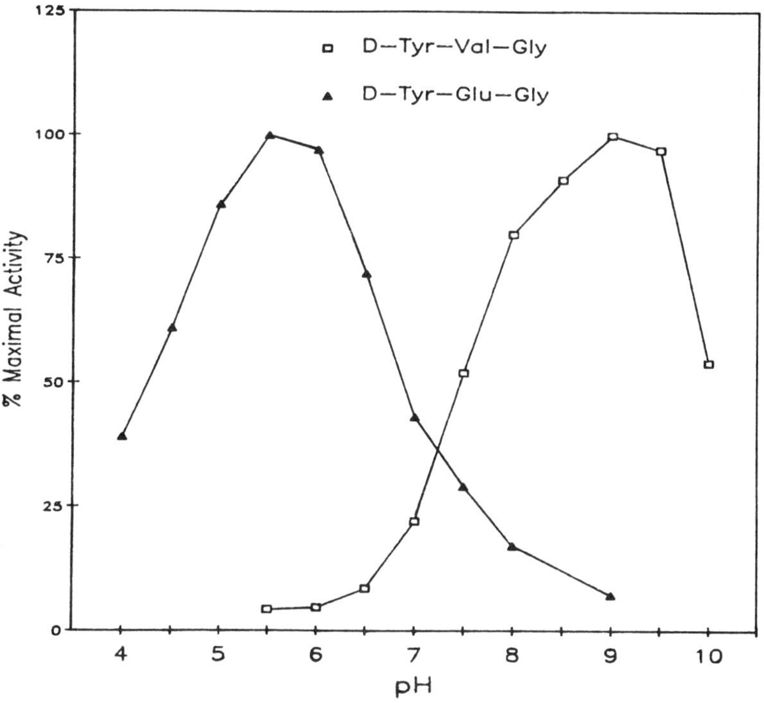

FIGURE 3. The pH dependence of PAM activity with two peptide substrates. The pH dependence of PAM activity, using synthetic D-Tyr-Val-Gly (*empty squares*) and D-Tyr-Glu-Gly (*filled triangles*) as substrates, is shown. Maximal activity has been normalized to 100% for each substrate. The buffers used and the details of the assays with D-Tyr-Val-Gly are given in ref. 38. For assays with D-Tyr-Glu-Gly, the ^{125}I-labeled substrate and amidated product were separated by reversed-phase high performance liquid chromatography (RP-HPLC) on a C-18 μBondapak column in 0.1% trifluoroacetic acid/4% acetonitrile eluted with a linear gradient to 20% acetonitrile over 48 min.

A number of processes must occur between the transport of ascorbate across the plasma membrane and the ascorbate-stimulated production of α-amidated peptides in secretory granules, but little is known of these steps. During the course of taking up ascorbate, intermediate pituitary cells were tested for their ability to α-amidate newly synthesized peptides. In the experiment shown in FIGURE 6 the extent of amidation of newly synthesized αMSH in cell extracts and in medium was determined as a function of the time after addition of ascorbate to the medium. The hormone in the

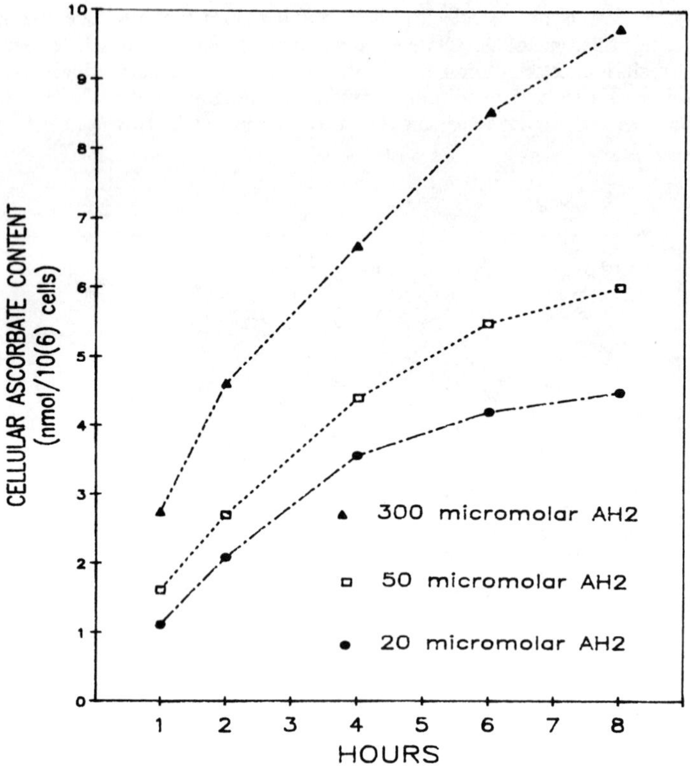

FIGURE 4. Ascorbate uptake into AtT-20 mouse pituitary tumor cells. The uptake of various concentrations of [^{14}C]ascorbate into AtT-20 cells was observed as a function of time. Cellular methods and techniques for biochemical analyses are given in reference 46. The label inside the cells was identified as reduced ascorbate by RP-HPLC with electrochemical detection and by treatment with ascorbate oxidase.

medium represents secretion from the cells under basal conditions. In the cell extracts (hatched bars) there was a greater than 10-fold increase in the ability of the cells to amidate peptides over the course of the 12-hour incubation in medium containing ascorbate. During this time, the newly synthesized αMSH secreted by the intermediate pituitary cells (solid bars) was amidated to the same extent as the corresponding peptide in the cell extracts. In general, peptides secreted by a tissue correspond rather precisely to the peptides stored in the tissue.

Given the results in FIGURE 3 on the disparate pH dependence of purified PAM acting on the synthetic tripeptide substrates corresponding to αMSH and joining peptide, it was important to compare the effects of addition of ascorbate to the culture medium on the amidation of these two peptides. FIGURE 7 demonstrates that the extent of amidation of both peptides declined to a similar level in the absence of added ascorbate. The addition of ascorbate to the culture medium stimulated amidation of both peptides to the same extent.

STUDIES ON THE EFFECT OF COPPER CHELATORS ON PEPTIDE AMIDATION IN ADULT RATS

Disulfiram (Antabuse) is the disulfide dimer of diethyldithiocarbamate (DDC). Disulfiram is used clinically as an alcohol deterrent. DDC reversibly inhibits the activity of PAM in the test tube and the inhibition of PAM by DDC is reversed by addition of an excess of copper ions. It is thought that the inhibitory actions of DDC on PAM are due to the relatively selective ability of DDC to chelate copper ions.[1,38] Earlier studies on the effects of dietary copper deficiency failed to show an effect on peptide α-amidation in pituitary or brain[40]; this lack of effect was thought to reflect the fact that these tissues retain copper better than the rest of the tissues in the body and are not readily depleted. Hence, the effects of disulfiram and DDC on peptide amidation were investigated in cultured pituitary cells and in adult rats. Previous studies on DBH had demonstrated the ability of disulfiram treatment *in vivo* to inhibit the production of NE from DA.[48] The ability of cultured pituitary cells to produce amidated peptides is inhibited in a dose-dependent manner by inclusion of micromolar amounts of DDC or disulfiram in the medium.[47]

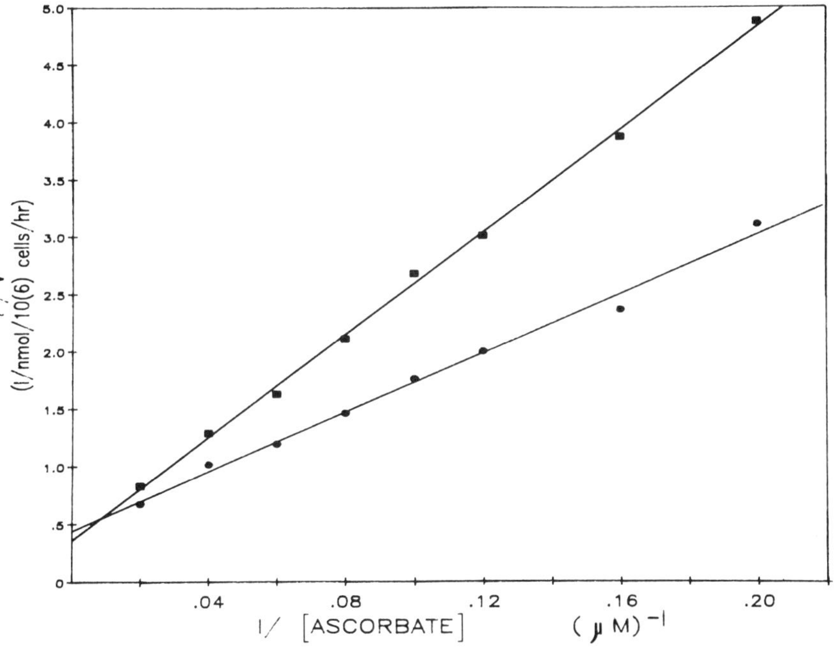

FIGURE 5. Concentration dependence of ascorbate uptake into AtT-20 cells. The Lineweaver-Burk transformation of the data from an uptake study using AtT-20 cells is displayed. Cells were incubated in medium containing [^{14}C]ascorbate for 30 min to determine the initial velocity of uptake. *Filled circles* show the direct concentration dependence of the uptake; data obtained in the presence of added 30 μM cold ascorbate are shown in the *filled squares.* The K_i for the added cold ascorbate is in good agreement with the K_m for [^{14}C]ascorbate.

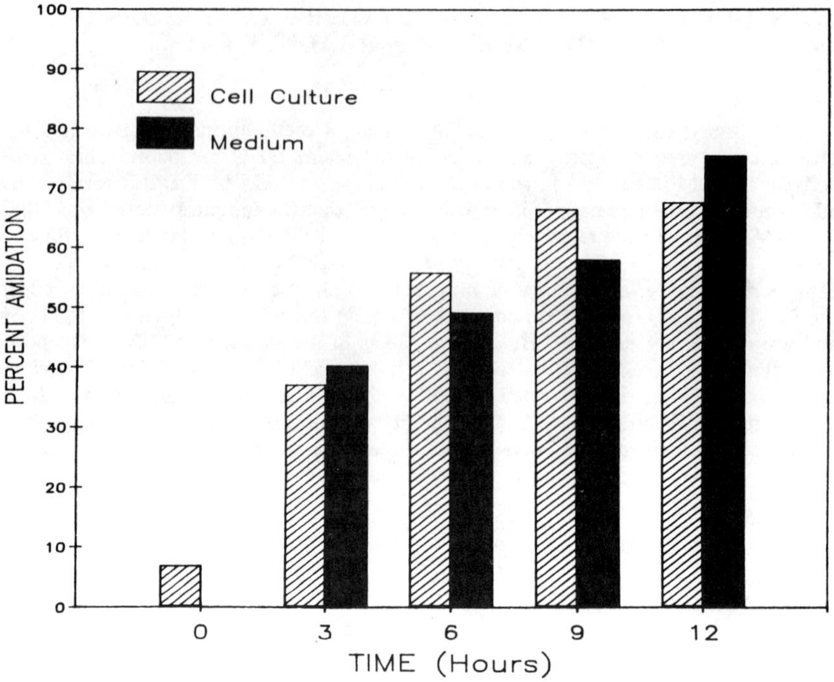

FIGURE 6. Time course of the effect of exogenous ascorbate on peptide amidation. Intermediate pituitary cells were maintained in complete serum-free medium[30] for 3 days in the absence of ascorbate, and then tested for their ability to synthesize amidated αMSH during a 5-hour incubation in medium containing 10 μM [^{3}H]Trp (11.3 Ci/mmol) ± 0.5 mM ascorbate. Cells were then incubated in medium containing the same concentration of ascorbate but lacking [^{3}H]Trp. At the times indicated, the extent of amidation of the newly synthesized αMSH in the cells and in the medium was determined as in May and Eipper.[43]

The ability of DDC and disulfiram to block peptide amidation in tissue culture having been demonstrated, similar effects were investigated *in vivo.* Initial studies used the high doses of drug commonly used in rat studies[47] (400 mg/kg/d disulfiram; 1000 mg/kg/d DDC) and demonstrated a massive loss of ability to α-amidate peptides *in vivo.* Doses of disulfiram in the range of the human therapeutic dose (250 mg/d or about 4 mg/kg/d) were given to rats for two days and the ability of the intact neurointermediate pituitary to α-amidate newly synthesized, radiolabeled peptides was assessed in culture in the absence of added drug (FIGURE 8). The ability of the melanotropes to amidate both joining peptide and αMSH changed in parallel. Rats given the human therapeutic dose of Antabuse retained very little of their normal ability to α-amidate peptides; the clinical implications of this effect of Antabuse remain to be determined. Despite the apparent differences in the two synthetic peptide substrates assayed in a test tube with purified PAM, both endogenous peptides were treated similarly by the enzyme in the secretory granules during experimental manipulation of copper and ascorbate.

One prediction from the data in FIGURE 8 is that the peptide-producing tissues of rats given disulfiram daily for many days would gradually be depleted of all their α-amidated peptides and would contain instead primarily Gly-extended peptides. This type of alteration would dramatically alter physiological function, since for most α-amidated peptides the corresponding Gly-extended peptide has greatly diminished biological potency. On the other hand, animals chronically given the therapeutic dose of disulfiram might adapt to the drug and correctly α-amidate peptides. The effects of chronic disulfiram treatment are shown in FIGURE 9, where the extent of α-amidation of αMSH and joining peptide stored in both lobes of the pituitary were determined. The animals did not develop an ability to overcome the effects of disulfiram and instead exhibited depletion of amidated peptides from both lobes of pituitary. Total pituitary content of pro-ATCH/endorphin-derived material was unaltered by drug treatment, with increased amount of Gly-extended peptides replacing the α-

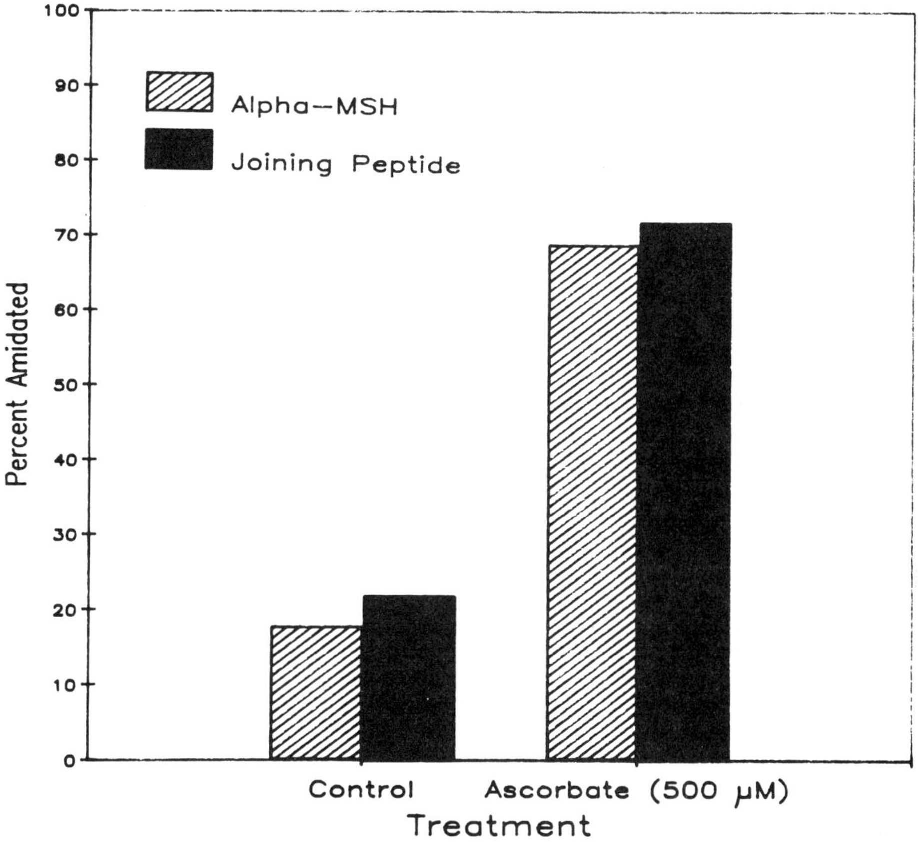

FIGURE 7. Acquisition of the ability to α-amidate peptides in ascorbate. Intermediate pituitary cells were maintained for 3 days as in FIGURE 6 and then tested for their ability to synthesize α-amidated αMSH and joining peptide during a 5-hour incubation in medium containing 140 μM [^{3}H]proline (21 Ci/mmol) in the presence or absence of 0.5 mM ascorbate. The extent of amidation of the newly synthesized peptides was determined as in refs. 43-45 and 47.

amidated ones. It is important to note that the ability to amidate αMSH and joining peptide was again lost roughly in parallel.

SUMMARY

There are many events in the posttranslational processing of bioactive peptides that occur in secretory granules and not to any great extent in other cellular organelles and that do not appear as modifications of the structure of many conventional neurotransmitters. In addition, at least two very important steps are unique to peptide-containing granules: (1) the peptides must begin their trek to the secretory granule in the RER as a larger precursor, rather than being taken up as a finished or nearly

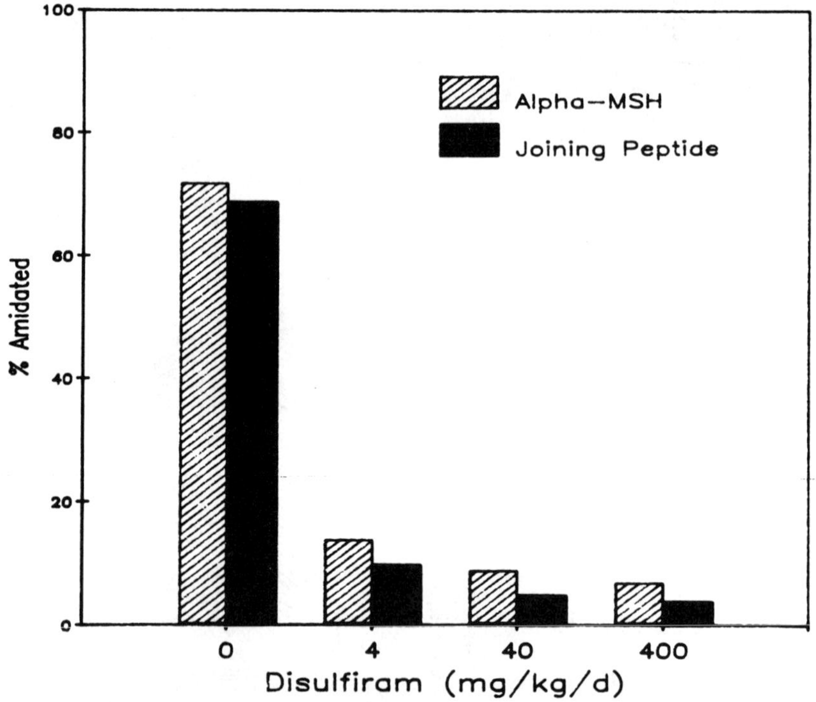

FIGURE 8. Inhibition of peptide amidation by acute disulfiram treatment. Male rats were treated for two days with the indicated doses of disulfiram given subcutaneously once daily in 0.9% NaCl, 0.5% Tween 80. Fourteen hours after the second injection, the animals were sacrificed and individual neurointermediate pituitaries were incubated for 6 h in 50 μl medium containing 100 μM [^{3}H]proline and 100 μM ascorbate and extracted as in Eipper *et al.*[45] Samples were fractionated by HPLC gel filtration into a pool of αMSH- and joining-peptide-sized material, and the extent of amidation of the newly synthesized joining peptide and αMSH was determined as in FIGURE 7. Data were averaged for two animals per treatment.

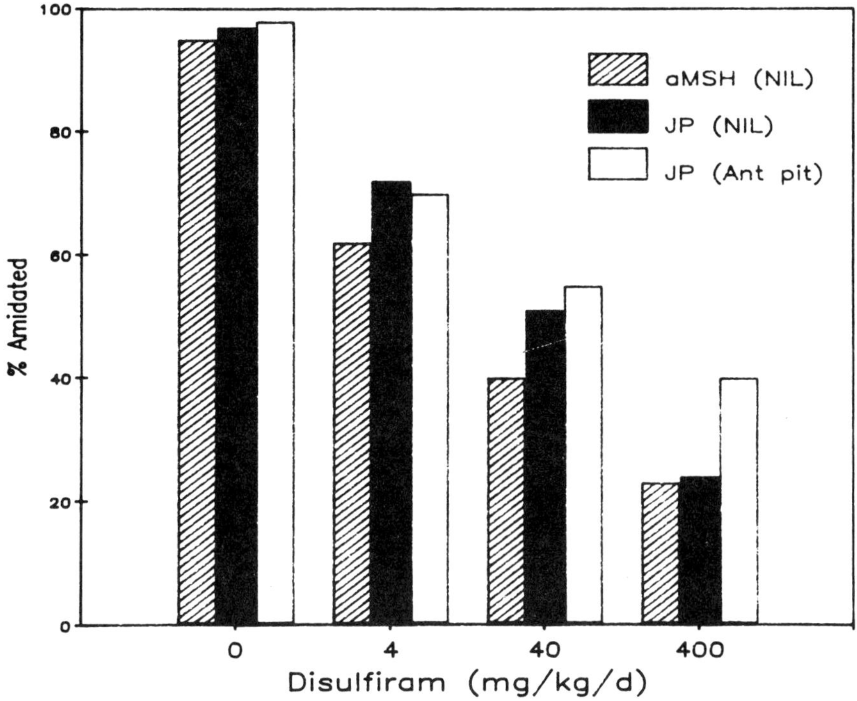

FIGURE 9. Inhibition of peptide amidation by chronic disulfiram treatment. Male rats (n = 10 per treatment) were treated with disulfiram for 11 days as described in FIGURE 8. Separated anterior pituitaries and neurointermediate pituitaries were pooled into groups of 3 or 4 pituitaries at the time of sacrifice. Aliquots of the extracts were fractionated by RP-HPLC and radioimmunoassayed to determine the extent of amidation of joining peptide[45,47]; the data have been corrected for the differential cross-reactivity of the general joining peptide antiserum with the amidated and Gly-extended forms of the peptide.[45] The extent of amidation of αMSH was determined by immunoassay after fractionation of another aliquot of each extract by cation exchange HPLC.[47] The amidation states of the peptides were confirmed using amide-specific antisera.[43–45,47]

finished product into a mature granule; (2) there is at least one crucial sorting step on the way from the RER to the secretory granule that must occur faithfully before the peptide correctly appears in the granule. As for small molecules such as the catecholamines, the posttranslational processing enzymes and any required cofactors must also be put into the granules if the final events of processing are to occur with fidelity.

Many of the posttranslational processing enzymes are only beginning to be identified. It is clear from these studies on purified PAM and peptide α-amidation as it occurs in cells that correlating test tube studies with the functioning of secretory granules is a worthwhile, if difficult, pursuit. The unique milieu inside the granule is difficult to mimic in a test tube. Transfection of peptide-producing cells with cDNAs encoding precursors with specific alterations in processing sites offers perhaps the best way to interface the studies of secretory granules and the posttranslational processing enzymes that mediate those functions.

REFERENCES

1. EIPPER, B. A., R. E. MAINS & C. C. GLEMBOTSKI. 1983. Proc. Natl. Acad. Sci. U.S.A. **80:** 5144-5148.
2. MAINS, R. E. & B. A. EIPPER. 1984. Endocrinology **115:** 1683-1690.
3. VON ZASTROW, M., T. R. TRITTON & J. D. CASTLE. 1986. Proc. Natl. Acad. Sci. U.S.A. **83:** 3297-3301.
4. LYNCH, D. R. & S. H. SNYDER. 1986. Ann. Rev. Biochem. In press.
5. BEERS, M. F., R. G. JOHNSON & A. SCARPA. 1986. J. Biol. Chem. **261:** 2529-2535.
6. LEVINE, M., K. MORITA & H. POLLARD. 1985. J. Biol. Chem. **260:** 12942-12947.
7. O'CONNOR, D. T. & L. J. DEFTOS. 1986. N. Engl. J. Med. **314:** 1145-1151.
8. COHN, D. V., J. J. ELTING, M. FRICKE & R. ELDE. 1982. Endocrinology **114:** 1963-1974.
9. KORNFELD, R. & S. KORNFELD. 1985. Ann. Rev. Biochem. **54:** 631-644.
10. SADLER, J. E. 1984. *In* Biology of Carbohydrates, Vol. 2. V. Ginsburg & P. W. Robbins, Eds.: 199-288. Wiley. New York.
11. DUNPHY, W. G. & J. E. ROTHMAN. 1985. Cell **42:** 13-21.
12. FARQUHAR, M. G. 1985. Ann. Rev. Cell Biol. **1:** 447-488.
13. KELLY, R. B. 1985. Science **230:** 25-32.
14. ORCI, L., P. HALBAN, M. AMHERDT, A. M. RAVAZZOLA, J. D. VASSALLI & A. PERRELET. 1985. J. Cell Biol. **99:** 2187-2192.
15. PALADE, G. 1975. Science **189:** 347-358.
16. HABENER, J. F., M. ROSENBLATT & J. T. POTTS. 1984. Physiol. Rev. **64:** 985-1053.
17. DOCHERTY, K. & D. F. STEINER. 1982. Ann. Rev. Physiol. **44:** 625-638.
18. LOH, Y. P., M. J. BROWNSTEIN & H. GAINER. 1984. Ann. Rev. Neurosci. **7:** 189-222.
19. GAINER, H., J. T. RUSSELL & Y. P. LOH. 1985. Neuroendocrinology **40:** 171-184.
20. GILMORE, R. & G. BLOBEL. 1985. *In* Current Communications in Molecular Biology. M. J. Gething, Ed.: 29-32. Cold Spring Harbor Laboratory. Cold Spring Harbor, N.Y.
21. WALTER, P., V. SIEGEL, L. LAUFFER, P. D. GARCIA, A. ULLRICH & R. HARKINS. 1985. *In* Current Communications in Molecular Biology. M. J. Gething, Ed.: 21-23, Cold Spring Harbor Laboratory. Cold Spring Harbor, N.Y.
22. SABATINI, D. D., G. KREIBICH, T. MORIMOTO & M. ADESNIK. 1982. J. Cell. Biol. **92:** 1-22.
23. FARQUHAR, M. G. & G. E. PALADE. 1981. J. Cell. Biol. **91:** 77s-103s.
24. MOORE, H.-P. H. & R. B. KELLY. 1986. Nature **321:** 443-446.
25. GLEMBOTSKI, C. C. 1982. J. Biol. Chem. **257:** 10493-10500.
26. MAINS, R. E., B. A. EIPPER, C. C. GLEMBOTSKI & R. M. DORES. 1983. Trends Neurosci. **6:** 229-285.
27. HERBERT, E., M. COMB, A. SEASHOLTZ, M. MARTIN & D. LISTON. 1986. *In* Neuropeptides in Neurologic and Psychiatric Disease. J. B. Martin & J. D. Barchas, Eds.: 33-46. Raven. New York.
28. NORDMANN, J. J. & S. J. NORRIS. 1984. Proc. Natl. Acad. Sci. U.S.A. **81:** 180-184.
29. EIPPER, B. A. & R. E. MAINS. 1980. Endocrine Rev. **1:** 1-27.
30. MAY, V. & B. A. EIPPER. 1986. Endocrinology **118:** 1284-1295.
31. JOHNSON, R. G., S. E. CARTY & A. SCARPA. 1981. J. Biol. Chem. **256:** 5773-5780.
32. LOH, Y. P., W. W. H. TAM & J. T. RUSSELL. 1984. J. Biol. Chem. **259:** 8238-8245.
33. ARVAN, P., G. RUDNICK & J. D. CASTLE. 1984. J. Biol. Chem. **259:** 13567-13572.
34. KOPPELL, W. N. & E. W. WESTHEAD. 1982. J. Biol. Chem. **257:** 5707-5710.
35. WINKLER, H. & S. W. CARMICHAEL. 1982. *In* The Secretory Granule. A. M. Poisner & J. M. Trifaro, Eds.: 4-79. Elsevier Biomedical Press. Amsterdam.
36. INGEBRETSEN, O. C., O. TERLAND & T. FLATMARK. 1980. Biochim. Biophys. Acta. **628:** 182-189.
37. TATEMOTO, K., M. CARLQUIST & V. MUTT. 1982. Nature **296:** 659-660.
38. MURTHY, A. S. N., R. E. MAINS & B. A. EIPPER. 1986. J. Biol. Chem. **261:** 1815-1822.
39. BRADBURY, A. F., M. D. A. FINNIE & D. G. SMYTH. 1982. Nature **298:** 686-688.
40. MAINS, R. E., A. C. MYERS & B. A. EIPPER. 1985. Endocrinology **116:** 2505-2515.
41. DILIBERTO, E. J. & P. L. ALLEN. 1981. J. Biol. Chem. **256:** 3385-3393.

42. EMESON, R. B. & B. A. EIPPER. 1986. J. Neurosci. **6:** 837-849.
43. MAY, V. & B. A. EIPPER. 1985. J. Biol. Chem. **260:** 16224-16231.
44. EIPPER, B. A., C. C. GLEMBOTSKI & R. E. MAINS. 1983. J. Biol. Chem. **258:** 7292-7298.
45. EIPPER, B. A., L. PARK, H. T. KEUTMANN & R. E. MAINS. 1986. J. Biol. Chem. **261**. In press.
46. CULLEN, E. I., V. MAY & B. A. EIPPER. 1986. In press.
47. MAINS, R. E., L. P. PARK & B. A. EIPPER. 1986. In press.
48. MUSACHIO, J. M., M. GOLDSTEIN, B. ANAGNOSTE, G. POCH & I. J. KOPIN. 1966. J. Pharm. Exp. Ther. **152:** 56-61.

Peptide Precursor Processing Enzymes within Secretory Vesicles

Y. PENG LOH

Section on Cellular Neurobiology
Laboratory of Neurochemistry and Neuroimmunology
National Institute of Child Health and Human Development
National Institutes of Health
Bethesda, Maryland 20892

INTRODUCTION

Studies on the biosynthesis of many neuropeptides and peptide hormones have shown that they are first cleaved from larger precursors and subsequently modified to yield biologically active peptides by a variety of posttranslational processing steps.[1-10] Recent advances in molecular cloning techniques have accelerated the determination of the primary structure of the peptide precursors and allowed the identification of proteolytic processing signals.[11-17] It is apparent from the amino acid sequence of the precursors that processing begins with selective proteolytic cleavages generally at paired basic residues that flank the potentially biologically active peptides (FIG. 1). This step is then followed by other enzymological events including proteolytic, acetylation, and amidation reactions (FIG. 2).

The biosynthetic process begins with the synthesis of the precursor (*i.e.*, the prepro-protein) on the rough endoplasmic reticulum (RER), followed by its transportation into the RER cisternae and subsequent translocation into the Golgi apparatus, where it is ultimately sorted and packaged into the secretory vesicle. Several posttranslational processing steps occur during this translocation of the precursor through the membrane systems of the cell. These include the removal of the signal peptide from the pre-pro-protein and asparagine-linked glycosylation of the precursor in the RER, disulfide bond formation in the RER, and more complex glycosylation and phosphorylation of the precursor in the Golgi.[2,4,6-8,10,18] These posttranslational events are used to complete the precursor's structure prior to the endoproteolytic and other enzymatic steps that produce the final peptide product. These latter processing events, in general, appear to be localized in the secretory vesicles.[10,19-22] Hence, the processing enzymes should be present in secretory vesicles.

Several key proteolytic enzymes are required for the processing of peptide precursors (FIG. 2). They include an enzyme(s) that cleaves at paired basic residues (prohormone converting enzyme, PCE), a carboxypeptidase B-like enzyme (CPB), and/or an aminopeptidase B-like enzyme (AP) for the removal of COOH- and NH_2-terminal extended basic residues from the cleaved peptides. Further modifications of the peptides at the NH_2 or COOH terminus, such as acetylation and amidation, are carried out by specific acetyltransferases (NAT) and a peptidyl-glycine α-amidating monooxygenase (PAM). The most extensively studied processing enzyme system has been that involved in the conversion of pro-opiomelanocortin (POMC) in intermediate and anterior lobe secretory vesicles. Enzymes within chromaffin granules, pituitary neural lobe secretory vesicles, and pancreatic islet cell vesicles that process proenkephalin, provasopressin/pro-oxytocin and precursors of insulin, glucagon, and somatostatin, have also been studied. In the following sections, processing enzymes that have been detected in various secretory vesicles will be discussed. Our primary focus will be on those enzymes involved in POMC conversion.

PROCESSING ENZYMES IN SECRETORY VESICLES

Pro-opiomelanocortin Processing Enzymes in Pituitary Anterior and Intermediate Lobe Secretory Vesicles

Pro-opiomelanocortin is synthesized and processed in the anterior and intermediate lobes of the pituitary. Our early work using rat intermediate lobe or anterior lobe secretory vesicles as an enzyme source and frog POMC as substrate revealed an acidic, paired basic residue-specific prohormone-converting enzyme (PCE) activity in these organelles.[23–26] We have subsequently purified such an enzyme from the soluble fraction of bovine intermediate lobe secretory vesicle lysate.[27] The enzyme was characterized as a ~70,000 Da glycoprotein and has an *in vitro* pH optimum of 4.0-5.0. It was inhibited by pepstatin A and diazoacetyl-norleucine methyl ester, both aspartyl protease inhibitors, but not by thiol or serine protease inhibitors.[28] PCE therefore appears to be an aspartyl protease and is quite different from other basic residue-specific enzymes such as pancreatic trypsin and lysosomal cathepsin B. An enzyme identical to soluble PCE has also been partially purified from the membrane fraction of bovine intermediate lobe secretory vesicles (Tuteja and Loh, unpublished data). PCE was shown to process mouse POMC at Lys-Arg pairs to yield 21-23 kDa ACTH, 13-kDa and 4.5-kDa ACTH (the glycosylated and nonglycosylated forms of $ACTH_{1-39}$, respectively, see FIG. 1), β-endorphin and a 16-kDa NH_2-terminal glycopeptide (see FIG. 3), products which are found in the anterior and intermediate pituitary *in situ.*[19,29] Interestingly, PCE did not cleave ACTH at the Lys-Lys-Arg-Arg residues to yield $ACTH_{1-14}$ and CLIP.[27] NH_2- and COOH-terminal analysis of the products indicated that cleavage occurred on the carboxyl side of the Lys-Arg pair, as well as between these basic residues (see FIG. 3). The specificity of PCE was further characterized using a simpler substrate, human β-lipotropin (see FIG. 4). PCE cleaved the Lys_{37}-Lys_{38} and the Lys_{57}-Arg_{58} pairs of β-lipotropin to yield β-MSH and β-endorphin$_{1-31}$ (see FIG. 4), but not the Lys_{86}-Lys_{87} residues to yield β-endorphin$_{1-27}$. The K_m for the cleavage of the two pairs of basic residues was very similar, suggesting that cleavage is not influenced by which basic amino acids (Lys or Arg) are within the pair.[28] HPLC analysis showed that 90% of the β-endorphin$_{1-31}$ cleaved from β-lipotropin contained an Arg at the NH_2 terminus.[28] These results indicate that PCE preferentially cleaved in between the basic residues at this site. However, cleavage at the Lys-Arg pair separating the ACTH from the β-lipotropin (see FIG. 1) occurred between and on the carboxyl site of the Arg with approximately equal frequency.[27] This is interesting since in the case of β-endorphin, the Lys-Arg pair cleaved was immediately followed by a Tyr, similar to the Arg-Lys pair preceding γ-MSH[16] (see FIGURE 1). In the latter case, the major endogenous form of this peptide found in the intermediate lobe was

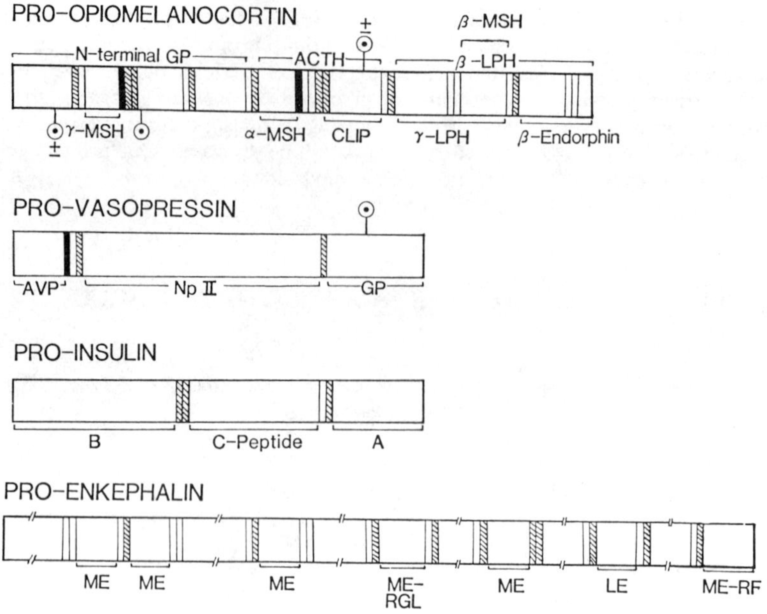

FIGURE 1. Diagrammatic representation of the structure of several prohormones and neuropeptide polyproteins, showing the paired basic amino acid residues (*open bars,* lysine; *hatched bars,* arginine), the glycine residues (*solid bars*) located at the NH_2 termini of the pairs of basic amino acids, and glycosylation sites (*circled dots*). In pro-opiomelanocortin, the circled-dot sites indicated by ± are not always glycosylated in different forms of this molecule. The lengths of the peptides and spacer regions have not been drawn to scale; regions shortened are indicated by *hatch marks; GP,* glycopeptide; *MSH,* melanotrophic-stimulating hormone; *ACTH,* adrenocorticotropin (1-39); *LPH,* lipotropin; *CLIP,* corticotropin-like intermediate lobe peptide; *ME,* Met-enkephalin; *ME-RGL,* Met-enkephalin-Arg-Gly-Leu; *ME-RF,* Met-enkephalin-Arg-Phe; *LE,* Leu-enkephalin; *AVP,* arginine vasopressin; *NpII,* neurophysin II. The positions of the paired basic amino acids and glycine were taken from the reported sequences for bovine pro-opiomelanocortin[16]; bovine proenkephalin,[12] bovine provasopressin,[15] and rat proinsulin.[17]

Lys-γ-MSH,[30,31] suggesting that cleavage may have occurred preferentially in between the paired basic residues *in situ.* It is possible that the amino acids flanking the paired basic residues may play a role in influencing the frequency of cleavage in between, or on the carboxyl side of, the dibasic pairs. Another feature of this enzyme is that it did not cleave all the pairs of basic residues within the POMC or β-lipotropin molecule. This may be due to conformational constraints imposed by the substrate with respect to PCE. It is apparent, however, that some of the paired basic residues which are resistant to PCE action are cleaved in some tissues, but not in others. For example, β-endorphin$_{1-27}$ and α-MSH plus CLIP (see FIG. 1) are present in the intermediate lobe, but not the anterior lobe of the pituitary.[19,29] It is likely that some POMC-synthesizing cells may contain other enzymes that can specifically carry out these additional cleavages, *i.e.,* cleavage at Lys_{86}-Lys_{87} of β-endorphin and the Arg_{17}-Arg_{18} in ACTH to form β-endorphin$_{1-27}$ and CLIP, respectively.

The pattern of cleavage of POMC by PCE indicates that two exopeptidases are necessary for the removal of the NH_2- and COOH-terminal extended basic residues from the cleaved peptides (see FIG. 2). A carboxypeptidase B-like enzyme (CPB) activity has been detected in rat anterior and intermediate lobe secretory vesicles.[32] This activity has been characterized as a thiol, metallopeptidase, and is stimulated by Co^{2+}, unlike lysosomal carboxypeptidase B. It has a pH optimum of 6.0 and cleaved the basic residues sequentially from the COOH terminus of $ACTH_{1-14}$ Lys_{15}-Lys_{16}-Arg_{17} to yield $ACTH_{1-14}$. Such a carboxypeptidase B-like enzyme has been purified to apparent homogeneity from bovine pituitary by Fricker and Snyder.[33] The purified enzyme is a glycoprotein and exists in a soluble and membrane-associated form. The soluble form has a molecular weight of 50,000,[33] the membrane form 52,500.[34] Recently this enzyme has been shown to be released together with ACTH in a coordinate manner from AtT-20 cells, a mouse anterior pituitary-derived tumor line.[35] This further supports a secretory vesicle localization of the carboxypeptidase B-like enzyme and its physiological role in peptide processing in anterior pituitary cells.

An aminopeptidase B-like (AP) activity (see FIG. 2) has been detected in bovine intermediate lobe secretory vesicles. The activity was primarily associated with secretory vesicle membranes. It has a pH optimum of 6.0 and is a metallopeptidase that is stimulated by Co^{2+} and Zn^{2+}, unlike other reported aminopeptidases. The enzyme

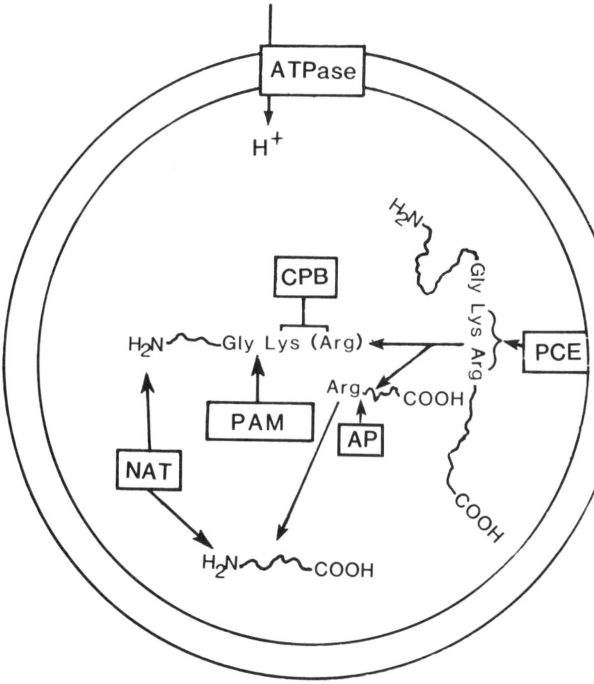

FIGURE 2. Diagrammatic representation of the secretory granule indicating proposed enzymatic steps involved in the processing of a prohormone. Processing begins with proteolytic cleavage at paired basic residues (as indicated by *arrows*) by a prohormone-converting enzyme (*PCE*), followed by the removal of the COOH-terminal and/or NH_2-terminal basic residues by a carboxypeptidase B-like enzyme (*CPB*) and an aminopeptidase B-like enzyme (*AP*), respectively. The cleaved peptides in some cases undergo further modification, such as *N*-acetylation by an acetylation enzyme (*NAT*) and COOH-terminal amidation by an amidation enzyme (*PAM*).

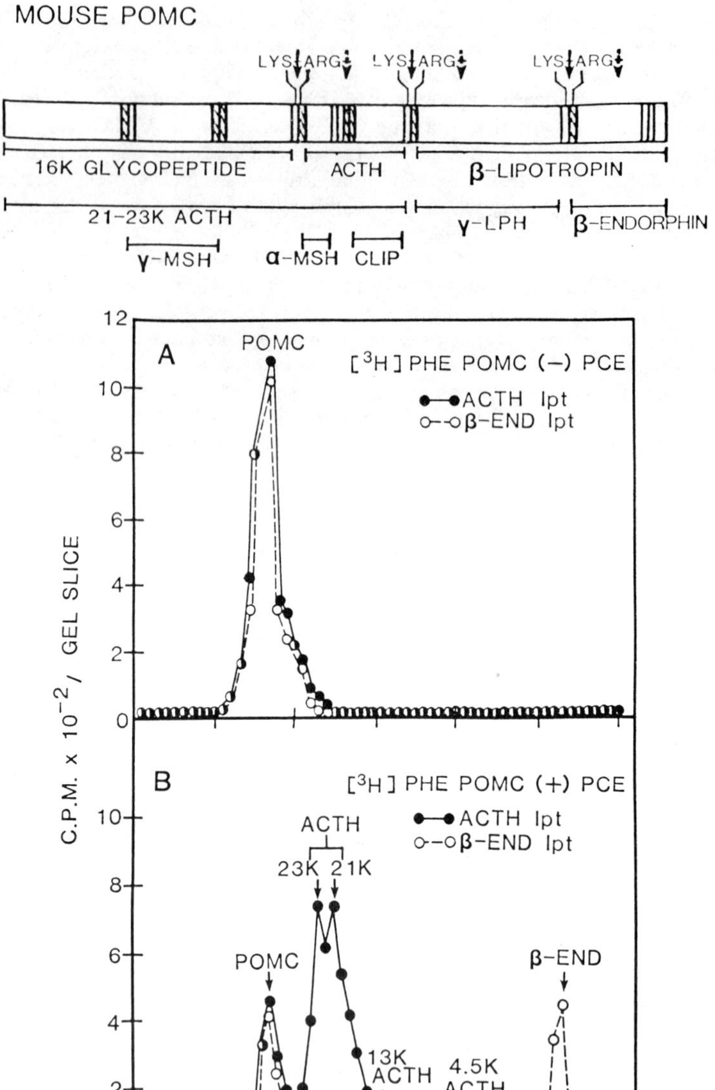

FIGURE 3. **(A)** Acid-urea gel profiles of [^{3}H]phenylalanine-labeled mouse POMC immunoprecipitated with ACTH and β-endorphin (β-*END*) antiserum after incubation for 5 h at 37°C in pH 4.0 buffer without pro-opiomelanocortin-converting enzyme (*PCE*). **(B)** Acid-urea gel profiles of anti-ACTH and anti-β-endorphin immunoprecipitated mouse POMC-derived products, generated by incubating [^{3}H]phenylalanine-labeled POMC with purified pro-opiomelanocortin-converting enzyme at pH 4.0 for 5 h at 37°C. The apparent molecular masses of the products are indicated. *Ipt,* immunoprecipitate.

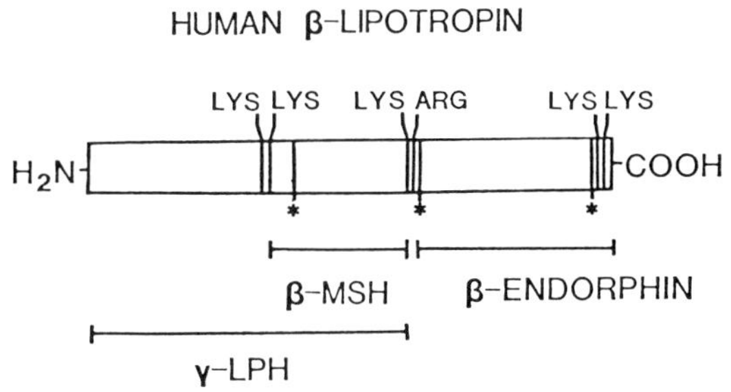

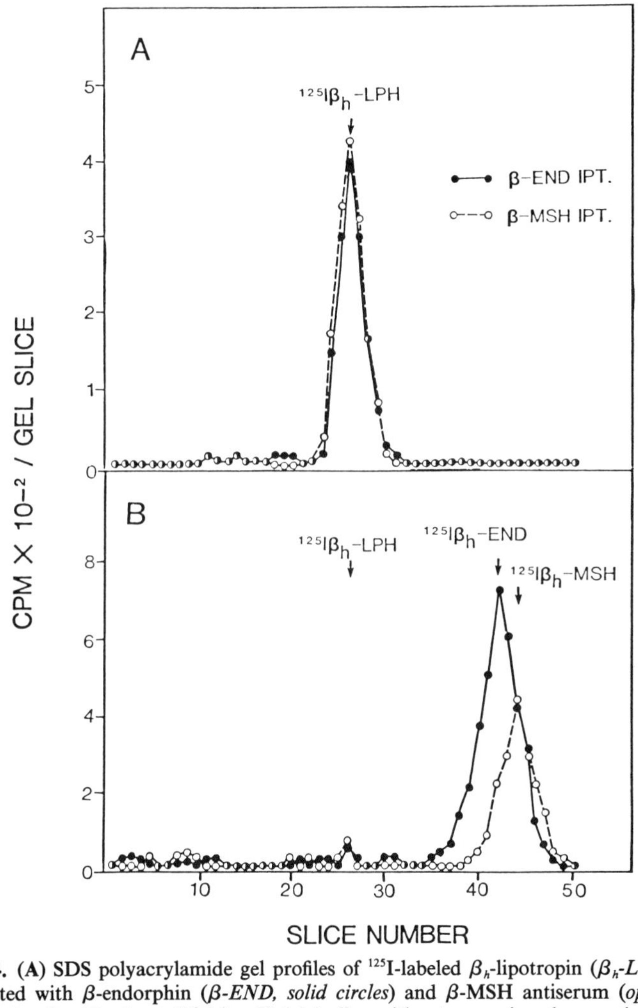

FIGURE 4. **(A)** SDS polyacrylamide gel profiles of ^{125}I-labeled β_h-lipotropin (*β_h-LPH*) immunoprecipitated with β-endorphin (*β-END, solid circles*) and β-MSH antiserum (*open circles*), after incubation for 30 min at 37°C in pH 4.0 buffer, without pro-opiomelanocortin-converting enzyme. **(B)** SDS-polyacrylamide-gel profiles of anti-β-END and anti-β-MSH immunoprecipitated products generated by incubating [^{125}I]β_h-LPH with pro-opiomelanocortin-converting enzyme in pH 4 buffer for 30 min at 37°C. The [^{125}I]β_h-LPH, [^{125}I]β_h-END, and [^{125}I]β_h-MSH marker are indicated. *IPT,* immunoprecipitate.

activity was shown to cleave the NH_2-terminal basic residue (Arg) from Arg-Met-enkephalin[36] and Arg-β-endorphin (Loh, unpublished data). It did not cleave the NH_2-terminal Tyr from Met-enkephalin, indicating the enzyme's specificity for a basic residue. Recently Dr. Nigel Birch, in our laboratory, has shown that the generation of Met-enkephalin from Arg-Met-enkephalin by the AP activity was approximately six times faster than from Lys-Arg Met-enkephalin (FIG. 5). Furthermore, incubation of Lys-γ_1-MSH with the membrane fraction of bovine intermediate lobe secretory

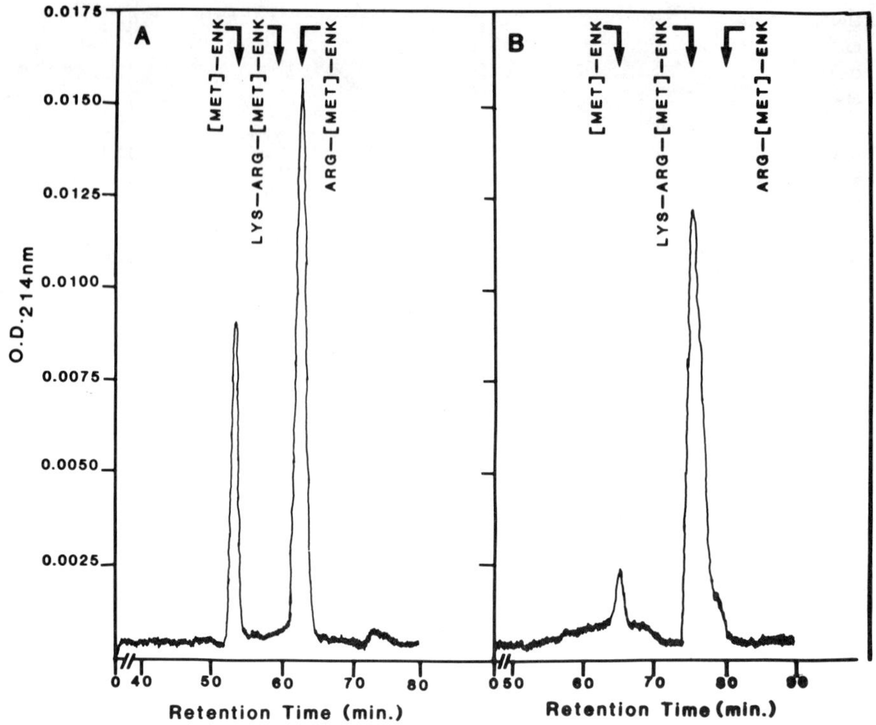

FIGURE 5. HPLC profiles showing Met-enkephalin formed from (**A**) Arg-Met-enkephalin or (**B**) Lys-Arg-Met-enkephalin when incubated with aminopeptidase B-like activity from intermediate lobe secretory vesicle membranes for 1 h at 37°C in pH 6.0 buffer. The HPLC conditions were isocratic 0.1% trifluoroacetic acid/14% acetonitrile, using a Biorad C_{18} reversed-phase column. Flow rate: 1.2 ml/min. The retention times of the different standards are indicated by the arrows. *ENK,* enkephalin.

vesicles did not result in the liberation of any detectable Lys. Thus, the enzyme has a greater affinity for NH_2-terminal Arg than Lys. Also, the secretory vesicle AP activity did not cleave an NH_2-terminal Arg if it was followed by a proline.[36] The specificity of this intermediate lobe AP enzyme probably accounts for the observation of NH_2-terminally extended Lys-γ-MSH as the major form of γ-MSH in bovine and rat intermediate pituitary,[30,31] as well as the existence of CLIP (see FIG. 1, $ACTH_{18-39}$), which has an NH_2-terminal extended Arg followed by a Pro in its amino acid sequence.[37]

In the intermediate lobe of the pituitary, two POMC-derived peptides, $ACTH_{1-14}$ and β-endorphin, undergo further modifications. $ACTH_{1-14}$ is acetylated at the NH_2 terminus and amidated at the COOH terminus to yield α-MSH. β-Endorphin is acetylated at the NH_2 terminus to yield Ac-β-endorphin, which loses its opiate activity upon such a modification.[38]

An acetyltransferase that is capable of acetylating the NH_2-terminal serine in $ACTH_{1-14}$ and tyrosine in β-endorphin has been detected in intermediate lobe secretory vesicles.[39–42] The enzyme is soluble and has a pH optimum of 7.0. Kinetic studies on the crude enzyme preparation showed similar K_m values for $ACTH_{1-13}NH_2$ and β-endorphin, but a higher K_m for the *O*-acetylation of α-MSH to form α-*N'O*-diacetyl α-MSH.[39] Radiation inactivation studies suggest that the molecular weight of the enzyme is 75,000.[43]

Enzymes with α-amidating activity have been reported in intermediate lobe and anterior lobe secretory vesicles.[44–47] Early work by Bradbury *et al.*[48] demonstrated an enzyme in secretory vesicles of porcine pituitary that amidates peptides with a COOH-terminal glycine. The enzyme activity was assayed using D-Tyr-Val-Gly (mimicking the COOH-terminal sequence of $ACTH_{1-14}$), and was characterized as a Cu^{2+}-dependent, neutral enzyme (pH optimum 7.0).[49] Glycine was found to be a mandatory amino acid in the COOH-terminal position of the peptide to be amidated, although the valine in the tripeptide substrate could be replaced by any neutral amino acid. This requirement for a COOH-terminal glycine in the peptide to be amidated was consistent with the fact that the amino acid sequences of amidated peptides invariably contain an extra COOH-terminal glycine in the sequences of their precursor molecules, often followed by paired basic residues. Subsequent studies on rat and bovine α-amidating enzyme revealed that it requires molecular oxygen and ascorbic acid, in addition to Cu^{2+}, and it was hence named peptidyl-glycine α-amidating monooxygenase (PAM).[45,46] The enzyme activity has also been shown to amidate naturally occurring substrates such as α-*N*-acetyl $ACTH_{1-14}$.[47] PAM was recently purified from bovine neurointermediate lobe and shown to exist in two forms, PAM-A (mol wt 48,000) and PAM-B (mol wt 42,000 and 37,000).[50] The properties of PAM-A and PAM-B are very similar to the PAM in the secretory vesicle crude extract. The physiological significance of the two forms of PAM remains to be determined.

Provasopressin and Pro-oxytocin Processing Enzymes in Neurosecretory Vesicles

Provasopressin and pro-oxytocin synthesized in the hypothalamo-neurohypophysial system are processed via an enzymatic cleavage at a Lys-Arg pair to liberate vasopressin and oxytocin, respectively. Recently Parish *et al.*[51] have purified a provasopressin-converting enzyme from bovine pituitary neural lobe secretory vesicles, the site of conversion of this prohormone *in vivo.*[52] The purified PCE in the neural lobe was found to have the same molecular weight, inhibitor profile, and specificity for paired basic residues as the intermediate lobe PCE. Neural lobe PCE cleaved human provasopressin primarily on the carboxyl side of the Lys_{11}-Arg_{12} to yield vasopressin-Gly_{10}-Lys_{11}-Arg_{12} (see FIG. 6). Moreover, neural lobe PCE also cleaved mouse POMC at a Lys-Arg pair to yield 21-23-kDa ACTH and β-lipotropin.[51] The

PCE, however, did not cleave the single Arg residue separating neurophysin from the glycopeptide (see FIG. 1). Another enzyme is likely involved in this cleavage step.

Recently, an enzyme activity that is capable of cleaving synthetic pro-oxytocin$_{1-18}$ at a pair of basic residues to liberate oxytocin has been reported in bovine neural lobe secretory vesicles.[53] Whether this enzyme activity will cleave intact pro-oxytocin or have properties similar to PCE awaits further characterization.

Although there have been reports of single Arg cleavage enzymes in the atrial gland of the mollusk, *Aplysia*,[54] and rat brain membranes,[55] to date no single Arg specific cleavage enzyme has been reported in neural lobe secretory vesicles.

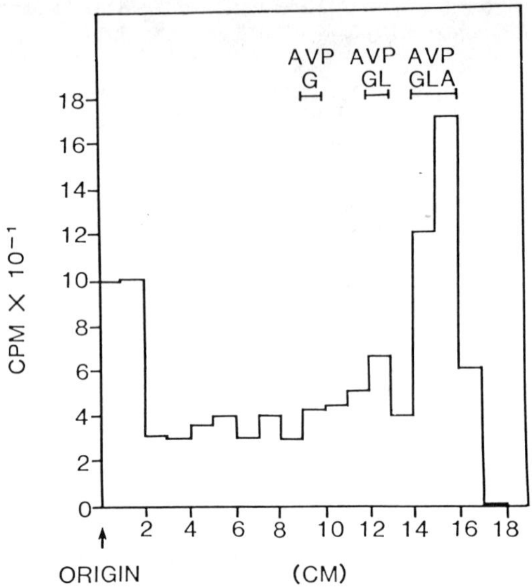

FIGURE 6. ^{35}S-labeled human provasopressin was incubated with purified prohormone-converting enzyme from bovine neural lobe secretory vesicles for 5 h at 37°C in pH 4.0 buffer. The incubate was immunoprecipitated with antibodies that recognize extended forms of arginine vasopressin. The generation of extended forms of immunoreactive arginine vasopressin (*AVP*) was assayed by high-voltage electrophoresis (*HVE*). The HVE profile (after subtracting background from a control incubate without enzyme) shows a major peak of [^{35}S]AVP-Gly$_{10}$-Lys$_{11}$-Arg$_{12}$ and minor peak of [^{35}S]AVP-Gly$_{10}$-Lys$_{11}$, indicating that the enzyme cleaved primarily on the carboxyl side of Lys$_{11}$-Arg$_{12}$ of provasopressin. The standards shown are AVP-GLA = AVP-Gly$_{10}$-Lys$_{11}$-Arg$_{12}$; AVP-GL = AVP-Gly$_{10}$-Lys$_{11}$, and AVP-G = AVP-Gly$_{10}$.

A carboxypeptidase B-like enzyme activity has been found in neural lobe secretory vesicles and shown to have the same properties as that described for the intermediate and anterior lobes. Rat neural lobe secretory vesicle CPB cleaved vasopressin-Gly$_{10}$-Lys$_{11}$-Arg$_{12}$ to yield vasopressin-Gly$_{10}$ by sequential removal of basic residues from the COOH terminus.[32] In another study, bovine neural lobe secretory vesicle CPB was shown to cleave oxytocin-Gly$_{10}$-Lys$_{11}$-Arg$_{12}$ to oxytocin-Gly$_{10}$.[53,56]

An aminopeptidase B-like enzyme activity was detected in bovine neural lobe secretory vesicles using Arg-met-enkephalin[36] and Arg$_{12}$-Ala-Leu-Asp-Tyr$_{18}$ (pro-oxytocin$_{12-18}$)[53] as substrates. The activity was shown to have properties identical to that described for intermediate lobe.[36] Although PCE cleaves provasopressin primarily on the carboxyl side of the Lys-Arg pair, a small percentage ($\leq 20\%$) of the cleavage does sometimes occur in between the basic residues (FIG. 6), thus necessitating an aminopeptidase B-like enzyme to remove the NH_2-terminal Arg from neurophysin (see FIG. 1). Furthermore, North *et al.*[57] have suggested that formation of neurophysin subsequent to the removal of vasopressin from the NH_2 terminus of provasopressin (see FIG. 1) may be carried out by an α-chymotrypsin-like enzyme found in bovine neural lobe secretory vesicles that could cleave between Val$_{107}$ and Arg$_{108}$ to yield neurophysin and the glycopeptide extended at the NH_2 terminus with an Arg. The Arg will then be removed by an aminopeptidase B-like enzyme.

Vasopressin and oxytocin are amidated at the COOH terminus. PAM activity identical in characteristics to that in intermediate and anterior lobes has been detected in neural lobe secretory vesicles using D-Tyr-Val-Gly as a substrate.[44]

Proenkephalin Processing Enzymes in Adrenal Medulla Chromaffin Granules

Proenkephalin is synthesized in adrenal medulla, and attempts have been made by numerous laboratories to search for a paired basic residue-specific proenkephalin-converting enzyme in chromaffin granules. To date, however, the paired basic residue specific activities found in bovine chromaffin granules have not yet been shown to cleave proenkephalin. Rather, the activities described have been demonstrated to cleave truncated forms of proenkephalin such as peptide E, peptide F, and BAM 12. These activities comprise serine and thiol proteases[58–62] (for a review see Loh *et al.*).[8] One of the serine protease activities has been purified 1000-fold and shown to cleave peptide E and peptide F, both in between and on the carboxyl side of the dibasic pair of residues.[58] The enzyme has a neutral pH optimum and a molecular weight of 20,000. The role of this and other reported paired basic residue specific chromaffin granule enzymes in proenkephalin processing remains to be determined.

A carboxypeptidase B-like enzyme is present in chromaffin granules and has been shown to cleave Arg from Met-enkephalin-Arg.[63,64] The enzyme that was first detected in these vesicles has been purified and shown to have properties identical to the enzyme purified from pituitary and brain.[33] Other posttranslational modification enzymes, *e.g.*, aminopeptidase-B, acetylation, and amidation enzymes, have not been described in chromaffin granules.

Proinsulin, Proglucagon, and Prosomatostatin Processing Enzymes in Pancreatic Islet Secretory Vesicles

An acid thiol protease activity capable of processing endogenously labeled proinsulin, proglucagon, and prosomatostatin to the respective hormones has been reported in anglerfish pancreatic islet secretory vesicles.[65,66] The enzyme activity appears to be distinct from lysosomal cathepsin B. In contrast, Steiner's group has reported the presence of cathepsin B and a precursor form of the enzyme in rat pancreatic islet secretory vesicles.[67] They propose that the precursor form of cathepsin B may play a role in proinsulin processing.

A carboxypeptidase B-like enzyme activity, similar to that described in the pituitary, has also been found in secretory vesicles of an insuloma, using hippuryl-L-arginine as substrate.[68] This enzyme presumably participates in the removal of COOH-terminal basic residues from peptide liberated from proinsulin by a paired basic residue-specific enzyme.

CONCLUDING REMARKS

TABLE 1 highlights some of the mammalian peptide precursor processing enzymes that have been detected in secretory vesicles of different tissues. Progress in the elucidation of paired basic residue-specific enzymes that can cleave the intact prohormone has been slow because of the difficulty in obtaining sufficient quantities of purified prohormone substrate, and the tedious and complex assay procedures required. So far only one secretory-vesicle-paired basic residue-specific enzyme (prohormone-converting enzyme, PCE) that is capable of cleaving intact peptide precursors (prohormones) has been described and purified to apparent homogeneity. This unique aspartyl protease (*i.e.,* PCE) is present in pituitary neural, intermediate, and anterior lobe secretory vesicles (TABLE 1), is functional at pHs between 4.0 and 5.0 consistent with the internal, acidic pH of the secretory vesicles (for review see Gainer *et al.*[10]), and processes a number of intact peptide precursors at paired basic residues in a highly specific manner. The enzyme does not cleave short synthetic peptides with paired basic residues[27,28,69]; perhaps these substrates lack the appropriate conformation. Interestingly, PCE from intermediate and neural pituitary secretory vesicles can also cleave bovine proinsulin[27,51] and human proenkephalin at paired basic residues (N. Birch, unpublished data). Processing of proenkephalin by PCE is perhaps not surprising since enkephalin is present in neural pituitary secretory vesicles.[70] These observations raise the possibility that PCE, or a similar aspartyl endopeptidase, may also be present in mammalian pancreatic islet secretory vesicles and adrenal medulla chromaffin granules.

We propose that for POMC processing PCE is the first enzyme to act on the prohormone in the secretory vesicle, cleaving it at paired basic residues to liberate various fragments which are secreted or further processed, depending on the tissue. For example, POMC is cleaved by PCE to yield ACTH, 16-kDa glycopeptide, β-lipotropin, and β-endorphin$_{1-31}$, and these products are secreted without further processing in the anterior pituitary. In the intermediate pituitary and various brain regions,[29,71,72] the ACTH is further processed to form α-MSH or desacetyl α-MSH plus CLIP, and the β-endorphin$_{1-31}$ to form β-endorphin$_{1-27}$ plus Gly-Gln.[73] These additional proteolytic processing steps do not appear to be carried out by PCE,[27,28] and therefore other highly specific enzymes in intermediate and brain secretory vesicles are required to cleave the paired basic residues of ACTH and β-endorphin$_{1-31}$. Such a proteolytic cascade mechanism involving a "core" enzyme (PCE) that is highly specific for certain dibasic residues within the intact prohormone, and additional proteases (present only in some tissues) that can specifically cleave the smaller liberated fragments to yield other biologically active products, could provide a very efficient means of regulating differential processing of POMC in different regions of the pituitary and brain. Differential proteolytic processing of other multivalent peptide precursors, *e.g.,* proenkephalin in brain and adrenal medulla,[74] may also be regulated by a similar multienzyme mechanism. Suffice it to say, for simple precursors such as pro-oxytocin and proinsulin that require only one or two cleavages at paired basic residues, one enzyme, *e.g.*, PCE, would be adequate for complete endoproteolytic processing of the precursor. Given what we have learned about proteolytic precursor processing enzymes so far, and the concept that there is a "core" prohormone converting endopeptidase that cleaves the intact precursor (but not small peptides), such as PCE, and other peptide endopeptidases that cleave only small fragments derived from complex precursors, it is apparent that one should always begin studies on prohormone-converting enzymes with the use of the endogenous, intact prohormone as substrate. The current ability to express peptide precursor genes in bacterial or mammalian systems should overcome the difficulty of obtaining sufficient precursor as substrate for such studies.

The purified prohormone converting enzyme (PCE) from neural lobe, intermediate lobe, and the activity from anterior lobe appear to be identical based on their inhibitor profile, size, and specificity, and I have therefore referred to them as a single enzyme. It is conceivable, however, that there may be subtle differences in their active sites that may become apparent only when their primary structures are known through cloning these enzymes, as in the case of the family of kallikreins.[75] The concept of a family of very closely related prohormone-converting endopeptidases the members of which are tissue- or cell-specific seems plausible, in view of the complexity involved in processing different intact prohormones. On the other hand, the exopeptidases are likely to be more general to all peptidergic cells, since the reactions catalyzed by these enzymes are relatively simple. Indeed, the wide distribution of an apparently identical carboxypeptidase B-like enzyme in the secretory vesicles of a number of endocrine and neural tissues[32–35,68] (see TABLE 1), and its ability to act on a variety of substrates, are consistent with the hypothesis that one universal, secretory-vesicle-specific, carboxypeptidase B-like enzyme performs the role of removing COOH-terminally extended basic residues from different prohormone-derived peptides in various peptide-synthesizing tissues. Although there is insufficient data at present to suggest that the same secretory vesicle aminopeptidase B-like enzyme operates in different peptidergic cells, this is likely to be the case.

Less is known about the tissue distribution of the intermediate pituitary secretory vesicle α-MSH/β-endorphin *N*-acetyltransferase.[39–43] Although *N*-acetyltransferases that can acetylate $ACTH_{1-13}NH_2$ have been reported in anterior pituitary vesicles,[40,42,76] hypothalamus[42] and lens,[43,77] these activities appear to be general serine *N*-acetyltransferases, and they do not *N*-acetylate the tyrosine of β-endorphin, unlike the intermediate lobe secretory vesicle *N*-acetyltransferase.

The α-amidating enzyme (PAM) appears to be fairly widely distributed. It is present in secretory vesicles of pituitary neural, intermediate and anterior lobes.[44] PAM activity has also been detected in hypothalamus, serum, thyroid, submandibular gland, and cerebrospinal fluid.[78–83] The activities from these sources, however, have varying properties and have not been sufficiently characterized to determine their similarity to the purified PAM from the neurointermediate lobe.[50]

Current studies on peptide precursor posttranslational modification enzymes therefore suggest that, perhaps with the exception of endoproteolysis, the same or very similar enzymes carry out each of the processing steps involved in the conversion of different precursors and peptides in various tissues. These secretory vesicle processing enzymes appear to be unique, however, and have properties distinct from other enzymes that share the same or related specificities, but are localized in other cellular compartments or nonendocrine or neuropeptide-producing tissues. It is also evident from this review that whereas the enzymes responsible for the late processing steps are well characterized, much more work needs to be done to elucidate and characterize the endoprotease(s) necessary for complete and differential processing of different peptide precursors.

TABLE 1. Properties of Some Mammalian Peptide Precursor Processing Enzymes Found in Secretory Vesicles

Enzyme	Reaction	Precursor and Peptide Substrates[b]	Tissue Distribution	Molecular Weight	Cofactors	pH Max *in Vitro*	Inhibitors	Refs.
Prohormone-converting[a] enzyme (paired basic residue-specific endopeptidase)	peptide¹-Lys-Arg-peptide² → peptide¹-Lys + Arg-peptide² or peptide¹-Lys-Arg + peptide²	pro-opiomelanocortin provasopressin proinsulin proenkephalin β-lipotropin	intermediate, neural and anterior pituitary	70,000	—	4.0–5.0	pepstatin A, diazoethylnorleucine methyl ester	27,28
Carboxypeptidase-B-like[a] enzyme (carboxypeptidase-E) (carboxypeptidase-H) (enkephalin convertase)	peptide-Lys-Arg → peptide + Lys + Arg	AVP-Gly-Lys-Arg OT-Gly-Lys-Arg Met-enkephalin-Arg $ACTH_{1-17}$	intermediate, neural and anterior pituitary adrenal medulla insuloma cells	50,000	Co^{2+}	5.4–5.8	Cu^{2+}, Cd^{2+} PCMB GEMSA	32–34, 68
Aminopeptidase-B-like enzyme	Arg-peptide → peptide + Arg	Arg-Met-enkephalin	intermediate and neural pituitary	?	Co^{2+}, Zn^{2+}	6.0	EDTA	36
Peptidyl glycine α-amidating monooxygenase (PAM)[a]	peptide-Gly → peptide-NH_2 + OHC-COOH	D-Tyr-Val-Gly α-*N*-acetyl-$ACTH_{1-14}$	intermediate, neural and anterior pituitary	48,000 (PAM A) 42,000 (PAM B) 37,000	Cu^{2+} ascorbic acid O_2	7.0–8.0	Diethyldithiocarbonate (reversed by $CuSO_4$)	44–50
Peptide acetyltransferase	NH_2-peptide → CH_3-C(=O)-NH-peptide	$ACTH_{1-13}NH_2$ β-endorphin	intermediate pituitary	75,000	AcetylCoA	7.0	—	39–43

[a] Purified to apparent homogeneity.
[b] AVP = arginine vasopressin; OT = oxytocin; PCMB = *p*-chloromercuribenzoate; GEMSA = guanidinoethylmercaptosuccinic acid.

REFERENCES

1. ZIMMERMAN, M., R. A. MUMFORD & D. F. STEINER, Eds. 1980. Ann. N.Y. Acad. Sci. **343:** 1-449.
2. DOCHERTY, K. & D. STEINER. 1982. Ann. Rev. Physiol. **44:** 625-638.
3. FREEDMAN, R. B. & H. C. HAWKINS, Eds. 1980. The Enzymology of Posttranslational Modification of Proteins, Vol. 1. Academic Press. London.
4. HABENER, J. F. 1981. Ann. Rev. Physiol. **43:** 211-223.
5. KOCH, G. & D. RICHTER, Eds. 1983. Biosynthesis, Modification and Processing of Cellular and Viral Polyproteins. Academic. New York.
6. CHRETIEN, M., S. BENJANNET, F. GOSSARD, C. GIANOULAKIS, P. CRINE, M. LIS & N. G. SEIDAH. 1979. Can. J. Biochem. **57:** 1111-1121.
7. LOH, Y. P. & H. GAINER. 1983. Biosynthesis and processing of neuropeptides *In* Brain Peptides, Vol. 1. D. Kreiger, M. J. Brownstein & J. Martin, Eds.: 76-116. Wiley. New York.
8. LOH, Y. P., M. J. BROWNSTEIN & H. GAINER. 1984. Ann. Rev. Neurosci. **7:** 189-222.
9. MAINS, R. E., B. A. EIPPER, C. C. GLEMBOTSKI & R. M. DORES. 1983. Trends Neurosci. **6:** 229-235.
10. GAINER, H., J. T. RUSSELL & Y. P. LOH. 1985. Neuroendocrinology **40:** 171-184.
11. COMB, M., P. H. SEEBURG, J. ADEMA, L. EIDEN & E. HERBERT. 1982. Nature **295:** 663-666.
12. GUBLER, U., P. H. SEEBURG, B. J. HOFFMAN, L. P. GAGE & S. UDENFRIEND. 1982. Nature **295:** 206-208.
13. JACOBS, J. W., R. H. GOODMAN, W. W. CHIN, P. C. DEE, J. F. HABENER, N. H. BELL & J. T. POTTS. 1981. Science **213:** 457-459.
14. KAKIDANI, H., Y. FURUTANI, H. TAKAHASHI, M. NODA, Y. MORIMOTO, T. HIROSE, M. ASAI, S. INAYAMA, S. NAKANISHI & S. NUMA. 1982. Nature **298:** 245-249.
15. LAND, H., G. SCHUTZ, H. SCHMALE & D. RICHTER. 1982. Nature **295:** 299-303.
16. NAKANISHI, S., A. INOUE, T. KITA, M. NAKAMURA, A. C. Y. CHANG, S. N. COHEN & S. NUMA. 1979. Nature **278:** 423-427.
17. VILLA-KOMAROFF, L., A. EFSTRATIADIS, S. BROOME, P. LOMEDICO, R. TIZARD, S. P. NABER, W. L. CHICK & W. GILBERT. 1978. Proc. Natl. Acad. Sci. U.S.A. **75:** 3727-3731.
18. CASTEL, M., H. GAINER & H. D. DELLMAN. 1984. Int. Rev. Cytol. **88:** 303-459.
19. LOH, Y. P. & H. A. GRITSCH. 1981. Eur. J. Cell Biol. **26:** 177-183.
20. STOECKEL, M. E., S. SCHIMCHOWITSCH, J. C. GARAND, G. SCHMITT, H. VAUDRY & A. PORTE. 1983. Cell Tiss. Res. **230:** 511-515.
21. MOORE, H.-P., M. WALKER, F. LEE & R. B. KELLY. 1983. Cell **35:** 531-538.
22. ORCI, L., M. RAVAZZOLA, M. AMHERDT, O. MADSEN, J.-D. VASALLI & A. PERRELET. 1985. Cell **42:** 671-681.
23. LOH, Y. P. & H. GAINER. 1982. Proc. Natl. Acad. Sci. U.S.A. **79:** 108-112.
24. LOH, Y. P. & T.-L. CHANG. 1982. FEBS Lett. **137:** 57-62.
25. CHANG, T.-L. & Y. P. LOH. 1983. Endocrinology **112:** 1832-1838.
26. CHANG, T.-L. & Y. P. LOH. 1984. Endocrinology **114:** 2092-2099.
27. LOH, Y. P., D. C. PARISH & R. TUTEJA. 1985. J. Biol. Chem. **260:** 7194-7205.
28. LOH, Y. P. 1986. J. Biol. Chem. **261:** 11949-11952.
29. EIPPER, B. A. & R. E. MAINS. 1980. Endocrinol. Rev. **1:** 1-27.
30. BROWNE, C. A., H. P. J. BENNETT & S. SOLOMON. 1981. Biochem. Biophys. Res. Commun. **100:** 336-343.
31. BOHLEN, P., F. ESCH, T. SHIBASAKI & A. BAIRD, N. LING & R. GUILLEMIN. 1981. FEBS Lett. **128:** 67-70.
32. HOOK, V. Y. H. & Y. P. LOH. 1984. Proc. Natl. Acad. Sci. **81:** 2777-2780.
33. FRICKER, L. D. & S. H. SNYDER. 1983. J. Biol. Chem. **258:** 10950-10955.
34. SUPATTAPONE, S., L. D. FRICKER & S. H. SNYDER. 1984. J. Neurochem. **42:** 1019-1023.
35. MAINS, R. E. & B. A. EIPPER. 1984. Endocrinology **115:** 1683-1690.
36. GAINER, H., J. T. RUSSELL & Y. P. LOH. 1984. FEBS Lett. **175:** 135-139.

37. SCOTT, A. P., J. G. RATCLIFFE, L. H. REES, J. LANDON, H. P. J. BENNETT, P. J. LOWRY & C. MCMARTIN. 1973. Nature New Biol. **244:** 65-67.
38. SMYTH, D. G., D. E. MASSEY, S. ZAKARIAN & M. D. A. FINNIE. 1979. Nature **279:** 252-254.
39. GLEMBOTSKI, C. C. 1982. J. Biol. Chem. **257:** 10501-10509.
40. CHAPPELL, M., Y. P. LOH & T. L. O'DONOHUE. 1982. Peptides **3:** 405-410.
41. O'DONOHUE, T. L. 1983. J. Biol. Chem. **258:** 2163-2167.
42. BARNEA, A. & G. CHO. 1983. Neuroendocrinology **37:** 434-439.
43. CHAPPELL, M. C., T. L. O'DONOHUE, W. R. MILLINGTON & E. S. KEMPNER. 1986. J. Biol. Chem. **261:** 1088-1091.
44. EIPPER, B. A., R. E. MAINS & C. C. GLEMBOTSKI. 1983. Proc. Natl. Acad. Sci. **80:** 5144-5148.
45. EIPPER, B. A., C. C. GLEMBOTSKI & R. E. MAINS. 1983. Peptides **4:** 921-928.
46. GLEMBOTSKI, C. C. 1984. J. Biol. Chem. **259:** 13041-13048.
47. GLEMBOTSKI, C. C. 1985. Archiv. Biochem. Biophys. **241:** 673-683.
48. BRADBURY, A. F., M. D. A. FINNIE & D. G. SMYTH. 1982. Nature **298:** 686-688.
49. BRADBURY, A. F. & D. G. SMYTH. 1983. Amidation of synthetic peptides by a pituitary enzyme: Specificity and mechanism of the reaction. *In* Peptides. K. Blaha & P. Mahlon, Eds.: 381-386. de Gruyter. Berlin.
50. MURTHY, A. S., R. E. MAINS & B. A. EIPPER. 1986. J. Biol. Chem. **261:** 1815-1822.
51. PARISH, D. C., R. TUTEJA, M. ALTSTEIN, H. GAINER & Y. P. LOH. 1986. J. Biol. Chem. **261:** 14392-14397.
52. GAINER, H., Y. SARNE & M. J. BROWNSTEIN. 1977. J. Cell. Biol. **73:** 366-381.
53. CLAMAGIRAND, C., M. CAMIER, H. BOUSSETTA, C. FAHY, A. MOREL, P. NICOLAS & P. COHEN. 1986. Biochem. Biophys. Res. Commun. **134:** 1190-1196.
54. WALLACE, E. F., E. WEBER, J. D. BARCHAS & C. J. EVANS. 1984. Biochem. Biophys. Res. Commun. **119:** 415-422.
55. DEVI, L. & A. GOLDSTEIN. 1985. Biochem. Biophys. Res. Commun. **130:** 1168-1176.
56. KANMERA, T. & I. M. CHAIKEN. 1985. J. Biol. Chem. **260:** 8474-8482.
57. NORTH, W. G., H. VALTIN, S. CHENG & G. R HARDY. 1983. Prog. Brain Res. **60:** 217-225.
58. LINDBERG, I., H. Y. T. YANG & E. COSTA. 1984. J. Neurochem. **42:** 1411-1419.
59. TROY, C. & J. M. MUSACCHIO. 1982. Life Sci. **31:** 1717-1720.
60. EVANGELISTA, R., P. RAY & R. V. LEWIS. 1982. Biochem. Biophys. Res. Commun. **106:** 895-902.
61. MIZUNO, K., A. MIYATA, K. KANEGAWA & R. H. MATSUO. 1982. Biochem. Biophys. Res. Commun. **106:** 186-193.
62. MIZUNO, K., M. KOJIMA & H. MATSUO. 1985. Biochem. Biophys. Res. Commun. **128:** 884-891.
63. FRICKER, L. D. & S. H. SNYDER. 1982. Proc. Natl. Acad. Sci. U.S.A. **79:** 3886-3890.
64. HOOK, V. Y. H. 1985. J. Neurochem. **45:** 987-989.
65. FLETCHER, D. J., B. D. NOE, G E. BAUER & J. P. QUIGLEY. 1980. Diabetes **29:** 593-599.
66. FLETCHER, D. J., J. P. QUIGLEY, G. E. BAUER & B. D. NOE. 1981. J. Cell. Biol. **90:** 312-322.
67. DOCHERTY, K., J. C. HUTTON & D. F. STEINER. 1984. J. Biol. Chem. **259:** 6041-6044.
68. DOCHERTY, K. & J. C. HUTTON. 1983. FEBS Lett. **162:** 137-141.
69. CHANG, T.-L., H. GAINER, J. T. RUSSELL & Y. P. LOH. 1982. Endocrinology **111:** 1607-1614.
70. VOIGT, K. H. & R. MARTIN. 1985. Coexistence of Unrelated Neuropeptides in Nerve Terminals. *In* Biogenetics of Neurohormonal Peptides. R. Hakanson & J. Thorell, Eds.: Academic Press, New York, N.Y.
71. ZAKARIAN, S. & D. G. SMYTH. 1982. Biochem. J. **202:** 561-571.
72. WATSON, S. J. & H. AKIL. 1980. Brain Res. **182:** 217-223.
73. PARISH, D. C., D. G. SMYTH, J. R. NORMANTON & J. H. WOLSTENCROFT. 1984. Nature **306:** 267-270.
74. LISTON, D. R., J. J. VANDERHAEGEN & J. ROSSIER. 1983. Nature **302:** 62-65.
75. MASON, A. J., B. A. EVANS, D. R. COX, J. SHINE & R. I. RICHARDS. 1983. Nature **303:** 300-307.

76. WOODFORD, T. A. & J. E. DIXON. 1979. J. Biol. Chem. **254:** 4993-4999.
77. GRANGER, M., G. E. TESSER, W. DEJONG & H. BLOEMENDAL. 1976. Proc. Natl. Acad. Sci. U.S.A. **73:** 3010-3014.
78. EMESON, R. B. 1984. J. Neurosci. **4:** 2604-2613.
79. WAND, G. S., R. L. NEY, R. E. MAINS & B. A. EIPPER. 1985. Neuroendocrinology **41:** 482-489.
80. KIZER, J. S., W. H. BUSBY, JR., C. COTTLE & W. W. YOUNGBLOOD. 1984. Proc. Natl. Acad. Sci. U.S.A. **81:** 3228-3232.
81. EIPPER, B. A., A. C. MYERS & R. E. MAINS. 1985. Endocrinology **116:** 2497-2504.
82. HUSAIN, I. & S. S. TATE. 1983. FEBS Lett. **152:** 277-281.
83. GOMEZ, S., C. DIBELLO, L. T. HUNG, R. GENET, J.-L. MORGAT, D. FROMAGEOT & P. COHEN. 1984. FEBS Lett. **167:** 160-164.

DISCUSSION OF THE PAPER

J. D. CASTLE (*Yale University School of Medicine, New Haven, Conn.*): What is the nature of association of POMC-converting enzyme within the granule membrane, and the extent of its release during exocytosis?

Y. P. LOH (*National Institutes of Health, Bethesda, Md.*): POMC converting enzyme is ionically associated with the vesicle membrane and is easily liberated from the membrane by 1 M NaCl. By our lysis procedure (freezing and thawing six times), approximately 80% of the activity in bovine intermediate lobe secretory vesicles is in the soluble fraction. However, it is possible that *in vivo,* more of it may be associated with the membrane. Interestingly, newly synthesized POMC is associated with the secretory vesicle membrane. At the present time all we know is that the enzyme is released. I have no quantitative data to determine the extent of the release from the bovine intermediate lobe secretory vesicles.

Regulation of Enkephalin, VIP, and Chromogranin Biosynthesis in Actively Secreting Chromaffin Cells

Multiple Strategies for Multiple Peptides

JAMES A. WASCHEK,[a] REBECCA M. PRUSS,[a]
RUTH E. SIEGEL,[a] LEE E. EIDEN,[a]
MARIE-FRANCE BADER,[b] AND DOMINIQUE AUNIS[b]

[a] *Laboratory of Cell Biology*
National Institute of Mental Health
National Institutes of Health
Bethesda, Maryland 20892

[b] *Unité INSERM U-44*
Centre de Neurochimie du CNRS
Strasbourg, France

Secretory cells must have a mechanism to replenish peptides that are secreted upon stimulation in order to maintain a continued ability to secrete in response to stimulation. Since peptide reuptake mechanisms are not known to exist in chromaffin cells, the stimuli for peptide release must in some way also produce a compensatory increase in the rate of peptide biosynthesis. We have studied the role of calcium in peptide biosynthesis, since it is well established that calcium acts as a second messenger in exocytotic secretion. A variety of pharmacologic agents have been employed to study the role of calcium in stimulus-secretion-synthesis coupling, and the results indicate that some, but not all, peptides secreted from chromaffin cells are under positive control by calcium. Reported here are studies in bovine chromaffin cells on the biosynthesis of two peptides, Met-enkephalin and vasoactive intestinal polypeptide (VIP), whose biosynthesis is tightly coupled to release, and one polypeptide, chromogranin A, whose biosynthesis is not.

MET-ENKEPHALIN

Met-enkephalin is secreted along with catecholamines from the adrenal medulla upon stimulation of nicotinic receptors with acetylcholine.[1,2] In primary cultures of chromaffin cells treated with nicotine, an increase in total Met-enkephalin biosynthesis is evident after 24 hours of stimulation (FIG. 1 and ref. 4). The induction of peptide is preceded by an increase in the amount of mRNA encoding the enkephalin polypeptide precursor, preproenkephalin A ($mRNA^{enk}$, FIG. 1). Thus, nicotine-stimulated secretion of Met-enkephalin is accompanied by an increase in Met-enkephalin biosynthesis, which occurs at a pretranslational locus.

If an enhanced rate of Met-enkephalin biosynthesis is coupled to its secretagogue-induced release, at what step do the two processes diverge? Exocytotic release of chromaffin granule contents after physiologic stimuli is believed to occur by a sequence of steps[6] including: binding of acetylcholine to the nicotinic receptor, opening of the receptor-operated sodium channel, depolarization of the plasma membrane, and opening of voltage-dependent calcium channels. Calcium influx then results in an increase in intracellular free calcium concentration and activation of proteins or factors involved in secretion. The pathway leading to enhanced biosynthesis could diverge from this secretory pathway at any point (FIG. 2). For example, the two pathways could diverge just after secretagogue binding if receptor occupancy activated adenylate cyclase directly to produce an enhanced rate of Met-enkephalin biosynthesis. If this were the case, it would be expected that elevation of cAMP would cause enhanced peptide biosynthesis, but stimulation of sodium or calcium influx would not. To systematically investigate the potential steps of pathway divergence, we used forskolin as an agent to increase intracellular cAMP, veratridine to cause sodium influx, elevated extracellular potassium to produce membrane depolarization independent of receptor-associated sodium influx, and barium to mimic the effect of elevated intracellular calcium. It was found that all of the above agents except forskolin elicited secretion of Met-enkephalin from chromaffin cells, while all agents, including forskolin, elicited an increase in Met-enkephalin biosynthesis at the level of mRNA (FIG. 3).

That veratridine, potassium, and barium were all able to produce an increase in biosynthesis similar to that caused by nicotine or acetylcholine suggests that pathways leading to Met-enkephalin secretion and biosynthesis diverge at or distal to calcium entry into the cell. Thus calcium influx or calcium activation of an intracellular factor appears to be the final step common to both pathways and ultimately couples these processes in chromaffin cells after nicotinic stimulation. Calcium may act to increase Met-enkephalin biosynthesis by activating adenylate cyclase. Moderate elevation of cAMP in bovine chromaffin cells is produced by nicotine,[7,8] and $mRNA^{enk}$ is increased by forskolin (FIG. 3) and other agents that elevate or mimic cAMP.[9] An alternative model for acetylcholine stimulation of Met-enkephalin biosynthesis is that receptor occupancy produces both calcium influx and activation of adenylate cyclase in parallel and that calcium and cAMP act cooperatively to increase $mRNA^{enk}$.

The role of calcium as a second messenger in mediating Met-enkephalin secretion and biosynthesis was studied further by examining the response to the agents discussed above in the presence of normal (1.8 mM) and reduced (100 μm) extracellular calcium. Treatment with reduced calcium inhibited secretion by nicotine, veratridine, and elevated potassium (FIG. 4), but barium-stimulated secretion was enhanced in medium containing reduced calcium. An inverse relationship between extracellular calcium and barium-induced secretion exists over a wide range of extracellular calcium concentration (Waschek and Eiden, in preparation), indicating that barium competes with calcium at a site involved in secretion, presumably the site of calcium entry into the cell. Inside the cell, barium appears to be able to act at all calcium sites actively involved in secretion. Stimulation of Met-enkephalin biosynthesis by all of the agents, except veratridine, was inhibited by reducing the concentration of extracellular calcium. Thus, in the presence of extracellular calcium, barium may act by depolarizing the cell, allowing calcium to enter the cell to stimulate biosynthesis at barium-insensitive sites. The step or steps at which only calcium may act are not involved in the

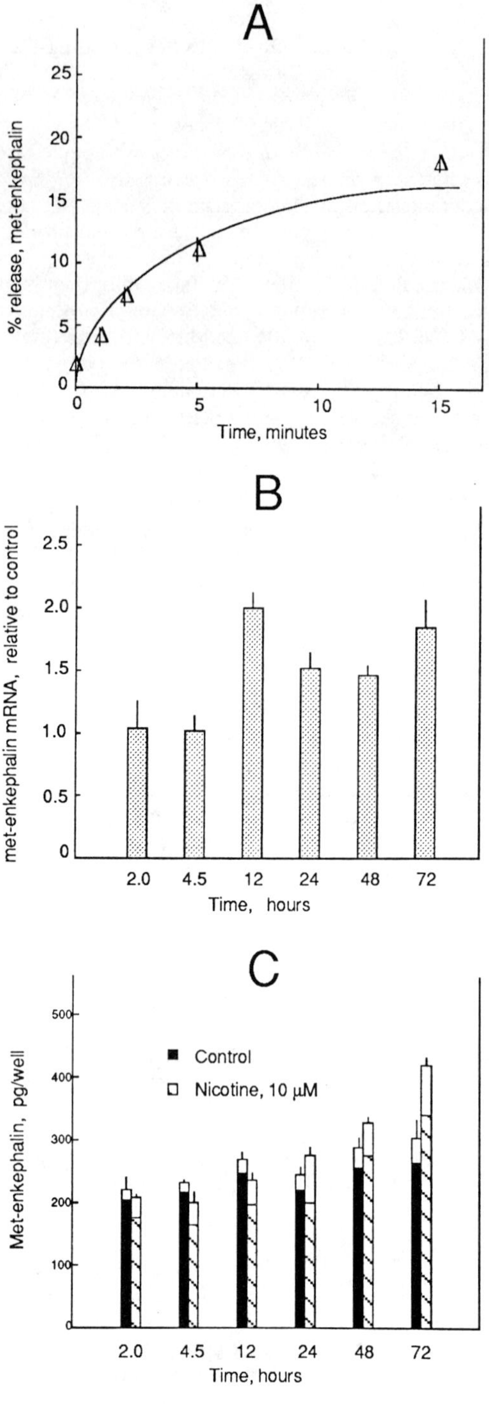
A
% release, met-enkephalin
25
20
15
10
5
0
0
5
10
15
Time, minutes
B
met-enkephalin mRNA, relative to control
2.5
2.0
1.5
1.0
0.5
0
2.0
4.5
12
24
48
72
Time, hours
C
Met-enkephalin, pg/well
500
400
300
200
100
0
Control
Nicotine, 10 μM
2.0
4.5
12
24
48
72
Time, hours

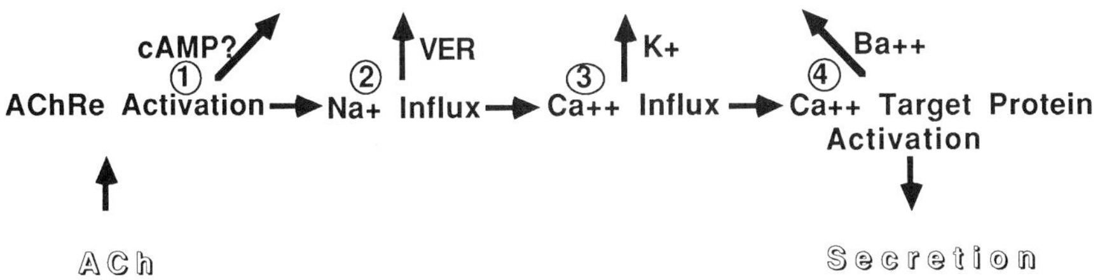

FIGURE 2. Possible sites of coupling of enkephalin secretion and activation of enkephalin biosynthesis in chromaffin cells. The signal for stimulation of enkephalin biosynthesis could be generated at any of the four shown sequential steps leading to peptide secretion. Agents that act independently at each of these steps are shown. If secretion and synthesis were coupled at step 1, veratridine, K^+, and Ba^{2+} would stimulate only secretion, and not synthesis. If coupling occurred at step 2, veratridine but not K^+ or Ba^{2+} would stimulate enkephalin biosynthesis. If coupling occurred at step 3 (depolarization per se), veratridine and K^+ would stimulate enkephalin synthesis. If coupling occurred at step 4, all three agents would stimulate both enkephalin secretion and biosynthesis.

cAMP stimulation of $mRNA^{enk}$ because the effect of forskolin to increase $mRNA^{enk}$ was not blocked in the medium containing reduced calcium (FIG. 4).

To further implicate a second-messenger role of calcium, the voltage-dependent calcium channel antagonist D600 was utilized to block pharmacologically the stimulated rise in intracellular calcium concentration. Release elicited by nicotine as well as veratridine, potassium, and barium was inhibited by 10 μM D600 as was the induction of $mRNA^{enk}$ produced by nicotine and potassium (FIG. 5). Higher doses of D600 (30 μM) were required to inhibit induction of $mRNA^{enk}$ by veratridine and

---

FIGURE 1. Nicotine-stimulated Met-enkephalin secretion and biosynthesis in bovine chromaffin cells. **(A)** Cumulative Met-enkephalin release (% of cell contents) after treatment with nicotine 10 μM for 15 minutes. Release after 15 minutes in untreated cells was about 2%. **(B)** Time course of increase in mRNA encoding the Met-enkephalin polypeptide precursor after treatment with nicotine 5 μM. **(C)** Time course of increase in immunoreactive Met-enkephalin in cells (*darkened or hatched portion of bars*) and medium (*open portion of bars*) after treatment with nicotine 10 μM. SEM bars refer to total peptide (cell plus medium). For all data points in **A**, **B**, and **C**, values are mean of 3 samples ± SEM. General methods are as follows: Chromaffin cells were obtained from fresh bovine adrenal medulla as described,[3] and were cultured in a humidified 95% air/5% CO_2 atmosphere in non-collagen-coated Costar® 24-well dishes (16 mm diameter) at a density of 0.3 to 1.0 × 10^6 cells per well in Basal Medium of Eagle with Earle's salts and Hepes, 25 mM, supplemented with 5% fetal bovine serum, penicillin, streptomycin, and cytosine arabinoside. After 2 to 4 days in culture, 30 to 60 min before adding drugs, medium was removed and replaced with fresh medium. Met-enkephalin immunoreactivity in medium and cell extracts was measured as described.[3] For Met-enkephalin mRNA quantitation,[4] RNA was extracted from cells, denatured, and electrophoresed on agarose gels as described. Total RNA extracted was estimated by densitometric scanning of the 28S band appearing on photographic negatives of UV illuminated gels stained with ethidium bromide. The RNA was transblotted from the gel to Gene Screen® and hybridized with a nick-translated 400 base pair *Pst*I fragment of the cDNA encoding the Met-enkephalin polypeptide precursor.[5] The mRNA was then quantified by densitometric scanning of film exposed to the hybridized blots. In some cases, after total RNA measurement, quantification of mRNA for enkephalin was carried out after direct slot blot RNA application, because hybridization to mRNA species other than enkephalin was negligible.

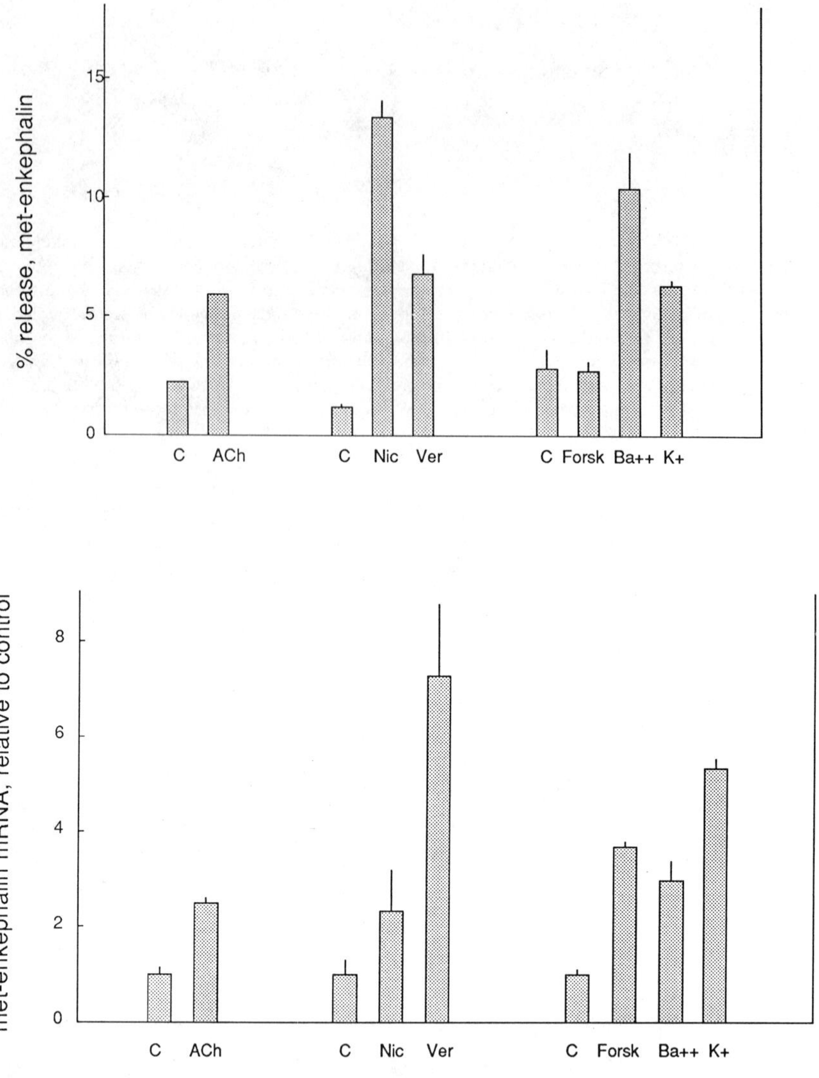

FIGURE 3. Secretion of Met-enkephalin and stimulation of mRNA encoding its polypeptide precursor after treatment with various reagents: C, control, ACh, acetylcholine 50 μM, Nic, nicotine 10 μM, Ver, veratridine 10 μM, Forsk, forskolin 25 μM, Ba^{2+}, barium chloride 1 mM, K^+, KCl 40 mM (replacing equimolar amount of NaCl). Data from separate experiments are grouped, with control values preceding drug treatment values. Methods are as indicated in FIGURE 1, except that release was measured over a 30 or 60 min time period. In studies utilizing barium, a 1:1 dilution of standard medium with a solution of chloride salts of sodium, potassium, calcium, and magnesium, at concentrations of 154, 5.4, 1.8 and 0.8 mM, respectively, was used. Replacement of sulfate, carbonate, and phosphate salts in this manner prevented precipitation of barium, which occurred in the standard medium at barium concentrations greater than 2 mM after 4 to 5 days.

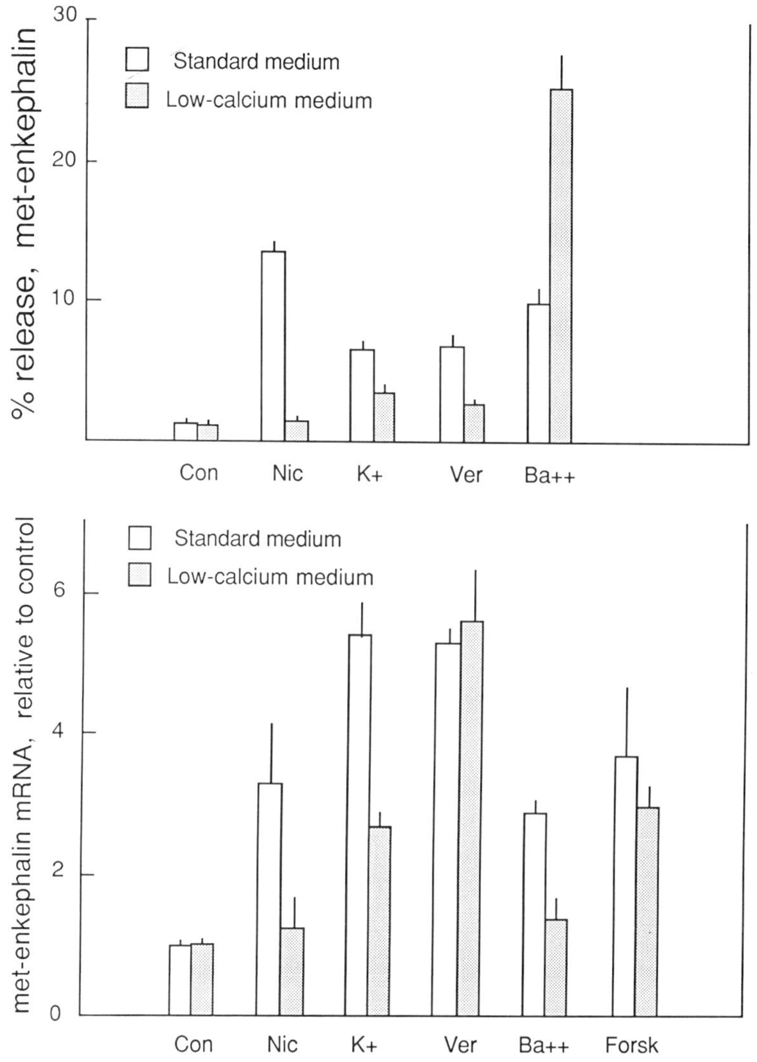

FIGURE 4. Met-enkephalin secretion and mRNA induction 30 to 60 min and 24 h, respectively, after stimulation with various agents in medium containing 1.8 mM calcium and reduced calcium (approximately 100 μM). Cells were preincubated with the respective medium for 30 to 60 min before drug addition. Drug doses are: nicotine, 10 μM; KCl, 40 mM (replacing equimolar amount of NaCl); veratridine, 10 μM; barium, 1 mM; and forskolin, 25 μM. Data are from two experiments having similar control values. SEM values are indicated, $n = 3$ to 6. Other details are as indicated in FIGURE 3.

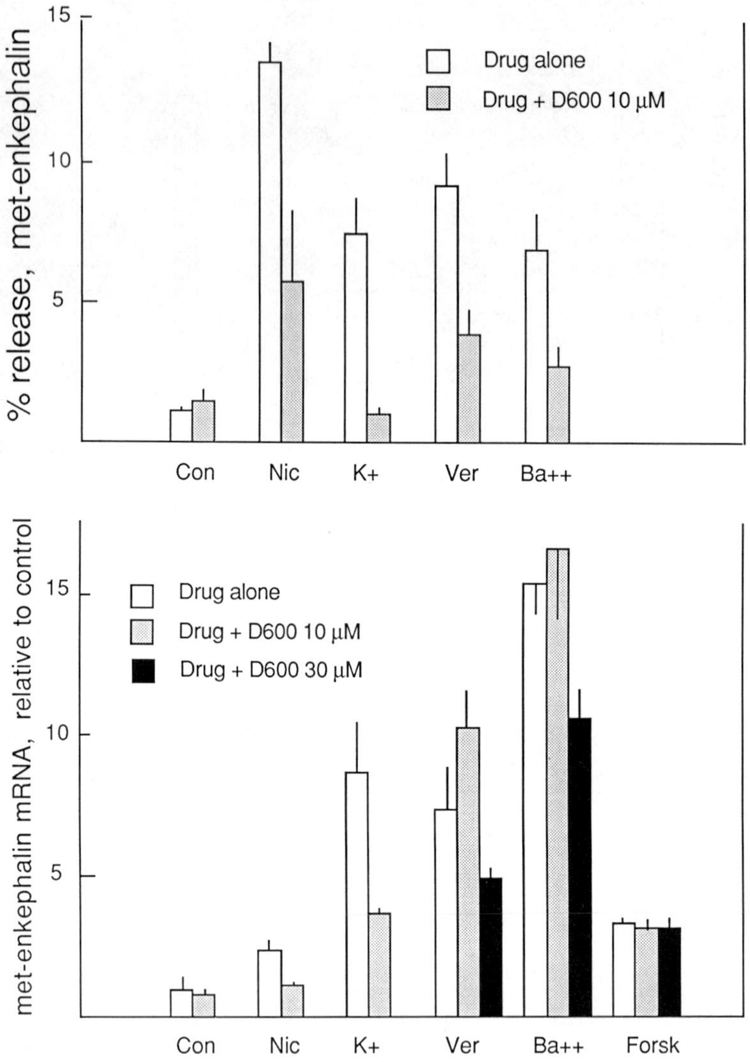

FIGURE 5. Met-enkephalin secretion and mRNA induction 30 to 60 min and 24 h, respectively, after stimulation with various agents in control and D600-containing medium. Cells were preincubated with the respective medium containing D600 for 30 to 60 min before drug addition. Data are from several experiments having similar control values. Nicotine doses are 10 μM (*upper panel*) and 5 μM (*lower panel*). Other drug doses (both panels) are KCl, 40 mM (replacing equimolar amount of NaCl); veratridine, 10 μM; barium, 1 mM; and forskolin, 25 μM. Doses of D600 are indicated. SEM values are indicated, n = 3 to 6.

barium, so it is apparent that calcium entry into the cell through the D600-sensitive channel is a prerequisite for stimulation of biosynthesis by these agents. The lack of sensitivity to lower doses of D600, however, distinguishes the regulation of this peptide from that of VIP (discussed below), perhaps due to differences in calcium channels on the individual cell types producing these peptides, or differences in calcium metabolism.

The above data allow construction of a model for nicotinic stimulation of enkephalin secretion and enhanced biosynthesis, in which both effects are driven by an influx of extracellular calcium (FIG. 6). Binding of acetylcholine to plasma membrane receptors causes opening of sodium channels, depolarization of the plasma membrane, and opening of voltage-dependent calcium channels. Calcium influx through these channels results in exocytotic release of chromaffin granule contents and stimulation of Met-enkephalin gene transcription by activation of an as yet unidentified intracellular factor or factors. Calmodulin may be one of these factors because introduction of calmodulin antibodies into chromaffin cells blocks secretion of catecholamines elicited by nicotine or a depolarizing concentration of potassium.[10] The calcium-activated factor(s) involved in biosynthesis induction may or may not subsequently act to stimulate adenylate cyclase. The small cAMP increase that occurs after nicotine treatment may act cooperatively with elevated intracellular calcium to increase peptide biosynthesis. Alternatively, cAMP is not involved at all in stimulus-secretion-synthesis coupling, but perhaps is involved in the stimulation of Met-enkephalin biosynthesis in phenotypic establishment during development, or in long-term regulation of biosynthesis by other hormones or neurotransmitters.

VASOACTIVE INTESTINAL POLYPEPTIDE (VIP)

In primary culture, bovine chromaffin cells synthesize VIP and release this peptide in response to the same agents that cause release of catecholamines and Met-enkephalin, including nicotine, veratridine, and elevated potassium.[3] The release of VIP, similar to that of Met-enkephalin, is accompanied by an increase in peptide biosynthesis (FIG. 7). A sensitive polynucleotide probe to quantify bovine VIP mRNA is not currently available, so it is not known for certain if the enhanced rate of biosynthesis of VIP occurs at a pretranslational locus. Cycloheximide blocks the increase in VIP peptide levels produced by nicotine, however, and the relative reduction in peptide levels by cycloheximide is greater in nicotine-treated cells than in untreated cells (FIG. 8), so it can be concluded that treatment with nicotine results in an increase in VIP by *de novo* synthesis of peptide. The effects of medium containing reduced calcium or 10 μM D600 on stimulation of VIP biosynthesis by nicotine, potassium, veratridine, and barium are qualitatively similar, but not identical to the effects of reduced calcium and D600 on stimulation of Met-enkephalin biosynthesis by these agents (FIG. 9). The difference is that, in a given experiment, induction of VIP biosynthesis by barium and veratridine is sensitive to blockade by lower doses of D600 than is biosynthesis of Met-enkephalin. VIP appears to be inducible in a subset of chromaffin cells distinct from those that produce Met-enkephalin (FIG. 10), so the dissimilarity may be due to differences in individual cell types, sensitivity to D600 at membrane channels, or to calcium at intracellular sites. Despite these differences in D600 sensitivity, secretagogue-stimulated Met-enkephalin and VIP secretion and biosynthesis appear to be coupled by calcium in very similar ways (FIG. 6).

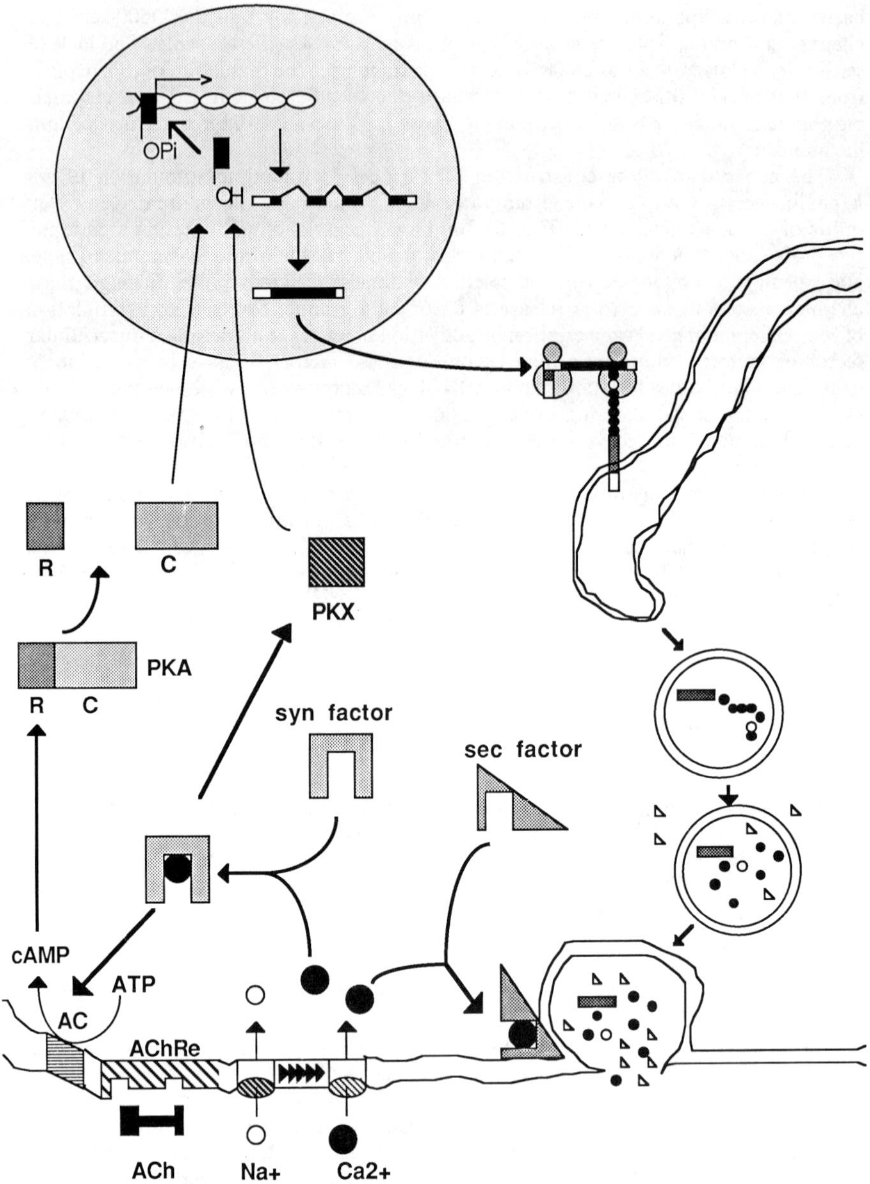

FIGURE 6. Model depicting possible events involved in secretagogue regulation of peptide release and biosynthesis. Acetylcholine (*ACh*) interacts with nicotinic receptors (*AChRe*) on the plasma membrane and opens the receptor-associated channel permeable to Na^+ (○). Influx of Na^+ through this channel causes depolarization of the plasma membrane, opening of voltage-dependent Ca^{2+} (●) channels, and a rise in intracellular free calcium. Calcium acts within the cell with one or more secretory (*sec*) factors to produce exocytotic release of granule contents and with one or more biosynthetic (*syn*) factors to produce a compensatory increase in biosynthesis

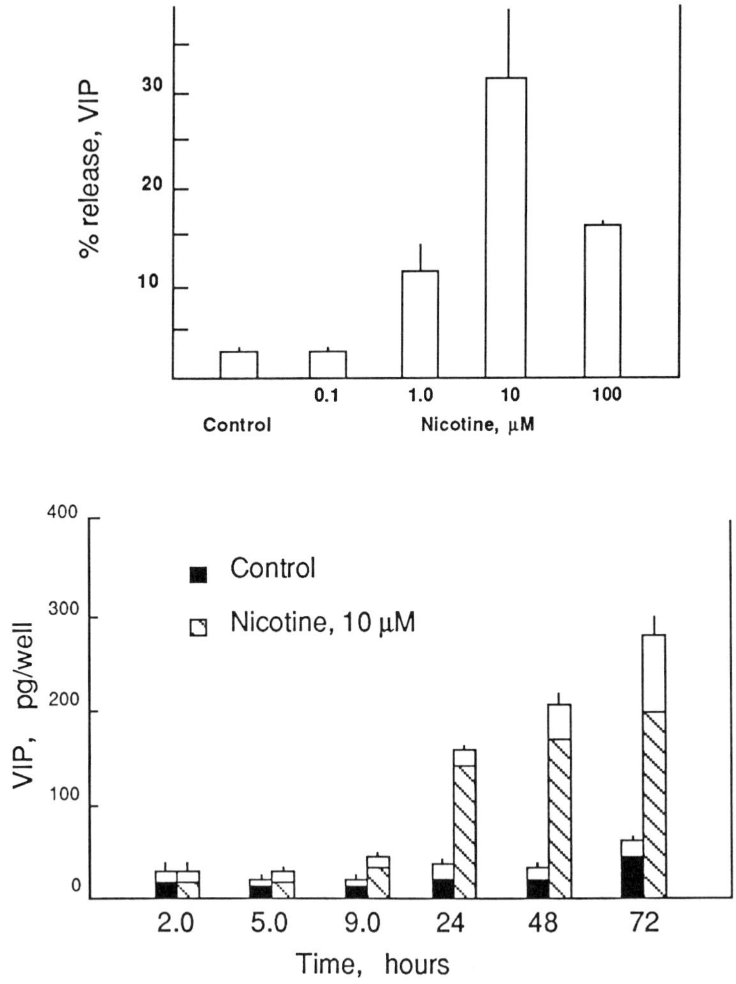

FIGURE 7. Secretion over 15 min of immunoreactive vasoactive intestinal polypeptide (*VIP*) after treatment with various doses of nicotine and the time course of increase in total VIP levels (cell extracts plus medium) after treatment with nicotine 10 μM. Data are mean of three samples ± SEM. Experimental details are as described in Eiden *et al.*[3]

of the secretory peptide. The calcium-stimulated increase in Met-enkephalin biosynthesis occurs in the nucleus at the level of gene transcription, and may be mediated by an undefined calcium-activated protein kinase (*PKX*), which upon activation phosphorylates a protein involved in the regulation of Met-enkephalin gene transcription. An alternate route of calcium action producing stimulation of the enkephalin gene involves activation of adenylate cyclase (*AC*), binding of the cAMP to protein kinase A (*PKA*), and dissociation of PKA into regulatory (*R*) and catalytic (*C*) subunits. The free catalytic subunit enters the nucleus and phosphorylates a protein involved in regulation of the enkephalin gene.

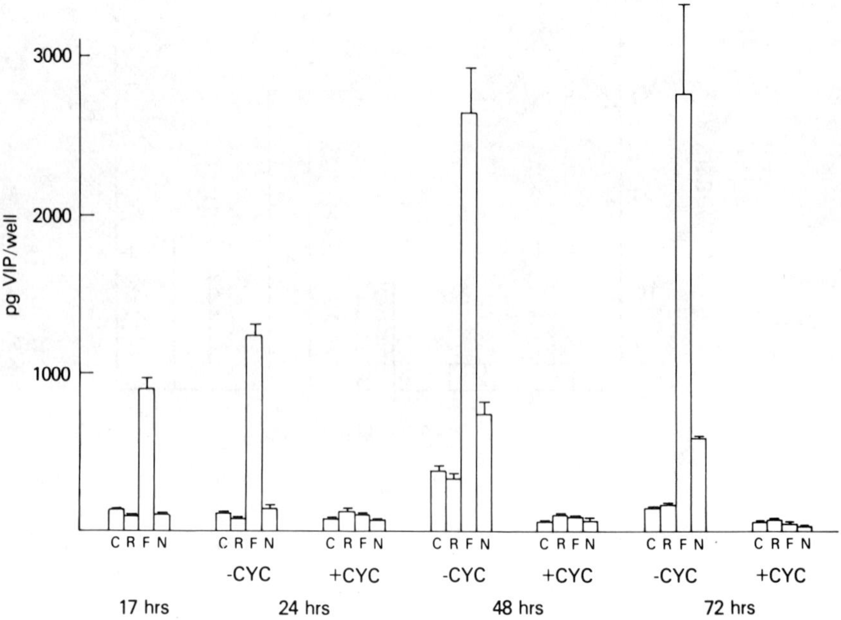

FIGURE 8. Effect of cycloheximide 500 nM (*CYC*) on induction of VIP by various agents in bovine chromaffin cells. *C,* control; *R,* reserpine 5 μM; *F,* forskolin 25 μM; and *N,* nicotine 5 μM. Values (mean ± SEM, n = 3) are total VIP in cell extracts and medium.[3] Drugs were added by removing 0.5 ml medium and replacing with 0.5 ml fresh medium containing the appropriate concentration of drug.

CHROMOGRANIN A

Chromogranin A is the most abundant constituent of the chromaffin vesicle.[12] There is about twenty times more chromogranin A than enkephalin, on a mass basis, in the chromaffin cell. Whereas chromogranin A, catecholamines, VIP, and enkephalins are released in parallel from the secretory vesicle in response to nicotinic stimulation (Eiden *et al.*[3] and Eiden, Iacangelo, Hsu, Hotchkiss, Bader, and Aunis, J. Neurochem., in press), chromogranin A biosynthesis seems to be regulated quite independently from that of the other neuropeptides. Long-term exposure to nicotine or forskolin, conditions that cause elevation of enkephalin and VIP as well as enkephalin mRNA, cause only a depletion of chromogranin A immunoreactivity from chromaffin cells in primary culture, with no evidence of a compensatory increase in biosynthesis (FIG. 11). It is possible that chromogranin A biosynthesis does not occur in culture at all, perhaps due to the absence of a factor that is required *in vivo.* This seems unlikely, since in fact chromogranin A mRNA levels are maintained at *in vivo* levels over several days in culture (Eiden, unpublished observations). A factor permissive for enhanced chromogranin synthesis that is present *in vivo* but absent in culture, however, may exist. Alternatively, chromogranin A may represent a class of vesicular constituents whose rate of synthesis is not coupled directly to secretion. If this is the case, then the ratio of enkephalin or VIP to chromogranin A contained within a chromaffin cell is a function of the secretory history of the cell, and a cell stimulated by repeated splanchnic firing would be expected to have a progressively lower ratio of chromogranin A to enkephalin with time, as shown in FIGURE 11 for chromaffin cells in culture continuously exposed to nicotine. This change in the ratio of secretory products within the cell may cause a corresponding change in the ratio of products actually secreted from the cell. A shift in the pattern of peptides secreted from an endocrine organ as a function of intensity and duration of stimulation may in some cases abet organismic homeostasis.

SUMMARY

Enkephalins, vasoactive intestinal polypeptide, and chromogranin A are all contained in the secretory vesicles of chromaffin cells in culture, and are all released from this compartment by secretagogues in a calcium-dependent way. The biosynthesis of each of these peptides, however, is under quite independent regulation. The synthesis and secretion of enkephalin is tightly coupled to acetylcholine and elevated potassium stimulation by calcium influx. Once calcium enters the cell, calcium acts at pharmacologically distinct sites to elicit secretion and enhanced biosynthesis of Met-enkephalin. This is demonstrated by the calcium-independent stimulation of enkephalin secretion by 1 mM barium, in contrast to the dependence on extracellular calcium of

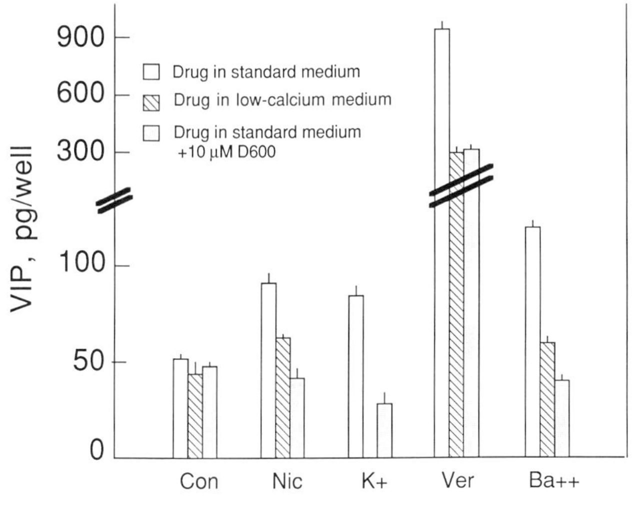

FIGURE 9. Effect of D600 (10 μM) and medium containing reduced calcium (approximately 100 μM versus 1.8 mM control) on VIP immunoreactivity (cell plus medium) stimulated by various agents for 72 hours. Data are pooled from several experiments having similar control values. Drug doses are: nicotine, 100 μM; KCl, 40 mM (replacing equimolar amount of NaCl); veratridine, 10 μM; and barium chloride, 1 mM. Other details are as indicated in FIGURE 3. SEM values are indicated, n = 3 to 6.

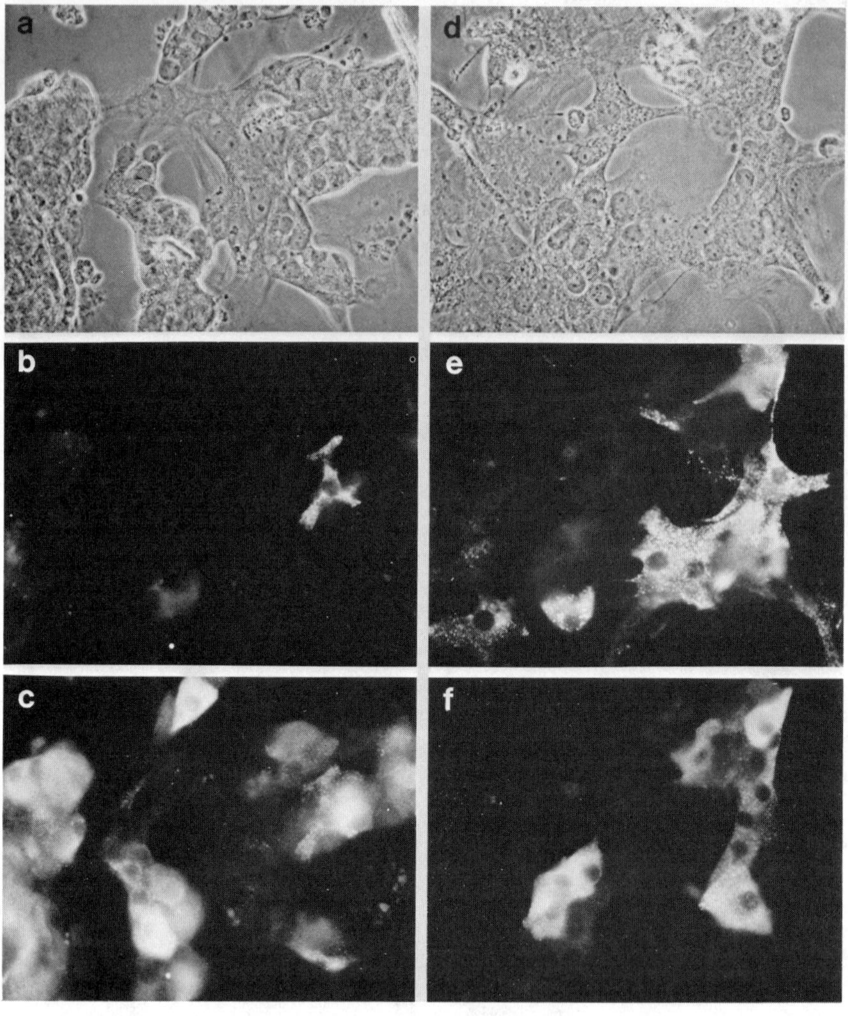

FIGURE 10. Immunocytochemical comparison of Met-enkephalin and VIP-positive cells in control cultures (*left*) and cultures treated with the combination of forskolin 25 μM, and the phorbol ester, 12-*O*-tetradecanoyl-phorbol-13-acetate (TPA), 5×10^{-8} M (*right*). Top photos are phase contrast. Middle and bottom photos are cells stained for VIP and Met-enkephalin, respectively. For details, see Pruss *et al.*[11]

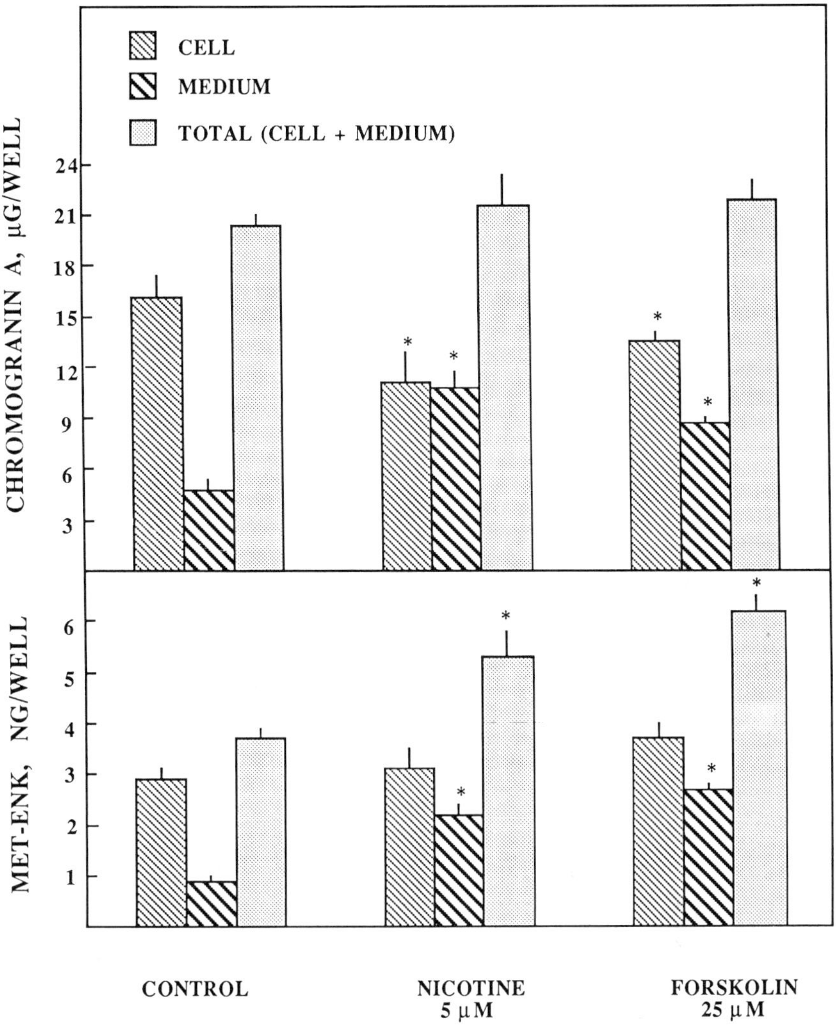

FIGURE 11. Effect of nicotine and forskolin on chromogranin A and Met-enkephalin biosynthesis in bovine chromaffin cells. Peptides were measured in cell extracts and medium by radioimmunoassay 72 hours after the addition of fresh medium, or forskolin (25 μM) or nicotine (5 μM) in fresh medium, to cultured cells plated at 300,000 cells per ml per well for 3 days.

barium-stimulated biosynthesis of this peptide. The synthesis and secretion of VIP is also coupled to acetylcholine and elevated potassium stimulation by calcium influx. Treatment with barium demonstrates that calcium acts at distinct sites to stimulate secretion and biosynthesis of this peptide; however induction of VIP by barium and veratridine shows greater sensitivity to the calcium channel blocker methoxyverapamil (D600) than does the induction of Met-enkephalin by these agents. These differences in D600 sensitivity may be due to differences in calcium metabolism or voltage-dependent calcium channels in enkephalin-producing and VIP-inducible subpopulations of chromaffin cells. Chromogranin A levels are essentially unaffected by any of the agents which increase enkephalin and VIP levels, although it is secreted in parallel with enkephalins and catecholamines from chromaffin cells in response to secretagogues. We suggest that peptide hormones such as VIP and enkephalins are regulated by calcium-dependent stimulus-secretion-synthesis coupling in the chromaffin cell. Cyclic AMP is a positive regulator of enkephalin and VIP biosynthesis, but does not affect acute release of these peptides. The cAMP/protein kinase A system may be a distal mediator of peptide biosynthesis stimulated by secretagogues. Alternatively, cAMP may be involved in early developmental establishment of phenotype or long-term regulation of peptide biosynthesis by other hormones or neurotransmitters. Chromogranin A may represent a class of intravesicular, soluble proteins that are expressed constitutively by the chromaffin cell in the presence or absence of positive regulators of other systems. The biosynthesis of chromogranin A may be coupled to the production or assembly of the secretory vesicle itself.

REFERENCES

1. VIVEROS, O. H., E. J. DILIBERTO, JR., E. HAZEM & K.-J. CHANG. 1979. Opiate-like materials in the adrenal medulla: Evidence for storage and secretion with catecholamines. Mol. Pharmacol. **16:** 1101-1108.
2. LIVETT, B. G., D. M. DEAN, L. G. WHELAN, S. UDENFRIEND & J. ROSSIER. 1981. Co-release of enkephalin and catecholamines from cultured chromaffin cells. Nature **289:** 317-319.
3. EIDEN, L. E., R. L. ESKAY, J. SCOTT, H. POLLARD & A. J. HOTCHKISS. 1983. Primary cultures of bovine chromaffin cells synthesize and secrete vasoactive intestinal polypeptide (VIP). Life Sci. **33:** 687-693.
4. EIDEN, L. E., P. GIRAUD, J. R. DAVE, A. J. HOTCHKISS & H.-U. AFFOLTER. 1984. Nicotinic receptor stimulation activates enkephalin release and biosynthesis in adrenal chromaffin cells. Nature **312:** 661-663.
5. GUBLER, U., P. SEEBURG, B. J. HOFFMAN, L. P. GAGE & S. UDENFRIEND. 1982. Molecular cloning establishes proenkephalin as precursor of enkephalin-containing peptides. Nature **295:** 206-208.
6. LIVETT, B. G. 1984. The secretory process in adrenal medullary cells. *In* Cell Biology of the Secretory Process. M. Cantin, Ed.: 309-358. Karger. Basel.
7. GUIDOTTI, A. & E. COSTA. 1979. Involvement of adenosine 3′5′-monophosphate in the activation of tyrosine hydroxylase elicited by drugs. Science **179:** 902-904.
8. AFFOLTER, H.-U., P. GIRAUD, A. J. HOTCHKISS & L. E. EIDEN. 1984. Stimulus-secretion-synthesis coupling: A model for cholinergic regulation of enkephalin secretion and gene transcription in adrenomedullary chromaffin cells. *In* Opiate Peptides in the Periphery, F. Fraioli, Ed.: 23-30. Elsevier. Amsterdam.
9. QUACH, T. T., F. TANG, H. KAGEYAMA, I. MOCHETTI, A. GUIDOTTI, J. MEEK, E. COSTA & J. SCHWARTZ. 1984. Enkephalin biosynthesis in the adrenal medulla. Mol. Pharmacol. **26:** 255-260.

10. KENIGSBERG, R. L. & J. M. TRIFARO. 1985. Microinjection of calmodulin antibodies into cultured chromaffin cells blocks catecholamine release in response to stimulation. Neuroscience **14:** 335-347.
11. PRUSS, R. M., J. R. MOSKAL, L. E. EIDEN & M. C. BEINFELD. 1985. Specific regulation of vasoactive intestinal polypeptide biosynthesis by phorbol ester in bovine chromaffin cells. Endocrinology **117:** 1020-1026.
12. WINKLER, H. 1976. The composition of adrenal chromaffin granules: An assessment of controversial results. Neuroscience **1:** 65-80.

DISCUSSION OF THE PAPER

A. GINTZLER: (*SUNY Downstate Medical Center, Brooklyn, N.Y.*) Since most agents that cause an increase in enkephalin mRNA content also cause release, do your think that release per se is the stimulus for increased synthesis?

L. EIDEN (*NIMH, National Institutes of Health, Bethesda, Md.*): We have eliminated the possibility that material once released feeds back on the cell to cause an increase in synthesis. Hexamethonium blocks nicotine-induced increase in enkephalin mRNA, even if added 60 min after the addition of nicotine, at which time nicotine-stimulated release has already occurred.

The Regulation of Enkephalin Levels in Adrenomedullary Cells and Its Relation to Chromaffin Vesicle Biogenesis and Functional Plasticity

O. H. VIVEROS,[a] E. J. DILIBERTO, JR.,[a] J.-H. HONG,[b]
J. S. KIZER,[c] C. D. UNSWORTH,[a,d] AND
T. KANAMATSU[b,e]

[a] *Laboratory of Cellular Neurochemistry*
Department of Medicinal Biochemistry
The Wellcome Research Laboratories
Research Triangle Park, North Carolina 27709

[b] *Laboratory of Behavioral and Neurological Toxicology*
National Institute of Environmental Health Sciences
Research Triangle Park, North Carolina 27709

[c] *Departments of Medicine and Pharmacology*
and
Biological Sciences Research Center
School of Medicine
University of North Carolina
Chapel Hill, North Carolina 27514

[d] *Howard Hughes Medical Institute*
Massachusetts General Hospital
Boston, Massachusetts 02114

[e] *Second Department of Physiology*
Toho University School of Medicine
Tokyo, Japan

A large number of different neuropeptides have been isolated from the chromaffin cells of the adrenal medulla of different vertebrates (for a review, see Unsworth and Viveros[1]). Several of these neuropeptides may coexist in the same adrenomedullary cell, but the possible combinations of different peptides coexisting in the same cell and their relative proportions of coexisting peptides show marked species variation. Nevertheless, in spite of these interspecies differences, enkephalins and other peptides derived from proenkephalin A (PEA) are always found in significant amounts.[2–5] Of the many adrenomedullary neuropeptides, the PEA-derived opioid peptides (OP) are currently the best studied and are the only ones where coexistence with catecholamines (CA) in chromaffin vesicles[3–7] and cosecretion with CA in the same proportion as they are contained in the storage vesicles[3,8–14] have been unequivocally demonstrated. After the cloning of preproenkephalin A mRNA (PPEA mRNA) from bovine adrenal medulla[15,16] and human pheochromocytoma,[17] significant efforts from several laboratories have been invested in the isolation and characterization of the enzymes involved in PPEA processing[18–24] and in determining the factors involved in the regulation of PPEA biosynthesis.[8,9,14,25–34] Our own efforts have concentrated on the regulation of PEA biosynthesis and processing in intact animals and in chromaffin cells in culture. We will review here some of our initial studies and present recent evidence for the involvement of multiple regulatory steps in the biosynthesis and processing of OP in adrenomedullary chromaffin cells.

The Effect of Increased Splanchnic Nerve Activity and Catecholamine-depleting Agents on Opioid Peptide Levels and Preproenkephalin A mRNA in the Adrenal Medulla in Situ

In early experiments,[8] it was found that a brief period of increased reflex splanchnic discharge in the cat, induced by insulin hypoglycemia, produced a proportional decrease in the content of CA and of native OP (NOP, peptides that without trypsin-carboxypeptidase B digestion displace ^{125}I-labeled [D-Ala^2, D-Leu^5]-enkephalin from rat brain membrane opiate receptors). This decrease in CA and OP results from the cosecretion of CA and OP. In the guinea pig,[8] this initial decrease is followed by an increase in OP that precedes the recovery of CA content and reaches levels twofold above control 6 days after insulin. A similar long-lasting increase in NOP and in total OP (TOP, OP activity measured with the radioreceptor assay in extracts previously digested with trypsin and carboxypeptidase B to release all the Leu- and Met-enkephalins contained in extended peptides) after a short period of increased splanchnic discharge has also been found in the rat.[30,34] In these latter experiments, the increase in cryptic OP (COP) (calculated as the difference of TOP-NOP) precedes the rise in NOP.

The early evidence that increased neurogenic stimulation enhanced the biosynthesis of PEA and its products has been further confirmed by studying PPEA mRNA levels after a brief period of stress. FIGURE 1 shows a Northern blot for PPEA mRNA from control rats and animals that have been through a 2-h period of hypoglycemia followed by 24 h of recovery before sacrifice. A single mRNA species of approximately 1.5 kilobases hybridizes with the ^{32}P-labeled cDNA PPEA probe in control and in insulin-pretreated animals. The intensity of the band from insulin-treated rats appears markedly increased. FIGURE 2 shows the time course of the changes in CA, NOP, COP, and PPEA mRNA after a 2-h period of insulin hypoglycemia. The levels of PPEA mRNA are already fourfold above control at the end of the stimulation period and continue to rise during the first 24 h of recovery. The increase in mRNA clearly precedes the rise in COP and NOP. A steady rise in COP is seen between 5 h and 96 h after insulin injection. The levels of NOP rise initially at a lower rate than COP, but later accelerate such that the ratio of COP/NOP shows a progressive decline between 24 and 96 h after insulin and stays well below control until the end of the experiment (FIG. 2B). The decay in PPEA mRNA has a half-life of approximately 4 days. If medullary PPEA mRNA is equally stable in control animals, a total suppression of mRNA degradation would give a doubling in mRNA levels after 24 h rather than the 16-fold change observed in FIGURE 2. These data strongly suggest that neurogenic stimulation enhances PPEA gene transcription. The initial increase in COP/NOP ratio is to be expected from a precursor-product relationship, but the increase in the rate of NOP accumulation with a marked decrease in COP/NOP

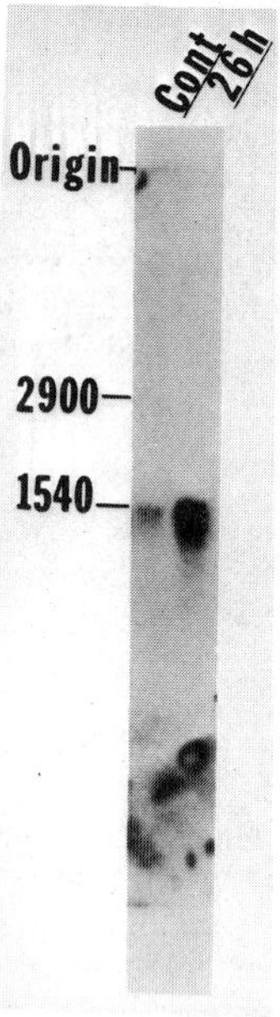

FIGURE 1. Northern blot of rat adrenal medulla preproenkephalin mRNA. Adrenal glands were rapidly removed from the animals following decapitation and the medullae freed of cortex under a dissecting microscope in the cold and frozen in liquid N_2. Eight pooled pairs of adrenal medullae were homogenized in 4 M guanidinium thiocyanate and RNA extracted as described.[56] Twenty μg of total RNA from control and treated rats were electrophoresed in a 1.3% agarose gel and transferred to a Gene Screen membrane. The membrane was hybridized with a ^{32}P-labeled cDNA probe coding for PPEA derived from rat brain. The left lane corresponds to total RNA from control animals. The right lane is an equivalent amount of total RNA extracted from adrenal medullae that had been reflexly stimulated by 2 h of insulin hypoglycemia followed by 24 h of recovery. Note that there is a single mRNA species of approximately 1.5 kilobases that hybridizes with the PPEA rat probe and that the intensity of the band is greatly increased after neurogenic stimulation of the gland.

ratio, while COP levels continue to rise, suggests that neurogenic stimulation results in a delayed increase in the rate of processing in addition to the increased rate of PEA biosynthesis (see following section).

Amidorphin and metorphamide (adrenophin), two amidated opioid peptides derived by posttranslational processing of PEA, have recently been identified in brain and in adrenal medulla.[35,38] These peptides are generated by the enzyme peptidylglycine α-amidating monooxygenase (PAM), which was first isolated from the pituitary[39,57] and is also present in brain.[40] Peptides of the general structure AA_n-Gly-COOH are substrates for PAM. There are two internal peptides in PEA where hydrolysis of paired basic amino acids would yield possible substrates for PAM. One corresponds to the first 26 amino acids on the amino-terminal side of peptide F and generates amidorphin and the second corresponds to peptide I^{15-24} and generates metorphamide. We have recently found PAM to be present in bovine and rat chromaffin vesicles (Diliberto, Kizer and Viveros, unpublished observations). In the rat, the enzyme activity distributes approximately in equal amounts between the membrane and soluble fractions. FIGURE 3 shows a steady rise in PAM specific activity in the large granular fraction that contains the intact chromaffin vesicles from the end of a 2-h period of insulin shock through 45 h following the termination of increased splanchnic discharge. The results indicate that neurogenic stimulation, in addition to increasing PEA biosynthesis, activates and/or induces PAM in the CA-enkephalin storage vesicles.

As shown in FIGURE 4, immobilization stress, which is also known to activate splanchnic nerve discharge,[41] also results in decreased CA content and increased PPEA mRNA. The earliest time after neurogenic stimulation in which an increase in PPEA mRNA is detected is 2 h (FIG. 5). Nevertheless, much shorter times of stimulation must be sufficient to increase transcription, as seen in FIGURE 6. Shortening the period of hypoglycemia to 45 min does not show an immediate rise in PPEA mRNA (FIG. 5), but when rats are in shock for 45 min followed by 1½ h of recovery, there is a fourfold increase in the levels of PPEA mRNA (FIG. 6). All of these changes are

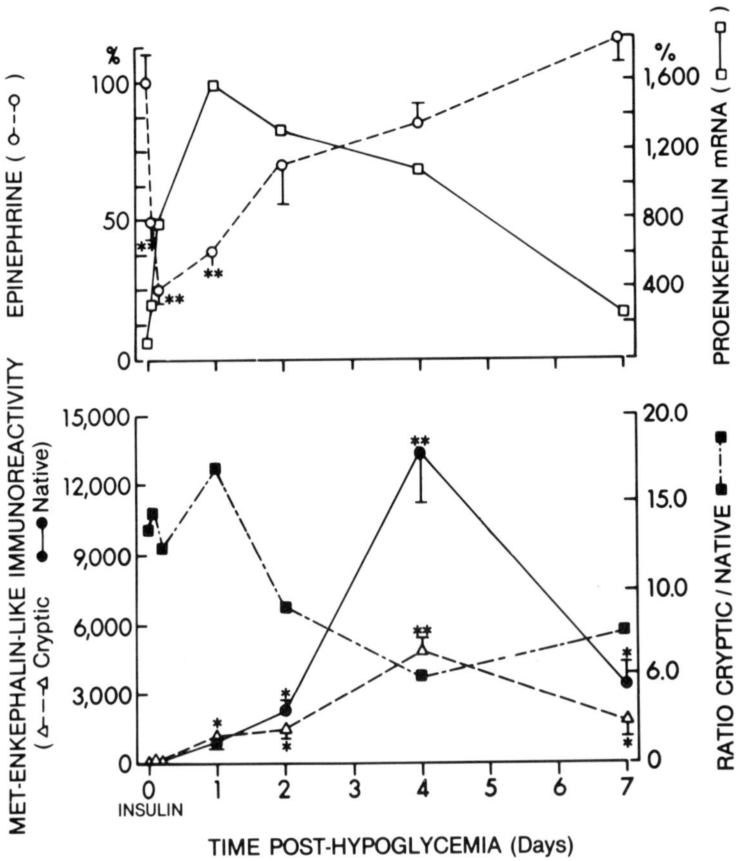

FIGURE 2. The effect of a 2-h period of increased splanchnic firing on the levels of epinephrine, Met-enkephalin, and preproenkephalin mRNA in the rat adrenal medulla. After 2 h of insulin (10 U/kg, ip) hypoglycemia, rats were recovered by administration of sucrose and food ad libitum and sacrificed at different times thereafter. For opioid peptide assay,[31] individual adrenal medulla pairs were homogenized in 0.5 M acetic acid; aliquots were treated with perchloric acid (final concentration, 0.5 M) for protein and catecholamine determinations. Native and cryptic opioid peptides are measured without or with previous partial digestion with trypsin and carboxypeptidase B, respectively. Unstimulated values were: epinephrine, 1.97 ± 0.21 μmol/μg protein; native opioid peptides, 0.9 ± 0.3 pmol/mg protein; cryptic opioid peptides, 12.1 pmol/mg protein (mRNA is expressed in relative densitometric units per μg of total RNA). For epinephrine and opioid peptides, $n = 8$, *$p < 0.02$, **$p < 0.005$. Each mRNA value obtained from a pool of eight medulla pairs.

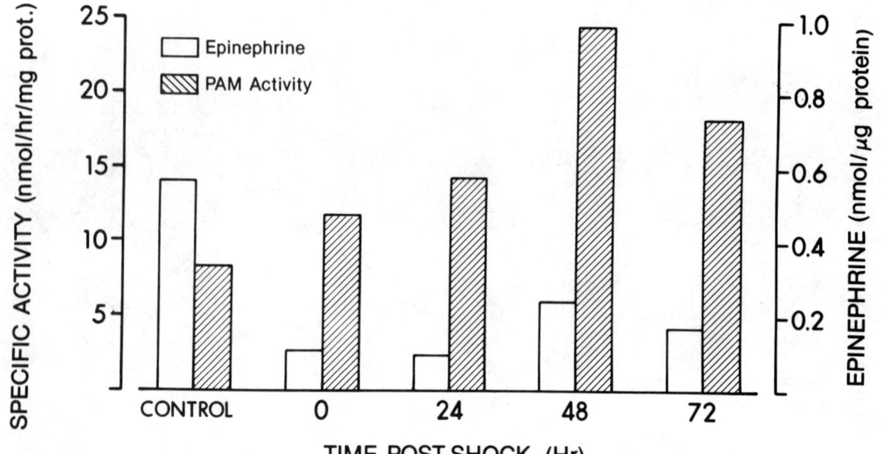

FIGURE 3. Peptidyl-glycine α-amidating monooxygenase activity in the rat adrenal medulla following 2 h of insulin-induced hypoglycemia. Animals were treated as described in legend to FIGURE 2. Adrenal medullae from five animals were homogenized in 300 μl of 0.3 M sucrose containing 10 mM Hepes and 1 mM EDTA, pH 7.2, and centrifuged at 800 ×g for 10 min. The pellet was rehomogenized in 200 μl buffered sucrose and centrifuged again at 800 ×g for 10 min. The combined supernatants were centrifuged at 26,000 ×g for 20 min to obtain a crude chromaffin vesicle (P_2) fraction and supernatant (S_2), Aliquots of each fraction were taken for catecholamine, proteins, and α-amidating monooxygenase activity. Enzyme activity in the P_2 fraction was estimated as previously described.[40]

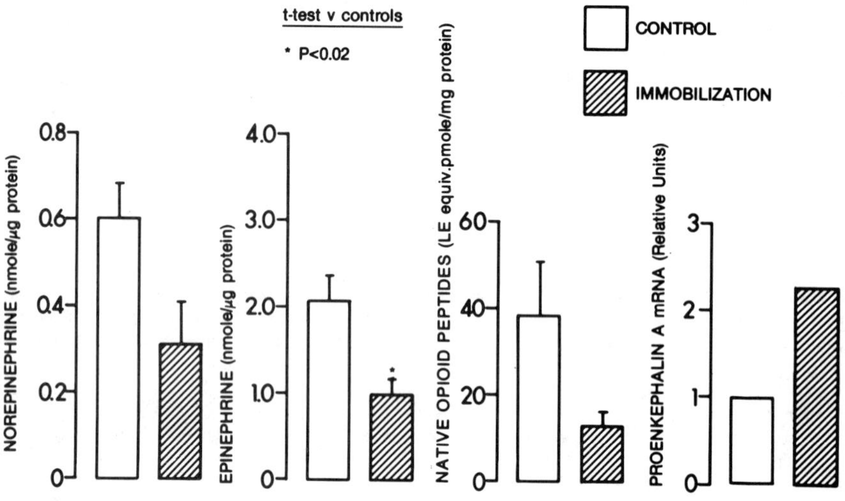

FIGURE 4. The effect of 2 h of immobilization stress on the catecholamine, opioid peptide, and proenkephalin mRNA levels in the rat adrenal medulla. Rats were sacrificed after 2 h of immobilization. For each mRNA determination, total RNA was extracted from eight pairs of pooled frozen glands. The mRNA values are the means of determinations on two sets of pooled medullae per point. For catecholamines and opioid peptides, $n = 5$.

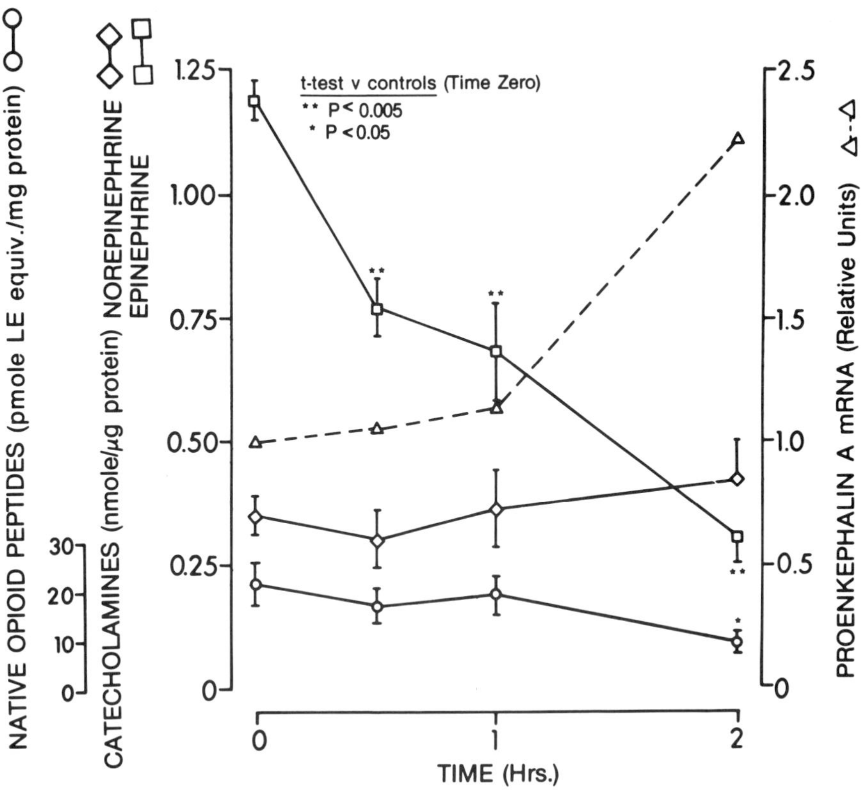

FIGURE 5. Short-term effects of insulin hypoglycemia on the levels of proenkephalin A mRNA and catecholamines in the rat adrenal medulla. Rats were injected with insulin (10 U/kg, sc) and were sacrificed 30, 60, and 120 min later; $n = 8$, $^{*}p < 0.05$, $^{**}p < 0.005$. Each value represents the $mRNA^{enk}$ from a pool of eight rats. The experiment was replicated once.

markedly blunted when the cholinergic splanchnic-medullary synaptic transmission is impaired by the combined treatment of the rats with chlorisondamine and atropine.

In apparent contrast to the above results, others have reported increased levels of PPEA mRNA and PEA-derived peptides after splanchnic denervation or after keeping adrenal medullae in organ culture.[2,42,43] However, those results may be explained, at least in part, by the surgical stress during the denervation procedures and by the initial changes that occur on explanting medullae into tissue culture. When care is taken to reduce anesthetic and surgical stress during splanchnic denervation to a minimum, we have not observed an early decrease in CA or in OP levels and there is no increase in OP later.[30] When more extensive and prolonged surgery is performed for unilateral splanchnicectomy, CA and OP show an initial decline on both the innervated and denervated glands. During recovery from surgery, there is a fast increase in NOP and COP in the decentralized gland. In the innervated gland, the levels of OP and CA stay below control for several days, probably an indication of protracted increase in splanchnic activity. Similarly, explanting rat adrenal medullae into culture dishes produces a massive release of CA and OP into the medium with decreased levels in the explant.[30] The decrease in OP levels is soon followed by a marked rise over the initial levels found in nonexplanted glands and may be a consequence of CA depletion rather than of denervation, as will be discussed below.

Large doses of reserpine deplete adrenomedullary catecholamines by blockade of the vesicular amine transport and by increased splanchnic discharge,[44,45] and conse-

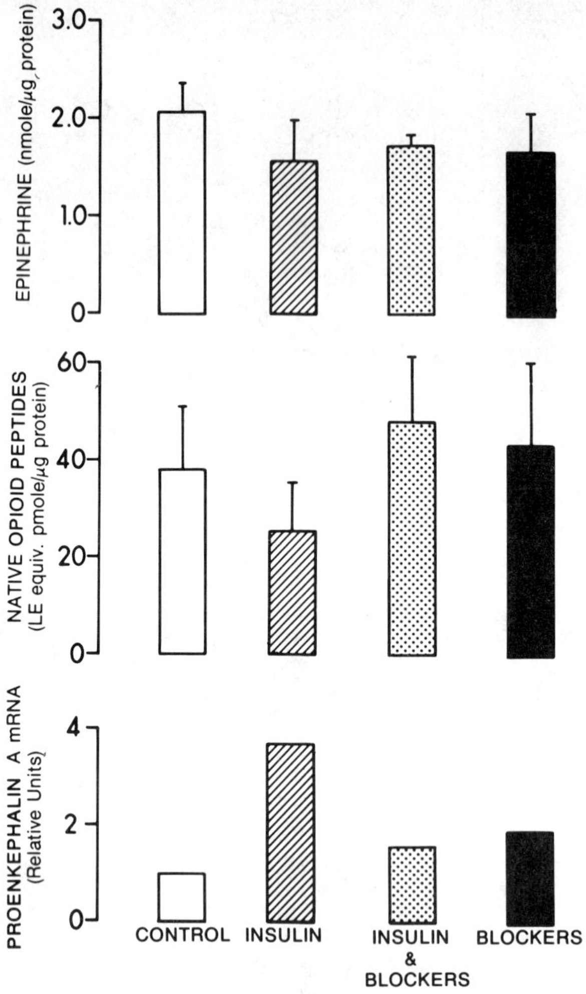

FIGURE 6. Epinephrine, opioid peptide, and proenkephalin mRNA levels after a brief period of increased splanchnic firing induced by insulin hypoglycemia: blockade by cholinergic antagonists. Fasted rats were injected with insulin (10 U/kg, ip); 45 min later, the hypoglycemia was terminated by administration of 2 ml of 40% sucrose by stomach tube and 1 ml of 20% dextrose, ip. Animals were sacrificed 135 min after the insulin injection. To block the splanchnic-adrenomedullary synapse, rats were treated with chlorisondamine (5 mg/kg, ip) and atropine (1 mg/kg, ip) 15 min prior to and again 45 min after the injection of insulin. For assay procedures, see legend to FIGURE 2.

quently deplete intravesicular soluble proteins, including soluble dopamine β-hydroxylase (DBH). Pharmacological blockade of the splanchnic adrenomedullary synapse suppresses DBH depletion induced by large doses of reserpine, but merely slows the rate of CA depletion.[45] By administering small doses of reserpine one day apart, it is possible to obtain CA depletion without concomitant secretion of vesicular components and initial maintenance of DBH levels.[44] Within a few days of two small doses of reserpine and preceding the recovery of vesicular CA uptake and tissue levels of CA, there is a marked increase in soluble and membrane-bound DBH, indicative of formation of new CA storage vesicles.[44,45] A similar study on the effect of CA depletion on medullary OP in cats and guinea pigs demonstrated that there was no initial depletion of OP but only a late increase in OP.[8] This increase in OP observed after reserpine *in vivo* is accompanied by a decrease[46] or no change (Unsworth, Hong, Diliberto and Viveros, unpublished) in PPEA mRNA. The increase in OP must then result from increased efficiency of translation and/or increased enkephalin precursor processing.

In summary, the experiments on regulation of enkephalin levels in the adrenal medulla *in situ* suggest that while CA depletion may increase enkephalin biosynthesis, acting mainly at a translational or posttranslational level, neurogenic stimulation, in addition to a possible action through CA depletion, also acts at the transcriptional level through the activation of cholinergic receptors. To further study the mechanisms involved in this regulation, several laboratories, including our own, have made use of bovine chromaffin cells in primary culture (BAMC).

Studies on Proenkephalin A Biosynthesis and Processing in Primary Cultures of Adrenomedullary Chromaffin Cells

Although reports from the various laboratories using BAMC in culture are generally in close agreement, there are a few differences that may arise from different culture conditions such as plating density, proportion of cortical and other cells in the cultures, addition of serum or other conditioning factors for plating efficiency, and survival and use of antibiotics and antimitotics.[47] The majority of our own results have been obtained from high-density cultures that are plated in the presence of serum to enhance cell attachment, but from which serum is subsequently removed within 48 to 72 h of plating and at least 24 h before the experiment is initiated. Under these conditions of absence of serum and mitotic inhibitors[47] the cells are viable, and they synthesize, store, secrete, and maintain the original levels of CA and OP for several weeks. Also, cellular protein content remains stable even in the absence of mitotic inhibitors and the cells maintain a rounded appearance with little or no extension of processes. When cells are plated at low density, process extension is enhanced and the presence of serum induces large increases of total cell protein (TABLE 1) even in the presence of mitotic inhibitors, and also markedly increases NOP and COP above initial levels when expressed on a 10^6 cell basis.

Wilson *et al.*[25] were the first to report that chromaffin cells synthesize Met- and Leu-enkephalin. In the same studies, they demonstrated that addition of reserpine to the cultures accelerated the *de novo* biosynthesis and increased the ratio of total and newly synthesized Met-enkephalin/Leu-enkephalin. These results suggest a concomitant acceleration of the rate of PEA biosynthesis and of the rate of processing of precursors to enkephalins. The increase in OP was markedly inhibited by cycloheximide and actinomycin D.[27] Other agents that deplete CA by interfering with synthesis or vesicular uptake were also found to increase the level of NOP and TOP in the cultures, while COP would not change[30,31] or would decrease, apparently depending on the culturing conditions (TABLE 1). Agents like forskolin and inhibitors of phosphodiesterase that increase endogenous levels of cAMP, or the direction addition of the cyclic nucleotide to these cultures produce a preferential increase in COP levels. High concentrations of insulin[48] or culturing of the cells in the presence of serum not only markedly increased TOP, COP, and NOP, but also increased total protein contents in the cultures (TABLE 1). When the enkephalins and the small- to mid-sized enkephalin-extended COP from cultures treated with CA-depleting drugs are separated by reversed-phase HPLC, there is a marked increase in Met- and Leu-enkephalin with a corresponding decrease in COP (FIG. 7), confirming that CA depletion accelerates PEA processing. Hook *et al.*[49] have recently reported that the V_{max} of carboxypeptidase H, an enzyme found in chromaffin vesicles that removes carboxyterminal basic amino acids from enkephalin hexapeptides,[22,23] is significantly increased and the substrate K_m decreased between 24 and 72 h of adding reserpine to the BAMC cultures. No increase in carboxypeptidase H immunoreactivity was observed.

Increasing endogenous levels of cAMP or its addition to BAMC cultures increases PPEA mRNA.[28,29,32] Of the agents that deplete CA, tetrabenazine consistently increases PPEA mRNA[34] (FIG. 8), inhibitors of CA synthesis produce no change, and reserpine produces no change (FIG. 8) or decreases mRNA,[32] suggesting that the increase in *de novo* synthesis of enkephalins induced by reserpine[25] may result from an improved efficiency of translation.[32]

Depolarization of BAMC in culture by high concentrations of potassium and activation of nicotinic receptors also increases PPEA mRNA levels and OP by a Ca^{2+}-dependent mechanism.[33,50] Thus, the experiments on cultured cells confirm the *in vivo* observations that regulation of enkephalin synthesis can be acted upon at the transcriptional, translational, and posttranslational levels. Furthermore, it becomes apparent that both cAMP-dependent and cAMP-independent mechanisms are involved in this regulation.

Biogenesis and Reuse of Chromaffin Vesicles in Bovine Chromaffin Cells in Culture

Studies with Dopamine β-Hydroxylase

In the bovine adrenal medulla, approximately 50% of the dopamine β-hydroxylase (DBH) activity is associated with the chromaffin vesicle membrane and the rest is soluble inside the vesicle.[44] By monitoring soluble- and membrane-bound DBH, CA uptake, and changes in the buoyant density of vesicles with different CA content, it was proposed that recovery of function of chromaffin cells *in vivo* after CA depletion by intense neurogenic stimulation[44,51] or by reserpine[44,45] requires resynthesis of both the soluble and membrane components of these organelles. Milder forms of stimulation may result in the reuse of the vesicle membranes that have been emptied through a previous cycle(s) of exocytosis.

Repeated injections of reserpine (once every other day for two weeks) result in an increase in total and in membrane-bound DBH activity but no apparent increase in the number and dimensions of storage vesicles in the rat adrenal medulla.[52] These

TABLE 1. Modulation of the Effect of Tetrabenazine and Forskolin on Opiod Peptide Levels of Bovine Chromaffin Cells in Culture by Fetal Calf Serum

	No Serum				Fetal Calf Serum			
Treatments	Protein ($\mu g/10^6$ Cells)	TOP (pmol/mg Protein)	COP (pmol/mg Protein)	NOP (pmol/mg Protein)	Protein ($\mu g/10^6$ Cells)	TOP (pmol/mg Protein)	COP (pmol/mg Protein)	NOP (pmol/mg Protein)
Control $n = 4$	132.7 ± 8.1^a	597.4 ± 35.2	435.1 ± 24.2	162.3 ± 12.8	278.8 ± 12.3^a	385.7 ± 10.9	298.4 ± 7.4	87.3 ± 3.9
Solvents $n = 8$	136.6 ± 2.1	557.9 ± 9.8	428.7 ± 9.1	129.2 ± 2.4	—	—	—	—
TBZ $n = 4$	142.5 ± 5.1^a	627.5 ± 31.9^b	361.1 ± 20.6^b	266.4 ± 14.0^c	286.3 ± 9.1^a	345.8 ± 13.9	209.2 ± 5.1^c	136.6 ± 8.9^b
Forskolin $n = 4$	152.9 ± 6.2^a	695.3 ± 26.4^c	565 ± 24.6^c	129.5 ± 6.2	263.2 ± 3.6^a	712.2 ± 17.0^c	557.0 ± 16.7^c	155.2 ± 4.1^c

Bovine adrenal medulla chromaffin cells were plated at 2×10^6 cells/plate in 35-mm plastic dishes in DMEM/F12 medium supplemented with 10% fetal calf serum.[47] Serum was removed from half of the cultures 72 h after plating. Treatments were initiated on day 4 of plating on cells harvested 3 days later as described.[31] The final concentrations of tetrabenazine (TBZ) and forskolin were 50 μM. TBZ was dissolved in 2 mM HCl and 20 μl added per 2 ml of medium. Four control plates received the same volume of acid. Forskolin was dissolved in 100% ethanol and 20 μl added per 2 ml of medium. Four control cultures received the same volume of ethanol. The acid and ethanol control plates did not differ in any of the parameters measured and have been pooled under "solvents." Mitotic inhibitors were not added to the cultures. Notice the large increase in protein content of plates maintained for the 7 days in the presence of serum as compared to those where serum was removed after 3 days of culturing. Also notice the marked decrease in the ratio of NOP/COP in the presence of serum. TBZ significantly increases TOP and NOP in the absence of serum, but only NOP in its presence; forskolin that increases TOP, NOP, and COP in the presence of serum increases only TOP and COP, but not NOP, in its absence.

[a] $p < 0.001$, when protein values are compared with and without serum.
[b] $p \leq 0.05$, when treatments are compared to its corresponding control.
[c] $p \leq 0.001$, when treatments are compared to its corresponding control.

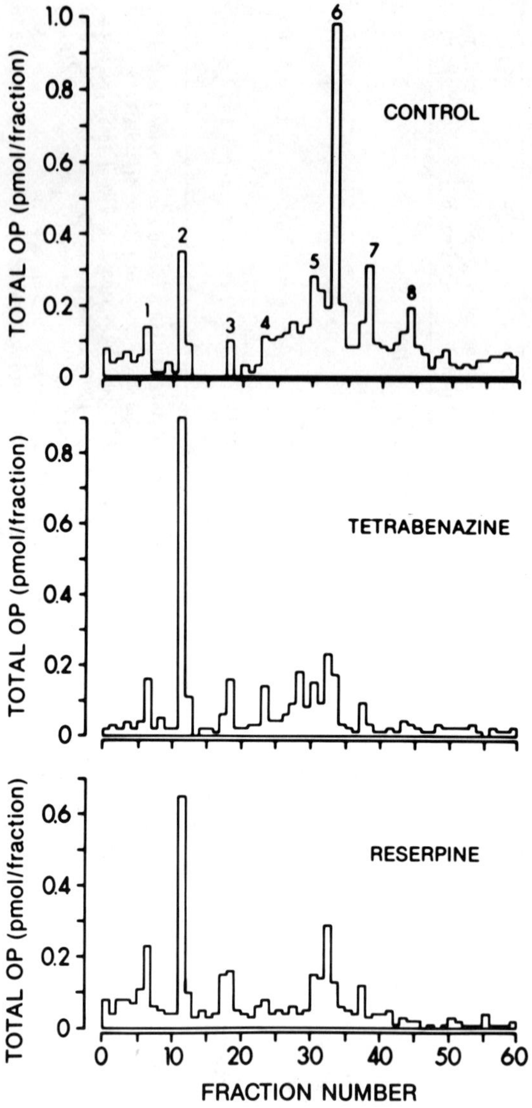

FIGURE 7. Catecholamine depletion by tetrabenazine and reserpine increases the processing of opioid peptides in bovine chromaffin cells in culture. Bovine adrenal medulla chromaffin cells were dispersed, plated, and maintained in a defined media as previously described.[47] After three days of exposure of the cultures to tetrabenazine (100 μM) or reserpine (0.2 μM), cells were extracted with 0.5 M acetic acid. After removing precipitated proteins, the acid extracts were dried, redissolved in HPLC buffer A, applied to a reversed-phase HPLC column (μBondapak, Waters Association) and fractionated with a nonlinear gradient from 16% acetonitrile, 5% tetrahydrofuran, and 0.05% trifluoroacetic acid (buffer A) to 60% acetonitrile, 5% tetrahydrofuran, and 0.05% trifluoroacetic acid (buffer B, Diliberto, Unsworth, and Viveros, in preparation). Each fraction was partially digested with trypsin and carboxypeptidase B prior to the radioreceptor assay to measure total opioid peptides eluted from the column. The different opioid

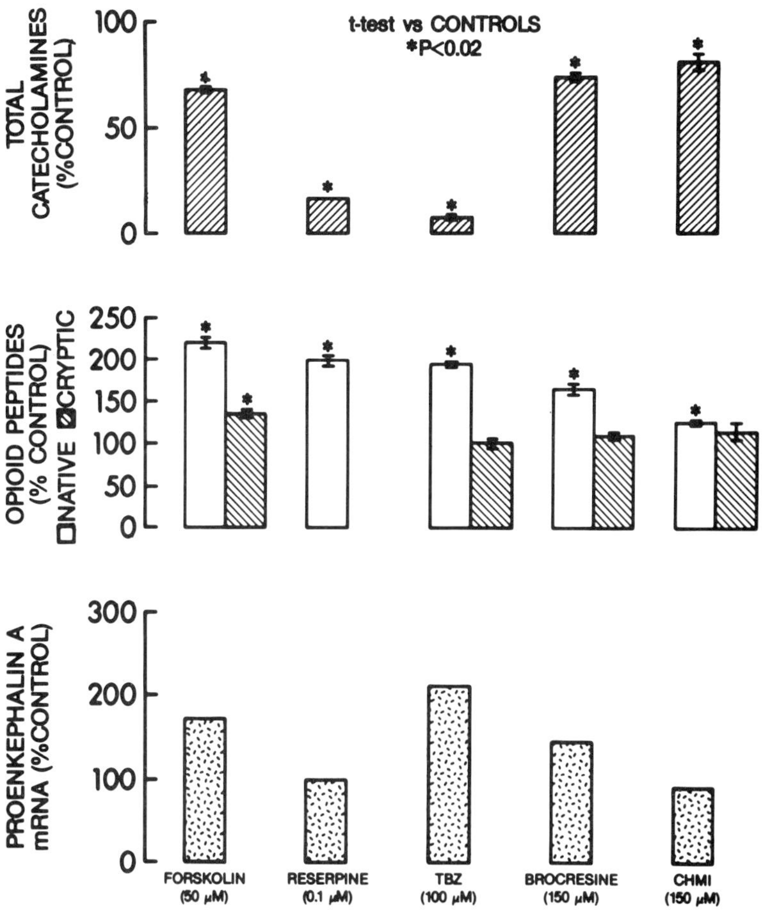

FIGURE 8. The effect of forskolin, catecholamine synthesis inhibitors, and blockers of vesicular catecholamine uptake on opioid peptides and preproenkephalin mRNA on chromaffin cells in culture. Cells were maintained in culture as described in FIGURE 7. Drugs were added to the cultures for three days and changes in catecholamines, opioid peptides, and PPEA mRNA are expressed as percentage of control plates treated with the corresponding vehicles. Assays as in legend to FIGURE 2.

---

peptides have been tentatively characterized by coelution with synthetic standards: peak 1, Met-enkephalin sulfoxide and/or Met-enkephalin-Arg^6; peak 2, Met-enkephalin; peak 3, Leu-enkephalin; peak 4, Met-enkephalin-Arg^6-Phe^7; peak 5, peptide B and/or amidorphin; peak 6, peptide E; peak 7, peptide F; and peak 8, unknown. Met-enkephalin-Arg^6-Gly^7-Leu^8 has a typical retention time of 21.3 min. Note the marked decrease in the intermediate-size opioid peptides with longer retention times (peaks 5-8) and increase in the enkephalin peaks (peaks 1-3) after tetrabenazine or reserpine treatment.

results suggest that the number of vesicles and the composition of the vesicle membrane and vesicle content can be independently regulated. Wilson *et al.*[48] found that most treatments that elevate TOP levels (tetrabenazine, forskolin, 8-bromo-cyclic AMP, and theophylline) failed to increase soluble and membrane-bound DBH measured by enzymatic activity or by immunotitration in bovine adrenomedullary chromaffin cells in culture. The increased amounts of OP induced by these treatments are stored in subcellular particles with properties identical to chromaffin vesicles on density-gradient centrifugation and the OP can be secreted when cells are stimulated by the usual secretagogues. These data are compatible with two interpretations: (1) BAMC in culture may be able to increase the rate of synthesis of new vesicles containing PEA-derived peptides but no DBH, and (2) newly synthesized OP may be inserted into preexisting vesicles. Though it is conceivable that BAMC in culture may make new vesicles with normal TOP content, but no soluble and membrane-bound DBH, this possibility appears to be unlikely since BAMC *ex vivo* (perfused glands) are able to synthesize membrane-bound DBH, soluble DBH, and chromogranins in parallel, and they incorporate these proteins into chromaffin vesicles.[53] Furthermore, BAMC in culture, when exposed to a high concentration of insulin, increase TOP levels and soluble and membrane-bound DBH in approximately the same proportion.[48] If continued experimentation successfully eliminates the first possibility, it may then be concluded that there is an unexpected degree of plasticity of otherwise "mature vesicles," wherein newly synthesized protein is transferred to these preexisting vesicles by an intracellular transport mechanism. Moreover, the failure to increase vesicular DBH by agents that deplete CA or increase intracellular cAMP levels in cultured cells is indicative of other regulatory factor(s) *in situ* that contribute to modulating vesicle biogenesis and composition but are being missed under current cell culture conditions.

Concluding Remarks

FIGURE 9 diagrammatically represents our current level of understanding of the physiological regulation of OP biosynthesis in relation to the biogenesis of other vesicle components. Increased splanchnic firing results in the secretion of acetylcholine at the splanchnic-adrenomedullary synapse. The activation of the cholinergic receptor activates adenylcyclase and initiates the chain of events that culminate in the Ca^{2+}-dependent quantal secretion of the whole soluble content of the chromaffin vesicle.[54] The cellular content of CA is decreased as a consequence of exocytosis and through a yet unknown feedback mechanism enhances PEA processing and biosynthesis at the translational and possibly also transcriptional levels. In parallel to the activation of the secretory pathway, activation of nicotinic receptors increases adenylcyclase activity[55] that seems to act predominantly by increasing PPEA mRNA, probably at the transcriptional level. Under culture conditions, the chromaffin cells may not respond to depletion of CA and increased levels of cAMP by making new chromaffin vesicles, but can modify the composition of preexisting vesicles by inserting into them increased amounts of COP and perhaps also already processed NOP (as may happen under the action of CA-depleting agents). *In situ* the chromaffin cells can make new vesicles to replace empty ones or those functionally damaged after reserpine. These new vesicles apparently can have a relative composition of their soluble components markedly different from preexisting vesicles. Since the new vesicles secrete their content

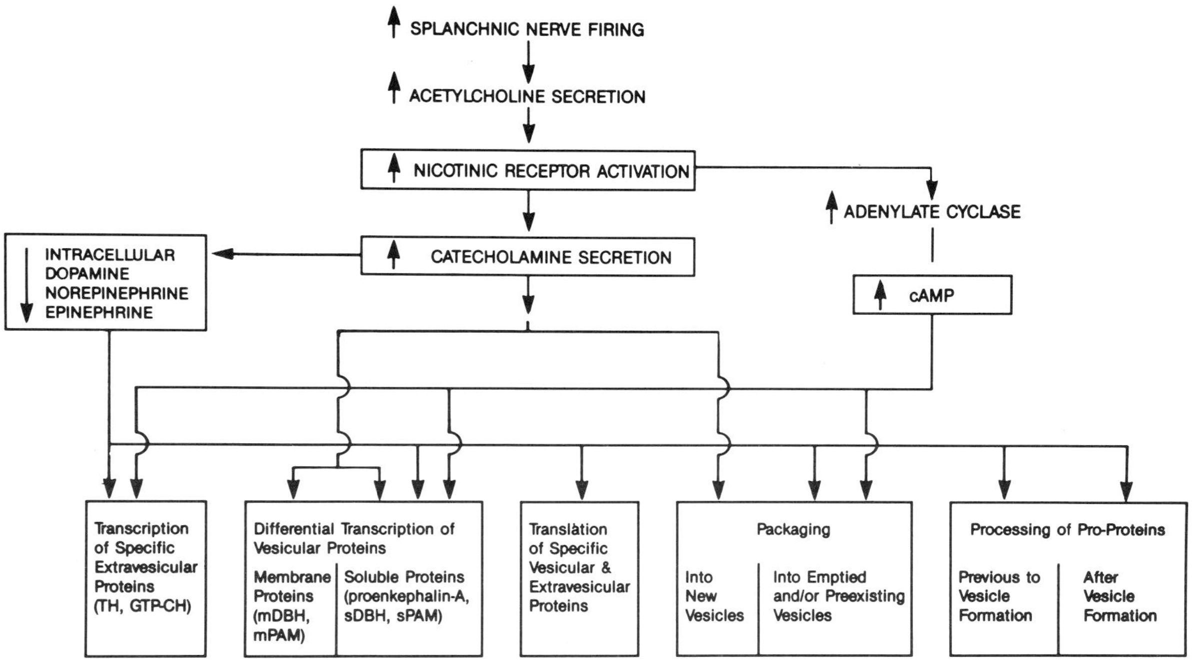

FIGURE 9. Multiple possible sites for the regulation of specific protein synthesis in adrenal medulla chromaffin cells. Splanchnic nerve firing by release of acetylcholine at the synaptic terminal may activate at least three pathways to condition the amount of catecholamines and neuropeptides synthesized, processed, stored, and secreted from chromaffin cells. Under culture conditions, activation of the cAMP-dependent pathway results in effects on OP biosynthesis and processing quite distinct from the effects of lowering intracellular catecholamine levels, and activation of either pathway does not result in increase in the activity or amount of soluble and/or membrane-associated dopamine β-hydroxylase (DBH). Neurogenic stimulation of the gland *in situ* has effects very similar to the activation of the cAMP- and catecholamine-dependent pathways, but in addition markedly increases the amount of soluble and membrane DBH. Abbreviations used are: *TH,* tyrosine hydroxylase; *GTP-CH,* GTP-cyclohydrolase; *PAM,* peptidyl-glycine α-amidating monooxygenase; *downward arrows* and *upward arrows,* mean decrease and increase in that particular element, respectively.

by exocytosis, the composition of the secreted materials and thus the actions of adrenomedullary stimulation on target cells may be importantly determined by the past functional history of the gland. Though it has been demonstrated that chromaffin cells can modify the rate of vesicle biogenesis *in situ,* it is possible that secretory cells may also constantly modify the composition of "mature storage/secretory vesicles" by inserting specific proteins and peptides according to functional demands. If this is a general phenomenon, it will markedly increase the plasticity of intercellular communications systems.

ACKNOWLEDGMENTS

The authors wish to thank Mrs. Cathy Auchter and Dr. Ricardo Rigual for their expert assistance with some of the unpublished experiments included in this review and Dr. Steven P. Wilson for many helpful discussions.

REFERENCES

1. Unsworth, C. D. & O. H. Viveros. 1986. Neuropeptides in the adrenal medulla. *In* Stimulus Secretion Coupling. K. Rosenbeck, Ed. CRC Press. Boca Raton, Fla. In press.
2. Schultzberg, M., J. M. Lundberg, T. Hökfelt, L. Terenius, J. Brandt, R. P. Elde & M. Goldstein. 1978. Enkephalin-like immunoreactivity in gland cells and nerve terminals of the adrenal medulla. Neuroscience **3:** 1169-1186.
3. Viveros, O. H., E. J. Diliberto, Jr., E. Hazum & K.-J. Chang. 1979. Opiate-like materials in the adrenal medulla: Evidence for storage and secretion with catecholamines. Mol. Pharmacol. **16:** 1101-1108.
4. Saria, A., S. P. Wilson, A. Molnar, O. H. Viveros & F. Lembeck. 1980. Substance P and opiate-like peptides in human adrenal medulla. Neurosci. Lett. **20:** 195-200.
5. Hexum, T. D., H.-Y. T. Yang & E. Costa. 1980. Biochemical characterization of enkephalin-like immunoreactive peptides of adrenal glands. Life Sci. **27:** 1211-1216.
6. Lewis, R. V., A. S. Stern, J. Rossier, S. Stein & S. Udenfriend. 1979. Putative enkephalin precursors in bovine adrenal medulla. Biochem. Biophys. Res. Commun. **89:** 822-829.
7. Varndell, I. M., F. J. Tapia, J. De Mey, R. A. Rush, S. R. Bloom & J. Polak. 1982. Electron immunocytochemical localization of enkephalin-like material in catecholamine-containing cells of the carotid body, the adrenal medulla and in pheochromocytomas of man and other mammals. J. Histochem. Cytochem. **30:** 682-690.
8. Viveros, O. H., E. J. Diliberto, Jr., E. Hazum & K.-J. Chang. 1980. Enkephalins as possible adrenomedullary hormones: Storage, secretion, and regulation of synthesis. *In* Neural Peptides and Neuronal Communication. E. Costa & M. Trabucchi, Eds.: 191-204. Raven Press. New York.
9. Viveros, O. H., S. P. Wilson, E. J. Diliberto, Jr., E. Hazum & K.-J. Chang. 1980. Enkephalins in adrenomedullary chromaffin cells and sympathetic nerves. *In* Advances in Physiological Sciences, Vol. 14. Endocrinology, Neuroendocrinology, Neuropeptides II. E. Stark, G. B. Makara, B. Halász & Gy. Rappay, Eds.: 349-353. Pergamon. New York.
10. Hexum, T. D., I. Hanbauer, S. Govoni, H.-Y. T. Yang & E. Costa. 1980. Secretion of enkephalin-like peptides from canine adrenal gland following splanchnic nerve stimulation. Neuropeptides **1:** 137-142.

11. KILPATRICK, D. L., R. V. LEWIS, S. STEIN & S. UDENFRIEND. 1980. Release of enkephalins and enkephalin-containing polypeptides from perfused beef adrenal glands. Proc. Natl. Acad. Sci. **77:** 7473-7475.
12. LIVETT, B. G., D. M. DEAN, L. G. WHELAN, S. UDENFRIEND & J. ROSSIER. 1981. Co-release of enkephalin and catecholamines from cultured adrenal chromaffin cells. Nature **289:** 317-319.
13. WILSON, S. P., K.-J. CHANG & O. H. VIVEROS. 1982. Proportional secretion of opioid peptides and catecholamines from adrenal chromaffin cells in culture. J. Neurosci. **2:** 1150-1156.
14. VIVEROS, O. H. & S. P. WILSON. 1983. The adrenal chromaffin cell as a model to study the co-secretion of enkephalins and catecholamines. J. Autonom. Nerv. Syst. **7:** 41-58.
15. NODA, M., Y. FURUTANI, H. TAKAHASHI, M. TOYOSATO, T. HIROSE, S. INAYAMA, S. NAKANISHI & S. NUMA. 1982. Cloning and sequence analysis of cDNA for bovine adrenal preproenkephalin. Nature **295:** 202-206.
16. GUBLER, U., P. SEEBURG, B. J. HOFFMAN, L. P. GAGE & S. UDENFRIEND. 1982. Molecular cloning establishes proenkephalin as precursor of enkephalin-containing peptides. Nature **295:** 206-208.
17. COMB, M., P. H. SEEBURG, J. ADELMAN, L. EIDEN & E. HERBERT. 1982. Primary structure of the human Met- and Leu-enkephalin precursor and its mRNA. Nature **295:** 663-666.
18. TROY, C. M. & J. M. MUSACCHIO. 1982. Processing of enkephalin precursors by chromaffin granule enzymes. Life Sci. **31:** 1717-1720.
19. LINDBERG, I., H.-Y. T. YANG & E. COSTA. 1982. An enkephalin-generating enzyme in bovine adrenal medulla. Biochem. Biophys. Res. Commun. **106:** 186-193.
20. EVANGELISTA, R., P. RAY & R. V. LEWIS. 1982. A "trypsin-like" enzyme in adrenal chromaffin granules: A proenkephalin processing enzyme. Biochem. Biophys. Res. Commun. **106:** 895-902.
21. MIZUNO, K., A. MIYATA, K. KANGAWA & H. MATSUO. 1982. A unique proenkephalin-converting enzyme purified from bovine adrenal chromaffin granules. Biochem. Biophys. Res. Commun. **108:** 1235-1242.
22. FRICKER, L. D. & S. H. SNYDER. 1982. Enkephalin convertase: Purification and characterization of a specific enkephalin-synthesizing carboxypeptidase localized to adrenal chromaffin granules. Proc. Natl. Acad. Sci. **79:** 3886-3890.
23. HOOK, V. Y. H., L. E. EIDEN & M. J. BROWNSTEIN. A carboxypeptidase processing enzyme for enkephalin precursors. Nature **295:** 341-342.
24. HOOK, V. Y. H. & L. E. EIDEN. 1984. Two peptidases that convert ^{125}I-Lys-Arg-(Met)enkephalin and ^{125}I-(Met)enkephalin-Arg^6, respectively, to ^{125}I-(Met) enkephalin in bovine adrenal medullary chromaffin granules. FEBS Lett. **172:** 212-218.
25. WILSON, S. P., K.-J. CHANG & O. H. VIVEROS. 1980. Synthesis of enkephalins by adrenal medullary chromaffin cells: Reserpine increases incorporation of radiolabeled amino acids. Proc. Natl. Acad. Sci. **77:** 4364-4368.
26. ROSSIER, J., J. M. TRIFARÓ, R. V. LEWIS, R. W. H. LEE, A. STERN, S. KIMURA, S. STEIN & S. UDENFRIEND. 1980. Studies with [^{35}S]methionine indicate that the 22,000-dalton [Met]enkephalin-containing protein in chromaffin cells is a precursor of [Met]enkephalin. Proc. Natl. Acad. Sci. **77:** 6889-6891.
27. WILSON, S. P., M. ABOU-DONIA, K.-J. CHANG & O. H. VIVEROS. 1981. Reserpine increases opiate-like peptide content and tyrosine hydroxylase activity in adrenal medullary chromaffin cells in culture. Neuroscience **6:** 71-79.
28. EIDEN, L. E. & A. J. HOTCHKISS. 1983. Cyclic adenosine monophosphate regulates vasoactive intestinal polypeptide and enkephalin biosynthesis in cultured bovine chromaffin cells. Neuropeptides **4:** 1-9.
29. QUACH, T. T., F. TANG, H. KAGEYAMA, I. MOCCHETTI, A. GUIDOTTI, J. L. MEEK, E. COSTA & J. P. SCHWARTZ. 1984. Enkephalin biosynthesis in adrenal medulla. Modulation of proenkephalin mRNA content of cultured chromaffin cells by 8-bromo-adenosine 3′,5′-monophosphate. Mol. Pharmacol. **26:** 255-260.
30. UNSWORTH, C. D., S. P. WILSON & O. H. VIVEROS. 1984. Enkephalins in the adrenal medulla: Regulation of synthesis and processing by catecholamines and cAMP in bovine chromaffin cells in primary culture and by the splanchnic innervation in the adrenal of the rat. *In* Endocrinology. F. Labrie & L. Proulx, Eds.: 993-998. Elsevier. New York.

31. WILSON, S. P., C. D. UNSWORTH & O. H. VIVEROS. 1984. Regulation of opioid peptide synthesis and processing in adrenal chromaffin cells by catecholamines and cyclic adenosine 3′:5′-monophosphate. J. Neurosci. **4:** 2993-3001.
32. EIDEN, L. E., P. GIRAUD, H.-U. AFFOLTER, E. HERBERT & A. J. HOTCHKISS. 1984. Alternative modes of enkephalin biosynthesis regulation by reserpine and cyclic AMP in cultured chromaffin cells. Proc. Natl. Acad. Sci. **81:** 3949-3953.
33. EIDEN, L. E., P. GIRAUD, J. R. DAVE, A. J. HOTCHKISS & H.-U. AFFOLTER. Nicotinic receptor stimulation activates enkephalin release and biosynthesis in adrenal chromaffin cells. Nature **312:** 661-663.
34. VIVEROS, O. H., C. D. UNSWORTH, T. KANAMATSU, J.-S. HONG & E. J. DILIBERTO, JR.. 1986. Nicotinic regulation of adrenomedullary opioid peptide synthesis and secretion: A model to study monoamine neuropeptide cotransmission. *In* Proceedings of the International Symposium on Tobacco Smoking and Health: A Neurobiological Approach. J. Humble & L. Davis, Eds. Plenum. New York. In press.
35. MATSUO, H., A. MIYATA & K. MIZUNO. 1983. Novel C-terminally amidated opioid peptide in human phaeochromocytoma tumour. Nature **305:** 721-723.
36. MIYATA, A., K. MIZUNO, N. MINAMINO & H. MATSUO. 1984. Regional distribution of adrenorphin in rat brain: Comparative study with PH-8P. Biochem. Biophys. Res. Commun. **120:** 1030-1036.
37. WEBER, E., F. S. ESCH, P. BÖHLEN, S. PATERSON, A. D. CORBETT, A. T. MCKNIGHT, H. W. KOSTERLITZ, J. D. BARCHAS & C. J. EVANS. 1983. Metorphamide: Isolation, structure, and biologic activity of an amidated opioid octapeptide from bovine brain. Proc. Natl. Acad. Sci. **80:** 7362-7366.
38. SEIZINGER, B. R., D. C. LIEBISCH, C. GRAMSCH, A. HERZ, E. WEBER, C. J. EVANS, F. S. ESCH & P. BÖHLEN. 1985. Isolation and structure of a novel C-terminally amidated opioid peptide, amidorphin, from bovine adrenal medulla. Nature **313:** 57-59.
39. BRADBURY, A. F., M. D. A. FINNIE & D. G. SMYTH. 1982. Mechanism of C-terminal amide formation by pituitary enzymes. Nature **298:** 686-688.
40. KIZER, J. S., W. H. BUSHBY, JR., C. COTTLE & W. W. YOUNGBLOOD. 1984. Glycine-directed peptide amidation: presence in rat brain of two enzymes that convert p-glu-his-pro-gly-OH into p-glu-his-pro-NH_2 (thyrotropin-releasing hormone). Proc. Natl. Acad. Sci. **81:** 3228.
41. KVETNANSKY, R. 1973. *In* Frontiers in Catecholamine Research. E. Usdin & S. Snyder, Eds.: 223. Pergamon. New York.
42. KILPATRICK, D. L., R. D. HOWELLS, G. FLEMINGER & S. UDENFRIEND. 1984. Denervation of rat adrenal glands markedly increases preproenkephalin mRNA. Proc. Natl. Acad. Sci. **81:** 7221-7223.
43. LAGAMMA, E. F., J. E. ADLER & I. B. BLACK. 1984. Impulse activity differentially regulates [Leu]enkephalin and catecholamine characters in the adrenal medulla. Science **224:** 1102-1104.
44. VIVEROS, O. H., L. ARQUEROS, R. J. CONNETT & N. KIRSHNER. 1969. Mechanism of secretion from the adrenal medulla. IV. The fate of the storage vesicles following insulin and reserpine administration. Mol. Pharmacol. **5:** 69.
45. VIVEROS, O. H., L. ARQUEROS & N. KIRSHNER. 1971. Mechanism of secretion from the adrenal medulla. VI. Effect of reserpine on the dopamine β-hydroxylase and catecholamine content and on the buoyant density of adrenal storage vesicles. Mol. Pharmacol. **7:** 434.
46. MOCCHETI, I., A. GUIDOTTI, J. P. SCHWARTZ & E. COSTA. 1985. Reserpine changes the dynamic state of enkephalin stores in rat striatum and adrenal medulla by different mechanism. J. Neurosci. **5:** 3379.
47. WILSON, S. P. & O. H. VIVEROS. 1981. Primary culture of adrenal medullary chromaffin cells in a chemically defined medium. Exp. Cell. Res. **133:** 159-169.
48. WILSON, S. P., O. H. VIVEROS & N. KIRSHNER. 1985. Relationship between the regulation of enkephalin-containing peptide and dopamine β-hydroxylase levels in cultured adrenal chromaffin cells. J. Neurochem. **45:** 1363-1370.
49. HOOK, V. Y. H., L. E. EIDEN & R. M. PRUSS. 1985. Selective regulation of carboxypeptidase peptide hormone-processing enzyme during enkephalin biosynthesis in cultured bovine adrenomedullary chromaffin cells. J. Biol. Chem. **260:** 5991-5997.

50. SIEGEL, R. E., L. E. EIDEN & H.-U. AFFOLTER. 1985. Elevated potassium stimulates enkephalin biosynthesis in bovine chromaffin cells. Neuropeptides **6:** 543-552.
51. VIVEROS, O. H., L. ARQUEROS & N. KIRSHNER. 1971. Mechanism of secretion from the adrenal medulla. VII. Effect of insulin administration on the buoyant density, dopamine β-hydroxylase, and catecholamine content of adrenal storage vesicles. Mol. Pharmacol. **7:** 444-454.
52. GAGNON, C., W. PFALLER, W. M. FISCHERA, M. SCHWAB, H. WINKLER & H. THOENEN. 1977. Increased specific activity of membrane-bound dopamine β-hydroxylase in chromaffin granules after reserpine treatment. J. Neurochem. **28:** 853-856.
53. LEDBETTER, F. H., D. KILPATRICK, H. L. SAGE & N. KIRSHNER. 1978. Synthesis of chromogranins and dopamine β-hydroxylase by perfused bovine adrenal glands. Am. J. Physiol. **235:** E475-E486.
54. VIVEROS, O. H., L. ARQUEROS & N. KIRSHNER. 1969. Quantal secretion from adrenal medulla: All-or-none release of storage vesicle content. Science **165:** 911-913.
55. GUIDOTTI, A. & E. COSTA. 1974. A role for nicotinic receptors in the regulation of the adenylate cyclase of adrenal medulla. J. Pharmacol. Exp. Ther. **189:** 665-675.
56. YOSHIKAWA, K., J.-H. HONG & S. L. SABOL. 1985. Electroconvulsive shock increases preproenkephalin messenger RNA abundance in rat hypothalamus. Proc. Natl. Acad. Sci. **82:** 589-593.
57. EIPPER, B., R. E. MAINS & C. C. GLEMBOTSKI. 1983. Identification on pituitary tissue of a peptide α-amidation activity that acts on glycine-extended peptides and requires molecular oxygen, copper and ascorbic acid. Proc. Natl. Acad. Sci. **80:** 5144-5148.

DISCUSSION OF THE PAPER

V. Y. H. HOOK (*Uniformed Services University of Health Sciences, Bethesda, Md.*): This morning we heard about the interactions of reserpine and tetrabenazine with the amine transporter. Does this imply that the amine transporter interferes with the peptide biosynthetic pathway? Or, how are reserpine and tetrabenazine envisioned to modify opiate peptide biosynthesis?

O. H. VIVEROS (*Wellcome Research Laboratories, Research Triangle Park, N.C.*): All drugs that deplete catecholamines (inhibitors of vesicular uptake and inhibitors of catecholamine synthesis at the tyrosine hydroxylase, DOPA decarboxylase, and dopamine β-hydroxylase level) have very similar effects that differ from the effects of increased cAMP levels. Thus, there is no relation to the amine transporter but to the decrease in catecholamine content in some intracellular pool that directly or indirectly signals for enhanced processing, and probably enhanced translation and increased levels of mRNA.

H. TAMIR (*Columbia University, New York, N.Y.*): Do PC_{12} cells have enkephalins?

O. H. VIVEROS: Yes they do, both in culture or in the transplanted tumors.

Dopamine β-Hydroxylase: Biochemistry and Molecular Biology

TONG H. JOH AND ONYOU HWANG

Laboratory of Molecular Neurobiology
Department of Neurology
Cornell University Medical College
New York, New York 10021

Dihydroxyphenylethylamine (dopamine, DA), norepinephrine (NE), and epinephrine (EP), collectively called catecholamines (CA), are synthesized from the amino acid tyrosine in adrenal medulla and in adrenergic neurons in the central nervous system. Biosynthesis of CA involves four enzymatic steps as shown in FIGURE 1: (1) tyrosine hydroxylase (TH) catalyzes the conversion of tyrosine to dihydroxyphenylalanine (DOPA); (2) DOPA is converted to DA by aromatic L-amino acid decarboxylase (AADC); (3) dopamine β-hydroxylase (DBH) catalyzes the enzymatic reaction of DA to NE; and (4) EP is synthesized from NE by phenylethanolamine *N*-methyltransferase (PNMT).

As early as 1962, Kirshner found that DA is transported into chromaffin granule where it is hydroxylated to become NE.[1] This confirmed the earlier belief that DBH, the enzyme responsible for the conversion of DA to NE, is located in the interior of chromaffin granule. All other enzymes in the CA biosynthetic pathway are known to be present in the cytoplasm. DBH is found even inside the granule of epinephrine cells, in which the terminal methylating enzyme, PNMT, is cytoplasmic. Thus, the sequence involved in biosynthesis of EP in chromaffin cells or adrenergic neurons is as follows: synthesis of DOPA and DA occur in cytoplasm; DA is transported into NE-containing vesicle where it is converted to NE; NE is then transported back to cytoplasm and converted to EP; and finally, EP is stored in EP-containing vesicle (FIG. 2). Whether these seemingly inefficient CA transport processes into and out of granules are involved in regulation of EP biosynthesis, and whether this unique location of DBH among the CA biosynthetic enzymes is related to phylogeny of CA system, remain unresolved.

Dopamine β-hydroxylase is known to exist in two forms: soluble (sDBH) and membrane bound (mDBH).[2,3] Whereas sDBH is stored in the granule interior and secreted,[4–9] mDBH is reinternalized upon exocytosis.[10–15] Thus, biochemically, sDBH is found in the soluble fraction upon granule lysis; mDBH remains associated with the granule membrane even after repeated washing with aqueous buffer and can be solubilized in active form only upon disruption of membrane with nonionic detergents.[16,17] DBH is recovered largely as a component of vesicular membrane such as that of synaptic vesicles[18,19] and chromaffin granules.[20,21] Studies on the release of DBH suggest that *in vivo* most of the enzyme may in fact exist as the membrane-bound form.[22,23]

The two forms have been shown to be similar in their immunoreactivity,[17,24–26] carbohydrate content,[27,28] mobilities on nondenaturing gels,[29] and binding affinities for various substrates.[30,31]

They are, however, apparently dissimilar in several properties. It has been reported that they differ in subunit structures; mDBH consists of two subunits. 70 kDa and 75 kDa in bovine adrenal medullary cells[32] and 73 kDa and 77 kDa in PC12 cells,[33] whereas sDBH has been reported to consist of one subunit of 70 kDa in bovine tissues[32] and 73 kDa in PC12 cells.[33] On the other hand, Sabban *et al.*[34] have suggested that in PC12 cells, the 77-kDa subunit is found in the membrane fraction and the 73-kDa subunit in the soluble fraction.

Furthermore, structural complexity of DBH is found within sDBH itself, which

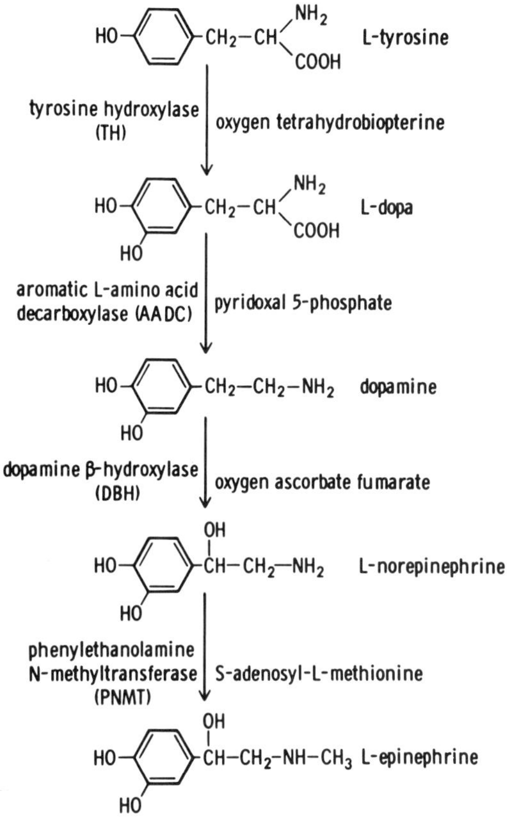

FIGURE 1. Catecholamine biosynthetic pathway. Enzymes and cofactors are indicated.

has been shown to run as three distinctive protein bands on SDS polyacrylamide gels.[35] This indicates that sDBH may consist of multiple subunits. This finding has been confirmed by our recent studies.[36] We have been successful in isolating sDBH from chromaffin granules of bovine adrenal medulla and purifying it to homogeneity. The purified sDBH runs as a single protein band on nondenaturing gel, as shown in FIGURE 3. When this sample is subjected to SDS-PAGE, the protein is separated into three distinct bands of 75, 72, and 69 kDa (FIG. 4a), all of which are positive in immunoblot analysis using our antibodies against DBH (FIG. 4b), the specificity of

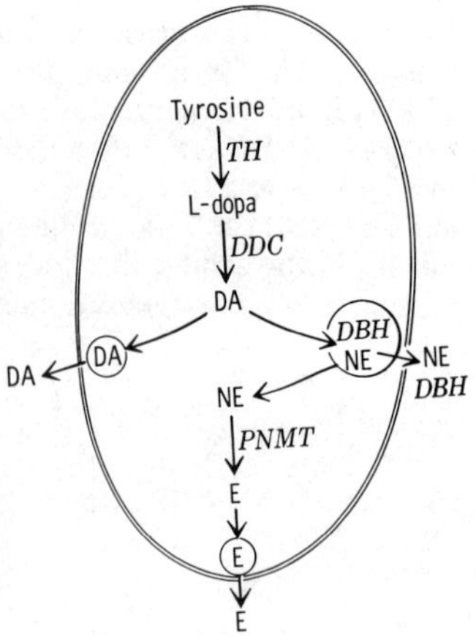

FIGURE 2. Catecholamine biosynthesis, storage, and release. In dopamine (DA) neurons, DA is stored in DA vesicles and released. In noradrenergic neurons, norepinephrine (NE) is synthesized from DA and stored inside the vesicles where dopamine β-hydroxylase (DBH) is located. Upon exocytosis, both NE and soluble DBH are released. In adrenergic neurons and adrenal chromaffin cells, NE in the vesicle is transported to cytoplasm where it is converted to epinephrine (EP), which is stored in and released from EP-containing vesicles. Unlike tyrosine hydroxylase (TH), aromatic L-amino acid decarboxylase (DDC), and phenylethanolamine *N*-methyltransferase (PNMT), which are cytoplasmic enzymes, DBH is located in the vesicle.

which has been demonstrated in our laboratory.[17] Individual bands are isolated separately and their purity determined by SDS-PAGE (FIG. 5).

The major 72-kDa protein band is eluted from the gel and subjected to gas phase sequencing. The NH_2-terminal amino acid sequence of this protein (72 kDa) shows that this single band on SDS gel consists of two polypeptide chains of equimolar concentration (FIG. 6). They differ in that one has three extra amino acids at its NH_2-terminus compared to the other. Although similar sequencing data have been reported earlier by Skotland *et al.*,[38] their results could not be further substantiated due to technical limitation at the time the studies were done.

We now have the corrected NH_2-terminal amino acid sequence of sDBH, obtained using more sensitive techniques (FIG. 6). We have determined the thirteenth amino acid, unidentified by Skotland *et al.*,[38] to be proline. Furthermore, the third amino acid, reported to be threonine by Skotland *et al.*,[38] has been corrected to proline. This change was based on the following: (1) Skotland *et al.*[38] identified amino acids as PTH derivatives by gas chromatography, and PTH-threonine derivative identified by gas chromatography coincides with that of proline; (2) our results show proline in the third position; (3) no threonine is detected in the entire sequence of the first 15 amino acids. Furthermore, the molecular structures and identity of the two additional proteins (75 kDa and 69 kDa) in sDBH fraction have not been established. They may represent DBH with different degrees of glycosylation, or cleavage products of DBH during purification. Thus the molecular structure of DBH is more complex than originally considered.

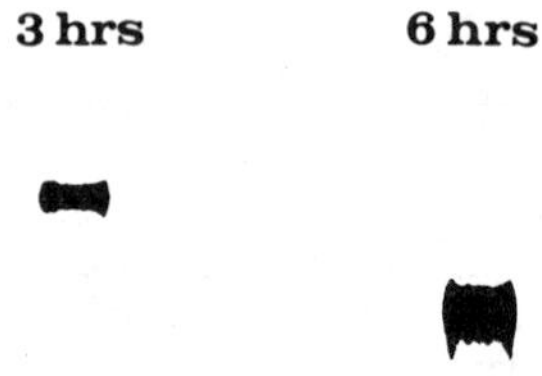

FIGURE 3. Electrophoretic patterns of purified soluble form of DBH on nondenaturing polyacrylamide gel, showing a single protein band at 3 and 6 hours of electrophoresis.

Hydrophilicity of the two forms also differ. Charge shift electrophoresis of detergent-solubilized membranes showed that mDBH is amphiphilic in nature, whereas sDBH displays hydrophilic behavior.[26,39] Slater *et al.* have reported that mDBH has a higher content of hydrophobic amino acids and peptides not found in sDBH.[17] Isotopically labeled phosphatidylcholine has been shown to bind to purified mDBH but not sDBH.[40] These findings have led to the belief that mDBH and sDBH are identical except that mDBH has a hydrophobic tail that anchors the protein to the membrane. In fact, Zaremba and Hogue Angeletti[41] have shown that mDBH is indeed a transmembrane protein that has its active site exclusively on the intragranular face.

However, the mechanism of biosynthesis of the two forms of DBH has been controversial. Two possibilities may be considered: (1) mDBH and sDBH are produced

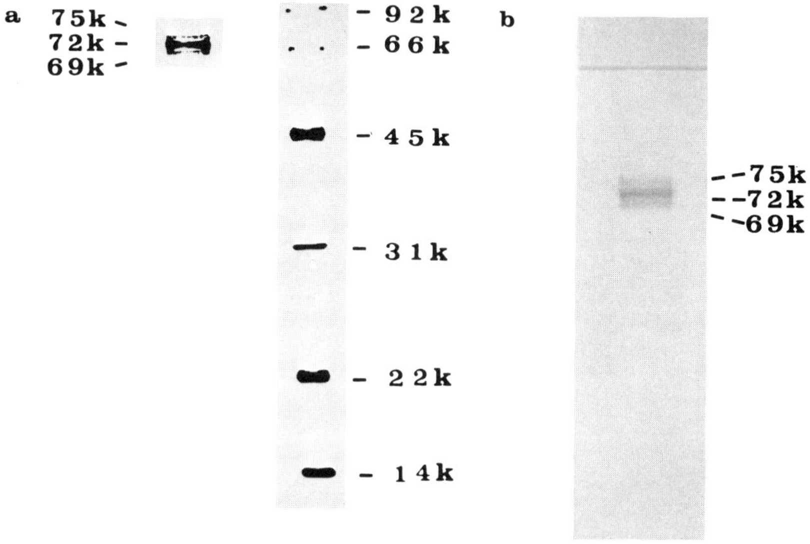

FIGURE 4. Electrophoretic patterns of soluble DBH. (**a**) Purified soluble DBH on SDS polyacrylamide gel showing three protein bands: 75 kDa, 72 kDa, and 69 kDa. (**b**) Immunoblot of crude soluble fraction of adrenal medulla using specific antisera against soluble DBH, demonstrating that all three bands are immunoreactive.

by posttranslational processing, and sDBH is a proteolytic cleavage product of mDBH; (2) mDBH and sDBH are synthesized independently from respective mRNAs, and these mRNAs may derive from the common separate gene(s).

Some studies have suggested the former. Helle and Serck-Hanssen[42] reported that newly synthesized mDBH can be converted to sDBH within the vesicle. Sabban *et al.*,[34] using pulse-chase experiments, suggested that mDBH is formed before sDBH. Helle *et al.* reported that incubation of the membrane fraction at 37°C converted the amphiphilic form to the hydrophilic form, suggesting that there is endogeneous proteolytic activity converting mDBH to sDBH. Other studies, however, have suggested the latter. Several early studies reported that mDBH and sDBH are synthesized separately.[43,44] Saxena and Fleming[32] also showed that there is no precursor-product

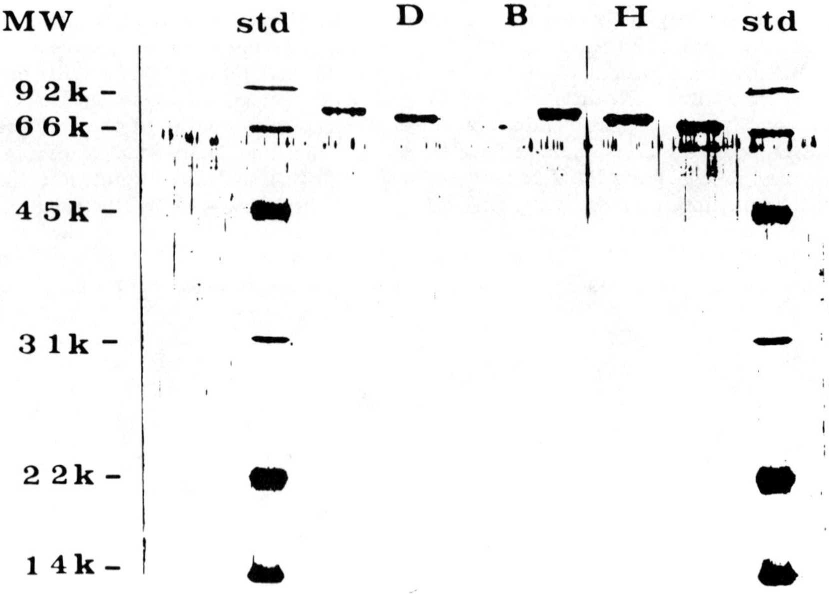

FIGURE 5. SDS-polyacrylamide electrophoretic pattern of three individual bands isolated from the SDS gel, demonstrating the purity of each band.

Hwang et al. (1986)

1	2	3	4	5	6	7	8	9	10	11	12	13	14	15 ..
( )	Glu	Ser	Pro	Phe	Pro	Phe	His	Pro	Phe	Leu	Ile	Pro	Leu	Asp..
		Pro	Ala	Glu	Ser	Pro	Phe	Ile	Pro	His	Asp	Arg	Glu	Gly..
									Leu					

Skotland et al. (1977) Biochem. Biophys. Res. Comm. 74, 1483.

1	2	3	4	5	6	7	8	9	10	11	12	13	14	15 ..
			Ala	Glu	Ser	Pro	Phe	Pro	Phe	His	Ile	(NI)	Leu	Asp..
Ser	Ala	Thr	Ala	Glu	Ser	Pro	Phe	Pro	Phe	His	Ile	(NI)	Leu	Asp..
													Ile	

Corrected Sequences

1	2	3	4	5	6	7	8	9	10	11	12	13	14	15 ...
			Ala	Glu	Ser	Pro	Phe	Pro	Phe	His	Ile	Pro	Leu	Asp...
Ser	Ala	Pro	Ala	Glu	Ser	Pro	Phe	Pro	Phe	His	Ile	Pro	Leu	Asp...

FIGURE 6. NH_2-terminal amino acid sequence of soluble form of DBH. The major protein band, 72-kDa protein, was eluted from the SDS gel and the NH_2-terminal amino acid sequence determined. Our results were compared with others,[38] and the corrected sequence determined.

relationship, and that the 70-kDa and 75-kDa subunits of mDBH generate identical peptide maps, which differ from that of the 70-kDa sDBH subunit.

In vitro translation of total mRNA followed by immunoprecipitation of DBH produces a single band on SDS polyacrylamide gel, although different authors have reported various molecular weights. Sabban and Goldstein[45] showed a 67-kDa translation product of bovine adrenal medulla mRNA and Benlot *et al.*[46] demonstrated that a 73-kDa protein band is produced from human pheochromocytoma. Our results show a single 72-kDa band of bovine adrenal medullary mRNA product (FIG. 7).

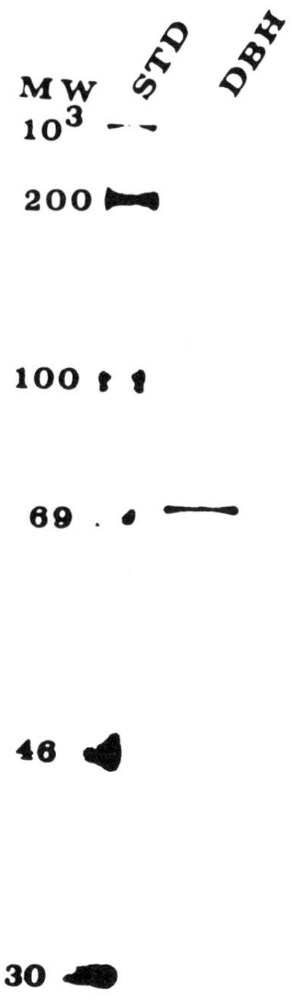

FIGURE 7. *In vitro* translation product of DBH mRNA. Total mRNA of bovine adrenal medulla was translated *in vitro* and DBH precipitated with specific antibody to bovine soluble DBH.

However, the possibility of the presence of multiple forms that are indistinguishable by molecular weight can not be eliminated, since native sDBH and mDBH are indistinguishable by many criteria.[31,47,48]

If it is true that sDBH is produced by the cleavage of a hydrophobic fragment of mDBH, it remains to be determined how a pool of mDBH is processed to become sDBH while another pool remains in the membrane in the same tissue. On the other hand, mDBH and sDBH may differ at the mRNA level and thus in their nascent protein structures. As membrane and secreted proteins, respectively, they are both expected to contain a leader peptide in their nascent form.[49–53] It is possible that a small sequence difference in their leader peptides allows that of mDBH to be protected from a specific protease whereas that of sDBH to be cleaved, thus removing the hydrophobic region. Alternatively, in addition to the leader peptide, mDBH may contain a hydrophobic tail not found in sDBH, which anchors the protein to the membrane. We hope that molecular biological techniques will help to resolve these questions.

SUMMARY

Dopamine β-hydroxylase (DBH) catalyzes the conversion of dopamine to norepinephrine (NE), and is known to exist in two forms: soluble and membrane-bound. It has been reported that the two forms are similar in their immunoreactivity, carbohydrate content, and binding affinities for various substrates, and are apparently dissimilar in subunit structures and hydrophilicity. Furthermore, added structural complexity is observed within sDBH itself. Our results indicate that purified sDBH, which runs a single band on a nondenaturing gel, exhibits three protein bands of 75 kDa, 72 kDa, and 69 kDa on SDS polyacrylamide gel. The majority of sDBH exists as a 72-kDa protein. The NH_2-terminal amino acid sequence of this 72-kDa protein indicates that it consists of two polypeptides of equimolar concentrations, where one differs from the other by three extra amino acids at its NH_2 terminus. Whether they are different proteolytic cleavage products is not known. Thus, the structure of DBH appears to be more complex than originally considered. *In vitro* translation of total mRNA of bovine adrenal medulla followed by immunoprecipitation of DBH produces a single 72-kDa band on SDS polyacrylamide gel. This suggests either that there is only one *in vitro* mRNA translation product, which is modified to become different forms of DBH, or that multiple translation products are present but are indistinguishable by molecular weight. These subjects have been discussed in detail in this paper.

REFERENCES

1. Kirshner, N. 1962. J. Biol. Chem. **237:** 2311-2317.
2. Smith, W.J. & N. Kirshner. 1967. Mol. Pharmacol. **3:** 52-62.
3. Winkler, H. 1976. Neuroscience **1:** 65-80.
4. Dutch, D. S., O. H. Viveros & N. Kirshner. 1968. Biochem. Pharmacol. **17:** 255-264.
5. Viveros, O. H., L. Arqueros, R. J. Connett & N. Kirshner. 1968. Mol. Pharmacol. **5:** 60-68.
6. Pollard, H., C. J. Pazoles, C. E. Creuts & O. Zindler. 1979. Intl. Rev. Cytol. **58:** 159-197.
7. Norman, T. C. 1976. Intl. Rev. Cytol. **46:** 1-77.
8. Liang, B. T. & R. L. Pearlman. 1979. J. Neurochem. **32:** 927-933.
9. Ledbetter, F. H. & N. Kirshner. 1981. Biochem. Pharmacol. **30:** 3246-3249.
10. Heuser, J. E. & T. S. Reese. 1973. J. Cell. Biol. **57:** 315-344.
11. Winkler, H., F. H. Schneider, G. Rufener, P. K. Nakane & H. Hornagl. 1974. Adv. Cytopharmacol. **2:** 127-139.

12. Fried, R. C. & M. P. Blaustein. 1978. J. Biol. Chem. **253:** 685-700.
13. Schneider, Y., P. Tulkers, C. DeDuve & A. Trouet. 1979. J. Cell Biol. **82:** 466-474.
14. Silver, M. A. & D. M. Jacobowitz. 1979. Brain Res. **167:** 65-75.
15. Herzog, V. 1981. Trends Biol. Sci. **6:** 315-344.
16. Friedman, S. & S. Kaufman. 1965. J. Biol. Chem. **240:** 4763-4773.
17. Slater, E. P., S. Zaremba & R A. Hogue-Angeletti. 1981. Arch. Biochem. Biophys. **211:** 288-296.
18. Klein, R. L., D. E. Kirksey, R. A. Rush & M. Goldstein. 1977. J. Neurochem. **28:** 81-86.
19. Kirksey, D. E., R. L. Klein, J. M. Baggett & M. S. Gasparis. 1978. Neuroscience **3:** 71-81.
20. Ciaranello, R. D., G. F. Wooten & J. Axelrod. 1975. J. Biol. Chem. **250:** 3204-3211.
21. Helle, K. B., M. Fillenz, C. Stanford, K. E. Phil & B. Srebro. 1979. J. Neurochem. **32:** 1351-1355.
22. Dixon, W. E., A. G. Garcia & S. M. Kirpekar. 1975. J. Physiol. **244:** 805-824.
23. Edwards, A. V., P. N. Furness & K. B. Helle. 1980. J. Physiol. **308:** 15-27.
24. Goldstein, M., E. Lauber & M. R. McKreghan. 1965. J. Biol. Chem. **240:** 2066-2072.
25. Rush, R. A., S. H. Kindler & S. Udenfriend. 1974. Biochem. Biophys. Res. Comm. **61:** 38-44.
26. Skotland, T. & T. Flatmark. 1979. J. Neurochem. **32:** 1861-1862.
27. Geissler, D., A. Martinek, R. U. Margolis, R. K. Margolis, J. A. Skrivanek, R. Ledeen, P. Konig & H. Winkler. 1977. Neuroscience **2:** 685-693.
28. Fischer-Colbrie, R., M. Davis, R. Zangerle & H. Winkler. 1982. J. Neurochem. **38:** 725-732.
29. Hortnagle, H., H. Winkler & H. Lochs. 1972. Biochem. J. **129:** 187-195.
30. Brodde, O. F., F. Arens, K. Huvermann & H. J. Schumann. 1976. Experientia **32:** 984-986.
31. Aunis, D., M. Bouclier, M. Pescheloche & P. Mandel. 1977. J. Neurochem. **29:** 439-447.
32. Saxena, A. & P. J. Fleming. 1983. J. Biol. Chem. **258:** 4147-4152.
33. McHugh, E. M., R. McGee & P. J. Fleming. 1985. J. Biol. Chem. **260:** 4409-4417.
34. Sabban, E. L., M. Goldstein & L. A. Greene. 1983. J. Biol. Chem. **258:** 7819-7823.
35. Speedie, M. K., D. L. Wong & R. D. Ciaranello. 1985. J. Chromatogr. **327:** 351-357.
36. Hwang, O., J. Smith & T. H. Joh. 1986. Neurosci. Abst. In press.
37. Joh, T. H. & M. E. Ross. 1983. *In* Immunohistochemistry. A. C. Cuello, Ed.: 121-138. IBRO Handbook Series.
38. Skotland, T., T. Ljones, T. Flatmark & K. Sletten. 1977. Biochem. Biophys. Res. Comm **74:** 1483-1489.
39. Bjerrum, O. J., K. B. Helle & E. Bock. 1979. Biochem. J. **181:** 231-237.
40. Albanesi, J. P., E. C. Robinson & N. Kirshner. 1980. Fed. Proc. **39:** 1655.
41. Zaremba, S. & R. A. Hogue-Angeletti. 1981. J. Biol. Chem. **256:** 12310-12315.
42. Helle, K. B. & G. Serck-Hanssen. 1981. *In* Chemical Neurotransmission. L. Stjarne, P. Hedqvist, H. Lagercrantz & A. Wennmalm, Eds.: 85-90. Academic. London.
43. Gagnon, C., R. Schatz, Y. Otten & H. Thoenen. 1976. J. Neurochem. **27:** 1083-1089.
44. Winkler, H. 1977. Neuroscience **2:** 657-638.
45. Sabban, E. L. & M. Goldstein. 1984. J. Neurochem. **43:** 1663-1668.
46. Benlot, C., J. Antreassian, J.-P. Henry, J. C. Lengrand, F. Gross & J. Thibault. 1985. Biochemie **67:** 589-595.
47. Ljones, T., T. Skotland & T. Flatmark. 1976. Eur. J. Biochem. **61:** 525-533.
48. Helle, K. B., G. Serck-Hanssen & K. L. Reichelt. 1977. Intl. J. Biochem. **8:** 693-704.
49. Blobel, G. 1977. *In* International Cell Biology, 1976-1977. B. R. Brinkley & K. E. Porter, Eds.: 318-328. Rockfeller Univ. Press. New York.
50. Blobel, G. 1982. Proc. Natl. Acad. Sci. U.S.A. **77:** 1496-1500.
51. Austin, B. M. 1979. FEBS Lett. **103:** 308-313.
52. Steiner, D. F., P. S. Quinn, S. J. Chan, J. Marsh & H. S. Tager. 1980. Ann. N.Y. Acad. Sci. **343:** 1.
53. Sabatini, D. D., G. Kreibichi, T. Morimoto & M. Adesnik. 1982. J. Cell. Biol. **92:** 1-22.

DISCUSSION OF THE PAPER

E. L. SABBAN (*New York Medical College, Valhalla, N.Y.*): Regarding the amino acid sequence of dopamine-β-hydroxylase, you report two sequences. Are these both for the soluble form, and if so, which of the "soluble forms" are these? There can also be microheterogeneity in the sugars of DBH, which can explain minor variations or somewhat different soluble forms.

T. H. JOH (*Cornell University Medical College, New York, N.Y.*): In regard to DBH in the rat, results from my lab and that of Dr. Patrick Fleming agree that in PC12 cells, soluble DBH consists of the 73-kDa subunit form. The membrane form of DBH contains 77-kDa subunit forms. There is no controversy now in this regard. Whether the 73-kDa form in some membrane fractions is a contaminant of the soluble form or an intrinsic component of membrane DBH remains to be seen. We often see small amounts of the 73-kDa forms in our isolation of membrane-bound DBH.

Chromogranin A: The Primary Structure Deduced from cDNA Clones Reveals the Presence of Pairs of Basic Amino Acids

MARK GRIMES,[a,b] ANNA IACANGELO,[c] LEE E. EIDEN,[c] BRUCE GODFREY,[d] AND EDWARD HERBERT[d]

[a] *Chemistry Department*
University of Oregon
Eugene, Oregon 97403

[c] *Laboratory of Cell Biology*
National Institute of Mental Health
National Institutes of Health
Bethesda, Maryland 20892

[d] *Institute for Advanced Biomedical Research*
The Oregon Health Sciences University
Portland, Oregon 97201

INTRODUCTION

Chromogranin A is the most abundant protein in bovine chromaffin granules. It is secreted into the bloodstream and levels have been found to increase during stress.[1] It was purified in 1967 by Smith and Winkler[2] and has been shown to be glycosylated.[3] The molecular mass range has been reported to be from 60-77 kDa.[2,4] Analysis of amino acid composition reveals an unusually acidic protein, containing 25% glutamic acid (and glutamine) and 8% aspartic acid (and asparagine, see TABLE 1). The chromogranin antigens in the brain and pituitary also have a molecular weight and pI similar to adrenal chromogranin A.[5] Secretory protein I (SP-I) of the parathyroid gland has been shown to be almost identical to chromogranin A in physical properties, amino acid composition, and NH_2-terminal sequence (see TABLE 1).[6,7] It is possible that the proteins in the adrenal medulla, brain, pituitary, and parathyroid represent the same gene product expressed in different tissues.

[b] Present address: Department of Biochemistry and Biophysics, University of California, San Francisco, Calif. 94143-0448.

Recently it has been shown that there are several species of chromogranins, all of which share a similar amino acid content, pI, and antigenic components.[5,8,9] There is, however, disagreement about the number of chromogranins and the biosynthetic relationship among them. Falkensammer *et al.*[10] have recently described two related chromogranins, a 71-kDa protein (chromogranin A) and a 96-kDa protein, which are precipitated with antibody raised against soluble proteins from chromaffin granules. Pulse-chase studies have failed to show conversion of the 96-kDa forms of chromogranins to the 71-kDa chromogranin A.[10]

Settleman *et al.*[11] have described four proteins of molecular masses 100, 85, 75 and 65 kDa that are recognized by antisera raised against the 75-kDa chromogranin A. O'Conner and Frigon[9] have described chromogranin A as a doublet, 67 kDa in size. Antisera against this protein recognizes a 52-kDa protein and several smaller

TABLE 1. Amino Acid Composition of Chromogranin A and SP-I

Amino Acid	Cgr A[a]	Cgr A[b]	SP-I[a]	SP-I[b]	Cgr A[c]
Ala	8.5	7.8	8.9	7.8	8.1
Asx	8.0	7.6	7.1	7.4	7.2
Arg	6.0	5.3	6.6	5.6	7.2
Cys	0.2	0.8	0.5	0.7	0.5
Glx	22.5	25.7	23.2	25.9	24.6
Gly	8.1	6.8	7.9	7.1	7.7
His	1.9	1.6	1.6	1.6	1.4
Ile	1.4	1.8	1.0	1.8	1.2
Leu	7.3	6.7	7.4	7.0	7.4
Lys	8.4	8.6	8.7	8.6	7.7
Met	1.4	1.8	1.6	1.8	1.7
Phe	1.7	1.6	1.5	1.5	1.4
Pro	9.2	11.6	8.7	10.7	8.6
Ser	7.7	7.3	7.6	7.3	7.7
Thr	2.6	2.4	2.1	2.4	2.1
Trp	1.4	0.8	1.1	0.9	1.4
Tyr	1.0	0.7	1.0	0.7	0.9
Val	3.9	3.6	3.5	3.5	3.5

[a] From reference 6.
[b] From reference 7.
[c] From the protein encoded by the cDNA clones, without the signal sequence. Asx = Asp + Asn and Glx = Glu + Gln.

ones. Hamilton *et al.*[7] have described chromogranin A and secretory protein I as being nearly identical proteins that occur as a doublet of 66 kDa and 72 kDa. Somogyi *et al.*[5] also describe chromogranins as a family of related but differently sized polypeptides, all having a very similar isoelectric point. Proteolytic cleavage of newly synthesized chromogranin A does occur, but is a very slow process, compatible with the 6-8 day half-life reported for the proenkephalin molecule in the same tissue.[10,12,13]

If the different chromogranins in the secretory vesicle are indeed related, there are several ways in which a family of proteins could arise during biosynthesis: (1) expression of several related genes could occur; (2) a single gene could express several mRNAs by an alternate splicing mechanism; (3) a single mRNA could give rise to several proteins by the use of multiple translation initiation points; or (4) a single protein could be posttranslationally processed into several different products. Possibility 4 probably cannot explain the origin of chromogranin A because of the pulse-chase experiments mentioned above,[10] but has not been rigorously ruled out.

In order to begin to answer some of the questions about the function, biosynthetic relationship, and expression of chromogranins, we employed the tools of recombinant DNA technology to obtain a cDNA clone for chromogranin A. This clone has allowed us to determine the primary structure of the chromogranin A protein, and it provides a probe with which to examine the relationship of different species of chromogranins to one another. Use of this probe has made it possible to detect several related mRNA species on Northern blots, which suggests that one large protein is not the source for all the chromogranins.

RESULTS

The oligonucleotide sequences that were used in these studies were designed to be complementary to the amino-terminal sequence of chromogranin A.[14] The region spanned by the two oligomers is shown in FIGURE 1A. These probes are designated the 5′ 14-mer and the 3′ 14-mer corresponding to their respective positions on the RNA template. A third probe was made, a 15-mer (FIG. 1A), complementary to the sequence of a cyanogen bromide fragment of chromogranin A (communicated before publication by Ruth Hogue Angeletti).[11] To determine the size of chromogranin A mRNA, Northern blots were done using poly-A^+ RNA from bovine adrenal medulla. FIGURE 1b shows the results of hybridization with all three probes. A predominant RNA species of approximately 2.1 kilobases (kb) was observed with all three probes. A 70-kDa protein has approximately 635 amino acid residues, so the coding region of the mRNA must be at least 1900 bases long plus the 5′ and 3′ noncoding regions and poly-A tail found on the typical message. Thus, 2.1 kb is a reasonable size for the chromogranin A mRNA. The predominant RNA produces a strong signal, hence it is not a rare species in this tissue, as would be expected since chromogranin A is an abundant protein. Other minor signals probably represent RNA species with sequence homology to one of the oligomers in the pool. Thus, it can be concluded from these data that (1) all the probes hybridize to one predominant RNA species, and (2) the size and amount of this RNA species could reasonably represent chromogranin A mRNA.

We attempted to use the oligomers to select a cDNA clone from clone banks derived from bovine adrenal medulla poly-A^+ RNA. We obtained many "false positives," that is, clones that hybridized to the oligonucleotides and contained partial or complete matches to one of the probe sequence, but did not represent chromogranin A mRNA, as determined by sequencing. In order to obtain the sequence of the 5′ end of chromogranin A mRNA and design a more specific probe for screening cDNA libraries, the probe pools were used to prime cDNA synthesis specifically from chromogranin A mRNA.

Primer Extension

The two 14-residue oligomer probe pools to the amino-terminal sequence of chromogranin A were 5′ labeled with ^{32}P and used as primers to synthesize a partial cDNA

A

N-TERMINUS

```
NH2 -  leu arg val asn ser pro met asn lys gly asp thr glu val met lys cys ile arg glu val ile ser asp  - COOH
                     [-----5'-14-mer----]             [-----3'-14-mer----]
                  5'-CCT/CTTA/GTTCATNGG-3'         5'-ATA/GCAT/CTTCATNAC-3'
```

CNBr FRAGMENT

```
NH2 -  asp glu leu pro phe val leu phe ala glu -X- -X- leu -X- -X- val glu gln glu glu glu leu  - COOH
                                                                 [-------15-mer------]
                                                            5'-T/CTCT/CTCT/CTCT/CTGT/CTC-3'
```

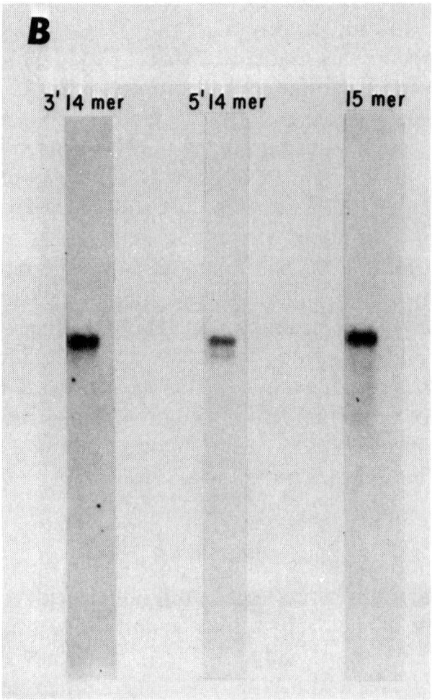

FIGURE 1. **(A)** Sequences of the oligomer pools. Both 14-mers are 16-fold degenerate and the 15-mer is 32-fold degenerate. **(B)** Hybridization of oligomers to chromogranin A mRNA. Bovine adrenal medulla poly-A$^+$ RNA (5 μg) was size fractionated on an agarose gel,[46] transferred to nitrocellulose, and hybridized to kinased oligomers (see legend to FIGURE 2A). The 3′-14-mer was hybridized at 25°C and the 5′-14-mer and 15-mer were hybridized at 30°C. The blot was washed at room temperature in 6 × SSC, 0.1% sark, 0.1% NaPPi, and exposed to X-ray film. After hybridization, the blot was washed in 10 mM EDTA, 0.1% SDS, 0.1% NaPPi at 60°C until no counts remained and reprobed. The same lane of the blot is shown hybridizing to all three probes. The order of probes hybridized to the blot was 5′-14-mer, 15-mer, 3′-14-mer.

copy of chromogranin A mRNA using the enzyme reverse transcriptase to elongate the primer on the mRNA template (FIG. 2A). FIGURE 2B illustrates the products of this primer extension reaction, fractionated on an 8% acrylamide gel and visualized by autoradiography. Each probe produced multiple primer extension products. We reasoned that the correct primer extension product should have the following properties. (1) It should be greater than 100 bases in length, to accommodate mRNA for a 15-30 amino acid signal sequence and 5′ untranslated region. (2) The transcript elongated from the 5′ 14-mer had to be 10 bases shorter than the transcript elongated from the 3′ 14-mer. The bands labeled A and B in FIGURE 2B met all these criteria.

The primer extension products A and B were cut out of the gel, eluted, and sequenced using the procedure developed by Maxam and Gilbert.[15] Sequencing of these primer extension products confirmed that they were primed from chromogranin A mRNA (see FIGURE 3). The first ATG at position 179 starts an open reading frame that predicts the same protein sequence for chromogranin A[14] with the exception of the second residue, which translates to proline rather than arginine. Interestingly, an 18-residue stretch of hydrophobic amino acids precedes the first amino acid that was determined by protein sequencing. This stretch of hydrophobic residues meets all the criteria for a signal sequence.[16] Since chromogranin A is a secreted protein, a signal sequence is not unexpected.

Screening

The sequence of the 5′ end of chromogranin A mRNA obtained by Maxam-Gilbert sequencing of specifically primed cDNAs was used to design new oligonucleotide probes exactly matching complementary sequences in DNA encoding chromogranin mRNA. The 18-residue oligonucleotide (18-mer) and the 30-residue oligonucleotide (30-mer) shown in FIGURE 3 were synthesized and used to screen cDNA libraries made from bovine adrenal medulla poly-A$^+$ RNA. We obtained a nearly full-length chromogranin A cDNA clone, pCHGR4A, using the 30-mer to screen a Berg-Okayama[17,18] bovine adrenal cDNA library constructed with the help of Hiroto Okayama. This clone contained an 1883 base insert that includes all but 37 bases of the 5′ end of chromogranin A mRNA. We also screened a cDNA library constructed from calf adrenal medulla RNA, also constructed using the Berg-Okayama cloning vector, kindly provided by Mike Brown, Joe Goldstein, and David Russel of the University of Texas Health Sciences Center at Dallas. The longest cloned insert found from this library, pAMCGR39, represented only about 2/3 of chromogranin A mRNA. We found no clones that hybridized to chromogranin A sequences that were not from the same mRNA.

Chromogranin A had been reported to be present in the pituitary, but appeared to have a different molecular weight,[19,20] so we wanted to determine whether a chromogranin A mRNA could be detected in this tissue, and if it was the same size as that in the adrenal medulla. We used the oligonucleotide (not shown) and cDNA probes to probe bovine pituitary poly-A$^+$ RNA. An RNA species of the same size as adrenal medulla chromogranin A mRNA is present in bovine pituitary (see FIG. 10). A Berg-Okayama library made from bovine pituitary mRNA was generously provided by Fred Esh from the Salk Institute in San Diego. The insert from pCHGR4A was used to screen the pituitary cDNA library to determine whether chromogranin A mRNA is identically expressed in bovine pituitary. The longest hybridization-

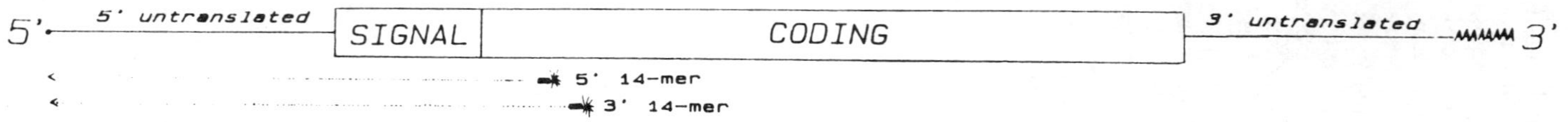

FIGURE 2A. Primer extension strategy. Primers were labeled at their 5′ ends by the following procedure. The oligonucleotide was added to a 10% molar excess of high specific activity γ-^{32}P-ATP (New England Nuclear) in 50 mM Tris pH 9.0, 10 mM $MgCl_2$, 1 mM dithiothreitol, 0.1 mM EDTA, 0.1 mM spermidine, and at least 1 unit per 10 picomoles primer of polynucleotide kinase (New England Biolabs) and was incubated for 30-60 minutes at 37°C. The reaction was stopped by the addition of EDTA to 20 mM and extracted with phenol/chloroform. An oligonucleotide primer that had been labeled at its 5′ end was coprecipitated in ethanol with 10 μg adrenal medulla polyadenylated RNA. The molar ratio of primer to chromogranin mRNA was calculated to be approximately 100:1 based on chromogranin mRNA being 0.05% of the total adrenal medulla mRNA and 2.1 kb in length. The pellet was resuspended in 10 μl 10 mM Hepes pH 7.4, 1 mM EDTA; 1 μl 1 M KCl was pipetted onto the side of the eppendorf tube (not mixed in). The RNA-primer mixture was heated to 90°C for 3 minutes and the KCl mixed in by a quick centrifugation. The temperature of the priming reaction was sequentially lowered in several steps, 5°-10°C each, over a total time period of 30 minutes, then brought to 37°C for 10 minutes. All the components necessary for a reverse transcriptase reaction were assembled on ice, then added to the RNA-primer mixture to give the following final concentrations: 50 mM Tris pH 8.6 at 42°C, 60 mM KCl, 8 mM $MgCl_2$, 0.4 mM DTT, 1 mM each dATP, dCTP, dGTP, and dTTP, 2 mM sodium pyrophosphate (to inhibit hairpin loop formation), RNA 0.1 mg/ml, 5 units/μg RNA RNAsin (Promega Biotech), and 5 units/μg RNA reverse transcriptase. The reaction was incubated at 37°C for 3 minutes, then 42°C for 30 minutes and terminated by adding EDTA to 10 mM, 1 μg tRNA (as a carrier) and extracting with one volume of 1:1 phenol/chloroform. The phenol/chloroform was reextracted with 1/5 volume 0.3M sodium acetate (to increase yield) and the aqueous phases were pooled and ethanol precipitated. Pellets were washed twice in 70% ethanol, dried, and resuspended in 4 μl 20 mM EDTA. 4μl 0.2 M NaOH was added, the sample heated to 55°C for 20 minutes and then run on an 8% acrylamide sequencing gel. Specific primer extension products were visualized by autoradiography for 2 hours at room temperature. The labeled cDNA fragments were cut out of the gel, eluted, and sequenced by the methods of Maxam and Gilbert[15]; 5 μg tRNA was added as a carrier to increase yield.

positive colony detected from this library, pPCGR8, has a 1842 base insert which includes all but 87 bases of the 5′ end of the mRNA.

Sequence Analysis

The clones pCHGR4A, from bovine adrenal medulla, and pPCGR8, from the bovine pituitary, were sequenced by the dideoxy termination method according to the strategies depicted in FIGURE 4. These two clones are identical in sequence except

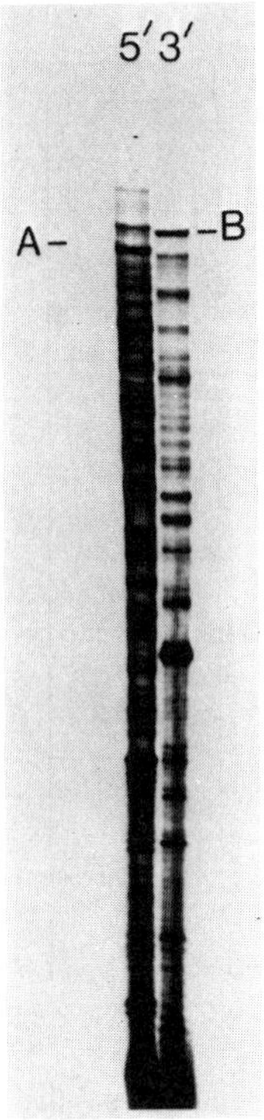

FIGURE 2B. Primer extension products. The 5′ and 3′ 14-mers were kinased and used to prime cDNA synthesis on poly-A^+ RNA derived from bovine adrenal medulla.

for the 40 base pair difference in length at the 5′ end. Thus, the same chromogranin A mRNA is expressed in the pituitary and the adrenal medulla. The size differences reported[19,20] between the two major chromogranins from these tissues are not due to translation of different mRNAs.

Clones pAMCGR7, pAMCGR15, pAMCGR39, pAMCGR54, and pAMCGR56 from the Brown, Goldstein, and Russel bovine adrenal medulla cDNA library were found to be incomplete cDNAs and were partially sequenced as shown in FIGURE 4. The poly-A tails were extraordinarily long in some of these clones (greater than 120 as estimated on sequencing gels), and were found to be progressively removed from

```
           10                                  30                                  50
5'-GCTCCTGGGATCTGCATCTGGTTGCTGGCGGCGTACACGGAGACCTTGCACCACCAAACC
   ^

           70                                  90                                  110
   CCATCCCCGCGCTGGTGTCGCCGCAGCTTGCCTGGAGCGAGCAGTCCAGCCGCCCCTCGC

          130                                 150                                 170
   CCGAGCGCGCGCCGTGGCCCGCCCCAGACCACCAGCTGCTCGGCGCCCCGGCTTCGCC
                                                       [-----18-mer-----]


              190                           210
ATG CGC TCC GCC GCG GTC CTG GCG CTT CTG CTC TGC GCG GGG CAA
 M   R   S   A   A   V   L   A   L   L   L   C   A   G   Q
{~~~~~~~~~~~~~~~~~~~~~~Signal sequence~~~~~~~~~~~~~~~~~~~~~~


                                    [----5'-14-mer---]
       230                          250
GTC ATT GCC CTG CCT GTG AAC AGC CCC ATG AAT AAA GGG GAC ACT
 V   I   A   L   P   V   N   S   P   M   N   K   G   D   T
~~~~~~~~~~~}    (R)             [---------------30-mer-----


     [----3'-14-mer---]
270
GAG GTG ATG --- --- --- --- --- --- --- --- --- -3'
 E   V   M   K   C   I   R   E   V   I   S   D
----------]
```

FIGURE 3. Sequence of the 5′ end of chromogranin A mRNA. The signal sequence and the positions of the oligonucleotides are identified. The two 14-mer probe pools were used as primers to specifically prime cDNA synthesis as described in the legend to FIGURE 2A. The 18-mer and 30-mer were designed from the specifically primed sequence data for screening cDNA libraries. The 5′ G residue was inferred to be the capped nucleotide found on mRNAs. The sequence shown here is derived both from the primer extension experiments and from sequence of chromogranin A cDNA clones pPCGR8 and pCHGR4A.

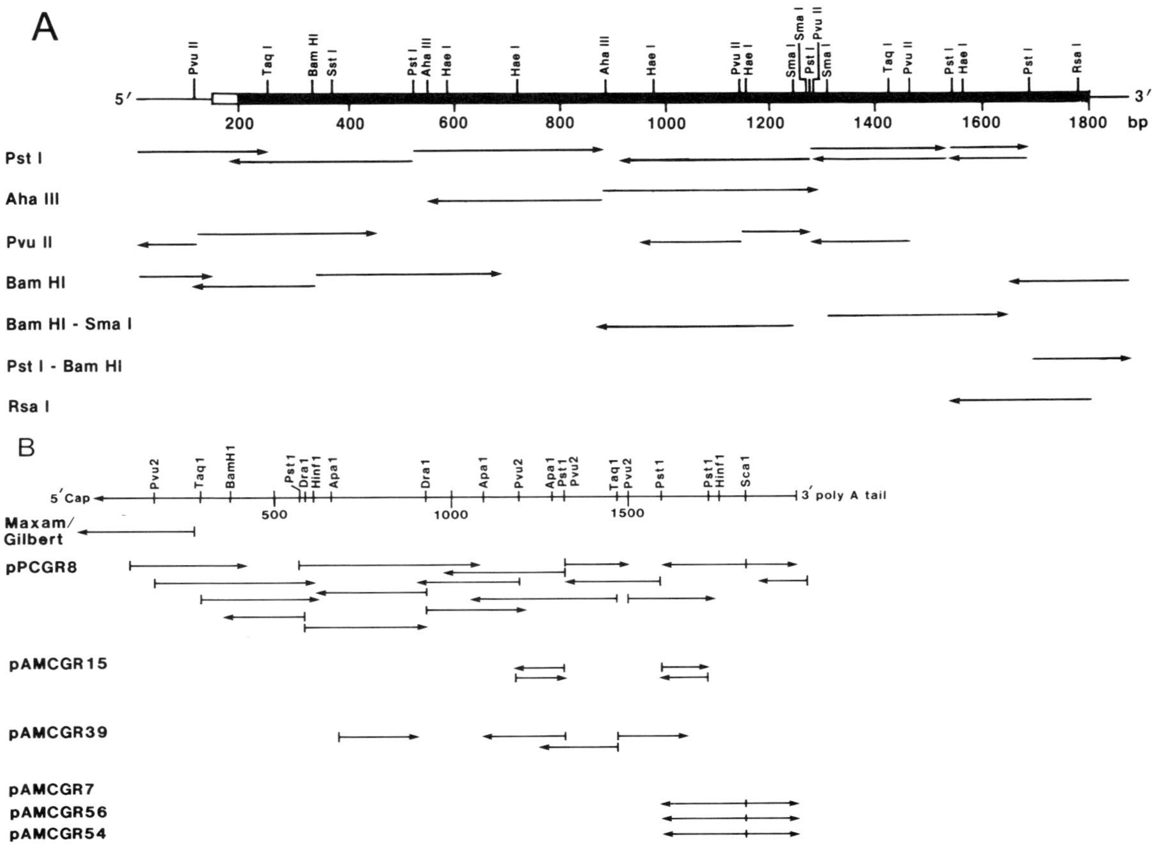

FIGURE 4. Sequencing strategy for cDNA clones. (**A**) Strategy for pCHGR4A. (**B**) Strategy for other cDNA clones. *Arrows* depict which strand was sequenced from each cDNA clone. Sequencing was done by the method of Sanger *et al.*[44] The region sequenced from the primer extension products is also shown.

the clones during replication of the plasmid DNA in the bacteria, causing the restriction fragments of the 3′ ends of these clones to appear heterogeneous in size. This artifact was troublesome when trying to determine which clones were related to one another by restriction mapping. Usually such long poly-A tails are not represented in cDNA clones. Perhaps the Berg-Okayama method of making cDNA (priming reverse transcriptase from a T-tailed plasmid) allows most of the poly-A tail to be included in some cDNA clones. In the pituitary and adrenal medulla the same mRNA is expressed, as represented by pPCGR8 and pCHGR4A. None of the other clones that were sequenced were found to be different from these, with the exception of pAMCGR39, which had a cytosine instead of a thymine residue substituted in the third position of a serine codon (nucleotide number 823). Presumably this clone represents an allelic variant that has no effect on the translated protein.

The Chromogranin A Protein

The complete sequence of chromogranin A mRNA and the deduced protein is shown in FIGURE 5. The total length of the mRNA without the poly-A tail is 1929 bases. Allowing for a poly-A tail of 100-200 nucleotides, this is consistent with the 2.1 kb messenger RNA detected by the oligomer and cDNA probes on Northern blots. The first ATG at position 179 starts an open reading frame that translates into a protein of 449 amino acids, including the signal sequence of 18 amino acids mentioned above. The amino acid composition of this protein is very similar to that reported for chromogranin A and secretory protein I (see TABLE 1), and is in partial agreement with the data of Hamilton *et al.*[7]

Because the molecular mass calculated from the protein sequence derived from cDNA is much smaller than that reported for the molecular weight of chromogranin A determined by gel electrophoresis of translation product (48 kDa versus 70-75 kDa), it was considered possible that the cDNA clones pPCGR8 and pCHGR4A encoded a member of the chromogranin family related to but not identical with chromogranin A. This appears very unlikely, however, because when adrenal medulla mRNA that had been enriched previously by hybrid selection with pCHGR4A is used in the cell-free translation experiment, a 75-kDa chromogranin A protein is produced (see FIG. 6A). The protein products produced by cell-free translation using selected and nonselected RNAs as templates were compared by cell-free translation and immunoprecipitation with chromogranin A antibody. The most abundant protein produced by the selected RNA had an apparent mobility of 75-kDa sodium dodecyl sulfate polyacrylamide gel electrophoresis, and was specifically immunoprecipitated by antibody to chromogranin A. Also, translation of an RNA produced by SP6 transcription of the coding region from pCHGR4A (from nucleotide 164 to the 3′ end) gives rise to a 75-kDa protein that reacts with chromogranin A antibody (FIG. 6B). Thus, chromogranin A is a 48-kDa protein that migrates with an apparent molecular mass of 75 kDa and is encoded by cDNA clones derived from the pituitary (pPCGR8) and adrenal medulla (pCHGR4A). Bovine chromogranin A has a molecular mass of 61 kDa when determined by filtration in 6 M guanidine HCl.[4] The discrepancy between the protein's mobility on SDS gels and its actual molecular mass could be due to a number of factors. It may have unusual SDS binding attributes, or the large number of charges on the molecule may cause it to appear to take up more space and have a higher coefficient of resistance during migration in the gel.

The chromogranin A protein is unlike any globular protein in its sequence. One unusual feature of this protein is the high content of acidic amino acids (19.3% glutamic acid + 5.1% aspartic acid = 24.4% acidic residues) dispersed throughout the protein and clustered in several regions (oligoglutamic acid regions). The three cysteine residues in the protein are in the first 60 amino acids: one at position 12 in the signal sequence and the others at positions 35 and 56. No *N*-linked glycosylation sites are present in chromogranin A.

Chromogranin A has been called a "random-coil polypeptide".[2, 23–28] Prediction of secondary structure-forming regions based on the primary structure of chromogranin A is consistent with this model. The protein has some possible α-helix-forming regions.[29] In all cases except in the signal sequence and the very NH_2-terminus (residues 31-40), however, the collection of residues that might form an α-helix structure are made up of many residues with the same charge (mostly glutamic acid), which would tend to destabilize formation of the α-helix due to charge-charge repulsion. The β-sheet score across the whole chromogranin A protein (except in the same region of the NH_2-terminus) is very low. Thus, analysis of the protein sequence in terms of its possible secondary structure-forming regions supports the idea that chromogranin A is, unlike most globular proteins, a "random coil polypeptide."

Except for the signal sequence, chromogranin A is an extremely hydrophilic protein. The rest of the protein scores very high on hydrophilicity tests, and very low on hydrophobicity. This is consistent with its behavior as a very soluble protein, and is inconsistent with a membrane-bound form of this protein as Settleman *et al.* have suggested.[11] The protein they purified from membrane fractions could be a related protein with a similar molecular weight and identical NH_2-terminus. We have not uncovered any cDNA clones that represent any mRNAs that encode such a membrane-bound chromogranin. Thus, the origin of the membrane bound chromogranin described by Settleman *et al.*[11] remains to be elucidated.

Related Proteins

FIGURE 7 compares the sequence of chromogranin A from the cDNA sequence with the sequences from other proteins with similar amino termini. Residues in common with chromogranin A are underlined. The NH_2-terminal sequence of chromogranin A deduced from the cDNA clone is identical to that of human and bovine chromogranin determined by protein sequencing by Kruggel *et al.*,[21] and differs by two amino acids from the protein sequence determination reported by Hogue-Angeletti[14] and Settleman *et al.*[11] for bovine chromogranin A. The latter may represent the NH_2-terminal sequence of another chromogranin A protein, or an allelic variant of the protein, or errors arising from the protein sequencing method.

The NH_2-terminal sequence of chromogranin A is identical to that of the NH_2 terminus of SP-I[6]; all the differences that were noted in Cohn *et al.*[6] are resolved by the sequence predicted from the primer extension product and cDNA clones (FIG. 7). Also, the determination of the position of leucine and methionine residues in the signal sequence of SP-I corresponds exactly to the positions of these residues in the 18 amino acid signal sequence of chromogranin A (see FIG. 7). Hamilton *et al.*[7] have compared SP-I and chromogranin A tryptic peptides and found that each protein exists in a 66-kDa and a 72-kDa form, which are resolved by HPLC. Subtle differences in the tryptic maps of the proteins derived from parathyroid and adrenal medulla

```
          10                  30                  50                  70                  90
GCTCCTGGGATCTGCATCTGGTTGCTGGCGGCGTACACGGAGACCTTGCACCACCAAACCCCATCCCCGCGCTGGTGTCGCCGCAGCTTG

                  110                 130                 150                 170
CCTGGAGCGAGCAGTCCAGCCGCCCCTCGCCCGAGCGCGCCGTGGCCCGCCCCAGACCACCAGCTGCTCGGCGCCCCGGCTTCGCC

             190                    210                            230
ATG CGC TCC GCC GCG GTC CTG GCG CTT CTG CTC TGC GCG GGG CAA GTC ATT GCC  CTG CCT GTG AAC
 M   R   S   A   A   V   L   A   L   L   L   C   A   G   Q   V   I   A    L   P   V   N
                                                                              20

     250                       270                           290                        310
AGC CCC ATG AAT AAA GGG GAC ACT GAG GTG ATG AAG TGT ATC GTC GAG GTC ATC TCT GAC ACA CTC
 S   P   M   N   K   G   D   T   E   V   M   K   C   I   V   E   V   I   S   D   T   L
                                                                     40

                           330                 350                      370
TCC AAG CCC AGC CCC ATG CCA GTC AGC AAG GAG TGT TTT GAG ACA CTC CGA GGA GAT GAA CGG ATC
 S   K   P   S   P   M   P   V   S   K   E   C   F   E   T   L   R   G   D   E   R   I
                                                                 60

                 390                     410                          430
CTC TCA ATC CTG CGA CAT CAG AAT TTG CTG AAA GAG CTC CAA GAC CTC GCT CTC CAA GGA GCC AAG
 L   S   I   L   R   H   Q   N   L   L   K   E   L   Q   D   L   A   L   Q   G   A   K
                                                     80

        450                           470                      490
GAG CGG ACA CAT CAG CAG AAG AAG CAC AGC AGT TAC GAG GAT GAA CTC TCA GAG GTG CTT GAG AAG
 E   R   T   H   Q   Q  [K   K]  H   S   S   Y   E   D   E   L   S   E   V   L   E   K
                                         100
```

```
510                          530                                          550                                          570
CCG  AAC  GAC  CAG  GCC  GAG  CCG  AAA  GAG  GTG  ACA  GAA  GAG  GTG  TCC  TCC  AAG  GAT  GCT  GCA  GAA  AAA
 P    N    D    Q    A    E    P    K    E    V    T    E    E    V    S    S    K    D    A    A    E   [K]
                                             120

                         590                                          610                                          630
AGA  GAC  GAT  TTT  AAA  GAG  GTG  GAG  GAG  AGT  GAT  GAA  GAC  TCG  GAC  GGA  GAC  AGG  CCT  CAG  GCT  CCC
[R]   D    D    F    K    E    V    E    E    S    D    E    D    S    D    G    D    R    P    Q    A    P
                                        140

               650                                          670                                          690
CAG  GGC  TTG  GGC  CGG  GGC  CCC  AAG  GTT  GAG  GAG  GAC  AAC  CAG  GCC  CCT  GGG  GAG  GAG  GAG  GAG  GCC
 Q    G    L    G    R    G    P    K    V    E    E    D    N    Q    A    P    G    E    E    E    E    A
                              160

   710                                    730                                          750                                          770
CCC  TCC  AAC  GCC  CAC  CCC  CTA  GCC  AGC  CTC  CCC  AGC  CCG  AAA  TAC  CCA  GGC  CCA  CAG  GCC  AAG  GAG
 P    S    N    A    H    P    L    A    S    L    P    S    P    K    Y    P    G    P    Q    A    K    E
                180

                              790                                          810                                          830
GAC  AGC  GAG  GGT  CCC  TCC  CAG  GGT  CCA  GCC  AGC  AGG  GAG  AAG  GGC  CTG  AGT  GCA  GAG  CAA  GGG  AGG
 D    S    E    G    P    S    Q    G    P    A    S    R    E    K    G    L    S    A    E    Q    G    R
   200                                                                                                  220

                    850                                          870                                          890
CAG  ACA  GAG  AGA  GAA  GAG  GAG  GAG  GAG  AAG  TGG  GAG  GAG  GCG  GAG  GCT  AGA  GAG  AAG  GCC  GTC  CCG
 Q    T    E    R    E    E    E    E    E    K    W    E    E    A    E    A    R    E    K    A    V    P
                                                                                                   240
```

FIGURE 5. (*Continued on following pages.*)

```
     910                        930                         950                     970
GAG GAA GAA AGC CCG CCC ACC GCA GCG TTT AAA CCC CCA CCG AGC CTC GGC AAC AAG GAG ACG CAG
 E   E   E   S   P   P   T   A   A   F   K   P   P   P   S   L   G   N   K   E   T   Q
                                                                     260

                          990                         1010                  1030
AGG GCT GCT CCA GGT TGG CCC GAG GAT GGA GCC GGG AAG ATG GGG GCT GAG GAG GCC AAG CCC CCT
 R   A   A   P   G   W   P   E   D   G   A   G   K   M   G   A   E   E   A   K   P   P
                                                             280

                1050                    1070                    1090
GAG GGG AAG GGG GAG TGG GCA CAC TCC CGG CAG GAG GAA GAG GAG ATG GCA AGG GCC CCT CAA GTC
 E   G   K   G   E   W   A   H   S   R   Q   E   E   E   E   M   A   R   A   P   Q   V
                                                   300

    1110                        1130                    1150
CTC TTC CGT GGT GGG AAG AGC GGG GAG CCC GAG CAG GAG GAG CAG CTC TCC AAG GAG TGG GAG GAC
 L   F   R   G   G   K   S   G   E   P   E   Q   E   E   Q   L   S   K   E   W   E   D
                                             320

1170                        1190                    1210                    1230
GCC AAG CGA TGG AGC AAG ATG GAC CAG CTG GCC AAG GAG CTG ACG GCC GAG AAG CGG CTG GAG GGG
 A  [K   R]  W   S   K   M   D   Q   L   A   K   E   L   T   A   E  [K   R]  L   E   G
                                       340

                    1250                    1270                    1290
GAG GAG GAG GAA GAG GAG GAC CCC GAC CGC TCC ATG AGG CTC TCC TTC CGG GCC CGG GGC TAC GGC
 E   E   E   E   E   E   D   P   D   R   S   M   R   L   S   F   R   A   R   G   Y   G
                             360

        1310                        1330                    1350
TTC AGG GGT CCC GGG CTG CAG CTG CGG CGA GGC TGG AGG CCG AAC TCC CGG GAG GAC AGC GTG GAG
 F   R   G   P   G   L   Q   L  [R   R]  G   W   R   P   N   S   R   E   D   S   V   E
                     380
```

```
 1370                                 1390                                 1410                                 1430
GCC GGC CTG CCC CTC CAG GTG CGC GGC TAC CCG GAA GAG AAG AAG GAG GAG GAG GGC AGC GCC AAC
 A   G   L   P   L   Q   V   R   G   Y   P   E   E  [K   K]  E   E   E   G   S   A   N
            400

                          1450                                 1470                                 1490
CGC AGA CCA GAG GAC CAG GAG CTG GAG AGC TTG TCA GCC ATC GAG GCA GAG CTG GAG AAG GTG GCC
[R   R]  P   E   D   Q   E   L   E   S   L   S   A   I   E   A   E   L   E   K   V   A
   420                                                                                440

                1510                                 1530                                 1560
CAC CAG CTG GAG GAG CTT CGG CGG GGC TGA GGCACTGACTGGCCCCACCAGCCAGGGCCCCGAGACACCTCGTGGTCCC
 H   Q   L   E   E   L  [R   R]  G  End

1580                     1600                     1620                     1640                     1660
 GGCTCTGTTGCCCCCTCTGCAGGTCCTGGCCAGACGGCCCCAGGCACTGCTTCCGGGAGGGAGGCCGCCCCCAGCCTGCCCAAGCCCAG

                 1680                     1700                     1720                     1740
 CCCACCCCATTGCCCCCACGCTCTTTTCCTCTTGCTCCTGACCCCTGCCCAGTGCGCCCCCCTGCAGGGCAGACCCCTTGCCTTTCAAC

  1760                     1780                     1800                     1820                     1840
 GATTATCCTGTCTCTGAACACAGGCAGCCTTCTCAAAGTTTCCCTTCCACCATTAGAGCCACTGGGCCTAACTGCAATAAGTACTGACC

                   1860                     1880                     1900                     1920
 TTTAGGTGAAAGCTGAGGACTCCTGACTATATTGTGTATGAAGTTTATCTAAGGAAAATAAATCTGCTCTGGGCTCTTTCCTGTTAAAAAAAA
```

FIGURE 5. (*Continued from preceding pages.*) Complete sequence of chromogranin A mRNA and protein. Numbers over the sequence refer to the nucleotides, starting at the 5′ cap. Numbers below the protein sequence refer to amino acid residues, starting with the first methionine of the signal sequence. Pairs of basic amino acids are *boxed*.

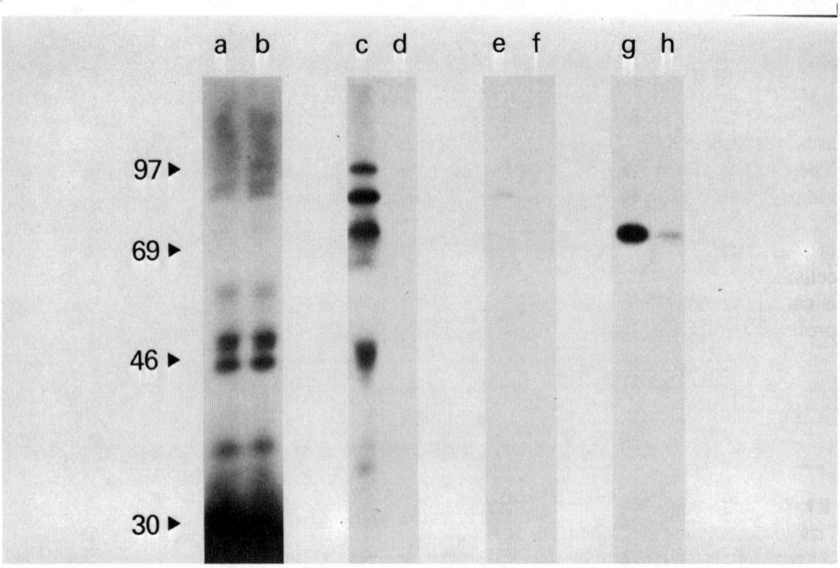

FIGURE 6A. Demonstration that pCHRG4A is a cDNA clone for chromogranin A messenger RNA. Immunologic identification of protein encoded by pCHRG4A as chromogranin A. Chromogranin-A-specific messenger RNA was obtained by hybridizing total bovine chromaffin cell RNA to pCHRG4A immobilized on nitrocellulose paper, washing away unbound RNA, and eluting specifically bound mRNA by boiling in water for 2 minutes. Hybrid-selected and hybrid-nonselected RNA were used to program rabbit reticulocyte lysate for cell-free translation with [^{35}S]methionine. Translation mixtures were immunoprecipitated in the presence or in the absence of excess unlabeled authentic bovine chromogranin A (courtesy of Dr. Claude Gagnon, Laval University, Quebec). Immunoprecipitation was carried out using the Europium antiserum whose immunological properties have been described elsewhere (Ehrhart, Grube, Bader, Aunis, and Gratzl, J. Histochem. Cytochem. **34:** 1673, 1986). Goat anti-rabbit gamma globulin antiserum was used to immunoprecipitate the labeled chromogranin A/antichromogranin A complex. The immune pellet was extensively washed and denatured in 1% SDS, 1% β-mercaptoethanol, 50 mM Tris pH 6.5 by heating for 120 min at 37°C and then 2 min at 95°C, and each sample electrophoresed on a 10% polyacrylamide gel, followed by fixation and fluorography. Lanes a and b, Cell-free translation of total chromaffin cell mRNA. Lane b shows all proteins present in the mixture, and lane a the proteins remaining after removal of chromogranin A by immunodepletion with the Europium antiserum. Lanes c and d, Immunoprecipitates from total chromaffin cell mRNA carried out in the presence (lane d) or the absence (lane c) of excess unlabeled antigen. The Europium antiserum precipitates three major protein bands, corresponding to chromogranins A (75 kDa), C (85 kDa), and B (100 kDa). Lanes e and f, Immunoprecipitates from pCHRG4A-nonselected RNA in the presence (lane f) and the absence (lane e) of excess unlabeled antigen. Lanes g and h, Immunoprecipitates from pCHRG4A-selected RNA in the presence (lane h) and the absence (lane g) of excess unlabeled antigen.

were noted. These differences could be ascribed to proteins with a common NH_2 terminus but different internal regions. However, it is more likely that these differences are due to different posttranslational modifications of the same proteins in the adrenal and parathyroid glands, since the proteins were purified directly from tissues rather than from cell-free translation systems. Thus, it is likely that these two proteins, chromogranin A from the adrenal medulla and SP-I from the parathyroid, represent the same gene product. Isolation and characterization of chromogranin A cDNA clones from the parathyroid should define the precise difference, if any, between SP-I and chromogranin A mRNAs.

Settleman *et al.*[11] have partially sequenced fragments of a 74-kDa chromogranin as well as higher and lower molecular mass proteins (100 kDa, 85 kDa, and 65 kDa). The latter chromogranins have NH_2-terminal sequences quite similar to, but not identical with, the reported sequence for chromogranin A (FIG. 7).

Chromogranin A differs by one out of its first 14 amino acids from the sequence of β-granin, a 21-kDa protein cosecreted with insulin from rat beta pancreatic islet cells.[22] Since this sequence is from rat β-granin, it remains to be seen if this difference is caused by tissue or species differences. It is possible that this 21-kDa protein is a proteolytic fragment of chromogranin A expressed in the pancreas.

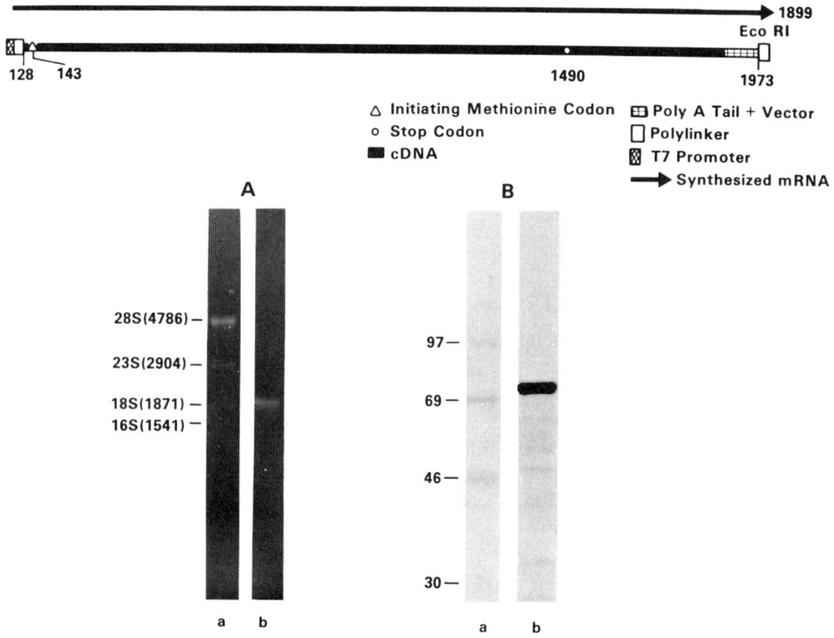

FIGURE 6B. Demonstration that the 48.3-kDa protein encoded by pCHRG4A has an apparent molecular mass on SDS PAGE of 73.5 kDa. The *Nar* 1-*Xho*1 fragment of pCHRG4A containing the entire protein coding region was subcloned into the GEM2 transcription vector, and transcribed from the T7 promoter after linearizing the plasmid with *Eco*R1. This RNA contains a single long open reading frame with initiating methionine at position 143 and stop codon at position 1490, thus coding for a protein of molecular mass 48.3 kDa. (**A**) Ribosomal RNA markers are shown in lane a and the *in vitro* transcribed mRNA in lane b. (**B**) About 400 mg of the mRNA shown in FIGURE 6A was translated in the rabbit reticulocyte lysate system as described in the legend to FIGURE 6A. Protein markers are shown in lane a, and the cell-free translation products of the mRNA synthesized from the pCHRG4A template in GEM2 are shown in lane b. The major product is a protein of apparent molecular mass 73.5 kDa.

	Signal Sequence: 1 5 10 15	N-terminal sequences: 20 25 30 35 40
Translated:	M R S A A V L A L L L C A G Q V I A	L P V N S P M N K G D T E V M K C I V E V I S D
SP-1(6,45):	M L L L L	L P V N S P M N K G D T E V M K x I V E
Bovine(21):		L P V N S P M N K G D T E V M K C I V E V I S D
Human(21):		L P V N S P M N K G D T E V M K C I V E V I S D
p100(11):		S P V N I H M N K G E V E V M K x I V E
p85(11):		M P V N I P M N K G E V E V M K I I V E
p75(11,14):		L R V N S P M N K G D T E V M K C I R E V I S D
p75, 32 kD peptide(11):		L P V N Q P G N K L D E E V L K L I V E
p65(11):		M P V D I N N H N E E V V T H K I L E V L x N A L
Rat B-granin(22):		L P V N S P M T x G D T x V

FIGURE 7. Amino termini of several chromogranin proteins. The residues that match those of chromogranin A are *underlined*.

Calcium-binding Regions

Chromogranin A has been reported to bind calcium.[30,31] In support of these data, chromogranin A shares regions of homology with S-100β, paravalbumin β, and intestinal calcium-binding proteins (FIG. 8). One class of calcium-binding pocket (or EF-hand region) in proteins consists of an α-helix, a β-turn about the calcium ion, and another α-helix.[32] FIGURE 8A summarizes sequence data from 26 EF-hand loops.[33] Four out of the seven strongly conserved amino acids are present in chromogranin A. This region has four of six ligand binding residues (the site designated $-Y$ is a carboxyl group from the peptide backbone). Chromogranin A, however, lacks the COOH-terminal ligand binding residues; they are replaced with prolines that would disrupt the COOH-terminal α-helix of the EF-hand. A second class of calcium-binding region can also be identified within chromogranin A as a region sharing 5-residue identity with intestinal calcium-binding proteins (FIG. 8B). A third class of calcium-binding region, identified recently by Geisow *et al.*[34] and Kretsinger and Creutz,[35] is incomplete within chromogranin A (FIG. 8C). It is not known whether the latter limited homology is significant. Other potential calcium-binding regions within chromogranin A include the oligoglutamic acid regions at positions 171-174, 224-228, 242-245, 298-301, and 353-358. The random-coil properties of chromogranin A may not allow it to form the α-helices of a calcium-binding pocket, but the regions identified by homologies to other calcium-binding proteins, and the oligoglutamic acid regions, are likely to be the source of the ability of chromogranin A to bind calcium with low affinity.

It is interesting to note that the two regions of chromogranin A identified by their homology to EF-hands are the two internal sequences most related to each other. Several repeating segments within chromogranin A protein sequence can be seen, although their relationship to one another is only apparent using a mutation-biased method for scoring amino acid relationships.[36] The best match of these repeated segments is only partial (the segments from amino acid 305 to 326 and from amino acid 153 to 175, see FIGURE 9). If the chromogranin A gene originated by a duplication of this unit, it must have occurred long ago in evolutionary time, for it has diverged considerably. Interestingly, the internal peptide sequences reported by Settleman *et al.*[11] are not contained within the chromogranin A molecule encoded by pPCGR8 and pCHGR4A, but they are related to the same family of repeating units (see FIG. 9). The single sequence that is most homologous to the peptides reported by Settelman *et al.* is the segment from amino acid 305 to 326. These data suggest that there may be several different species of chromogranins of the same apparent molecular mass (75 kDa), derived from the same common ancestral repeating unit, the EF-hand.

Possible Cleavage Products

The chromogranin A protein contains eight pairs of basic amino acid residues at positions 95 (KK), 132 (KR), 332 (KR), 348 (KR), 383 (RR), 410 (KK), and 419 (RR), and one at the COOH terminus 447 (RR). Since pairs of basic residues serve as recognition signals for endoproteolytic processing enzymes in many neuropeptide precursors, this protein may be cleaved into smaller peptides. If all the Lys-Arg or Lys-Lys pairs are recognized as cleavage sites the sizes of the peptides arising from

A

EF-hand: Loop Region

Ligand binding:	+X		+Y		+Z		-Y		-X			-Z
Conserved residues:	D	pos	D N	G	D N	G	X	I V	neg pol	pos F, Y	neg	neg
% Conservation:	96	46	100	65	81	100	0	96	92	69	62	100
Chromogranin A 143:	D	E	D	S	D	G	D	R	P	Q	A	P

Examples

Chromogranin A - Bovine 134:	DDFKEVEESDE DSDGD RPQAPQGLGRGPKVEEDNQ
S-100B protein - Bovine:	EVVDKVMETL DSDGD GECDFQEFMA
Parvalbumin beta - Carp 90:	GETKTFLKAG DSDGD GKIGVDEFTA
Troponin C - rabbit:	EMIAEFKAAFDMF DADGG CDISVKELGTVMRMLGQT

B

EF-Hand with Double Insertions

Chromogranin A - Bovine	312:	GGKSGEPEQEE QLSKE WEDAKRWSKM
Calcium-binding protein, intestinal - Pig	25:	KYAAKEGDPN QLSKE ELKQLIQAEF
Calcium-binding protein, intestinal - Bovine	22:	KYAAKEGDPN QLSKE ELKLLLQTEF
Calcium-binding protein, intestinal - Rat (fragment)	16:	KYAAKEGDPN QLSKE ELKLLIQSEF

C

Calcium-Dependent Membrane Binding Proteins

	----Loop------- [~~~alpha-helix~~~]
Consensus (Geisow et.al.):	K G h G T D E x x L I p I L A p R
Chromogranin A 95:	Q Q K K H S S Y E D E L S E V L E K P N D
Lipocortin 128-144:	K G L G T D E D T L I E I L A S R

FIGURE 8. Calcium binding regions within chromogranin A. **(A)** EF-hand loop region and examples of homology. **(B)** Examples of homology to EF-hands with double insertions.[33] **(C)** A crude homology to calcium-dependent membrane binding proteins.[34,35] Protein sequences in **A** and **B** were obtained from the National Biomedical Research Foundation Protein Identification Resource, release 7.0.

Internal fragment sequences:

```
CGRA Translated:     298 - E E E E M A R A P Q V L F R G G K S G E P E Q E E Q L S K E W E - 329
     p75, 30 kD peptide:             D E L P F V L F A E I x L x x V E Q E E E L
     p75, 15 kD peptide:             P D E P Q V L F E G K K E G A P E Q E E Q L S K
     p75, 12 kD peptide:             A D A P E V L F R G D K S G D P E Q E E
```

```
 CGRA    135 - D F K E V E E S D E D S D G D R P Q A P Q G L G R G P K V E E D N Q A P G E E E E A P S N A H P L A S L P S P
p75, 30 kD peptide:- - - - - - - - - - - - - - - D E L P F V L F A E I x L x - - - - - - x V E Q E E E L - - - - - - - - - -
p75, 15 kD pep - - - - - - - - - - - - - - - - - P D E P Q V L F E G K K - - - - - - - - - - - - E G A P E Q E E Q L S K - - - -
p75, 12 kD pep - - - - - - - - - - - - - - - - - A D A P E V L F R G D K S G D - - - - P E Q E E - - - - - - - - - - - - - -
 CGRA    198 - - - - - - - - - - - E D S E G - P S Q G P A S R E K G L S A E Q G R Q T E R E E E E E - - - - - - - - - - - - -
 CGRA    290 - E W A H S R Q E E E E - - - - M A R A P Q V L F R G G K S G E - - - - P E Q E E Q - - - - - - - - - - L S K
```

FIGURE 9. Internal protein sequences of chromogranin A. Residues in common with chromogranin A 298 or 135 are *underlined.*

proteolytic processing of this protein would be 8.3 kDa (residues 19-94), 3.7 kDa (97-131), 22 kDa (133-331), 1.4 kDa (334-347), 6.5 kDa (350-409), and 4.3 kDa (412-449). However, the KK at position 410 is preceded by a proline three residues upstream (PEEKK), which may mask its ability to be cleaved. The COOH-terminal residue 449 is a glycine, which often contributes its nitrogen to form an amide on the adjacent residue during posttranslational processing. It is noteworthy that three out of the five Lys-Lys or Lys-Arg pairs are preceded by a glutamic acid residue (positions 131, 347, and 409), and one is preceded by glutamine (position 94). It is difficult to determine if the peptides found in chromaffin granules (which other workers have suggested are derived from chromogranins) are in fact the same size as these possible cleavage products of chromogranin A. If chromogranin A were processed like a peptide hormone, the resulting peptides are likely to have anomalous mobilities on SDS gels, as chromogranin A does, so the apparent size of each peptide is not enough information to determine its origin.

In addition, it is not possible to predict the glycosylation sites in chromogranin A because they are all *O*-linked.[3] Any of the peptides arising from cleavage of chromogranin A might be glycosylated, since potential *O*-linked glycosylation sites (serine or threonine residues) exist throughout the protein. Thus, possible glycosylated peptides and electrophoretic mobilities of these fragments would not be predictable. The generation of antibodies against chemically synthesized peptides contained within the chromogranin A sequence should allow us to determine whether the pairs of basic amino acids are used as cleavage sites in chromogranin A.

Tissue Distribution of Chromogranin A mRNA

Since chromogranin immunoreactivity has been reported in a number of neuroendocrine and brain tissues, we determined where chromogranin A mRNA is expressed by Northern blot analysis. In addition, because related proteins have been described, it is of interest to determine whether related mRNAs can be detected by this method.

The distribution of chromogranin A mRNA in endocrine tissues is consistent with the reported distribution of chromogranin A and SP-I, demonstrating that the pituitary, brain (caudate nucleus), and parathyroid and adrenal medulla are sites of synthesis of this protein (FIG. 10A). It is present in abundance in human pheochromocytoma as well, and has about the same size, demonstrating a homology between species. Chromogranin A mRNA was abundant in the caudate nucleus (FIG. 10A), but could not be detected in total RNA from cerebellum or hypothalamus (data not shown), confirming previous reports that this protein is not uniformly distributed in the brain.[5] The caudate nucleus is also rich in mRNA encoding proenkephalin. Thus, chromogranin A and Met-enkephalin appear to be coexpressed in at least two nervous tissues, the adrenal medulla and the caudate nucleus.

Chromogranin A mRNA is expressed in the pituitary and adrenal medulla. Chromogranin A mRNA is more abundant in the neurointermediate lobe than in the anterior lobe of the pituitary (FIG. 10). SP-I and chromogranin A are also identical or nearly identical gene products. An mRNA of the same size as chromogranin A mRNA is an extremely abundant species in bovine parathyroid (see FIG. 10A), much more abundant in fact than in the adrenal medulla. This is indicated by the finding that 0.1 μg of parathyroid poly-A$^+$ RNA gives a 5-10-fold stronger signal than 1 μg of adrenal medulla poly-A$^+$ RNA. The accumulation of data about their common properties strongly suggests that these proteins represent the same gene product, and

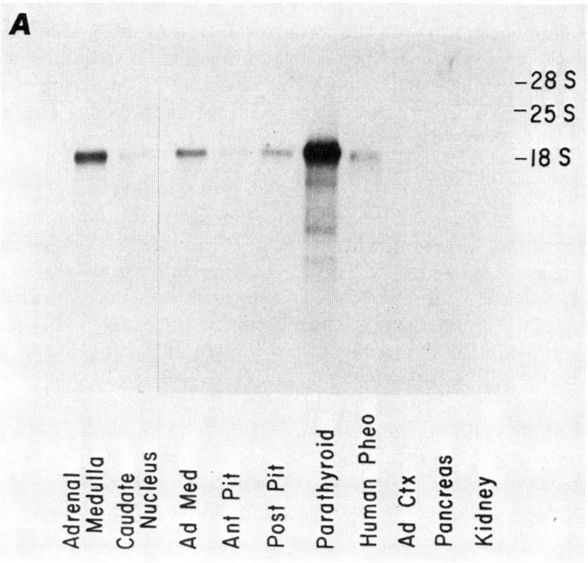

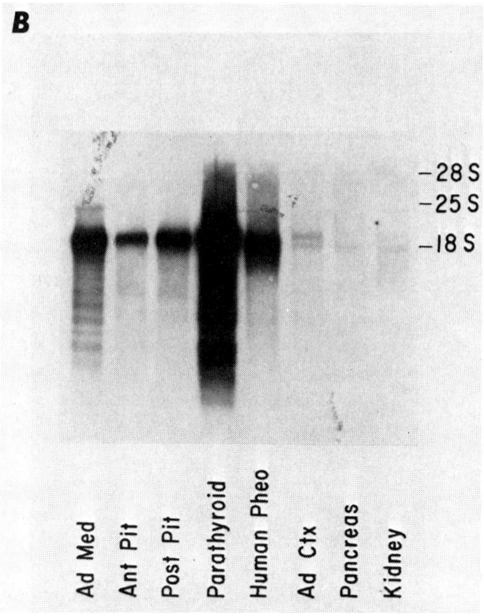

FIGURE 10. Tissue distribution of chromogranin A mRNA; 1 μg poly-A$^+$ RNA from all the tissues except parathyroid, 0.1 μg, was size fractionated on agarose gels and transferred to nylon membrane[43] (Schleicher and Schuell). The blot was hybridized at 60°C with an SP6 RNA transcript encoding the 5′ end (antisense) of chromogranin A (*Pst*1-*Pst*1 fragment). (**A**) The left two lanes, adrenal medulla and caudate nucleus, are from a separate blot that had been exposed 4 hours at room temperature to X-ray film. The rest of the blot was exposed 2 hours at room temperature. (**B**) Two-day room temperature exposure of the same blot shown in **A.**

that minor differences in peptides reported by Hamilton *et al.*[7] are due to differences in posttranslational processing. The minor differences in distribution of chromogranin A and SP-I reported previously[5,37] may also be derived from tissue-specific posttranslational processing, which is reflected in the epitopes of antibodies used in these studies.

After longer exposures of the same Northern blot, hybridizing bands about the same size as ribosomal 18S and 28S RNA appear in the adrenal cortex, caudate nucleus, pancreas, and kidney (FIG. 10B). These RNAs have been through one round of poly-A^+ selection, but probably contain some residual 18S and 28S RNA, which is nonspecifically hybridizing to the probe in this experiment. Chromogranin immunoreactivity has not been detected in the kidney, and chromogranin A cDNA does not specifically hybridize to any RNA species in this tissue.

Chromogranin A antigen has not been reported to be present in the adrenal cortex. This tissue contains a small amount of chromogranin A mRNA, which can be seen under long exposures of the autoradiogram (FIG. 10B). Chromogranin A may be a minor component in this tissue. We think it is likely, however, that a small amount of adrenal medulla contaminated the adrenal cortex tissue during the dissection. The chromogranin A mRNA in the adrenal cortex is probably a result of this contamination rather than a reflection of the presence of a small amount of chromogranin A mRNA in adrenal cortical cells. *In situ* hybridization should help to determine if the latter is the case.

β-Granin from the rat pancreas has been reported to have an NH_2-terminal sequence very similar to chromogranin A. FIGURE 10B shows bovine pancreas poly-A^+ RNA that was transferred to a nylon membrane and probed with a 500-base SP6 transcript derived from the 5′ end (*Pst*1 fragment) of the chromogranin A cDNA clone pPCGR8. This probe contains the region encoding the NH_2 terminus of the protein, yet does not specifically hybridize to any mRNA in the pancreas (FIG. 10B). The reasons for this could be: (1) the mRNA for bovine β-granin is not homologous enough to chromogranin A to hybridize at this degree of stringency; (2) β-granin is expressed only in rat pancreas, not in bovine pancreas; or (3) during the dissection, the part of the pancreas that contains β-granin (islet cells) was not taken. The latter is perhaps more likely because only a small portion of the pancreas, a large diffuse tissue, was used to prepare RNA. Further experiments, using more pancreas tissue to isolate RNA, are needed to resolve the issue of the relationship between β-granin and chromogranin A.

Other RNA Species

In the adrenal medulla a large species of mRNA (3.3 kb) hybridizes to all regions of chromogranin A mRNA. FIGURE 10 shows the result with the 5′ end used as a probe. Smaller bands can be seen as well; these differ with the probe used in the hybridization. The 3.3-kb RNA species in the adrenal medulla could be an precursor form of chromogranin A mRNA (unspliced transcript), or it could be a mRNA encoding one of the larger chromogranins. Interestingly, the 3.3-kb band is much less abundant in the parathyroid and not visible in the pituitary or caudate nucleus, even under long exposures (FIG. 10B). It is also of interest that minor proteins of molecular mass 87 and 100 kDa are partially enriched by hybrid selection with pCHRG4A, and are also immunoprecipitated with antichromogranin A antiserum (Europium 81), which has been demonstrated to react equally well with chromogranins of apparent molecular masses 100, 87, and 75 kDa as primary protein translates.[38] The latter two proteins probably correspond to the so-called chromogranins B and C identified and characterized by Fischer-Colbrie and Frischenshlager[12] and Falkensammer *et al.*[10] Possibly the mRNAs for the 87- and 100-kDa chromogranins share a limited homology with the cDNA clone. Thus, the larger and smaller RNAs are likely to be related but distinct gene products encoding other chromogranin proteins.

Genomic Southern blots probed at high stringency suggest that there is one gene for chromogranin A in the cow (not shown). Various segments of the cDNA clone were used as probes. Each of these probes hybridized to a single band on genomic DNA cut with enzymes that do not cut the cDNA clone (*Bgl*2, *Eco*R1, *Hind*2, and *Hind*3). There is at least one intron in the bovine gene, as the 5′ end (*Pst*1 fragment) and the 3′ end (*Pvu*2 fragment) hybridize to different bands in genomic DNA cut with the above enzymes (data not shown). The full structure of the gene will have to await characterization of a genomic clone.

DISCUSSION

One of the goals of this research is to provide a basis for future work on the function of chromogranins. A critical first step in this direction is to determine the sites (tissues) where these proteins are expressed. If, as appears to be the case for chromogranins, these proteins are expressed only in neurosecretory cells in association with neurotransmitters and neuropeptides,[5] then a role is suggested for chromogranins in secretion of these compounds. A broader distribution would suggest a more general function. Up to this point, studies of the tissue distribution of chromogranins have been done with antibodies. A major problem in such studies is that antibodies often cross-react with other proteins or react with molecular groups added posttranslationally to a protein. Thus, an independent measure of the sites of synthesis is needed. The isolation of a cDNA clone for chromogranin A provides an excellent tool for a much more precise localization of sites of chromogranin A synthesis.

The origin of the related chromogranin proteins is a very interesting issue at this point. We find no evidence for multiple initiation points on chromogranin A mRNA. These proteins could be encoded by a single gene that gives rise to alternately spliced mRNAs, which contain some exons in common and some that are different. Alternatively, each protein could be transcribed from a separate gene that evolved through duplication and subsequent divergence from a common ancestor. It is unlikely that the related RNAs, which are sufficiently homologous to hybridize to chromogranin A probes, have genes that do not hybridize to the same sequences. Thus, it can be argued that the most likely relationship among the various chromogranin proteins is that they originate from a single gene which is alternately spliced to produce different sized mRNAs. An alternative explanation is that the 3.3-kb RNA detected on Northern blots in the adrenal medulla is an unspliced primary transcript and the smaller RNA species are degradation products. We can now hope to obtain cDNA probes specific for each of the related mRNAs to determine whether they come from one or several genes.

The function of chromogranins is an important unresolved question. The occurrence of chromogranin immunoreactivity and mRNA expression suggests that chromogranins may play a general role in neuropeptide secretion or processing. Since chromogranins are in the same cells as neuropeptides, they also may be generally in the same secretory vesicles.[4,5,20,37,39–41] The presence of pairs of basic amino acid residues within chromogranin A suggests that specific processing of chromogranin A could occur. This could account for the smaller chromogranin peptides contained in secretory vesicles. Thus, a hormonal function can be imagined for the chromogranins. The Lys-Lys or Lys-Arg sequences of chromogranin A would then at least compete for neuropeptide processing enzymes, even if they are not efficiently cleaved, so chromogranins may affect processing. The generation of antibodies specific to synthetic peptides will tell us whether the pairs of basic residues in chromogranin A are used as processing signals.

Chromogranin A binds calcium with high capacity and low affinity,[30,31] and contains regions similar to other calcium-binding proteins. It has been suggested that chromogranin A may be a storage protein that regulates osmotic pressure inside the secretory vesicle.[42] Perhaps the calcium-binding properties of this protein do confer some stability on the secretory vesicle. Clearly, much exciting work on the biology of chromogranins lies ahead.

ACKNOWLEDGMENTS

We thank Hans-Urs Affolter for technical assistance and Jeff Bell for helpful discussions about calcium-binding proteins. Thanks also to Fred Esh and Mike Brown for sending bovine pituitary and calf adrenal cDNA libraries. M. G. is supported by the Oregon Affiliate of the American Heart Association.

REFERENCES

1. O'Conner, D. T. & K. N. Bernstein. 1984. Radioimmunoassay of chromogranin A in plasma as a measure of exocytotic sympathoadrenal activity in normal subjects and patients with pheochromocytoma. N. Eng. J. Med. **311:** 764.
2. Smith, A. D. & H. Winkler. 1967. Purification and properties of an acidic protein from chromaffin granules of bovine adrenal medulla. Biochem. J. **103:** 483.
3. Kiang, W. L., T. Krusius, J. Finne, R. U. Margolis & R. K. Margolis. 1982. Glycoproteins and proteoglycans of the chromaffin granule matrix. J. Biol. Chem. **257:** 1651.
4. O'Conner, D. T., R. J. Parmer & L. J. Deftos. 1984. Chromogranin A: Studies in the endocrine system. Trans. Assoc. Am. Physicians **97:** 242.
5. Somogyi, P., A. J. Hodgson, R. W. DePotter, R. Fischer-Colbrie, M. Schober, H. Winkler & I. W. Chubb. 1984. Chromogranin immunoreactivity in the central nervous system: Immunochemical characterization, distribution and relationship to catecholamine and enkephalin pathways. Brain Res. Rev. **8:** 193.
6. Cohn, D. V., R. Zangerle, R. Fisher-Colbrie, L. L. H. Chu, J. J. Elting, J. W. Hamilton & H. Winkler. 1982. Similarity of secretory protein I from parathyroid gland to chromogranin A from adrenal medulla. Proc. Natl. Acad. Sci. U.S.A. **79:** 6056.
7. Hamilton, J. W., L. L. H. Chu, J. B. Rouse, K. Reddig & R. R. MacGregor. 1986. Structural characterization of adrenal chromogranin A and parathyroid secretory protein-I as homologs. Arch. Biochem. Biophys. **244:** 16.
8. Kilpatrick, L., F. Gavine, D. Apps & J. Phillips. 1983. Biosynthetic relationship between the major matrix proteins of adrenal chromaffin granules. FEBS Lett. **164:** 383.

9. O'CONNOR, D. T. & R. P. FRIGON. 1984. Chromogranin A: The major catecholamine storage vesicle soluble protein. J. Biol. Chem. **259:** 3237.
10. FALKENSAMMER, G., R. FISCHER-COLBRIE, K. RICHTER & H. WINKLER. 1985. Cell-free and cellular synthesis of chromogranin A and B of bovine adrenal medulla. Neuroscience **14:** 735.
11. SETTLEMAN, J., R. FONSECA, J. NOLAN & R. HOGUE-ANGELETTI. 1985. Relationship of multiple forms of chromogranin. J. Biol. Chem. **260:** 1645.
12. FLEMINGER, G., E. EZRA, D. L. KILPATRICK & S. UDENFRIEND. 1983. Processing of enkephalin-containing peptides in isolated bovine adrenal chromaffin cultures. Proc. Natl. Acad. Sci. U.S.A. **80:** 6418.
13. FISCHER-COLBRIE, R. & I. FRISCHENSCHLAGER. 1985. Immunological characterization of secretory proteins of chromaffin granules: Chromogranins A, chromogranins B and enkephalin-containing peptides. J. Neurochem. **44:** 1854.
14. HOGUE-ANGELETTI, R. 1977. Non-identity of chromogranin A and dopamine-β-monooxygenase. Arch. Biochem. Biophys. **181:** 364.
15. MAXAM, A. M. & W. GILBERT. 1980. DNA sequencing by chemical cleavage. *In* Methods in Enzymology, Vol. 65. L. Grossman & K. Moldave, Eds.: 499. Academic. New York.
16. BLOBEL, G. & B. DOBBERSTEIN. 1975. Transfer of proteins across membranes. I. Presence of proteolytically processed and unprocessed nascent immunoglobulin light chains on membrane-bound ribosomes of murine myeloma. J. Cell Biol. **67:** 835.
17. OKAYAMA, H. & P. BERG. 1982. High efficiency cloning of full-length cDNA. Mol. Cell. Biol. **2:** 161.
18. OKAYAMA, H. & P. BERG. 1983. A cDNA cloning vector that permits expression of cDNA inserts in mammalian cells. Mol. Cell. Biol. **3:** 280.
19. ROSA, P., G. FUMAFALLI, A. ZANINI & W. B. HUTTNER. 1985. The major tyrosine-sulfated protein of the bovine anterior pituitary is a secretory protein present in gonadotrophs, thyrotrophs, mammotrophs and corticotrophs. J. Cell Biol. **100:** 928.
20. O'CONNER, D. T. 1983. Chromogranin: Widespread immunoreactivity in polypeptide hormone producing tissues and in serum. Reg. Peptides. **6:** 263.
21. KRUGGEL, W., D. T. O'CONNER & R. V. LEWIS. 1985. The amino terminal sequences of bovine and human chromogranin A and secretory protein I are identical. Biochem. Res. Comm. **127:** 380.
22. HUTTON, J. C., F. HANSEN & M. PESHAVARIA. 1985. β-Granins: 21 kDa co-secreted peptides of the insulin granule closely related to adrenal medullary chromogranin A. FEBS Lett. **188:** 336.
23. PHILLIPS, J. H. 1982. Commentary: Dynamic aspects of chromaffin granule structure. Neuroscience **7:** 1595.
24. SHARP, R. R. & R. SEN. 1982. Molecular mobilities and the lowered osmolality of the chromaffin granule aqueous phase. Bioch. Biophys. Acta **721:** 70.
25. RICHARDS, E. P. & R. R. SHARP. 1977. Analysis of the carbon-13 and proton NMR spectra of bovine chromaffin granules. Bioch. Biophys. Acta **497:** 14.
26. RICHARDS, E. P. & R. R. SHARP. 1977. Molecular mobilities of soluble components in the aqueous phase of chromaffin granules. Bioch. Biophys. Acta **497:** 260.
27. SEN, R., R. R. SHARP, L. E. DOMINO & E. F. DOMINO. 1979. Composition of the aqueous phase of chromaffin granules. Bioch. Biophys. Acta **587:** 75.
28. DANIELS, A. J., R. J. P. WILLIAMS & P. E. WRIGHT. 1978. The character of the stored molecules in chromaffin granules of the adrenal medulla: A nuclear magnetic resonance study. Neuroscience **3:** 573.
29. SCHELLMAN, C. (Personal communication of unpublished data on chromogranin localization.) Am. J. Path. **116:** 464.
30. FELICITAS, U. R. & M. GRATZL. 1986. Chromogranins, widespread in endocrine and nervous tissue, bind calcium. FEBS Lett. **195:** 327.
31. ABORG, C. H. & B. UVNAS. 1977. The ability of ATP-free granule material from bovine adrenal medulla to bind inorganic cations and biogenic amines. Acta. Physiol. Scand. **99:** 476.
32. TUFTY, R. M. & R. H. KRETSINGER. 1975. Troponin and parvalbumin calcium binding regions predicted in myosin light chain and T4 lysozyme. Science **187:** 167.

33. GARIEPY, J. & S. HODGES. 1983. Primary sequence analysis and folding behavior of EF hands in relation to the mechanism of action of troponin C and calmodulin. FEBS Lett. **160:** 1.
34. GEISOW, M. J., U. FRITSCHE, J. M. HEXHAM, B. DASH & T. JOHNSON. 1986. A consensus amino-acid sequence repeat in Torpedo and mammalian calcium-dependent membrane-binding proteins. Nature **320:** 636.
35. KRETSINGER, R. H. & C. E. CREUTZ. 1986. Cell biology: Consensus in exocytosis. Nature **320:** 573.
36. STADEN, R. 1982. An interactive graphics program for comparing and aligning nucleic acid homologies and amino acid sequences. Nuc. Acids Res. **10:** 2951.
37. COHN, D. V., J. J. ELTING, M. FRICK & R. ELDE. 1984. Selective localization of the parathyroid secretory protein-I/adrenal medulla chromogranin A protein family in a wide variety of endocrine cells of the rat. Endocrinology **114:** 1963.
38. EIDEN, L. E. *et al.* 1986. In press.
39. O'CONNOR, D. T., D. W. BURTON, R. J. PARMER & L. J. DEFTOS. 1984. Human chromogranin A: Detection by immunohistochemistry in C cells and diverse polypeptide hormone producing tumors. *In* Endocrine Control of Bone and Calcium Metabolism. D. V. Cohn, T. Fufita, J. T. Potts & R. V. Talmage, Eds.: 187. Elsevier Science Publishers. Amsterdam.
40. WILSON, B. S. & R. V. LLOYD. 1984. Detection of chromogranin in neuroendocrine cells with a monoclonal antibody. Am. J. Pathol. **115:** 458.
41. DESTEPHANO, D. B., R. V. LLOYD, A. M. PIKE & B. S. WILSON. 1984. Pituitary adenomas. An immunohistochemical study of hormone production and chromogranin localization. Am. J. Path. **116:** 464.
42. HELLE, K. B., R. K. REED, K. E. PIHL & G. SERCK-HANSSEN. 1985. Osmotic properties of the chromogranins and relation to osmotic pressure in catecholamine storage granules. Acta. Physiol. Scand. **123:** 21.
43. LEHRACH, H., D. DIAMOND, J. M. WOZNEY & H. BOEDTKER. 1977. RNA molecular weight determinations by gel electrophoresis under denaturing conditions, a critical reexamination. Biochemistry **16:** 4743.
44. SANGER, F., S. NICKLEN & A. R. COULSON. 1977. Sequencing of DNA by dideoxy chain termination. Proc. Natl. Acad. Sci. U.S.A. **74:** 546.
45. MAJZOUB, J. A., H. M. KRONENBERG, J. T. POTTS, JR., A. RICH & J. F. HABENER. 1979. Identification and cell-free translation of mRNA coding for a precursor of parathyroid secretory protein. J. Biol. Chem. **254:** 7449.

DISCUSSION OF THE PAPER

A. SCARPA (*Case Western Reserve University, Cleveland, Ohio*): Do you have any measurement of the affinity of these EF putative binding Ca^{2+} sites on chromogranin A? Second, do you have any idea on whether or not calcium binding will promote a major tertiary structure change of chromogranin A?

M. GRIMES (*University of Oregon, Eugene, Ore.*): We have not done Ca^{2+} binding studies on chromogranin A. I referred to the work of Felicitas and Gratzl[30] and Aborg and Unvas.[31] It's possible that Ca^{2+} binding might promote secondary and tertiary structure. However, the studies of Winkler's, Sharp's and Phillips's groups do not support this idea. They did ORD and NMR studies on intact secretory vesicles, which have an internal calcium concentration of 10-20 mM, and deduced that most proteins (*i.e.,* chromogranin A which is 50% of soluble protein) are in a random coil state.

How Sensitive and Specific Is Measurement of Plasma Chromogranin A for the Diagnosis of Neuroendocrine Neoplasia?[a]

DANIEL T. O'CONNOR AND LEONARD J. DEFTOS

Departments of Medicine
University of California
San Diego, California 92161
and
Veterans Administration Medical Center
San Diego, California 92161

INTRODUCTION

Although chromogranin A was originally found in adrenal medullary catecholamine storage vesicles, from which it is coreleased with catecholamines, we and others have recently found it in secretory vesicles of a variety of peptide-secreting endocrine tissues,[1-9] suggesting the possibility of its costorage and corelease with many peptide hormones.

Since chromogranin A is found in peptide-producing endocrine neoplasia,[4,8,9] we wondered whether its release into plasma would provide a useful diagnostic tool for a variety of neuroendocrine neoplasia. These studies explore the diagnostic sensitivity and specificity of such measurements.

MATERIALS AND METHODS

Chromogranin A Radioimmunoassay

Human chromogranin A was isolated and characterized from chromaffin granules of human pheochromocytoma, as previously described.[10-12] Human chromogranin A was measured by a soluble phase, double antibody radioimmunoassay as previously described.[10,13]

[a]This work was supported by the Veterans Administration, the National Institutes of Health (grant nos. AM-36,400 and AM-15,888), the American Heart Association and the American Cancer Society. Dr. O'Connor is an Established Investigator of the American Heart Association.

Human Subject Samples

Serum or plasma samples were obtained from 54 healthy controls as well as subjects with a variety of peptide-producing endocrine neoplasia ($n = 69$): pheochromocytoma ($n = 15$), aortic body tumor ($n = 1$), gut carcinoid tumor ($n = 11$), pancreatic islet cell tumor ($n = 5$), oat cell lung carcinoma ($n = 12$), medullary thyroid carcinoma ($n = 13$), thyroidal C cell hyperplasia ($n = 3$), parathyroid adenoma ($n = 7$), and primary parathyroid hyperplasia ($n = 2$). All diagnoses were made histologically.

Samples were also obtained from 68 other subjects with a variety of other medical illnesses that are not associated with peptide secretion. The illnesses were both benign and malignant, and of both the endocrine and the nonendocrine variety.

To assess the day-to-day stability of plasma chromogranin A immunoreactivity, plasma samples were obtained from a group of 44 healthy male adult controls on three occasions separated by a mean of 7 days each.

Other Radioimmunoassays

Plasma parathyroid hormone and calcitonin were measured by established procedures.[14,15]

RESULTS

The circulating immunoreactive chromogranin A concentration was stable within normal individuals over a 2-week time course (FIG. 1), with an overall mean day-to-day coefficient of variation of 18.0%.

TABLE 1 displays plasma chromogranin A values in normal controls as well as subjects with a variety of neuroendocrine neoplasia. Several features bear mentioning. First, tumors other than pheochromocytoma had elevations of plasma chromogranin A, even though the original antigen was isolated from a pheochromocytoma.[5] In fact, the very highest values were observed in subjects with carcinoid tumor. Second, plasma chromogranin A values were quite heterogeneous (some normal, some elevated) in subjects with oat cell lung carcinoma; those subjects with elevated chromogranin A also tended to have elevated plasma calcitonin (TABLE 1), suggesting that chromogranin A elevations reflect peptide production by endocrine neoplasia, rather than being an indicator for neoplasia in general.

Correlations (TABLE 2) between plasma chromogranin A and the plasma concentration of the tumors' usual resident hormones were generally poor, though there was some correspondence between plasma chromogranin A and plasma calcitonin in oat cell lung carcinoma subjects ($r = 0.51$, $n = 12$, $p = 0.09$).

In the pheochromocytoma subjects, there was a significant correlation (FIG. 2) between plasma chromogranin A and tumor mass ($r = 0.83$, $n = 15$, $p < 0.01$), though the correlation was heavily dependent on an outlier with the largest tumor.

In 68 other "control" subjects selected to span a variety of medical illnesses—both benign and malignant, both endocrine (not peptide producing) and nonendocrine—the highest plasma chromogranin A value seen was 151 ng/ml, modestly more than the highest value of 103 ng/ml seen in the 54 healthy normal controls.

From an analysis of plasma chromogranin A values in subjects with peptide-producing endocrine neoplasia versus the various "control" groups (54 healthy normal controls and 68 controls with various medical illnesses), true and false positive and negative plasma chromogranin A values may be calculated for the diagnosis of neuroendocrine neoplasia, based on a cutoff of 151 ng/ml as the upper value in the

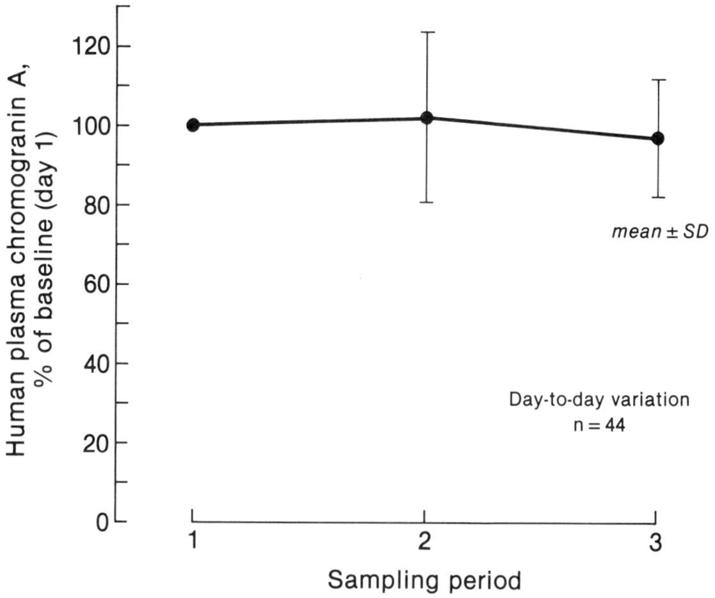

FIGURE 1. Plasma chromogranin A in 44 healthy adult male controls, measured in three separate samples obtained at intervals of 7 ± 1 days. The results are plotted as percentage of the initial value, mean ± SD. At the second sampling, the mean coefficient of variation was 21.1%; at the third sampling, it was 14.9%. The overall mean day-to-day coefficient of variation was 18.0%.

various control groups (TABLE 3). The results suggest a sensitivity of 81% and a specificity of 100% for plasma chromogranin A in the recognition of neuroendocrine neoplasia.

Gel filtration of tumor plasma samples (FIG. 3), coupled to the chromogranin A radioimmunoassay, showed a variety of size forms of chromogranin A immunoreactivity, both larger and smaller than the ^{125}I-labeled purified human chromogranin A molecule. No clear pattern has yet emerged linking a given tumor site to a given gel filtration/radioimmunoassay pattern (TABLE 4).

TABLE 1. Plasma Chromogranin A, Calcitonin, and Parathyroid Hormone in Subjects with Neuroendocrine Neoplasia[a,b]

Group, Subgroup	n	Plasma Chromogranin A, ng/ml	Plasma Calcitonin, pg/ml	Plasma Parathyroid Hormone pg/ml
Normal controls	54	57 ± 2	<100	<200
Males	27	60 ± 3	<100	<200
Females	27	57 ± 3	<100	<200
Pheochromocytoma	15	736 ± 399[c]	—	—
Aortic body tumor	1	215	—	—
Carcinoid tumor	11	35,400 ± 17,400[c]	—	—
Pancreatic islet cell tumor	5	4,960 ± 4,680[c]	—	—
Oat cell lung carcinoma	12	221 ± 51[c]	193 ± 77	—
Stratified by plasma calcitonin				
Normal calcitonin	7	151 ± 46	25 ± 11	—
Elevated calcitonin	5	319 ± 77[c]	428 ± 124	—
Stratified by plasma chromogranin A				
Normal chromogranin A	3	35 ± 10	40 ± 26	—
Elevated chromogranin A	9	283 ± 52[c]	243 ± 98	—
C cell disorders				
Medullary thyroid carcinoma	13	564 ± 184[c]	13,200 ± 6,320	—
C cell hyperplasia	3	276 ± 67[c]	366 ± 249	—
Hyperparathyroidism				
Primary				
Adenoma	7	218 ± 13[c]	—	699 ± 173
Hyperplasia	2	1,680 ± 1,450[c]	—	683 ± 285

[a] See O'Connor and Deftos.[11]
[b] The values are shown as mean ± SEM.
[c] $p < 0.01$, comparing plasma concentration of chromogranin A to that in the control group.

TABLE 2. Basal Correlations between Plasma Chromogranin A and Other Peptide Hormones

Group	n	Independent Variable	Dependent Variable	r	p
Oat cell lung carcinoma	12	plasma chromogranin A	plasma calcitonin	0.51	0.09
Medullary thyroid carcinoma	13	plasma chromogranin A	plasma calcitonin	−0.15	>0.1
Parathyroid adenoma	7	plasma chromogranin A	plasma parathyroid hormone	−0.54	>0.1

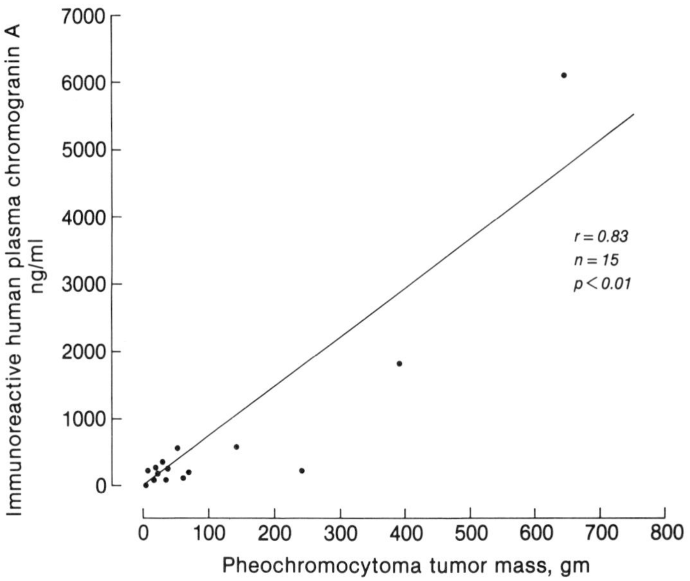

FIGURE 2. Plasma chromogranin A in 15 preoperative subjects with pheochromocytoma, plotted as a function of tumor mass in grams (determined after removal). There is a significant correspondence ($r = 0.83$, $n = 15$, $p < 0.01$).

TABLE 3. Sensitivity and Specificity of Chromogranin A Elevations in the Diagnosis of Peptide-Producing Endocrine Neoplasia

True and False Positives and Negatives in All Subjects

	Polypeptide-hormone-producing endocrine tumor	
	no	yes
normal chromogranin A (< 151 ng/ml)	122	11
elevated chromogranin A (> 151 ng/ml)	0	46

Overall Sensitivity and Specificity

- sensitivity = (true positives)/(true positives + false negatives) = (46)/11 + 46) = 0.81
- specificity = (true negatives)/(true negatives + false positives) = (122)/(122 + 0) = 1.0

For the purpose of this table, peptide-producing endocrine neoplasia are defined as histologically documented tumors of the following varieties: pheochromocytoma, aortic body tumor, carcinoid tumor, pancreatic islet cell tumor, medullary thyroid carcinoma, thyroidal C cell hyperplasia, parathyroid adenoma, and primary parathyroid hyperplasia. Oat cell lung carcinoma is excluded from the analysis because the endocrine nature of all such tumors is not established. The 151 ng/ml plasma chromogranin A cutoff is the highest value observed in the several "control" groups.

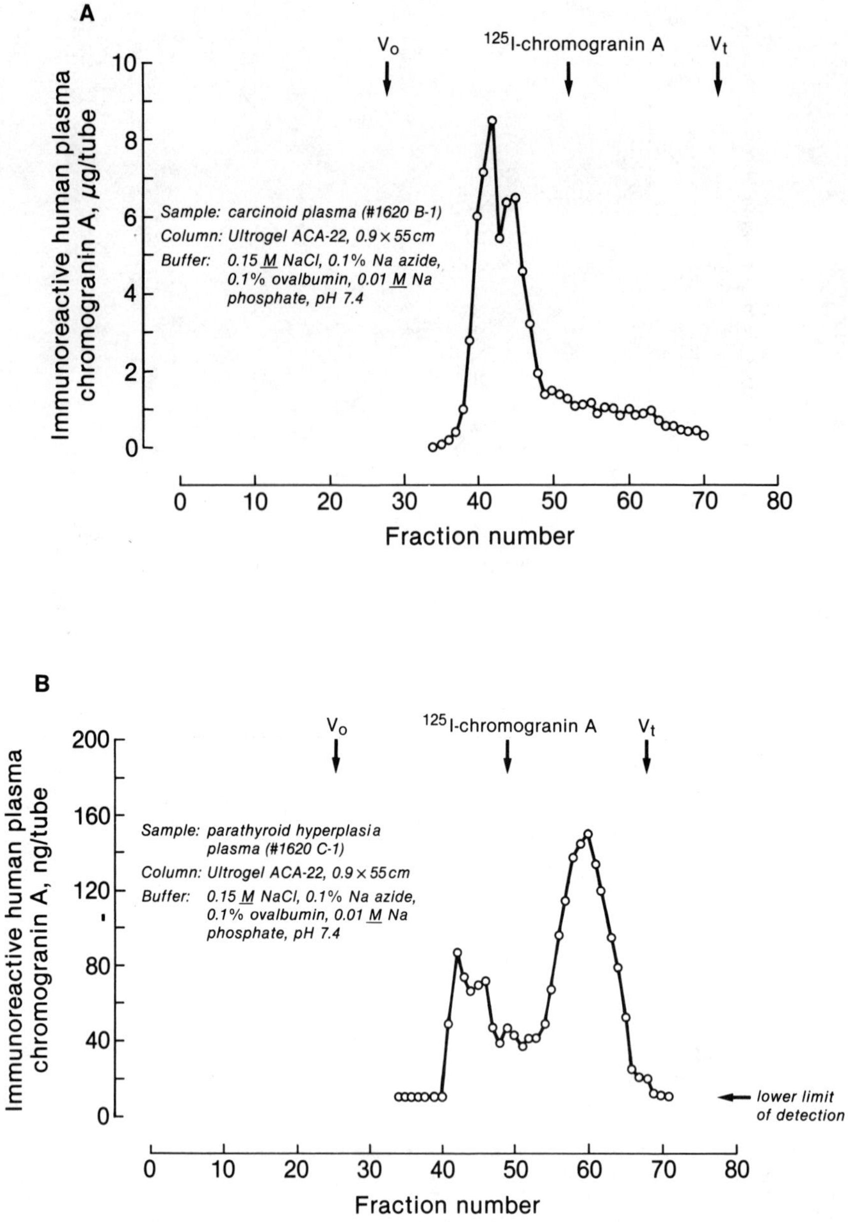

FIGURE 3. Immunoreactive plasma chromogranin A size characterization by gel filtration. The tumor plasma source and the specifications of the column are noted in each panel. The column was standardized for void volume (V_o, by elution of blue dextran), total internal volume (v_t, by elution of KCl), and the elution position of purified ^{125}I-labeled human chromogranin A. **(A)** Plasma from a patient with a gut carcinoid tumor. **(B)** Plasma from a patient with primary parathyroid hyperplasia.

TABLE 4. Relative Gel Filtration Elution Positions (Sizes) of Human Plasma Chromogranin A Immunoreactivity

Sample	K_d Values	Major K_d
^{125}I-labeled, purified human chromogranin A	0.55	+
Pheochromocytoma chromaffin vesicle lysate (no. 1630-1A)	0.31	+
	0.36	
	0.53	
	0.86	
Pheochromocytoma plasma (no. 1630L-1)	0.50	+
	0.86	
Pancreatic islet cell tumor plasma (no. 1630L-3)	0.45	+
	0.67	
	0.83	
Carcinoid tumor plasma (no. 1620D-1)	0.32	+
	0.39	
Medullary thyroid carcinoma plasma (no. 1630D-1)	0.47	+
	0.79	
	0.98	
Parathyroid hyperplasia plasma (no. 1620C-1)	0.39	
	0.48	
	0.55	
	0.81	+

The column K_d, for any given peak, was calculated as $K_d = V_e - V_o/V_t - V_o$, where V_e = the elution volume for any given peak, V_o = the column void volume (defined by the elution of blue dextran), and V_t = the column total internal volume (defined by the elution of KCl). The "+" indicates the K_d for the quantitatively major immunoreactive peak on each run. The gel filtration conditions are as given in FIGURES 3A and 3B.

DISCUSSION

The results indicate that chromogranin A secretion into plasma is found in a variety of neuroendocrine neoplasia, not limited to pheochromocytoma. The plasma chromogranin elevations seemed to be specific for peptide-producing tumors rather than neoplasia in general. Thus, these neuroendocrine neoplasia might well be characterized as "chromograninomas." Finally, the high sensitivity (81%) and specificity (100%) of plasma chromogranin A suggest that it may find potential use in screening patients with suspected endocrine neoplasia.

REFERENCES

1. O'CONNOR, D. T. 1983. Chromogranin: Widespread immunoreactivity in polypeptide hormone producing tissues and in serum. Regulatory Peptides **6:** 263-280.
2. O'CONNOR, D. T., D. W. BURTON, R. J. PARMER & L. J. DEFTOS. 1984. Human chromogranin A: Detection by immunohistochemistry in C cells and diverse polypeptide hormone

producing tumors. *In* Endocrine Control of Bone and Calcium Metabolism. D. V. Cohn, T. Fujita, J. T. Potts & R. V. Talmadge, Eds.: 187-190. Elsevier Science Publishers. Amsterdam.

3. O'Connor, D. T., D. W. Burton & L. J. Deftos. 1983. Chromogranin A: Immunohistology reveals its universal occurrence in normal polypeptide hormone producing endocrine glands. Life Sci. **33:** 1657-1663.
4. O'Connor, D. T., D. W. Burton & L. J. Deftos. 1983. Immunoreactive human chromogranin A in diverse polypeptide hormone producing human tumors and normal endocrine tissues. J. Clin. Endocrinol. Metabol. **57:** 1084-1086.
5. Cohn, D. V., R. Zangerle, R. Fischer-Colbrie, L. L. H. Chu, J. J. Elting, J. W. Hamilton & H. Winkler. 1982. Similarity of secretory protein I from parathyroid gland to chromogranin A from the adrenal medulla. Proc. Natl. Acad. Sci. U.S.A. **79:** 6056-6059.
6. Cohn, D. V., J. J. Elting, M. Frick & R. Elde. 1984. Selective localization of parathyroid secretory protein I/adrenal medulla chromogranin A in a wide variety of endocrine cells of the rat. Endocrinology **114:** 1963-1984.
7. Kruggel, W., D. T. O'Connor & R. V. Lewis. 1985. The amino terminal sequences of bovine and human chromogranin A and secretory protein I are identical. Biochem. Biophys. Res. Commun. **127:** 380-383.
8. Lloyd, R. V. & B. S. Wilson. 1983. Specific endocrine tissue marker defined by a monoclonal antibody. Science **222:** 628-630.
9. Wilson, B. S. & R. V. Lloyd. 1984. Detection of chromogranin in neuroendocrine cells with a monoclonal antibody. Am. J. Pathol. **115:** 458-468.
10. O'Connor, D. T., R. P. Frigon & R. L. Sokoloff. 1984. Human chromogranin A: Purification and characterization from catecholamine storage vesicles of human pheochromocytoma. Hypertension **6:** 2-12.
11. O'Connor, D. T. & Deftos, L. J. 1986. Secretion of chromogranin A by peptide producing endocrine neoplasms. N. Engl. J. Med. **314:** 1145-1151.
12. O'Connor, D. T. 1986. Radioimmunoassays of chromogranin. *In* Quantitative Analysis of Catecholamines and Related Compounds. A. M. Krstulovic, Ed.: 302-315. Ellis Horwood, Ltd., Chichester, U.K.
13. O'Connor, D. T. & K. N. Bernstein. 1984. Radioimmunoassay of chromogranin A in plasma as a measure of exocytotic sympathoadrenal activity in normal subjects and patients with pheochromocytoma. N. Engl. J. Med. **311:** 764-770.
14. Roos, B. A. & L. J. Deftos. 1979. Parathyroid hormone. *In* Methods of Hormone Radioimmunoassay. 2d Ed. B. M. Joffe, Ed.: 401-418. Academic. New York.
15. O'Connor, D. T., R. P. Frigon & L. J. Deftos. 1983. Immunoreactive calcitonin in catecholamine storage vesicles of human pheochromocytoma. J. Clin. Endocrinol. Metabol. **56:** 582-585.

Transport of Ascorbic Acid into Pituitary Cultures

E. I. CULLEN, V. MAY, AND B. A. EIPPER

Department of Neuroscience
The Johns Hopkins University School of Medicine
Baltimore, Maryland 21205

In 1982 Bradbury, Finnie, and Smyth described an activity in porcine pituitaries that converted the synthetic substrate D-Tyr-Val-Gly to the amidated product D-Tyr-Val-NH_2. Glyoxylate was detected in the reaction mixture, and mass spectra indicated that the nitrogen of the glycine residue was the source of the amide nitrogen.[1] Later work showed that amidating enzyme purified from bovine pituitaries and activity from several tissues required molecular oxygen and was stimulated by copper and ascorbic acid.[2–5] Thus, the enzyme is most likely a monooxygenase and has been called peptidylglycine α-amidating monooxygenase, or PAM. PAM activity has been localized to secretory granules in rat hypothalamus and mouse AtT-20 pituitary tumor cells.[4,5]

Primary cultures of intermediate pituitary cells lose the ability to α-amidate αMSH, but the loss can be reversed by incubating the cells in ascorbic acid.[6,7] To continue investigating the relationship among cellular ascorbic acid concentration, amidating ability, and PAM activity, we studied ascorbic acid transport in primary cultures of rat anterior and intermediate pituitary and mouse AtT-20 tumor cells. When incubated for 7 to 9 hours in 50 μM ^{14}C-labeled ascorbic acid, all three cell preparations concentrated ascorbic acid 20- to 40-fold, producing intracellular ascorbate concentrations of 1 to 2 mM, based on experimentally determined cell volumes.[8] All three cell preparations also displayed saturable ascorbic acid uptake; half-maximal rates of ascorbic acid uptake occurred between 9 and 18 μM ascorbate. In primary cultures of intermediate pituitary, uptake of ascorbate in medium containing 5 mM sodium was less than 20% of the uptake in medium containing 156 mM sodium. When primary anterior pituitary and AtT-20 cells were incubated in medium containing 5 mM sodium, the initial rate of ascorbic acid transport was less than 10% of that measured in medium containing 156 mM sodium. In contrast to systems that transport dehydroascorbic acid, pituitary ascorbate transport was not inhibited by glucose[8] (FIG. 1). HPLC analyses of extracts from cells incubated in ^{14}C-labeled ascorbic acid showed that over 90% of the intracellular label comigrated with authentic ascorbic acid.[8] Although ascorbic acid was oxidized rapidly in culture media in the absence of cells, incubation of ascorbate in the presence of cells stabilized the ascorbic acid substantially.[8] ^{14}C-labeled ascorbic acid was stable in physiological buffer containing 1 mM thiourea.[8] In the presence of 1 mM thiourea, unlabeled ascorbic acid competitively inhibited the uptake of ^{14}C-labeled ascorbic acid with a K_i that compared well with the K_m for transport (FIG. 2). This result supports the conclusion that reduced ascorbic acid is transported into pituitary cells by a sodium-dependent active transport system.

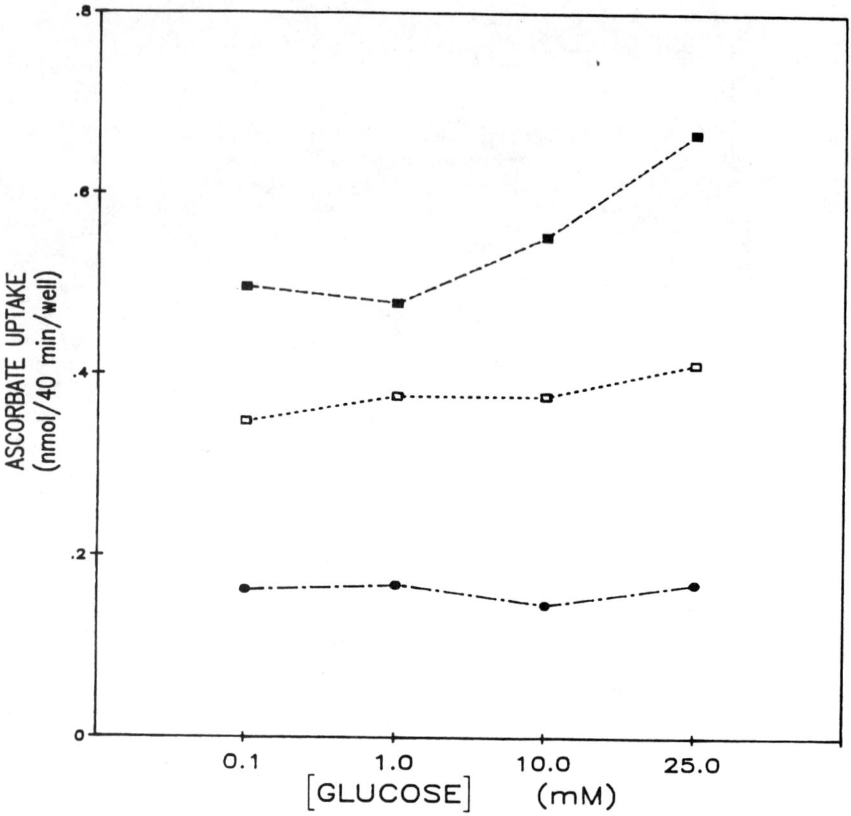

FIGURE 1. Effect of glucose on ^{14}C-labeled ascorbic acid uptake in AtT-20 cells. AtT-20 cells were incubated at 36° in either 5 (●), 15 (□), or 50 (■) μM ^{14}C-labeled ascorbic acid dissolved in physiological buffer containing 1 mM thiourea and the indicated amounts of D-glucose. After 40 minutes of incubation, the medium was removed, the cells were washed three times, and then extracted with 2 mg/ml BSA in 5 N acetic acid. After thrice freezing and thawing the extract it was centrifuged, and the radioactivity in the supernatant was determined. Each point represents the mean of three wells. Standard deviations for all points were less than 9% of the mean.

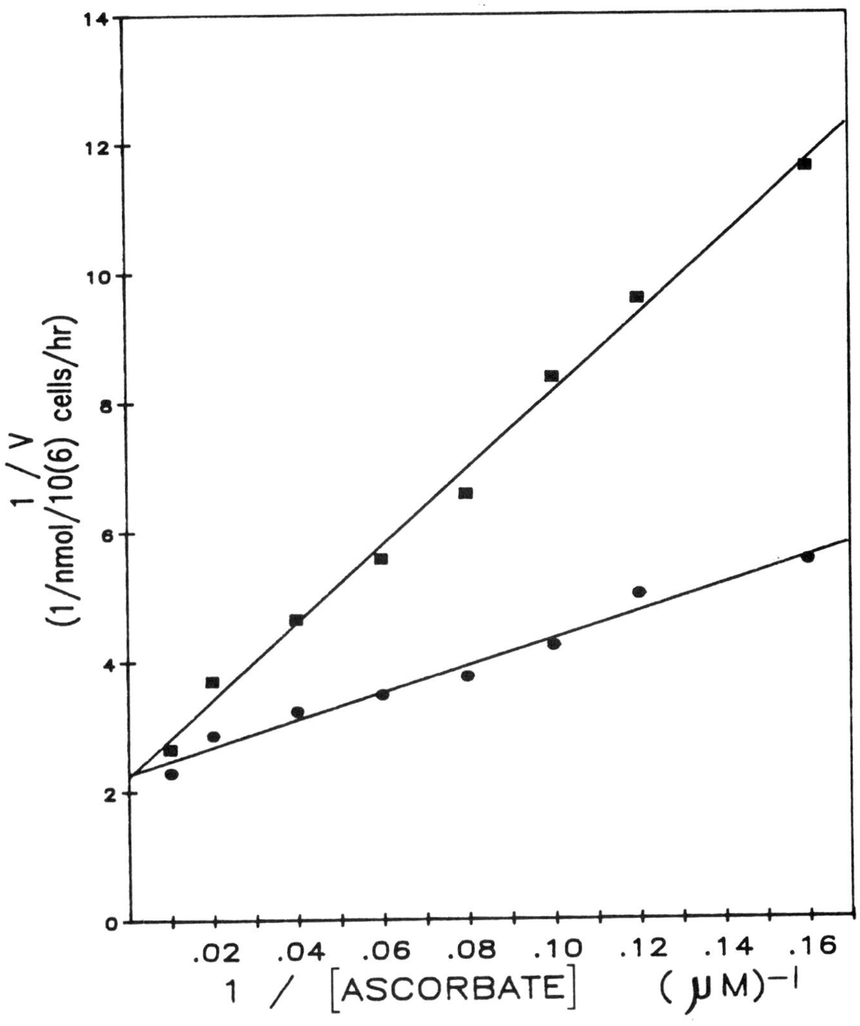

FIGURE 2. Inhibition of ^{14}C-labeled ascorbic acid uptake by unlabeled ascorbic acid. Primary cultures of rat anterior pituitary were incubated for 30 minutes at 36°C at the indicated concentrations of ^{14}C-labeled ascorbic acid in the absence (●) or presence (■) of 30 μM unlabeled ascorbic acid. Uptake was determined as in FIGURE 1. Each point represents the mean of duplicate wells. Correlation coefficients from linear regression analyses were both greater than 0.98.

REFERENCES

1. BRADBURY, A. F., M. D. FINNIE & D. G. SMYTH. 1982. Nature **298:** 686-688.
2. MURTHY, A. S. N., R. E. MAINS & B. A. EIPPER. 1986. J. Biol. Chem. **261:** 1815-1822.
3. EIPPER, B. A., A. C. MYERS & R. E. MAINS. 1985. Endocrinology **116:** 2497-2504.
4. EMESON, R. B. 1984. J. Neurosci. **4:** 2604-2613.
5. MAINS, R. E., C. C. GLEMBOTSKI & B. A. EIPPER. 1984. Endocrinology **114:** 1522-1530.
6. MAY, V. & B. A. EIPPER. 1985. J. Biol. Chem. **260:** 16224-16231.
7. GLEMBOTSKI, C. C. 1984. J. Biol. Chem. **259:** 13041-13048.
8. CULLEN, E. I., V. MAY & B. A. EIPPER. 1986. Mol. Cell. Endocrinol. **48:** 239-250.

Molecular Biology of Carboxypeptidase E (Enkephalin Convertase), a Neuropeptide-synthesizing Enzyme

L. D. FRICKER[a] AND E. HERBERT

Institute for Advanced Biomedical Research
The Oregon Health Sciences University
Portland, Oregon 97201

Carboxypeptidase E (CPE) is the carboxypeptidase-B-like enzyme involved in the biosynthesis of many peptide hormones and neurotransmitters.[1] This enzyme has been purified to homogeneity from bovine pituitary, and a partial amino acid sequence has been determined. Oligonucleotide probes corresponding to the amino acid sequence have been used to isolate a cDNA clone from a bovine pituitary cDNA library.[2] The cDNA encoding CPE hybridizes primarily to mRNA of approximately 3.3 kb in bovine brain, pituitary, and adrenal medulla. In these tissues, the CPE cDNA probes also hybridize to 2.1 and 2.6 kb mRNA, although these species represent less than 5% of the 3.3 kb mRNA. Restriction fragment analysis of bovine genomic DNA indicates a single gene encodes CPE.

The predicted amino acid sequence of the cDNA clone contains several pairs of basic amino acids, and displays some homology with carboxypeptidases A (CPA) and B (CPB). Although the overall homology is low, all of the amino acids thought to be essential for the catalytic activity of CPA and CPB[3] have been conserved in CPE (FIG. 1). These amino acids include Tyr-227, Tyr-278, and Glu-300 of CPE. Amino acids that bind Zn^{2+} in CPA and CPB correspond to His-72, Glu-75, and His-225 in CPE. The arginine in CPA and CPB that binds to the COOH terminus of the substrate has also been conserved in CPE (Arg-147). The amino acid in CPA and CPB that binds to the side of the substrate's COOH-terminal amino acid has not been conserved in CPE. This amino acid is an aspartate in CPB and an isoleucine in CPA. The corresponding amino acid in CPE is a glutamine, although there are many other differences between the substrate binding regions of CPE and the other carboxypeptidases.

CPE contains an additional 100 amino acids at the COOH terminus and two short internal segments that are not found in CPA or CPB. If CPE has a three-dimensional structure similar to the other carboxypeptidases, these extra portions of CPE would be located near the substrate binding region. This may account for the differences in substrate specificity between the carboxypeptidases: CPE is very specific for COOH-terminal basic amino acids, whereas CPA and CPB both hydrolyze a wide variety of COOH-terminal amino acids.

[a] Current address: Department of Molecular Pharmacology, Albert Einstein College of Medicine, Bronx, New York 10461.

```
RESIDUE      1                  10                       20                     30                      40                      50                      59

BOVINE E     R P Q E D G I S F E - Y H R Y P E L R E A L V S V W L Q C A A - V S R I Y T V G R S F E G R E L L V L E L S D N P G V H E
BOVINE B           T T G H S Y E K Y N N W E T I E A W T E Q V A S E N P D L I S R S A - I G T T F L G N T I Y L L K V G K - P G S N K
CRAYFISH B               M D W T S Y H D Y D E I N A W L D S L A T D Y P E L A S V E D - V G L S Y E G R T M K L L K L G K - G G A D K
BOVINE A     A R S T N T F N Y A T Y H T L D E I Y D F M D L L V A E H P Q L V S K L Q - I G R S Y E G R P I Y V L K F S T - G G S N R
RAT    A     A L S T D S F N Y A T Y H T L D E I Y E F M D L L V A E H P Q L V S K I Q - I G N T F E G R P I H V L K F S T - G G T N R

RESIDUE      60                    70                    80                    90                    100                    110                  120

BOVINE E     P G E P E F K Y I G N M H G N E A V G R E L L I F L A Q Y L C N E Y Q K G N E T I V Q L I H N T R I H I M P S L N P D G F
BOVINE B     P A - - V F M D C G - F H A R E W I S P A F C Q W F V R E A V R I Y G R E I H M T E F L D - K L D F Y V L P V V N I D G Y
CRAY.  B     P I - - I F I D G G - I H A R E W I A P S T V T Y I V N E F V S N - - - S A T Y D D I L S - N V N F Y V M P T I N P D G Y
BOVINE A     P A - - I W I D L G - I H S R E W I T Q A T G V W F A K K F T E N Y G Q N P S F T A I L D - S M D I F L E I V T N P N G F
RAT    A     P A - - I W I D T G - I H S R E W V T Q A S G V W F A K K V T K D Y G Q D P T F T A V L D - N M D I F L E I V T N P D G F

RESIDUE     121                 130                    140                    150                    160                    170                178

BOVINE E     E K A A S Q L G E L K D W F V G R S N A Q - - - G I D L N R N F P D L D R I V Y I N E K E G G P N N H L L K N L K K I V D
BOVINE B     I Y T W T T N R M W R K T R S T R A G S S - C T G T D L N R N F D A G W C S I G A S N N P C S E T - - - - - - - - - - - -
CRAY.  B     A Y T F T D D R L W R K T R S E T G S V L G C K G A D P N R N W S F H W D E V G A S D S P C S D I - - - - - - - - - - - -
BOVINE A     A F T H S E N R L W R K T R S V T S S S L - C V G V D A N R N W D A G F G K A G A S S S P C S E T - - - - - - - - - - - -
RAT    A     A Y T H K T N R M W R K T R S H T Q G S L - C V G V D P N R N W D A G L G K A G A S S N P C S E T - - - - - - - - - - - -

RESIDUE      180                 190                    200                    210                    220                228

BOVINE E     Q N T K L A P E T K A V I H W I M D I P F V L S A N L H G G D L V A N Y P Y D E T R S G S A H E Y S - - - - - - - - - - -
BOVINE B     - - - - - - - - - - - - Y C G S A A E S E K E S K A V - - A D F I R N H L S S I K A Y L T I H S Y S Q M M L Y P Y S Y D -
CRAY.  B     - - - - - - - - - - - - Y A G P E P F S E V E M R N V - - R D Q I L E Y A A N I K V Y L T F H S Y S Q L W M Y P W G F T -
BOVINE A     - - - - - - - - - - - - Y H G K Y A N S E V E V K S I - - V D F V K N H - G N F K A F L S I H S Y S Q L L L Y P Y G Y T T
RAT    A     - - - - - - - - - - - - Y R G K F P N S E V E V K S I - - V D F V T S H - G N I K A F I S I H S Y S Q L L L Y P Y G Y T S

RESIDUE        230                    240                    250                    260                    270                280

BOVINE E     - S C P D D - - D I F Q S L A R A Y S S F N P P M S D P D R P P C R K N D D D S S F V E G T T N G A A W Y S V P G G M Q D
BOVINE B     Y K L P K N N V E L - N T L A K G A V K K L A S L H - - - - - - - - - - - - - G T T Y S Y G P G A - T T I Y P A S G G S D D
CRAY.  B     S D L P D D W Q D L - D T L A T N A V D A L T A V H - - - - - - - - - - - - - G T R Y E I G S S T - N T I Y A A A G G S D D
BOVINE A     Q S I P D K I - E L N Q - V A K S A V A A L K S L Y - - - - - - - - - - - - - G T S Y K Y G S I I - T T I Y Q A S G G S I D
RAT    A     E P A P D Q A - E L D Q - L A K S A V T A L T S L H - - - - - - - - - - - - - G T E F K Y G S I I - D T I Y Q A S G S T I D

RESIDUE        290                    300                    310                    320                    330                340

BOVINE E     F N Y L S S - N C F E I T V E L - - - S C E K F P - P E E T L K N Y W E D N K N S L I S Y I Q Q I H R G V K G F V R D ...
BOVINE B     W A Y D Q G - I K Y S F T F E L R D K G R Y G F V L P E S Q I Q P T C E E T M L A I K Y V T - - - - S Y V L E H L
CRAY.  B     W A K G E G G V K Y A Y T I E L R D T G N Y G F L L P E N Q I I P T G E E T F E G V K V V A - - - - N F V K D T Y S
BOVINE A     W S Y N Q G - I K Y S F T F E L R D T G R Y G F L L P A S Q I I P T A Q E T W L G V L T I M E - - - H T V N N
RAT    A     W T Y S Q G - I K Y S F T F E L R D T G L R G F L L P A S Q I I P T A E E T W L A L L T I M D - - - H T V K H P Y
```

FIGURE 1. The amino acid sequence of crayfish CPB and the alignment of this sequence with bovine CPB, bovine CPA, and rat CPA was obtained from Titani *et al.*[4] The amino acid sequence of bovine CPE was derived from the nucleic acid sequence of a cDNA clone encoding this enzyme.[2] Only the first 340 amino acids from the NH_2-terminal of the enzymatically active forms of CPE are shown, and an additional COOH-terminal 94 amino acids are predicted from the cDNA sequence. Regions of homology between bovine CPE and the other carboxypeptidases are indicated by *boxes. Dashes* within the amino acid sequences indicate gaps, which have been introduced to optimize the alignment of homologous regions. Amino acids thought to be important in substrate binding and catalytic activity of CPA and CPB[3] are indicated by *asterisks.*

Of the 90 amino acids that have been conserved between bovine CPB, crayfish CPB, bovine CPA, and rat CPA,[4] 35 are present in bovine CPE (39% homology). The overall homologies between bovine CPE and bovine CPB (17%), crayfish CPB (20%), bovine CPA (20%), and rat CPA (20%) are similar, and are considerably lower than the homology between bovine CPA and bovine CPB (48%). This suggests that CPE diverged from an ancestral carboxypeptidase before CPA and CPB diverged.

REFERENCES

1. Fricker, L. D. 1985. Trends Neurosci. **8:** 210-214.
2. Fricker, L. D., C. J. Evans, F. S. Esch & E. Herbert. 1986. Nature. **323:** 461-464.
3. Schmid, M. F. & J. R. Herriott. 1976. J. Mol. Biol. **103:** 175-190.
4. Titani, K., L. H. Ericsson, S. Kumar, F. Jacob, H. Neurath & R. Zwillig. 1984. Biochemistry **23:** 1245-1250.

Biosynthesis and Maturation of Carboxypeptidase H[a]

VIVIAN Y. H. HOOK, MIKLOS PALKOVITS,[b] AND HANS-URS AFFOLTER[b]

Department of Biochemistry
Uniformed Services University of the Health Sciences
Bethesda, Maryland 20814

[b] *Laboratory of Cell Biology*
National Institute of Mental Health
Bethesda, Maryland 20902

Carboxypeptidase H (CPH)[c] is one of several processing enzymes required for the conversion of peptide hormone precursors such as proenkephalin,[1,2] provasopressin,[3] and others[4] into their smaller biologically active forms. CPH, which cleaves COOH-terminal lysine or arginine residues, is a unique carboxypeptidase that is specifically involved in the processing of a variety of peptides.

Recent studies[5] have suggested that CPH may be more active in mature than in immature granules. To test this hypothesis directly, the state of activation of CPH was compared in pituitary stalk and posterior pituitary of the rat hypothalamo-neurohypophysial system (FIG. 1). To determine if CPH in nerve terminal regions may be more active than CPH in axonal regions, rat pituitary stalk and posterior pituitary were dissected and the ratio of CPH activity to immunoreactivity (i.r.) was measured and calculated for each region. This ratio serves as an index of the level of activity per unit number of enzyme molecules and can differentiate between highly active and inactive populations of CPH.[6] Comparison of these ratios (0.3/0.05 = 6.0) shows that CPH molecules in posterior pituitary were six times more active than those in pituitary stalk (FIG. 2). When CPH activity was measured after pituitary stalk transection, the enzyme activity was nearly abolished after nerve terminal degeneration (data not shown). This result suggests that CPH synthesized in the cell bodies of the hypothalamic paraventricular and supraoptic nuclei is axonally transported to nerve terminals in posterior pituitary. These studies show that CPH enzyme molecules may be activated during axonal transport (pituitary stalk) to nerve terminals (posterior pituitary).

The hypothesis that CPH may be activated implies that it may be synthesized as an inactive precursor. In preliminary studies, poly-A^+ mRNA isolated from bovine

[a] The opinions or assertions contained herein are the private ones of the authors and are not to be construed as official or reflecting the views of the Department of Defense or the Uniformed Services University of the Health Sciences.

[c] Carboxypeptidase H (CPH) has been referred to as "carboxypeptidase B-like"[1,3] and "enkephalin convertase."[2] The Enzyme Nomenclature Committee of IUB has recently designated the name "carboxpeptidase H" to distinguish this enzyme from other known carboxypeptidases. (Eur. J. Biochem. **157:** 16. 1986.)

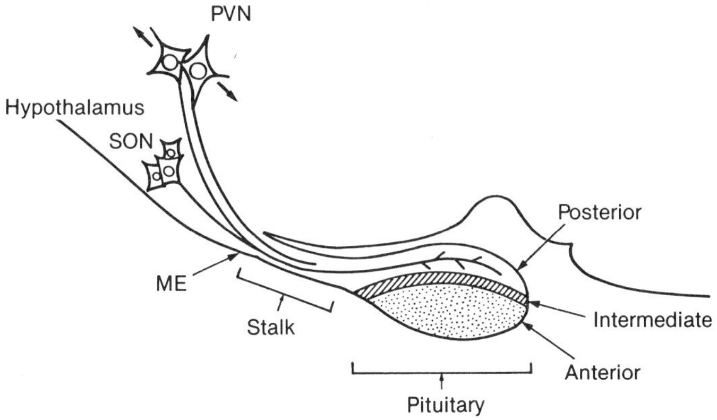

FIGURE 1. Schematic illustration of the rat hypothalamo-neurohypophysial system. *PVN* = paraventricular nucleus; *SON* = supraoptic nucleus; *ME* = median eminence. Cell bodies originating in the PVN and SON of the hypothalamus project axons via the pituitary stalk that terminate as nerve endings in posterior pituitary.

adrenal medulla was translated in a cell-free translation system with [^{35}S]methionine. Immunoprecipitation of ^{35}S-labeled proteins with specific CPH antiserum[7] and analysis by SDS-PAGE showed that CPH itself appears to be first synthesized as a precursor that would then require posttranslational processing and activation to form enzymatically active CPH.

These results suggest that maturation of relatively inactive CPH to form the active enzyme occurs as the secretory vesicles are transported along the axon to nerve terminals. Thus, as the secretory vesicle matures during axonal transport, regulatory mechanisms must be present to activate CPH. In addition to CPH, "trypsin-like" processing enzyme(s) are also necessary for complete peptide precursor processing. Previous studies that have identified precursor forms of provasopressin[8] and proenkephalin[9] in the hypothalamo-neurohypophysial system have shown that large precursor forms are present in the cell bodies of the supraoptic nucleus, intermediate-

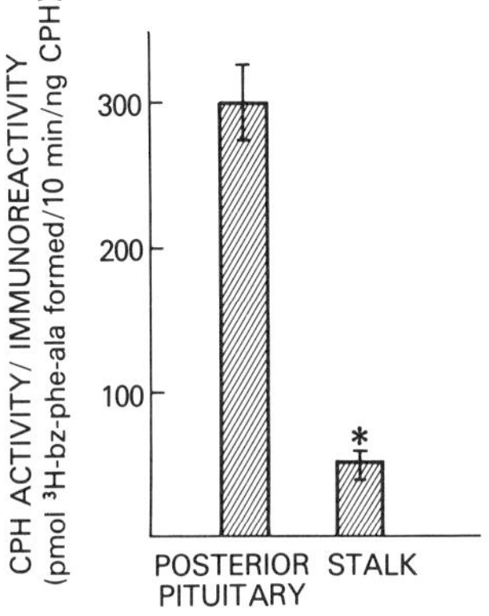

FIGURE 2. Activation state of carboxypeptidase H in rat pituitary stalk and posterior pituitary. Pituitary stalk and posterior pituitary were dissected, immediately frozen on dry ice, and homogenized in 0.1 M Na-acetate pH 5.7. CPH enzyme activity[11] and immunoreactivity[8] were measured and the ratio of activity and immunoreactivity was calculated.

sized precursor fragments are found in pituitary stalk, and primarily small peptides are found in posterior pituitary. These findings suggest that the trypsin-like enzyme(s) that cleave provasopressin and proenkephalin may also be activated as the secretory vesicles are axonally transported from neuronal cell bodies to terminal regions. It will be important in future studies to investigate the mechanisms that regulate the coordinate activation of the multiple processing enzymes required for the synthesis of peptide neurotransmitters and peptide hormones.

REFERENCES

1. HOOK, V. Y. H., L. E. EIDEN & M. J. BROWNSTEIN. 1982. A carboxypeptidase processing enzyme for enkephalin precursors. Nature **295:** 341-342.
2. FRICKER, L. D. & S. H. SNYDER. 1982. Enkephalin convertase: Purification and characterization of a specific enkephalin-synthesizing carboxypeptidase localized to adrenal chromaffin granules. Proc. Natl. Acad. Sci. U.S.A. **79:** 3886-3890.
3. HOOK, V. Y. H. & P. Y. LOH. Carboxypeptidase B-like converting enzyme activity in secretory granules of rat pituitary. 1984. Proc. Natl. Acad. Sci. U.S.A. **81:** 2776-2780.
4. GAINER, H., J. T. RUSSELL & Y. P. LOH. 1985. The enzymology and intracellular organization of peptide precursor processing: The secretory vesicle hypothesis. Progr. Neuroendocrinol. **40:** 171-184.
5. HOOK, V. Y. H. & L. E. EIDEN. 1985. Nicotine stimulates release of carboxypeptidase peptide hormone processing enzyme and (Met)enkephalin from cultured chromaffin cells. Biochem. Biophys. Res. Comm. **128:** 563-570.
6. HOOK, V. Y. H. 1985. Differential distribution of carboxypeptidase processing enzyme activity and immunoreactivity in membrane and soluble components of chromaffin granules. J. Neurochem. **45:** 987-989.
7. HOOK, V. Y. H., E. MEZEY, L. D. FRICKER, R. M. PRUSS, R. SIEGEL & M. J. BROWNSTEIN. 1985. Immunochemical characterization of carboxypeptidase B-like peptide hormone processing enzyme. Proc. Natl. Acad. Sci. U.S.A. **82:** 4745-4749.
8. GAINER, H., Y. SARNE & M. J. BROWNSTEIN. 1977. Biosynthesis and axonal transport of rat neurohypophysial proteins and peptides. J. Cell Biol. **73:** 366-381. 1984.
9. LISTON, D., G. PATEY, J. ROSSIER, P. VERBANCK & J.-J. VANDERHAEGHEN. Processing of proenkephalin is tissue-specific. 1984. Science **225:** 734-737.
10. STACK, G., L. D. FRICKER & S. H. SNYDER. 1984. A sensitive radiometric assay for enkephalin convertase and other carboxypeptidase-B-like enzymes. Life Sci. **34:** 113-121.

The Secretogranins/Chromogranins: What Biochemistry, Cell Biology and Molecular Biology Tell Us about Their Possible Functions

W. B. HUTTNER, U. M. BENEDUM, A. HILLE,
AND P. ROSA

European Molecular Biology Laboratory
D-6900 Heidelberg, Federal Republic of Germany

Secretogranin I (chromogranin B) and secretogranin II, two tyrosine-sulfated secretory proteins first characterized in PC12 cells[1] and anterior pituitary,[2,3] respectively, have recently been shown to occur in a wide variety of endocrine and neuronal cells[4] (collaborative study with P. De Camilli and A. Zanini). Such a widespread distribution is also known for chromogranin A (O'Connor *et al.*[5] and Somogyi *et al.*[6] and refs. therein). Although secretogranins I and II and chromogranin A are clearly distinct proteins, they have several properties in common and may therefore be regarded as one class of proteins that probably exert similar functions (see Rosa *et al.*[4] and refs. therein). All three proteins are selectively expressed in cells that (1) concentrate the secretory product, store it and release it upon stimulation, (2) secrete into the internal milieu, and (3) secrete peptides with a rapid regulatory action after exocytosis. All three proteins are sorted, together with the regulatory peptides, to secretory storage granules and undergo, at least in some cells, partial proteolytic processing before exocytosis. All three proteins are rich in acidic amino acids and are posttranslationally modified by *O*-glycosylation, phosphorylation, and sulfation. As a result, they are strongly negatively charged and very soluble proteins at neutral pH.

In order to elucidate the function of the secretogranins/chromogranins, we decided to clone cDNAs coding for these proteins. We have determined the primary structure of bovine chromogranin A[7] (collaborative study with R. Frank and J. Mallet). Chromogranin A consists of 431 amino acid residues plus an NH_2-terminal cleaved signal peptide of 18 residues. Interesting features of the chromogranin A sequence include (1) a pronounced hydrophilicity, in particular repeated clusters of glutamic acid residues, (2) the occurrence of eight potential dibasic cleavage sites, which may point to a role as a peptide precursor, and (3) the presence of -Arg-Gly-Asp-, a three amino acid sequence involved in the receptor-mediated binding of several constitutively secreted proteins to cell membranes.

We have previously suggested[4] that the properties common to the secretogranins/chromogranins may provide us with essential clues as to their as yet unknown function. Specifically, it has been discussed that these proteins, by means of their large number of glutamic acid residues, may function as helper proteins in the packaging of regulatory peptides into secretory granules.[4,8] Additional roles as helper proteins in the secretion of regulatory peptides may be: (1) secretogranins/chromogranins keep the content of secretory granules in a state that allows proteolytic processing of peptide precursors, and (2) secretogranins/chromogranins ensure immediate solubility of regulatory peptides after exocytosis. Such roles would be consistent with the biochemical properties as well as the cellular and subcellular distribution of these proteins (see Rosa *et al.*[4] and refs. therein) and with the fact that they are secreted from the acidic milieu of secretory granules into fluids with neutral pH at which these proteins are highly soluble. Both the proposed ability to allow peptide processing and the capacity to ensure immediate solubility of regulatory peptides after exocytosis would make the secretogranins/chromogranins important for the proper functioning of the endocrine and nervous system. For chromogranin A, these hypotheses can now be investigated using molecular genetics.

REFERENCES

1. Lee, R. W. H. & W. B. Huttner. 1983. J. Biol. Chem. **258:** 11326-11334.
2. Zanini, A. & P. Rosa. 1981. Mol. Cell. Endocrinol. **24:** 165-179.
3. Rosa, P. & A. Zanini. 1981. Mol. Cell. Endocrinol. **24:** 181-193.
4. Rosa, P., A. Hille, R. W. H. Lee, A. Zanini, P. De Camilli & W. B. Huttner. 1985. J. Cell Biol. **101:** 1999-2011.
5. O'Connor, D. T., D. Burton & L. J. Deftos. 1983. Life Sci. **33:** 1657-1663.
6. Somogyi, P., A. J. Hodgson, R. W. Potter, R. Fischer-Colbrie, M. Schober, H. Winkler & I. W. Chubb. 1984. Brain Res. Rev. **8:** 193-230.
7. Benedum, U. M., P. A. Baeuerle, D. S. Konecki, R. Frank, J. Powell, J. Mallet & W. B. Huttner. 1986. EMBO J. **5:** 1495-1502.
8. Rosa, P., G. Fumagalli, A. Zanini & W. B. Huttner. 1985. J. Cell Biol. **100:** 928-937.

Chloroquine and Monensin Alter the Posttranslational Processing and Secretion of Dopamine β-Hydroxylase and Other Proteins from PC12 Cells[a]

ESTHER L. SABBAN, LORRAINE J. KUHN, AND MAHESH SARMALKAR

Department of Biochemistry
New York Medical College
Valhalla, New York 10595

The subcellular pathway for the posttranslational processing and secretion of dopamine β-hydroxylase was examined in PC12 pheochromocytoma cells. Dopamine β-hydroxylase, the enzyme which catalyzes the formation of norepinephrine from dopamine, is present as a 77-kDa membrane-bound subunit form and a 73-kDa soluble subunit form in near equal amounts[1,2] and the 77-kDa form appears to be a precursor of the 73-kDa form. Treatment of PC12 cells with a secretagogue results in release of DBH (apparent M_r = 73,000).[2,3]

To study the processing and secretory pathway, we used monensin and chloroquine. Monensin, a monovalent cationic ionophore, has been used in a number of other cell systems to block exit of secretory proteins from the Golgi apparatus.[4] Chloroquine, a lysomotropic agent, is a weak base, and probably acts by raising intravesicular pH. Chloroquine has been proposed to divert secretion from a regulated to a constitutive pathway.[5]

Treatment of PC12 cells with increasing concentrations of chloroquine resulted in an increased proportion of the 73-kDa subunit form of DBH (FIG. 1). At 200 μM chloroquine, DBH consists almost exclusively of the 73-kDa subunit form. In contrast to this specific effect on DBH biosynthesis, the profiles of total proteins synthesized in the presence or absence of various concentrations of chloroquine did not differ markedly. These results are similar to those obtained after treatment of PC12 cells with 200 nM monensin.[3] These findings suggest that posttranslational processing of DBH probably occurs prior to acidification of the secretory granules. Elevation of intravesicular pH as a consequence of chloroquine treatment may allow processing to resume in the altered granules. Alternatively, chloroquine treatment could be disrupting exit of newly synthesized DBH from the Golgi apparatus.

[a] This study was supported by NIH grant no. NS 20440. L. J. Kuhn is a recipient of the American Association of University Women Educational Foundation Fellowship.

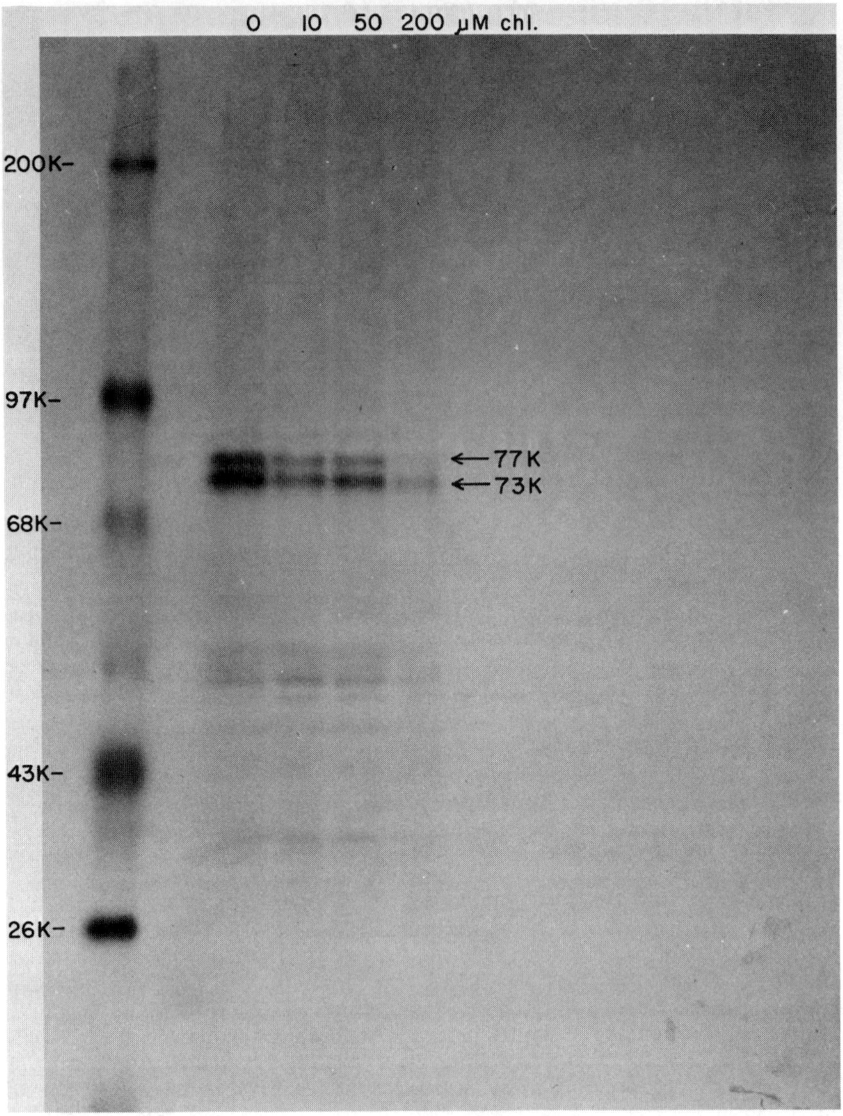

FIGURE 1. Effect of chloroquine concentration on subunit forms of dopamine β-hydroxylase. PC12 cells were treated with the specified concentration of chloroquine for 1 hour and labeled with [^{35}S]methionine for 3 hours in the presence of these additives. Dopamine β-hydroxylase was immunoprecipitated and analyzed by gel electrophoresis and fluorography.

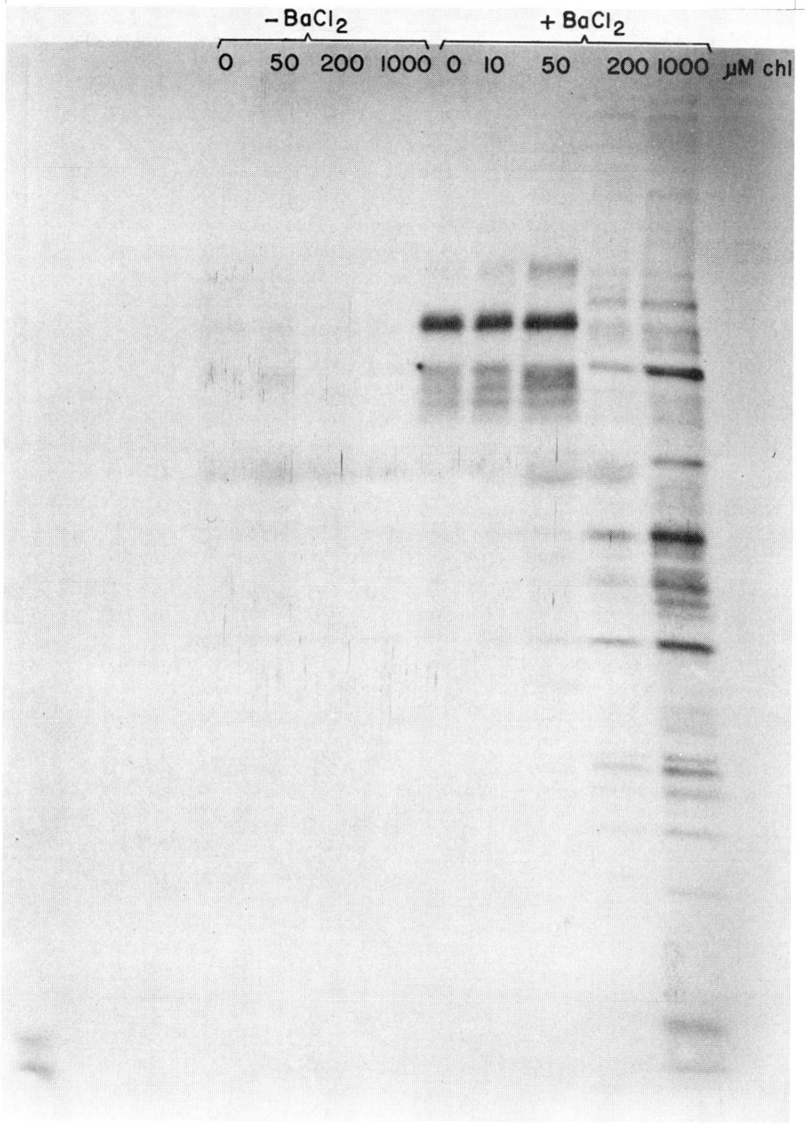

FIGURE 2. Effect of chloroquine concentration on stimulated secretion from PC12 cells. PC12 cells were treated with the specified concentrations of chloroquine for 1 hour, then labeled with [^{35}S]methionine in the presence of these additives. The cells were incubated for 10 min with 4 mM $BaCl_2$ in PBS or with PBS alone. The proteins released were concentrated using an Amicon Minicon B-15 concentrator and analyzed by gel electrophoresis and fluorography.

We have previously found that monensin treatment altered the profile of [^{35}S]Met-labeled proteins released in response to the secretagogue $BaCl_2$. We investigated the effect of chloroquine on the stimulated secretion of proteins from PC12 cells. At 200 μM chloroquine, release of the higher molecular weight proteins (M_r = 105,000-110,000, 90,000, 80,000, 76,000, and 73,000), normally observed in control preparations, appeared to be impeded. In addition, a subset of lower molecular weight proteins was now released (FIG. 2). In the absence of secretagogue, very little protein was released under these conditions. We cannot rule out whether some proteolytic event or disruption of the normal subcellular pathway allows release of different proteins. Nevertheless, based on these results, it is possible that chloroquine causes a disruption of the proton gradient in the vesicles. This may lead to secretion of a new group of proteins that may have been loosely attached to the granules.

ACKNOWLEDGMENTS

We thank Dr. Menek Goldstein for antisera to dopamine β-hydroxylase.

REFERENCES

1. SABBAN, E. L., L. A. GREENE & M. GOLDSTEIN. 1983. J. Biol. Chem. **258:** 7812-7818.
2. MCHUGH, E. M., R. J. MCGREE & P. J. FLEMING. 1985. J. Biol. Chem. **260:** 4409-4417.
3. KUHN, L., M. HADMAN & E. L. SABBAN. 1986. J. Biol. Chem. **261:** 3816-3838.
4. TARTAKOFF, A. M. 1983. Cell **32:** 1026-1028.
5. MOORE, H. P., B. GUMBINER & R. B. KELLY. 1983. Nature **302:** 434-436.

IRCM-Serine Protease #1 from Pituitary and Heart

A Common Prohormone Maturation Enzyme

NABIL G. SEIDAH, JAMES A. CROMLISH,
G. THIBAULT, AND M. CHRÉTIEN

Clinical Research Institute of Montreal
Montreal, Quebec H2W 1R7, Canada

A novel proteinase, which we have called IRCM-serine protease #1 (IRCM-SP1), was purified from both porcine neurointermediate and anterior pituitary lobes. This enzyme displayed "trypsin-like" specificity toward a number of tripeptide-coumarin-containing substrates, with k_{cat}/K_m values ranging from 10^4-10^6 $M^{-1}s^{-1}$. The best substrate was *Z*-Ala-Lys-Arg-AMC with a k_{cat}/K_m of 2.27×10^6 $M^{-1}s^{-1}$. IRCM-SP1 appears to be a homologous dimer of M_r 169,000-190,000 determined by gel filtration and gradient gel electrophoresis. The monomeric subunits of the enzyme are composed of a 38,000 catalytic chain disulfide linked to another polypeptide resulting in a subunit molecular mass of 88,000 (88 kDa/38 kDa). A similar enzyme was also isolated from whole human pituitaries and from rat heart atria and ventricles. The rat enzyme migrates on SDS-PAGE with an apparent molecular mass 96 kDa/45 kDa and the human enzyme as 91 kDa/41 kDa. The final enzymatic activity capable of cleaving COOH terminus to Arg in the substrate *Z*-Ala-Lys-Arg-AMC was found to be 244, 55, and 6 nmol/min/g starting protein in porcine neurointermediate lobe, rat heart atrium, and ventricle respectively. The nine times enrichment in atria versus ventricles suggest an enrichment of this enzyme in endocrine tissues.

The cleavage specificity of IRCM-SP1 from the different species and endocrine tissues was further probed with a number of synthetic and natural substrates derived from human pro-opiomelanocortin (POMC) and from proatrial natriuretic factor (pro-ANF 1-126). By using HPLC, amino acid analysis, and microsequencing it was possible to show that IRCM-SP1 cleaved COOH-terminally to all pairs of basic residues known to be cleaved *in vivo* with these substrates (FIG. 1). In addition, this enzyme cleaved in between pairs of basic amino acids found in the POMC NH_2-terminal glycopeptide fragment NT 1-76, also known to be cleaved *in vivo.* In pro-ANF 1-126 cleavage post Arg-Arg↓ was found to be favored over that post Pro-Arg↓ or in between Arg↓Arg. All those three bonds are known to be cleaved *in vivo,* albeit the Pro-Arg↓ cleavage was favored in atria whereas the Arg-Arg↓ and Arg↓Arg represent the favored sites in whole brain and hypothalamus. Therefore, the fact that IRCM-SP1 from either pituitary or heart and from either human, porcine, or rat cleaves the POMC-related and pro-ANF polypeptides at the expected pairs of basic residues and COOH-terminal to some single basic residues suggests that both tissues could utilize the same prohormone maturation enzyme.

TABLE 1

Subcellular[a] Fractions	Protein mg	Protein %	IR-7B2[b] μg	IR-7B2[b] ng/mg Protein	IR-7B2[b] %	IR-NT[b] mg	IR-NT[b] μg/mg Protein	IR-NT[b] %	IRCM-SP#1[c] mU[f]	IRCM-SP#1[c] mU/mg Protein[g]	IRCM-SP#1[c] %	IRCM-SP#1[d] mU[f]	IRCM-SP#1[d] mU/mg Protein[g]	IRCM-SP#1[d] %	IRCM-SP#1[e] mU[f]	IRCM-SP#1[e] mU/mg Protein[g]	IRCM-SP#1[e] %
P_1	284	13.3	1.1	3.9	3.2	1.45	5.1	4.6	708	2.5	0.8	13,500	47.5	3.1	4,240	14.9	2.7
P_2	305	14.4	8.0	26.2	14.8	3.51	11.5	11.2	987	3.2	1.1	1,400	4.6	0.3	5,010	16.4	3.2
P_3	698	32.7	19.7	28.2	36.2	9.40	13.5	30.0	920	1.3	1.0	23,000	33.0	5.3	31,400	45.0	20.3
P_4	133	6.22	0.6	4.5	1.0	0.7	5.3	2.1	74,500	560	81.3	122,000	917	28.2	18,700	141	12.1
S_4	715	33.5	24.3	34.0	44.8	16.3	22.8	52.1	14,500	20.3	15.8	272,000	380	63.0	95,500	134	61.7

[a] Subcellular fractionation performed with 18 g wet weight of porcine anterior lobes.
[b] 7B2 and NT represent secretory polypeptide markers for gonadotrophs and corticotrophs respectively.
[c] Proteolytic activity was measured in subcellular fractions without further purification.
[d] Proteolytic activity measured after purification on a phenyl boronate column and dialysis against 10,000 molecular weight cut-off membranes.
[e] Proteolytic activity was measured after concentration of the phenyl boronate purified fractions to 0.5 ml.
[f] One milliunit of activity (mU) is equivalent to the amount of enzyme that hydrolyzes 1 pmole of *Z*-Ala-Lys-Arg-AMC in one minute (pmol/minute) under standard assay conditions.
[g] In each case mg of protein refers to the amount of protein initially present in each subcellular fraction, and not the amount of protein present in the purified fractions.

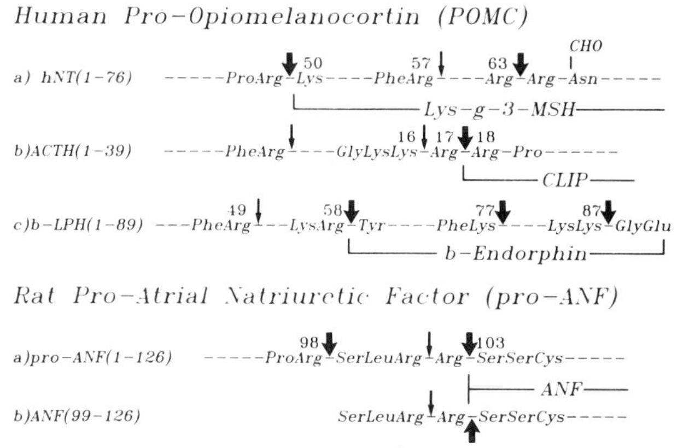

FIGURE 1

Results of subcellular localization studies of IRCM-SP1 in porcine anterior pituitary are shown in TABLE 1. Differential centrifugation of anterior pituitary gave five subcellular fractions including four pellets: P1 (484 $\times g/10$ min), P2 (4070 $\times g/$ 15 min), P3 (27,000 $\times g/15$ min), and P4 (100,000 $\times g/5$ h) and a final supernatant S4. Extraction of each subcellular fraction and partial purification on a phenyl boronate column revealed a complex pattern of IRCM-SP1 distribution. The majority of active enzyme is found in the high-speed pellet P4. A latent, partially activable form was also found in P3, which contains the majority of mature secretory granules (see TABLE 1). This distribution suggests a tentative model where the active enzyme is found in small vesicles, possibly clathrin-coated vesicles. These vesicles may be immature secretory granules containing prohormone and active enzyme or granules containing only the active enzyme which would later fuse with prohormone-containing granules. Upon cleavage of the prohormone and granule maturation the enzyme would be inactivated (possibly by metal ions and low pH) and hence would be found in a latent form in mature secretory granules. Our model suggests that tissue-specific processing could result from the time of contact of the substrate and active enzyme and their relative concentrations.

Opioid Peptides of the Electric Ray *Narcine braziliensis*

CHRISTOPHER D. UNSWORTH AND
ROBERT G. JOHNSON, JR.

Howard Hughes Medical Institute
and
Departments of Medicine and Neurology
Massachusetts General Hospital
Boston, Massachusetts 02114

In recent years it has become apparent that colocalization of multiple messengers within secretory vesicles may be a common phenomenon in neurotransmission. While the innervation of the electric organ of a variety of marine species has been extensively studied as a model for cholinergic neurotransmission, the presence of enkephalin-like peptides has also been reported in this tissue.[1] This therefore represents a model in which interactions between cholinergic and peptidergic systems might be investigated.

By using a combination of reversed-phase HPLC and specific radioimmunoassays for Met^5- and Leu^5-enkephalin (ME, LE), the presence of these peptides was demonstrated in the electric organ of *Narcine braziliensis.* A synaptosomal preparation from this tissue, known to be rich in acetylcholine and ATP, was also shown to contain both of these enkephalins with a ME:LE ratio of approximately 12:1 (FIG. 1). This is in contrast to previous reports where only LE was detected in *Torpedo,*[1] but does agree with reports of ME as the predominant form in the central nervous system of other species of fish.[2] Cell bodies of the neurons which innervate the electric organ are located in a pair of distinct nuclei (electromotor nuclei) on the dorsal surface of the brain stem. These nuclei were also shown to contain ME immunoreactive material (ME-IR), which coeluted with authentic ME on reversed-phase HPLC. The presence of precursor forms or "cryptic" ME-IR in this tissue was revealed by sequential digestion with trypsin and carboxypeptidase B. Investigation of a number of other brain areas revealed the widespread presence of ME-IR in both fully processed and precursor forms (TABLE 1). As with the synaptosomal preparation from the electric organ, a high ME:LE ratio was found in some of these brain areas (*e.g.,* olfactory bulb and lower brain stem).

These data clearly indicate that the enkephalins, particularly ME, are widely employed neurotransmitters in the central nervous system of this species. The high ME:LE ratio seen in the electric organ and in several brain areas raises an interesting question as to the molecular origin of these enkephalins. Further studies of the precursor forms may indicate whether ME and LE are derived from common or distinct precursor proteins in this species. The presence of ME-IR in both the electromotor nuclei and a synaptosomal preparation from the electric organ suggests that this neuropeptide may be colocalized in this classically cholinergic system. Although the cellular and subcellular localization of the enkephalins in the electric organ still requires definitive characterization, this system potentially represents an excellent

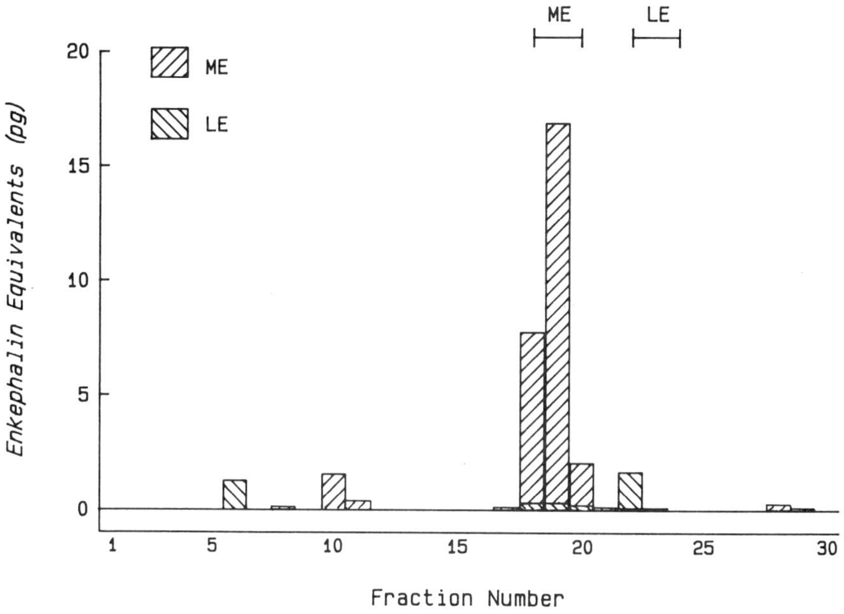

FIGURE 1. A seppak-treated acetic acid extract of synaptosomes prepared from the electric organ was chromatographed on C-18 reversed-phase HPLC with a linear gradient of acetonitrile from 16% to 36% over 30 min at a flow rate of 1 ml/min. One-minute fractions were collected, lyophilized, and aliquots assayed for Met^5- and Leu^5-enkephalin immunoreactivity.

TABLE 1. Met^5-Enkephalin Levels in the Central Nervous System and Electric Organ of *Narcine braziliensis*

Tissue	Met^5-Enkephalin	Cryptic Met^5-Enkephalin
Optic tectum	27.30 ± 2.48	68.80 ± 16.09
Olfactory bulb	12.51 ± 0.87	10.33 ± 0.80
Cerebellum	7.31 ± 1.10	11.44 ± 2.97
Upper brainstem	12.93 ± 1.92	18.93 ± 3.03
Lower brainstem	9.49 ± 1.06	13.16 ± 1.13
Electromotor nuclei	1.36 ± 0.27	2.67 ± 0.51
Electric organ	0.031 ± 0.006	nd

The brain areas and electric organs were dissected from anaesthetized rays on ice. The tissues were homogenized in 1 M acetic acid and stored at −70°C. Following centrifugation, aliquots of the supernatant were assayed for Met^5-enkephalin immunoreactivity (immunoassay kit, Immunonuclear Corp.). Levels of cryptic Met^5-enkephalin were calculated as the increase in immunoreactivity following sequential digestion with trypsin (10 μg/ml) and carboxypeptidase B (1 μg/ml). These values are from rays approximately 1 year of age: mean weight = 44.17 g, mean length = 148 mm. They are mean ± SEM, n = 10. Units are ng/mg protein; nd = not detectable.

model to study the interaction among colocalized multiple messengers. Such interactions may occur following release where pre- or postsynaptic modulation of the cholinergic effects is a potential function of the enkephalins.[3] In addition, there may be intraneuronal regulation of the relative amounts of the various components of a multiple messenger system as has been reported for the adrenergic-peptidergic systems of the adrenal medulla. The electromotor system of *Narcine braziliensis* allows study of neurotransmitter biochemistry and function at many levels. Precursor biosynthesis including mRNA transcription, protein translation, and vesicular packaging can be studied in a clearly defined population of cell bodies. The long axons and abundance of nerve terminals allow investigation of peptide processing, vesicular storage, axonal transport, transmitter release, and postsynaptic events. This therefore represents a valuable system that may add to our understanding of interactions between cholinergic and enkephalinergic systems colocalized in a single neuronal population.

REFERENCES

1. MICHAELSON, D. M., G. MCDOWALL & Y. SARNE. 1984. The "Torpedo" electric organ is a model for opiate regulation of acetylcholine release. Brain Res. **305:** 173-176.
2. LINDBERG, I. On the evolution of proenkephalin. Trends Pharm. Sci. p. 216. (June 1986).
3. MICHAELSON, D. M., G. MCDOWALL & Y. SARNE. 1984. Opiates inhibit acetylcholine release from "Torpedo" nerve terminals by blocking Ca^{2+} influx. J. Neurochem. **43:** 614-618.

Movements of Vesicles on Microtubules[a]

MICHAEL P. SHEETZ,[b] RONALD VALE,[c]
BRUCE SCHNAPP,[c] TRINA SCHROER,[b] AND
THOMAS REESE[c]

[b] *Department of Cell Biology and Physiology*
Washington University Medical School
St. Louis, Missouri 63110

[c] *NINCDS at Woods Hole*
Marine Biological Laboratory
Woods Hole, Massachusetts 02543

INTRODUCTION

Many observations of organelle movements within different cells under a wide variety of conditions have been linked to microtubule integrity.[1] Fast axoplasmic transport within the squid giant axon provides an excellent model system both for viewing vesicle transport and for obtaining sufficient quantities of material for biochemical measurements. The application of new video technologies to light microscopy has made it possible to view directly the abundant bidirectional organelle traffic within the axon.[2] Previous microscopic studies had revealed only a fraction of the organelle traffic measured in biochemical studies.[3] The abundant movements of organelles upon "single filaments" in both directions were seen not only within axons and isolated axoplasm, but also in regions of dissociated axoplasm where single filaments had become disentangled from the bulk of the axoplasm.[4]

With a reproducible method of preparing active dissociated axoplasm it was possible to explore the exact nature of the "transport filaments," which might have been microtubules, actin filament bundles, or a complex of both.[5] Sequential examination of the same "transport filaments" in the light and electron microscopes revealed that a "transport filament" was a single microtubule,[6] and no actin was found associated with the filaments.[7] Further proof that organelles could move on microtubules came from the reconstitution of organelle movement with semipurified organelles and purified microtubules,[8] or axonemal microtubules.[9]

The observation that purified microtubules would move on glass coated with an axoplasmic supernatant fraction free of organelles[10] provided a simple assay for purification of a microtubule motor. A major affinity purification step for the isolation of the motor was suggested by the observation that the nonhydrolyzable analogue of

[a] This work was supported in part by NIH grants GM-33352 and NS-14576 to M. Sheetz. M. Sheetz is an established investigator of the American Heart Association. T. Schroer is supported by a Muscular Dystrophy Association Postdoctoral Fellowship.

ATP, AMP-PNP, would not only inhibit organelle movement but also would cause microtubules to become decorated with organelles.[11] The possibility that the motor was trapped in a high-affinity state for microtubules was supported by the finding that AMP-PNP caused the motor activity to pellet with microtubules. Subsequent purification steps yielded a protein which was a complex of 110-, 65-, and 70-kDa polypeptides in a 4:1:1 ratio and chromatographed on a gel filtration column with an apparent native molecular mass of 660 kDa. The subunit molecular weights and relative insensitivities to both vanadate ion and the sulfhydryl reagent, *N*-ethylmaleimide, differentiated this protein from the other known microtubule-based motor, dynein.[12] It was, therefore, named "kinesin" from the Greek verb "to move" because of its ability to power the movement of beads on microtubules and microtubules on glass, and to increase vesicle movement on microtubules.[10]

Microtubules supported the movement of organelles in both directions in axons and dissociated axoplasm. With purified kinesin, movement in only one direction was observed using beads or organelles. An assay system was needed to determine the direction of kinesin movement relative to the microtubule polarity, which is defined by the direction of preferred microtubule polymerization. In the chosen assay centrosomes were used to seed microtubule growth and provided a radial array with all of the fast growing (plus) ends away from the center.[13] When kinesin-coated beads were added to the centrosomal microtubule arrays, they always moved away from the center toward the plus end of the microtubule and at the glass surface microtubules were pushed toward the centrosome (FIG. 1).[14]

RETROGRADE FACTOR

If crude supernatants from axoplasm were used to coat plastic beads, the beads were found to move toward the minus as well as the plus ends of microtubules.[8] Thus, a component was lost in the purification of kinesin that enabled beads to move toward the minus end of the microtubule. To determine if kinesin was also involved in movements toward the minus end, an anti-kinesin antibody bound to agarose beads was used to extract the kinesin from the crude axoplasmic supernatant that contained both motile activities. When kinesin was removed, the supernatant-coated beads would only move in the minus direction. Further indications that the retrograde moving component was distinct from kinesin came from the finding that treatment with the sulfhydryl reagent *N*-ethylmaleimide or low concentrations of vanadate would inactivate the minus moving factor but not kinesin. There is thus another motor in axoplasm besides kinesin that will move objects in the minus direction on microtubules.

Within axons the polarity of microtubules is such that the plus ends are oriented toward the synapse.[15,16] Kinesin would be expected to drive organelle movements in the anterograde direction from the cell body to the synapse, whereas the minus end motile factor would be responsible for moving organelles in the retrograde direction. There are many questions which remain to be answered, including: How do the motors become attached to the organelle surface since they can exist in a soluble form? How does the retrograde factor reach the synapse? What happens to kinesin when it reaches the synapse? The central question is, Do kinesin and the retrograde factor actually power vesicle movements *in vivo?* In this paper we will summarize the findings supporting the hypothesis that they are the motors and discuss some of the major difficulties with that hypothesis.

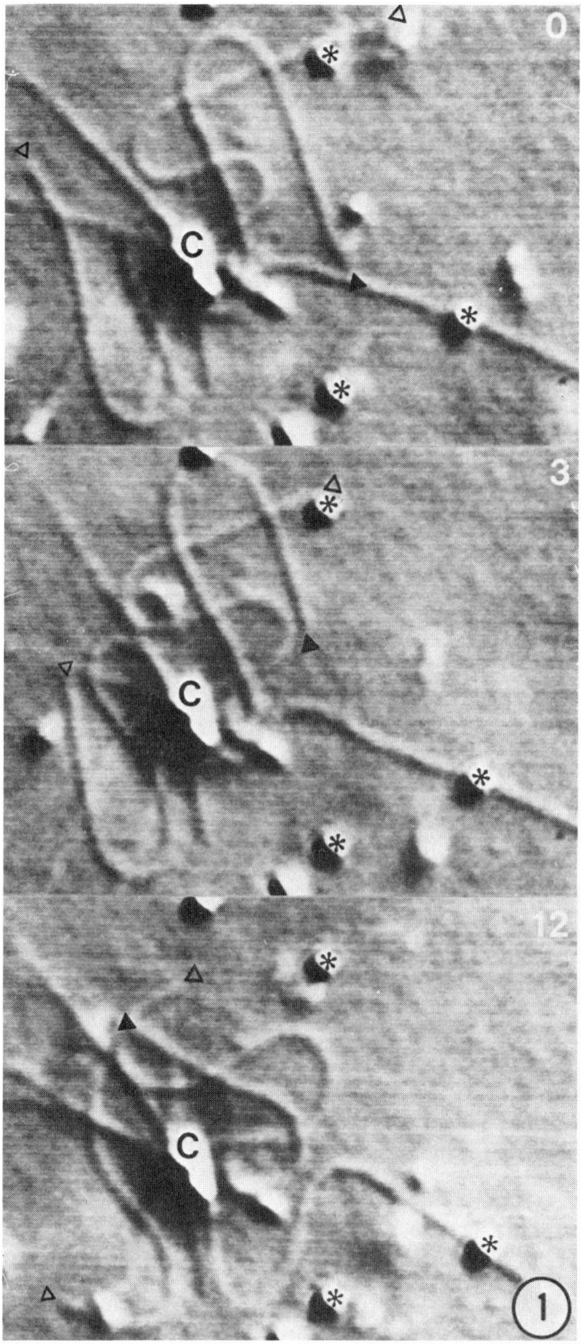

FIGURE 1. Movement of centrosomal microtubules on a glass coverslip occurs when kinesin and ATP are added. Movements of three microtubules are indicated by *arrowheads* at their trailing (+) ends. The direction of these kinesin-induced movements is toward the centrosome (*C*). *Asterisks* indicate stationary beads attached to glass coverslip; elapsed times in seconds are indicated at right. Video DIC image; magnification 8000×.

VESICLES CONTAIN DIRECTIONALITY SIGNALS AND MOTORS ON THEIR SURFACE

Movement of vesicles from the cell body to the axon must be programmed onto the vesicle surface, and a different program must exist for vesicles moving from the synapse to the cell. If this is the case, then the moving vesicles derived from axoplasm which are moving should move unidirectionally on microtubules. In studies of dissociated axoplasm, which maintains activity for several hours, the movement of smaller vesicles with the apparent sizes of 0.2-0.6 μm on single microtubules was unidirectional with only three observations of direction reversals out of several thousand vesicle movements.

In dissociated axoplasm preparations an important question remained: Are the microtubules coated with motor proteins? Or, do the vesicles contain the motors? The latter seems more likely. When vesicles were purified from axoplasm by sucrose density gradient centrifugation, some vesicles did move on purified microtubules[10] and as well on axonemal tubules.[9] The number of moving vesicles per micron of microtubule length was significantly lower with purified vesicles than with similar concentrations of vesicles in dissociated axoplasm. There was also considerable variability in the percentage of organelles that moved toward the plus end versus the minus end of microtubules in a centrosome assay from one experiment to the next.

In order to determine if the vesicles carried motor proteins into the gradient, the sucrose gradient was fractionated and each fraction was analyzed for its ability to promote bead movement on microtubules. Triton X-100 was added to each fraction prior to assay to remove vesicle membranes. When the gradient fractions were analyzed for the amount of motile activity, they exhibited a peak of motile activity in a dense gradient fraction (FIG. 2). Since the centrifugation time and *g*-force were such that only very large particles could travel this far into the gradient, the motors must have been attached to larger structures such as vesicles. Vesicles isolated from these fractions were capable of moving on purified microtubules. This data supports the concept that motors can remain bound to vesicles and copurify with them.

If motor proteins were bound to vesicles and kinesin was one of the motors, then it should be possible to find kinesin on the purified vesicles. Purification of organelles from squid optic lobes was accomplished by initial density gradient sedimentation followed by addition of sucrose and flotation through a second density gradient. The final vesicle fraction from a 12% sucrose layer contained kinesin as determined by Western immunoblot analysis with an anti-kinesin antibody. Further, when intact vesicles (12-18% sucrose layers) were incubated with monoclonal anti-kinesin antibody bound to agarose, approximately 50% of the vesicle kinesin was bound to the resin. Thus, one of the cytoplasmic microtubule motors, kinesin, copurifies with vesicles and is available externally for binding to antibodies.

MOTOR BINDING TO VESICLES APPEARS REVERSIBLE

Purification of vesicles by sucrose gradient sedimentation may cause decreased motility on microtubules *in vitro* for a variety of reasons. One possibility is that motor binding to the vesicles is reversible and motors come off the vesicles in the gradient. If this were the case, then the addition of purified motors should increase vesicle motility. Increases in vesicle motility were observed after the addition of purified kinesin in four out of five experiments (TABLE 1). Stimulation of purified organelle movement was always observed when a crude supernatant was added. In experiments where vesicles were further purified, the addition of supernatant was required for vesicle motility (T. Schroer and M. Sheetz, in preparation). The possibility exists that a supernatant enzyme activity, which is a contaminant of the kinesin preparations, causes the stimulation of vesicle movement. These findings, however, are consistent with the hypothesis that kinesin is reversibly bound to the vesicle surface.

TABLE 1. Movements Supported by Squid and Bovine Translocators

Sample	Organelle Movement[b]
Buffer	+
S2 axoplasmic supernatant[a]	+ + + to + + + +
Squid gel filtration	+ + to + + + +
Squid hydroxyapatite	+ + to + + + +
Bovine gel filtration	+ to + +

Organelle movements promoted by S2 axoplasmic supernatant or purified squid or bovine translocator; movement occurred consistently in preparations. Squid gel filtration peak fractions ($n = 7$; equivalent to lane 30 in Fig. 4 in Vale *et al.*,[8]) and hydroxyapatite peak fractions ($n = 3$; equivalent to lanes 40, 41 in Fig. 5) were tested in motility buffer or 100 mM KC1, 50 mM Tris (pH 7.6), 5 mM $MgC1_2$, 0.5 mM EDTA, 2 mM ATP at final protein concentrations between 10 and 150 μg/ml. Bovine gel filtration peak fractions ($n = 3$) were tested in KC1/Tris buffer at protein concentrations between 20 and 50 μg/ml. Organelle movement was assessed by viewing a 20 μm × 20 μm field of microtubules for 2.5–15 min and counting the number of different organelles that made directed movements along microtubules. The following rating system was employed: −, no movement in all preparations; +, 0–3 movements/min; + +, 3–7 movements/min; + + +, 7–15 movements/min; + + + +, 15–34 movements/min. The range of organelle movement in different preparations is reported. Microtubule bead, or organelle movement assays are described in Experimental Procedures in Vale *et al.*[8]

[a] Values from Vale *et al.*[5]

[b] The velocity of organelle movement in all samples was approximately 1.64 ± 0.24 μm/sec.

VESICLE MOVEMENTS AND KINESIN

Although the circumstantial evidence strongly suggests that kinesin moves vesicles toward the plus ends of microtubules (anterograde direction) *in vivo,* there are a number of problems which remain unexplained. The majority of kinesin is found in a soluble form in dissociated axoplasm and not on vesicle surfaces. What prevents it *in vivo* from attaching to retrograde vesicles or other cytoplasmic constituents and moving them all away from the cell body? Preliminary studies of the kinesin vesicle interaction have shown that the binding of kinesin to vesicles can occur, and yet those vesicles will not move on microtubules (Schroer *et al.,* unpublished results). This finding suggests that in addition to attachment to a vesicle an activation of bound kinesin may occur or that vesicle binding can occur with fewer kinesins than needed for motility. The anionic beads require a critical density of myosin on their surface to move at all, and they move at the maximal velocity when they do move, which is consistent with the latter possibility.

An implication of the large amount of soluble kinesin and the findings in TABLE 1 is that kinesin may be binding to the vesicles reversibly. If the ratio of free to bound kinesin in FIGURE 2 reflects the true affinity constant, then there is obviously a low-affinity binding of kinesin with the vesicles. A weak interaction will have difficulty transmitting force from kinesin to the vesicle, since the force of movement will be likely to break the kinesin-vesicle interaction. An obvious alternative explanation is that there are both strong and weak interactions of kinesin with vesicles and that the strong are saturated with a small fraction of the total kinesin.

There is a problem of what to do with kinesin when it reaches the synapse. Is it degraded there or transported back to the cell body? In smaller cells kinesin can easily diffuse from the periphery back to the cell center as a soluble molecule. Because axonal processes are so long, a diffusion mechanism is impractical but a weak binding to retrograde vesicles might allow kinesin to be moved toward the cell body without inhibiting retrograde movement.

There are many other practical problems to be understood through experimentation before it is possible to state unequivocally that kinesin powers vesicle movements on microtubules *in vivo.* The role of the retrograde axoplasmic motor in retrograde vesicle movements is even less clear.

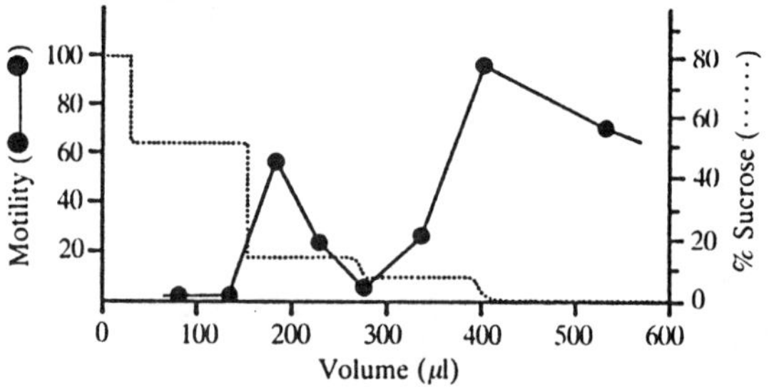

FIGURE 2. The motility of fractions of a sucrose gradient sedimentation of dissociated axoplasm reveal a peak of motile activity at the 50–15% sucrose interace. The axoplasm dissociation and sucrose gradient procedures are described in Vale *et al.*[14] Gradients were centrifuged at 135,000 × g for 2 hours at 4°C, and the graph shows the data from a typical experiment. Motility (number of bead movements per 10 minutes) was assayed with a standard density of microtubules using carboxylated latex beads and Triton as described in Vale *et al.*[14]

KINESIN-DEPENDENT MOVEMENTS IN OTHER SYSTEMS

Kinesin has been isolated from a variety of sources, including sea urchin eggs,[17] rabbit liver and kidney, and cultured cells. In addition, polyclonal antibodies to squid kinesin (110 kDa) cross-react on Western blots with a 95–140-kDa component in every species and cell type tested. It is, therefore, likely that kinesin or a close analogue might act similarly in all cell systems to move objects along microtubules. One of the important controlling elements in cell form and polarity is the radial microtubule array emanating from the centrosome. Since kinesin would be expected to move objects to the periphery of the cell on those microtubules, it could be very useful in building and maintaining polar distributions of materials within cells.

SUMMARY

Many cytoplasmic vesicles are observed to move along microtubules. Often, bidirectional movement of particles is observed on a single microtubule. We have isolated one cytoplasmic motor, kinesin, and defined another, the axoplasmic retrograde factor, which are capable of powering anionic latex beads toward the plus and minus ends of microtubules, respectively. Observations of vesicle movements show that vesicles have a defined direction of movement and that vesicles copurify with a kinesin motor activity. Current evidence suggests the hypothesis that kinesin and the retrograde motors power vesicle movements *in vivo* by attachment to the appropriate vesicle.

REFERENCES

1. SCHLIWA, M. 1984. Mechanisms of intracellular organelle transport. *In* Cell Muscle Motility, Vol. 5. J. W. Shaw, Ed.: 1-81. Plenum. New York.
2. ALLEN, R. D., J. METUZALS, I. TASAKI, S. T. BRADY & S. P. GILBERT. 1982. Fast axonal transport in squid giant axon. Science **218:** 1127-1129.
3. GRAFSTEIN, B. & D. S. FORMAN. 1980. Intracellular transport in neurons. Physiol. Rev. **60:** 1167-1283.
4. ALLEN, R. D., D. G. WEISS, J. H. HAYDEN, D. T. BROWN, H. FUJIWAKE & M. SIMPSON. 1985. Gliding movement of and bidirectional organelle transport along single native microtubules from squid axoplasm: Evidence for an active role of microtubules in cytoplasmic transport. J. Cell Biol. **100:** 1736-1752.
5. VALE, R. D., B. J. SCHNAPP, T. S. REESE & M. P. SHEETZ. 1985. Movement of organelles along filaments dissociated from the axoplasm of the squid giant axon. Cell **40:** 449-454.
6. SCHNAPP, B. J., R. D. VALE, M. P. SHEETZ & T. S. REESE. 1985. Single microtubules from squid axoplasm support bidirectional movement of organelles. Cell **40:** 455-462.
7. SCHNAPP, B. J., M. P. SHEETZ, R. D. VALE & T. S. REESE. 1984. Filamentous actin is not a component of transport filaments isolated from squid axoplasm. J. Cell Biol. **99:** 351a.
8. VALE, R. D., T. S. REESE & M. P. SHEETZ. 1985. Identification of a novel force generating protein (kinesin) involved in microtubule-based motility. Cell **42:** 39-50.
9. GILBERT, S. P., R. D. ALLEN & R. D. SLOBODA. 1985. Translocation of vesicles from squid axoplasm on flagellar microtubules. Nature **315:** 245-248.
10. VALE, R. D., B. J. SCHNAPP, T. S. REESE & M. P. SHEETZ. 1985. Organelle, bead and microtubule translocations promoted by soluble factors from the squid giant axon. Cell **40:** 559-569.
11. LASEK, R. J. & S. T. BRADY. 1985. Attachment of transported vesicles to microtubules in axonal microtubules in axoplasm is facilitated by AMP-PNP. Nature **316:** 645-647.
12. GIBBONS, I. R. 1981. Cilia and flagella of eukaryotes. J. Cell Biol. **91:** 107-124.
13. MITCHISON, T. & M. KIRSCHNER. 1984. Microtubule assembly nucleated by isolated centrosomes. Nature **312:** 232-237.
14. VALE, R. D., B. J. SCHNAPP, T. MITCHISON, E. STEUER, T. S. REESE & M. P. SHEETZ. 1985. Different axoplasmic proteins generate movement in opposite directions along microtubules in vitro. Cell **43:** 623-632.

15. BURTON, R. R. & J. L. PAIGE. 1981. Polarity of axoplasmic microtubules in the olfactory nerve of the frog. Proc. Natl. Acad. Sci. U.S.A. **78:** 3269-3273.
16. HEIDEMANN, S. R., J. M. LANDERS & M. A. HAMBORG. 1981. Polarity orientation of axonal microtubules. J. Cell Biol. **91:** 661-665.
17. SCHOLEY, J. M., M. E. PORTER, P. M. GRISSOM & J. R. MCINTOSH. 1985. Identification of kinesin in sea urchin eggs, and evidence for its localization in the mitotic spindle. Nature **318:** 483-486.

DISCUSSION OF THE PAPER

S. H. DEVOTO (*Rockefeller University, New York, N.Y.*): There seems to be something puzzling about the receptor for kinesin, namely it is present on glass beads but absent from synaptic vesicles.

M. P. SHEETZ (*Washington University Medical School, St. Louis, Mo.*): Synaptic vesicles wouldn't be expected to possess the receptor because they don't have access to microtubules.

G. RUDNICK (*Yale University School of Medicine, New Haven, Conn.*): What is known about the way that kinesin or the retrograde factor hydrolyze ATP? Are there any similarities to myosin?

M. P. SHEETZ: We have not found conditions yet to measure the hydrolysis of ATP by kinesin nor the retrograde factor (which has not yet been purified). Studies by Shahid Khan (at Albert Einstein College of Medicine) and Bruce Schnapp (at the National Institutes of Health) show that kinesin-dependent movement of microtubules on glass is inhibited by ADP and phosphate and is half-maximal at 50 μM ATP.

J. D. CASTLE (*Yale University School of Medicine, New Haven, Conn.*): First, what are the determinants of kinesin-organelle interaction? Second, rotary shadowing of kinesin indicates 700 Å length, what is the relation to the 110-kDa protein?

M. P. SHEETZ: In response to your first question, we are currently studying what is necessary for kinesin to produce organelle movements and have no answer yet. In preliminary studies in collaboration with T. Schroer and R. Kelly (at the University of California at San Francisco) we have found that synaptic vesicles will not move on microtubules when mixed with kinesin and ATP.

In response to your second question, the size of the major kinesin subunit is sufficient to form the long (700 Å) stalk seen by rotary shadowing and preliminary studies of antibody decoration support this hypothesis.

R. HOLZ (*University of Michigan Medical School, Ann Arbor, Mich.*): What is the role of neurofilaments in organelle movement?

M. P. SHEETZ: At present there is no evidence to implicate neurofilaments in any organelle movements. In addition, studies of the structure of neurofilaments (K. Weber and colleagues) indicate that they are composed of an antiparallel array of subunits, which suggests that they may not be polar filaments. If they are not polar, then they could not support directed movement of organelles.

Calmodulin and the Secretory Vesicle[a]

JOSÉ-MARÍA TRIFARÓ

Department of Pharmacology
University of Ottawa
Ottawa, Ontario K1H 8M5, Canada

SUSAN FOURNIER

Department of Pharmacology
McGill University
Montreal, Quebec, Canada

INTRODUCTION

The chromaffin cells of the adrenal medulla store their secretory products in membrane-bound organelles, the chromaffin granules. On stimulation by acetylcholine, the chromaffin cell depolarizes and Ca^{2+} channels open up, resulting in an increase in Ca^{2+} influx. When the intracellular concentration of Ca^{2+} reaches a critical level (10^{-6}-10^{-5} M), the release mechanism is triggered and the chromaffin granules release their contents to the cell exterior by exocytosis.[1-3] Therefore, Ca^{2+} plays a key role in the secretory process of the chromaffin cell and in other such systems that store their secretory products in subcellular granules.[3] The question of where exactly Ca^{2+} plays a role, however, still awaits elucidation. Ca^{2+} may be involved either in membrane fusion, in a contractile-like event leading to the release of the chromaffin cell secretion products,[4,5] or in secretory granule mobility as a result of the regulation of cytoplasm viscosity.[6,7] It has been postulated that some type of contractile event might be involved in the process of secretion by the chromaffin cells.[4,5,8] The presence of contractile proteins in these cells and the association of some of these proteins with the chromaffin granules help support this hypothesis.[9-18] Moreover, it has also been demonstrated that the activation of myosin by actin in both smooth muscle and in nonmuscle cells requires the intervention of the Ca^{2+} regulatory protein calmodulin (CM).[19] In addition, CM has been shown to regulate other cytoskeletal components through specific binding proteins such as tubulin kinase, caldesmon, spectrins, and tau factor.[20] CM is also involved in the cyclic nucleotide and glycogen metabolisms and regulates calcium transport systems.[20]

In this paper we discuss the properties of the chromaffin cell CM, its subcellular distribution, and its interaction with chromaffin granule and plasma membrane pro-

[a] This work was supported by MRC grant no. PG-20.

teins. The effects of the treatment with CM antagonists and of the intracellular delivery of CM antibodies on secretion are also analyzed.

SUBCELLULAR DISTRIBUTION OF CALMODULIN

It has been known for several years that the total cellular levels of CM remain constant[21] regardless of the functional state of the cell (resting or stimulated). It is, therefore, possible that changes in the subcellular distribution of CM (free versus membrane-bound) are important for its modulating effects. In this regard, it has been shown recently that during the release of luteinizing hormone from the anterior pituitary stimulated by the gonadotropin-releasing hormone, the concentration of CM in the target cell cytosol decreased with a concomitant increase in the amount of CM bound to the plasma membrane.[22] Termination of target cell stimulation was accompanied by re-establishment of the CM distribution observed in the resting cell.

Immunocytochemical Evidence

Using an affinity chromatography purified CM antibody, and the indirect immunofluorescence technique, we have observed that CM was uniformly distributed throughout the entire cultured chromaffin cell, including processes and growth cones. This was an expected finding, since most of the published literature suggests that CM is a cytosolic protein.[23] As indicated above, however, there is some evidence in favor of membrane-bound CM.[22] Therefore, it was decided that one way to visualize the chromaffin cell membrane-bound CM was to permeabilize the cells by treatment with 20 μM digitonin, a procedure that allows many cytosolic components to leave the cell. Permeabilized chromaffin cells were then fixed and immunostained for CM. Cells thus treated showed a particulate CM fluorescence localized in cell bodies and their processes (FIG. 1). Interestingly enough, this punctate fluorescence was similar to that observed when cells were immunostained for dopamine β-hydroxylase.[24] The similarity between these two antibody (anti-DBH and anti-CM) staining patterns suggested that, perhaps, part of the chromaffin cell CM was associated with the chromaffin granules.

Biochemical Evidence

The cytochemical evidence described above prompted us to study the subcellular distribution of CM in chromaffin cells, and in addition, to test the possibility that the Ca^{2+} concentration would affect the subcellular distribution of this Ca^{2+}-binding protein.[25] The results obtained clearly demonstrated that when a chelating agent (EGTA) was present during homogenization and subcellular fractionation of the chromaffin cell, the amount of free CM present in the cytosol (67% of total cellular CM) was much higher than that recovered in the same fraction (43% of total cellular CM) when the chelating agent was omitted from the entire procedure.[25] More membrane-bound CM (nuclear, microsomal, and "crude granule" fractions) was recovered under this latter condition; however, the total cellular content of CM remained unchanged.[25]

The traces of Ca^{2+} that might be present in the solutions prepared in the absence of chelating agent together with the Ca^{2+} that might have leaked out from cellular depots (mitochondria, endoplasmic reticulum, etc.) during homogenization and subcellular fractionation might have been responsible for the decrease in the free form and the increase in the membrane-bound CM detected in our experiments.[25] Further resolution of the "crude granule fraction" by sucrose density gradient centrifugation demonstrated that the distribution of CM in the density gradient was similar to the distribution of chromaffin granules rather than to that of mitochondria, Golgi elements, and lysosomes.[25] In this case, there was also more CM bound to chromaffin granules when EGTA was omitted from the density gradient (FIG. 2).

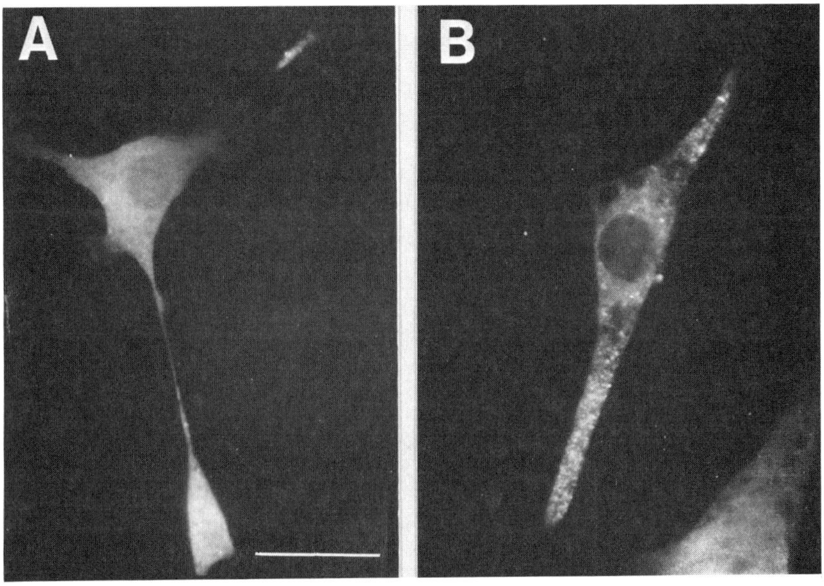

FIGURE 1. Cultured chromaffin cells as they appear after staining with antibodies against CM. Chromaffin cells cultured for 10 days were (**A**) fixed with 3.7% formaldehyde for 20 min or (**B**) permeabilized by treatment with digitonin (digitonin, 20 μM; potassium glutamate, 140 mM; Pipes, 20 mM; glucose, 10 mM; and $CaCl_2$, 1 μM) for 10 min prior to fixation. Both preparations were dehydrated as previously described[18] and stained by affinity-chromatography-purified antibodies against CM. The preparations were examined by incident light fluorescence microscopy.[18] A fine granular fluorescence is only observed in the preparation treated with digitonin (**B**). The *bar* in **A** indicates 50 μM. (From Trifaró *et al.*[20])

The observations described here demonstrate that the subcellular distribution of CM in the chromaffin cell depends on the presence or the absence of Ca^{2+}, suggesting that changes in the intracellular levels of Ca^{2+} might control the subcellular distribution of CM in the chromaffin cell. Moreover, the finding that the density gradient distri-

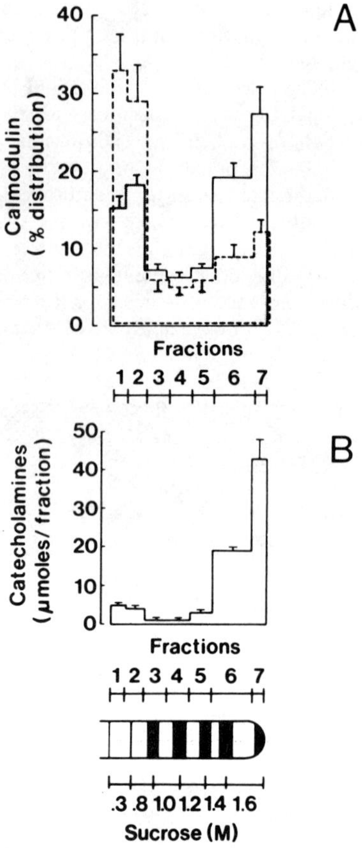

FIGURE 2. Effect of a chelating agent on the distribution of calmodulin in different subcellular fractions obtained by sucrose density gradient centrifugation of "crude granule fractions." Bovine adrenal medullae were homogenized in the presence or in the absence of 2 mM EGTA and "crude granule fractions" (20,000 × g sediments) were prepared by differential centrifugation. The "crude granule" sediments were resuspended in 0.3 M sucrose and layered onto sucrose density gradients prepared in the presence (- - - - -) or in the absence (——) of the chelating agent. The tubes were centrifuged at 113,000 ×g for 70 min and seven fractions were collected from each tube. (**A**) The calmodulin content of the fractions was determined by RIA.[25] (**B**) Catecholamines were assayed as previously described.[25] (Modified From Hikita *et al.*[25])

bution of membrane-bound CM followed that of chromaffin granules suggested that CM may play an important role in secretory granule function.

CALMODULIN-BINDING PROTEINS IN CHROMAFFIN GRANULE MEMBRANES

In order to further characterize the CM-chromaffin granule association, the binding of ^{125}I-calmodulin to granule membranes has been studied by two groups of investigators.[25–28] The results obtained in our laboratory have shown that in the presence of 10^{-4} M free Ca^{2+}, saturable and high-affinity CM-binding sites (K_d = 9.8 nM; B_{max} = 25 pmol/mg protein) were present in chromaffin granule membranes.[28] A second, nonsaturable, CM-binding site activity could also be detected only when the free Ca^{2+} concentration was between 10^{-7} M and 2.5×10^{-7} M.[28] No binding occurred at lower Ca^{2+} levels. When chromaffin granule membranes were delipidated by solvent extraction, CM binding was observed between 10^{-6} and 10^{-4} M free Ca^{2+}. However, no binding was detected at lower Ca^{2+} concentrations. Thus, it appears that Ca^{2+} concentrations between 10^{-7} M and 2.5×10^{-7} M promote the binding of CM to some solvent-soluble components of the chromaffin granule membrane. It is possible that, under these experimental conditions (10^{-7} M-2.5×10^{-7} M free Ca^{2+}), one or two molecules of Ca^{2+} are bound per molecule of CM, and this Ca^{2+} binding results in the exposure of a hydrophobic surface, which may explain the binding of CM to some hydrophobic (solvent-extractable) components of the granule membrane.[28]

Calcium-dependent CM binding has been detected in a variety of secretory granules.[29–32] Burgoyne and Geisow[26] have also reported the presence of Ca^{2+}-dependent CM-binding sites in chromaffin granules. However, the values (binding characteristics) reported by us[25,28] are quite different from those reported by Burgoyne and Geisow.[26] We do not have at the present time an explanation for this discrepancy. Nevertheless, it is interesting to point out that a recent paper describing the binding of CM to synaptic vesicles[32] indicated binding parameters that were in agreement with the values obtained with granule membranes in our laboratory.[28]

Experiments were also carried out to characterize the CM acceptor component(s) of the chromaffin granule membranes. Thus, CM-binding proteins associated with the granule membrane were identified first by photoaffinity cross-linking.[28] These experiments allowed the identification of a CM-binding protein complex, of molecular weight 82,000, which was formed in the presence of 10^{-6}-10^{-4} M free Ca^{2+} (FIG. 3). This cross-linked product was specific because it was not detected either in the absence of calcium, in the presence of nonlabeled CM, or in the absence of cross-linked activation.[28] When solvent-treated membranes were used, a second and specific CM-binding protein complex (M_r = 70,000) was formed (FIG. 3). Since the apparent molecular weight of CM in our electrophoresis systems was 17,000, these experiments suggested the presence of two CM-binding proteins, of molecular weights 65,000 and 53,000, in the chromaffin granule membrane. This result was confirmed by the use of CM-affinity chromatography. When detergent-solubilized granule membranes were applied to the affinity column in the presence of Ca^{2+}, two polypeptides of apparent molecular weights of 65,000 and 53,000 were specifically eluted by EGTA-containing buffer.[28] Since detergent treatments or solvent extractions were necessary to detect the 53,000 CM-binding protein, it can be concluded from the experiments described above that only the M_r 65,000 CM-binding polypeptide may play a role in the interaction between

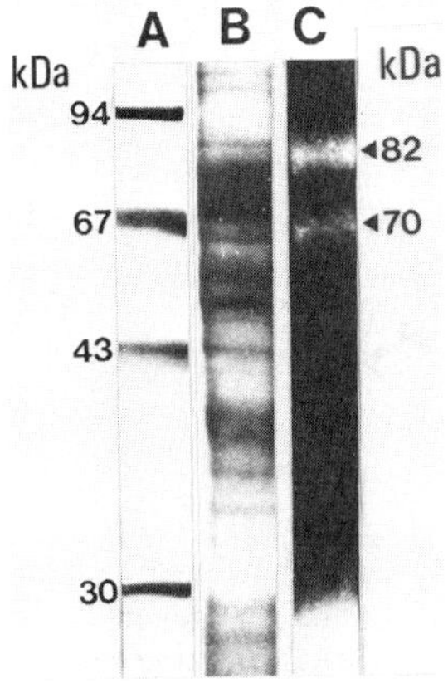

FIGURE 3. Identification of calmodulin-binding proteins present in chromaffin granule membranes after delipidation with acetone-ethanol. Delipidation of the membranes did not produce modifications in the protein composition, since the Coomassie blue staining pattern of nontreated membranes was similar to the pattern obtained with membranes treated with acetone-ethanol (lane B). Delipidated membranes (120 μg protein) were incubated in the presence of ^{125}I-labeled calmodulin, washed, and subsequently incubated with cross-linker and irradiated (lane B). The autoradiograph (lane C) showed the presence of two cross-linked products of molecular masses 82 kDa and 70 kDa. (Modified from Bader *et al.*[28])

CM and secretory granules in chromaffin cells. It is, therefore, possible that these two polypeptides are two subunits of a large protein, with one subunit (M_r = 53,000) deeply embedded in the membrane and stabilizing the second (M_r = 65,000) subunit (CM acceptor). Furthermore, the presence of a CM-dependent kinase in chromaffin granule membranes has been reported.[33] These authors have indicated a K_d value for CM-dependent phosphorylation of 400 nM (at 10^{-4} M free Ca^{2+}) that is almost 40 times higher than the K_d of the chromaffin binding site for CM. Thus, we consider it unlikely that the M_r 65,000 CM-binding polypeptide is the protein kinase described by Treiman *et al.*[33]

Similar experiments were also performed to characterize the CM acceptor components of chromaffin plasma membranes.[34,35] Purified plasma membranes obtained on polycationic Cytodex 1 beads (Pharmacia) were detergent-solubilized and applied to a CM affinity column. Polypeptides of apparent molecular weights 65,000 and 53,000 were eluted with EGTA-containing buffer. This result was confirmed using I^{125}-labeled CM overlay on nitrocellulose sheets containing transferred plasma membrane proteins. A major polypeptide band of apparent molecular weight 65,000 was labeled in the presence of 10^{-4} M Ca^{2+}; no labeling was seen in the presence of an EGTA-containing buffer.

Experiments carried out in our laboratory suggest that the CM binding polypeptides eluted from both granule and plasma membrane proteins are identical. Several lines of evidence support this hypothesis. Firstly, simultaneous two-dimensional SDS polyacrylamide gel electrophoresis show an identical pattern in terms of molecular weight and isoelectric focusing points (M_r = 65,000, pI = 6.3; M_r = 53,000, pI = 5.6).[35] Secondly, I^{125}-labeled CM overlay experiments on nitrocellulose show identical labeling patterns for both granule and plasma membrane proteins.[35] Finally, Matthew *et al.*[36] have described two different monoclonal antibodies that react with antigenic determinants of a 65,000 polypeptide present on the surface of brain synaptic vesicles. Furthermore, they showed that this antigenic component was widely distributed in neuronal and secretory tissues of neural crest origin, including the adrenal medulla.[36] Immunocytochemical experiments on adrenal medullary slices showed the antibodies reacting with granule membranes as well as with the inner surface of plasma membranes.[36] Using these two antibodies, we have recently performed immunoblotting experiments on granule and plasma membranes as well as calmodulin-binding proteins from isolated granule membranes (FIG. 4). In all cases, the antibody reacted with the 65,000 molecular weight CM-binding polypeptide.

Therefore, it appears that the 65,000 molecular weight polypeptide isolated from chromaffin granule and plasma membranes is the same protein as the synaptic vesicle surface antigen described by Matthew *et al.*[36] One possibility is that this protein may serve as the common link (docking protein) between the granule and the plasma membrane.

CALMODULIN IN SECRETION

The Use of Calmodulin Antagonists

Calmodulin's implication in the regulation of numerous Ca^{2+}-dependent cellular functions has led to the application of pharmacological agents to modify CM-mediated processes. This has become a potentially useful approach to the control of these cellular processes. Although phenothiazine antipsychotics were the first class of established CM antagonists,[20] at present a wide range of chemically unrelated substances such as butyrophenones, naphthalene sulfonamides, vinca alkaloids, local anesthetics, calmidazolium, and compound 48/80 have been found to antagonize CM as well.[20] With the use of these antagonists, CM has been implicated in cellular secretion from a number of cell systems, including the adrenal medulla.

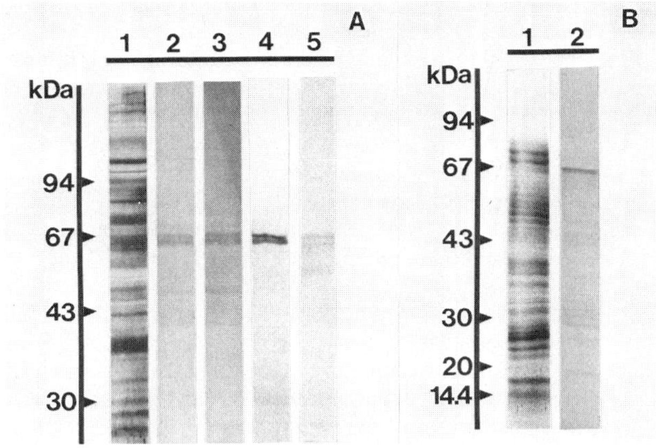

FIGURE 4. Immunoblots of chromaffin granule and plasma membrane proteins and isolated chromaffin granule calmodulin-binding proteins (CMBP) using monoclonal antibodies (mAb 30, mAb 48) against rat brain synaptic vesicle protein.[36] **(A)** Chromaffin granule membrane proteins and isolated CMBP were resolved on SDS polyacrylamide gel (7.5-15%) and transferred electrophoretically to nitrocellulose sheets (3 hours, 60 volts). The nitrocellulose blots were incubated 1 hour in phosphate-buffered saline (PBS) containing 3% BSA and 10% normal goat serum at room temperature. Blots were then incubated 2 hours at room temperature with mAb 30 or mAb 48 at a final dilution of 1:250. After a 1 hour PBS wash, the blots were incubated for 2 hours with a 1:1000 dilution of horseradish peroxidase labeled anti-mouse immunoglobulins (Miles). Immunoreactive bands were visualized by incubation in PBS containing 0.18 mg/ml 2-chloronapthol and 0.01% H_2O_2. Lane 1 represents transferred granule membrane proteins stained with Amido Black. The mAb 30 (lane 2) and mAb 48 (lane 3) detected the same proteins of molecular weight 65,000. Lane 4 represents a Coomassie-blue-stained gel of CMBP isolated from granule membranes. The mAb 30 detected a protein of molecular weight 65,000 (lane 5). The same results were obtained using mAb 48. **(B)** The conditions described above for blotting and staining of granule membrane proteins were followed for the plasma membrane proteins. Lane 1 represents a Coomassie-blue-stained polyacrylamide gel of plasma membrane proteins. When the blot of these proteins was incubated with mAb 30 (lane 2), the 65,000 molecular weight protein was detected. In all cases, the 65,000 molecular weight protein was resolved into a doublet on nitrocellulose blots. This doublet could be seen on SDS polyacrylamide gels (**A:** lane 4) and could be resolved into 2 subunits on two-dimensional gel electrophoresis.[28,35]

We were the first to use the CM antagonist trifluoperazine (TFP) to test for CM's involvement in adrenal chromaffin cell secretion.[37] In these experiments, we demonstrated TFP to be capable of inhibiting, in a dose-dependent fashion, catecholamine secretion from the chromaffin cells in response to two different types of stimulation (*i.e.*, cholinergic and K^+-induced depolarization).[37] Additional evidence has subsequently been provided from several other laboratories demonstrating that both TFP[20] and W7[20] were capable of inhibiting both nicotinic receptor-mediated and high K^+-induced release of catecholamines from adrenal chromaffin cells. In agreement with our findings, these other investigators found the CM antagonists much more potent in inhibiting receptor-mediated release than high K^+-induced release (*i.e.,* the ID_{50}s for this effect of TFP were 2×10^{-7} M for ACh-induced output and 2.2×10^{-6} M for K^+-induced release),[37] suggesting a partial manifestation of this inhibition at the level of the cholinergic receptor.

In an attempt to examine more carefully the inhibitory effect of TFP, we proceeded to test for its effects on Ca^{2+} movements in the chromaffin cell. We found that TFP at concentrations (10^{-6}-10^{-5} M) which produce significant inhibition of the secretory response was ineffective in blocking Ca^{2+} influx into the chromaffin cell[37] (FIG. 5). Furthermore, the antagonist did not inhibit the (Ca^{2+}-Ca^{2+}) exchange mechanism operating in these cells.[37] Consequently, we have, with the use of TFP, provided circumstantial evidence for a role for CM in chromaffin cell secretion, as the effects of the CM antagonist were found to be exerted distal to Ca^{2+} entry. In agreement with our findings, several other CM antagonists have been found to inhibit cellular secretion in other cell systems at a step distal to Ca^{2+} entry.[38–40]

In support of our findings are the earlier observations of Baker and Knight, who demonstrated that micromolar concentrations of TFP blocked Ca^{2+}-activated secretion from chromaffin cells rendered leaky by current pulses (an experimental situation where all ion fluxes associated with membrane depolarization are bypassed), suggesting that Ca^{2+} may trigger secretion via an intracellular CM-dependent reaction.[41]

Electrophysiological studies on this cell system (chromaffin cells) employing the patch-clamp technique have also suggested an intracellular effect of TFP.[42] In support of such a suggestion are the *in vitro* studies which demonstrate that the binding of labeled CM to chromaffin granule membranes is inhibited by TFP.[28]

Microinjection of Calmodulin Antibodies

Calmodulin's involvement in cellular secretion in various secretory systems has been suggested indirectly with the use of antagonists. More direct evidence for its involvement can only be obtained through the use of more highly selective probes. In this regard, the effect of an antigen (in this case CM) should be most selectively antagonized by its appropriate antibody, provided that the antigen (intracellular CM) is readily accessible to its antibody (a large macromolecule whose entry into the cell is restricted by the plasma membrane). Steinhardt and Alderton[43] attempted to overcome this restriction by using a cell free system prepared from sea urchin eggs. They demonstrated that the fusion of cortical granules to isolated cortical surfaces was blocked in the presence of antibodies raised against CM, thus providing the first bit of direct evidence for a role for CM in cellular secretion.

Polyclonal monospecific antibodies raised in sheep against native rat testis CM demonstrated cross-reactivity with bovine adrenal medullary chromaffin cell CM as verified by immunoprecipitation.[44,45] Consequently, this antibody appeared quite appropriate for use in the cultured chromaffin cell system.

In our laboratory, we have developed a technique for the intracellular injection of macromolecules into chromaffin cells in monolayer culture.[46] This technique employs erythrocyte ghosts as the vesicles for microinjection, phytohemagglutinin, a plant lectin, as an attachment agent, and polyethylene glycol as the fusagen.[46] With the use of this technique, high erythrocyte ghost-chromaffin cell fusion indices have been

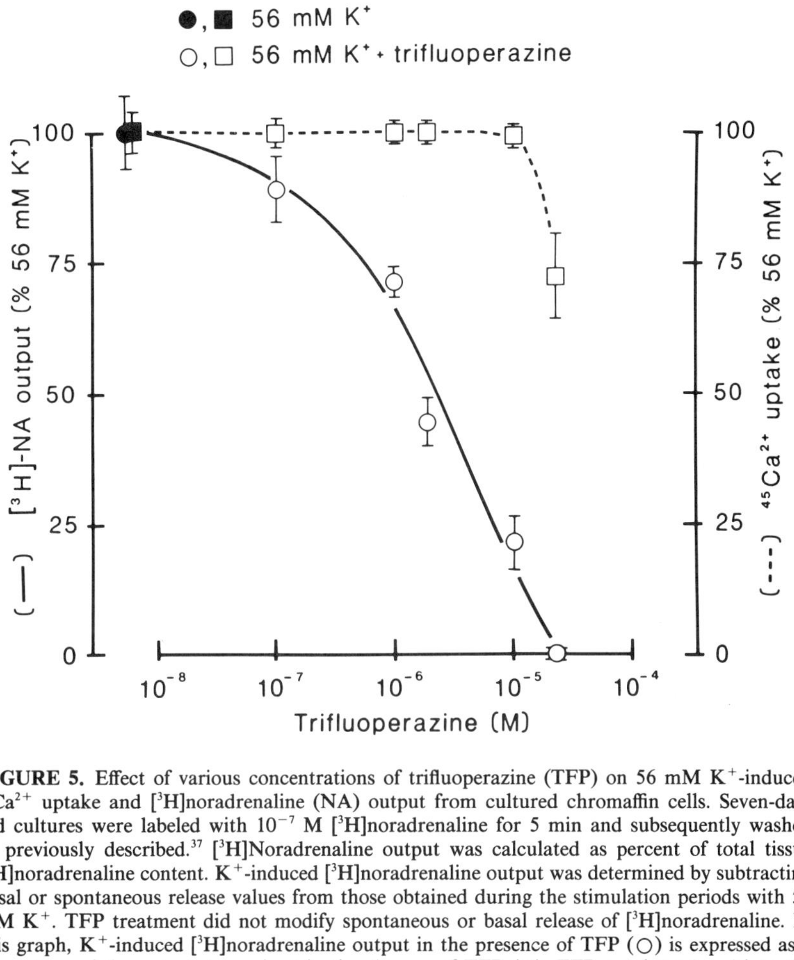

FIGURE 5. Effect of various concentrations of trifluoperazine (TFP) on 56 mM K^+-induced $^{45}Ca^{2+}$ uptake and [^{3}H]noradrenaline (NA) output from cultured chromaffin cells. Seven-day-old cultures were labeled with 10^{-7} M [^{3}H]noradrenaline for 5 min and subsequently washed as previously described.[37] [^{3}H]Noradrenaline output was calculated as percent of total tissue [^{3}H]noradrenaline content. K^+-induced [^{3}H]noradrenaline output was determined by subtracting basal or spontaneous release values from those obtained during the stimulation periods with 56 mM K^+. TFP treatment did not modify spontaneous or basal release of [^{3}H]noradrenaline. In this graph, K^+-induced [^{3}H]noradrenaline output in the presence of TFP (○) is expressed as a percentage of the output determined in the absence of TFP (●). TFP was introduced into the medium 20 min prior to and was present during the period within which spontaneous and K^+-induced release of [^{3}H]noradrenaline was monitored. The cell cultures were also exposed to $^{45}Ca^{2+}$ (specific activity: 2.25 μCi/μmol, 1 Ci = 37 GBq) during K^+-stimulation in the presence or in the absence of TFP. $^{45}Ca^{2+}$ uptake was determined from the desaturation curves plotted with the values obtained. For further experimental details see Kenigsberg *et al.*[37] The K^+-induced $^{45}Ca^{2+}$ uptake in the presence of TFP (□) is expressed as a percentage of the uptake determined in the absence of TFP (■). Each value represents the means ± SE of four separate culture dishes.[24]

obtained (FIGS. 6A, B) with an average of 46.2 ± 1.1% (n = 14) of the total chromaffin cell population efficiently injected.[44-46] Chromaffin cell viability was found to be well maintained during and after fusion with the average cell loss 12 ± 0.4% (n = 14) of the total cell population.[44-46] In addition, the functional parameters that characterize the chromaffin cells in culture were found to be unaltered after fusion-induced microinjection as the endogenous catecholamine stores, the uptake of exogenous catecholamines via the high-affinity uptake mechanism for catecholamines, as well as the cells' response to various secretagogues remained unchanged.[44-46] The delivery of intact antibodies raised against CM directly into the cytoplasm of cultured chromaffin cells by this erythrocyte-ghost-mediated microinjection technique was found to inhibit catecholamine release from the chromaffin cells during periods of stimulation with either acetylcholine (10^{-4} M) or a depolarizing concentration (56 mM) of potassium (FIG. 7).[44,45] However, the intracellular delivery of the CM antibodies into the cultured chromaffin cells did not alter the cell's ability to accumulate exogenous amines via its high-affinity uptake system. It therefore appears that CM is directly involved in the process of stimulus-secretion coupling in the chromaffin cell while it is of little significance to the high-affinity amine uptake system.[44,45] Although the results obtained with the use of these monospecific antibodies raised against CM strongly suggest a role for CM in chromaffin cell secretion and complement the indirect evidence obtained with the use of CM antagonists,[20,37] it still remains possible that the antibody-antigen interaction might nonselectively inhibit catecholamine secretion. In addition, from the aforementioned evidence, indirect and direct, one cannot distinguish with any degree of certainty which CM-dependent process(es) is of significance to stimulus-secretion coupling in the chromaffin cell. *In vitro* studies carried out in our laboratory have demonstrated that the Ca^{2+}-dependent specific binding of radiolabeled CM to isolated chromaffin granule membranes is inhibited in the presence of these monospecific CM antibodies.[28] In view of the fact that both the plasma and chromaffin granule membranes are central to all facets of stimulus-secretion coupling, it is possible that the specific Ca^{2+}-dependent binding of CM to the granule membrane may be essential to the secretory process of the chromaffin cell.

Finally, with the advent of this microinjection technique and suitable monoclonal antibodies, many of these questions pertaining to the underlying molecular mechanisms involved in cellular secretion may be answered in the near future.

CYTOSKELETON PROTEINS AND CALMODULIN IN STIMULUS-SECRETION COUPLING

Adrenal medullary chromaffin cells are among those nonmuscle cells from which cytoskeleton proteins have also been isolated and characterized.[4,5,24]

On the basis of their well-established role in muscle contraction, it is generally assumed that the prime function of contractile proteins (cytoskeleton) in nonmuscle cells is to generate the force required in many of the expressions of cell motility. Cells move and subcellular organelles (mitochondria, lysosomes, secretory granules, etc.) move within cells, and some of these movements might require the intervention of contractile proteins.[5]

Catecholamines are released from the chromaffin cell by exocytosis, a mechanism that follows the movement of secretory vesicles to the plasma membrane and involves the fusion of secretory granule membranes with the cell plasma membrane and the subsequent extrusion of the soluble granular contents to the cell exterior.[3,47]

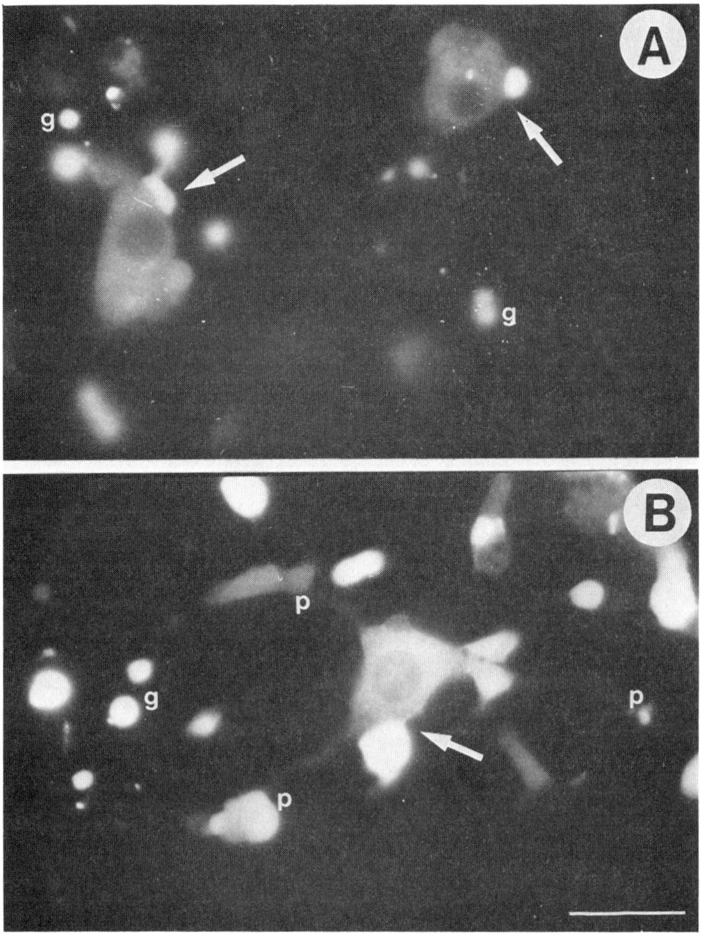

FIGURE 6. Chromaffin cells grown on collagen-coated glass coverslips in monolayer cultures were fused with FITC-BSA-loaded ghosts, then mounted on glass slides in regular Locke's solution (**A** and **B**). These unfixed cell preparations were observed by epifluorescence immediately after mounting. Ghost to cell fusion sites are indicated by *single straight arrows.* The injected substances were evenly distributed throughout the chromaffin cell cytoplasm and processes (*p*). Some of the intact ghosts (*g*) remained attached to the collagen substrate. Photographs (**A** and **B**) were taken with a 35-mm Kodak Tri-X pan film (ASA 400). The *horizontal bar* represents 50 μM.

By reason of the similarities between *stimulus-secretion coupling* and *excitation-contraction coupling* in muscle, it has been proposed that the secretory process of the adrenal chromaffin cell might be mediated by contractile elements associated with the secretory granule or present elsewhere in the cell.[4,5,24] The association of some of these proteins with chromaffin granules[5,24] as well as with secretory vesicles present in other secretory systems[48] suggest that contractile and regulatory proteins might play a role in the cellular transport of secretory granules and/or in the other steps involved in the secretory process.

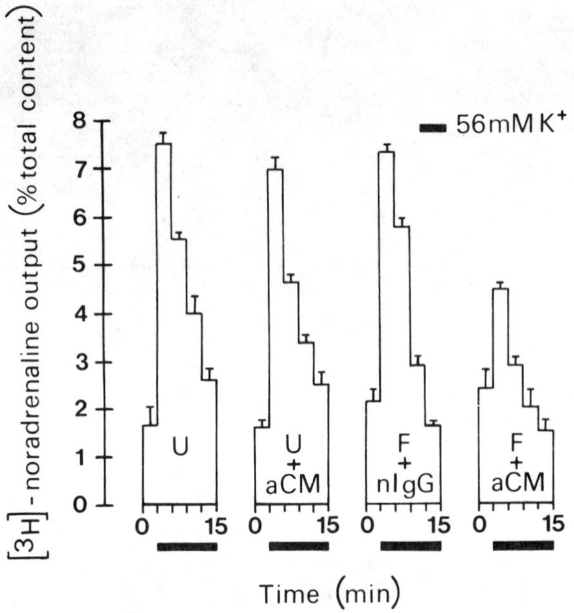

FIGURE 7. Pattern of release of [^{3}H]noradrenaline from 9-day-old cultured chromaffin cells at rest (0-3 min) and during periods of stimulation (3-15 min) with 56 mM K^+. From left to right: unfused cells (U), unfused cells incubated with exogenous calmodulin antibodies (aCM) at a concentration of 9.3 μg/ml (U + aCM) and chromaffin cells that had undergone fusion with either normal IgG-loaded erythrocyte ghosts (F + nIgG) or calmodulin antibody-loaded erythrocyte ghosts (F + aCM). Each *bar* represents the mean ± SE of the values obtained from four culture dishes from the same culture preparations. (Modified from Kenigsberg and Trifaró.[45,46])

Neurofilaments and microtubules are major elements of the neuronal cytoskeleton.[49,50] The staining obtained in cultured chromaffin cells using neurofilament antibodies is very similar to the staining observed in cultured fibroblasts using vimentin antibodies or in cultured epithelial cells using keratin antibodies.[50] This similarity in the localization might be an indication of similar roles for the different intermediate filaments.

In contrast with the distribution of neurofilaments, a very thin and highly ramified network of microtubules is observed in the cell bodies, neurites, and growth cones of chromaffin cells.[49] The effects observed after colchicine treatment (*i.e.,* retraction of neurites, movement of the secretory granules back to the cell bodies, and a diffuse anti-tubulin staining pattern) suggest that microtubules are probably involved in the transport of chromaffin granules to the cell periphery.[49,51] However, the microtubules do not seem to be involved in the final steps of secretion, since the exocytotic release of catecholamines induced by high K^+ is not inhibited by either colchicine or vinblastine.[51,52]

Many of the Ca^{2+}-binding events in the eukaryotic cell are mediated by the intracellular Ca^{2+}-binding protein CM.[53–55] The process of secretion is one of the Ca^{2+}-mediated events in which CM seems to be involved. How CM might be involved in secretion remains to be determined. However, the presence of CM-binding proteins in secretory granules and plasma membranes may suggest that CM is involved either with the transport of granules to release sites or with the process of interaction between granule and plasma membranes during exocytosis. Alternatively, CM might be involved with the process of retrieval of granule membranes after exocytosis. In this regard, high-affinity binding sites for CM have been found in brain coated vesicles.[56]

If exocytosis is a true contractile event, cytoskeleton proteins and their regulatory ones might be involved with the mechanism of transport of secretory granules (microtubules, microfilaments?) to release sites on the plasma membrane, with the fusion process (CM, CM-binding proteins?), with the extrusion phenomenon itself, or with the control of ion influx through the plasma membrane.[20] In any case, we will discuss here two possible molecular mechanisms of action for the cytoskeleton proteins.

The first possible mechanism would involve actin, α-actinin, myosin, and the regulatory protein, CM. In this case, a sliding filament mechanism similar to that found in muscle would operate in chromaffin cells. Myosin should be arranged in bipolar filaments. In this regard, we have demonstrated that chromaffin cell myosin can form bipolar filaments.[10] The α-actinin present in granule membranes would provide binding sites for actin. Actin-binding sites (spectrin) are also present in plasma membranes. It is known that stimulation depolarizes the cell and Ca^{2+} entry is activated with a subsequent increase in intracellular Ca^{2+}. These events, in turn, would activate the sliding filament mechanism through a CM-dependent action. At the molecular level, this implies the activation of myosin ATPase by actin, a mechanism that requires, in all nonmuscle tissues so far investigated, the phosphorylation of the 20,000-Da myosin light chain. This phosphorylation step is regulated by CM.[19] Our experiments with trifluoperazine have shown that this CM antagonist does inhibit catecholamine release in response to either acetylcholine or high K^+ stimulation at a step distal from Ca^{2+} entry, thus suggesting that the drug is antagonizing CM. Moreover, more direct evidence for the role of CM was obtained from the experiments on the intracellular delivery of CM antibodies. In this case, catecholamine release in response to stimulation was inhibited subsequent to the microinjection of the antibodies.[44,45] Although the results obtained with the CM antibodies provide us with a more direct evidence for CM involvement in the process of secretion of the chromaffin cells, one cannot conclude with certainty at this point that the role of CM in secretion is through the modulation of myosin light chain phosphorylation. Therefore, other CM-dependent processes might be involved in secretion, and these are discussed below.

A second possible mechanism in which contractile proteins may play a role in secretion is based on the viscosity properties of actin and does not necessarily require the intervention of myosin (FIG. 8). In this particular case, actin would control chromaffin cell cytosol viscosity through the formation of a mesh of microfilaments which would be cross-linked and stabilized by the chromaffin granule alpha-actinin and spectrin (fodrin).[5,24,48] In this regard, it has been shown recently in *in vitro* experiments that chromaffin granules will induce actin polymerization and gel formation, effects blocked by raising the concentration of Ca^{2+} in the medium.[48,57] There-

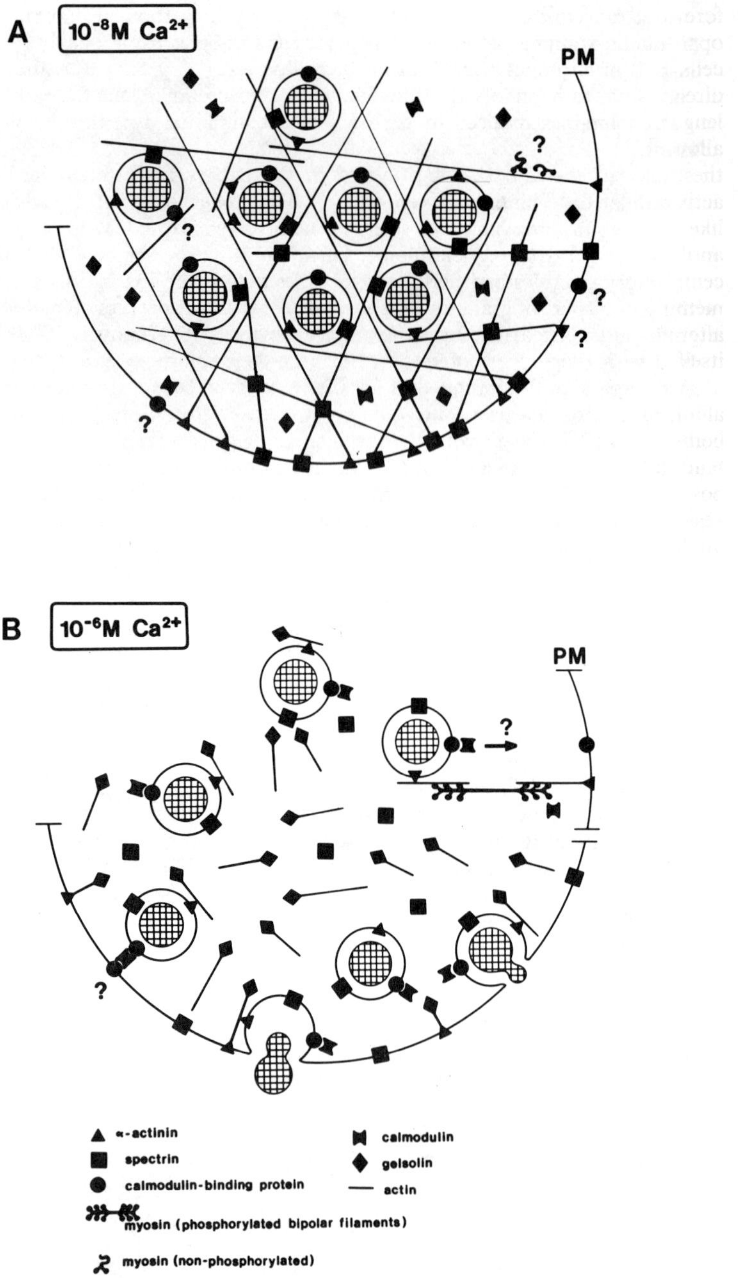

FIGURE 8. Schematic representation of a possible mechanism in which cytoskeleton proteins and their regulatory counterparts may play a role in chromaffin cell secretion. (**A**) Under resting conditions (intracellular Ca^{2+} of about 10^{-8} M), actin would control chromaffin cell cytosol

fore, under resting conditions (10^{-8} M free Ca^{2+}), the cytosolic actin network would oppose the movement of secretory granules towards release sites (FIG. 7A). When the cells are stimulated, Ca^{2+} enters the cell (10^{-6} M free Ca^{2+}). This would produce, directly or through the activation of a specific protein (gelsolin?), a shortening in the length of the actin microfilaments. This would decrease the viscosity of the cytosol, allowing the movement of granules towards release sites (FIG. 8B). A protein with these properties first isolated from rabbit lung macrophages and named gelsolin is activated when the concentration of Ca^{2+} is increased to 10^{-6} M.[58] Similarly, a gelsolin-like protein has been isolated from chromaffin cells and found to cross-react with an antibody raised against macrophage gelsolin.[6,7] Moreover, increasing the Ca^{2+} concentration also dissociates actin filaments from the chromaffin granule and plasma membrane spectrins.[48] CM might be involved in the modulation of these processes; alternatively, it might be involved in the granule/plasma membrane fusion process itself.

We have presented here data on the presence, in granule membranes, of high-affinity binding sites for CM. Moreover, the observation that two monoclonal antibodies against a M_r 65,000 synaptic vesicle protein[36] recognize the M_r 65,000 CM-binding proteins present in chromaffin granules and vesicle antigens from anterior and posterior pituitaries[36] suggest the presence of CM-binding proteins in all secretory vesicles. Recent morphological observations have also suggested that trifluoperazine might inhibit carbamylcholine-induced catecholamine secretion by interfering with the process of fusion between chromaffin granules and plasma membranes.[59] This could also be the site of inhibition for the CM antibody when delivered intracellularly by erythrocyte-ghost-mediated microinjection.[44,45]

ACKNOWLEDGMENT

We wish to thank Dr. R. Kelly and Dr. L. Reichardt for their generous supply of monoclonal antibodies.

---

viscosity through the formation of a mesh of microfilaments which would also be cross-linked and stabilized by (1) spectrin present in the cytosol and in chromaffin granule and plasma membranes and (2) α-actinin present in granule membranes and probably in plasma membranes. The secretory granule membranes also contain a calmodulin-binding protein. A similar calmodulin-binding protein is also present in plasma membranes. At the Ca^{2+} concentration found in the resting cell, calmodulin and gelsolin are not activated. In addition, there is probably a large percentage of nonfilamentous, nonphosphorylated myosin. (**B**) When the cell is stimulated, Ca^{2+} enters the cell (intracellular Ca^{2+} concentration of about 10^{-6} M) and produces (1) a dissociation of actin from spectrin and (2) activation of gelsolin with the consequent capping and shortening of the actin microfilaments. Calcium ions do not seem to affect the binding of actin to chromaffin granule membrane α-actinin. As a result of both 1 and 2, the cytosol viscosity should decrease, allowing the movement of secretory granules towards the plasma membrane releasing sites. Whether an actin-myosin interaction (sliding mechanism) is also involved in granule movement remains to be determined. The intracellular Ca^{2+} concentration reached during stimulation is sufficient to activate calmodulin-dependent processes, including the binding of calmodulin to granule membranes and phosphorylation of myosin light chains, a condition required for myosin activation and bipolar filament formation. (From Trifaró *et al.*[6])

REFERENCES

1. SMITH, A. D. & H. WINKLER. 1972. Handbook of Experimental Pharmacology, Vol. 33. H. Blaschko & E. Muscholl, Eds.: 538. Springer. Berlin.
2. VIVEROS, O. H. 1975. Handbook of Physiology: Endocrinology. H. Blaschko, G. Sayers & A. D. Smith, Eds.: 389. American Physiological Society. Washington, D.C.
3. TRIFARÓ, J.-M. 1977. Annu. Rev. Pharmacol. Toxicol. **17:** 27.
4. TRIFARÓ, J.-M. 1978. Neuroscience **3:** 1.
5. TRIFARÓ, J.-M., R. W. H. LEE, R. L. KENIGSBERG & A. CÔTÉ. 1982. *In* Synthesis, Storage and Secretion of Adrenal Catecholamines. Advances in the Biosciences, Vol. 36. F. Izumi & M. Oka, Eds.: 151. Pergamon Press. Oxford.
6. TRIFARÓ, J.-M., M.-F. BADER & J.-P. DOUCET. 1985. Can. J. Biochem. Cell Biol. **63:** 661.
7. BADER, M.-F., J.-M. TRIFARÓ, O. K. LANGLEY, D. THIERSE & D. AUNIS. 1986. J. Cell Biol. **102**(2): 636.
8. POISNER, A. M. & J.-M. TRIFARÓ. 1967. Mol. Pharmacol. **3:** 561.
9. TRIFARÓ, J.-M. & C. ULPIAN. 1975. FEBS Lett. **57:** 198.
10. TRIFARÓ, J.-M. & C. ULPIAN. 1976. Neuroscience **1:** 483.
11. GABBIANI, G., M. DA PRADA, G. RICHARDS & A. PLETSCHER. 1976. Proc. Soc. Exp. Biol. Med. **152:** 135.
12. CREUTZ, C. E. 1977. Cell Tissue Res. **178:** 17.
13. HESKETH, J. E., D. AUNIS, P. MANDEL & G. DEVILLIERS. 1978. Biol. Cell. **33:** 199.
14. TRIFARÓ, J.-M., C. ULPIAN & H. PREIKSAITIS. 1978. Experientia **34:** 1568.
15. LEE, R. W. H., W. E. MUSHYNSKI & J.-M. TRIFARÓ. 1979. Neuroscience **4:** 843.
16. TRIFARÓ, J.-M. & R. W. H. LEE. 1978. *In* Catecholamines: Basic and Clinical Frontiers. Proceedings of the Fourth International Catecholamine Symposium. E. Usdin, J. J. Kopin & J. Barchas, Eds.: 358. Pergamon. New York.
17. AUNIS, D., B. GUEROLD, M. F. BADER & J. CIESELSKI-TRESKA. 1980. Neuroscience **5:** 2261.
18. LEE, R. W. H. & J.-M. TRIFARÓ. 1981. Neuroscience **6:** 2087.
19. ADELSTEIN, R. S., M. A. CONTI & M. D. PATO. 1980. Ann. N.Y. Acad. Sci. **356:** 142.
20. TRIFARÓ, J.-M. & R. L. KENIGSBERG. 1986. *In* Stimulus-Secretion Coupling in Chromaffin Cells. K. Rosenheck, Ed. CRC Press. Boca Raton, Fla. In press.
21. MEANS, A. R. & J. G. CHAFOULEAS. 1982. Cold Spring Harbor Symposia on Quantitative Biology, Vol. XLVI. Cold Spring Harbor Laboratory. Cold Spring Harbor, N.Y. pp. 903.
22. CONN, P. M., J. G. CHAFOULEAS, D. C. ROGERS & A. R. MEANS. 1981. Nature **292:** 264.
23. OHNO, S., Y. FUJII, N. USUDA, T. NAGATA, T. ENDO, T. TANAKA & H. HIDAKA. 1981. *In* Calmodulin and Intracellular Ca^{++} Receptors. S. Kakiuchi, H. Hidaka & A. R. Means, Eds.: 35. Plenum. New York.
24. TRIFARÓ, J.-M., R. L. KENIGSBERG, A. CÔTÉ, R. W. H. LEE & T. HIKITA. 1984. Can. J. Physiol. Pharmacol. **62:** 493.
25. HIKITA, T., M. F. BADER & J.-M. TRIFARÓ. 1984. J. Neurochem. **43:** 1087.
26. BURGOYNE, R. D. & M. J. GEISOW. 1981. FEBS Lett. **131:** 127.
27. GEISOW, M. J., R. D. BURGOYNE & A. HARRIS. 1982. FEBS Lett. **143:** 69.
28. BADER, M.-F., T. HIKITA & J.-M. TRIFARÓ. 1985. J. Neurochem. **44:** 526.
29. GRINSTEIN, S. & W. FURUYA. 1982. FEBS Lett. **140:** 49.
30. MOSKOWITZ, N., W. SCHOOK, K. BECKENSTEIN & S. PUSZKIN. 1983. Brain Res. **263:** 243.
31. OLSEN, S. F., J. SLANINOVA, M. TREIMAN, T. SAERMARK & N. A. THORN. 1983. Acta Physiol. Scand. **118:** 355.
32. HOOPER, J. E. & R. B. KELLY. 1984. J. Biol. Chem. **259:** 141.
33. TREIMAN, M. R., W. WEBER & M. GRATZL. 1983. J. Neurochem. **40:** 661.
34. FOURNIER, S. & J.-M. TRIFARÓ. 1985. Can. Fed. Biol. Soc. Toronto. **28:** 116.
35. FOURNIER, S. & J.-M. TRIFARÓ. 1986. Third International Symposium on Chromaffin Cell Biology. Coolfont, W. Va. **41:** 47.
36. MATTHEW, W. D., L. TSAVALER & L. F. REICHARDT. 1981. J. Cell Biol. **91:** 257.
37. KENIGSBERG, R. L., A. CÔTÉ & J.-M. TRIFARÓ. 1982. Neuroscience **7:** 2277.

38. NACCACHE, P. H., F. P. MOLSKI, T. ALOBAIDI, E. L. BECKER, H. J. SHOWELL & R. I. SHA'AFI. 1980. Biochem. Biophys. Res. Commun. **97:** 62.
39. SCHETTINI, G., M. J. CRONIN & R. M. MACLEOD. 1983. Endocrinology **112:** 1801.
40. OKUMURA-NOJI, K. & R. TANAKA. 1984. Neurochem. Int. **6:** 91.
41. BAKER, P. F. & D. E. KNIGHT. 1981. Phil. Trans. R. Soc. Lond. **B296:** 83.
42. CLAPMAN, D. E. & E. NEHER. 1984. J. Physiol. **353:** 541.
43. STEINHARDT, R. A. & J. M. ALDERTON. 1982. Nature **295:** 154.
44. TRIFARÓ, J.-M. & R. L. KENIGSBERG. 1983. Fed. Proc. Feds. Amer. Soc. Exp. Biol. **42:** 456.
45. KENIGSBERG, R. L. & J.-M. TRIFARÓ. 1985. Neuroscience **14:** 335.
46. KENIGSBERG, R. L. & J.-M. TRIFARÓ. 1985. J. Neuroscience Methods **13:** 103.
47. TRIFARÓ, J.-M. & X. L. CUBEDDU. 1979. *In* Trends in Autonomic Pharmacology. S. Kalsner, Ed.: 195. Urban & Schwarzenberg. Baltimore.
48. AUNIS, D. & D. PERRIN. 1984. J. Neurochem. **42:** 1558.
49. BADER, M. -F., F. BERNIER-VALENTIN, B. ROUSSET & D. AUNIS. 1984. Can. J. Physiol. Pharmacol. **62:** 502.
50. LAZARIDES, E. 1980. Nature **283:** 249.
51. TRIFARÓ, J.-M., M.-F. BADER, A. CÔTÉ, R. L. KENIGSBERG, T. HIKITA & R. W. H. LEE. 1985. *In* Contractile Proteins in Muscle and Non-Muscle Systems. E. E. Alia, N. Arena & M. Russo, Eds.: 459 Praeger. New York.
52. TRIFARÓ, J.-M., B. COLLIER, A. LASTOWECKA & D. STERN. 1972. Mol. Pharmacol. **8:** 264.
53. CHEUNG, W. Y. 1980. Science **207:** 19.
54. MEANS, A. R. & J. R. DEDMAN. 1980. Nature **285:** 73.
55. BROSTROM, C. O. & D. J. WOLFF. 1981. Biochem. Pharmacol. **30:** 1395.
56. MOSKOWITZ, N., W. SCHOOK, M. LISANTI, E. HUA & S. PUSZKIN. 1982. J. Neurochem. **38:** 1742.
57. FOWLER, V. M. & H. B. POLLARD. 1982. Nature **295:** 336.
58. STENDAHL, O. I. & T. P. STOSSEL. 1980. Biochem. Biophys. Res. Commun. **92:** 675.
59. BURGOYNE, R. D., M. J. GEISOW & J. BARRON. 1982. Proc. R. Soc. Lond. **B216:** 111.

DISCUSSION OF THE PAPER

N. KIRSHNER (*Duke University Medical Center, Durham, N.C.*): In experiments that we have reported we found that TFP and pimozide caused a parallel inhibition of calcium uptake and catecholamine secretion. We also found that neither TFP nor pimozide had any effect on calcium-evoked secretion in digitonin-permeabilized cells.

J.-M. TRIFARÓ (*University of Ottawa, Ottawa, Ont.*): As you can see, our experiments clearly indicate that at the concentrations of TFP used, the effect of the drug in blocking stimulation-induced secretion is distal to Ca^{2+} entry. Moreover, experiments from Neher's laboratory in Germany have indicated, using patch clamp techniques on chromaffin cells, that TFP has at least three types of effects: one effect at the level of the cholinergic receptor, another effect at the level of the Ca^{2+} channel but at concentrations, I believe, larger than 10 μM, and a third effect at an intracellular site.

UNIDENTIFIED SPEAKER: Do you know which lipids are involved in binding calmodulin to vesicles?

J. M. TRIFARÓ: As I have said, the low-affinity binding of calmodulin observed at 2×10^{-7} M Ca^{2+} is due to a lipid component of the secretory vesicle membrane. We have not determined the nature of the lipid involved in this interaction.

L. E. EIDEN (*NIMH, National Institutes of Health, Bethesda, Md.*): Does the possibility exist that anticalmodulin antibodies inhibit secretion only because the immune precipitate makes a physical barrier to secretion within the cell? Can you mention any other controls you have done to address this possibility?

J.-M. TRIFARÓ: The microinjection of Fab fragments prepared from anticalmodulin antibodies also blocks the release of catecholamines from chromaffin cells stimulated by either acetylcholine or a depolarizing concentration of K^+.

T. R. SODERLING (*Howard Hughes Medical Institute, Nashville, Tenn.*): Do you have any evidence that either the 53-kDa or 65-kDa calmodulin-binding proteins are also phosphoproteins?

J.-M. TRIFARÓ: We still have to test whether these peptides are subunits of a calmodulin-dependent kinase, and whether or not they are autophosphorylated. Future experiments will be directed to test this possibility.

Tubulin- and Actin-binding Proteins in Chromaffin Cells[a]

DOMINIQUE AUNIS, MARIE-FRANCE BADER,
O. KEITH LANGLEY, AND DOMINIQUE PERRIN

Unité INSERM U-44
Centre de Neurochimie du CNRS
F-67084 Strasbourg, France

INTRODUCTION

Extensive biochemical, immunochemical, and pharmacological studies have shown that secretion from chromaffin cells occurs by exocytosis.[1-6] Exocytosis is a membrane process, which requires calcium and ATP.[7-9] Early observations showing that the events involved in the secretory process have many features in common with the process of "excitation-contraction coupling" in muscle led to the concept of "stimulus-secretion coupling."[10] This concept was supported by the discovery that chromaffin cells do contain contractile proteins. There is not yet a sufficient basis for regarding the secretory cell as a close analogue of the muscle cell and for viewing actomyosin as the force-generating system involved in the secretory process.

The purpose of the present chapter is to reconsider the popular idea that cytoskeletal proteins act as "contractile machinery" in secretory cells.[11] Actin-binding proteins have recently been described in chromaffin cells, and their calcium-dependent activity results in the alteration of actin filaments. In addition, granule membrane has been demonstrated to contain tubulin-binding as well as actin-binding activities.[12-14] Data are also now accumulating favoring the idea that the cytoskeleton could primarily act as a barrier preventing granules from undergoing exocytosis. For exocytosis to occur, this barrier has to be disrupted, thus introducing a regulatory exocytotic mechanism. In addition the disruption step may result in a dramatic rearrangement of cell organization, thus preparing the cell for subsequent exocytosis.

THE CHROMAFFIN CELL CYTOSKELETON

Conventional electron microscope images of epoxy-embedded preparations show that chromaffin cells have a prominent cytoskeleton consisting of abundant cytoplasmic microtubules and a subplasmalemmal network of microtubules and microfilaments. Because of electron scattering of the resin and insufficient contrast, microtubules are

[a] A part of the work described in this paper was supported by grants from the Ministère de l'Industrie et de la Recherche (no. 82-E-1195).

the only cytoskeletal structures visible in the cytoplasm, while microfilaments are not detectable. These conventional techniques have revealed chromaffin granules to be associated with microtubules in the intact gland,[15] which are abundant in the Golgi domain of cultured chromaffin cells, from which they seem to radiate, and in the subplasmalemmal space (FIG. 1a,b). In contrast to these classical techniques, cytoskeletal microfilaments are clearly demonstrated after embedding in polyethyleneglycol[16] or LR White resin. As shown in FIGURE 1c, the electron microscopic image of LR-White-processed rat adrenal medulla reveals a highly organized, three-dimensional network composed of filaments interconnecting membrane-limited organelles. These electron-dense organelles are chromaffin granules that seem to be linked to the filaments, suggesting a continuum between secretory organelles, mitochondria, and plasma membrane. Such a three-dimensional network has also been described in the axoplasm of nerve cells using a variety of electron microscopical techniques,[17–21] including the network of fine cross-connections between microtubules and neurofilaments, and also the cross-bridges linking longitudinal structures with membrane-limited organelles such as vesicles and mitochondria revealed by the quick-freezing, deep-etching technique.[22]

The possibility of a relationship between the granule membrane and filamentous structures poses the following questions: (1) Does the granule membrane possess sites able to bind tubulin, the main component of microtubules? (2) Similarly, does the granule membrane possess actin-binding sites? (3) Is the three-dimensional network, which probably forms a viscoelastic gel, susceptible to modification by the intervention of regulatory proteins?

GRANULE MEMBRANE CYTOSKELETON BINDING SITES

Tubulin-binding sites

Tubulin-chromaffin granule membrane interaction has been demonstrated by using a radiolabeled ligand receptor assay and an antibody retention assay.[12] The properties of the interaction are summarized in TABLE 1. It is time- and temperature-dependent, reversible, specific, and saturable. The apparent dissociation constant of the tubulin-chromaffin granule membrane was found to be 30 nM, a value that is within the expected intracellular free tubulin concentration range.

Tubulin was estimated in purified granule membrane preparation by using a radioimmunoassay procedure[12,23] to represent 2.2 μg/mg membrane protein. At this stage, it was difficult to conclude whether tubulin is a component of the granule membrane. This was further checked by subcellular fractionation on continuous sucrose gradients. The crude chromaffin granule P_2 fraction was isolated, resuspended, and deposited on sucrose continuous gradients. After centrifugation, gradients were fractionated and organelle-specific enzymes and tubulin content were measured. As indicated in FIGURE 2, granules sedimented near the bottom of the tube as shown from dopamine-β-hydroxylase activity measurements, and were well separated from mitochondria and plasma membrane. The profile of endogenous tubulin showed sedimentable tubulin with low activity in the plasma-membrane-containing fractions, while a large activity was recovered in the mitochondrial and the chromaffin granule fractions. The tubulin peak did not, however, coincide exactly with the granule peak,

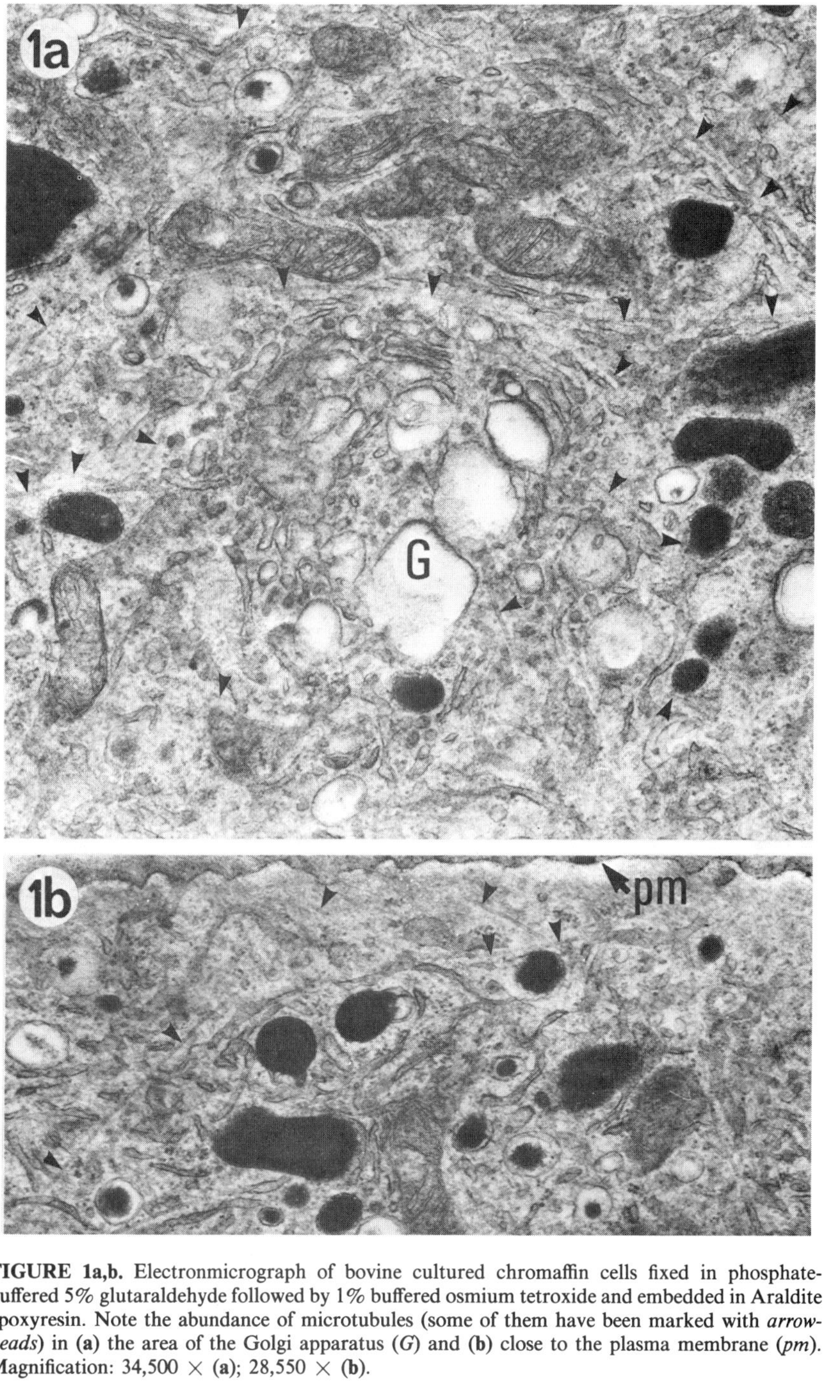

FIGURE 1a,b. Electronmicrograph of bovine cultured chromaffin cells fixed in phosphate-buffered 5% glutaraldehyde followed by 1% buffered osmium tetroxide and embedded in Araldite epoxyresin. Note the abundance of microtubules (some of them have been marked with *arrowheads*) in (**a**) the area of the Golgi apparatus (*G*) and (**b**) close to the plasma membrane (*pm*). Magnification: 34,500 × (**a**); 28,550 × (**b**).

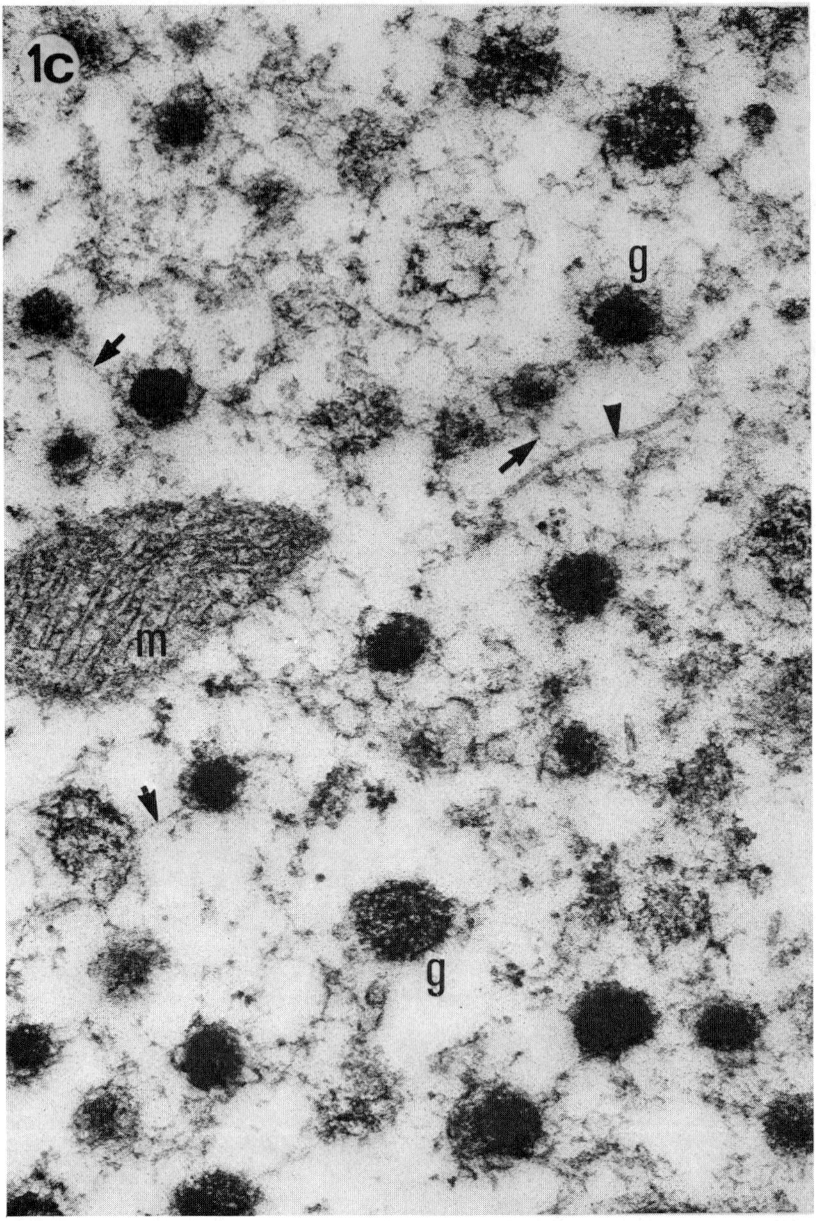

FIGURE 1c. Electronmicrograph of chromaffin cell from rat adrenal medulla fixed in phosphate-buffered 4% *p*-formaldehyde, postfixed in 1% osmium tetroxide and embedded in LR White (London Resin Company), a hydrophilic acrylamide-based resin which, due to its low inherent electron density, affords a better visualization of microfilaments. Note the extensive network of thin filaments (*arrows*) interconnecting secretory granules (*g*), microtubules (*arrowhead*), and mitochondria (*m*). Magnification: 47,000 ×.

TABLE 1. Tubulin-Binding Properties of Chromaffin Granule Membrane[a]

Binding parameters	
Apparent dissociation constant	$K_d = 3.0 \pm 1.3 \times 10^{-8}$ M
Hill coefficient	$h = 1.02 \pm 0.01$
Maximal binding capacity	$B_{max} = 0.4 \pm 0.1$ nmol/mg protein
Binding properties	
Temperature	37°C: association 0°C: dissociation
Colchicine (10^{-4} M)	no effect
Buffer composition	MES 0.1 M: induction phosphate 0.1 M: inhibition

[a] Based on Bernier-Valentin, Aunis, and Rousset.[12]

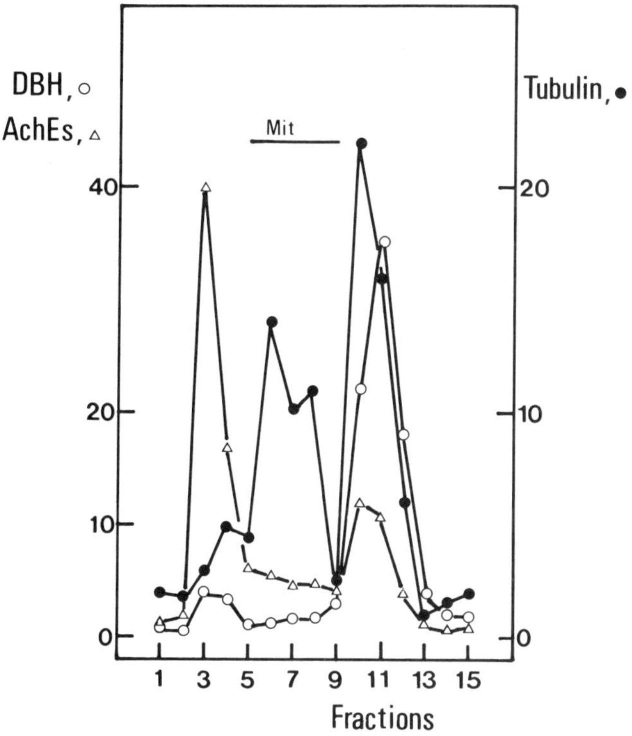

FIGURE 2. Distribution of marker enzymes and tubulin in a sucrose density gradient after centrifugation of the crude P_2 chromaffin granule fraction. Acetylcholinesterase (AchEs, marker of the plasma membrane) and dopamine β-hydroxylase (DBH, the marker of chromaffin granules) were measured and activities are given as percentage of total recovered activity. Location of mitochondria is indicated in the figure (*Mit*) from monoamine oxidase activity measurements. Highest activity was found in fractions 6-7. Tubulin was estimated in each fraction by a radioimmunoassay (see Bernier-Valentin *et al.*[12]). The content in each fraction was calculated as percentage of total tubulin content recovered in the gradient. Top of the gradient (fraction 1) corresponds to the lowest sucrose density (Aunis, Bernier-Valentin, and Rousset, unpublished data).

suggesting a possible granule heterogeneity. To examine this possibility, radioiodinated tubulin was added to adrenal medullary homogenate and subcellular fractionation was performed. As shown in FIGURE 3, the exogenous radioactive tubulin profile differed slightly from that of the endogenous tubulin shown in FIGURE 2. A first significant peak was found at the top of the gradient where plasma membranes sedimented; no radioactivity was observed with the mitochondria but significant activity

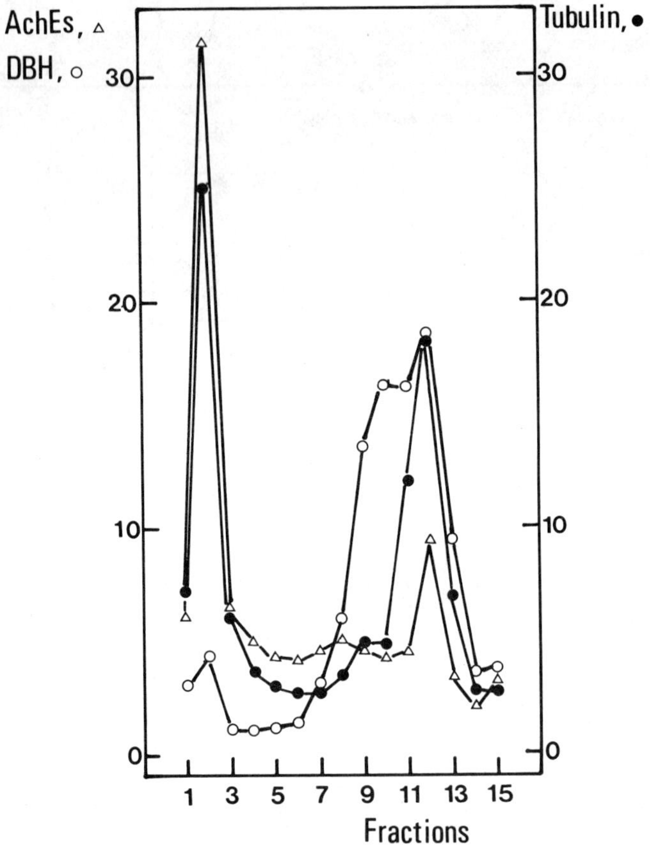

FIGURE 3. Distribution of marker enzymes and exogenous radioiodinated tubulin in sucrose density gradient after centrifugation of the crude P_2 chromaffin granule fraction. ^{125}I-tubulin (prepared as in ref. 12) was added to adrenal medulla homogenate in the presence of GTP and EGTA. Suspension was left at 4°C for 10 min (conditions of microtubule depolymerization) and P_2 fraction was isolated; 9% of initial radioactive tubulin was recovered in the P_2 fraction, while 63% and 24% were recovered in the 100,000 g soluble fraction and in the microsomal fraction respectively. The P_2 fraction was centrifuged; the gradient was then fractionated and each fraction was analyzed for marker enzymes (acetylcholinesterase [AchEs]: plasma membrane; dopamine-β-hydroxylase [DBH]: chromaffin granules) and for tubulin content from radioactivity estimation. The small AchEs peak in fraction 12 corresponds either to plasma-membrane-derived vesicles sedimenting in high-density fractions or to activity associated with secretory granules.[38,39]

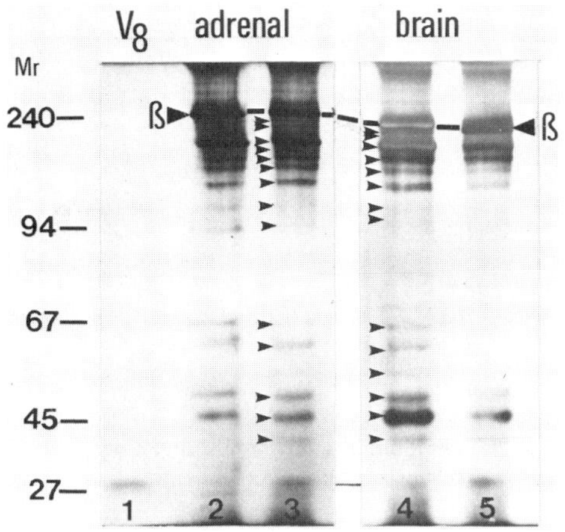

FIGURE 4. One-dimensional peptide map of peptides generated by mild proteolysis with *Staphylococcus* V8 protease of adrenal β-fodrin chain[2,3] and brain β-fodrin chain.[4,5] Digestion was performed at two protease concentrations. β indicates the native undigested proteins, and *arrowheads* breakdown peptides. Peptide profiles were very similar, indicating that adrenal β-fodrin and brain β-fodrin are structurally related.

was found associated with some granule-containing fractions, mostly with the densest granules, thus confirming granule heterogeneity either in terms of occupancy of tubulin sites or of their capacity to exchange tubulin. At this stage, we can conclude that the granule membrane has the capacity to bind tubulin. The significance of such a binding remains speculative.[12] It is possible that such a binding is related to the transport of the granules from their site of assembly, the Golgi domain, to the cell periphery. However, the nature of the transport in chromaffin cells is still unknown. The possible involvement of proteins that have been described in nonendocrine tissues, such as kinesine,[24] has yet to be demonstrated.

Actin-binding Sites

Several reports have described the presence of actin[13,14] and of α-actinin[25,26] in the chromaffin granule membrane. α-Actinin stabilizes actin nuclei in membrane, which can promote the assembly of F-actin from G-actin.[27] In addition, chromaffin granule membranes, when mixed with skeletal muscle F-actin, dramatically increase the apparent viscosity of F-actin solutions.[13,14] We demonstrated that the actin-cross-linking property of the chromaffin granule membrane was due to the presence of a spectrin-like molecule.[14] The α-subunit of spectrin (240 kDa) was identified in chromaffin granule membrane using electroblots with an antibody directed against human erythrocyte spectrin, while the β-subunit (235 kDa) was identified using an antibody against bovine brain fodrin.[14,28] Thus the granule membrane protein is more closely related to brain spectrin (fodrin) than to erythrocyte spectrin on the basis of both the molecular weight and breakdown peptides obtained by mild digestion with V8 protease (FIG. 4).

Incubation of granule membranes in low ionic strength buffers provokes the detachment of fodrin from membranes and thus this binding was studied in more detail. Fodrin was incubated with granule membranes and the membranes were purified through a sucrose cushion; proteins were separated by polyacrylamide gel electrophoresis and fodrin was estimated by laser scanning densitometry. The properties of the fodrin binding to granule membrane are summarized in TABLE 2.

TABLE 2. Fodrin-binding Properties of Chromaffin Granule Membrane

Binding parameters ($n = 3$)	
Michaelis constant	3.3×10^{-8} M
Maximum binding capacity	0.13 ± 0.01 nmol/mg protein
Dissociation constant	$K_d = 1.36 \pm 0.13 \times 10^{-8}$ M
Number of fodrin binding sites per granule	19 ± 4
Binding properties	
Reversibility	$T_{1/2} = \infty$ at 4°C
	$T_{1/2} = 60$ min at 20°C
Ca^{2+}	no effect
Trypsin	75% inhibition after 30 min digestion at trypsin/granule protein ratio of 1:100.
Properties	
Viscosity	Dramatic increase of F-actin solution viscosity of protein-coated granule membrane

By using immunoaffinity-purified anti-α-fodrin antibody, fodrin was localized at the ultrastructural level in the subplasmalemmal region of chromaffin cells.[29] Fodrin was not detectable in the cytoplasm or in the nucleus, but was confined to the region extending 230 ± 70 nm (approximately the length of the fodrin molecule) from the cell membrane towards the cell interior. This result indicated that although the granule membrane has the capacity to bind fodrin, this binding occurs in the cell only where fodrin is present, that is, the subplasmalemmal region. What is the significance of such a binding in the subplasmalemmal space?

ORGANIZATION OF THE SUBPLASMALEMMAL REGION

The subplasmalemmal region is characterized by the abundance of microfilaments connecting the plasma membrane, secretory granules, and microtubules. Fodrin is exclusively present in this region and could be short filaments connecting granules to actin filaments. Recently, we have shown, using the immunofluorescent technique, that the labeling pattern of α-fodrin, forming a continuous ring at the periphery of resting cells, was dramatically altered in nicotine-stimulated or high potassium-stimulated chromaffin cells in culture; after stimulation numerous patches of fodrin were produced.[28]

Identifying secreting cells with a marker of sites of exocytosis (*i.e.*, the transient presence on the plasma membrane of an inner granule membrane protein), it was possible to correlate the number of cells with a disrupted fodrin ring with nicotine concentration and catecholamine release in the external medium (FIG. 5). In addition, this process was found to be a slow phenomenon. The absence of calcium in the external medium blocked both the secretion and the α-fodrin rearrangement (FIG. 5). The main stimulus activating α-fodrin movement results from the rise of intracellular calcium concentration following cell stimulation.[28] Calmodulin seems to be also involved in the process of both secretion and fodrin rearrangement.[28] It is interesting to correlate this observation with recent data showing that the injection of anticalmodulin antibody into living chromaffin cells resulted in inhibition of catecholamine release in response to acetylcholine or high potassium.[30]

The redistribution of α-fodrin in the subplasmalemmal region of cultured chromaffin cells suggests that the subplasmalemmal cytoskeletal organization is affected

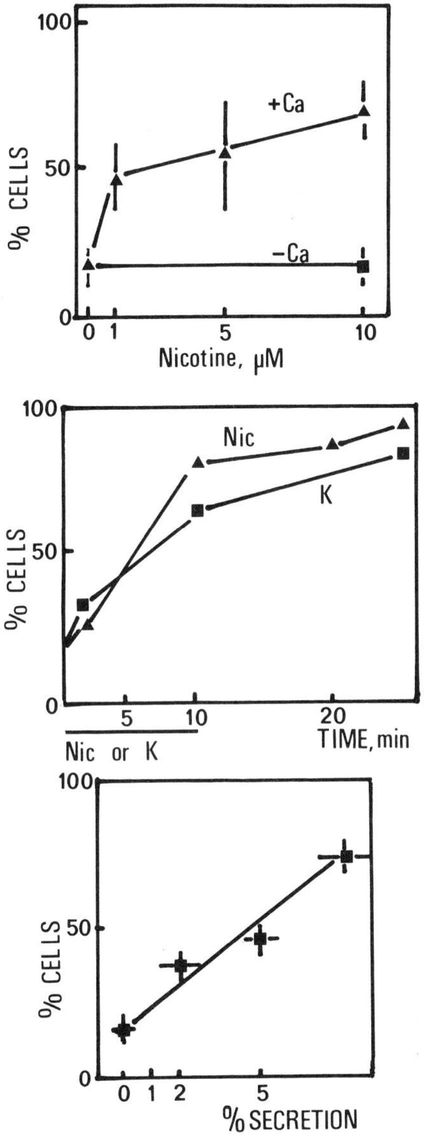

FIGURE 5. Effect of stimulation on the number of cultured chromaffin cells showing discontinuous α-fodrin staining (from ref. 28). *Top panel:* Chromaffin cells in culture were stimulated with nicotine at indicated concentrations for 10 min in the presence of 2.2 mM Ca^{2+} or in the absence of Ca^{2+} in the external medium; *% cells* indicate the percentage of chromaffin cells showing altered distribution of fodrin. *Middle panel:* Kinetics of α-fodrin patch formation in chromaffin cells. At indicated times after stimulation with 10 µM nicotine or 56 mM K^+ in the presence of 2.2 mM Ca^{2+}, cells were decorated with anti-α-fodrin and counted. *Bottom panel:* Relationship between the number of chromaffin cells showing discontinuous α-fodrin staining after treatment with 5 µM nicotine for 10 min in Ca^{2+}-containing solution and released catecholamine. Calculated correlation coefficient was $r = 0.93$.

by internal calcium concentration, calmodulin, and external stimuli. Fodrin clearance induced by the rise of internal calcium concentrations following cell stimulation may reflect the calcium-promoted disconnection of secretory granules from the actin filament lattice, which results in a complete reorganization of the granule population within the actin lattice. In order to decrease the viscoelasticity of the actin gel and to allow granule movement, the presence of proteins able to shorten actin filaments can be predicted. Such satellite proteins must be actin-binding proteins, which are activated by calcium. This property was used to purify these proteins.

Bovine adrenal medullary tissue was homogenized, and the 100,000 g soluble fraction was fractionated on a DNAse I-Sepharose column in the presence of calcium.[31] When calcium concentration was diminished to 10^{-8} M, three proteins of molecular mass 93, 91, and 85 kDa were eluted from the column (FIG. 6). The 93-kDa component was shown to be brevin resulting from blood contamination, whereas the 91-kDa

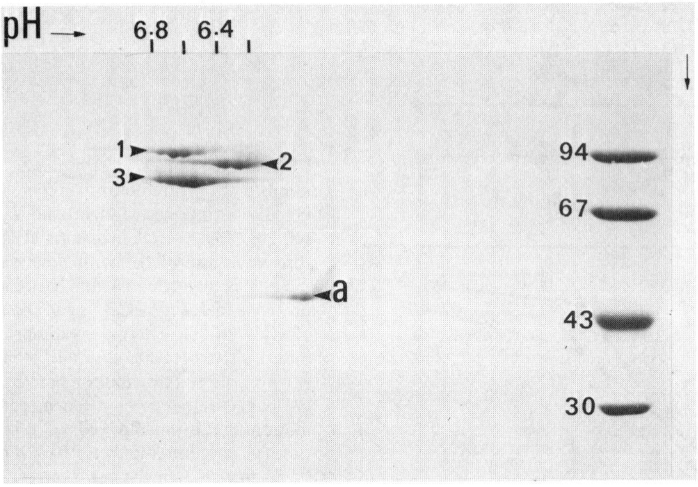

FIGURE 6. Two-dimensional gel electrophoresis of actin-binding proteins purified on DNAse-I-Sepharose affinity column. Four proteins were detected: *1* indicates brevin-like protein (93 kDa, pH_I 6.7-6.5); *2* shows the gelsoline-like protein (91 kDa, pH_I 6.4-6.2); *3* is the 85 kDa component (85 kDa, pH_I 6.7-6.3) and *a* indicates actin. Note the presence of multiple spots due to isoforms.

protein was demonstrated to be a gelsolin-like protein.[31] Gelsolin has primarily been described in macrophages as a protein characterized by a Ca^{2+}-dependent ability to sever actin filaments and to nucleate actin polymerization. Whereas gelsolin is inactive at nM Ca^{2+} concentrations, at μM Ca^{2+} concentrations it is able to shorten actin filaments.[32] A similar property was found for chromaffin cell gelsolin: in the presence of μM Ca^{2+} concentrations, it shortened the length of actin filaments by 90% with a concomitant dramatic decrease of viscosity of F-actin solutions.[31]

Thus, in the subplasmalemmal space, calcium controls the activity of both fodrin and gelsolin. The latter protein is capable of reducing the viscosity and consequently may be responsible with fodrin for modulating the cytoskeletal organization in the region where exocytosis occurs.

Recently, Burgoyne *et al.*[33] have described a new cytosolic actin-binding protein with a molecular mass of 70 kDa which is related to caldesmon, a calmodulin-regulated protein. This protein is exclusively localized at the periphery of chromaffin cells in culture. In other systems, including smooth muscle[34] and platelets,[35] caldesmon has been purified, characterized, and shown to bind reversibly to F-actin and calmodulin. This binding of caldesmon to calmodulin is Ca^{2+}-dependent, and calcium acts as a "flip-flop" switch between caldesmon-calmodulin and caldesmon-F-actin.[34,36] In the micromolar-range of concentrations the caldesmon-calmodulin complex increases; in the nanomolar range of concentrations the complex caldesmon-F-actin is formed. The binding of caldesmon to F-actin prevents the actin-myosin interaction in smooth muscle, whereas the caldesmon-calmodulin complex allows the binding of myosin to F-actin. Myosin has been described to be present in secretory cells, however its function has not yet been elucidated. The presence of caldesmon, which promotes the actin-myosin interaction in presence of calcium, suggests that an actomyosin complex could generate forces in stimulated cells.

CONCLUSIONS

From all these considerations, it can be pointed out that the Ca^{2+}-dependent activation of fodrin and gelsolin results in solation at some points in the subplasmalemmal space where exocytosis has to take place. This solation results in the facilitation of granule movements during reorganization of subplasmalemmal region following cell stimulation. Participation of other proteins, such as caldesmon, myosin, and others yet to be characterized is probable, but remains to be demonstrated in order to understand molecular mechanisms underlying secretion.

ACKNOWLEDGMENTS

We thank Mr. J. C. Artault for printing electron micrographs and Ms. S. Ott for carefully typing the manuscript.

REFERENCES

1. DOUGLAS, W. W. 1968. Brit. J. Pharmacol. **34:** 451-474.
2. VIVEROS, O. H. 1975. Mechanism of secretion of catecholamines from adrenal medulla. *In* Handbook of Physiology: Endocrinology, Vol. 6. H. Blaschko, G. Sayers & A. D. Smith, Eds.: 389-426. American Physiological Society. Washington, D.C.
3. SMITH, A. D. 1972. Brit. Med. Bull. **29:** 123-129.
4. SMITH, A. D. & H. WINKLER. 1972. Handb. Exp. Pharmacol. **33:** 538-617.
5. RUBIN, R. P. 1984. Stimulus-secretion coupling. *In* Cell Biology of the Secretory Process. M. Cantin, Ed.: 52-72. Karger. Basel.
6. DINER, O. 1967. C. R. Acad. Sci. D **265:** 616-619.
7. KNIGHT, D. E. & P. F. BAKER. 1982. J. Membr. Biol. **68:** 107-140.

8. BADER, M. F., D. THIERSE, D. AUNIS, G. AHNERT-HILGER & M. GRATZL. 1986. J. Biol. Chem. **261:** 5777-5783.
9. BURGOYNE, R. D. 1984. Biochim. Biophys. Acta **779:** 201-216.
10. DOUGLAS, W. W. 1975. Secretomotor control of adrenal medullary secretion: Synaptic, membrane and ionic events in stimulus-secretion coupling. *In* Handbook of Physiology: Endocrinology, Vol. 6. H. Blaschko, G. Sayers & A. D. Smith, Eds.: 367-388. American Physiological Society. Washington, D.C.
11. TRIFARÓ, J.-M. 1978. Neuroscience **3:** 1-24.
12. BERNIER-VALENTIN, F., D. AUNIS & B. ROUSSET. 1983. J. Cell Biol. **97:** 209-216.
13. FOWLER, V. & H. B. POLLARD. 1982. Nature **295:** 336-339.
14. AUNIS, D. & D. PERRIN. 1984. J. Neurochem. **42:** 1558-1569.
15. POISNER, A. M. & P. COOKE. 1975. Ann. N.Y. Acad. Sci. **253:** 653-669.
16. KONDO, H., J. J. WOLOSEWICK & G. D. PAPPAS. 1982. J. Neurosci. **2:** 57-65.
17. METUZALS, J. & I. TASAKI. 1978. J. Cell Biol. **78:** 597-621.
18. ELLISMAN, M. H. & K. R. PORTER. 1980. J. Cell Biol. **87:** 464-479.
19. SCHNAPP, B. J. & T. S. REESE. 1982. J. Cell Biol. **94:** 667-679.
20. LEBEUX, Y. S. & J. WILLEMOT. 1975. Cell Tiss. Res. **160:** 1-68.
21. WANG, Y. & H. R. MAHLER. 1976. J. Cell Biol. **71:** 639-658.
22. HIROKAWA, N. 1982. J. Cell Biol. **94:** 129-142.
23. BADER, M. F., J. CIESIELSKI-TRESKA, D. THIERSE, J. E. HESKETH & D. AUNIS. 1981. J. Neurochem. **37:** 917-933.
24. SCHNAPP, B. J. & T. S. REESE. 1986. Trends Neurosci. **9:** 155-162.
25. JOCKUSCH, B. M., M. M. BURGER, M. DA PRADA, J. G. RICHARDS, C. CHAPONNIER & G. GABBIANI. 1977. Nature **270:** 628-629.
26. BADER, M. F. & D. AUNIS. 1983. Neuroscience **8:** 165-181.
27. WILKINS, J. A. & S. LIN. 1981. Biochim. Biophys. Acta **642:** 55-66.
28. PERRIN, D. & D. AUNIS. 1985. Nature **315:** 589-592.
29. LANGLEY, O. K., D. PERRIN & D. AUNIS. 1986. J. Histochem. Cytochem. **34:** 517-525.
30. KENIGSBERG, R. L. & J.-M. TRIFARÓ. 1984. Neuroscience **14:** 335-347.
31. BADER, M. F., J.-M. TRIFARÓ, O. K. LANGLEY, D. THIERSE & D. AUNIS. 1986. J. Cell Biol. **102:** 636-646.
32. YIN, H. L. & T. P. STOSSEL. 1979. Nature **281:** 583-585.
33. BURGOYNE, R. D., T. R. CHEEK & K. M. NORMAN. 1986. Nature **319:** 68-70.
34. SOBUE, K., Y. MURAMOTO, M. FUJITA & S. KAKIUCHI. 1981. Proc. Natl. Acad. Sci. U.S.A. **78:** 5652-5655.
35. LEWIS, W. G., G. P. COTE, A. S. MAK & L. B. SMILLIE. 1983. FEBS Lett. **156:** 269-273.
36. KAKIUCHI, S. & K. SOBUE. 1983. Trends Biochem. Sci. **8:** 59-62.
37. GRATZL, M. 1984. Anal. Biochem. **142:** 148-154.
38. BURGUN, C., D. MARTINEZ DE MUNOZ & D. AUNIS. 1985. Biochim. Biophys. Acta **839:** 219-227.

DISCUSSION OF THE PAPER

A. SCHNEIDER (*Albany Medical College, Albany, N.Y.*): The kinetics of fodrin patching appears to be slower than the kinetics of the secretory response (10 minutes versus less than one minute). Does this suggest a role for fodrin patching in the recycling of secretory granules rather than in the release process?

D. AUNIS (*Centre de Neurochimie du CNRS, Strasbourg, France*): Fodrin rearrangement in chromaffin cells seems to require several minutes. I think that this fodrin reorganization is a long-term process, related to the reorganization of cytoskeleton in inner regions of the secretory cell, thus preparing granules to undergo exocytosis. However, the involvement of fodrin in the endocytotic process cannot yet be excluded.

M. J. GEISOW (*National Institute for Medical Research, London, England*): What happens to the distribution of fodrin in the digitonin permeabilized, anti-fodrin treated cells? Is there a difference before and after the Ca^{2+} induction of secretion?

D. AUNIS: This is no real difference in the localization of fodrin; one would need careful electron microscopy.

V. P. WHITTAKER (*Max-Planck-Institut für Biophysikalische Chemie, Göttingen, Federal Republic of Germany*): If I understand your position correctly, you regard exocytosis of granules less as an activity-driven process than as a spontaneous process which needs to be inhibited by a layer of protein on the cytoplasmic surface of the plasma membrane. The removal of this layer then leads to spontaneous exocytosis. Perhaps we should take this concept one step further and suggest that a sudden contraction of the cell, or nerve terminal, could move storage granules or vesicles against cleared patches of plasma membrane, thus inducing exocytosis without there being a positive mechanism driving granule fusion.

D. AUNIS: The cytoplasmic area just beneath the cell membrane is composed of a complicated network of actin filaments. A population of secretory granules seems to be embedded into this network, extending on 200 nm from the cell membrane. This network has to be considered as a barrier which must be dissolved in order to have granules undergo exocytosis. However, it may happen that on stimulation the few granules trapped into this network undergo osmotic changes leading to the exocytotic process, and the clearance of fodrin occurs after exocytosis. This promotes intracellular reorganization of fodrin of cytoskeleton in order to induce granule movement from inner layers to cell periphery, thus preparing inner granules for the next-coming exocytosis. This granule movement may require a force-generating system similar to the one involved in axonal transport.

B. WIEDENMANN (*Institute of Cell and Tumor Biology, Heidelberg, Federal Republic of Germany*): Wouldn't you expect to see a punctuate pattern, using immunofluorescence microscopy combined with a fodrin antibody in your cells, since you showed that fodrin binds to chromaffin granules *in vitro*? In this context, it is worthwhile remembering that tubulin has been found to be associated with coated vesicles *in vitro;* still, using specific antibodies to tubulin, no punctuate patten typical for vesicle staining was observed.

D. AUNIS: Fodrin is not exclusively bound to the secretory granules present beneath the cell membrane. In this region, actin filaments are formed in abundance, and fodrin can bind to these filaments, to plasma membranes, and to other structures in addition to granules. Therefore, electron microscopy is necessary to look for membrane-protein relationships in cells.

Similarities and Differences among Neuroendocrine, Exocrine, and Endocytic Vesicles[a]

J. DAVID CASTLE,[b] RICHARD S. CAMERON,[b]
PETER ARVAN,[c] MARK VON ZASTROW,[b] AND
GARY RUDNICK[d]

[b] *Department of Cell Biology*
[c] *Department of Internal Medicine*
and
[d] *Department of Pharmacology*
Yale School of Medicine
New Haven, Connecticut 06510

INTRODUCTION: THE GENERAL ROLE OF SECRETORY AND ENDOCYTIC CARRIERS

In eukaryotic cells, macromolecular export and import activities are mediated by vesicular shuttles that serve to interconnect internal membrane-bounded compartments with the cell surface. In the export (secretory) pathway the shuttle has an origin in association with the *trans* aspect of the Golgi complex. It conveys a subset of the total synthetic products centrifugally for discharge, is conserved during the release process of exocytosis,[1] and undergoes (in large part) compensatory recycling to its origin for reutilization.[2,3] By contrast, the import (endocytic) pathway has as its primary base the plasma membrane, operates centripetally to deliver the entire spectrum of internalized cargo (ligand-receptor complexes plus components of the fluid phase) to endosomes,[4,5] a prelysosomal dispatch center that uses a reduced internal pH to segregate cargo from carrier. The shuttle recycles with efficiency to the cell surface.[6] Export and import activities are ubiquitous cellular functions, yet current evidence suggests that the pathways within the same cell may remain distinct with, at most, limited coupling. Thus endocytosis that is not compensatory to secretion does not relocate membrane to the Golgi compartment,[7] and that which is linked to secretion serves to clear the cell surface selectively of the export carrier[8–10] and may be targeted to lysosomes only in cases of normal or induced organelle turnover.[11,12] Further, in multicellular organisms, functional specialization for either pathway tends to occur in distinct cell types that are professional for secretory or absorptive activities.

Within the secretory pathway, the most general exit route from the Golgi complex involves no appreciable intracellular accumulation of exportable macromolecules. The

[a]The research on parotid granules reviewed herein was supported by grant nos. GM26524 and AM29868 from the National Institutes of Health.

shuttles are relatively fast-moving and the products are conveyed and expelled continuously without storage. Certain cell types in which intermittent secretion is a functional specialization supplement this operation with a second pathway involving concentration and packaging of selected exportable products for intracellular storage and subsequent conditional mobilization.[13,14] These operations are characteristically initiated in Golgi-associated condensing vacuoles,[15,16] and the discharge of storage granules is amplified by external stimuli via processes mediated by intracellular second messengers.[17] The storage alternative involves only a subset of the total spectrum of macromolecules that enter the intracellular transport pathway. It is currently assumed that segregation of the stored species may involve specific class-wide (and possibly pH-dependent) sorting mechanisms[14] (analogous to those that sequester lysosomal hydrolases[18,19]) that operate in late- or post-Golgi compartments. These general features are summarized in the schematic diagram shown in FIGURE 1.

In epithelial cells that mediate molecular exchanges with the external environment, the issue of cell surface polarity and presence of plasmalemmal domains of distinct composition and function introduce an additional order of complexity to the traffic of vesicular carriers. Thus export traffic can be targeted to distinct cell surfaces and in exocrine cells favors a selective route to the apical cell surface, whereas import traffic, favoring an origin from either cell surface depending on the cell type, can include a transcellular component that operates as a shuttle to the opposite cell surface. Interestingly, in both the export and import carrier systems, the pathway leads to analogous vesicular cisternal structures with flattened lamellar (or even tubular) radial extensions.[20–24] Here compartmental volume and surface area are distributed differentially[7] as if undergoing a partial segregation. The electron micrograph shown in FIGURE 2 depicts this structure, which in the secretory pathway is located at the *trans* aspect of the Golgi complex and has been defined as a kinetic intermediate leading to storage granule formation.[16] Any dynamic role that it might serve in segregating and distributing secretory polypeptides to various destinations remains obscure at present, although the endocytic counterpart may serve as a model in this regard since it is associated with differentiating cargo and carrier traffic in both import and transcellular activities.[22–24] Thus in the most general sense, the vesicular carrier systems to and from the cell surface operate in an analogous manner to conserve and recycle shuttle membrane from a particular home base while bringing about a net and specific relocation of internal volume.

A LIMITED COMPARISON OF THE PROPERTIES OF EXOCRINE AND ENDOCRINE STORAGE GRANULES

In exocrine and endocrine secretory cells the major export pathway involves intracellular storage and stimulus-amplified discharge. The same series of intracellular compartments are used to synthesize, process, concentrate, and store macromolecules destined for release in both cell types; however, the secretory compositions differ markedly, reflecting distinct primary roles in mediating exchanges with the external environment and in coordinating the function of other cell types. Further, exocrine discharge is strictly a function of polarized epithelial cells, occurring only at a specific apical domain of the cell surface that is segregated from the adjoining basolateral surfaces by junctional complexes. Thus the storage granules of exocrine and endocrine cells must retain and concentrate different products and fuse with plasmalemmal domains of different composition. Evidently, a comparative analysis of the properties of these compartments should aid in delineating the compositional and functional features that may be general to vesicular carriers and those that may reflect a specific form of secretory activity.

Most of our efforts have been focused on understanding the composition and properties of the storage granules of the rat parotid gland. Through the early efforts of Schramm and collaborators, this tissue was shown to be especially favorable for

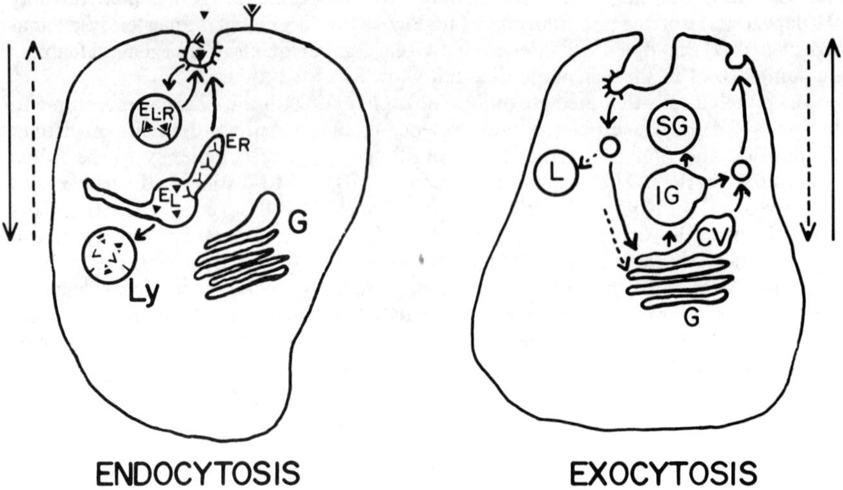

FIGURE 1. Schematic diagram depicting the general operation of endocytotic and exocytotic shuttling operations. Endocytosis has its origin at the cell surface, functioning centripetally (*solid vertical arrow*) to deliver cargo to the endosomal compartment (*E*) where a number of ligand-receptor complexes ($L \cdot R$) dissociate at acidic internal pH. Receptor-containing membrane (in E_R) and accumulated unbound ligand within the cisterna, E_L, become segregated with membrane recycling for further shuttling and ligand destined to enter secondary lysosomes (*Ly*). Exocytosis constitutes the terminal portion of the secretory pathway and has its origin in the Golgi region. Vesicles exit from the *trans* Golgi cisternae for centrifugal (*solid vertical arrow*) constitutive shuttling to the cell surface, or alternatively, storage granules form at this same location involving the progressive accumulation and concentration of macromolecular secretory products in condensing vacuoles (*CV*), immature granules (*IG*), and mature secretory granules (*SG*). The latter are released in response to extracellular secretory stimuli. Vesicular volume is exported from the cell whereas membrane is conserved and selectively recycled, probably mostly for reutilization with only a fraction undergoing lysosomal degradation.

investigating stimulus-dependent exocrine secretion and for studying isolated cell organelles that are nominally free of major proteolytic and lipolytic activities.[25] Parotid acinar cells normally devote most of their protein synthetic capability to manufacturing more than a dozen major proteins for export, and our recent studies suggest that ~85% of this production is destined for intracellular storage in granules as a reserve for stimulus-dependent release. The 1 μm diameter granules occupy nearly 30% of cytoplasmic volume,[26] and in response to β-adrenergic agonists undergo massive discharge by exocytosis[27] at the apical lumen (~12% of the cell surface[28]). Compensatory

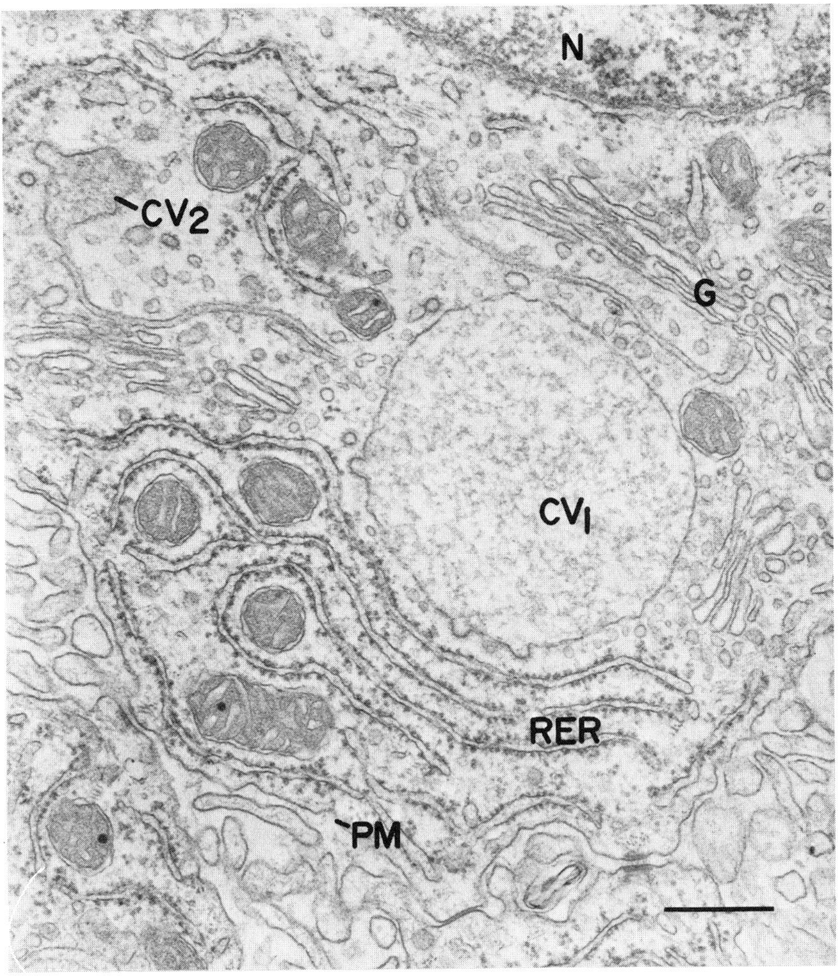

FIGURE 2. Electron micrograph of the Golgi region of a parotid acinar cell showing a very large spherical condensing vacuole (CV_1) and a separate condensing vacuole (CV_2) having a vesicular cisterna and an extended lamellar arm with a high surface:volume ratio. Both structures are located at the *trans* aspect of the Golgi complex, and the presence of coated evaginations on their periphery provides static evidence of membrane shuttling, consistent with these structures serving in the formation of the exocrine storage compartment. Other structures, nucleus (*N*), rough endoplasmic reticulum (*RER*), Golgi cisternae (*G*), and plasma membrane (*PM*) are marked. *Bar* = 0.5 μm.

reinternalization following discharge is rapid,[29] selective,[8] and involves recycling to the Golgi complex for at least partial reutilization.[13] The secretion granules are readily isolated in either isosmotic media for biophysical studies[25,30] or in hyperosmotic sucrose/Ficoll where exceptional purity is required for subfractionation and subsequent analysis of membrane composition.[31]

Studies of isolated exocrine storage granules have tended to emphasize the spectrum and diverse roles of secretory polypeptides,[32] whereas comparable analyses of neuroendocrine granules, particularly adrenal chromaffin granules, have focused most frequently on the biophysical and biogenic amine sequestering properties.[33] In order to provide a limited basis for more direct comparison and contrast of the two types of granules, we have constructed TABLE 1, which comprises accumulated data for isolated rat parotid granules (mostly from our own studies and those of Schramm

TABLE 1. Compositional Comparison of Parotid and Chromaffin Granules

Content	Parotid	Chromaffin
Protein	300 mg/ml (10-20 mM)	170 mg/ml
Principal component	α-amylase ~40% (pI 6.8)	chromogranins (pI ~ 4.5)
Other	8 other major species	dopamine β-hydroxylase ~5%
Catecholamine	no evidence	550 mM
ATP	≤ 10 μM	122 mM
Ascorbic Acid	3 mM	22 mM
Na^+	~ 4 mM	
K^+	2 mM	2 mM
Ca^{2+}	~ 8 mM	17 mM
Mg^{2+}	~ 5 mM	5 mM
Total solutes	~ 40 mM	~ 750 mM
Internal pH	6.8	5.7
Buffering capacity	85-140 mM	70-210 mM

The data on rat parotid granules has been compiled primarily from studies carried out in the authors' laboratory.[30,35,41] The data on divalent ions of parotid granules was initially reported by Wallach & Schramm,[39] and related studies have been reviewed recently.[32] For chromaffin granules, much of the data shown has been tabulated previously by Winkler[34] and by Njus *et al.*[33] The data concerning chromaffin granule internal pH and buffering capacity have been reviewed recently.[33]

and co-workers) and for bovine adrenal chromaffin granules (taken from tables that have appeared in earlier reviews[33,34]). Both types of storage organelles have very high internal protein content; however, there is a clear difference in the diversity of polypeptides stored—more than eight major species dominated by α-amylase in parotid granules versus a much more limited spectrum emphasizing the acidic chromogranins (secretogranins) in chromaffin granules. The most striking contrast concerns catecholamine and ATP content, which are >0.5 M and 0.1 M, respectively, in chromaffin granules but are not observed at levels that exceed possible contributions by contaminating organelles in isolated parotid granules. Thus the exocrine granules are fundamentally different in that they do not supplement input into the storage compartment by a local accumulation of high concentrations of small molecular secretory products. Both types of granules contain and secrete ascorbic acid,[35,36] which is known to function as a cofactor for dopamine β-hydroxylase in chromaffin granules and for peptide α-amidating monooxygenase possibly in both types of granules.[37,38] A variety of inorganic cations appear to be present at comparable levels in both types of granules[33,39]; in the case of parotid granules the levels of Na^+ and K^+ observed may be underestimates since our recent observations suggest that cation permeability of the exocrine granule membranes may not be quite as restricted as that observed in chromaffin granule membranes.[40] Thus for the content species that have been identified so far, there is a very large difference in the total solute concentration maintained in the two types of granules; chromaffin granules store more than twice the isosmotic levels of total solutes, probably as a nonideal solution,[33] whereas parotid granules apparently achieve transmembrane osmotic equilibrium with much lower internal solute concentrations. This may contribute to a largely different basis for stability during storage.

The limiting membranes of the two types of granules seem to share similar anion permeability properties, being relatively impermeable to sulfate and increasingly permeated by isethionate, acetate, chloride, and thiocyanate.[30,41–43] Further, both types of granule membranes appear to have a permeability to H^+ that is much lower than in the presence of protonophores.[41,44,45] There is, however, a striking contrast in both internal pH and level of the membrane-associated H^+ translocating ATPase activity between the two types of granules. Chromaffin and most other types of endocrine secretion granules have an acidic internal pH that appears to be well-buffered, probably by components such as ATP and the acidic content proteins.[33] In spite of the internal buffering capacity, it is possible to demonstrate ATP-driven intragranular acidification of $\leq$ 0.5 pH unit. The electrogenic H^+ pump responsible for this action has been shown under appropriate conditions to drive a transmembrane H^+ electrical potential of >70 mV[44–46] and is known to play a major role in the sustained capacity for chromaffin granules to accumulate and store catecholamines (reviewed in Njus *et al.*[33]). Parotid granules have a nearly neutral internal pH ($\sim$6.8) that does not appear to differ substantially from that of the surrounding cytoplasm.[30] This pH appears to be buffered to about the same extent as that observed in chromaffin granules.[30] However, the major contributions to buffer capacity are likely to be made by the principal stored secretory proteins such as amylase (pI $\sim$ 6.8), since appropriate smaller molecular buffering species are not known to be present. Further, most of the evidence we have obtained suggests that parotid granule membranes possess relatively little H^+-translocating ATPase activity. Thus, in contrast to results obtained with chromaffin granules, it has not been possible to observe ATP-dependent internal acidification or chemical inhibition of ATPase activity under conditions where effective inhibition of chromaffin granule H^+ ATPase is observed.[41] Membrane potential measurements suggest at most low levels of H^+ ATPase; an ATP-dependent inside-positive potential of ≤ 6 mV was measured (as compared to >70 mV observed for chromaffin granules examined in parallel under identical experimental conditions[41]).

For at least two reasons, the elevated intragranular pH and relative lack of H^+ ATPase observed for normal parotid granules may not be generalizable as characteristics of exocrine granules. First, the recent report of Chander *et al.*[47] on lamellar bodies isolated from type II epithelial cells of lung and our own preliminary measurements on zymogen granules from the exocrine pancreas have identified intragranular pH values in the range 6.1-6.3, and even though both we and DeLisle and Hopfer[48] have not obtained evidence for H^+ ATPase associated with zymogen granules, this activity seems to be present in the isolated lamellar body preparations.[47] Second, the near absence of acidifying ATPase activity in parotid granules appears to be conditional, and may depend on the properties of the secretory content being stored. This consideration was revealed through studies of parotid secretion granules that were isolated from rats that had received chronic administration of the β-adrenergic agonist, isoproterenol. This treatment induces parotid acinar cell hyperplasia, hypertrophy of the organelles composing the secretory pathway (especially secretion granules), and modifies the transcriptional program such that an altered spectrum of secretory proteins is produced, stored, and discharged.[49,51] Secretion granules isolated following treatment were found to possess a slightly alkaline internal pH relative to the surrounding cytoplasm (probably reflecting the large proportion of basic proteins being stored), yet exhibit readily detectable H^+-translocating ATPase activity as witnessed by the ability to observe ATP-driven internal acidification and changes in membrane potential (as well as inhibition of both acidification and ATPase activity by known pump inhibitors and protonophores[52]). Although these granules with modified internal composition are not analogues for the condensing granules of normal parotid,[52] it may be worth considering the possibility that the forming exocrine storage compartment possesses H^+-ATPase activity for supporting its sorting, processing, and concentrating operations. In this case, the granules of treated animals may maintain this activity, which is normally lost during parotid granule maturation. Speculation aside, however, the results serve to reinforce the observation made above that exocrine granules can exhibit H^+-ATPase activity just as other vesicular shuttles do, and further, show that acidifying activity and an acidic granule interior are not obligatorily coupled properties.

Thus with the limited comparisons made to date, it seems clear that many of the differences identified between exocrine and biogenic amine-storing endocrine granules relate to the functions of uptake and sustained accumulation of high concentrations of positively charged biogenic amines.

THE COMPOSITION OF SHUTTLE MEMBRANES USED FOR EXOCRINE AND ENDOCRINE STORAGE

A number of high-quality granule membrane fractions prepared from both exocrine and endocrine tissues are now available and make possible certain generalizations concerning the composition of membranes used for secretory storage. Characteristically these granule membranes have a phospholipid:protein weight ratio that is higher than that of most other cellular membranes and for preparations of documented high purity, is in the range of 2-6 mg/mg.[31] Further, cholesterol levels (33-40 mole percent of total phospholipid[34,53]) are higher than those observed in other intracellular membranes but lower than generally reported for plasma membranes.[54] Thus as a compartment, the membrane containers of both endocrine and exocrine granules possess overall compositional characteristics that are distinct from other cellular membranes. With the exception of membrane components involved in catecholamine accumulation and metabolism,[55–57] however, rather little is known about the activities of individual membrane polypeptides in either endocrine or exocrine storage granules. Catalogs of the polypeptides from both sources have been developed primarily by two-dimensional (isoelectric focusing SDS) polyacrylamide gel electrophoresis,[58–60] and in the absence of functional information, we have used this strategy to compare the patterns obtained for granule membrane preparations from different tissue sources with the intention of searching for compositional overlap. Common polypeptides could then become an object of focus for functional studies since their involvement in general secretory activities such as polypeptide storage, discharge, or membrane retrieval seems quite likely. In our efforts using this approach, we have viewed the highly purified parotid granule membrane fraction[31] as a compositional standard and have focused initially on other types of exocrine granules and subsequently on endocrine carriers of two varieties—the Golgi-derived vesicles of hepatocytes that export newly synthesized components of the blood plasma without appreciable intracellular storage and the chromaffin granule membranes of bovine adrenal medulla. Because of difficulties encountered in obtaining and adequately resolving membrane polypeptides at levels that are detectable by chemical staining, we have examined electrophoretograms of samples that have been radioiodinated (using either chloramine T[61] or lactoperoxidase[62]) prior to separation. Thus the relative levels of different polypeptides observed may be more representative of tyrosine content than amount of protein.

The reproducible pattern observed for radioiodinated (lactoperoxidase technique) parotid granule membrane polypeptides resolved by isoelectric focusing in the horizontal dimension and by SDS-PAGE in the vertical dimension containing 25-30 distinct focused species is shown in FIGURE 3b. The labeled polypeptides all have acidic isoelectric points and fall into three general size classes: a high molecular mass set (80- > 110 kDa that contains many of the membrane glycoproteins; a group of intermediate sizes (40-65 kDa); and a prominent lower molecular mass group (20-35 kDa). Most of the species being observed are likely to be integral membrane components, because either saponin-sulfate[31] or high pH carbonate[63] treatments have been used prior to solubilization and electrophoresis to reduce contamination by residual secretory protein to negligible levels. When the patterns for purified granule membranes from exocrine pancreatic, submandibular, and exorbital lacrimal granules were examined, a similar three-group pattern was observed, but an even more striking observation was the extent of apparent compositional overlap with the parotid pattern.[60] This was particularly evident for the 20-35-kDa group of species where at least eight polypeptides could be recognized as having common isoelectric points and apparent molecular weights, but it also applied to a number of the 40-65-kDa polypeptides and to a lesser extent to the highest molecular weight group. The analysis has been extended one step further by carrying out comparative peptide mapping[64] of individual corresponding parotid and pancreatic membrane components using both tryptic and chymotryptic digests. The results verified the identity of components originating from different granule membranes and further served to rule out the possibility that the prominently labeled lower molecular weight species represented degradation products of higher molecular weight polypeptides.

When this comparative analysis of membrane proteins was extended to the Golgi-derived vesicles of rat hepatocytes that are responsible for endocrine secretion of plasma proteins, three groups of radiolabeled polypeptides similar to those observed for parotid were identified. Further, the resemblance, but not identity, in the two-dimensional (IEF-SDS) pattern of low molecular mass species (20-35 kDa) to that observed for parotid suggests the presence of related membrane components.[65] This notion is supported by observations of considerable peptide overlap when chymotryptic digests of individual parotid granule membrane and liver vesicle membrane polypeptides were mapped by two-dimensional thin-layer electrophoresis and chromatography. Although the complexity and resolution of the present analysis does not permit strong conclusions to be drawn, the comprehensive view suggests that carrier membranes of post-Golgi export traffic to the two different epithelial cell surfaces may share a spectrum of similar but not structurally identical polypeptide components. Evidently further analyses must focus on the primary structure of different domains of such polypeptides.

The two-dimensional pattern of radioiodinated polypeptides observed for adrenal chromaffin granule membranes is shown in FIGURE 3a. Comparison to FIGURE 3b immediately indicates that the major polypeptide components differ from those observed in parotid and other exocrine granules. Evidently a large part of the difference reflects the absence in parotid membranes of the machinery involved in catecholamine metabolism, particularly prominent species like dopamine β-hydroxylase. If one, however, compares the minor polypeptides of the chromaffin granule membrane pattern to the parotid pattern, then it is possible to identify a number of species that have similar isoelectric points and apparent molecular weights even in view of the different

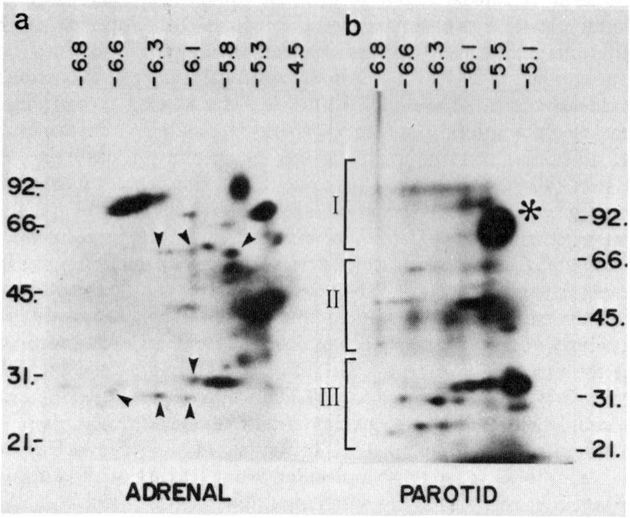

FIGURE 3. Autoradiograms of two-dimensional (IEF-SDS) polyacrylamide gel profiles of radioiodinated (lactoperoxidase technique) membrane polypeptides from (**a**) bovine adrenal medullary chromaffin granules and (**b**) rat parotid secretion granules. For the parotid pattern, the three groups of polypeptides distinguished in the text on the basis of molecular weight are indicated by brackets (I-III). The *asterisk* identifies residual lactoperoxidase. Many of the polypeptides that overlap the profile observed for other membrane preparations involve the species in group III. Group II is more variable, showing numerous identities with certain types of granule membranes but not with others, whereas group I is enriched in the membrane glycoprotein components. Even though group III polypeptides are a prominent radioiodinated species, they may not be major constituents of the granule membrane, as signified by their minor contributions to chemically stained polypeptide profiles. For chromaffin granule membranes (**a**) the contributions of polypeptides known to function in biogenic amine processing are prominent and may obscure possible compositional similarities to the profile observed for exocrine storage granule membranes; potential candidates in the latter regard are indicated by *arrowheads.*

animal sources of the two tissues that were fractionated. Some of these polypeptides (marked in FIG. 3a) fall in the same low molecular weight range as the polypeptides identified as common components among the exocrine granules examined, and as in the case of the liver Golgi vesicle analogues, deserve further structural study as possible elements involved in the basic shuttling function of vesicular carriers. It is hoped that it will be possible to extend this general analysis to the membranes of other types of endocrine storage granules, *e.g.,* those of the posterior pituitary, where the absence of machinery for biogenic amine processing and storage might facilitate a more straightforward comparison for similar quantities of radiolabeled membrane polypeptides.

CONCLUSIONS

The vesicular carrier systems that operate between the cell surface and internal compartments seem to be organized according to a common plan even though the points of origin and ultimate destinations differ for shuttles of specific function. Each of the systems relocate internal volume into which selected components have been segregated for transport, and the surface comprising carrier membrane is largely conserved and undergoes substantial compensatory recycling and reutilization. An inwardly directed electrogenic H^+ translocating ATPase activity may be a common element to all shuttles at least in their formative stages and may serve a central role in the segregation and concentration functions. Secretion granules are a specialized form of export shuttle in which a subset (usually very large) of total secretory products is stored at high concentration prior to discharge. The exocrine and amine-storing endocrine granules that have been compared both contain internal contents of stored protein at high concentration yet maintain very different internal concentrations of total solutes, a difference that may be established due to secondary uptake of small molecular species by already formed granules. The latter process appears to require the sustained presence of H^+ pumping activity. In spite of these functional differences for the storage compartments, the initial attempts to examine the spectrum of membrane polypeptides comparatively suggest that there are underlying elements of compositional overlap that may involve a family of species already shown to be common components of exocrine storage membranes and possibly representing basic elements of vesicular carrier function. It is hoped that further study of these polypeptides individually will reveal the extent to which they are general components of cell-surface-interacting carriers and will lead to some insight into their role in shuttle function.

SUMMARY

Secretory and endocytic vesicles have analogous functions as cyclic carriers between specific cellular compartments. The centrifugally functioning secretory system operates from the Golgi complex, whereas the centripetally functioning endocytic system operates from the cell surface. Further, within polarized epithelial cells the export traffic can be directed to a distinct plasmalemmal domain which distinguishes exocrine from endocrine secretion and import traffic can be directed transcellularly. These shuttle operations involve a special class of lipid-rich, protein-poor membranes that appear to use an inwardly directed H^+-translocase activity to varying extents for pH-dependent sorting and for accumulation and concentration of transported molecules. Comparative analyses of purified membrane preparations from exocrine and endocrine sources identify compositional overlap between different types of shuttle membrane. However, the structural elements that specify a particular origin or destination for a given carrier or determine function in storage and stimulus-dependent shuttling remain unknown.

ACKNOWLEDGMENT

The authors are grateful for the valuable participation of Patricia Cameron in many of the studies discussed.

REFERENCES

1. PALADE, G. E. 1975. Science **189:** 347-358.
2. PATZAK, A. & H. WINKLER. 1986. J. Cell Biol. **102:** 510-515.
3. HERZOG, V. & M. G. FARQUHAR. 1977. Proc. Natl. Acad. Sci. U.S.A. **74:** 5073-5077.
4. HELENIUS, A., I. MELLMAN, D. WALL & A. HUBBARD. 1983. Trends Biochem. Sci. **8:** 245-250.
5. STEINMAN, R. M., I. S. MELLMAN, W. A. MULLER & Z. A. COHN. 1983. J. Cell Biol. **96:** 1-27.
6. MULLER, W. A., R. M. STEINMAN & Z. A. COHN. 1980. J. Cell Biol. **86:** 292-303.
7. MARSH, M., G. GRIFFITHS, G. E. DEAN, I. MELLMAN & A. HELENIUS. 1986. Proc. Natl. Acad. Sci. U.S.A. **83:** 2899-2903.
8. DE CAMILLI, P., D. PELUCHETTI & J. MELDOLESI. 1976. J. Cell Biol. **70:** 59-74.
9. LINGG, C., R. FISCHER-COLBIE, W. SCHMIDT & H. WINKLER. 1983. Nature **301:** 610-611.
10. PHILLIPS, J. H., K. BURRIDGE, S. P. WILSON & N. KIRSHNER. 1983. J. Cell Biol. **97:** 1906-1917.
11. HERZOG, V. & H. REGGIO. 1980. Eur. J. Cell Biol. **21:** 141-150.
12. FARQUHAR, M. G. & G. E. PALADE. 1981. J. Cell Biol. **91:** 77s-103s.
13. TARTAKOFF, A. & P. VASSALI. 1978. J. Cell Biol. **79:** 694-707.
14. KELLY, R. B. 1985. Science **230:** 25-32.
15. JAMIESON, J. D. & G. E. PALADE. 1967. J. Cell Biol. **34:** 577-596.
16. JAMIESON, J. D. & G. E. PALADE. 1971. J. Cell Biol. **48:** 503-522.
17. RASMUSSEN, H. & P. Q. BARRETT. 1984. Physiol. Rev. **64:** 938-984.
18. SAHAGIAN, G. G., J. DISTLER & G. W. JOURDIAN. 1981. Proc. Natl. Acad. Sci. U.S.A. **78:** 4289-4293.
19. HOFLACK, B. & S. KORNFELD. 1985. Proc. Natl. Acad. Sci. U.S.A. **82:** 4428-4432.
20. NOVIKOFF, A. B., M. MORI, N. QUINTANA & A. YAM. 1977. J. Cell Biol. **75:** 148-165.
21. CLAUDE, A. 1970. J. Cell Biol. **47:** 745-766.
22. ABRAHAMSON, D. R. & R. RODEWALD. 1981. J. Cell Biol. **91:** 270-280.
23. HOPKINS, C. R. 1983. Cell **35:** 321-330.
24. GEUZE, H. J., J. W. SLOT, G. J. A. M. STROUS, H. F. LODISH & A. L. SCHWARTZ. 1983. Cell **32:** 277-287.
25. SCHRAMM, M. & D. DANON. 1961. Biochim. Biophys. Acta **50:** 102-112.
26. BLOOM, G. D., B. CARLSOO, A. DANIELSSON, H. GUSTAFSSON & R. HENRIKSSON. 1979. Med. Biol. **57:** 224-233.
27. AMSTERDAM, A., I. OHAD & M. SCHRAMM. 1969. J. Cell Biol. **41:** 753-773.
28. COPE, G. H. 1983. J. Microsc. **131:** 187-202.
29. KOIKE, H. & J. MELDOLESI. 1981. Exp. Cell Res. **134:** 377-388.
30. ARVAN, P., G. RUDNICK & J. D. CASTLE. 1984. J. Biol. Chem **259:** 13567-13572.
31. CAMERON, R. S. & J. D. CASTLE. 1984. J. Membr. Biol. **79:** 127-144.
32. WALLACH, D. 1982. *In* The Secretory Granule. A. Poisner & J. M. Trifaro, Eds.: 247-276. Elsevier. New York.
33. NJUS, D., J. KNOTH & M. ZALLAKIAN. 1981. Curr. Top. Bioenerg. **11:** 107-147.
34. WINKLER, H. 1970. Neuroscience. **1:** 65-80.
35. ZASTROW, M. V., T. R. TRITTON & J. D. CASTLE. 1984. J. Biol. Chem. **259:** 11746-11750.
36. TERLAND, O. & T. FLATMARK. 1975. FEBS Lett. **59:** 52-56.
37. ZASTROW, M. V., T. R. TRITTON & J. D. CASTLE. 1986. Proc. Natl. Acad. Sci. U.S.A. **83:** 3297-3301.

38. GLEMBOTSKI, C. C., B. A. EIPPER & R. E. MAINS. 1984. J. Biol. Chem. **259:** 6385-6392.
39. WALLACH, D. & M. SCHRAMM. 1971. Eur. J. Biochem. **21:** 433-437.
40. JOHNSON, R. G. & A. SCARPA. 1976. J. Gen. Physiol. **68:** 601-631.
41. ARVAN, P., G. RUDNICK & J. D. CASTLE. 1985. J. Biol. Chem. **260:** 14945-14952.
42. DOLAIS-KITABGI, J. & R. L. PERLMAN. 1975. Mol. Pharmacol. **11:** 745-750.
43. PHILLIPS, J. H. 1977. Biochem. J. **168:** 289-297.
44. JOHNSON, R. G. & A. SCARPA. 1979. J. Biol. Chem. **254:** 3750-3760.
45. HOLZ, R. W. 1979. J. Biol. Chem. **254:** 6703-6709.
46. HOLZ, R. W. 1978. Proc. Natl. Acad. Sci. U.S.A. **75:** 5190-5194.
47. CHANDER, A., R. G. JOHNSON, J. REICHERT & A. B. FISHER. 1986. J. Biol. Chem. **261:** 6126-6131.
48. DELISLE, R. C. & U. HOPFER. 1986. Am. J. Physiol. **250:** G489-G496.
49. SCHNEYER, C. A. 1962. Am. J. Physiol. **203:** 232-236.
50. MUENZER, J., C. BILDSTEIN, M. GLEASON & D. M. CARLSON. 1979. J. Biol. Chem. **255:** 5623-5628.
51. BENNICK, A. 1982. Molec. Cell. Biochem. **45:** 83-99.
52. ARVAN, P. & J. D. CASTLE. 1986. J. Biol. Chem. In press.
53. MELDOLESI, J., P. DECAMILLI & D. PELUCHETTI. 1974. *In* Secretory Mechanisms of Exocrine Glands. N. A. Thorn & O. H. Peterson, Eds.: 137-148. Munksgaard. Copenhagen.
54. ROUSER, G., G. J. NELSON, S. FLEISCHER & G. SIMON. 1968. *In* Biological Membranes. D. Chapman, Ed.: 5-70. Academic. New York.
55. GABIZON, R. & S. SCHULDINER. 1985. J. Biol. Chem. **260:** 3001-3005.
56. RUSSELL, J. T., M. LEVINE & D. NJUS. 1985. J. Biol. Chem. **260:** 226-231.
57. DUONG, L. T., P. J. FLEMING & J. T. RUSSELL. 1984. J. Biol. Chem. **259:** 4885-4889.
58. FISCHER-COLBRIE, R., R. ZANGERLE, I. FRISCHENSCHLAGER, A. WEBER & H. WINKLER. 1984. J. Neurochem. **42:** 1008-1016.
59. BADER, M. F. & D. AUNIS. 1983. Neuroscience **8:** 165-181.
60. CAMERON, R. S., P. L. CAMERON & J. D. CASTLE. 1986. J. Cell Biol. In press.
61. GREENWOOD, F. C., M. M. HUNTER & J. S. GLOVER. 1963. Biochem. J. **89:** 114-123.
62. HUBBARD, A. L. & Z. A. COHN. 1976. *In* Biochemical Analysis of Membranes. A. H. Maddy, Ed.: 427-501. Wiley. New York.
63. FUJIKI, Y., A. L. HUBBARD, S. FOWLER & P. B. LAZAROW. 1982. J. Cell Biol. **93:** 97-102.
64. ELDER, J. A., R. A. PICKETT, J. HAMPTON & R. A. LERNER. 1977. J. Biol. Chem. **252:** 6510-6515.
65. CAMERON, R. S., P. ARVAN & J. D. CASTLE. 1986. *In* Handbook of Physiology: Gastrointestinal Tract. J. G. Forte, Ed. In press. American Physiological Society. Washington, D.C.

DISCUSSION OF THE PAPER

J. H. PHILLIPS (*University of Edinburgh, Edinburgh, Scotland*): I think you need to be very cautious about making analyses of major proteins in secretory granule membranes. In the case of the chromaffin granule membrane there are two proteins which are overwhelmingly more abundant than other polypeptides. These are dopamine β-hydroxylase and cytochrome b_{561}, composing together about 40-50% of total membrane protein, and these both have a highly specialized function in the adrenal medulla. Those polypeptides that we know are found in very many neuroendocrine granule membranes are all present at an abundance of 1% or less of total membrane protein. These are the subunits of the H^+-translocating ATPase and three antigens identified by monoclonal antibodies (SV-2, p65, and synaptophysin).

J. D. CASTLE (*Yale University School of Medicine, New Haven, Conn.*): I agree with your comment and may have misled you during the talk. The prominence of the common proteins in the granule types we have examined is emphasized by radioiodination of these apparently integral membrane proteins. These actually are minor species by protein staining (silver or Coomassie blue). The high molecular weight glycoproteins which are major bands on SDS gels are poorly iodinatable.

Regulated Secretory Pathways of Neurons and Their Relation to the Regulated Secretory Pathway of Endocrine Cells

PIETRO DE CAMILLI AND FRANCESCA NAVONE

CNR Center of Cytopharmacology
and
Department of Medical Pharmacology
University of Milan
20129 Milan, Italy

Neurons are highly specialized secretory cells that can secrete in a regulated way at least two types of neurotransmitter substances: small nonpeptide molecules and peptides.[1,2] Evidence is accumulating that the release of these two classes of substances involves two types of secretory vesicles. One type, the small synaptic vesicle (SSV), is the "typical" synaptic vesicle, *i.e.*, a small (40-60 nm in diameter) vesicle which, after standard fixation conditions, exhibits by electron microscopy a clear core. The other type, the large dense-core vesicle (LDCV), is a larger vesicle (diameter greater than 60 nm and variable from neuron to neuron) that has an electron-dense core.[a,1-3] Small synaptic vesicles are thought to contain classical neurotransmitters only, whereas increasing evidence indicates that peptide neurotransmitters are stored in LDCVs.[2,3] Large dense-core vesicles may also contain nonpeptide molecules such as amines.[3] The segregation of different types of neurotransmitters in two classes of vesicles is related to the different mechanisms involved in their biosynthesis. All the machinery necessary to synthesize and load classical neurotransmitters into vesicles is present in nerve terminals, where SSVs undergo a continued local exo-endocytotic recycling. Thus, at each of these cycles, SSVs can be refilled with classical neurotransmitters.[4,5] In contrast, peptides can be synthetized, loaded, and concentrated into granules only in perikarya, which are therefore the only sites at which LDCVs can be assembled.[5,6]

Recently we have found that the synaptic-vesicle-associated protein synapsin I, a protein thought to play an important regulatory role in the control of neurotransmitter release,[7] is selectively associated with the surface of SSVs.[8] These findings have indicated that, even though some features may be common to regulated exocytosis of all secretory organelles of all cell types,[5,9] neuronal secretion via SSVs might be characterized by some special type of regulation. In fact, evidence is accumulating

[a] The term "dense-core vesicle" has been used to define two types of vesicles in nerve endings.[3] One type, the small dense-core vesicle, acquires an electron-opaque core under certain fixation conditions, and has a diameter of 40-60 nm. The dense core is thought to represent a catecholamine-containing precipitate. We include this type of vesicle in the SSV population. The other type, the LDCV, is larger and has an electron-opaque core, irrespective of the fixation procedure used.

that release from SSVs and from LDCVs can be, at least partially, independently regulated.[2,10] For example, the relative proportion of a peptide and a classical neurotransmitter released from the same nerve terminal in response to electrical stimulation appears to vary depending upon the frequency of stimulation.[2,10]

In order to further understand features of neuronal secretion mediated by LDCVs and by SSVs respectively, and the relationship of these two types of neuronal secretion to secretion from nonneuronal cells, it is important to establish similarities and differences between SSVs and LDCVs and between these organelles and the secretory granules of nonneuronal cells. In this short review we will summarize studies aimed at characterizing such relationships, focusing on work carried out in our own laboratory.

Synapsin I: A Nerve-Terminal-Specific Protein Selectively Associated with SSVs

Synapsin I was first identified as a major endogenous substrate for cAMP-dependent phosphorylation present in mammalian brain.[11] Synapsin I is the collective name for two closely related peptides, synapsin Ia and synapsin Ib (M_r around 80,000 in SDS gels), which have very similar properties.[12] For a review of these properties see De Camilli and Greengard.[7] Briefly, synapsin I is an extremely basic, acid-soluble, asymmetric protein, which undergoes multisite phosphorylation by cAMP-dependent protein kinase and by Ca^{2+} calmodulin-dependent protein kinases type I and type II.[13,14] It is a neuron-specific protein, present in virtually all neurons.[15,16] It represents approximately 0.3-0.4% of the total brain protein.[17] Evidence obtained by subcellular fractionation and by light and electron microscopy immunocytochemistry has indicated that, by far, the largest fraction of brain synapsin I is associated with the cytoplasmic surface of synaptic vesicles in nerve endings[15,16] (FIG. 1a, 2b, 3, and 4). More recent work has shown that synapsin I is associated selectively with SSVs and not with LDCVs[8] (FIGS. 5a and 5b). Synapsin I represents approximately 6% of the total protein in a highly purified SSV fraction.[17] It is a peripheral membrane protein of the vesicle membrane, which can be dissociated from the membrane by raising the ionic strength of the suspending media.[12,17] Binding of synapsin I to the vesicle surface is specific, saturable, of high affinity (K_d in the nM range), and appears to be mediated by a protein of the vesicle surface.[17,18] Since synapsin I is not an intrinsic membrane protein, and since it is known to be synthetized on free ribosomes,[19] a pool of nonvesicular synapsin I is likely to be present in nerve cells. The precise site at which synapsin I becomes associated with the vesicle surface, however, remains to be determined.

Phosphorylation-dephosphorylation of synapsin I is thought to play an important role in the regulation of neurotransmitter release. A variety of physiological and pharmacological manipulations of intact neurons or of isolated nerve endings, which induce release of classical neurotransmitters, or potentiate the release evoked by other stimuli, also produce an increase in the state of phosphorylation of the molecule (see De Camilli and Greengard[7] for a review). The sites of the molecule that become phosphorylated depend upon the type of second messenger (cAMP or Ca^{2+}) involved in the response to the stimulus. Recently Llinas *et al.* have shown that injection in the axon terminal of the squid giant axon of Ca^{2+}/calmodulin-dependent protein kinase type II potentiates several parameters of evoked neurotransmitter release.[20] Since synapsin I is a very good substrate for the kinase, the phosphorylation of synapsin

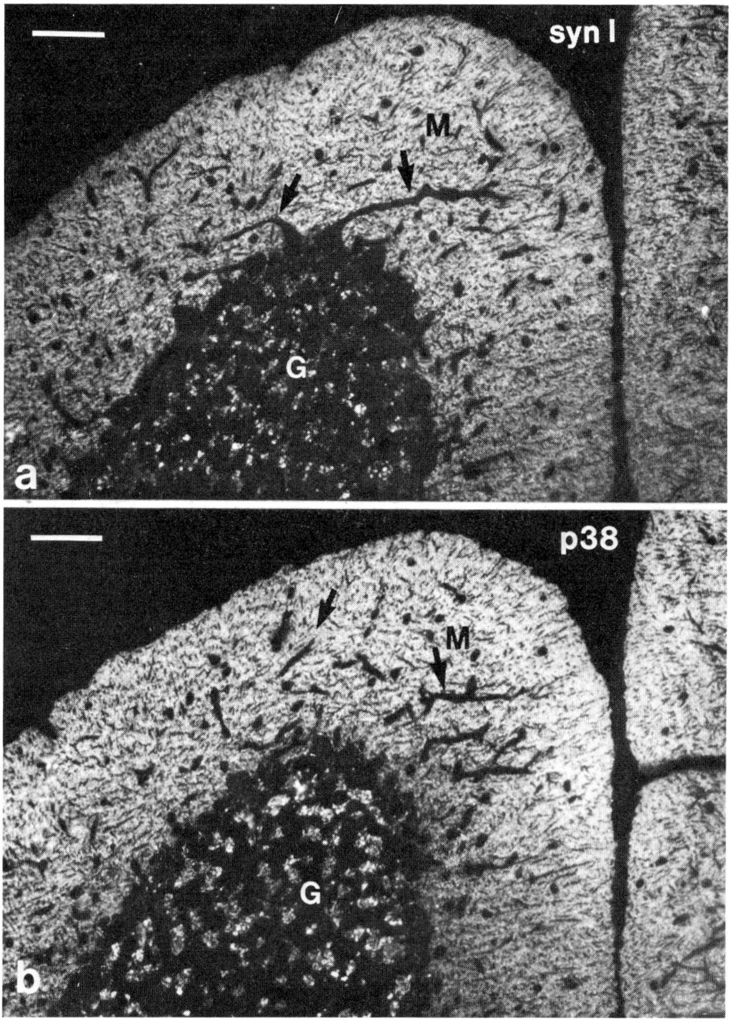

FIGURE 1. Distribution of (**a**) synapsin I and (**b**) p38 immunoreactivities in the rat cerebellar cortex visualized by immunofluorescence in 1-μm-thick plastic sections. Immunoreactivity appears white. The two sections are spaced a few micrometers apart. Synapsin I and p38 exhibit an almost identical distribution, which closely parallels the known distribution of nerve terminals. These are densely packed in the molecular layer (*M*), and arranged in clusters in the granule cell layer (*G*). *Arrows* point to dendrites of Purkinje cells, which appear as negative images outlined by immunoreactive axon endings. Calibration bars = 100 μm. (F. Navone, R. Jahn, P. Greengard, and P. De Camilli, unpublished observations.)

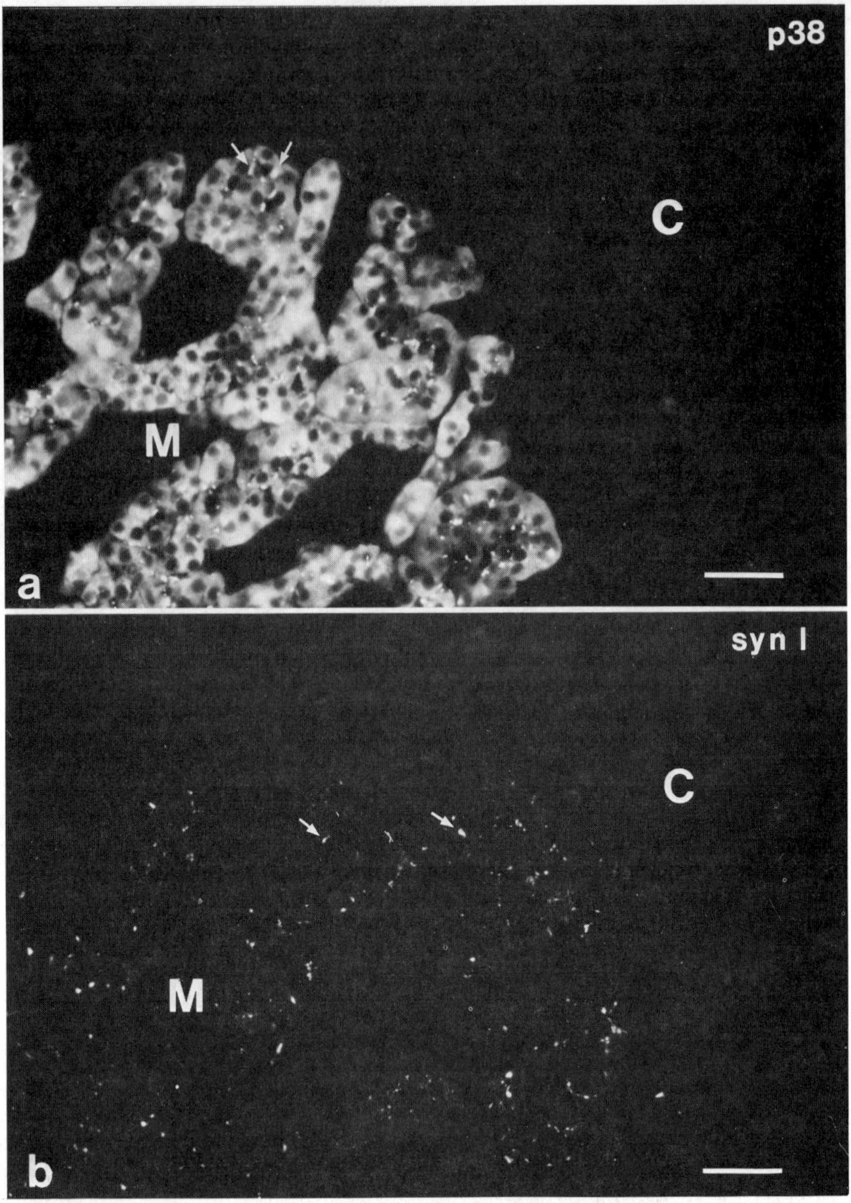

FIGURE 2. Localization of (**a**) p38 and (**b**) synapsin I immunoreactivities in the rat adrenal gland visualized by immunofluorescence in frozen sections. The two sections are spaced a few micrometers apart; p38 is present both in the nerve terminals that innervate the medulla (bright dots indicated by *arrows*), as well as in chromaffin cells themselves. In contrast, synapsin I immunoreactivity is present only in nerve terminals (*arrows*). *C,* cortex; *M,* medulla. Calibration bars=40μm. (F. Navone, R. Jahn, G. Di Gioia, P. Greengard, and P. De Camilli, unpublished observations.)

I, or of a synapsin-I-like protein of the squid giant terminal, might be involved in these effects.

The precise function of synapsin I is still unknown. Clearly such function has to be searched for in those properties that are specific for neuronal secretion. In this respect the finding that synapsin I is associated selectively with SSVs represents an important advancement, since SSVs are thought to be neuron-specific organelles. Thus, it is important to identify the properties that are common to all SSVs, and in particular, those in which SSVs differ from LDCVs.

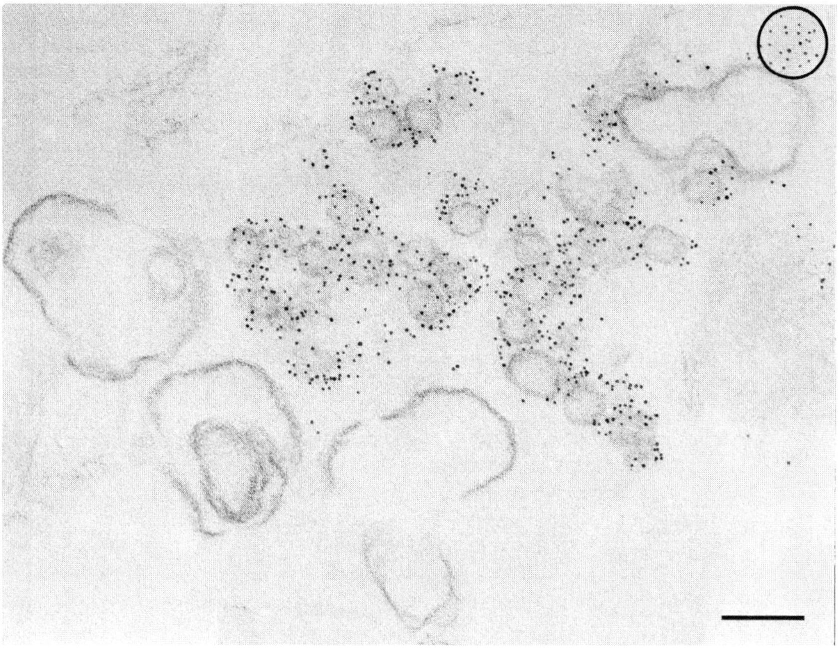

FIGURE 3. Localization of synapsin I immunoreactivity in isolated subcellular particles of bovine brain demonstrated by immunogold labeling. Bovine hypothalami were homogenized in a hypotonic buffer; the homogenate was fixed, agarose-embedded, and immunolabeled (see refs. 8 and 16 for experimental procedures). The figure shows remnants of a lysed nerve ending. In spite of the lack of an intact plasma membrane, SSVs are still clustered together, thus revealing the existence of a connecting cytoplasmic matrix. Synapsin I, as demonstrated by the distribution of gold particles, is selectively localized on SSVs. Other vesicular profiles are unlabeled. A black circle encloses a tangentially cut SSV, where the regular and dense apposition of synapsin I molecules on the vesicle surface can be seen. Calibration bar = 120 nm.

One set of properties in which SSVs are likely to differ from LDCVs are those that relate to their interactions with the cytoskeleton. SSVs in nerve terminals are often organized in clusters (FIG. 3) which, at classical synapses, are localized in close register with the synaptic cleft. LDCVs are in general excluded from such clusters and have a more scattered distribution.[8] Furthermore, SSVs, but not LDCVs, undergo a local exo-endocytotic recycling in nerve terminals (FIG. 4). These different functional characteristics of SSVs and LDCVs imply different interactions of their surfaces with the cytoskeletal matrix of nerve endings.

Synapsin I might act as a link protein between the SSV surface and some constituent(s) of the cytoskeletal matrix.[8] Like well-characterized proteins that link membranes to the cytoskeleton,[21] synapsin I is an extrinsic membrane protein and is present at high density on the membrane. In addition, it has been shown to bind to cytoskeletal proteins *in vitro* (namely to spectrin,[22] tubulin,[23] neurofilaments,[24] and actin—see M. Bahler and P. Greengard as quoted in ref. 7), although the physiological significance of these observed bindings remains to be established.

It has been proposed[22] that synapsin I might be the brain form of protein 4.1, a protein involved in anchoring the erythrocyte plasma membrane to the submembranous cytoskeleton.[21] Protein 4.1 and synapsin I have many similar properties and show a weak cross-reactivity with certain antibodies,[22] but they also exhibit substantial differences.[7] Furthermore, immunofluorescence patterns produced in the brain by antibodies directed against erythrocyte protein 4.1 are quite different from immunofluorescence patterns produced by antibodies directed against synapsin I (FIG. 6 and P. De Camilli, T. Petrucci, G. Di Gioia, and T. Leto, unpublished observations),

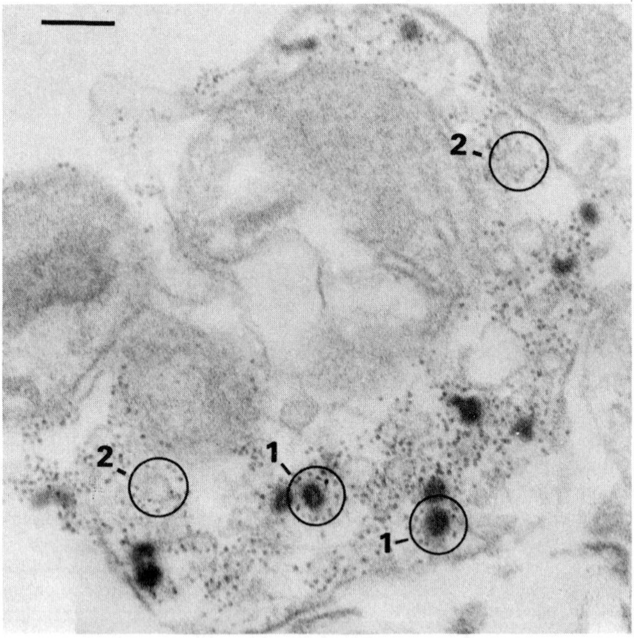

FIGURE 4. Immunoferritin localization of synapsin I immunoreactivity in an isolated nerve ending, demonstrating that synapsin I is present also on SSVs that have already undergone at least one exo-endocytotic cycle. Rat brain synaptosomes were incubated in a medium containing high K^+ (to stimulate exocytosis)[4] and an extracellular tracer (to label endocytic vesicles), then fixed under mild lytic conditions, agarose-embedded, and finally immunolabeled.[16] Type VI peroxidase has been used as the extracellular tracer. Vesicles that have already undergone an exo-endocytotic cycle have a black core, due to presence of peroxidase reaction product. Immunoferritin labels both vesicles labeled (*1*) and unlabeled (*2*) by peroxidase reaction product. Calibration bar = 120 nm. (S. M. Harris and P. De Camilli, unpublished experiments.)

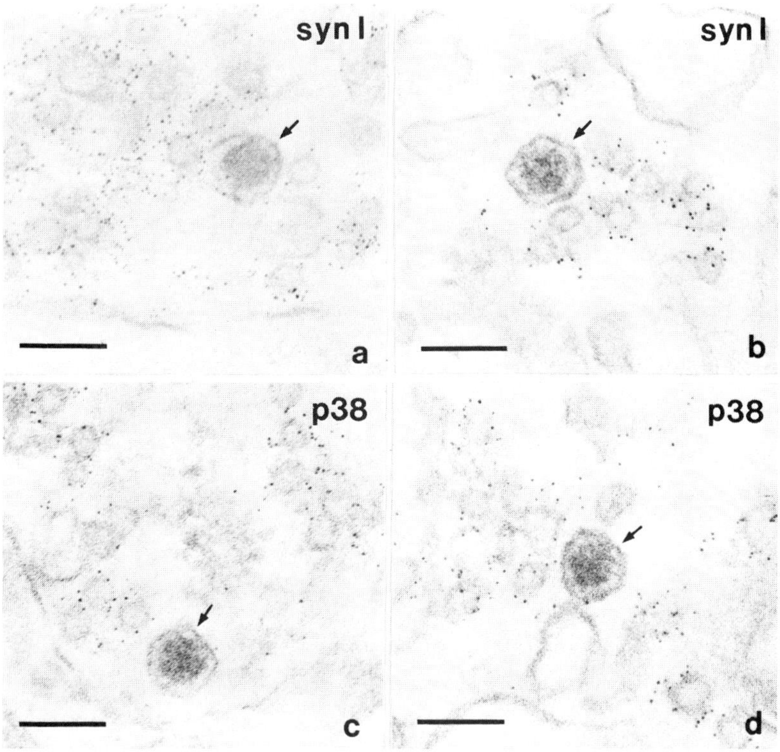

FIGURE 5. Selective association of synapsin I (**a** and **b**) and p38 (**c** and **d**) with SSVs demonstrated by immunogold labeling of isolated, lysed, hypothalamic nerve endings (for experimental procedures see Navone *et al.*[8]). Immunogold labels only SSVs. LDCVs, which are indicated by *arrows*, are unlabeled. Also unlabeled are other membrane profiles. Calibration bars = 120 nm. (F. Navone, R. Jahn, P. Greengard, and P. De Camilli, unpublished observations.)

suggesting that other types of 4.1 related proteins exist in brain. Thus, even though the two proteins might belong to the same class of proteins (link proteins between membranes and cytoskeletal elements), their physiological function is likely to be quite distinct.

Phosphorylation-dephosphorylation of the various phosphorylation sites of synapsin I might modify the interaction of SSVs with the cytoskeletal matrix of the terminal (via changes in the binding affinity of the protein for the cytoskeleton and for the vesicle membrane, respectively) and therefore control in some way the availability of SSVs for exocytosis. An inhibitory effect of the phosphorylation of the tail region of the molecule (which is mediated by Ca^{2+}/calmodulin) on its binding to the vesicle membrane has been demonstrated at least under certain *in vitro* conditions.[17,18] A possible model of how phosphorylation-dephosphorylation of synapsin I might regulate the exo-endocytotic cycles of SSVs is discussed in ref. 7.

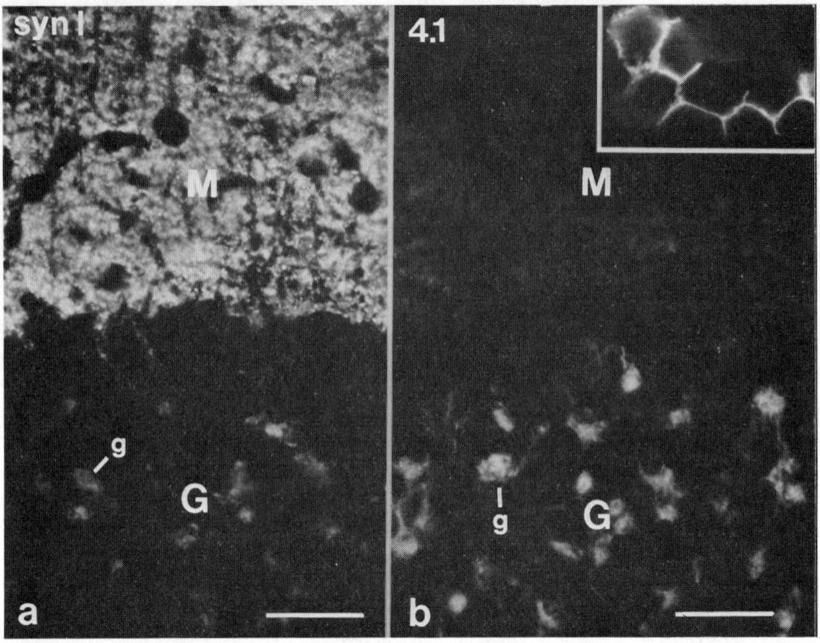

FIGURE 6. Comparison of the distribution of (**a**) synapsin I and (**b**) protein 4.1 immunoreactivities in the rat cerebellar cortex. 10-μm-thick frozen sections were stained by immunofluorescence. Antibodies raised against bovine synapsin I and human erythrocyte protein 4.1 have been used for the comparison. The fluorescence patterns produced by the two antibodies are very different. In the molecular layer (*M*) synapsin I immunoreactivity is very intense, while protein 4.1 immunoreactivity is almost undetectable. In contrast, in the glomeruli (*g*) of the granule cell layer (*G*) protein 4.1 immunoreactivity is more intense than synapsin I immunoreactivity. Furthermore, high power observation of the glomeruli by light microscopy revealed that the fluorescence pattern produced by the two antibodies in the glomeruli was different. The *inset* of *b* shows a portion of the choroid plexus adjacent to the cerebellar cortex shown in the main field. The basolateral surface of the epithelial cells of the plexus is very rich in protein 4.1 immunoreactivity and serves as a positive control for the specificity of the staining. Corresponding fields were completely negative in sections stained for synapsin I. Calibration bars = 50 μm. (P. De Camilli, T. Petrucci, G. Di Gioia, and T. Leto, unpublished observations.)

Protein p38, an Integral Membrane Protein Present on SSVs of Neurons and on Small Clear Vesicles of Neuroendocrine Cells[b]

Protein p38, Like Synapsin I, Is Present at All Nerve Terminals and Is Selectively Associated with SSVs

SSVs can be prepared from mammalian brain with a high degree of purity.[17] When highly purified fractions of SSVs are solubilized and analyzed by SDS polyacrylamide gel electrophoresis they appear to be primarily constituted of a rather small number of major polypeptides.[17] These major polypeptides are probably those proteins that are common to all SSVs irrespective of the specific classical neurotransmitter they store. If SSVs are prepared in low ionic strength media, one such protein is synapsin I.[17] Synapsin I is the only protein that can be quantitatively removed from synaptic vesicles by raising the ionic strength of the suspending media.[17] Some of the other major polypeptides are likely to be intrinsic membrane proteins. One such protein, protein p38 (referred to as p36 in Huttner *et al.*[17]) has been recently purified and characterized.[25–28]

Protein p38 (also called synaptophysin[26]) is an acidic intrinsic transmembrane glycoprotein. It exhibits a M_r of 38,000 in reduced SDS gels, but it appears to occur primarily as a dimer (M_r 76,000) in its native state[27] (R. Jahn, personal communication). It is an highly immunogenic protein. Sera obtained by injecting in rabbits highly purified SSV preparations, which had been previously stripped of synapsin I, are directed primarily against this molecule.[29] Its distribution in the various regions of the brain, as established by a sensitive RIA, closely parallels the distribution of synapsin I, and in all regions the two proteins are always present in roughly equimolar amounts.[30,31] Furthermore, light microscopy immunostaining patterns for p38 and for synapsin I are very similar in nearly all regions of the central and peripheral nervous system, and always indicate a predominant concentration of the two antigens in nerve terminals[30,31] (FIG. 1). As established by electron microscopy immunocytochemistry (immunogold labeling of lysed nerve endings), p38, like synapsin I, is selectively localized on the membrane on SSVs[31] (FIGS. 5c and 5d). Minor differences in the intracellular distribution of the two proteins, however, have also been observed by both light and electron microscopy immunocytochemistry. Thus, in contrast to synapsin I, p38 is also detectable in the Golgi area, occasionally in vesicles in dendrites, and in some large vacuoles and in nerve endings.[31] These localizations of p38, which involve only a very minor fraction of total neuronal p38, are probably related to the different type of association of p38 and synapsin I with membranes, and are consistent with the idea of a dynamic, rather than permanent, interaction of synapsin I with synaptic vesicle membranes.[7,16,17]

The presence in nearly equal amounts of synapsin I and p38 in all brain regions, and the colocalization of the predominant pool of both proteins on SSVs, suggest that they are present at a stoichiometric ratio close to 1 at the surface of SSVs.[31] Consistent

[b]The term "neuroendocrine cell" is used here to mean nonneuronal cells of the neuroendocrine system. The term does not encompass endocrine cells specialized for the secretion of nonpeptide molecules such as steroids.

with this idea, in a highly purified SSV fraction prepared in low ionic strength they are both enriched about 20-fold over the homogenate.[25]

Protein p38 is Also Present in Neuroendocrine Cells on a Population of Small Vesicles Distinct from Secretory Granules

In contrast to synapsin I, which is a neuron-specific protein, p38 is also present in a variety of neuroendocrine cells[b]; p38 immunoreactivity has been found in chromaffin cells of the adrenal (FIG. 2a), endocrine cells of the pancreas, all cell types of the anterior pituitary (FIG. 7a), C cells of the thyroid, cells of the diffuse neuroendocrine system of the gut, and in tumor and cell lines derived from such tissues.[25,26,30–32] p38 immunoreactivity in neuroendocrine cells is accounted for by a protein that has a slightly slower mobility in SDS gels than brain p38.[26,28,31] Such difference has been shown to be attributable to the sugar moiety of the molecule. In particular, when neuronal and endocrine p38 were chemically deglycosylated, they yielded a peptide of identical mobility.[28,31]

The presence of p38 in nonneuronal neuroendocrine cells was at first puzzling, because secretory organelles of these cells can be seen as related to neuronal LDCVs rather than to SSVs (see below). An investigation of the subcellular localization of p38 in endocrine cells has provided an explanation to such finding. By two different and complementary electron microscopy immunocytochemical approaches (labeling of intact tissue, ultrathin frozen sections,[33] and of dispersed subcellular particles, agarose-embedded tissue homogenates,[8,16]), we have demonstrated that p38 in endocrine cells is localized on a population of small vesicles with clear content that are scattered among secretory granules and are concentrated in the area of the Golgi complex (*trans* side)[31] (FIGS. 8 and 9). These vesicles have variable shapes, ranging from small round or oval vesicles to irregularly shaped vacuoles and short tubules. No p38 was detectable on typical secretory granules, *i.e.,* the granules where peptides (and in some cases amines[3]) are stored. The bulk of the plasma membrane was p38-negative, but a few small plasma membrane invaginations positive for p38 were seen.[31]

These electron microscopical results were supported by a comparison of the fluorescence patterns obtained in endocrine cells after immunostaining for p38 or for a "marker" (secretogranin II, see below) of secretory granules (FIG. 7). As can be seen from FIGURE 7, p38 immunoreactivity does not colocalize with secretory granules. Finally, these observations are consistent with the report that a fraction of secretory granules purified from the adrenal medulla does not contain a detectable amount of p38.[26]

The function of p38 is not known. Its high concentration in vesicle membranes, as suggested by biochemical and immunocytochemical analysis in nerve cells, and by immunocytochemistry in endocrine cells[31] (FIGS. 8 and 9), is consistent with the possibility that it might act as an anchoring site for the cytoskeleton on the vesicle membrane. In fact, its presence on SSVs at a stoichiometric ratio to synapsin I close to 1 raises the possibility that p38 might be the binding site of SSVs for synapsin I. This possibility, however, remains to be tested. It is also of interest that p38 has some similarities (molecular weight, ability to form dimers, acidic isoelectric point, presence at high density in membranes) with glycophorin, the transmembrane glycoprotein that acts as the binding site for protein 4.1 in the erythrocyte plasmalemma.[21] So far, however, we have been unable to detect any cross-reactivity between p38 and glycophorin.

SSVs and LDCVs Are Organelles of Two Distinct Regulated Secretory Pathways of Neurons Related to Two Distinct Pathways of Neuroendocrine Cells

Two Distinct Regulated Secretory Pathways Common to All Neurons

The selective localization of both synapsin I and p38 on SSVs in nerve endings adds further evidence to the idea that SSVs and LDCVs are organelles of two distinct

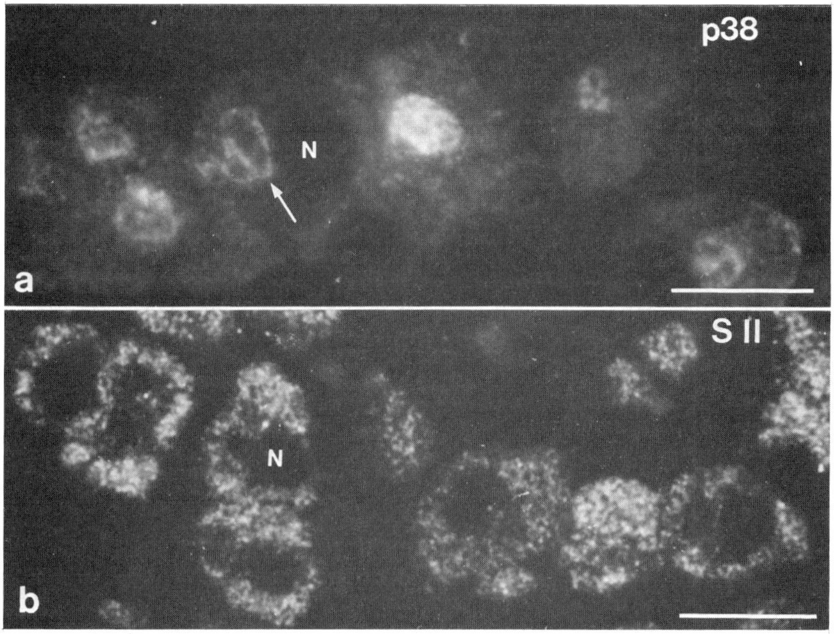

FIGURE 7. Localization of (**a**) p38 and (**b**) secretogranin II immunoreactivity in cells of the anterior pituitary (1-μm-thick plastic sections of a rat pituitary gland) demonstrated by immunofluorescence. The two antigens have a very different intracellular distribution. An intense accumulation of p38 immunoreactivity is visible (*arrow*) on a structure adjacent to the nucleus (*N*), which has the morphology and the localization expected for the Golgi complex in pituicytes. A lower level of finely granular p38 immunoreactivity is visible throughout the cell cytoplasm. Secretogranin II immunoreactivity occurs in the form of bright puncta scattered throughout the cytoplasm. The distribution of such puncta corresponds to that of secretory granules. Calibration bars = 20 μm. (G. Di Gioia, F. Navone, R. Jahn, P. Greengard, and P. De Camilli, unpublished observations.)

and independently regulated secretory pathways. We hypothesize that these two pathways to exocytosis coexist in all nerve endings.

Until a few years ago it was thought that a given neuron could only secrete either peptides or a classical neurotransmitter, and that only a few highly specialized neurons

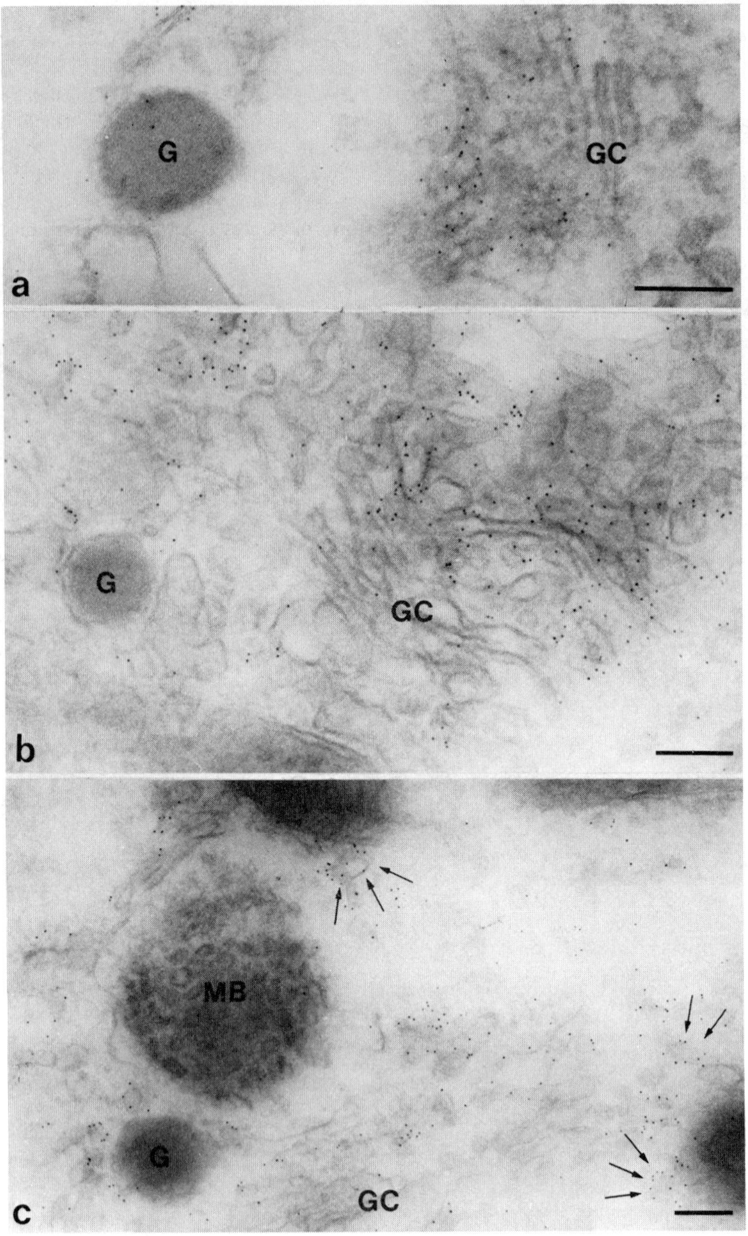

FIGURE 8. Intracellular localization of p38 in cells of the anterior pituitary demonstrated by immunogold labeling of ultrathin frozen sections.[33] In all fields gold particles are concentrated on areas rich in vesicles and tubules with clear content. The association of gold particles with individual small vesicles is clearly seen in field **c** (*arrows*). Secretory granules (*G*) are unlabeled. *GC,* cisternae of the Golgi complex; *MB,* multivesicular body. Calibration bars = 150 nm. (F. Navone, R. Jahn, H. Stukenbrok, P. Greengard, and P. De Camilli, unpublished observations.)

were peptidergic. The presence of regulatory peptides destined to secretion via regulated exocytosis has now been documented in a very large number of neurons.[1] Even motor end plates (both mammalian[34] and amphibian end plates, P. De Camilli, A. Greco, M. Matteoli, C. Haimann, and J. Polak, unpublished observations), previously thought to be clear examples of nerve endings secreting only classical neurotransmitter, have been shown to contain a secretory peptide, CGRP. In addition, the identification of proteins common to a large number of LDCVs, irrespective of the specific regulatory peptides they contain (see below the example of chromogranin and secretogranins), has allowed the demonstration of a distribution of peptidergic pathways much more widespread than previously thought.

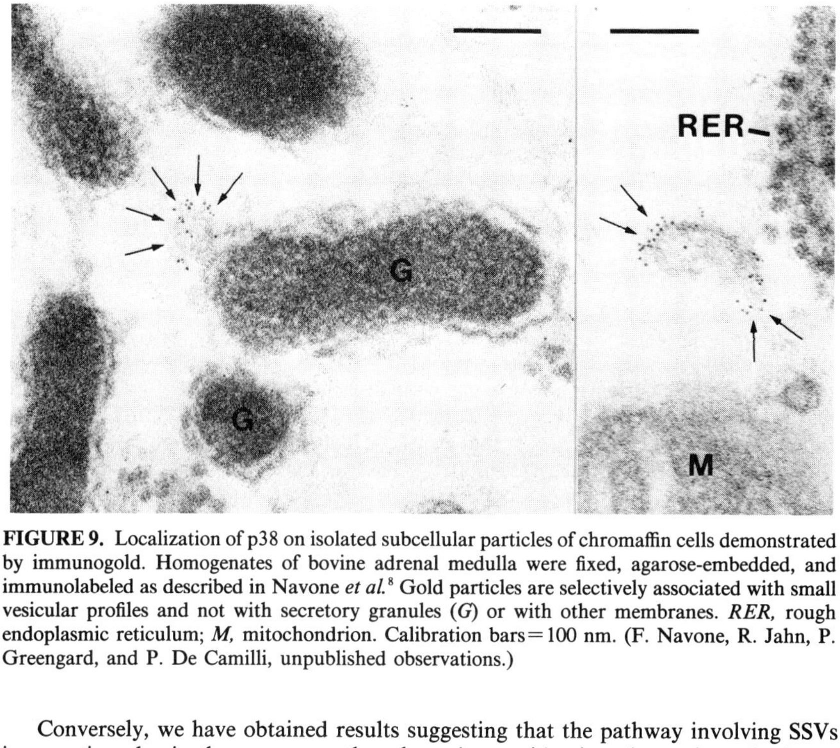

FIGURE 9. Localization of p38 on isolated subcellular particles of chromaffin cells demonstrated by immunogold. Homogenates of bovine adrenal medulla were fixed, agarose-embedded, and immunolabeled as described in Navone *et al.*[8] Gold particles are selectively associated with small vesicular profiles and not with secretory granules (*G*) or with other membranes. *RER,* rough endoplasmic reticulum; *M,* mitochondrion. Calibration bars = 100 nm. (F. Navone, R. Jahn, P. Greengard, and P. De Camilli, unpublished observations.)

Conversely, we have obtained results suggesting that the pathway involving SSVs is operating also in those neurons thought to be peptidergic only, such as the hypothalamic neurons projecting to the posterior pituitary. Terminals of the posterior pituitary contain clusters of small vesicles with a clear core, similar in size to SSVs of other neurons.[35] Their function is unknown and it has been proposed that such vesicles might represent endocytic vesicles involved in retrieving granule membrane after exocytosis of neurosecretory granules.[35] We have now found that both synapsin I and p38 are present on the membrane of small vesicles but not on that of neurosecretory granules in terminals of the neurohypophysis[8] (F. Navone, G. Di Gioia, R. Jahn, P. Greengard & P. De Camilli, unpublished observations). These findings clearly indicate that small vesicles of the posterior pituitary, rather than being a by product of peptide-containing granules,[35] are a functionally distinct organelle biochemically similar to typical SSVs.

The Neuronal Secretory Pathway Involving LDCVs Is Equivalent to the Regulated Secretory Pathway of Neuroendocrine Cells

LDCVs have many characteristics in common with secretory granules of the "regulated"[9] secretory pathway of neuroendocrine cells. They share the same mechanism of assembly in the Golgi area, they have a dense core when seen by electron microscopy, they undergo a regulated interaction with the plasmalemma, and, in particular, they have the same type of content: they contain regulatory peptides (hormones and neurohormones) and they also may contain amines[3] (see also other papers in this volume).

Recently it has become clear that secretory granules of a variety of neuroendocrine cells contain common proteins, irrespective of the specific type of regulatory hormones they store. Possibly, these proteins are involved in some coadjuvant role in peptide secretion. For example, they might be involved in processing, sorting, or packaging prohormones and hormones. Among these proteins are chromogranin (chromogranin A[36–38]), secretogranin I[39,40] (also referred to as chromogranin B[37,41,42]), and secretogranin II[39,40,43] (also referred to as chromogranin C[37,42]). These three proteins have several important characteristics in common and are therefore thought to represent a class of proteins.[39] They all have an acidic amino-acid composition, are further acidified posttranslationally by covalent modification, are highly soluble at physiological pH, and are heat-stable. Although they are present in granules of a variety of neuroendocrine tissues[37,39–43] (FIGS. 7 and 10), their relative ratio varies in granules of different endocrine cells.[37,39,42]

Chromogranin and secretogranins are also present in neurons, where they appear to be localized in LDCVs[39,44] (FIG. 10). Also in the nervous tissue, as in endocrine tissues, their distribution is much more widespread than that of any known peptide neurotransmitter[39,44] (A. Greco, P. Rosa, W. Huttner, A. Zanini, and De Camilli, unpublished observations). In fact they can be a tool to explore peptidergic pathways for which the specific regulatory peptide has not yet been identified. The presence of these proteins both in neuroendocrine secretory granules and in neuronal LDCVs provides additional evidence for the idea that the regulated pathways for peptide secretion are highly related in the two cell types (FIG. 11).

The Neuronal Secretory Pathway Involving SSVs Is Related to a Previously Unidentified Pathway of Neuroendocrine Cells

The presence and the subcellular distribution of p38 in neuroendocrine cells suggest that neuronal SSVs are related to an organelle of neuroendocrine cells but that this organelle is not the secretory granule. It is represented, instead, by a population of pleiomorphic small vesicles and tubules with clear content (see schematic drawing in FIG. 11). The concentration of these vesicular elements in the Golgi area, and the presence of patches of p38 in the plasmalemma (see above), suggest that they might connect functionally the Golgi area and the cell surface. The physiological function of this previously uncharacterized organelle, however, remains to be defined. As in the case of SSVs, these endocrine vesicles might be able to accumulate substances to be secreted. Alternatively, they might not have a secretory role. For example, they might be transport vesicles primarily involved in the traffic of membrane components. It should be noted that these vesicles do not appear to be clustered at subplasma-

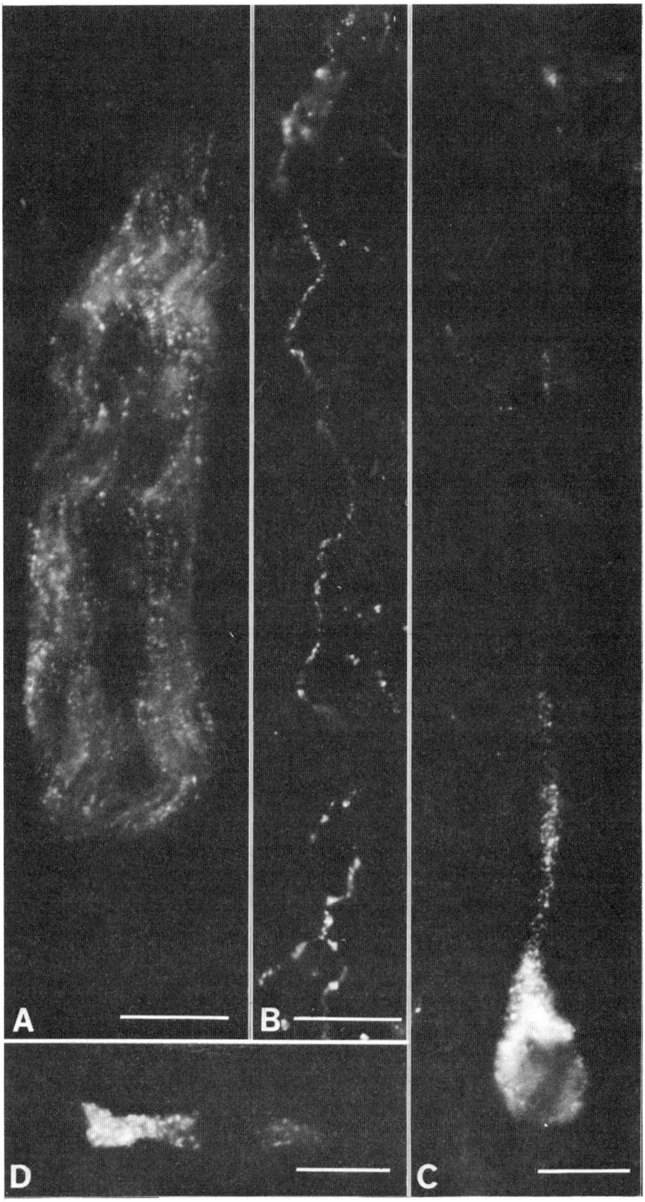

FIGURE 10. Localization of secretogranin II by immunofluorescence in neurons and endocrine cells of 10 μm-thick frozen sections. (**A**) Small nerve in the connective tissue surrounding the vas deferens. (**B**) Varicose fiber in the cerebellar cortex. (**C**) Neurons in the CA2 region of the hippocampus. (**D**) Endocrine cell in a duodenal gland. At all these locations secretogranin II immunoreactivity occurs in the form of discrete puncta that represent individual LDCVs or endocrine secretory granules. Calibration bars: **A** and **B**=30 μm; **C** and **D**=20 μm. (P. De Camilli, F. Navone, P. Rosa, A. Zanini, and W. Huttner, unpublished observations.)

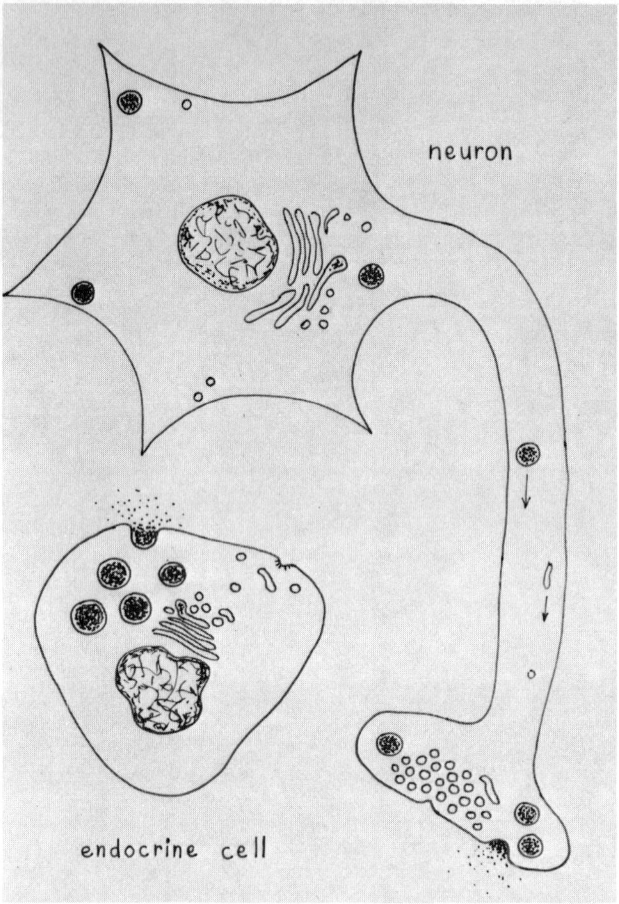

FIGURE 11. Schematic drawing illustrating the relation between neuronal and endocrine vesicular secretion. Nerve endings contain two types of secretory organelles, SSVs and LDCVs. The two types of organelles are involved in the secretion of different classes of neurotransmitters, have different life cycles in the terminal, and their membranes have a distinct composition. Since the exocytosis of both types of organelles is regulated, these can be seen as elements of two different "regulated" secretory pathways.[9] LDCVs are the equivalent organelles in neurons of secretory granules of endocrine cells. Thus, the neuronal secretory pathway involving LDCVs can be considered equivalent to the "regulated" secretory pathway of endocrine cells. SSVs are biochemically related to an as yet uncharacterized endomembrane system of endocrine cells. This is composed of pleiomorphic vesicles and tubules with clear content which are highly concentrated at the area of the Golgi complex (*trans* side) and which appear to be functionally connected to the plasma membrane. The function of this endomembrane system is still unknown.

membrane sites as is the case for SSVs in nerve endings. This observation raises the possibility that in contrast to what is the case for SSVs, their interaction with the plasma membrane might be nonregulated. In this respect it may be of interest that synapsin I is absent from these vesicles.

Identifying the function of p38-positive endocrine vesicles will be very important, not only to learn more about the physiology of endocrine cells, but potentially also to disclose previously unknown properties of neuronal SSVs. The possibility cannot be excluded that SSVs might have some other important function in addition to that of storing secretory substances. One piece of data in favor of this idea is our recent finding that both synapsin I[45,46] and p38 (P. De Camilli and R. Jahn, unpublished observations) are present in sensory endings, which are known to contain small vesicles morphologically similar to SSVs of efferent nerve endings. Thus, small vesicles in efferent and afferent nerve endings might share important biochemical similarities. This result challenges the obligatory secretory function of SSVs, since no secretion of classical neurotransmitters has been documented at sensory endings.

We plan in the future to compare further the molecular composition of p38-positive endocrine vesicles with that of neuronal SSVs. Thus far, two other intrinsic membrane proteins of SSVs (p65 and SV2 respectively) have also been detected in neuroendocrine cells.[47,48] It has been inferred that these proteins in endocrine cells are localized in the membrane of peptide granules,[47,48] but this has not been demonstrated. Furthermore, it has not been determined whether these two proteins are present also in LDCVs in nerve terminals. Published electron micrographs of chromaffin cells stained by immunoperoxidase for protein p65[47] are consistent with the possibility that p65 is present on small vesicles rather than on secretory granules. Thus, it is possible that further similarities might exist between neuronal SSVs and p38-positive vesicles of endocrine cells.

Neuroendocrine cells and neurons are embryologically, phylogenetically, and functionally related cells specialized for intercellular communication. Neurons may be seen as endocrine cells that have specialized for fast intercellular communication. Two key aspects of this specialization are the ability to form long processes and the development, at the end of such processes, of a secretory system partially independent from the cell body (the one based on locally recycling SSVs). Our findings suggest that such a secretory system does not appear ex novo in neurons but is the adaptation of an endomembrane system already present in endocrine cells (see FIG. 11). An important step in the adaptation of p38-positive vesicles to neuronal function seems to be the appearance of synapsin I.

An important question that will have to be answered in the future is whether p38-positive organelles of endocrine cells are in turn related to vesicular organelles of nonneuronal, nonendocrine cells. The undetectability of SSV antigens in cells other than neuroendocrine cells does not exclude this possibility, because it could be explained by tissue-specific differences of these proteins.

REFERENCES

1. HÖKFELT, T., O. JOHANSSON & M. GOLDSTEIN. 1984. Science **225:** 1326-1334.
2. LUNDBERG, J. M. & T. HÖKFELT. 1983. Trends Neurosci. **6:** 325-333.
3. KLEIN, R. L., H. LAGERCRANTZ & H. ZIMMERMANN, Eds. 1985. Neurotransmitter Vesicles. Academic Press. pp. 1-384.
4. CECCARELLI, B. & W. P. HURLBUT. 1980. Physiol. Rev. **60:** 396-441.
5. REICHARDT, L. F. & R. B. KELLY. 1983. Ann. Rev. Biochem. **52:** 871-926.

6. BROWNSTEIN, M. J., J. T. RUSSEL & H. GAINER. 1980. Science **207:** 373-378.
7. DE CAMILLI, P. & P. GREENGARD. 1986. Biochem. Pharmacol. (commentary). **35:** 4349-4357.
8. NAVONE, F., P. GREENGARD & P. DE CAMILLI. 1984. Science **226:** 1209-1211.
9. KELLY, R. B. 1985. Science **230:** 25-32.
10. ANDERSSON, P. O., S. R. BLOOM, A. V. EDWARDS & J. JARHULT. 1982. J. Physiol. **322:** 469-483.
11. JOHNSON, E. M., T. UEDA, H. MAENO & P. GREENGARD. 1972. J. Biol. Chem. **247:** 5650-5652.
12. UEDA, T. & P. GREENGARD. 1977. J. Biol. Chem. **252:** 5155-5163.
13. HUTTNER, W. B., L. J. DE GENNARO & P. GREENGARD. 1981. J. Biol. Chem. **256:** 1482-1488.
14. KENNEDY, M. B. & P. GREENGARD. 1981. Proc. Natl. Acad. Sci. U.S.A. **78:** 1293-1297.
15. DE CAMILLI, P., R. CAMERON & P. GREENGARD. 1983. J. Cell Biol. **96:** 1337-1354.
16. DE CAMILLI, P., S. M. HARRIS, W. HUTTNER & P. GREENGARD. 1983. J. Cell Biol. **96:** 1355-1373.
17. HUTTNER, W. B., W. SCHIEBLER, P. GREENGARD & P. DE CAMILLI. 1983. J. Cell Biol. **96:** 1374-1388.
18. SCHIEBLER, W. B., R. JAHN, J. P. DOUCET, J. ROTHLEIN & P. GREENGARD. 1986. J. Biol. Chem. **261:** 8383-8390.
19. DE GENNARO, L. J., S. KANAZIR, W. C. WALLACE, R. M. LEWIS & P. GREENGARD. 1983. Cold Spring Harbor Symposia on Quantitative Biology: Molecular Neurobiology, Vol. 48. Cold Spring Harbor, N.Y. pp. 337-345.
20. LLINAS, R., T. MCGUINNESS, C. S. LEONARD, M. SUGIMORI & P. GREENGARD. 1985. Proc. Natl. Acad. Sci. U.S.A. **82:** 3035-3039.
21. MARCHESI, V. T. 1985. Annu. Rev. Cell Biol. **1:** 531-562.
22. BAINES, A. & V. BENNETT. 1985. Nature **315:** 410-413.
23. BAINES, A. & V. BENNETT. 1986. Nature **319:** 145-147.
24. GOLDENRING, J. R., R. S. LASHER, M. L. VALLANO, T. UEDA, S. NAITO, N. H. STERNBERGER, L. A. STERNBERGER & R. J. DE LORENZO. 1986. J. Biol. Chem. **261:** 8495-8504.
25. JAHN, R., W. SCHIEBLER, C. OUIMET & P. GREENGARD. 1985. Proc. Natl. Acad. Sci. U.S.A. **82:** 4137-4141.
26. WIEDENMANN, B. & W. W. FRANKE. 1985. Cell **41:** 1017-1028.
27. REHM, H., B. WIEDENMANN & H. BETZ. 1986. EMBO J. **5:** 535-541.
28. JAHN, R., F. NAVONE, P. GREENGARD & P. DE CAMILLI. 1987. Biochemical and immunocytochemical characterization of p38, an integral membrane glycoprotein of small synaptic vesicles. Ann. N.Y. Acad. Sci. This volume.
29. JAHN, R., W. SCHIEBLER & P. GREENGARD. 1984. Proc. Natl. Acad. Sci. U.S.A. **81:** 1684-1687.
30. NAVONE, F., R. JAHN, P. GREENGARD & P. DE CAMILLI. 1985. Neurosci. Lett. (Suppl.) **22:** 5231.
31. NAVONE, F., R. JAHN, G. DI GIOIA, H. STUKENBROK, P. GREENGARD & P. DE CAMILLI. 1986. J. Cell Biol. **103:** 2511-2527.
32. WIEDENMANN, B., W. W. FRANKE, C. KUHN, R. MOLL & V. E. GOULD. 1986. Proc. Natl. Acad. Sci. U.S.A. **83:** 3500-3504.
33. GRIFFITH, G., K. SIMONS, G. WARREN & K. T. TOKUJASU. 1983. Methods Enzymol. **96:** 466-485.
34. RODRIGO, J., J. M. POLAK, G. TERENGHI, C. CERVANTES, M. A. GHATEI, P. K. MULDERRY & S. R. BLOOM. 1985. Histochemistry **82:** 67-74.
35. DOUGLAS, W. W., J. NAGASAWA & R. SCHULZ. 1971. Mem. Soc. Endocrinol. **19:** 353-378.
36. WINKLER, H. 1976. Neuroscience **1:** 65-80.
37. WINKLER, H. 1987. The life cycle of catecholamine-storing vesicles. Ann. N.Y. Acad. Sci. This volume.
38. BENEDUM, U. M., P. A. BAEUERLE, D. S. KONECKI, R. FRANK, J. POWELL, J. MALLET & W. HUTTNER. 1986. EMBO J. **5:** 1495-1502.
39. ROSA, P., A. HILLE, R. W. H. LEE, A. ZANINI, P. DE CAMILLI & W. B. HUTTNER. 1985. J. Cell Biol. **101:** 1999-2011.

40. LEE, R. W. H. & W. B. HUTTNER. 1983. J. Biol. Chem. **258:** 11326-11334.
41. FALKENSAMMER, G., R. FISCHER-COLBRIE, K. RICHTER & H. WINKLER. 1985. Neuroscience **14:** 735-746.
42. FISCHER-COLBRIE, R., C. HAGN & M. SCHOBER. 1987. Chromogranins A, B, and C: Widespread constituents of secretory vesicles. Ann. N.Y. Acad. Sci. This volume.
43. ZANINI, A. & P. ROSA. 1981. Mol. Cell. Endocrinol. **24:** 165-179.
44. SOMOGYI, P., A. J. HODGSON, R. W. DEPOTTER, R. FISCHER-COLBRIE, M. SCHOBER, H. WINKLER & I. W. CHUBB. 1984. Brain Res. Rev. **8:** 193-230.
45. DE CAMILLI, P., M. P. CANEVINI, R. ZANONI, M. VITTADELLO, C. TRIBAN & A. GORIO. 1985. Soc. Neurosci. Abstr. **11:** 1131.
46. FAVRE, D., E. SCARFONE, G. DI GIOIA, P. DE CAMILLI & D. DEMEMES. 1986. Brain Res. **384:** 379-382.
47. MATTHEW, W. D., L. TSAVALER & L. F. REICHARDT. 1981. J. Cell Biol. **91:** 257-269.
48. BUCKLEY, K. & R. B. KELLY. 1985. J. Cell Biol. **100:** 1284-1294.

Ascorbic Acid Release from Adrenomedullary Chromaffin Cells: Characteristics and Subcellular Origin

JANE KNOTH, O. HUMBERTO VIVEROS, AND
EMANUEL J. DILIBERTO, JR.

Wellcome Research Laboratories
Research Triangle Park, North Carolina 27709

INTRODUCTION

Primary cultures of bovine adrenomedullary cells actively take up and concentrate ascorbic acid.[1,2] Following a brief labeling period with L-[^{14}C]ascorbic acid, cells stimulated with either nicotinic agonists or depolarized by high K^+ or veratridine release the newly acquired ascorbic acid (NA-ASC).[3–5] Daniels *et al.*[4] demonstrated that NA-ASC was not labeling the catecholamine storage vesicles, but rather the postmicrosomal supernatant compartment(s). Additionally, they found that NA-ASC release was receptor mediated and Ca^{2+} dependent. NA-ASC and endogenous catecholamine (CA) release did, however, exhibit differential sensitivities to various Ca^{2+} channel blockers as well as differential rates of release in response to various secretagogues. Reported here are some of our recent observations on: (1) the differential release of NA-ASC and endogenous CA, (2) the requirements for maximal NA-ASC release, and (3) the compartmentalization of NA-ASC within chromaffin cells.

METHODS

Chromaffin cells were isolated from bovine adrenal medullae and maintained in primary cultures according to Wilson and Viveros.[6] Cells were plated in multiwell (16 mm/well) plates at a density of 5.2×10^5 cells/cm^2. After 3 to 7 days in culture, the experiments were carried out as described in the figure and table legends. The acetylcholinesterase activity was assayed radiometrically by measuring the amount of [^{14}C]acetate liberated from the hydrolysis of ^{14}C-labeled acetylcholine.[7] Separation of the [^{14}C]acetate product from unhydrolyzed [^{14}C]acetylcholine was accomplished by cation exchange chromatography (Biorad 50W-X8, Na^+ form) and the [^{14}C]acetate in the eluate assayed by liquid scintillation spectrometry.

RESULTS

Catecholamine secretion from cultured chromaffin cells is modified by the osmolality of the incubation medium.[8] FIGURE 1 depicts the 1,1-dimethyl-4-piperazinium (DMPP) induced release profiles of CA, NA-ASC, and α-[methyl-^{3}H]-aminoisobutyric acid (AIB) from chromaffin cells exposed to different osmolalities. Under hyperosmotic conditions, a progressive inhibition of release for all three molecules is obtained, although the degree of inhibition is different for CA as compared to NA-ASC and AIB. Release of NA-ASC and AIB is maximal under hyposmotic conditions, whereas CA release is greatest in an isotonic medium.

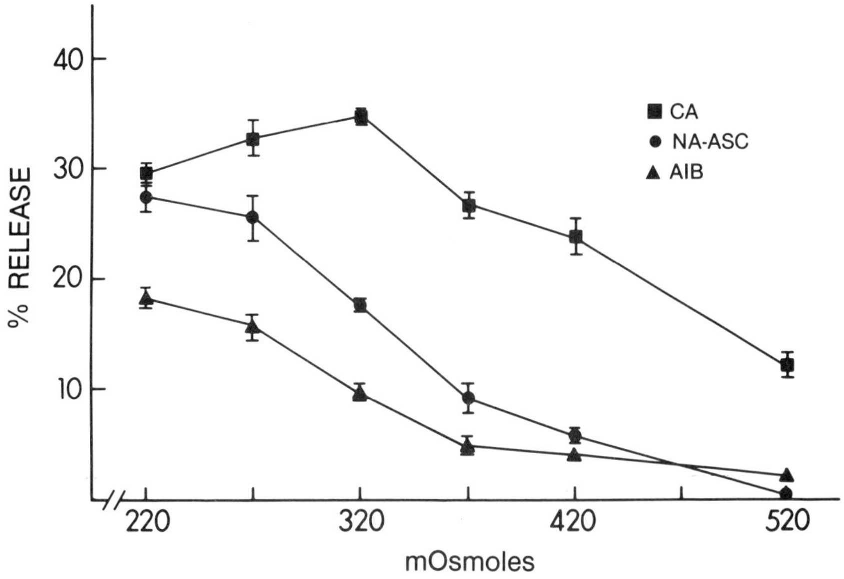

FIGURE 1. Effect of changes in the osmolality of the incubation medium on catecholamine (CA), newly acquired [^{14}C]ascorbate (NA-ASC), and α-[methyl-^{3}H]-aminoisobutyric acid (AIB) release. After three days in culture chromaffin cells (1 × 10^6 cells/well) were washed three times with 0.5 ml balanced salt solution (BSS) and then incubated with 0.5 ml BSS containing 200 μM ^{14}C-labeled ascorbic acid (9.9 mCi/mmol) and 0.25 μM α-[methyl-^{3}H]-aminoisobutyric acid (20 Ci/mmol) for 30 min at 37°. The cells were washed with 0.75 ml BSS for two 5-min intervals at 25°. For a final 10-min wash, the cells were incubated with 0.75 ml experimental buffer (100 mM NaCl, 4.2 mM KCl, 1.0 mM NaH_2PO_4, 11.2 mM glucose, 10 mM Hepes, 0.7 mM $MgCl_2$, 2 mM $CaCl_2$, pH 7.4) to which different amounts of sucrose were added to obtain the desired osmolality. The cells were stimulated for 15 min at room temperature by the addition of 0.45 ml experimental buffer containing 10 μM DMPP. The media and cell extracts were assayed for [^{14}C]ascorbate and α-[methyl-^{3}H]-aminoisobutyric acid by liquid scintillation counting and endogenous catecholamine content by the fluorometric trihydroxyindole method.[9] Release is expressed as percentage release of total cell content minus unstimulated release. Results are expressed as the mean ± SEM (n = 3).

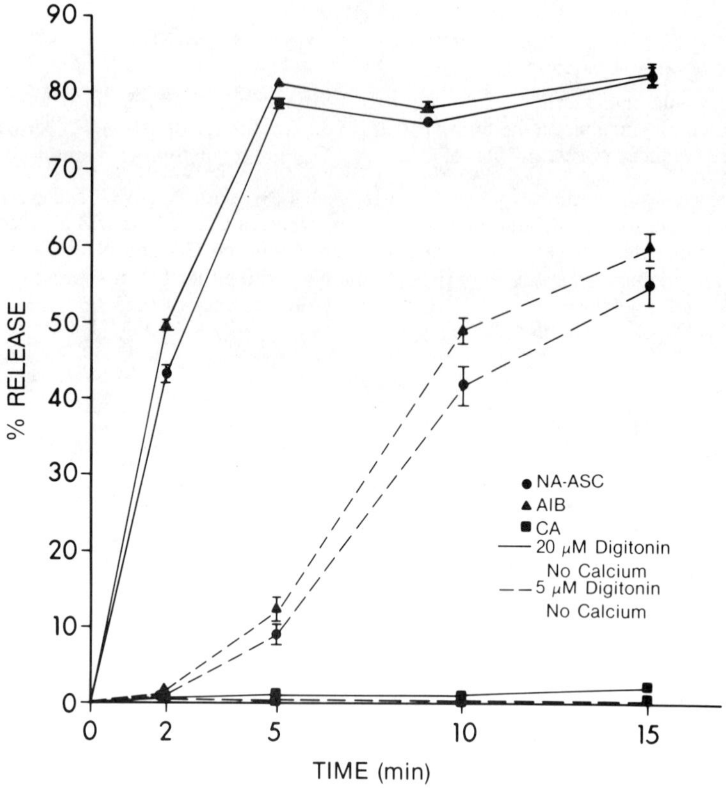

FIGURE 2. Catecholamine (CA), newly acquired [^{14}C]ascorbate (NA-ASC), and α-[methyl-^{3}H]-aminoisobutyric acid (AIB) release from digitonin-permeabilized chromaffin cells. Chromaffin cells were labeled with [^{14}C]ascorbate (200 μM) and α-[methyl-^{3}H]-aminoisobutyric acid (0.25 μM) for 30 min at 37°. Cells were washed and then incubated with the experimental buffer (139 mM glutamic acid, 20 mM Pipes, 5 mM $MgSO_4$, 5 mM ATP, 5 mM glucose, 5 mM EGTA, 4.43 mM $CaCl_2$, pH 6.6) containing 5 or 20 μM digitonin for the indicated times at room temperature. The media and cell extracts were assayed as described in FIGURE 1. Release is expressed as percentage of total cell content minus basal release (mean ± SEM).

Permeabilization of chromaffin cells has been a useful tool to study the exocytotic process.[10–13] Cells can be permeabilized by the addition of detergents such as digitonin and saponin or by exposure to an intense electric field. Exocytosis of catecholamines occurs in these cells only if extracellular Mg-ATP and Ca^{2+} are present. In contrast, release of NA-ASC from digitonin-permeabilized cells occurs in both the presence of absence of Ca^{2+} and release is detergent concentration and time dependent (FIG. 2).

NA-ASC and AIB were probably released from the same intracellular compartment, since the release profiles for NA-ASC and AIB were nearly identical under various osmotic conditions and in digitonin-permeabilized cells. To determine if NA-ASC and AIB are labeling the cisternae of the endoplasmic reticulum, release of endogenous soluble acetylcholinesterase (AchE)[14] was monitored from cells that had been subjected to changes in the osmolality of the external medium. The osmolality profile for AchE release more closely resembled the profile for CA release than that of AIB release (FIG. 3). In these experiments, AIB rather than NA-ASC release was monitored simultaneously with AchE and CA because of the use of [^{14}C]acetylcholine as the AchE substrate. When the cells were stimulated with DMPP at various osmolalities, AchE release was maximal under isosmotic conditions. Moreover, release of AchE in saponin-permeabilized cells displayed a similar Ca^{2+} dependency to CA release (data not shown).

To determine the ionic requirements for maximal NA-ASC release, the effects of Cl^- and Na^+ substitution were investigated. Release of both NA-ASC and AIB were quite sensitive to various anionic substitutions, whereas CA release was affected to a much lesser extent (TABLE 1). A clear difference between CA and NA-ASC release was observed upon substitution of the external Na^+ with either Li^+ or sucrose (TABLE 2). NA-ASC release was drastically reduced in the presence of sucrose, whereas CA release was only minimally affected.

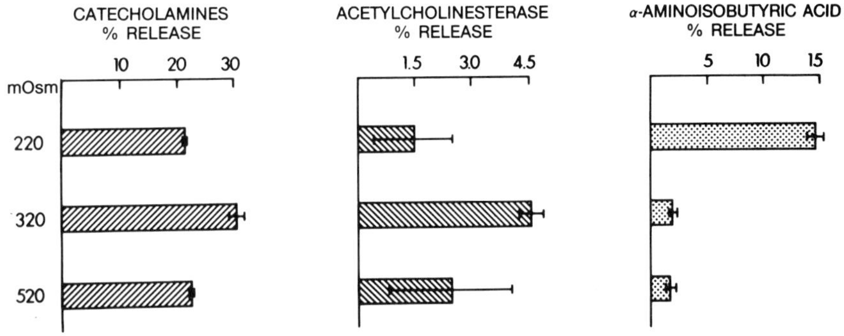

FIGURE 3. The effect of medium osmolality on DMPP-induced release of endogenous catecholamines, soluble acetylcholinesterase, and α-[methyl-^{3}H]-aminoisobutyric acid from cultured chromaffin cells. Chromaffin cells were labeled with α-[methyl-^{3}H]-aminoisobutyric acid (0.25 μM) for 30 min at 37°. Cells were washed and then stimulated with 10 μM DMPP for 15 min. The osmolality of the buffer was varied by increasing the concentration of NaCl (at 220 mOsM [NaCl] = 100 mM). Cells were washed and assayed as described in FIGURE 1. Enzyme activity was measured as described in METHODS. Release is expressed as percentage of total cell content minus unstimulated release (mean ± SEM; n = 3).

The effects of various inhibitors were investigated to explore the possibility that NA-ASC was being released from the cell via an anion channel or a transporter (TABLE 3, experiments I-III). 4,4-Diisothiocyano-2,2′-disulfonic acid stilbene (DIDS), an anion channel blocker, inhibited NA-ASC release about 40% and was without effect on CA release. Phloridzin, an inhibitor of ascorbate uptake into the cells,[1] inhibited NA-ASC about 70%, however, CA release was inhibited to nearly the same extent. Addition of ouabain, an inhibitor of the $(Na^+ + K^+)$-ATPase, nearly doubled basal release and increased veratridine-induced release of NA-ASC, CA, and AIB by 83%, 60%, and 192%, respectively. NA-ASC release is apparently dependent upon metabolic energy since the inhibitors 2-deoxyglucose, KCN, and carbonyl cyanide-(trifluoromethoxy)-phenyl hydrazone (FCCP) greatly attenuated its release (TABLE 3, experiment IV).

TABLE 1. Effect of Anion Replacement on Veratridine-induced Release of Newly Acquired [^{14}C]Ascorbate (NA-ASC), α-[methyl-^{3}H]-Aminoisobutyric Acid (AIB), and Endogenous Catecholamines (CA)

Anion Replacement	CA	NA-ASC	AIB
		% Maximal Release	
Cl	100	100	100
Glutamate	93	67	64
Br	62	90	88
Isethionate	73	47	53
NO_3	71	22	45
SO_4	63	32	35

Primary cultures of bovine adrenomedullary chromaffin cells were labeled with [^{14}C]ascorbate (200 μM) and α-[methyl-^{3}H]-aminoisobutyric acid (0.25 μM) for 30 min at 37°C, washed, and then stimulated for 20 min with 100 μM veratridine. The various anion replacements were made by complete substitution of Cl^-. Values for release were normalized according to 100% release in a 270 mOsM buffer containing 125 mM NaCl. Actual values for percentage release in NaCl medium were CA, 28.24 ± 2.6; NA-ASC, 20.82 ± 3.4; and AIB, 14.50 ± 2.0 (percentage release of total cell content minus unstimulated release; mean ± SEM). Cells were labeled, washed, and assayed as described in FIGURE 1.

DISCUSSION

Ascorbic acid is actively taken up and concentrated within adrenomedullary chromaffin cells.[1,2] Following a brief exposure to ascorbic acid, a small fraction of the total cell content is further concentrated in the catecholamine storage vesicles.[1,4] Upon CA secretion, vesicular ascorbic acid is released along with other soluble components of the vesicles via an exocytotic mechanism.[15] In addition to release of ascorbic acid from chromaffin vesicles, however, the major component of newly acquired ascorbic

TABLE 2. Effect of Na^+ Replacement on DMPP-induced Release of Newly Acquired [^{14}C]Ascorbate (NA-ASC), α-[methyl-^{3}H]-Aminoisobutyric Acid (AIB), and Endogenous Catecholamines (CA)

NaCl Replacement	CA	NA-ASC	AIB
		% Maximal Release	
NaCl	100	100	100
LiCl	53	42	55
Sucrose	69	16	8

Cultured chromaffin cells were labeled for 30 min at 37°C with [^{14}C]ascorbate (200 μM) and α-[methyl-^{3}H]-aminoisobutyric acid (0.25 μM). Cells were washed and then stimulated for 15 min with 10 μM DMPP. NaCl in the medium was replaced with either equiosmolar LiCl or sucrose. Values for release were normalized according to 100% release in a 320 mOsM buffer containing 150 mM NaCl. Actual values for percentage release in NaCl medium were CA, 30.79 ± 1.90; NA-ASC, 5.32 ± 0.43; and AIB, 3.27 ± 0.67 (percentage release of total cell content minus unstimulated release; mean ± SEM, n = 3). Cells were labeled, washed, and assayed as described in FIGURE 1.

TABLE 3. Effects of Various Additions on Newly Acquired [^{14}C]Ascorbate (NA-ASC), α-[methyl-^{3}H]-Aminoisobutyric Acid (AIB), and Endogenous Catecholamine (CA) Release

	Additions	CA	NA-ASC	AIB
			% Release	
Experiment I	veratridine	23.6 ± 0.8	18.2 ± 2.0	16.9 ± 0.9
	DIDS + veratridine	25.2 ± 0.8	10.4 ± 1.3	10.7 ± 0.5
Experiment II	veratridine	17.4 ± 1.4	14.0 ± 0.3	—
	phloridzin + veratridine	6.4 ± 1.9	5.7 ± 2.1	—
Experiment III	ouabain	4.2 ± 0.5	4.4 ± 0.5	11.1 ± 0.4
	veratridine	32.3 ± 0.3	10.8 ± 0.5	10.9 ± 0.4
	ouabain + veratridine	51.7 ± 3.2	19.8 ± 0.5	31.7 ± 0.4
Experiment IV	veratridine	33.6 ± 6.0	31.1 ± 0.9	17.0 ± 1.9
	veratridine + 2-deoxyglucose + KCN	16.9 ± 1.5	1.2 ± 0.4	3.3 ± 0.5
	veratridine + 2-deoxyglucose + FCCP	0.5 ± 1.5	0.6 ± 0.1	2.5 ± 0.3

Chromaffin cells were labeled with [^{14}C]ascorbate (200 μM) and α-[methyl-^{3}H]-aminoisobutyric acid (0.25 μM) for 30 min at 37°. Cells were washed and then stimulated for 60 min (Experiment III) or 20 min (Experiments I, II, and IV) at 25° in the presence of 100 μM veratridine. The inhibitors (100 μM ouabain, 100 μM DIDS, 2 mM phloridzin, 10 μM FCCP, 11.2 mM 2-deoxyglucose, or 2 mM KCN) were present during the last 10-min wash and also during stimulation. Release is expressed as percentage of total cell content. Nonstimulated release has been subtracted under each condition and results are expressed as the mean ± SEM (n = 3).

acid release was from a separate compartment(s).[4,5] Daniels *et al.*[3] found that if the cells were stimulated following a brief incubation with ^{14}C-labeled ascorbic acid, NA-ASC was released in a receptor-mediated, Ca^{2+}-dependent manner even though NA-ASC had not yet accumulated within the chromaffin vesicle.[4] Consequently, NA-ASC was being coreleased, but not costored with the catecholamines. We have further investigated NA-ASC release to understand better the mechanism of NA-ASC release as well as its compartmentalization within these cells.

Differential release of NA-ASC and CA was clearly evident when cultured-chromaffin cells were subjected to changes in the osmolality of the external medium, or when the cells were permeabilized by the detergent digitonin. Under both conditions, the release profiles for NA-ASC and CA were very different. In contrast, release of NA-ASC and AIB were nearly identical under various osmotic conditions as well as in digitonin-permeabilized cells, suggesting corelease from the same or very similar compartment. AIB (a nonmetabolizable amino acid) has frequently been used as a cytosolic marker and its release from neuronal tissue has previously been described.[16,17] AIB's assignment as a cytosolic marker was determined from its distribution into the postmicrosomal supernatant fraction. The techniques employed for distribution studies do not distinguish fragile subcellular organelles that are disrupted during sample processing. Therefore, the possibility that AIB labels a mechanically fragile compartment (such as the endoplasmic reticulum, ER) cannot be excluded. A soluble isoenzyme of AchE, which was localized to the cisternae of the endoplasmic reticulum, was shown to be released with CA upon stimulation of the adrenal medulla.[18,19] In recent studies, Mizobe and Livett[14] and Mizobe *et al.*[20] demonstrated nicotinic-receptor-mediated exocytotic release of AchE from chromaffin cells in culture. AchE release was Ca^{2+} dependent and proportional to, but independent of, CA release. These authors also suggest that the origin of the AchE is from the cisternae of the endoplasmic reticulum. This was based on biochemical studies[21,22] as well as cytochemical studies[19,23,24] demonstrating the localization of the enzyme to the cisternae of the endoplasmic reticulum and the plasma membrane. To explore the possibility that AIB and NA-ASC were labeling the endoplasmic reticulum, release of soluble acetylcholinesterase (AchE) from the cisternae of the endoplasmic reticulum was examined under various osmotic conditions. Release of CA and AIB was also measured. Release of AIB was monitored simultaneously with AchE and CA release to substitute for NA-ASC release under these conditions. The release profile for AchE closely resembled the CA release profile and maximal release was obtained under isosmotic conditions. Additionally, AchE release in saponin-permeabilized cells was also very similar to Ca^{2+}-dependent CA release (data not shown). Soluble AchE was in fact released from an intracellular compartment rather than released from the plasma membrane since pretreatment of the cells with the nonpenetrable AchE inhibitor, echothiophate iodide, was without effect.

The subcellular distribution and differential release studies support the contention that NA-ASC and AIB are costored within a cytosolic compartment and are subsequently coreleased upon stimulation of the cell. The precise nature of this compartmentalization within the cytosol, however, is not known. Since NA-ASC and AIB may be colocalized within the cytosol, the mechanism by which NA-ASC and AIB are released was examined. As previously noted, NA-ASC release is receptor mediated and Ca^{2+} dependent. To determine the requirements for maximal NA-ASC release, the effects of various inhibitors as well as ionic conditions were tested.

NA-ASC release is apparently dependent on both the ionic environment and metabolic energy. Both Cl^- and Na^+ substitution markedly attenuated NA-ASC release, whereas ouabain enhanced both basal and veratridine-induced NA-ASC and AIB release. Addition of ouabain should effectively increase the Na^+ concentration inside the cell. As previously described, ouabain increased the basal release and potentiated veratridine-induced CA release.[25,26] AIB release was particularly sensitive to ouabain addition (for both basal and stimulated release). AIB is presumed to be actively taken up into the cell via a Na^+-dependent amino acid transporter. Since AIB release was very sensitive to the Na^+ concentration and/or gradient, AIB may be released via reversal of its uptake transporter. Additionally, inhibition of both glycolysis (by 2-deoxyglucose) and oxidative phosphorylation (by KCN or FCCP) led to a marked inhibition of AIB and NA-ASC release. Since release of NA-ASC is very similar to AIB release, it is tempting to speculate that a similar type of mechanism may also be operating for NA-ASC release (*i.e.,* reversal of its uptake transporter). We are currently investigating this possibility as well as the possibility that it is simply being released via an anion channel or another transporter.

Finally, although ascorbic acid has long been recognized for its putative role as cofactor in β-hydroxylation, only recently has it been appreciated for its potential as a regulator of neuronal activity. Ascorbic acid has been found to modify neurotransmitter receptor binding to neostriatal membranes[27,28] and to alter the firing rate in nigrostriatal neurons.[29] There may be additional regulatory functions for ascorbic acid in the CNS, as suggested by its release from various neuronal tissues upon electrical stimulation or K^+-depolarization.[30] In order to better understand the functional roles of ascorbic acid, it is important to have a good understanding of its compartmentalization within neuronal cells and also the mechanism by which it is released. The chromaffin cell affords an excellent model system for studying both the compartmentalization and release of ascorbic acid.

REFERENCES

1. DILIBERTO, E. J., G. D. HECKMAN & A. J. DANIELS. 1983. J. Biol. Chem. **258:** 12886-12894.
2. LEVINE, M. & H. B. POLLARD. 1983. FEBS Lett. **158:** 134-138.
3. DANIELS, A. J., G. DEAN, O. H. VIVEROS & E. J. DILIBERTO. 1982. Science **216:** 737-739.
4. DANIELS, A. J., G. DEAN, O. H. VIVEROS & E. J. DILIBERTO. 1983. Mol. Pharmacol. **23:** 437-444.
5. LEVINE, M., A. ASHER, H. POLLARD & O. ZINDER. 1983. J. Biol. Chem. **258:** 13111-13115.
6. WILSON, S. P. & O. H. VIVEROS. 1981. Exp. Cell. Res. **133:** 159-169.
7. JOHNSON, C.D. & R. L. RUSSELL. 1975. Anal. Biochem. **64:** 229-238.
8. HAMPTON, R. Y. & R. W. HOLZ. 1983. J. Cell Biol. **96:** 1082-1088.
9. ANTON, A. H. & D. F. SAYRE. 1962. J. Pharmacol. Exp. Ther. **138:** 360-375.
10. DUNN, L. A. & R. W. HOLZ. 1983. J. Biol. Chem. **258:** 4989-4993.
11. WILSON, S. P. & N. KIRSHNER. 1983. J. Biol. Chem. **258:** 4994-5000.
12. BROOKS, J. C. & S. TREML. 1983. J. Neurochem. **40:** 468-473.
13. KNIGHT, D. E. & P. F. BAKER. 1982. Membr. Biol. **68:** 107-140.
14. MIZOBE, F. & B. G. LIVETT. 1983. J. Neurosci. **3:** 871-876.
15. VIVEROS, O. H., A. J. DANIELS & E. J. DILIBERTO. 1984. *In* Coexistence of Neuroactive Substances in Neurons, Vol. 19. V. Chan-Palay & S. Palay, Eds.: 305-323.
16. ORREGO, F., J. JANKELEVICH, L. CERUTI & E. FERRERA. 1974. Nature **251:** 55-56.
17. VARGAS, O., & F. ORREGO. 1976. J. Neurochem. **26:** 31-34.
18. CHUBB, I. W. & A. D. SMITH. 1975. Proc. R. Soc. London, Ser. B. **191:** 245-261.
19. SOMOGYI, P., I. W. CHUBB & A. D. SMITH. 1975. Proc. R. Soc. Lond. (Biol.) **191:** 271-283.
20. MIZOBE, G., M. IWAMOTO & B. G. LIVETT. 1984. J. Neurochem. **42:** 1433-1438.
21. WILSON, S. P. & N. KIRSHNER. 1977. Mol. Pharmacol. **13:** 382-385.
22. MIZOBE, F. & M. IWAMOTO. 1983. Biomedical Research. **4:** 543-548.
23. MILLER, T. J. & K. UNSICKER. 1981. Cell Tiss. Res. **4:** 543-548.

24. CARMICHAEL, S. W. 1982. Soc. Neurosci. Abstr. **8:** 409.
25. POCOCK, G. 1983. Mol. Pharmacol. **23:** 671-680.
26. POCOCK, G. 1983. Mol. Pharmacol. **23:** 681-697.
27. HEIKKILA, R. E., F. S. CABBAT & L. MANZINO. 1982. J. Neurochem. **38:** 1000-1006.
28. HEIKKILA, R. E. & F. S. CABBAT. 1983. J. Neurochem. **41:** 1384-1392.
29. GARDINER, T. W., M. ARMSTRONG-JAMES, A. WOODBURN CAAN, R. M. WIGHTMAN & G. V. REBEC. 1985. Brain Res. **344:** 181-185.
30. MILBY, K. H., I. N. MEFFORD, W. CHEY & R. N. ADAMS. 1981. Brain Res. Bull. **7:** 237-242.

Characterization of Calcium-Dependent Chromaffin Granule Binding Proteins

CARL E. CREUTZ, WILLIAM H. MARTIN, WILLIAM J. ZAKS, DEBRA S. DRUST, AND HELEN C. HAMMAN

Department of Pharmacology
University of Virginia
Charlottesville, Virginia 22908

Approximately 20 soluble proteins, called chromobindins, bind to chromaffin granule membranes in the presence of Ca^{2+}.[1,2] We have been characterizing these proteins in order to understand the roles they may play in exocytosis in the chromaffin cell. We report here several recent observations on (1) the ATP dependence of a subgroup of the chromobindins, (2) modulation of the Ca^{2+} dependence of synexin and calelectrin by fatty acids, (3) the aggregation of chromaffin granules by a protein derived from p36, a substrate for the *src* kinase, and (4) the cloning of a cDNA coding for synexin.

Seven chromobindins, ranging in mass from 53 to 59 kDa, appear to form a multisubunit complex that is regulated by ATP as well as Ca^{2+}. This complex, which we have named chromobindin A, is eluted from a Superose 6 gel filtration column as a 725-kDa protein. Electron micrographs of negatively stained chromobindin A indicate the complex is toroidal with an outer diameter of 150 Å and a central hole of 50 Å diameter. The binding of chromobindin A to granule membranes is stimulated by Ca^{2+}, however, Ca^{2+} is not required to maintain binding. The complex is released from the membrane only upon simultaneous treatment with EGTA and ATP. Chromobindin A appears to bind to a specific protein receptor on the membrane. The complex does not bind to liposomes or mitochondria; furthermore, pronase treatment of granule membranes inhibits binding of the complex. At present the function of chromobindin A *in vivo* is not clear, however, the properties of this complex may be partially responsible for the ATP dependence of exocytosis.

We have been comparing the properties of the chromaffin granule aggregating proteins synexin[3] and calelectrin.[4] Although these proteins bind to membranes at low concentrations of Ca^{2+} (1 to 10 μM), they promote membrane aggregation and fusion at seemingly supraphysiological levels of Ca^{2+} (> 10 μM). We have recently found, however, that the addition of free, *cis*-unsaturated fatty acids to model systems involving these proteins and chromaffin granules or granule membranes results in enhanced Ca^{2+} sensitivity of protein binding (FIG. 1A), membrane aggregation (FIG. 1C), or granule fusion (FIG. 1B). Therefore, model systems that include these naturally occurring fatty acids come close to reproducing the Ca^{2+} dependency of exocytosis seen in permeabilized cell models.[5]

One of the granule-binding proteins cross-reacts with antisera to p36, a prominent substrate for the *src* kinase in transformed cells.[6] We isolated the 33-kDa proteolytic

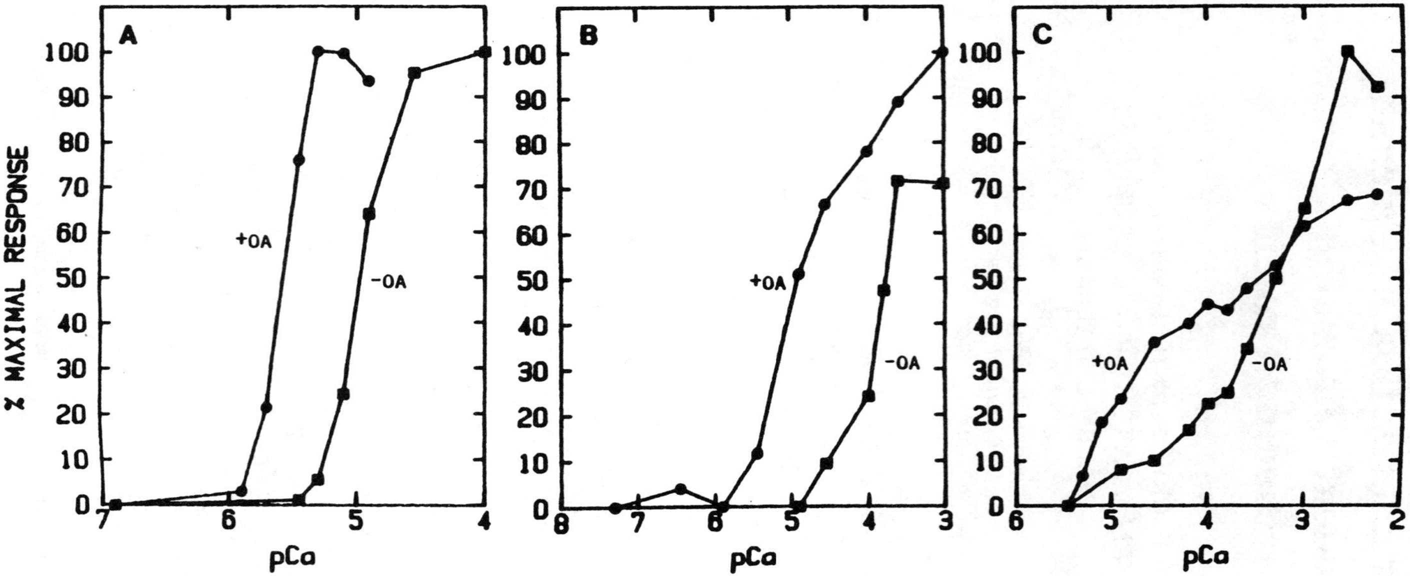

FIGURE 1. Effect of oleic acid (1 μg/μg granule membrane protein) on the Ca^{2+} dependencies of three activities of 32-kDa calelectrin. (**A**) Binding of ^{125}I-labeled calelectrin to chromaffin granule membranes in the presence (+ OA, *closed circles*) or absence (− OA, *closed squares*) of oleic acid. (**B**) Fusion of intact chromaffin granules induced by calelectrin in the presence or absence of oleic acid. Fusion was monitored by measuring the dequenching of the fluorescence probe octadecyl rhodamine after an 11.7 min incubation (Hoekstra *et al.* 1984. Biochemistry **23:** 5675-5681). (**C**) Aggregation of chromaffin granule membranes induced by calelectrin in the presence or absence of oleic acid. Aggregation was monitored by measuring the turbidity (A540) increase in a membrane suspension after a 6-min incubation.

"core"[7] of p36 from bovine intestinal epithelia in order to examine its interaction with chromaffin granules. This core protein exhibited a synexin-like activity, *i.e.*, it aggregated isolated chromaffin granules in a Ca^{2+}-dependent manner (FIG. 2). At pH 6.0, half-maximal granule aggregation was promoted by 8.7 μM Ca^{2+}, a considerably lower concentration of Ca^{2+} than the 200 μM required to promote half-maximal granule aggregation by synexin or calelectrin in the absence of free fatty acids.

We have recently initiated a study of the primary structure of synexin. The purified protein was degraded by cyanogen bromide cleavage and the resulting peptides isolated by HPLC. A partial sequence of 44 amino acids was obtained by automated Edman degradation. Two 29-base oligonucleotide probes were constructed that code for portions of the sequence and were used to screen 200,000 plaques from a bovine adrenal

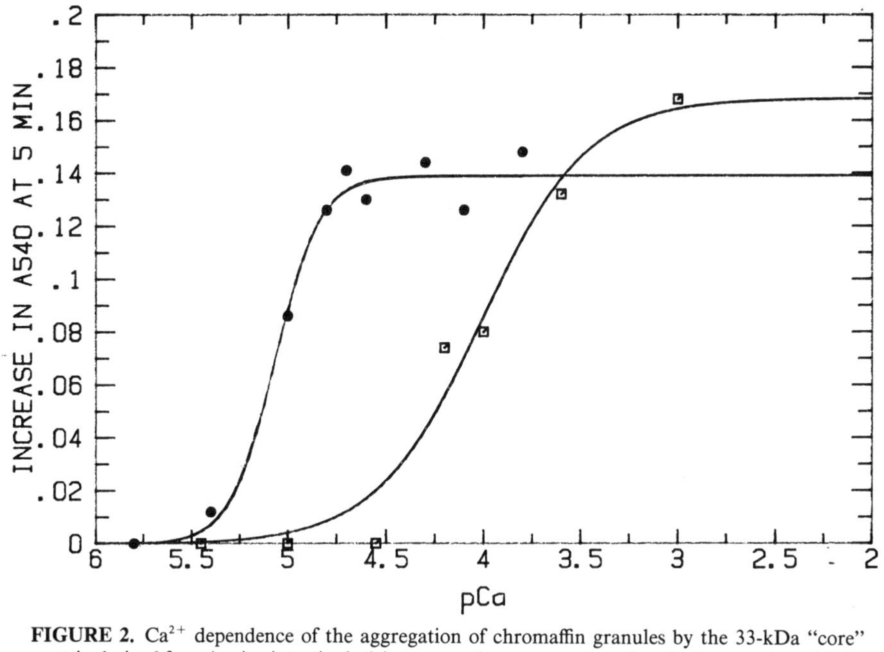

FIGURE 2. Ca^{2+} dependence of the aggregation of chromaffin granules by the 33-kDa "core" protein derived from bovine intestinal p36. Aggregation was monitored as the increase in turbidity (A540) of a granule suspension after a 5-min incubation.[3] *Solid circles,* incubation at pH 6.0; *open squares,* incubation at pH 7.0.

medullary cDNA library in λgt10. Four clones were purified that hybridize with both probes. All four were found to contain a six-kilobase cDNA insert, suggesting that the mRNA for synexin has approximately four times the coding capacity necessary for this protein of mass 47 kDa.

REFERENCES

1. CREUTZ, C. E., L. G. DOWLING, J. J. SANDO, C. VILLAR-PALASI, J. H. WHIPPLE & W. J. ZAKS. 1983. J. Biol. Chem. **258:** 14664-14674.

2. GEISOW, M. J. & R. D. BURGOYNE. 1982. J. Neurochem. **38:** 1735-1741.
3. CREUTZ, C. E., C. J. PAZOLES & H. B. POLLARD. 1978. J. Biol. Chem. **253:** 2858-2866.
4. SUDHOF, T. C., U. EBBECKE, J. H. WALKER, V. FRITSCHE & C. BOUSTEAD. 1984. Biochemistry **23:** 1103-1109.
5. KNIGHT, D. E. & P. F. BAKER. 1982. J. Membr. Biol. **68:** 107-140.
6. GEISOW, M., J. CHILDS, B. DASH, A. HARRIS, G. PANAYOTOU, T. SUDHOF & J. H. WALKER. 1984. EMBO J. **3:** 2969-2974.
7. GLENNEY, J. 1986. J. Biol. Chem. **261:** 7247-7252.

SVP38: A Synaptic Vesicle Protein Whose Appearance Correlates Closely with Synaptogenesis in the Rat Nervous System

STEPHEN H. DEVOTO AND COLIN J. BARNSTABLE

The Rockefeller University
New York, New York 10021

Molecules specifically associated with synaptic vesicles are obvious candidates to carry out some of the specialized functions of synaptic vesicles. Several proteins of the synaptic vesicle have been identified, however, none of them have a proven function. Synapsin I was identified as a major phosphoprotein of the synaptic vesicle and may regulate the availability of vesicles for release.[1] Other vesicle proteins have been identified with the aid of monoclonal antibodies, including a 65,000-Da transmembrane protein (p65),[2] and a 100,000-Da transmembrane glycoprotein.[3] We and others have characterized a 38,000-Da synaptic vesicle protein that has been named p38,[4] synaptophysin,[5] and SVP38.[6]

The monoclonal antibody we produced against this protein stained synaptic structures throughout the rat nervous system. Subcellular fractionation of homogenized rat cerebral cortex, based on established procedures[7] (FIG. 1A), was undertaken to identify the location of SVP38 within the cell. The results of this fractionation showed the antigen to be consistently enriched in the fractions known to contain synaptic vesicles (FIG. 1B). The fractions from the controlled pore glass chromatography, the last step in the isolation of synaptic vesicles, were analyzed by immunoblotting for SVP38 and synapsin I. Both proteins showed the same elution profile (FIG. 1C), demonstrating that SVP38 is also a synaptic vesicle protein.

One approach to understanding the function of synaptic vesicle proteins is to compare their developmental appearance with the appearance of other features of presynaptic function. Molecules expressed only subsequent to synapse formation can be expected to carry out different functions from those also expressed in the growth cone prior to synapse formation.

We have used a dot immunobinding assay to analyze the developmental expression in the cerebral cortex of three synaptic vesicle proteins: synapsin I, p65, and SVP38. These data are shown in FIGURE 2, along with comparable information on synapse formation in the cortex, as measured using the electron microscope.[8] SVP38 levels increase in parallel with the formation of synapses, while synapsin I and p65 are expressed earlier and do not increase to the same extent during synaptogenesis. Since the electron microscope analysis of synapse formation relied on the appearance of the electron-dense postsynaptic density, it is possible that the initial stages of synapse formation precede the expression of SVP38.

The difference in developmental expression between synapsin I, p65, and SVP38

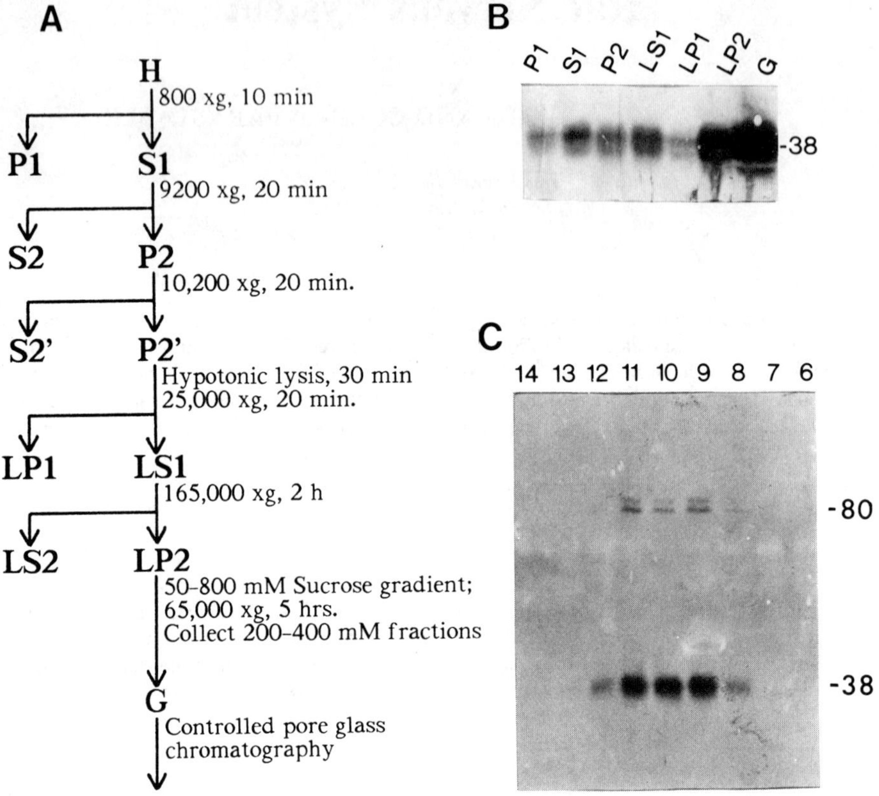

FIGURE 1. **(A)** Schematic illustration of procedure used for subcellular fractionation. **(B)** Various of the fractions were separated electrophoretically on a 0.1% SDS, 10% polyacrylamide gel, transblotted onto nitrocellulose paper and probed for SVP38 with a monoclonal antibody, followed by a peroxidase conjugated secondary antibody and visualization with 3,3′ diaminobenzidine. **(C)** Fractions from the controlled pore glass column, numbered at the top, separated on a gel as in **B**, and immunoblotted for both synapsin I and for SVP38.

implies that the molecular composition of vesicles in the brain changes during development. Vesicles within the growth cone seem to utilize synapsin I and p65 without requiring SVP38, whereas synaptic vesicles utilize all three proteins. SVP38 may be the molecule conferring on a synaptic vesicle the properties that distinguish it from membrane vesicles found at other stages of development.

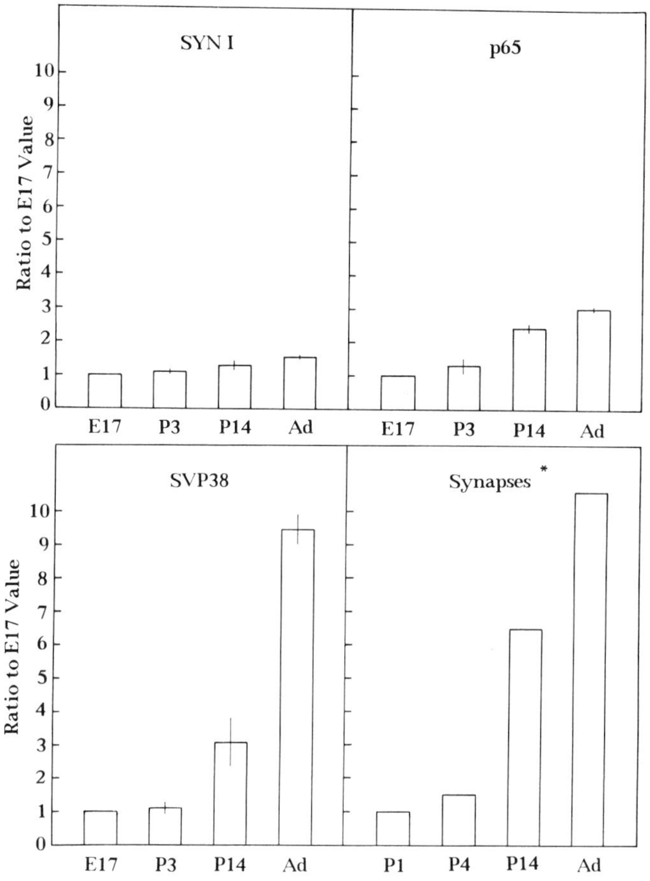

FIGURE 2. Each protein was monitored with a dot-immunobinding assay.[4] Antibodies to synapsin I and p65 were kindly provided by Dr. P. Greengard (Rockefeller University) and Dr. L. Reichardt (UCSF), respectively. The data are expressed as the ratio of the amount found in equal amounts of the S1 fraction from cortex on postnatal day 3 (P3), P14, and adult (Ad) to the amount found in forebrain on embryonic day 17 (E17). *Synapses** refers to the number of electron-dense postsynaptic densities, as measured using the electron microscope.[8] Their data are redrawn as the ratio of the number of synapses at P4, P14, and adult to the number at P1, in order to compare them to the figures for the three synaptic vesicle proteins.

REFERENCES

1. Llinas, R., T. L. McGuinness, C. S. Leonard, M. Sugimori & P. Greengard. 1985. Proc. Natl. Acad. Sci. U.S.A. **82:** 3035-3039.
2. Matthew, W. D., L. Tsavaler & L. F. Reichardt. 1981. J. Cell Biol. **91:** 257-269.
3. Buckley, K. & R. B. Kelly. 1985. J. Cell Biol. **100:** 1284-1294.
4. Jahn, R., W. Schiebler, C. Ouimet & P. Greengard. 1985. Proc. Natl. Acad. Sci. U.S.A. **82:** 4137-4141.
5. Wiedenmann, B. & W. W. Franke. 1985. Cell **41:** 1017-1028.
6. Devoto, S. H., K. A. Akagawa, D. Hicks & C. J. Barnstable. Submitted for publication.
7. Huttner, W. B., W. Schiebler, P. Greengard & P. De Camilli. 1983. J. Cell Biol. **96:** 1374-1388.
8. Armstrong-James, M. & R. Johnson. 1970. Z. Zellforsch. **110:** 559-568.

Biochemical and Immunocytochemical Characterization of p38, an Integral Membrane Glycoprotein of Small Synaptic Vesicles

R. JAHN,[a] F. NAVONE,[a] P. GREENGARD, AND P. DE CAMILLI[b]

[a]*The Rockefeller University, New York, New York 10021*

[b]*CNR Center of Cytopharmacology and Department of Medical Pharmacology University of Milan, Milan, Italy*

A major protein component of small synaptic vesicles, p38, has recently been partially characterized.[1-3] We have now carried out a more detailed analysis of this protein. We have purified p38 from rat brain homogenate 313-fold to virtual homogeneity using affinity chromatography involving an immobilized monoclonal antibody (TABLE 1). Typically, yields of 3-4 mg of protein were obtained from 30 rat brains. It was determined that p38 constitutes 0.3% of the total protein of rat forebrain homogenate and 6-8% of the total protein of highly purified synaptic vesicles. In a PAS reaction p38 was positive, but did not bind Concanavalin A, indicating that it contains carbohydrate side chains of the complex type. After deglycosylation with trifluoromethanesulfonic acid, p38 exhibited a reduced molecular weight of approximately 34,000. A survey of several species by an immunoblotting procedure using various monoclonal and polyclonal antibodies revealed cross-reacting proteins of similar molecular weight in the nervous system of all vertebrate classes (FIG. 1) and in some invertebrate species (not shown). Not all of the antibodies, however, reacted with each species tested.

Immunocytochemistry at the light and electron microscopic level revealed that p38 is present at virtually all synapses throughout the central and peripheral nervous system, where it is selectively associated with small synaptic vesicles. Large dense-core vesicles were consistently unlabeled. The findings that at least two major components, one extrinsic (synapsin I[4]) and one intrinsic (p38, present study), of the membrane of small synaptic vesicles are shared by small synaptic vesicles of virtually all neurons, but not by large dense-core vesicles, support the idea that small synaptic vesicles and large dense-core vesicles are organelles of two distinct pathways for regulated neuronal secretion.

A tissue survey revealed that in addition to the nervous system, p38 is present in a variety of endocrine cells and cell lines (chromaffin cells of the adrenal, endocrine

TABLE 1. Purification of p38 from 15 Rat Brains

Fraction	Total Protein (mg)	p38 (mg)	Enrichment
Homogenate	1995	6.38	1
Low-speed supernatant	1100	2.99	0.9
High-speed pellet	822	3.01	1.2
Triton pellet	284	0.77	0.8
Triton extract	497	1.78	1.1
Affinity chromatography flow-through	484	0.17	0.1
Affinity chromatography eluate	1.62	1.62	313

Assay of p38 was as described.[1] The homogenate was centrifuged at 800 ×g for 10 min. The resulting "low-speed supernatant" was spun at 100,000 ×g for 90 min. The resulting "high-speed pellet" was extracted with 100 ml of phosphate-buffered saline containing 1% Triton X-100. After another spin at 100,000 ×g for 90 min, the resulting pellet ("Triton pellet") was discarded, and the supernatant ("Triton extract") was used for affinity chromatography. As can be seen from the table, virtually all of the purification was achieved by the affinity chromatography step.

pancreas, pituicytes, neuroendocrine cells of the gut, C cells of the thyroid, PC 12 cells, GH_4 cells) where it is localized on smooth-surfaced clear vesicles of variable size that are scattered throughout the cytoplasm and concentrated in the Golgi area. No labeling on secretory granules could be found. In all endocrine systems, p38 exhibited a somewhat higher apparent molecular weight than in the brain. In order to analyze whether this difference is due to a difference in the protein core or in the

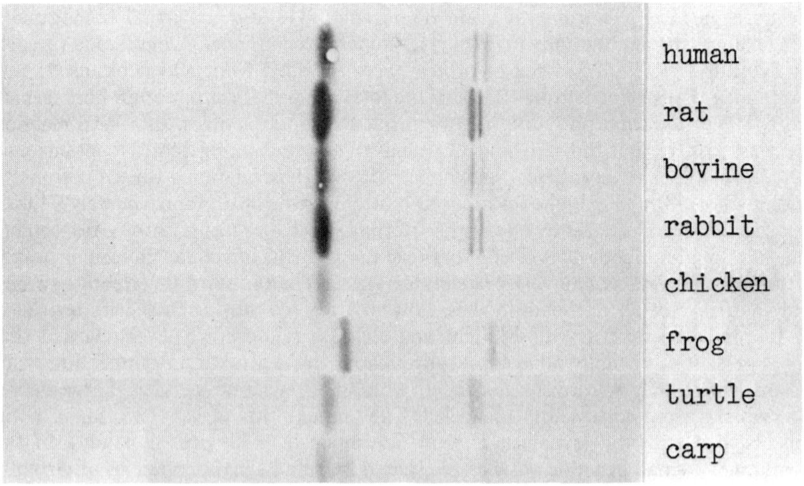

FIGURE 1. Distribution of p38 in the nervous system of different species. Brain homogenates were separated on a 7-15% linear gradient SDS polyacrylamide gel (10 μg/lane) and assayed for the presence of p38 by immunoblot analysis. The blot was labeled with monoclonal antibody C 7.2[1] and subsequently labeled with a monoclonal antibody against synapsin I (M 10.22) for comparison.

sugar part of the molecule, p38 was partially purified from PC 12 cells and deglycosylated with trifluoromethanesulfonic acid in parallel with purified brain p38. At the end of the deglycosylation reaction the two proteins exhibited apparently identical molecular weights, indicating that the difference between the two proteins is due to a difference in the sugar part of the molecule. Our results indicate that an endomembrane system related to small synaptic vesicles of neurons is present in a variety of endocrine cells.

REFERENCES

1. JAHN, R., W. SCHIEBLER, C. OUIMET & P. GREENGARD. 1985. Proc. Natl. Acad. Sci. U.S.A. **82:** 4137-4141.
2. WIEDENMANN, B. & W. W. FRANKE. 1985. Cell **41:** 1017-1028.
3. REHM, H., B. WIEDENMANN & H. BETZ. 1986. EMBO J. **5:** 535-541.
4. NAVONE, F., P. GREENGARD & P. DE CAMILLI. 1984. Science **226:** 1209-1211.

Synaptophysin, an Integral Membrane Protein of Vesicles Present in Normal and Neoplastic Neuroendocrine Cells

BERTRAM WIEDENMANN,[a] HUBERT REHM,[b] AND WERNER W. FRANKE[c]

[a]*Department of Internal Medicine*
[b]*Center of Molecular Biology*
University of Heidelberg
and
[c]*Institute of Cell and Tumor Biology*
German Cancer Research Center
D-6900 Heidelberg, Federal Republic of Germany

Synaptophysin is a glycosylated polypeptide of M_r 38,000 that is an integral component of presynaptic vesicles of a variety of mammalian species detected by immunocytochemistry and immunoblotting using murine monoclonal antibody SY 38[1] or other antibodies specific for this protein.[2] This *N*-glycosylated protein, which has been reported to present a Ca^{2+}-binding site on its cytoplasmic domain[3] also exists in a variety of normal neuroendocrine tissues of both the neuronal and epithelial differentiation lines (*e.g.,* adrenal medulla, inner and outer plexiform layer of the retina, various brain structures, enterochromaffin cells and pancreatic islets, for details see Wiedenmann and Franke[1]). Correspondingly, synaptophysin was also found in various endocrine tumors of both neural and epithelial derivation, including pheochromocytomas, paragangliomas and neuroblastomas, bronchial and gastrointestinal carcinoids, islet cell carcinomas, medullary thyroid carcinomas, and at least some small cell carcinomas of the lung.[4] Metastases of such tumors have also been found to express synaptophysin as determined by immunofluorescence microscopy and immunoblotting of proteins from cell fractions or total tissue (unpublished data). For example, FIGURE 1 presents the specific detection of synaptophysin in a lymph node metastasis of a medullary carcinoma of the thyroid by immunofluorescence microscopy with SY 38. These findings indicate that synaptophysin is a marker of neuroendocrine differentiation which continues to be expressed during malignant transformation and growth as well as during metastasis.

Synaptophysin is also present in a certain type of neuroendocrine vesicles present in permanently growing neuroendocrine cells such as those of the rat cell line PC 12, which is believed to be derived from a rat pheochromocytoma, but is unusual in that it coexpresses intermediate filaments of both the neuronal and the epithelial type, *i.e.,* neurofilaments and cytokeratin filaments.[5] In such cell cultures, synaptophysin is seen in a typical punctate pattern throughout most of the cytoplasm (FIG. 2a). Immunoelectron microscopy (FIG. 2b) with SY 38, using colloidal gold labeling, identifies

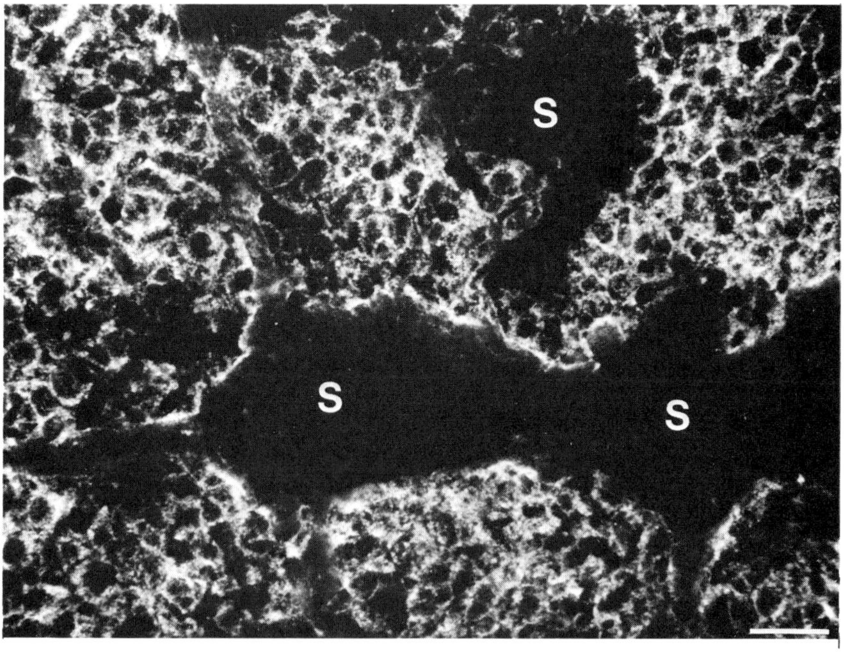

FIGURE 1. Identification of synaptophysin in a frozen section of a lymph node metastasis of a medullary thyroid carcinoma, using immunofluorescence microscopy with monoclonal antibody SY 38 (for methods see Wiedenmann & Franke[1]). Tumor cells are intensely stained; stromal (*S*) cells are unstained. *Bar*=25 μm.

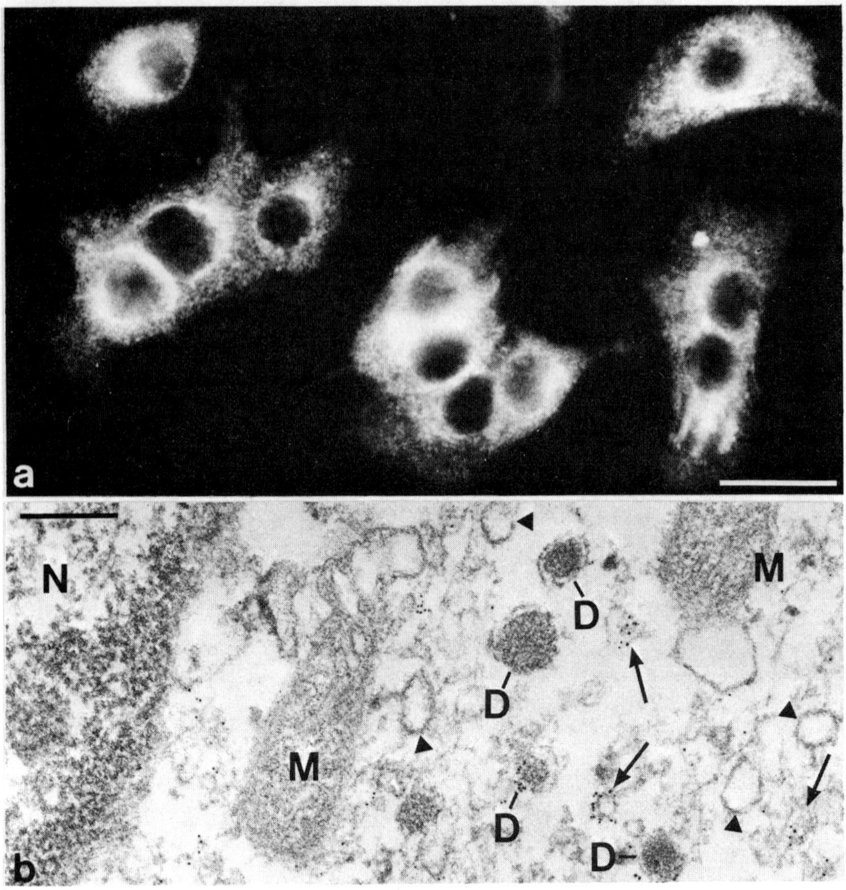

FIGURE 2. Identification of synaptophysin in cultured rat PC 12 cells. (**a**) Immunofluorescence microscopy with antibody SY 38 (for details see Wiedenmann *et al.*[4]). *Bars*=20 μm. (**b**) Immunoelectron microscopy with SY 38. Specific antibody complexes were visualized with a secondary goat anti-mouse antibody coupled to colloidal gold particles of 5 nm diameter. *N,* nucleus; *D,* dense core vesicles; *M,* mitochondria; *arrowheads* denote a type of larger vesicles with an empty-appearing content. *Arrows* denote small, electron-translucent, *i.e.,* "empty-looking" vesicles that are labeled with colloidal gold particles. *Bar*=0.25 μm.

synaptophysin in association with frequent small vesicles, which are very similar in transparency and size to the presynaptic vesicles.[1] Synaptophysin labeling was not detected in another type of neuroendocrine vesicles, the "dense core" vesicles (FIG. 2b).

In summary, our data show that synaptophysin is expressed in a certain category of vesicles that are present in a wide variety of normal and neoplastic cells and can also be formed in cultured cells. Isolation and biochemical characterization of these synaptophysin-containing vesicles should help in our understanding of the pathway of neuroendocrine storage and secretion marked by synaptophysin.

REFERENCES

1. WIEDENMANN, B. & W. W. FRANKE. 1985. Cell **45:** 1017-1028.
2. JAHN, B., W. SCHIEBLER, C. OUIMET & P. GREENGARD. 1985. Proc. Natl. Acad. Sci. U.S.A. **82:** 4137-4141.
3. REHM, H., B. WIEDENMANN & H. BETZ. 1986. EMBO J. **55:** 535-541.
4. WIEDENMANN, B., W. W. FRANKE, C. KUHN, R. MOLL & V. E. GOULD. 1986. Proc. Natl. Acad. Sci. U.S.A. **83:** 3500-3504.
5. FRANKE, W. W., CH. GRUND & T. ACHTSTÄTTER. 1986. J. Cell Biol. **103:** 933-943.

PART VI. EXOCYTOSIS FROM THE PERSPECTIVE OF THE SECRETORY VESICLE

Exocytosis from the Vesicle Viewpoint: An Overview

D. E. KNIGHT AND P. F. BAKER

Medical Research Council
Secretory Mechanisms Group, Department of Physiology
Kings College
Strand, London WC2 R2LS, United Kingdom

Exocytosis, the process by which intracellular vesicles fuse with the inner surface of the plasma membrane, is generally thought to follow two pathways—one that is apparently unregulated and thus occurs continuously, and another that is triggered by a second messenger.[1] In each case, exocytosis serves not only to release into the extracellular fluid substances previously stored within vesicles, but also to alter the composition, and hence properties, of the plasma membrane by the incorporation of vesicular membrane. Thus continuous exocytosis provides for a steady turnover of membrane components and release into the extracellular fluid of a variety of proteins including immunoglobulins and albumin. Triggered exocytosis, however, leads to a rapid change in membrane composition, as in the case when antidiuretic hormone acts to increase the water permeability of the distal nephron, or a sudden release into the extracellular fluid of many hormones, enzymes, and neuronal transmitter substances. The existence of both pathways in the same cell raises the question of how the different proteins are sorted.

The evidence so far strongly suggests that proteins destined for exocytosis carry with them an address that determines whether they enter the continuous or triggered pathway. The sorting probably occurs at the level of the Golgi and leads not only to cases where a vesicle can contain several secretory proteins, *e.g.*, chromogranins, dopamine β-hydroxylase, and enkephalins in chromaffin cell vesicles, but also to the cases where the sorted proteins are segregated and packaged into chemically distinct vesicles which, although destined for the triggered pathway, can be selectively released in response to suitable stimuli.[2–4] Smaller molecular weight molecules destined for export are accumulated within vesicles as a result of specific transporters incorporated into the vesicular membrane (*e.g.*, catecholamines, acetylcholine, serotonin, ATP, etc.). The release of transmitter substances in the nervous system may be a particularly interesting example of selective release. There is evidence that many nerves contain two types of vesicle, one containing a conventional transmitter and the other a peptide,[5,6] and it appears that the pattern of nervous impulses may determine which subpopulation of secretory vesicles and their chemically distinct contents is released.

THE EVIDENCE FOR EXOCYTOSIS

The weight of evidence supporting the idea that exocytosis is the mechanism by which secretion occurs seems overwhelming. On the biochemical front there is the observation that all the soluble components of the secretory vesicles are released together into the extracellular fluid,[7] and the correlation between the incorporation of vesicle membrane antigens into the plasma membrane and secretion.[8,9] Electron microscopy, and more recently image-enhanced microscopy, on a variety of preparations including chromaffin cells, mast cells, sea urchin eggs, the neuromuscular junction, and *Paramecium* have captured the various stages of the fusion event and shown the rapid disappearance of the secretory products from the vesicles.[10–13] The most striking of the pharmacological evidence supporting a vesicular role in secretion comes from studies using false transmitters whereby chemically modified transmitter substances, although able to be transported into the cell from the extracellular fluid, fail to be secreted if they are not also accumulated within secretory vesicles.[14] Finally on the electrophysiological front there are the early classical studies of Katz[15] demonstrating unit sizes of miniature end-plate potentials and thus the quantal nature of release of acetylcholine, and more recently electrical capacitance measurements of patches of membrane, which under conditions favoring secretion show signals consistent with addition of new membrane into the patches, the sizes of the additions corresponding to secretory vesicles.[16,17]

In spite of this evidence, however, there still seems to be some debate about whether exocytosis is the whole explanation for acetylcholine secretion. An alternative suggestion is that transmitter from a cytosolic pool can escape directly across the plasma membrane by means of pumps or channels, the pool being maintained by virtue of rapid equilibration with the stores of vesicular acetylcholine.[18] Part of the evidence for this comes from the observation that newly synthesized transmitter (the synthetic machinery for which is thought to be cytosolic) is released in preference to the older transmitter contained within vesicles. Furthermore the existence of some sort of pump or channel in the plasma membrane, as predicted by this model, is strongly supported by a component of acetylcholine release that is not obviously quantal.

Returning to the origin of quanta, according to the channel hypothesis the amount of acetylcholine released into the extracellular fluid should be dependent on the driving force, *i.e.,* the electrochemical gradient. The movement of transmitter through a channel would be expected to be passive, and hence the model predicts that the amount of positively charged acetylcholine passing through the channel would increase at more positive intracellular potentials, *i.e.,* conditions that would increase the electrochemical gradient and so favor acetylcholine efflux. When the intracellular potential is raised so as to trigger transmitter release and the preparation cooled down to slow the rate of secretion, however, the quantal components of the end-plate potential can be seen to be of the same size as the miniature end-plate potentials observed at resting potentials. Furthermore the nonquantal noise component also seems independent of the presynaptic membrane potential.[19] Hence the amount of acetylcholine released does not appear to be dependent on the electrochemical gradient and the data therefore offer no support for the involvement of channels in the release of acetylcholine.

Such data of course cannot be used to argue against the active transport of acetylcholine by pumps, and nonquantal release may well reflect the insertion of pumps into the plasma membrane. Accumulation of acetylcholine into the intracellular vesicles utilizes acetylcholine transporters in their limiting membranes. These transporters act to accumulate acetylcholine from low cytosolic concentrations into high concentrations within the vesicle.[20] Immediately after exocytosis the vesicular membrane, including these pumps, is incorporated into the plasma membrane. The continued operation of these pumps after insertion by exocytosis may give rise to the observed nonquantal component of acetylcholine release. Support for this idea comes from the finding that agents that inhibit the transport of acetylcholine into vesicles also inhibit the nonquantal release.[21] Nevertheless it seems likely that the debate will continue until electrical capacitance measurements are made on acetylcholine-secreting tissues that give an unequivocal measure of the increase in membrane area during secretion, or until measurements are made, convincing to all, of the size of the cytosolic pool of acetylcholine, as the existence of a sizable free pool is central to the non-exocytotic hypothesis.

SECOND MESSENGERS

Continuous or nonregulated exocytosis is, as its name implies, not subject to obvious control, but the experiments designed to investigate this question are still in their infancy.

Triggered secretion, on the other hand, has been more widely studied. Early studies demonstrating the dependence of secretion on extracellular calcium led to the hypothesis that a rise in intracellular calcium triggered exocytosis,[15,22] and this was tested directly both by introducing Ca^{2+} directly into cells and by measuring Ca^{2+} changes associated with secretion by the use of suitable Ca^{2+} sensors, *e.g.*, aequorin, quin-2 and fura-2.[23–28] Although such data suggest a role for Ca^{2+}, other studies reveal an apparent independence of Ca^{2+} on secretion,[17,29,30] or even that a decrease in cytosolic Ca^{2+} is the trigger for secretion.[30,31] But even in the experiments where an increase in Ca^{2+} is shown to activate secretion, the measured Ca^{2+} levels associated with agonist-evoked secretion are often smaller than the levels needed to trigger the same level of secretion when triggered by introducing calcium directly into the cell, *e.g.*, by ionophores. This discrepancy may in part be explained by assuming that agonists cause only a localized increase in cytosolic Ca^{2+} and that the measured average $Ca^{2+}{}_i$ rise is much smaller than the Ca^{2+} level at the site of exocytosis, or perhaps ionophores generate localized high concentrations of Ca^{2+} distant from the exocytotic site. Such explanations seem inadequate, however, when considering the $Ca^{2+}{}_i$ measurements made in platelets.[25,27] Secretion is triggered by ionophores, *e.g.*, ionomycin, when $Ca^{2+}{}_i$ levels are raised into the micromolar range, but is not triggered by submicromolar levels (FIGS. 1a and 1b). The natural agonist thrombin, however, triggers secretion under conditions when Ca^{2+} increases transiently only up to 0.1 μM (FIG. 1c), and in the case of collagen secretion can occur with no detectable increase in $Ca^{2+}{}_i$ (FIG. 1d). Such data therefore strongly suggest that second messengers other than Ca^{2+} may play a role in stimulus-secretion coupling. Likely candidates are cyclic nucleotides, products of PI metabolism, guanosine nucleotides, or molecules as yet undefined.

One approach to determining the intracellular factors controlling secretion is to manipulate experimentally, in a defined way, the chemical composition at the exocytotic site. Triggering of exocytosis by bringing together isolated secretory vesicles and plasma membranes in a defined chemical environment has on the whole proved unsatisfactory. In some cases when the secretory vesicles are fixed to the plasma membrane, the cells can be broken open, and the vesicle plasma membrane fragments exposed to chemical manipulation. In such cases, for instance the cortical granule reaction of the sea urchin egg, exocytosis is triggered by micromolar levels of Ca^{2+}.[32]

Access to the interior of cells can also be achieved by diffusing Ca^{2+} and other solutes into the cytosol, through patch pipettes or from the extracellular fluid after making the plasma membrane permeable. Various techniques have been used to render cells leaky,[33,34] including toxins such as α-toxin and streptolysin O, divalent ion-chelating agents, sendai virus, detergents, ATP, and electropermeabilization. This latter technique[35–38] has some advantages insofar as it is controlled, quick, chemically clean, and is equally effective on all cells in a homogeneous population. Also, as the voltage difference imposed by the electric field across the wall of a sphere is, in the simple case, proportional to the diameter of the sphere, the larger the sphere the smaller the field required to effect dielectric breakdown. It is therefore possible to break down the plasma membrane of a cell while leaving undisturbed the much smaller intra-cellular-membrane-bound organelles. Where measurements and calculations have been made, the pore sizes induced in the plasma membrane by this method are typically 4 nm in diameter.[35]

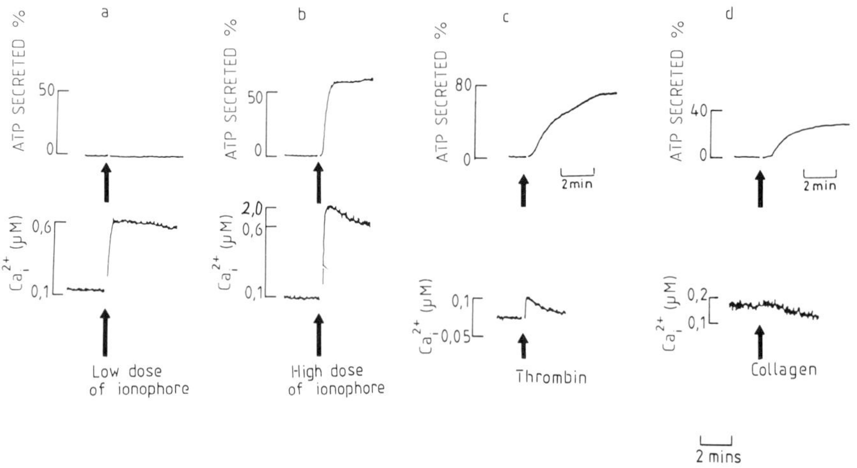

FIGURE 1. Experiments showing the various Ca^{2+} levels necessary to trigger ATP secretion from platelets. **(a,b)** Micromolar levels of Ca^{2+} are necessary when ionophore is used to raise intracellular Ca^{2+}. (From Rink and Hallam.[25]) **(c,d)** The secretory response is triggered at Ca^{2+} levels that are over an order of magnitude smaller when the agonists thrombin or collagen are used. (From Rink *et al.*[27])

By using this technique, calcium buffers can be introduced directly into the cytosol, and FIGURE 2a shows that the secretory products from endocrine, exocrine, neuronal, and paracrine cells are all triggered by micromolar levels of Ca^{2+}, the sensitivity to calcium being constant over the pH range 6.6 to 7.4 and being in close agreement with measurements of Ca^{2+} required to activate secretion induced by ionophores in intact cells. There is, however, no direct evidence that the release of these secretory products into the extracellular fluid is by exocytosis. For example, it could be argued that calcium buffers in leaky cells lead to a destabilization of the secretory vesicles and an unloading of vesicle contents into the cytosol where they would be free to diffuse out into the extracellular fluid through the pores in the plasma membrane. In the case of the chromaffin cells, however, this is unlikely, as using the same biochemical criteria as for exocytosis from intact cells, both the small molecular weight catechol-

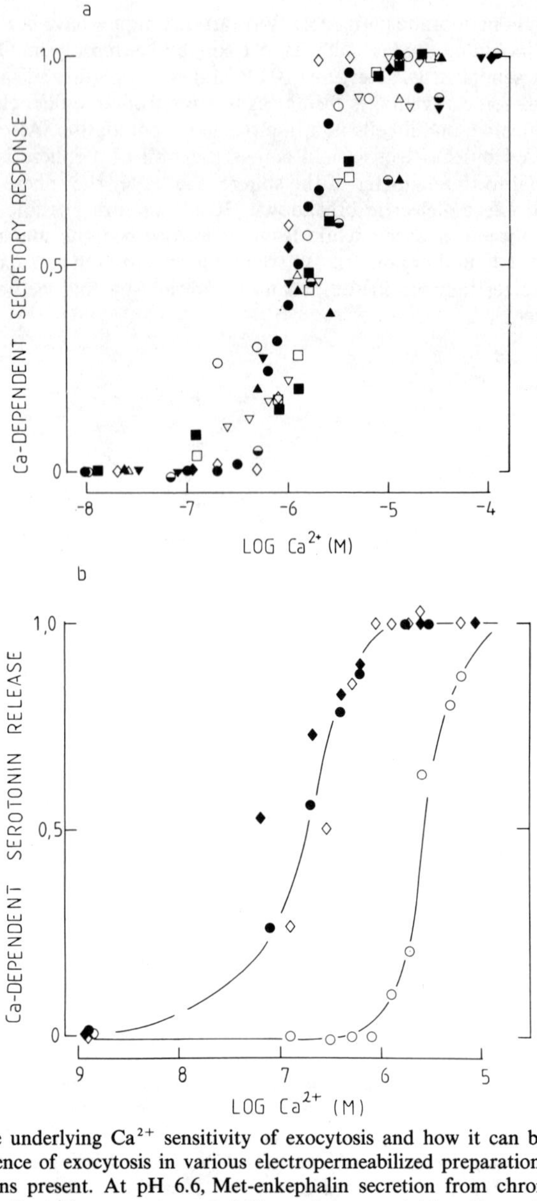

FIGURE 2. The underlying Ca^{2+} sensitivity of exocytosis and how it can be modulated. (**a**) Calcium dependence of exocytosis in various electropermeabilized preparations measured with no other additions present. At pH 6.6, Met-enkephalin secretion from chromaffin cells (◇), insulin from B cells (◆), serotonin and β-N-acetylglucosaminidase from platelets (■, □), acetylcholine and ATP from *Torpedo* synaptosomes (△, ▲), amylase from acinar cells (▽), cortical granule reaction from sea urchin eggs (○), and catecholamine secretion from chromaffin cells measured at pH 6.6 (●), 7.0 (▼), and 7.4 (⊖). (From Knight.[78]) (**b**) The sensitivity to calcium for serotonin release from electropermeabilized platelets is increased by about an order of magnitude when either diacylglycerol (20 μM, ●), GTP-γ-S (50 μM, ◆), or thrombin with GTP (0.6 U/ml and 10 μM, ◇) are present. Platelets in the absence of additions (○). (From Knight and Scrutton[44] and Knight.[78])

amines and the much larger vesicular molecule dopamine β-hydroxylase are released at the same time in response to a calcium challenge and in the same relative proportions as found in the soluble pool of the secretory vesicles, whereas the cytosolic enzyme lactate dehydrogenase escapes from the cell at a much slower rate and in a calcium-independent manner. The same criterion may also be used as evidence for secretion of lysosomal enzyme *N*-acetyl β-glucosaminidase from electropermeabilized platelets, as this enzyme is larger than the cytosolic enzyme lactate dehydrogenase which does not escape from electropermeabilized platelets.

With a few notable exceptions[17,30] other messenger systems introduced into the leaky cell have so far not been effective at triggering secretion in a calcium-independent manner, and it may be that where cyclic nucleotides have been implicated as second messengers, their role may be one of altering the calcium homeostasis of the cell, or perhaps regulating the production of other messenger systems. These other messenger systems seem in the main to modulate the calcium sensitivity of the exocytotic machinery rather than to operate in a calcium-independent manner.

In the case of the electropermeabilized platelet, for example,[38] thrombin increases the apparent affinity of exocytosis for Ca^{2+} to such an extent that secretion takes place at or near to the resting $Ca^{2+}{}_i$ level of the cell.[39,40] Thrombin and collagen trigger phospholipase C activity, so generating diacylglycerol, and the finding that diacylglycerol shifts the calcium activation curve in much the same way as thrombin[41] might explain the secretory response at or near the resting $Ca^{2+}{}_i$ level from intact cells induced by collagen or thrombin. Such data implicate diacylglycerol as a second messenger for secretion (FIG. 2b).

GUANINE NUCLEOTIDE BINDING PROTEINS

A role for GTP has also been implicated in secretion. In leaky platelets GTP enhances the effect of thrombin to increase the Ca^{2+} sensitivity for secretion, and the nonhydrolyzable analogues of GTP increase the calcium sensitivity of secretion in the absence of the agonist, and in much the same way as does diacylglycerol.[42–44] It is likely therefore that the role of guanine nucleotides is via diacylglycerol, as there is evidence of phospholipase C activity being under guanine nucleotide binding protein control in much the same way as the adenylate cyclase system.[45–47] Here therefore, guanine nucleotides can be regarded as permissive rather than as true second messengers.

There is, however, also evidence from chromaffin cells,[48] neutrophils,[29] and platelets[44] that suggest guanine nucleotides might regulate secretion by acting at a site other than at the level of the agonist-induced phospholipase C site, and it has even been suggested that they might act as far down the stimulus secretory pathway as to be at or near the exocytotic site itself.[29,48] In the case of the leaky nicotinic bovine chromaffin cell, the nonhydrolyzable analogue, GTP-γ-S, inhibits calcium-dependent exocytosis—unlike its stimulatory effect on the leaky muscarinic chicken chromaffin cell.[48] The inhibitory effect seems not to be a consequence of inhibiting basal phospholipase activity and hence decreasing endogenous diacylglycerol levels, as exogenously added diacylglycerol or phorbol ester (both of which shift the Ca^{2+} activation curve) fail to overcome the inhibitory effect of GTP-γ-S (FIG. 7b). Furthermore, GTP-γ-S acts to reduce the extent of secretion rather than alter the Ca^{2+} sensitivity or ATP requirement.[48] In the case of the sendai virus permeabilized neutrophil where

GTP-γ-S stimulates secretion, the guanine nucleotide binding protein controlled phospholipase C activity can be selectively blocked by pertussis toxin leaving the other GTP-γ-S stimulatory site unaltered.[29] Lastly in the case of the platelet the data indicates that there may be a target for guanine nucleotides other than at the agonist-mediated phospholipase C site, the evidence being that two quite separate effects of GTP-γ-S occur, *i.e.,* one that shows synergism with the effect of the agonist thrombin, and another that seems to be simply additive and independent of the effect of the agonist.[44] In the case of the adenylate cyclase system or phospholipase C systems, it is easy to understand that the G (or N) proteins are part of a signal transduction mechanism linking an extracellular signal to the control of an intracellular enzyme.[49] If the exocytotic site proves to be under G protein control (with the adenylate cyclase system in mind), it is not clear what the nature of the physiological signal is that modulates secretion through this transducer; but it may prove that extracellular agonists utilize this system to exert direct effects on exocytosis.

SITE OF ACTION OF SECOND MESSENGERS

We still know very little about what mechanisms are involved in the membrane fusion step of exocytosis or, at the molecular level, how calcium and diacylglycerol alter the rate and extent of secretion. It is therefore perhaps worthwhile to search for clues from the membrane fusion processes involving viruses, for which something is known at the molecular level. Some viruses have a pH-sensitive spike protein projecting from their membranes. When the viruses are internalized into endocytotic vacuoles and the pH of the endosome decreases, the spike protein undergoes a conformational change revealing a hydrophobic sequence that buries itself into the neighboring wall of the endosome, and this apparently leads to fusion between the viral and endosome membranes.[50] It may be that calcium and diacylglycerol serve to trigger a similar process, revealing a hydrophobic part of a different protein that leads to fusion between vesicular and plasma membranes. Ca^{2+} and H^+ are known to compete at the same sites on a number of calcium-binding proteins,[51] and so it may be no coincidence therefore that when the proton concentration inside leaky cells also approaches 10 μM, secretory products appear in the extracellular fluid. It may be of course simply that at pH 5, proteins are being altered to an extent that we are simply observing a nonspecific release, but it is also worthwhile to explore the possibility that it is exocytosis, and to see if conditions that alter Ca^{2+} dependent secretion also alter H^+-induced release.

THE CYTOSKELETON AND EXOCYTOSIS

The electropermeabilized cell preparation lends itself very well to controlling the chemical environment of the cytosol and thus studying the mechanism of exocytosis. One rather surprising result from such investigations, in view of the popular debate about an involvement of microfilaments, or microtubules, in exocytosis, is that exposure of leaky chromaffin cells to agents that either disrupt the cytoskeleton or prevent its disassembly do not alter the calcium-dependent secretory response.[35] This might of course be because the agents do not affect microfilaments or microtubules *in situ* in the same way they do with isolated filaments or tubules, or alternatively it might be what it seems, *i.e.*, that the cytoskeleton is not intimately involved with exocytosis.

DEPENDENCE OF EXOCYTOSIS ON THE NATURE OF THE MAJOR ANION AND CATION

Another rather interesting finding is that if electropermeabilized cells are exposed to extracellular concentrations of the physiological anion chloride, the calcium-dependent secretory response is inhibited.[35] The effect is not peculiar to chloride as some other anions have similar inhibitory effects, the order of potency following the Hofmeister lyotropic series. This suggests that these anions might be interacting and perturbing a protein or protein matrix essential for secretion. Unlike the effect of certain anions, the secretory response seems independent of the nature of the major cation, the calcium-dependence being unaltered when cells are incubated in either a K- or Na-based medium, or even one based on sucrose and containing only contaminating levels (less than 0.5 mM) of K and Na. This finding offers no support for the idea that Ca-activated monovalent cation channels in the vesicle membrane (similar to those found in the plasma membrane) are involved in the process of exocytosis.

OSMOTIC FORCES AND EXOCYTOSIS

The finding that increasing the osmolarity with sucrose inhibits secretion from electropermeabilized cells[35] supports the idea that fusion of vesicles with the plasma membrane is aided by an osmotic imbalance across the vesicle membrane.[52] Similar findings have been reported for cortical granule reactions in sea urchin eggs.[53] In adrenal medullary cells rendered permeable with digitonin, however,[54] the cells are not inhibited by an increase in osmolarity with sucrose. One possibility here, though, is that whereas sucrose might reduce the osmotic pressure differences across the vesicle wall, the detergent used to permeabilize the cell could also be affecting the surface properties of the vesicle in such a way as to offset the sucrose effect. The idea that the sucrose effect on electropermeabilized cells is a consequence of cell shrinkage due to the sucrose not equilibrating across the plasma membrane[54] is unlikely, as inhibition occurs when sucrose and calcium buffers are incubated over a twenty-five minute period, the time for equilibration of sucrose across the leaky plasma wall being the same as that for the calcium buffers, *i.e.*, about 3 to 5 minutes. Raising the osmolarity with smaller molecular weight solutes, *e.g.*, potassium glutamate or glycine, also inhibits secretion. In the case of potassium glutamate, however, the inhibitory effect may be due to a high ionic strength rather than simply an osmotic effect. In the case of glycine there should be no increase in ionic strength, and, using the same electropermeabilization procedure, the solute should equilibrate across the leaky plasma membrane in much less than a minute. If glycine is not acting at a specific inhibitory site, these data strongly suggest that the inhibition of secretion is a consequence of increasing the osmotic pressure.

If the role of Ca^{2+} in exocytosis is to mediate some essential chemical reaction, then it would be expected that a submaximal Ca^{2+} level would alter the rate of

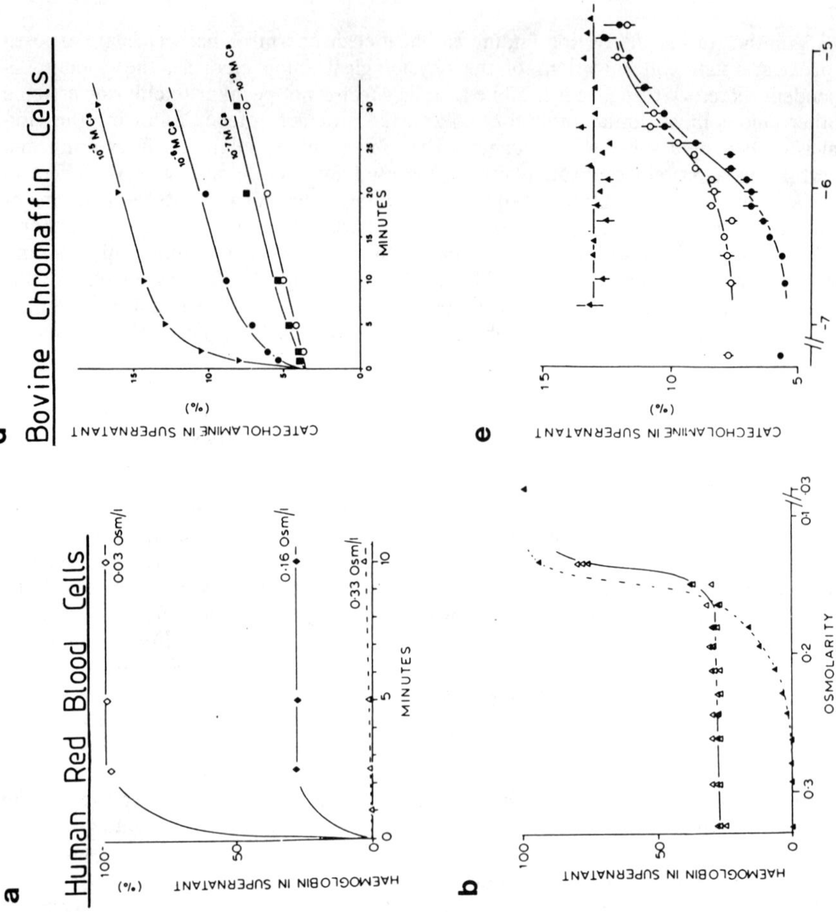
a
Human Red Blood Cells
HAEMOGLOBIN IN SUPERNATANT (%)
0·03 Osm/l
0·16 Osm/l
0·33 Osm/l
MINUTES
b
HAEMOGLOBIN IN SUPERNATANT
OSMOLARITY
d
Bovine Chromaffin Cells
CATECHOLAMINE IN SUPERNATANT (%)
MINUTES
e
CATECHOLAMINE IN SUPERNATANT (%)
LOG Ca2+ (M)

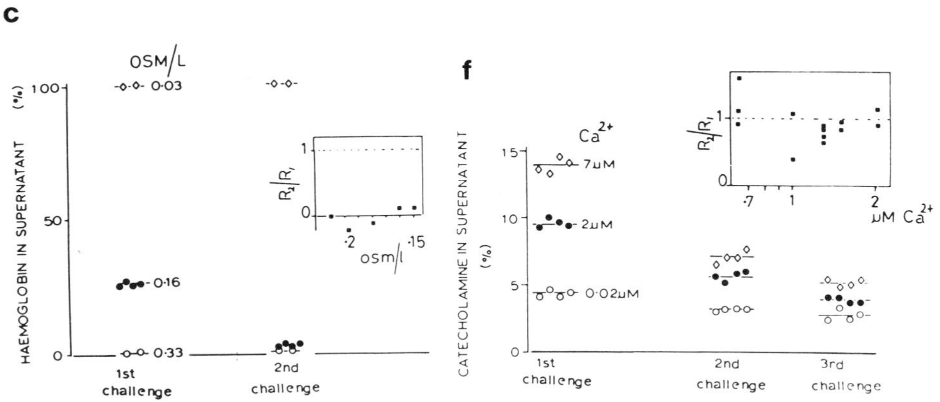

FIGURE 3. Similarities and dissimilarities between the characteristics of hemolysis and secretion. **(a,d)** In both systems a submaximal stimulus leads to a reduction in the extent of hemolysis and secretion rather than simply an effect on the rate. **(b)** Hemolysis curve for red blood cells previously exposed to 0.16 osmolar solutions. Cells previously exposed to hypotonic solutions resulting in about 30% hemolysis were returned to isotonic solution, and then again challenged with hypotonic solutions (△). The osmolarity has to be decreased to at least the level in the first hypotonic challenge before any further hemolysis occurs. Hemolysis curve for control cells not prechallenged with hypotonic solution (*dotted line*). **(c)** Cells initially challenged with 0.16 osmolar solutions (●) giving about 30% hemolysis were washed into isotonic solution of 0.33 osmolar (○), and then given a second 0.16 osmolar challenge (●). The amount hemolysed on this second challenge expressed as a ratio of the amount lysed on the first challenge is shown in the *inset*, and for the case of several hypotonicities. **(e)** Calcium dependence of catecholamine release from cells previously challenged with micromolar levels of Ca^{2+}. Chromaffin cells that had previously been exposed to 1 μM Ca^{2+} releasing about 30% of the available pool of catecholamine were returned to a low Ca^{2+} concentration (10^{-8} M) and then again challenged with various Ca^{2+} concentrations (○). Calcium dependence for control cells not preincubated with micromolar levels of Ca^{2+} (●). The dose response curves for the two conditions are very similar, unlike the case for the curves shown in **b.** **(f)** Cells initially challenged with 2 μM Ca^{2+} (●), giving about 50% release of the available catecholamine pool in the cells, were washed into a solution to reduce the Ca^{2+} down to 0.01 μM and then challenged again with the same Ca^{2+} level (●). The percentage of the available pool released by this second challenge (R_2) is expressed as a ratio of the amount in the first challenge (R_1) and is shown in the *inset* for a range of Ca^{2+} challenges used. (From Knight and Baker.[35])

secretion rather than its extent. This is not observed, however, in two cases examined: the leaky platelet and adrenal medullary cell, where submaximal Ca^{2+} affects the extent of secretion (FIG. 3d). If such a phenomenon also holds true for intact cells, the physiological significance is that a sustained elevated $Ca^{2+}{}_i$ would only give rise to a transient secretory response rather than a sustained one. At a Ca^{2+} concentration that is 50% effective therefore, i.e., approximately 1 μM, only half the vesicles capable of exocytosis do so. Raising the Ca^{2+} further to somewhere in the range of 2 μM allows a few more vesicles to fuse with the plasma membrane. Such data might be explained if a population of vesicles capable of exocytosis had different and distinct thresholds. A sustained submaximal Ca^{2+} challenge would therefore only recruit the part of this population whose thresholds were exceeded. A similar argument could be made for the case where each cell within a suspension of cells had different and distinct thresholds for secretion. Electron microscopy studies, however, do not support this last idea. It is tempting to link these observations with the possible osmotic involvement, and to consider the analogy with red blood cell hemolysis where the cells in suspension have distinct and different osmotic thresholds. A submaximal hypotonic shock causes a fraction of the cells in suspension to lyse, the effect being seen on the extent of hemolysis rather than simply on the rate (FIG. 3a). If the tonicity of the suspension is restored to isotonicity, and then returned to the same hypotonic state as before, no further hemolysis occurs as a result of this second hypotonic challenge (FIGS. 3b and 3c). This is because all the cells with a threshold for hemolysis lower than that corresponding to the first challenge would have already lysed. The relative amount of hemolysis due to the second challenge, compared with that of the first challenge, is always zero (FIG. 3c, inset). Here, however, the similarity between the osmotic fragility system and secretory systems seems to disappear, as chromaffin cells that secrete 50% of their releasable pool in response to a 1 μM Ca^{2+} challenge release 50% of what is left on the second 1 μM Ca^{2+} challenge (FIGS. 3e and 3f). The relative amounts secreted by a second or even third Ca^{2+} challenge is very similar to the relative amounts secreted at the first challenge, the ratio being close to 1 (FIG. 3f, inset).

POSSIBLE ROLES OF MgATP IN EXOCYTOSIS

In several preparations there appears to be general agreement that in addition to exocytosis being activated by micromolar levels of Ca^{2+}, millimolar levels of MgATP are required. In the case of catecholamine secretion from chromaffin cells the requirement for MgATP is specific, the effect being mainly on the extent of calcium-dependent exocytosis rather than on the calcium sensitivity (FIG. 4a). As nonhydrolyzable analogues of ATP are ineffective, the role of MgATP may be one of phosphorylation or hydrolysis.

Phosphorylation

The simplest model to link Ca, ATP, and secretion is one in which Ca alters the rate at which some key protein associated with exocytosis is phosphorylated. In

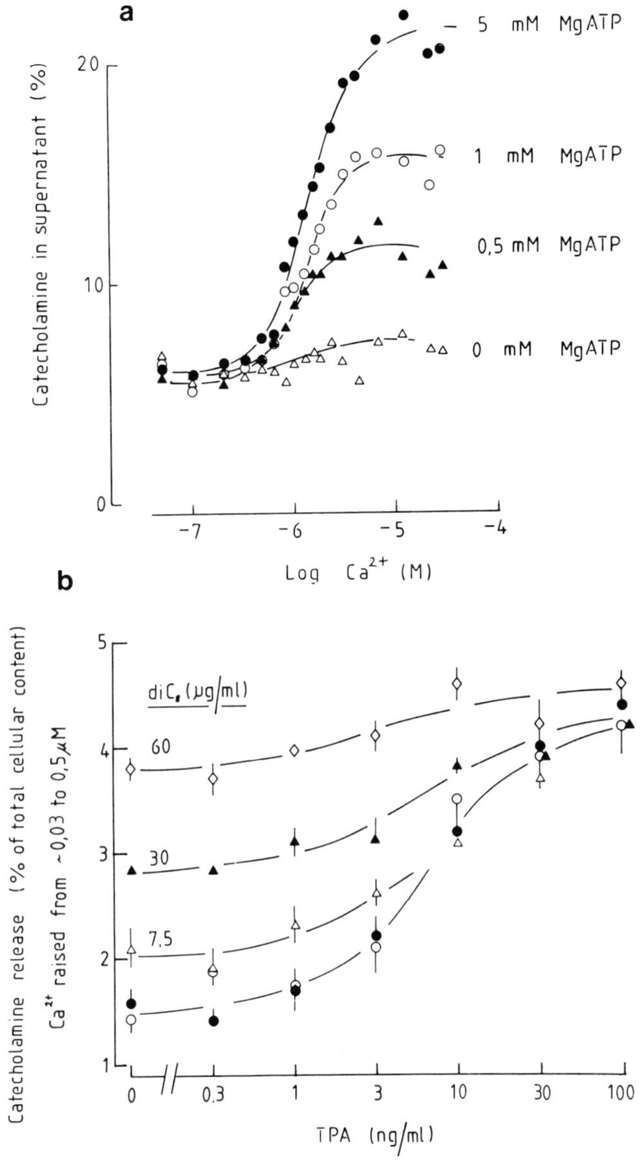

FIGURE 4. **(a)** MgATP dependence of catecholamine secretion from electropermeabilized bovine chromaffin cells. (From Knight and Baker.[35]) **(b)** The effect of the phorbol ester TPA, and the diacylglycerol diotanoylglycerol (diC_8) on the calcium sensitivity of catecholamine secretion. The amount of catecholamine secreted from electropermeabilized bovine chromaffin cells as a result of raising the Ca^{2+} from 0.08 to 0.5 μM increases in the presence of TPA; diC_8 alone has much the same effect but does not potentiate the effect seen by saturating levels of TPA.

electropermeabilized cells, however, where the Ca^{2+} and ATP levels can be defined accurately, phosphorylation studies do not reveal a single phosphorylated protein uniquely associated with secretion, but rather 10 or more easily visible bands—any one of which could be significant. If the data showing the dependence of secretion on ATP is to be fitted to a simple model, it is best fitted to a model in which one ATP molecule is involved in a chemical reaction directly related to exocytosis. The data therefore provide no evidence for cooperative ATP sites or of multiple phosphorylation sites necessary for secretion. If in fact only one phosphorylated step was involved in an exocytotic event, the amount of phosphorylated protein associated with the average level of secretion would probably be too small to be detected by simple experimental techniques. Furthermore, the finding that submaximal Ca^{2+} levels or MgATP levels affect mainly the extent of secretion rather than simply the rate (FIG. 3d) implies, on this simple model, that the rate of change of phosphorylation would be the controlling factor for secretion, rather than the absolute amount of protein phosphorylated. There is evidence that one phosphoprotein, synapsin 1, which seems to be generally distributed in the nervous system (although apparently not so in the chromaffin cell), may play some role in exocytosis.[55,56]

ATPases and the Proton Pump

The role of MgATP may also be one involving hydrolysis. The inwardly directed ATPase proton pump in the vesicle membranes acidifies the interior of the secretory vesicle and generates a positive internal potential with respect to the cytosol.[57] The sole role of ATP in secretion, however, is unlikely to be through operation of this ATPase pump[58] because (1) the pump can operate with a fairly broad nucleotide specificity, whereas secretion seems to be specific for ATP, and (2) the amount of ATP needed to operate the pump maximally is over an order of magnitude less than that needed for secretion. Furthermore, various agents can be used either to block the pump or to enhance or diminish the pump's effect on pH or potential gradient across the vesicle membrane. Under these conditions, the measured potential and pH gradient across the vesicle membrane show little correlation with calcium-dependent secretion[59,60] (FIG. 5).

Protein Kinase C

Another role for MgATP in secretion is as a substrate for the enzyme protein kinase C.[61] This enzyme is activated by calcium and diacylglycerol, the two primary messengers for exocytosis. The involvement of protein kinase C in secretion is further supported by the finding that those phorbol esters, which activate protein kinase C in the same way as do diacylglycerols, can also substitute for diacylglycerol in enhancing the secretory response.[42,62] The effect of a maximal dose of phorbol ester is not further increased by addition of diacylglycerol (FIG. 4b), suggesting that the two chemicals act along the same pathway, possibly at the same site, *i.e.,* protein kinase C. The attraction for the protein kinase C hypothesis is that one single molecule serves as a receptor for Ca^{2+}, diacylglycerol, and ATP. Protein kinase C has been shown to bind to chromaffin granules in a calcium-dependent manner and to plasma mem-

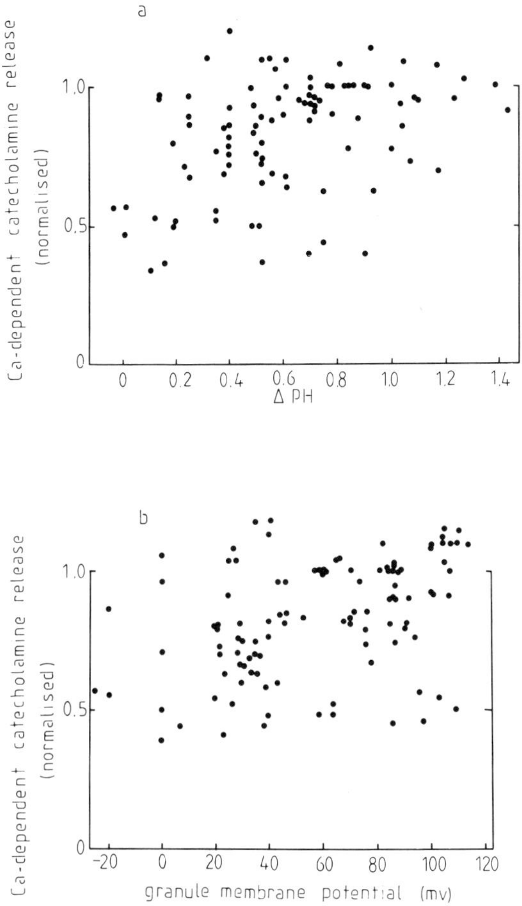

FIGURE 5. Ca^{2+}-evoked secretion from electropermeabilized chromaffin cells is not dependent on (**a**) the pH gradient across the vesicle membrane (inside acidic), or (**b**) the potential difference across the vesicle membrane (inside positive). The pH and potential gradients were measured *in situ* in the electropermeabilized cell. (From Knight and Baker.[60])

branes in a phorbol-ester- and diacylglycerol-dependent manner.[63,64] The activated protein kinase C may therefore operate in a way analogous to a viral spike protein. It seems more likely, however, that it acts through phosphorylation. One such site of phosphorylation is the 40-kDa protein lipocortin found in platelets.[65] In the dephosphorylated form, lipocortin acts to inhibit phospholipase A_2. Phosphorylation removes the inhibition, thereby increasing the products arachidonic acid and lysolecithin. It has been suggested that the liberation of one of these products, *i.e.*, arachidonic acid, may be involved in the mechanism of exocytosis. We have no evidence to support this postulate, however, as (1) under some conditions phosphorylation of the 40-kDa protein is not paralleled with secretion,[66] and (2) addition of arachidonic acid to leaky platelets or adrenal cells in the presence or absence of aspirin or indomethacin fail to alter the Ca activation curve in the presence or absence of diacylglycerol. The data do not rule out, however, the possibility of an involvement of the other product, *i.e.*, lysolecithin, in the mechanism of exocytosis.

SELECTIVE RELEASE

Secretory systems do not all respond to activators of protein kinase C in the same way. In the platelet, for example, they have quite different effects on secretory responses from two distinct populations of vesicles.[66] Thrombin, diacylglycerol, and phorbol esters increase the sensitivity of release of serotonin from amine storage vesicles, whereas they mainly alter the extent of release of acid hydrolases from the lysosomes without markedly altering the calcium sensitivity (FIG. 6). The mechanisms giving rise to these two different effects are intriguing; the observation also offers an explanation for the selective release seen with some agonists. For example, low levels of thrombin activate serotonin release from intact platelets without significant acid hydrolase release, whereas higher levels that cause near maximal serotonin release begin to activate appreciable acid hydrolase release.[3] This can be explained in terms of low concentrations of thrombin producing small amounts of inositoltrisphosphate and diacylglycerol, the intracellular Ca^{2+} being increased slightly by calcium mobilization and the calcium sensitivity for serotonin release being increased so as to activate appreciable secretion at this Ca^{2+} level. The acid hydrolase secretory machinery would not be expected to operate until the $Ca^{2+}{}_i$ was raised further by exposure to higher levels of thrombin.

Although the emphasis at the moment is on a protein kinase C involvement with secretion, other calcium receptors such as calmodulin,[67] calectrins,[68] and synexin[69] are still very much in the running and should not be ignored. Furthermore, the requirement for MgATP in all secretory systems is by no means clear. In toxin permeabilized transformed chromaffin cells, for example, calcium-dependent secretion has been reported to occur in the virtual absence of nucleotides,[70] and the cortical granule reaction from sea urchin egg plaques can be triggered by solutions containing micromolar Ca^{2+} but lacking MgATP. When leaky adrenal medullary cells are incubated in a medium lacking MgATP, the decrease in the calcium-dependent response occurs over minutes and can be correlated with the expected efflux of MgATP from the cytosol. With platelets, however, where the efflux of MgATP occurs in less than a minute, the secretory response still decreases over minutes. At first sight, therefore, these data suggest that calcium-dependent exocytosis can proceed in the absence of cytosolic MgATP and that either there is a bound pool of MgATP or a stable phosphorylated

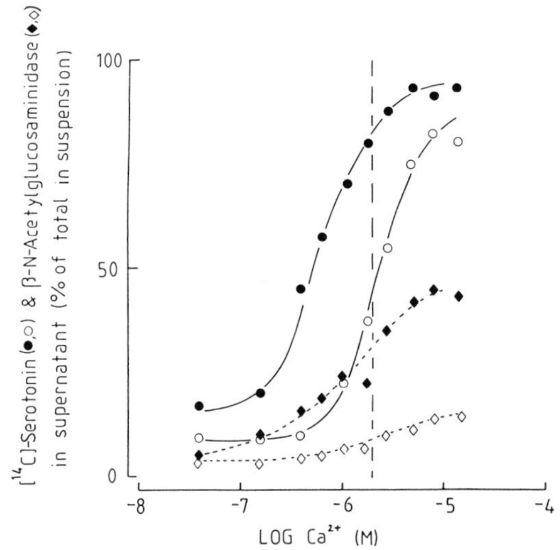

FIGURE 6. The effect of thrombin on the Ca^{2+} dependence of amine storage vesicle (○, ●) and lysosomal (◇, ◆) secretion from electropermeabilized platelets in the presence (*closed symbols*) or absence (*open symbols*) of 0.6 U/ml thrombin. The Ca^{2+} concentration, when added alone, needed to activate half-maximal secretion of both lysosomal and amine secretion is close to 2 μM and is shown by the *dotted line.* In the presence of thrombin, diacylglycerol, or TPA, however, the Ca^{2+} sensitivity for lysosomal secretion does not alter appreciably, unlike that for amine release. (From Knight *et al.*[66])

intermediary, or that maybe the role of MgATP is not one of phosphorylation or hydrolysis, but simply of conferring stability to the system.

TOXINS AND EXOCYTOSIS

In many biological systems, toxins have been used as powerful tools to elucidate mechanism, and it may also be the case with exocytosis that toxins might direct us to physiologically relevant molecules. α-Latrotoxin from the venom of black widow spider causes a dramatic stimulation of exocytosis from cholinergic synapses,[71] whereas Botulinum toxins from *Clostridium botulinus* bacteria block cholinergic transmission.[72] The recent finding that one of the botulinum toxins, botulinum D, also blocks catecholamine secretion from bovine chromaffin cells by acting downstream of the Ca^{2+} transient at or near the site of exocytosis (FIG. 7b) promises to be a very useful tool.[73] It may turn out that the inhibitory effects of botulinum toxin D and GTP-γ-S take place by a similar mechanism, as do cholera toxin and GTP-γ-S effects on adenylate cyclase. To date, however, we have found no evidence of ADPribosylation induced by botulinum toxin type D in chromaffin cells.

Although the electropermeabilized cell preparations have allowed some detailed studies on secretion and can be likened to the skinned muscle preparation for studies of contraction, it has some limitations that should not be overlooked. First, by virtue

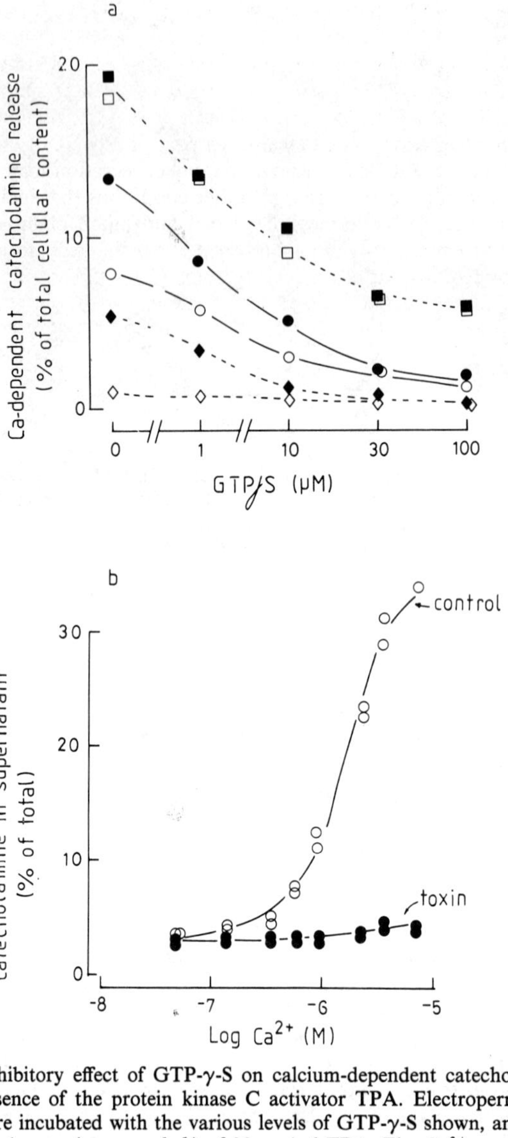

FIGURE 7. (**a**) Inhibitory effect of GTP-γ-S on calcium-dependent catecholamine release in the presence or absence of the protein kinase C activator TPA. Electropermeabilized bovine chromaffin cells were incubated with the various levels of GTP-γ-S shown, and in the presence (*closed symbols*) or absence (*open symbols*) of 20 ng/ml TPA. The Ca^{2+} was then raised from 0.01 μM to 0.4 μM (◇,◆), to 1 μM (○, ●), or to 10 μM (□, ■), and the catecholamine secreted as a result of this Ca^{2+} increase measured. (**b**) Inhibition of calcium-dependent catecholamine secretion from electropermeabilized bovine chromaffin cells pretreated with botulinum toxin type D. (From Knight *et al.*[73])

of accessing the interior of the cell by diffusion through holes in the plasma membrane, the technique unavoidably makes the solutions inside and outside the cell the same. Second, the plasma membrane potential is collapsed. Lastly, whereas we can define what small molecular weight substances go into cells, we cannot control what leaves. It is quite possible therefore than an electropermeabilized cell is not as sophisticated as an intact one. As such we may be characterizing a rather simplified system, and thus at times be distant from physiological conditions. Furthermore, if the secretory event not only involves exocytosis but also, as part of a cycle of events, endocytosis, then studies of the intracellular conditions that affect secretion may not be exclusively the conditions that control exocytosis but rather conditions that include endocytosis.[74] A combination of the electropermeabilized cell and patch clamp techniques might overcome some of these shortcomings and thus allow the exocytotic and endocytotic events to be dissected, and to allow investigation of the effects of plasma membrane potential and of different solutions across the plasma membrane on exocytosis, *e.g.,* investigation of a possible specific role for extracellular calcium in the secretory response.

Finally, other promising avenues of study include (1) extension of the existing technology that allows synthesis within a model cell of foreign secretory proteins[75] and foreign membrane receptors,[76] including the full range of components of the exocytotic apparatus, and (2) the study of secretory mutants.[75] These approaches may begin to provide new insight into the relation between protein structure and the secretory process.

REFERENCES

1. Kelly, R. B. 1985. Science **230:** 25-31.
2. Holmsen, H., C. A. Dangelmaier & H.-K. Holmsen. 1981. J. Biol. Chem. **256:** 9393-9396.
3. Knight, D. E., T. J. Hallam & M. C. Scrutton. 1982. Nature **296:** 256-257.
4. Theoharides, T. C., P. K. Bondy, N. D. Tsakalos & P. W. Askenase. 1982. Nature **297:** 229-231.
5. Lundberg, J. M. & T. Hokfelt. 1983. Trends Neurosci. **6:** 325-333.
6. Bartfai, T. 1985. Trends Pharmacol. Sci. **6:** 331-334.
7. Viveros, O. H., L. Arqueros & N. Kirshner. 1968. Life Sci. **7:** 609-618.
8. Lingg, C., R. Fischer-Colbrie, W. Schmidt & H. Winkler. 1983. Nature **301:** 610-611.
9. Phillips, J. H., K. Burridge, S. P. Wilson & N. Kirschner. 1983. J. Cell Biol. **97:** 1906-1917.
10. Heuser, J. E., T. S. Reese, M. J. Dennis, Y. Jan, L. Jan & L. Evans. 1979. J. Cell Biol. **81:** 275-300.
11. Chandler, D. E. & J. E. Heuser. 1980. J. Cell Biol. **86:** 666-674.
12. Ornberg, R. L. & T. S. Reese. 1981. J. Cell Biol. **90:** 40-54.
13. Schmidt, W., A. Patzak, G. Lingg, H. Winkler & H. Plattner. Eur. J. Cell Biol. **32:** 31-37.
14. Smith, A. D. 1972. Biochem. Soc. Symp. **36:** 103-131.
15. Katz, B. 1969. The Release of Neuronal Transmitter Substances. Liverpool University Press. Liverpool.
16. Neher, E. & A. Marty. 1982. Proc. Natl. Acad. Sci. U.S.A. **79:** 6712-6716.
17. Fernandez, J. M., E. Neher & B. D. Gomperts. 1984. Nature **312:** 453-455.
18. Tauc, L. 1982. Physiol. Rev. **62:** 857-893.
19. Katz, B. & R. Miledi. 1981. Proc. R. Soc. Lond., Ser. B. **212:** 131-137.
20. Parsons, S. M. & R. Koenigsberger. 1980. Proc. Natl. Acad. Sci. U.S.A. **77:** 6234-6238.
21. Edwards, C., V. Dolezal, S. Tucek, H. Zeinkova & F. Vyskocil. 1985. Proc. Natl. Acad. Sci. U.S.A. **82:** 3514-3518.

22. Douglas, W. W. 1968. Brit. J. Pharmacol. **34:** 451-474.
23. Baker, P. F. 1974. *In* Recent Advances in Physiology, Vol. 9. Churchill-Livingstone. London. pp. 51-86.
24. Llinas, R., J. R. Blinks & C. Nicholson. 1972. Science 1127-1129.
25. Rink, T. J. & T. J. Hallam. 1984. Trends Biochem. **9:** 215-219.
26. Knight, D. E. & N. T. Kesteven. 1983. Proc. R. Soc. Lond., Ser. B **218:** 177-199.
27. Rink, T. J., A. Sanchez & T. J. Hallam. 1983. Nature **305:** 317-319.
28. Tsien, R. Y. 1981. Nature **290:** 527-528.
29. DiVirgilio, F., D. P. Lew & T. Pozzan. 1984. Nature **310:** 691-693.
30. Barrowman, M. M., S. Cockcroft & B. D. Gomperts. 1986. Nature **319:** 504-507.
31. Shoback, D. M., J. Thatcher, R. Leombruno & E. M. Brown. 1984. Proc. Natl. Acad. Sci. U.S.A. **81:** 3113-3117.
32. Whitaker, M. J. & P. F. Baker. 1983. Proc. R. Soc. Lond., Ser. B. **218:** 397-413.
33. Baker, P. F., D. E. Knight & J. A. Umback. 1985. Cell Calcium **6:** 5-14.
34. Gomperts, B. D. & J. M. Fernandez. 1985. Trends Biochem. Sci. **10:** 414-417.
35. Knight, D. E. & P. F. Baker. 1982. J. Membr. Biol. **68:** 107-140.
36. Baker, P. F. & D. E. Knight. 1978. Nature **276:** 620-622.
37. Knight, D. E. & M. C. Scrutton. 1986. Biochem. J. **234:** 497-506.
38. Knight, D. E. & M. C. Scrutton. 1980. Thrombosis Res. **20:** 437-446.
39. Knight, D. E. & M. C. Scrutton. 1983. Thromb. Haemostas. **50:** 93.
40. Haslam, R. J. & M. M. L. Davidson. 1984. Biochem. J. **222:** 351-361.
41. Knight, D. E. & M. C. Scrutton. 1984. Nature **309:** 66-68.
42. Haslam, R. J. & M. M. L. Davidson. 1984. FEBS Lett. **174:** 90-95.
43. Knight, D. E. & M. C. Scrutton. 1985. FEBS Lett. **183:** 417-422.
44. Knight, D. E. & M. C. Scrutton. 1986. Eur. J. Biochem. **160:** 183-190.
45. Haslam, R. J. & M. M. L. Davidson. 1984. J. Recept. Res. **4:** 605-629.
46. Cockcroft, S. & B. D. Gomperts. 1985. Nature **314:** 534-536.
47. Berridge, M. J. 1984. Biochem. J. **220:** 345-360.
48. Knight, D. E. & P. F. Baker. 1985. FEBS Lett. **189:** 345-349.
49. Birnbaumer, L. *et al.* 1985. Recent Progress in Hormone Research, Vol. 49. pp. 41-99.
50. White, J., M. Kielan & A. Helenius. 1983. Q. Rev. Biophys. **16:** 151-195.
51. Cox, J. A., M. Comte, A. Malonoe, D. Burger & E. A. Stein. 1984. *In* Metal Ions in Biological Systems, Vol. 17. H. Sigel, Ed.: 215-273. Marcel Dekker. New York.
52. Zimmerberg, J., F. S. Cohen & A. Finkelstein. 1980. J. Gen. Physiol. **75:** 241-270.
53. Zimmerberg, J. & M. J. Whitaker. 1985. Nature **315:** 581-584.
54. Holz, R. W. 1986. Ann. Rev. Physiol. **48:** 175-189.
55. Navone, F., P. Greengard & P. DeCamillo. 1984. Science **226:** 1209-1211.
56. Llinas, R., T. L. M. McGuiness, C. S. Leonard, M. Sugimori & P. Greengard. 1985. Proc. Natl. Acad. Sci. U.S.A. **82:** 3035-3039.
57. Salama, G., R. G. Johnson & A. Scarpa. 1980. J. Gen. Physiol. **75:** 109-140.
58. Pollard, H. B., C. J. Pazoles, C. E. Creutz & O. Zinder. 1979. Int. Rev. Cytol. **58:** 159-197.
59. Holtz, R. W., R. A. Senter & R. R. Sharp. 1983. J. Biol. Chem. **258:** 7506-7513.
60. Knight, D. E. & P. F. Baker. 1985. J. Memb. Biol. **83:** 147-156.
61. Nishizuka, Y. 1984. Nature **308:** 693-698.
62. Knight, D. E. & P. F. Baker. 1983. FEBS Lett. **160:** 98-100.
63. Wolf, M., H. Levine, S. May, P. Cuatrecasas & N. Sahyoun. 1985. Nature **317:** 546-549.
64. Kraft, A. S. & W. B. Anderson. 1983. Nature **301:** 621-623.
65. Touqui, L., B. Rothhut, A. M. Shaw, A. Fradin, B. B. Vargaftig & F. Russo-Marie. 1986. Nature **321:** 177-180.
66. Knight, D. E., V. Niggli & M. C. Scrutton. 1984. Eur. J. Biochem. **143:** 437-446.
67. Trifaro, J. M. & R. L. Konigsberg. 1983. Fed. Proc. **42:** 456.
68. Sudhof, T. C., J. H. Walker & U. Fritsche. 1985. J. Neurochem. **44:** 1302-1307.
69. Creutz, C. E., C. J. Pazoles & H. B. Pollard. 1978. J. Biol. Chem. **253:** 2858-2866.
70. Ahnert-Hilger, G., S. Bhakdi & M. Gratzl. 1985. J. Biol. Chem. **260:** 12730-12734.
71. Meldolesi, J., H. Scheer, L. Madeddu & E. Wank. 1986. Trends Pharmacol. Sci. **80:** 151-155.

72. Simpson, L. L. 1986. Ann. Rev. Pharmacol. Toxicol. **26:** 427-453.
73. Knight, D. E., D. A. Tonge & P. F. Baker. 1985. Nature **317:** 719-721.
74. Baker, P. F. & D. E. Knight. 1981. Phil. Trans. R. Soc. Lond., Ser. B., Biol. Sci. **296:** 83-103.
75. Colman, A. & J. Morser. 1979. Cell **17:** 517-526.
76. Sakmann, B. *et al.* 1985. Nature **318:** 538-543.
77. Novic, P., C. Field & R. Schekman. 1980. Cell **21:** 205-215.
78. Knight, D. E. 1986. CIBA Symposia, Vol. 122. D. Evered & J. Whelen, Eds.: 250-270. Wiley. Chichester.

DISCUSSION OF THE PAPER

A. Scarpa (*Case Western Reserve University, Cleveland, Ohio*): You suggest that phosphorylation of a protein might be involved in exocytosis and that Ca^{2+} could be involved in its regulation. Is it not equally possible that dephosphorylation might be the key step, with Ca^{2+} regulating the phosphatase?

D. E. Knight (*Kings College London, London, England*): You are right, and we have no hard evidence to exclude the possibility that dephosphorylation is involved in exocytosis. The reason why I mentioned phosphorylation and exocytosis was that it seemed to follow the two main observations, *i.e.,* we need Ca^{2+} and MgATP for secretion from chromaffin cells, and we can see numerous Ca^{2+}-activated phosphorylations in leaky chromaffin cells and platelets. In the same preparations we are unable to resolve any Ca^{2+}-activated dephosphorylation.

Also, if phosphatases were involved then you would expect agents that inhibit phosphatases to also inhibit exocytosis. Low levels of F^-, (1 mM), however, have no effect on the calcium-dependence of secretion. Higher levels (10 mM) do have an effect. They inhibit calcium-dependent secretion from leaky chromaffin cells, but enhance calcium-dependent secretion from leaky platelets. In these two cases the effects of F^- mimic the effects of GTP-γ-S. It is therefore likely that any F^- effects are through G binding proteins, rather than through phosphatases.

Synexin and Chromaffin Granule Membrane Fusion

A Novel "Hydrophobic Bridge" Hypothesis for the Driving and Directing of the Fusion Process[a]

HARVEY B. POLLARD, EDUARDO ROJAS, AND A. LEE BURNS

Laboratory of Cell Biology and Genetics
National Institute of Diabetes, Digestive and Kidney Diseases
National Institutes of Health
Bethesda, Maryland 20205

INTRODUCTION

Membranes of secretory vesicles fuse with each other and with plasma membranes during exocytosis in many different cell types. Although much intellectual and physical effort has been expended in the search, the mechanism for this fusion process remains unknown. The missing knowledge seems to involve those parts of the process that specifically drive and direct fusion. After all, fusion only occurs at specific times and places in the cell in spite of close and continuous proximity of secretory vesicles to each other and to plasma membranes.

One answer to this problem may involve a new class of proteins, of which synexin is the primary example. Synexin is a widely distributed 47,000-Da calcium-binding protein, initially discovered in the bovine adrenal medulla, which causes isolated chromaffin granules to aggregate by close juxtaposition of granule membranes.[1] Further addition of small amounts of arachidonic acid, or any other fatty acid with *cis*-unsaturated double bonds, then causes the aggregated granules to fuse into continuous structures of 1-10 μm diameter.[2,3] These fusion products formed *in vitro* by synexin, calcium, chromaffin granules, and arachidonic acid are remarkably similar to the compound exocytotic structures observed by electron microscopy in secreting chromaffin cells.[4]

The likely relevance of the synexin reaction to exocytosis from chromaffin cells is also manifest in other ways. Arachidonic acid, for example, is released in substantial amounts by secreting chromaffin cells, in close temporal approximation to secretion.[5,6]

[a]This work was supported in part by the NIDDK-Cystic Fibrosis Foundation Joint Fellowship Program.

However, synexin alone has recently been found to induce the direct fusion of chromaffin granule ghosts[8] in a manner partially sensitive to calcium. Synexin also causes direct fusion of phosphatidylserine or phosphatidylethanolamine liposomes in a manner exclusively dependent on calcium.[9,10] Finally, the synexin reaction itself is also blocked by phenothiazine drugs, such as trifluoperazine and promethazine. Interestingly, both drugs block exocytosis from isolated chromaffin cells with ID_{50}s of 1-2 μM.[7] However, whereas trifluoperazine also acts in this dose range on calmodulin and exhibits local anesthetic actions, promethazine only exhibits these other effects at about 100-fold higher concentrations. Thus the drug studies do not exclude synexin from involvement in the secretion process.

For these and other reasons we would anticipate that a detailed understanding of how synexin causes chromaffin granule membranes to contact and fuse would prove valuable. In the course of our recent analysis of synexin action and of the molecular biology of synexin, enough new details have become available to enable us to develop a detailed hypothesis for how synexin both drives and directs the fusion process. In the remaining parts of this paper we plan to present these data and to develop a "hydrophobic bridge" hypothesis for synexin action.

SYNEXIN INTERACTS WITH ACIDIC PHOSPHOLIPIDS

The standard assay for synexin activity is its ability to cause chromaffin granules to aggregate. The process absolutely requires calcium, and ultrastructural data show that the points of contact between adjacent granule membranes are punctate pentalaminar complexes, reminiscent of structures seen in secreting cells. Most of the evidence seems to indicate that the site of calcium action on the system is on the synexin molecule, rather than on the membrane per se. The action of calcium on synexin is to induce polymerization of the protein, and current evidence indicates that the active form of synexin is the synexin polymer.[11]

The initial stage of granule aggregation involves binding of the synexin polymer to the granule membrane. Knowing more about this interaction would likely prove important since synexin must first interact with membrane surface components, perhaps including the phospholipid head groups, to fuse membranes. In this context, the receptor(s) for synexin probably include phospholipids, since synexin can aggregate and fuse liposomes composed of phosphatidylserine and some other types of lipids.[10] Furthermore, synexin can bind to a column composed of chromaffin granule lipids in a calcium-dependent manner.[12] However, the nature of the selectivity of synexin for different phospholipids, has received little detailed study.

More recently we have had the opportunity to investigate whether different types of lipids have different affinities for synexin. As iterated above, studies of synexin-lipid interaction have been performed by others with liposomes. Different phospholipids do adopt distinct conformations and long-range organizations, however, when placed in liposome form. In addition, nonpolar lipids can not be studied in isolation in this format. To bypass these problems we deposited small amounts of highly purified lipids dissolved in ethanol in wells of 96-position microtiter plates, constructed from highly hydrophobic plastic. After evaporation of the solvent we added buffer, presuming that the phospholipids would form oriented monolayers at the water-hydrophobic plastic interface (see FIG. 1). We also presumed that nonpolar lipids would simply adhere

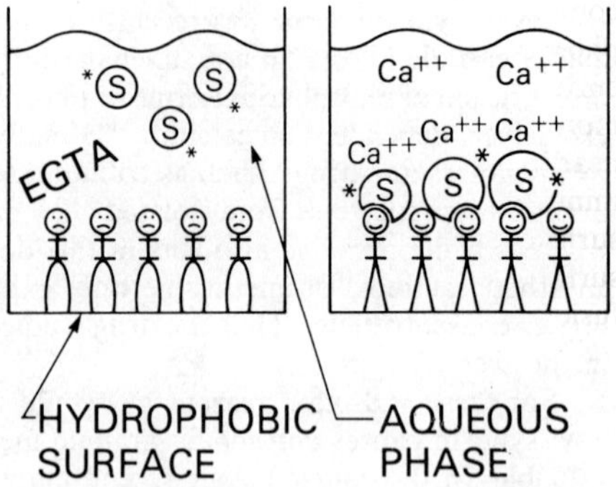

FIGURE 1. Description of the solid-phase ^{125}I-synexin binding assay to lipids. In the absence of free calcium the synexin does not stick to the lipids on the bottom of the well. In the presence of calcium, however, synexin associates with the lipids, and can be detected by the label found attached to the lipid-rich part of the well.

to the bottom of the well. To these wells we then added ^{125}I-synexin, labeled with Bolton-Hunter reagent, in the presence or absence of calcium.

As shown in TABLE 1, ^{125}I-synexin was found to bind differentially to different phospholipids in a calcium-dependent manner. Phosphatidylserine and phosphatidylethanolamine had the highest "affinity," while that of phosphatidic acid was slightly lower. Phosphatidylinositol had a substantially lower apparent affinity, but more synexin bound to the latter phospholipid than to the others. By contrast, synexin did not bind to phosphatidylcholine, nor did it bind to cholesterol or diglyceride. These nonreactive lipids also had no modifying effects on synexin reactivity with the acidic phospholipids. Whether the affinity of synexin for these acidic phospholipids is related to calcium complexes with the acidic headgroup, or to the charged headgroup per se, may eventually prove to be of some interest.

Regardless of the exact mechanism, however, these results seem to indicate that synexin has affinity for acidic phospholipids rather than basic ones. The possible biological relevance of this finding is that the membrane surfaces facing the cytosolic

TABLE 1. Calcium-Dependent Binding of ^{125}I-Synexin to Phospholipids

Phospholipid Species	Phospholipid for Half-Maximum Synexin Binding, pmol/well	Synexin Bound at Half-Maximum Lipid, pmol/well
PE	29	4.8 (8%)
PS	36	6.0 (11%)
PA	90	3.0 (6%)
PI	400	13.5 (20%)
PC	$\gg$2,000	none (0%)

Synexin was labeled with ^{125}I-Bolton-Hunter reagent to a specific activity of 526 cpm/pmol. The lipids (Avanti) were added in 20 μl ethanol to individual wells, vacuum-dried, and rehydrated with buffer. Following blocking with BSA, ^{125}I-synexin (110 pmol or about 60,000 cpm) was added to each well in a final volume of 20 μl of sucrose-histidine buffer, pH 6.0, in either 1 mM EGTA or 1 mM $CaCl_2$, and incubated at room temperature for 3 hours. After a rapid washing procedure, in either EGTA or $CaCl_2$, as appropriate, individual wells were cut out and counted. In the particular case of phosphatidylethanolamine, for example, radioactivity observed in EGTA was about 200 cpm, whereas that in $CaCl_2$ was about 11,000 cpm. Binding to phosphatidylcholine or other nonreactive lipids was quite similar to the former value.

compartment are highly enriched in these acidic phospholipids, including the inositol phospholipids. Indeed, nonreactive phosphatidylcholine is widely presumed to be enriched on the extracellular monolayer of the plasma membrane and on the interior monolayer of the chromaffin granule membrane. Thus when calcium concentration near the plasma membrane rises in the stimulated chromaffin cell, the synexin polymer immediately forms and binds to acidic phospholipids, either on the plasma membrane surface or on a nearby chromaffin granule membrane surface, or on both membrane surfaces simultaneously. Certainly the last possibility could be a prelude to membrane fusion and exocytosis.

SYNEXIN ALSO PENETRATES INTO THE HYDROPHOBIC INTERIOR OF THE PHOSPHOLIPID BILAYER

The differential affinity of synexin for phospholipids indicates that the polar head groups are making a strong contribution to directing the interaction. Indeed, if synexin were eventually to mediate membrane fusion events then synexin would obviously have to interact first with charged elements of membrane surface molecules, including phospholipids. However, it has been difficult to perceive in molecular terms how simple binding of synexin to the membrane surface could lead to membrane fusion.

One possibility is that after binding to the membrane surface synexin then proceeds to penetrate into the substrate of the bilayer. If this were true, substantial rearrangements of two juxtaposed membranes would have to ensue. To test this hypothesis we have developed a technique for measuring the capacitance of a phospholipid bilayer formed at the tip of a patch pipette. We predicted that if synexin did penetrate into the bilayer the capacitance of the bilayer would be increased. The reason for this is as follows. The capacitance of a membrane, C, is the product of a measure of the dielectric constant, k, and the ratio of the area, A, of the capacitor to the distance, d, between parallel plates. The equation is

$$C = k(A/d).$$

Adding synexin molecules to the low dielectric region of the bilayer would raise the apparent value of the dielectric constant, k, and thus raise the capacitance of the membrane.

As shown in FIGURE 2, we were able to prepare pure bilayers of phosphatidylserine at the tip of a patch pipette by a double dip method, and measure the capacitance of the membrane. To make this measurement we put a voltage pulse (top traces, FIG. 2) across the membrane and determined the change in the charge, dQ/dt (current), as a function of time (middle traces, FIG. 2). Upon application of a voltage, current flowed into the capacitor and then stopped flowing when the capacitor was fully charged. Upon turning off the voltage, the current flowed out in the opposite direction and then stopped flowing (middle traces, FIG. 2). This explains the two biphasic curves coincident with on- and off-phases of the voltage pulse.

The charge across the capacitor, the time integral of the current, was then measured as shown in FIGURE 2 (lower traces). The pure phosphatidylserine bilayer was found to have a capacitance (response shown by dashed lines in FIG. 2) of 5.2 picofarads, of which only 2% (104 femtofarads) represents the capacitance of the bilayer, the remainder being just the capacitance of the pipette. This membrane was stable for up

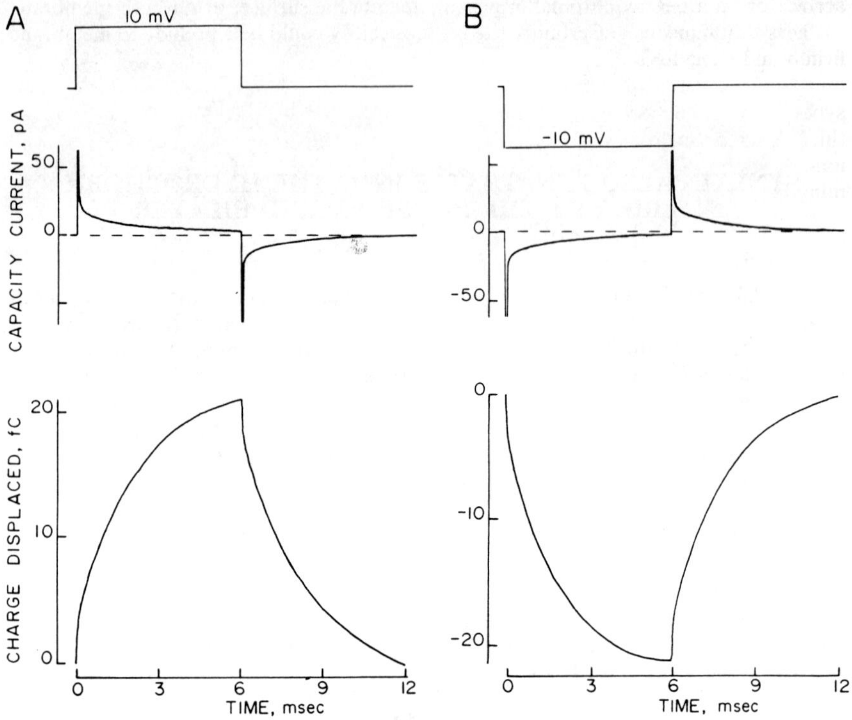

FIGURE 2. Synexin increases the capacitance of phosphatidylserine bilayers formed in the tip of a patch pipette. Pure phosphatidylserine (Avanti) was used to generate a monolayer at an air-water interface and a bilayer of this phospholipid formed at the tip of a patch pipette by a double dip method. The capacitance of the bilayer in the presence or absence of synexin was then measured. In both **A** and **B** the *dashed lines* represent controls in the absence of synexin. The conditions for the experiment were 150 mM KCl, 1 mM $CaCl_2$, 10 mM Na Hepes buffer, pH 7.2, and temperature of 32°C. The pH value of 7.2 was chosen to approximate the isoelectric point of synexin and thus minimize electrical driving forces, if any. Synexin was added to a final concentration of 85 nM in the bath. The initial concentration of synexin was 1.3 μM, however, and was initially applied in close proximity to the tip of the patch clamp pipette. The response to added synexin was apparent in less than one minute. **(A)** The *top trace* represents the potential across the bilayer (+10 mV). The *middle trace* is the capacitative current, during and after the pulse. The *lower trace* is the time integral of the current record shown in the middle trace (*i.e.*, the charge placed on the membrane capacitance). **(B)** The same representation as in **A**, except that the voltage pulse is negative (−10 mV).

to 3 hours. Furthermore, addition of either $CaCl_2$ (1 mM) or EDTA (1-10 mM), or both in combination, had no effect on the capacitance parameters or stability of the bilayer. Upon adding synexin in the presence of calcium, however, the capacitance of the pure phosphatidylserine bilayer rose rapidly to a value nearly 27-fold greater (2.8 picofarads) than with the pure bilayer alone. Neither bovine serum albumin (1 mg/ml) nor anilinonaphthaline sulfonate (10 μM) affected capacitance. The synexin effect absolutely required the additional presence of calcium in the system.

It was of course critical to this analysis that there be substantial and unambiguous evidence for the existence of a bilayer at the tip of the pipette. We determined this on the basis of three independent criteria. The capacity of the putative bilayer without synexin was 104 femtofarads, as noted above, consistent with an appropriate membrane thickness. The membrane was also found capable of incorporating gramicidin A, an ion-selective channel, and supporting characteristic single channel events. Finally, minute negative or positive pressures (by mouth) always broke the membranes at the end of an experiment.

THE PRIMARY SEQUENCE OF A SYNEXIN CLONE HAS THE HYDROPHOBICITY CHARACTERISTICS OF AN INTRINSIC MEMBRANE PROTEIN

The picture we have of synexin at this point is that in the absence of calcium it behaves as a "soluble" protein, albeit atypical in various aspects. In the presence of calcium, however, the protein polymerizes, binds to the charged regions of acidic phospholipids, and then penetrates into the substance of the phospholipid bilayer. Finally, these processes are apparently reversible since they are readily reversed by EGTA, generating "soluble" synexin. How can synexin express such disparate properties?

We decided that these properties could be best understood in terms of the chemical structure of synexin, and we have therefore been engaged in cloning and sequencing the gene for this protein. For this purpose we have used the λgt11 expression vector system containing a library of cDNA molecules generated from mRNA preparations from bovine adrenal medulla. The cloned sequences express peptides that react with two independently derived monoclonal antibodies to synexin. Thus the sequences derived are indeed those of authentic synexin, and we have at present sequenced about 30% of the molecule.

Although limited, these data have revealed to us some facts that may serve to illuminate how synexin works. One simple way of analyzing linear sequence data is to estimate the relative hydrophobicity, neutrality, or charged character of each amino acid, taking into account neighboring amino acids. We did this with our synexin clone 547P, comprising about 180 amino acids. From this analysis we determined that there were lengthy regions of hydrophobicity, punctuated by shorter neutral and charged regions.

These data are summarized in TABLE 2, where it is evident that about 43% of the clone has strong hydrophobic character. Furthermore, the hydrophobic regions occur in segments that average approximately 15 residues in length. Of the remaining sequence, 31% has neutral character whereas only 26% has charged character. The segmental character is still observed for neutral and charged regions, but the average length of charged regions is about half that for hydrophobic regions. The average length of neutral regions seems intermediate between the two others.

TABLE 2. Summary of Hydrophobicity Indices

Sequence	SYN Clone: 547P		BSA-p		AChR-αp	
	% Total Length	Average Length	% Total Length	Average Length	% Total Length	Average Length
Charged	26.1	7.8 (6)	54.1	11.3 (29)	28.7	8.7 (15)
Neutral	31.1	9.3 (6)	24.6	7.2 (20)	17.9	6.3 (13)
Hydrophobic	42.7	15.4 (5)	20.2	7.6 (16)	48.1	13.8 (16)

The decision on whether to call a region hydrophobic, neutral, or charged was made as follows. According to the paradigm, highly hydrophobic regions are given a value of −4. Neutral regions are 0. Charged regions are +4. We decided that any two residues sequentially within the zone between −0.2 and +0.2 would be termed neutral. Any two residues sequentially more negative or more positive would be termed hydrophobic or charged, respectively. This approach generated remarkably few ambiguities (ca. 1%), and seemed to make the fewest assumptions.

The significance of these data, however, becomes manifest when compared with a typical "soluble" protein such as bovine serum albumin or a typical "intrinsic membrane" protein such as the acetylcholine receptor α-subunit. We analyzed these sequences exactly as we did the synexin clone, and found the BSA hydrophobicity pattern to be quite the reverse of that for synexin. The charged segments made up more than half of the total sequence, and were nearly twice as long as the less frequent neutral and hydrophobic segments.

By contrast the acetylcholine receptor α-subunit was more nearly like the synexin clone 547P. These data are summarized in TABLE 2. Indeed, further analysis reveals that the synexin clone and acetylcholine receptor hydrophobicity patterns are more like most intrinsic membrane proteins.

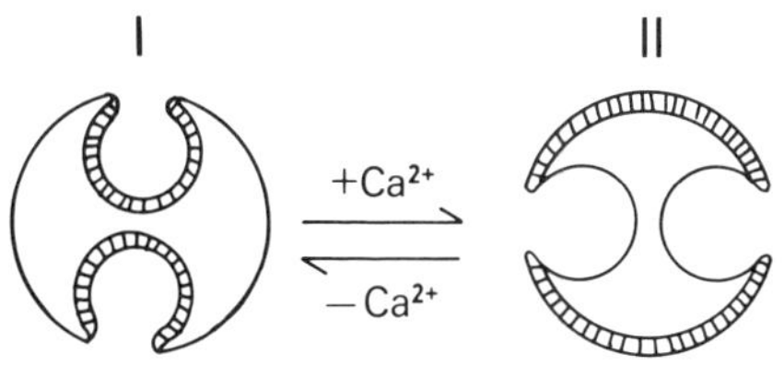

FIGURE 3. Two-dimensional model for calcium-dependent conformational transitions in synexin. The *hatched areas* represent hydrophobic sequences. The *solid lines* represent charged and neutral sequences. In the presence of calcium the water-soluble synexin undergoes a conformational change, exposing its hydrophobic domains and shielding its charged and neutral domains within the structure.

From this analysis one could imagine that synexin is able to penetrate the membrane bilayer because of its ability to assume a conformational character similar to typical intrinsic membrane proteins. The "soluble" character of synexin in low calcium must therefore be by a mechanism in which the hydrophobic sequences are hidden from the aqueous phase, again by a conformational strategy. Therefore, one could imagine that synexin in the absence of calcium would be water soluble, keeping its hydrophobic sequences buried under a hydrophilic shield. Upon the addition of calcium, however, a conformational change would occur, exposing the hydrophobic sequences. This would provide the hydrophobic forces favoring polymerization, interaction with phospholipids, and insertion into the bilayer. The concept that interaction of calcium with synexin promotes exposure of hydrophobic sites is supported by at least one piece of experimental data with anilino-naphthalene-sulphonate (ANS) (C. E. Creutz and H. B. Pollard, unpublished results).

These concepts can be summarized graphically in the sketch in FIGURE 3. This is purposefully a quite general two-dimensional model drawn to represent calcium-dependent conformational states that a synexin type sequence could assume. In state I the conformational organization places charged or neutral sequences outwards and places hydrophobic sequences towards the interior. Interestingly, about 50% of the total perimeter of the model structure is "hydrophobic." Upon adding calcium the model changes to state II, in which the hydrophobic surface is on the outside and the charged and neutral surfaces are towards the interior. The protein in state II could then easily polymerize and/or insert into a nearby membrane using hydrophobic driving forces.

The polymerization properties of this two-dimensional model are of interest because we have previously had the opportunity to study calcium-induced synexin polymers by electron microscopy. These polymers consist of cigar-shaped rods of 100 Å × 50 Å, which themselves associate side-by-side into ribbons and clusters. We have previously estimated that each rod is composed of at least four synexin monomers. In terms of the two-dimensional model of synexin conformation states, we could imagine that four state II molecules could polymerize into a rod, as shown in FIGURE 4A,

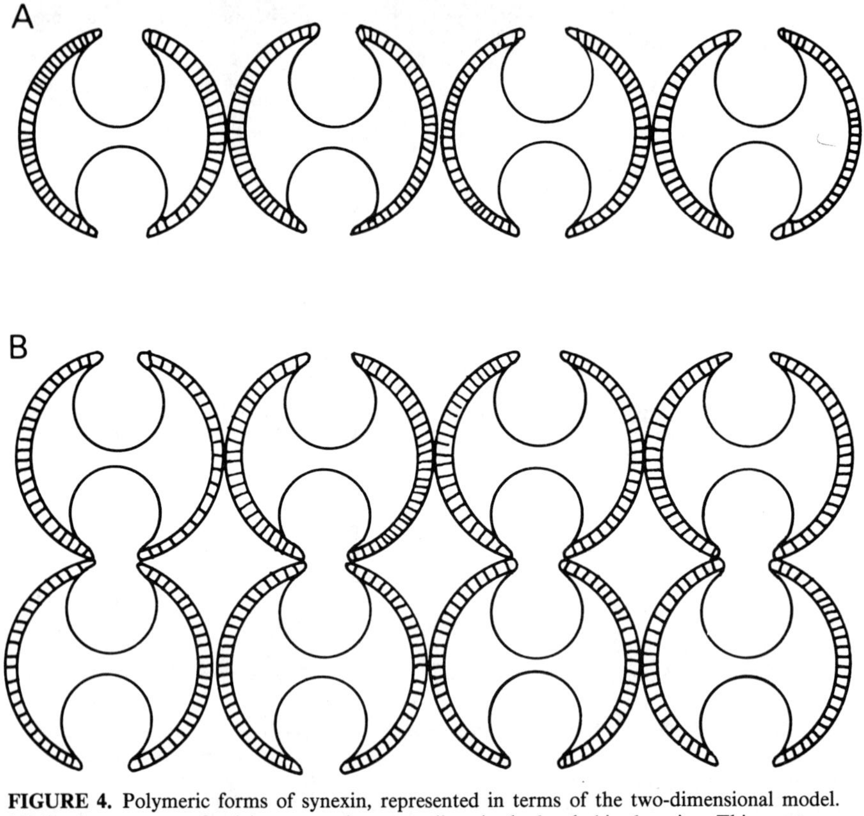

FIGURE 4. Polymeric forms of synexin, represented in terms of the two-dimensional model. (**A**) In the presence of calcium, synexin externalizes its hydrophobic domains. This creates a driving force to cause polymerization. The elementary polymeric structure observed by EM is a 50 Å × 100 Å cigar-shaped rod, which could contain as many as four synexin monomers. (**B**) The cigar-shaped polymer can associate side-by-side into two-dimensional ribbons or three-dimensional stacks. The *hatched lines* and *solid lines* represent hydrophobic and charged or neutral domains, as described in FIGURE 3.

and that two or more such rods could polymerize into a ribbon, as depicted in FIGURE 4B. At present the model as shown does not explain why the rod in FIGURE 4A is limited in length, but more detailed information on synexin structure will undoubtedly lead to further refinements.

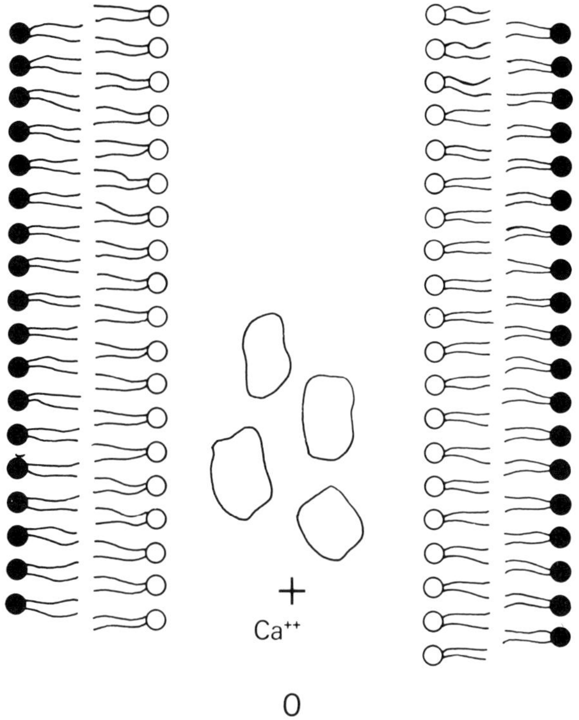

FIGURE 5. Two membranes are poised for fusion (stage 0) as monomeric synexin molecules (structures between membranes) are exposed to a pulse of calcium. The two membranes could be a chromaffin granule membrane and a plasma membrane, or two chromaffin granule membranes. *Small circles* represent charged head groups of phospholipids. *Unfilled circles* represent the head groups of phospholipids oriented towards cytoplasm, and include the acidic phospholipids having affinity for synexin. *Filled circles* represent the head groups of phospholipids oriented away from the cytoplasm, and include phosphatidylcholine, which has little affinity for synexin. The two *wavy lines* attached to the head groups represent fatty acid chains on the phospholipids.

A "HYDROPHOBIC BRIDGE" HYPOTHESIS FOR THE DRIVING AND DIRECTING OF THE FUSION PROCESS

We now intend to summarize these data in terms of a model for membrane fusion. The model is based on the fundamental concept that synexin molecules, in the presence of calcium, form hydrophobic polymers capable of binding to specific membranes and inserting into the bilayer. To understand how such a process could lead to membrane fusion, suppose that one synexin polymer could contact and be inserted into two neighboring membranes simultaneously. The result would be a "hydrophobic bridge" linking two membranes, as shown sequentially in FIGURES 5 and 6. This step is not

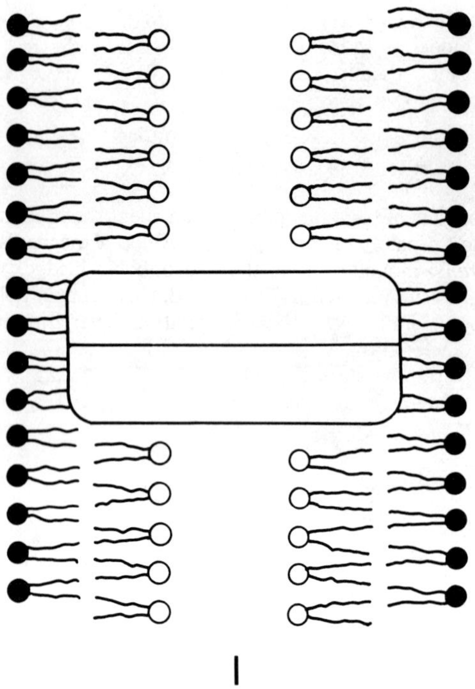

FIGURE 6. Synexin polymerizes in the presence of calcium and the resulting hydrophobic structures bind to acidic phospholipids. Subsequently the synexin polymer penetrates into the bilayer of each membrane, causing them to adhere (stage I). Membrane adhesion is, of course, a necessary preliminary step to eventual fusion. The line down the middle of the synexin polymer represents the fact that the polymer may be several synexin molecules thick. The open space between the membranes serves to emphasize the bridge-like character of the connection between membranes, not that hydrophobic domains are exposed to water for any length of time. In fact, electron micrographs of synexin-aggregated granules indicate very little space between adhering membranes.

so extraordinary inasmuch as a necessary preliminary to any fusion process must be that the membranes make contact. The specificity of interaction, according to this model, is the requirement for calcium and the selective affinity of acidic phospholipids for the synexin polymer.

Once having formed the hydrophobic synexin bridge, we anticipate that the neighboring phospholipids would not be able to distinguish easily the hydrophobic protein from the hydrophobic bilayer interior. The result would be that the phospholipids on the adjoined surfaces could simply move across the bridge, as illustrated in FIGURE 7. Evidence for movement of phospholipids from one membrane to another via a synexin-related bridge is, in fact, readily available. Chromaffin granule membranes can be labeled with the fluorescent probe octadecylrhodamine (R18) at self-quenching concentrations on the external leaflet. Upon adding synexin and unlabeled chromaffin granule membranes, the fluorescence of the R18 rises. This occurs presumably by transfer of the R18 molecules to the unlabeled membrane and resultant dequenching. Under the same conditions, true fusion of chromaffin granule ghosts can also be observed. By "true fusion" we mean the mixing of included volumes. This can be followed by loading granule ghosts with self-quenching concentrations of FITC-dextran and observing the dequenching of the FITC-dextran as the loaded ghosts fuse with empty ghosts in a synexin-dependent manner. We have performed the membrane mixing and volume mixing experiments in concert and have determined that membrane mixing occurs 2-5 times faster than volume mixing.[8]

The hydrophobic bridge hypothesis thus provides a mechanism for connecting adjoining (*cis*) monolayers of fusing membranes, an event that definitely occurs during biological membrane fusion. It remains now to develop a mechanism for connecting the contralateral (*trans*) monolayers via the hydrophobic bridge. We have achieved this end by recalling, as summarized in FIGURE 4B, that synexin polymers themselves form higher order polymers by side-to-side association. Therefore, it is likely that the hydrophobic bridge depicted in FIGURES 6 and 7 might be several monomers thick. Within the low dielectric domain of the membrane bilayer and under pressure from the phospholipids clustered on the *trans* monolayer, the synexin molecules making up the hydrophobic bridge might dissociate into two neighboring bridges. As shown in FIGURE 8, the regions of the new hydrophobic bridge unoccupied by phospholipids could provide an isoenergetic, hydrophobic space for the *trans* monolayers to enter.

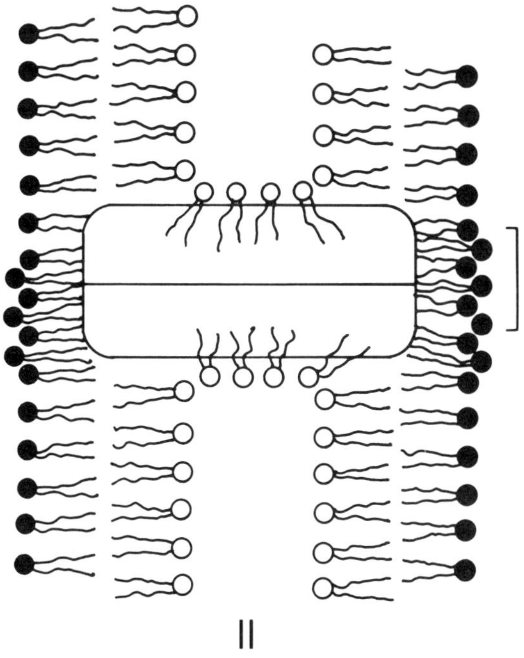

FIGURE 7. Phospholipids on the *cis* monolayers of the adherent membranes cross the synexin bridge towards each other (stage II). The nonpolar fatty acid tails remain oriented either towards the interior of the bilayer or the hydrophobic bridge formed by polymerized synexin. We do not rule out continued interactions of synexin with the polar head group or indeed continued interaction with both parts of the phospholipid. The phospholipids on the *trans* bilayer may become crowded (*bracketed area*) as phospholipids are removed from the *cis* monolayers.

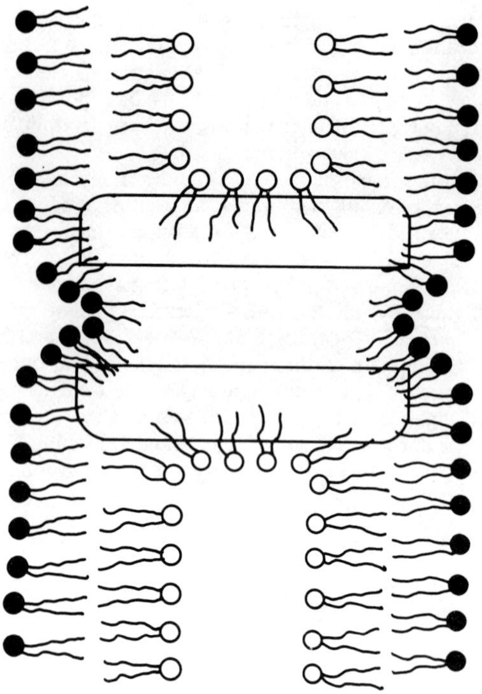

FIGURE 8. The synexin polymer bridge, immersed in a nonpolar environment, may undergo dissociation (stage III). The *trans* monolayer phospholipids may intrude into the nonpolar space left between synexin molecules. This may be an isoenergetic event.

The phospholipids in the *trans* monolayers, depicted with opaque head groups, could loop in completely without energetic constraint, guided by the newly dissociated hydrophobic bridge. The hydrophobic tails of the most central phospholipids would meet on the bridge as shown in FIGURE 9, in the region marked with an asterisk. A modest, isoenergetic rearrangement of these tails, to align completely with the hydrophobic bridge, as shown in FIGURE 10, would lead to morphologic fusion with the *cis* monolayers and the *trans* monolayers coincident, as in true biological fusion processes.

The fusion structure shown in FIGURE 10 contains a remnant of the hydrophobic bridge. At present we do not know whether remnants of synexin are left in fused membranes. However, our present data are consistent with the concept that if synexin were indeed the hydrophobic bridge material, it might not remain within the membrane if the concentration of calcium in the aqueous phase were lowered. Measurement of capacitance of phosphatidylserine bilayers in the tip of patch pipettes, immersed in synexin and calcium, show elevated capacitance, as shown in FIGURE 2. Chelation of the calcium, however, can cause breakage of the bilayer. One reason for the breakage could be that synexin within the bilayer reverts to its water-soluble conformation, a structure incompatible with a low dielectric environment. The apparent fragility of the pure phosphatidylserine bilayer may not obtain in more biological systems, in-

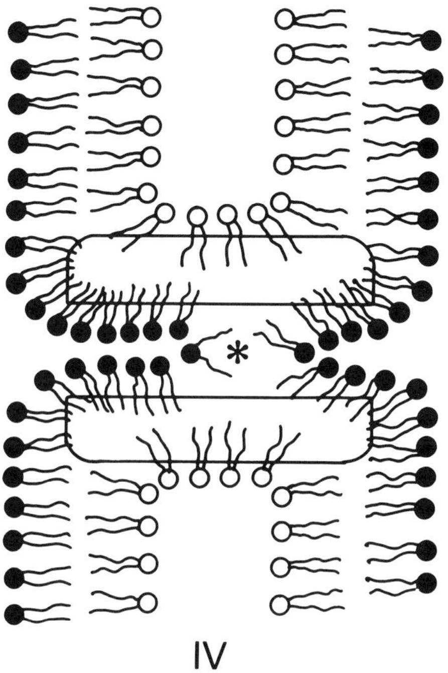

FIGURE 9. The *trans* monolayer phospholipids continue to intrude into the nonpolar space between synexin molecules until the fatty acid chains of the leading phospholipids make contact (stage IV). The *asterisk* represents the fact that the fatty acid chains must decide whether to remain cohesive with each other or to express a preference for interaction with the substance of the hydrophobic bridge. In order to get the membranes to fuse we anticipate they make the latter decision.

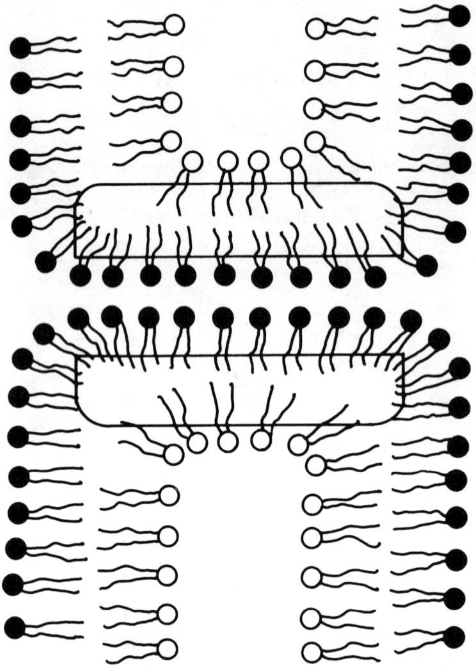

FIGURE 10. The *cis* monolayers and the *trans* monolayers of the two membranes are now respectively continuous, thus generating a genuinely fused structure (stage V). The previously separated volumes are now continuous. Bits of the hydrophobic bridge material, however, are left in the substance of the fused bilayers.

asmuch as chromaffin granules aggregated by synexin do not break when the calcium is subsequently chelated. It is therefore possible that the final fused membrane structure may extrude its hydrophobic bridge remnants, as shown in FIGURE 11. This point, as indeed others in the model, are explicit enough to generate simple experimental tests.

ANALYSIS AND CONCLUSIONS

The problem of how membranes fuse is at the crux of many events of interest to modern cell biology.These include not only exocytosis, as emphasized in this paper, but also organelle biogenesis, membrane recycling, membrane synthesis, endocytosis and phagocytosis, cell division, viral invasion and escape, fertilization, and many others. It has been long appreciated that the *cis* and *trans* monolayers of fusing

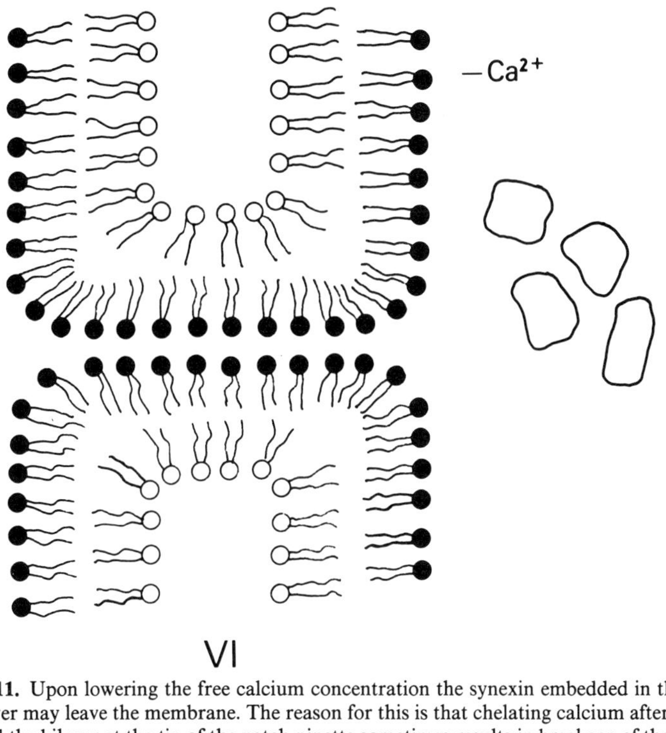

FIGURE 11. Upon lowering the free calcium concentration the synexin embedded in the membrane bilayer may leave the membrane. The reason for this is that chelating calcium after synexin has entered the bilayer at the tip of the patch pipette sometimes results in breakage of the bilayer. Perhaps the reason is that synexin within the membrane undergoes a conformational transition to its water-soluble form, a form unsuitable for continued existence within the bilayer. The latter result also means that at least some of the synexin molecule, including the calcium detector, remains exposed to the aqueous environment as the majority of the synexin mass becomes enveloped within the bilayer.

membranes eventually become continuous, respectively. But the driving and directing forces for this process and the spatial and temporal specificity within the cell have remained obscure. The hydrophobic bridge hypothesis described here, however, does provide not only a spatial and temporal specificity but also a driving and directing force for fusion processes occurring at least during exocytosis. The strength of this hypothesis is that it is built on the bedrock of known properties of synexin.

An additional advantage of a hydrophobic bridge, whether built of synexin or any other operationally equivalent organization of molecules, is that it provides a common connection perpendicular to the plane of adjoining membranes destined for fusion. This connection is the armature for the 90° change in orientation of the hydrophobic bilayer that occurs during fusion. This connecting element is absent in many previous theories of membrane fusion, missing perhaps because of an operational concentration on fusion of pure lipid systems. With the hydrophobic bridge hypothesis in mind, however, it is even possible that fusion of pure liposomes may prove to be interpretable in terms of an operational equivalent to the hydrophobic bridge described above.

Finally, we should emphasize the importance of the ability of synexin to exist not only in aqueous solution but also within the membrane bilayer. The problem of movement of proteins into and through membranes, as in the rough endoplasmic reticulum during biosynthesis, has been solved in some cases by leading the protein through the membrane with a nonpolar leader sequence. For other proteins entry has been by means of a nonpolar loop. Posttranslational modification with fatty acids has also been viewed as a potential technique for membrane entry, or at least association. In the case of synexin, the data from the patch clamp studies indicate that as much as 80% of the protein may become incorporated into the bilayer. The continued sensitivity of the otherwise incorporated synexin to calcium, however, indicates that the calcium sensor sequences, at least, remain exposed to the aqueous environment. This means that the drawings in the various figures of fusion should be viewed as strictly caricatures, with the two-dimensional, general character heavily stressed.

The reason for the ability of synexin to assume a membrane-incorporated form appears to be related to its similarity in hydrophobic sequence organization to that of some standard intrinsic membrane proteins such as the acetylcholine receptor α-subunit. Yet to be discovered, however, is the mechanism by which synexin, but apparently not authentic intrinsic membrane proteins, can also contrive to be water soluble when necessary. We anticipate that synexin may prove to be a prototype for a new class of such proteins. Indeed, nearly a dozen have now been reported,[13] with more certain to come.

ACKNOWLEDGMENTS

The authors wish to thank Dr. Shlomo Nir for a critical analysis of the model, Dr. Peter Lelkes for useful discussion, Dr. Andres Stutzin for help with measurements of membrane capacitance, Ms. Judith Heldman and Dr. Claudio Parra for help with measurements of synexin binding to lipids, Dr. Mira Srivastava for assistance in the molecular biology work, and Ms. Diane Johnson-Seaton and Dr. George Lee for preparation of synexin.

REFERENCES

1. CREUTZ, C. E., C. J. PAZOLES & H. B. POLLARD. 1978. J. Biol. Chem. **253:** 2858-2866.
2. CREUTZ, C. E. & H. B. POLLARD. 1981. Biophys. J. **37:** 119-120.
3. CREUTZ, C. E. 1981. J. Cell Biol. **91:** 247-256.
4. POLLARD, H. B., C. E. CREUTZ, V. M. FOWLER, J. H. SCOTT & C. J. PAZOLES. 1982. Cold Spring Harbor Symp. Quant. Biol. **46:** 819-833.
5. HOTCHKISS, A., H. B. POLLARD, J. H. SCOTT & J. AXELROD. 1981. Fed. Proc. **40:** 256.
6. FRYE, R. A. & R. N. HOLZ. 1984. J. Neurochem. **43:** 146-150.
7. POLLARD, H. B., J. H. SCOTT & C. E. CREUTZ. 1983. Biochem. Biophys. Res. Comm. **113:** 908-915.
8. STUTZIN, A. 1986. FEBS Lett. **197:** 274-280.
9. HONG, K., H. DUZGUNES & D. PAPAHADJOPOULOS. 1981. J. Biol. Chem. **256:** 3641-3644.
10. HONG, K., H. DUZGUNES & D. PAPAHADJOPOULOS. 1982. Biophys. J. **37:** 297-305.
11. CREUTZ, C. E., C. J. PAZOLES & H. B. POLLARD. 1979. J. Biol. Chem. **254:** 553-558.
12. CREUTZ, C. E., L. G. DOWLING, J. J. SANDO, C. VILLAR-PALASI, J. H. WHIPPLE & W. ZAKS. 1983. J. Biol. Chem. **258:** 14664-14674.
13. POLLARD, H. B. *et al.* 1985. Vitamins and Hormones **42:** 109-195.

Are Changes in Intracellular Free Calcium Necessary for Regulating Secretion in Parathyroid Cells?

E. F. NEMETH AND A. SCARPA

Department of Physiology and Biophysics
Case Western Reserve University
Cleveland, Ohio 44106

INTRODUCTION

Among the various endocrine factors contributing to the regulation of extracellular Ca^{2+} homeostasis is parathyroid hormone (PTH), which plays a major role in increasing serum and extracellular Ca^{2+} by actions in the kidney and in bone. In turn, the rate of secretion of PTH by the parathyroid gland is sensitive to the concentration of Ca^{2+} present in extracellular fluids, and it is now well established that increases or decreases in extracellular Ca^{2+} suppress or enhance PTH secretion, respectively.[1] Although it is clear that extracellular Ca^{2+} acts as the principal physiological stimulus modulating PTH secretion, both *in vivo* and *in vitro,*[1,2] it is unknown how a change in the extracellular Ca^{2+} concentration is detected by the parathyroid cell and how this recognition event is transformed into an intracellular signal(s) that controls secretion. The identity of this intracellular signal is likewise far from clear.

A salient feature of stimulus-secretion coupling in various other cells is an increase in the concentration of intracellular free Ca^{2+}, $[Ca^{2+}]_i$. And there is now a vast body of literature supporting the view that Ca^{2+} acts as an intracellular signal activating exocytotic secretion in diverse cell types.[3,4] It can be readily appreciated, however, that the parathyroid cell, if it is to conform to this general calcium hypothesis of stimulus-secretion coupling, must somehow be equipped with mechanisms that effect a reciprocal relationship between the concentration of extracellular Ca^{2+} and $[Ca^{2+}]_i$. Previous studies using the fluorescent indicator quin-2 to assess $[Ca^{2+}]_i$ have repeatedly shown that the parathyroid cell simply does not behave in this manner, and that increases in the concentration of extracellular Ca^{2+} are paralleled by increases in $[Ca^{2+}]_i$.[5–8] Increases in $[Ca^{2+}]_i$ are thus associated with an inhibition of secretion of PTH, suggesting that, at least in this cell, intracellular free Ca^{2+} may act as an inhibitory signal regulating secretion.

Alternatively, it can be postulated that changes in $[Ca^{2+}]_i$, although associated with alterations in the rate of secretion, are not causally involved in the regulation of secretion in parathyroid cells.

In this report, we summarize some of our recent studies centering on the role of cytosolic Ca^{2+} in stimulus-secretion coupling in parathyroid cells, using quin-2 or fura-2 to assess $[Ca^{2+}]_i$. The measurements obtained with the latter indicator reveal that an increase in the concentration of extracellular Ca^{2+} elicits two mechanistically distinct changes in $[Ca^{2+}]_i$ in parathyroid cells. One of these responses is associated with the generation or modulation of an intracellular signal(s) that regulates PTH secretion, although we suspect that this signal may not be cytosolic Ca^{2+}.

METHODS

Dissociated bovine parathyroid cells were prepared by collagenase digestion of minced parathyroid glands and purified on Percoll gradients as described previously.[8] The standard cell buffer contained (mM): NaCl, 126; KCI, 4; $MgSO_4$, 1; KH_2PO_4, 0.7; Na Hepes, 20 (pH 7.45); $CaCl_2$, 0.5-2. It was supplemented with 1 mg/ml glucose and 1% or 2% bovine serum albumin (BSA). Phosphate was omitted from this buffer in those experiments where the effects of $LaCl_3$ were examined.

Parathyroid cells were loaded with quin-2 or fura-2 by incubating in standard buffer (2mM $CaCl_2$ and 2% BSA) containing 20 μM or 1 μM of the acetoxymethylester of quin-2 or fura-2, respectively. Incubation at 37°C was for 15 min, after which the cells were washed and incubated a further 15 min (37°C) in ester-free buffer. The cells were subsequently washed twice in buffer containing 0.5 mM $CaCl_2$ and 1% BSA before use in experiments to measure $[Ca^{2+}]_i$ or secretion of PTH.

The fluorescence of quin-2- or fura-2-loaded cells was measured as previously described using excitation and emission wavelengths of 339 and 499 nm, respectively.[8] Fluorescent signals were calibrated after correction for dilution artifacts and extracellular leakage of indicator using K_ds for quin-2 and for fura-2 of 115 and 224 nM, respectively.

For secretion studies, control- or indicator-loaded cells were incubated (37°C) for various times as indicated in the figure legends. Following incubation with or without test substances, the cells were rapidly sedimented by centrifugation and the amount of PTH in samples of the supernatant was determined by radioimmunoassay.[9]

RESULTS AND DISCUSSION

Increase in Steady-State Levels of Ca^{2+} and Inhibition of Secretion

FIGURE 1 shows the time-dependent changes in fluorescence of the indicators quin-2 and fura-2 trapped within dissociated bovine parathyroid cells when the concentration of extracellular Ca^{2+} in the suspending buffer is increased.

There are obvious qualitative differences in the pattern of change in $[Ca^{2+}]_i$ reported by these two Ca^{2+} indicators. In cells loaded with quin-2, increases in the concentration of extracellular Ca^{2+} elicit monophasic increases in $[Ca^{2+}]_i$ that reach steady-state levels in about 5 min (FIG. 1, upper panel). In contrast, fura-2 reports a rapid, transient increase in $[Ca^{2+}]_i$ followed by a lower level that attains steady-state within 1-2 min (FIG. 1, lower panel). The different results provided by quin-2 and by fura-2 likely result from the high intracellular concentrations of quin-2 that are necessary to obtain a fluorescent signal and that create an artificially high Ca^{2+}-buffering capacity within the cell. Since fura-2 has approximately 30 times more Ca^{2+}-dependent fluorescence than quin-2, much lower intracellular concentrations of indicator are sufficient to obtain usable fluorescent signals. Incubation with the ester of fura-2 typically achieves intracellular concentrations of 75-150 μM fura-2. Slow, monophasic increases in $[Ca^{2+}]_i$, as depicted in FIGURE 1, are seen when the intracellular concentration of quin-2 is 0.8-1 mM. However, when much lower concentrations of quin-2 are used (about 200 μM intracellularly), rapid and transient increases in $[Ca^{2+}]_i$ that mimic those seen in fura-2-loaded cells are observed.[10] The estimated values for steady-state cytosolic Ca^{2+} are routinely lower in quin-2-loaded cells (see FIG. 1), and this too appears to result from the greater Ca^{2+} buffering capacity imposed upon the cell by

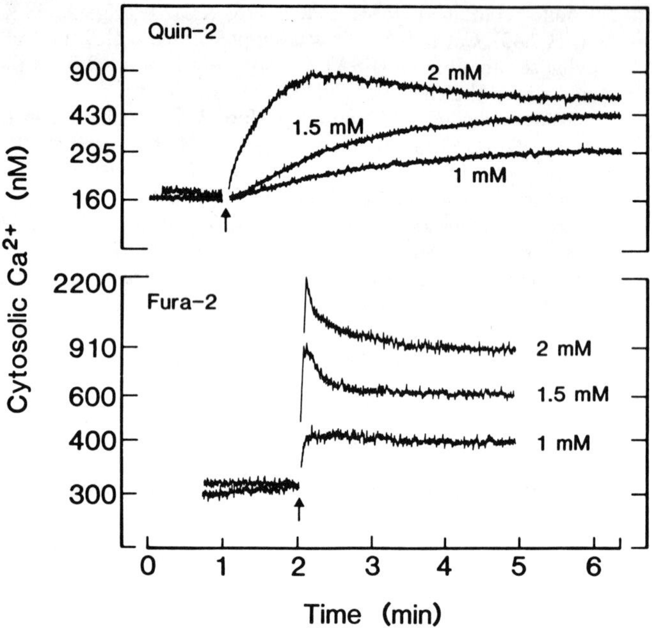

FIGURE 1. Effects of increases in the concentration of extracellular Ca^{2+} on $[Ca^{2+}]_i$ in parathyroid cells as reported by quin-2 (*top panel*) or fura-2 (*bottom panel*). Cells were initially equilibrated (37°C) in buffer containing 0.5 mM $CaCl_2$. At the *arrow*, a small volume (3-9 μl) of $CaCl_2$ was added to achieve the final concentration of extracellular Ca^{2+}, which is indicated near the respective trace.

quin-2 rather than from some intrinsic property of either indicator. Quin-2, at concentrations that are necessary to measure cytosolic Ca^{2+}, thus provides only a partially correct view of parathyroid cell physiology and masks, by buffering action, rapid and transient increases in $[Ca^{2+}]_i$ elicited by increases in extracellular Ca^{2+}.

Despite the different patterns of change in $[Ca^{2+}]_i$ reported by quin-2 and by fura-2, it is clear, using either indicator, that increases in the concentration of extracellular Ca^{2+} are paralleled by increases in $[Ca^{2+}]_i$, and that these increases in $[Ca^{2+}]_i$ are accompanied by corresponding decreases in secretion of PTH (FIG. 2). The more pronounced changes in $[Ca^{2+}]_i$ and in secretion in response to extracellular Ca^{2+} are noted at physiological concentrations between 1 and 2 mM.

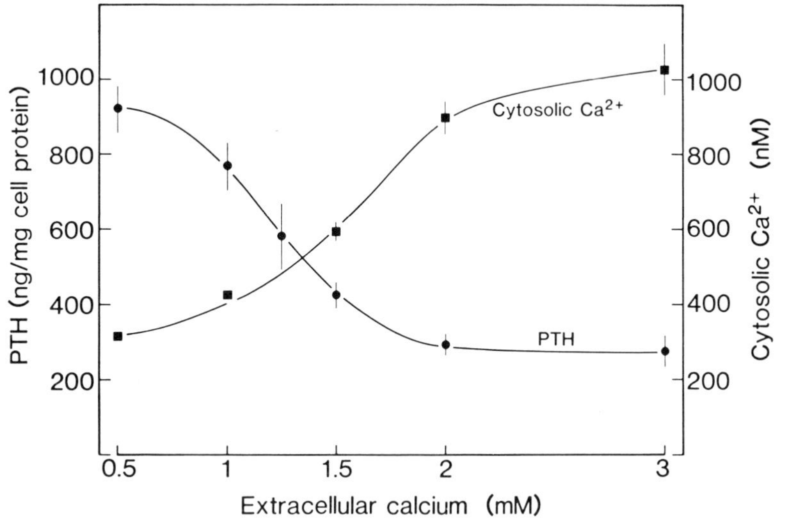

FIGURE 2. Reciprocal relationship between the secretion of PTH and steady-state $[Ca^{2+}]_i$ in parathyroid cells loaded with fura-2. Cells were incubated (37°C) for 20 min in the presence of the indicated concentration of extracellular Ca^{2+} before samples were taken and assessed for their content of PTH (●). In parallel experiments, the cells were equilibrated in buffer containing the indicated concentration of $CaCl_2$ to obtain a steady-state value of cytosolic Ca^{2+} (■). Values for PTH secretion are the mean ± SEM of five separate cell preparations, whereas those for steady-state $[Ca^{2+}]_i$ are from seven different preparations.

Taken together, the results are consistent with the notion of a "reversed intracellular Ca^{2+} signal" in parathyroid cells, *i.e.,* an increase in $[Ca^{2+}]_i$ produces an inhibition of secretion.

Modulation of PTH Secretion without Changes in Steady-State $[Ca^{2+}]_i$

Despite the obvious reciprocal relationship between $[Ca^{2+}]_i$ and secretion of PTH, there are conditions where these two parameters can be dissociated. FIGURE 3 shows the result of an experiment where the $[Ca^{2+}]_i$ was measured before and after the addition of various secretogogues and correlated with parallel measurements of PTH secretion. The addition of extracellular Ca^{2+} to parathyroid cells to raise the extracellular Ca^{2+} concentration from 0.5 to 2 mM results in a rapid transient increase, which is followed by a steady-state increase in $[Ca^{2+}]_i$ from 300 to 900 nM. These increases in steady-state $[Ca^{2+}]_i$ are accompanied by a decrease in PTH secretion from 700 to 360 ng/mg cell protein. The subsequent addition of 10 μM dopamine or 5 μM norepinephrine (not shown) returns the secretory rate to values obtained when $[Ca^{2+}]_i$ is low, yet causes no detectable changes in $[Ca^{2+}]_i$. In an additional experiment, the addition of 20 mM LiCl to cells suspended in 2 mM $CaCl_2$ results in an approximately twofold increase in rate of PTH secretion as shown previously.[8] This change in secretion likewise occurs in the absence of any change in $[Ca^{2+}]_i$.

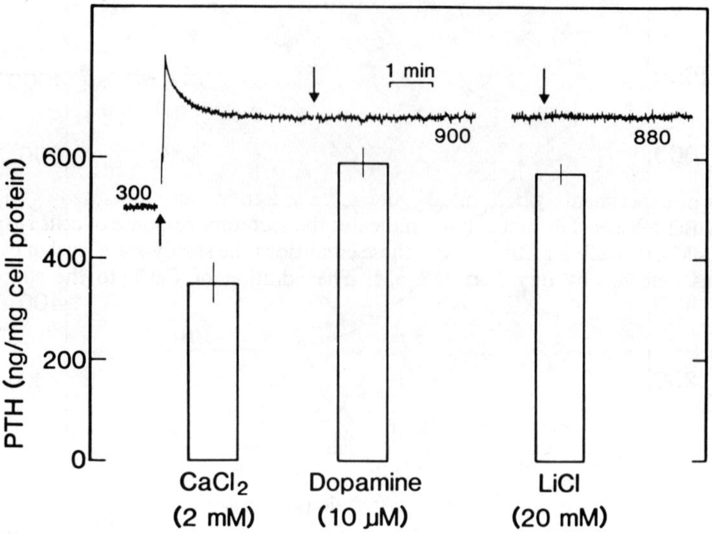

FIGURE 3. Potentiation of PTH secretion unaccompanied by changes in steady-state $[Ca^{2+}]_i$. Cells were loaded with fura-2, washed, and resuspended in buffer containing 2 mM $CaCl_2$. Incubation at 37°C in the presence or absence of dopamine or LiCl was for 5 min before samples were taken and assessed for PTH. Above the columns representing PTH secretion are the corresponding fluorescent traces obtained from the same cell preparation. At the *arrows,* the concentration of extracellular $CaCl_2$ was increased from 0.5 to 2 mM or dopamine or LiCl was added at the indicated final concentration. The numbers accompanying the traces are $[Ca^{2+}]_i$, in nM.

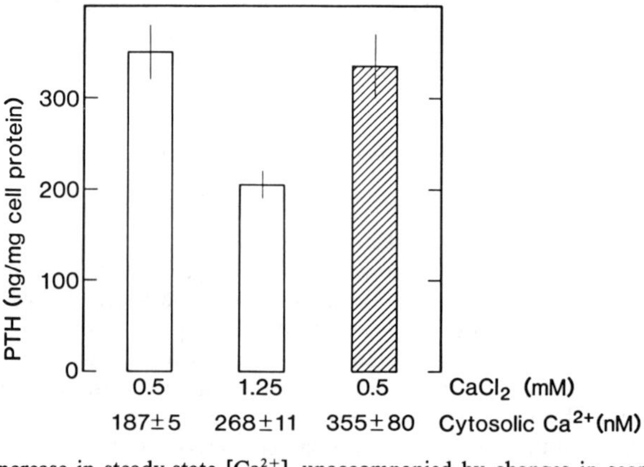

FIGURE 4. Increase in steady-state $[Ca^{2+}]_i$ unaccompanied by changes in secretion of PTH. Cells were loaded with quin-2 and incubated for 5 min in buffer containing 0.5 or 1.25 mM $CaCl_2$ in the presence (*hatched column*) or absence (*open columns*) of 1 μM ionomycin. In parallel experiments, steady-state $[Ca^{2+}]_i$ was determined at both concentrations of extracellular $CaCl_2$ and after the addition of ionomycin; these values are given below the appropiate columns. In fura-2-loaded cells, ionomycin likewise increases $[Ca^{2+}]_i$ and fails to depress secretion in buffer containing 0.5 mM $CaCl_2$.

Changes in Steady-State $[Ca^{2+}]_i$ Unaccompanied by Changes in Secretion of PTH

To further explore the relationship between steady-state $[Ca^{2+}]_i$ and PTH secretion, the effect of ionomycin, a divalent cation ionophore that would be expected to alter $[Ca^{2+}]_i$ at a constant value of extracellular Ca^{2+}, was examined. FIGURE 4 shows the results of experiments where $[Ca^{2+}]_i$ and PTH secretion were measured in the same population of cells. The first column indicates the secretory response of cells suspended in 0.5 mM extracellular Ca^{2+}. Under these conditions the steady-state level of cytosolic Ca^{2+} as detected by quin-2 is 187 nM. The addition of $CaCl_2$ to the suspending

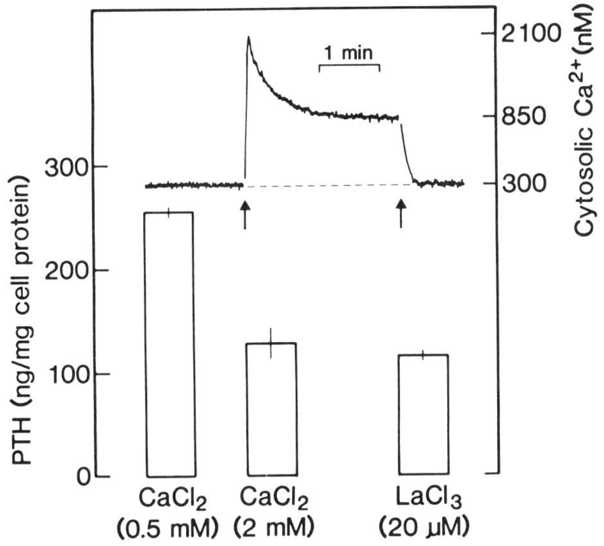

FIGURE 5. Decrease in steady-state $[Ca^{2+}]_i$ unaccompanied by changes in secretion of PTH. Cells were loaded with fura-2 and incubated 20 min in buffer containing 0.5 or 2 mM $CaCl_2$, the latter with or without 20 μM $LaCl_3$. The columns represent the secretory response, and above them is a representative fluorescent recording showing the corresponding changes in $[Ca^{2+}]_i$. At the *arrows,* the concentration of extracellular $CaCl_2$ was increased from 0.5 to 2 mM, or 20 μM $LaCl_3$ was added to cells equilibrated in 2 mM Ca^{2+}-containing buffer.

medium to achieve a final concentration of extracellular Ca^{2+} of 1.25 mM raises $[Ca^{2+}]_i$ to 268 nM. As expected, PTH secretion is depressed by approximately 50% (middle column). On the other hand, the addition of ionomycin to cells bathed in 0.5 mM extracellular Ca^{2+} causes no significant change in PTH secretion (hatched column). In this condition, however, $[Ca^{2+}]_i$ is increased above the level observed in 1.25 mM Ca^{2+}, where the rate of secretion is nearly half of the control rate. These observations indicate that the rate of PTH secretion can be readily dissociated from the steady-state level of $[Ca^{2+}]_i$ rendered artificially high with ionomycin.

The experiment of FIGURE 5 shows the reverse of this condition, where the $[Ca^{2+}]_i$ was artificially lowered without altering the concentration of extracellular Ca^{2+}. This was achieved by adding a low concentration of La^{3+}, which specifically depresses steady-state $[Ca^{2+}]_i$.[10-12] The level of cytosolic Ca^{2+} in parathyroid cells suspended initially in buffer containing 0.5 mM $CaCl_2$ is typically around 300 nM. Upon increasing extracellular Ca^{2+} to 2 mM, $[Ca^{2+}]_i$ rapidly peaks to about 2 μM and then settles into a new steady-state level around 850 nM. This higher steady-state level of cytosolic Ca^{2+} is accompanied by a 50% decrease in secretion of PTH. The subsequent addition of $LaCl_3$ (20 μM) to cells bathed in high extracellular Ca^{2+} provokes a rapid decrease in $[Ca^{2+}]_i$ to levels comparable to those obtained in the presence of low extracellular Ca_2^{+} (300 nM). Under these conditions, however, PTH secretion remains suppressed to an extent equal to that observed in the absence of La^{3+}, despite a threefold reduction in $[Ca^{2+}]_i$. These observations, taken together with those of FIGURE 4, show that steady-state $[Ca^{2+}]_i$ can be increased or decreased without an accompanying change in the secretory response of parathyroid cells. In each instance, the level of extracellular Ca^{2+} remains constant, and this might explain the absence of change in the secretory response despite changes in steady-state $[Ca^{2+}]_i$.

Rapid Ca^{2+} Transients and PTH Secretion

If cytosolic Ca^{2+} is to play some role in regulating secretion of PTH, then our attention must clearly shift to the initial cytosolic Ca^{2+} transients induced by extracellular Ca^{2+} since changes in steady-state $[Ca^{2+}]_i$ can be readily dissociated from changes in secretion.

Cytosolic Ca^{2+} transients arise from the rapid mobilization of cellular stores of Ca^{2+} in constrast to the steady-state $[Ca^{2+}]_i$, which is dependent on the presence of extracellular Ca^{2+} and likely reflects transmembrane Ca^{2+} influx.[10-12] A comparison of FIGURES 1 and 2 shows that cytosolic Ca^{2+} transients are observed in fura-2-loaded

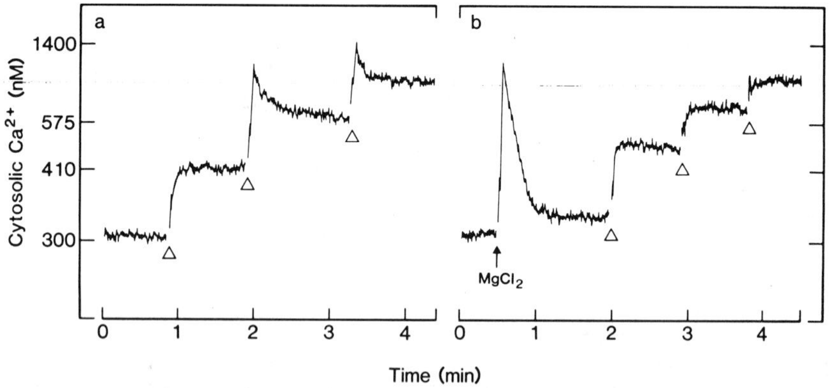

FIGURE 6. Mg^{2+} evokes transient increases in $[Ca^{2+}]_i$ and blocks cytosolic Ca^{2+} transients elicited by extracellular Ca^{2+}. Cells were loaded with fura-2 and equilibrated at 37°C in buffer containing 0.5 mM $CaCl_2$. At each of the *arrowheads,* the concentration of extracellular $CaCl_2$ was increased by 0.5 mM increments to a final of 2 mM. Panel **a** is a control tracing and shows the transient and steady-state increases in $[Ca^{2+}]_i$ elicited by stepwise increases in the concentration of extracellular Ca^{2+}. Panel **b** shows that increasing the concentration of extracellular $MgCl_2$ from 1 to 8 mM (in the presence of 0.5 mM $CaCl_2$) evokes only a cytosolic Ca^{2+} transient. In the presence of 8 mM $MgCl_2$, transient but not steady-state increases in $[Ca^{2+}]_i$ evoked by extracellular Ca^{2+} are abolished.

cells at concentrations of extracellular Ca^{2+} (1-2 mM) that produce the largest decrements in PTH secretion. Alterations in the level of extracellular Ca^{2+} outside this range fail to elicit cytosolic Ca^{2+} transients. Other divalent cations, such as Mg^{2+}, also elicit rapid and transient increases in $[Ca^{2+}]_i$ (FIG. 6). Unlike cytosolic Ca^{2+} transients evoked by extracellular Ca^{2+}, however, those elicited by Mg^{2+} (or Sr^{2+}) are not followed by large increases in steady-state $[Ca^{2+}]_i$. Moreover, increasing the concentration of extracellular Mg^{2+}, or adding Sr^{2+} (not shown), inhibits cytosolic Ca^{2+} transients without affecting increases in steady-state $[Ca^{2+}]_i$ elicited by extracellular Ca^{2+} (FIG. 6). This finding, together with various others,[11] shows that all these cations act at a similar site to cause the rapid mobilization of Ca^{2+} from some cellular store.

FIGURE 7 shows that various divalent cations (including Ba^{2+}, not shown) all cause an inhibition of PTH secretion at concentrations that elicit cytosolic Ca^{2+} transients. Because the level of steady-state intracellular Ca^{2+} changes very little upon increasing Mg^{2+} or adding Sr^{2+}, but secretion is inhibited maximally, these results are yet further evidence for the view that steady-state $[Ca^{2+}]_i$ is not an important intracellular signal regulating secretion of PTH. They also focus attention on the cytosolic Ca^{2+} transient as a possible event necessary to produce inhibition of secretion.

To examine the possible role of these rapid changes in $[Ca^{2+}]_i$ in regulating PTH secretion, we took advantage of the ability of quin-2 to buffer, and thereby suppress, cytosolic Ca^{2+} transients. FIGURE 8 shows that PTH secretion at different extracellular Ca^{2+} concentrations is virtually identical in controls and in cells where the intracellular concentration of quin-2 is 1.2 mM. Parathyroid cells loaded with this concentration of quin-2 do not show rapid cytosolic Ca^{2+} transients in response to increasing extracellular Ca^{2+} concentrations (see FIG. 1). FIGURE 8b shows that in fura-2-loaded cells, the addition of 8 mM Mg^{2+} inhibits PTH release by 56%, and evokes a transient, 8.5-fold increase in $[Ca^{2+}]_i$. Such Ca^{2+} transients are inhibited by 97% in cells loaded with quin-2, yet this buffering action fails to affect the inhibitory action of Mg^{2+}, which depresses secretion by 56%. These observations compel us to suppose that cytosolic Ca^{2+} per se may not be the intracellular signal acting to suppress secretion of PTH.

CONCLUSIONS

The use of fluorescent indicators to measure $[Ca^{2+}]_i$ has revealed some intriguing aspects of stimulus-secretion coupling in parathyroid cells. Small changes in the concentration of extracellular Ca^{2+}, which is the major physiological secretagogue, result in large changes in $[Ca^{2+}]_i$. Increasing the concentration of extracellular Ca^{2+} evokes two mechanistically distinct responses: a rapid and transient increase in $[Ca^{2+}]_i$ that arises from the mobilization of cellular Ca^{2+} and an increase in the steady-state $[Ca^{2+}]_i$ which possibly results from influx of Ca^{2+} across the plasma membrane.[10-12] Unlike other secretory cells, these increases in $[Ca^{2+}]_i$ are associated with an inhibition of secretion. Nonetheless, it is clear from the results reported here that changes in $[Ca^{2+}]_i$, of both a transient and a sustained nature, can be dissociated from changes in secretion. Thus, when the concentration of extracellular Ca^{2+} is low, the secretory rate of PTH is high, despite artificially induced increases in $[Ca^{2+}]_i$ (as with ionomycin). Conversely, PTH secretion remains depressed in the presence of high extracellular Ca^{2+} concentrations, in spite of net decreases in $[Ca^{2+}]_i$ (as with La^{3+}). Likewise, blocking transient increases in $[Ca^{2+}]_i$ by loading cells with high concentrations of quin-2 fails to block the inhibitory effects of increased extracellular Ca^{2+} or Mg^{2+}.

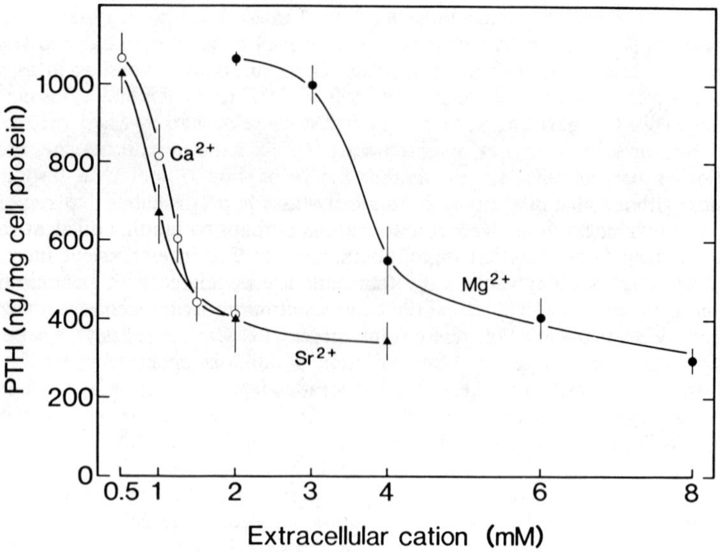

FIGURE 7. Inhibition of PTH secretion by various divalent cations. Parathyroid cells were loaded with fura-2 and incubated (37°C) for 30 min in buffer containing the indicted concentration of $CaCl_2$ (○) or in buffer containing 0.5 mM $CaCl_2$ and the indicated concentration of $SrCl_2$ (▲) or $MgCl_2$ (●).

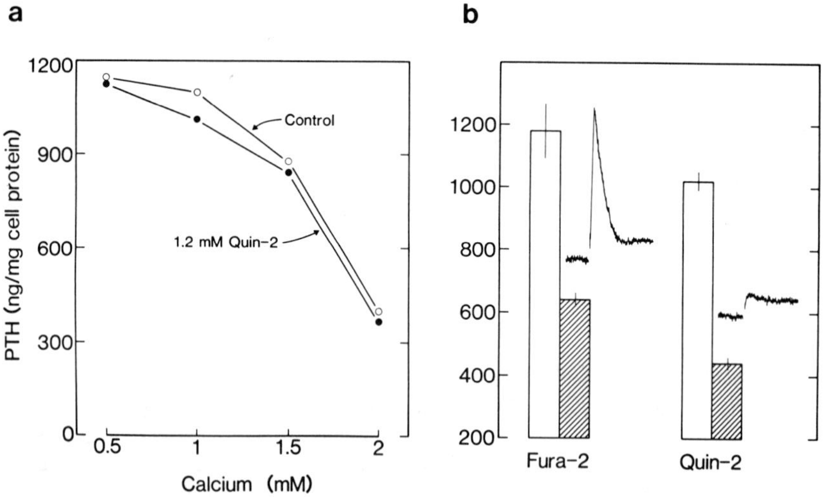

FIGURE 8. Regulation of PTH secretion by extracellular Ca^{2+} or Mg^{2+} in the absence of transient increases in $[Ca^{2+}]_i$. **(a)** Parathyroid cells were loaded, or not (control), with a high concentration of quin-2 and incubated for 15 min in buffer containing the indicated concentration of $CaCl_2$. Cytosolic Ca^{2+} transients are effectively buffered at this concentration of intracellular quin-2 (see FIG. 1). **(b)** Parathyroid cells were loaded with quin-2 (about 0.8 mM intracellularly) or fura-2 (about 100 μM intracellularly) and incubated in buffer containing 0.5 mM $CaCl_2$ with (*hatched columns*) or without (*open columns*) 8 mM $MgCl_2$ for 15 min at 37°C. The accompanying traces, obtained from the same preparations of indicator-loaded cells, show that quin-2 effectively blocks the cytosolic Ca^{2+} transients evoked by Mg^{2+}.

The common denominator underlying the regulation of PTH secretion in all these studies is not $[Ca^{2+}]_i$ but rather the concentration of extracellular Ca^{2+}, or more properly, the concentration of extracellular divalent cations. Extracellular cations must therefore produce changes in some intracellular signal besides Ca^{2+}.

This realization, together with the evidence for a "divalent cation receptor" on parathyroid cells,[10,12] allows us to propose a tentative model for the regulation of secretion in parathyroid cells. The model holds, in its initial form, that there exists a receptor on the surface of the parathyroid cell which is promiscuous and responds to a variety of divalent cations. Receptor activation, perhaps by binding divalent cations, leads to the rapid mobilization of cellular Ca^{2+} resulting in a transient increase in $[Ca^{2+}]_i$. Although such cytosolic Ca^{2+} transients are associated with an inhibition of secretion, they do not appear to be the causal factor coupling receptor activation to the control of secretion. Rather, these transients appear to provide just one convenient index of receptor activation and the results suggest that some other, as yet unidentified, intracellular signal plays a pivotal role in regulating secretion of PTH. It remains to be determined if receptor activation (or occupancy) results in the formation of an inhibitory intracellular signal or the loss of a stimulatory one.

REFERENCES

1. HABENER, J. F., M. ROSENBLATT & J. T. POTTS. 1984. Parathyroid hormone: Biochemical aspects of biosynthesis, secretion, action, and metabolism. Physiol. Rev. **64:** 985-1053.
2. BROWN, E. M. 1982. PTH secretion in vivo and in vitro. Regulation by calcium and other secretagogues. Mineral Electrolyte Metab. **8:** 130-150.
3. DOUGLAS, W. W. 1974. Involvement of calcium in exocytosis and the exocytosis-vesiculation sequence. Biochem. Soc. Symp. **39:** 1-28.
4. RUBIN, R. P. 1982. Calcium and Cellular Secretion. Plenum. New York.
5. SHOBACK, D. M., J. THATCHER, R. LEOMBRUNO & E. M. BROWN. 1984. Relationship between parathyroid hormone secretion and cytosolic calcium concentration in dispersed bovine parathyroid cells. Proc. Natl. Acad. Sci. U.S.A. **81:** 3113-3117.
6. LARSSON, R., G. AKERSTROM, E. GYLFE, H. JOHANSSON, S. LJUNGHALL, J. RASTAD & C. WALLFELT. 1985. Paradoxical effects of K^+ and D-600 on parathyroid hormone secretion and cytoplasmic Ca^{2+} in normal bovine and pathological human parathyroid cells. Biochem. Biophys. Acta **847:** 263-269.
7. MUFF, R. & J. A. FISCHER. 1986. Stimulation of parathyroid hormone secretion by phorbol esters is associated with a decrease of cytosolic calcium. FEBS Lett. **194:** 215-218.
8. NEMETH, E. F., J. WALLACE & A. SCARPA. 1986. Stimulus-secretion coupling in bovine parathyroid cells. Dissociation between secretion and net changes in cytosolic Ca^{2+}. J. Biol. Chem. **261:** 2668-2674.
9. WALLACE, J. & A. SCARPA. 1982. Regulation of parathyroid hormone secretion *in vitro* by divalent cations and cellular metabolism. J. Biol. Chem. **257:** 10613-10616.
10. NEMETH, E. F. & A. SCARPA. 1986. Cytosolic Ca^{2+} and the regulation of secretion in parathyroid cells. FEBS Lett. **203:** 15-19.
11. NEMETH, E. F. & A. SCARPA. 1986. Regulation of cytosolic Ca^{2+} in parathyroid cells by extracellular divalent cations: Contributions from cellular and extracellular sources of Ca^{2+}. *In* Calcium Ion: Membrane Transport and Cellular Regulation. F. Carpenedo, P. Debetto, M. Floreani & S. Luciani, Eds.: 27-33. University of Padua Press. Padua, Italy.
12. NEMETH, E. F. 1986. A receptor for extracellular cations on parathyroid cells. This volume.

Inositol Phospholipid Turnover and Protein Phosphorylation in Secretory Responses[a]

MASAYOSHI GO, HIDEAKI NOMURA, TATSURO KITANO, JUNKO KOUMOTO, USHIO KIKKAWA, NAOAKI SAITO, CHIKAKO TANAKA, AND YASUTOMI NISHIZUKA

Departments of Biochemistry and Pharmacology
Kobe University School of Medicine
Kobe 650, Japan

Receptor-mediated hydrolysis of inositol phospholipids was recently discovered to be a common mechanism for transducing various extracellular signals into the cell.[1,2] At an early phase of cellular responses, inositol-1,4,5-trisphosphate mobilizes Ca^{2+}, whereas 1,2-diacylglycerol activates protein kinase C. These two intracellular mediators are generated from the hydrolysis of a single molecule, phosphatidylinositol-4,5-bisphosphate (PIP_2).[3–5] Evidence available to date suggests that protein kinase C plays crucial roles in the relay of information from a group of hormones, some neurotransmitters, secretagogues, and other biologically active substances across the membrane to regulate many intracellular Ca^{2+}-dependent processes. The role of this enzyme in such signal transduction was first demonstrated for the release of serotonin from platelets,[6] and subsequently shown for secretion and exocytosis from a variety of endocrine and exocrine tissues as well as neuronal and inflammatory cells. This article will briefly summarize some aspects of the role of protein kinase C in secretory responses.

PIP_2, a minor component of inositol phospholipids, is produced from phosphatidylinositol (PI) via phosphatidylinositol-4-phosphate (PIP) by sequential phosphorylation of the *myo*-inositol moiety. Although PIP_2 has been recently regarded as a primary target, PI and PIP also disappear when cell surface receptors are stimulated.[3–5] It is plausible at present that in most tissues including platelets the three inositol phospholipids are broken down, probably at different rates at different times, resulting in the formation of diacylglycerol and three inositol phosphates.

The early evidence for the link between the diacylglycerol formation and protein kinase C activation came primarily from the experiments with platelets.[7] Inositol phospholipids in mammalian tissues frequently contain a 1-stearoyl-2-arachidonyl backbone. FIGURE 1 shows that protein kinase C is in fact activated effectively by 1-

[a] This research has been supported by the Ministry of Education, Science and Culture, Japan; Muscular Dystrophy Association; Yamanouchi Foundation for Research on Metabolic Disorders; Merck Sharp & Dohme Research Laboratories; Central Research Laboratories; Ajinomoto Company; Biotechnology Laboratories; Takeda Chemical Industries; and Meiji Institute of Health Sciences.

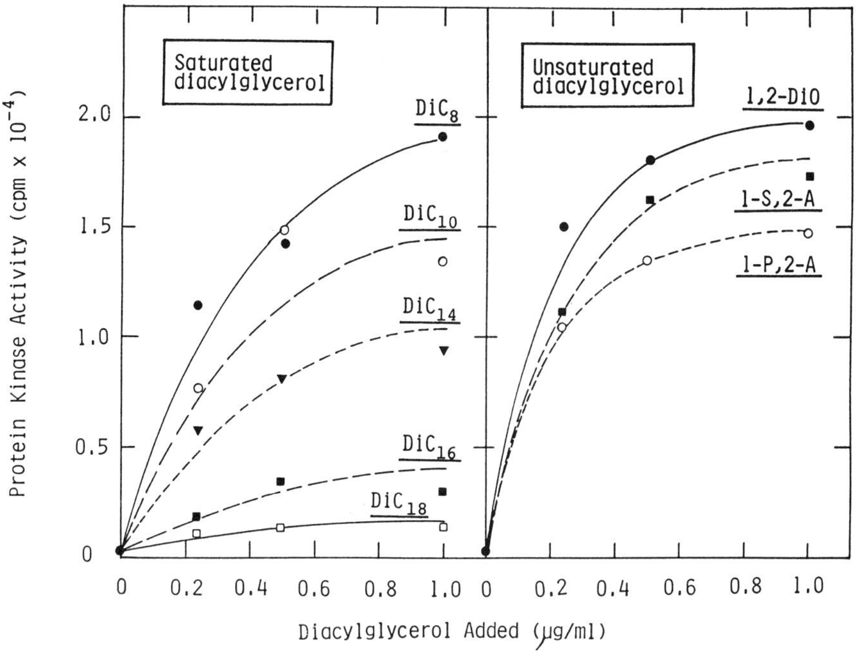

FIGURE 1. Activation of protein kinase C by diacylglycerols with various fatty acyl moieties. Homogeneous protein kinase C purified from rat brain was assayed in the presence of phosphatidylserine, Ca^{2+}, and diacylglycerol indicated. The detailed conditions will be described elsewhere. *DiC_8*, 1,2-dioctanoylglycerol; *DiC_{10}*, 1,2-didecanoylglycerol; *DiC_{14}*, 1,2-dimyristylglycerol; *DiC_{16}*, 1,2-dipalmitoylglycerol; *DiC_{18}*, 1,2-distearoylglycerol; *1,2-DiO,* 1,2-dioleoylglycerol; *1-S,2-A,* 1-stearoyl-2-arachidonylglycerol; and *1-P,2-A,* 1-palmitoyl-2-arachidonylglycerol.

stearoyl-2-arachidonylglycerol. In *in vitro* reactions, other diacylglycerols containing at least one unsaturated fatty acid are active, and those having saturated fatty acids with short chains appear to also be active. All diacylglycerols active in this role have a 1,2-*sn*-configuration, and neither 2,3-*sn*- nor 1,3-diacylglycerol supports the enzyme activation.[8] In the experiments shown in FIGURE 2, this stereospecificity of diacylglycerol is shown with both purified enzyme and intact cell systems. When human platelets were stimulated by three stereoisomers of permeable diacylglycerol, dioctanoylglycerol, 1,2-*sn*- but not 2,3-*sn*-enantiomer was found to be readily intercalated into the membrane, and activated protein kinase C directly, as judged by the phosphorylation of its specific endogeneous phosphate acceptor protein that has an approximate molecular weight of 47,000. In the presence of Ca^{2+} ionophore, A23187, 1,2-*sn*-enantiomer caused full activation of the release of serotonin. 1,3-Dioctanoylglycerol was inactive. Among these isomers only 1,2-*sn*-dioctanoylglycerol was converted in situ very rapidly to the corresponding phosphatidic acid in both intact and broken platelet preparations. Thus, the diacylglycerol which functions in the stimulus-response coupling possesses a 1,2-*sn*-glycerol backbone, and other isomers are not involved in the signal transduction through inositol phospholipid breakdown.[9] In cells and tissues, several diacylglycerols may be produced as intermediates in triacylglycerol metabolism as well as in phospholipid biosynthesis. As pointed out elsewhere,[9] however, the diacylglycerol derived from triacylglycerol by the action of lipoprotein lipase

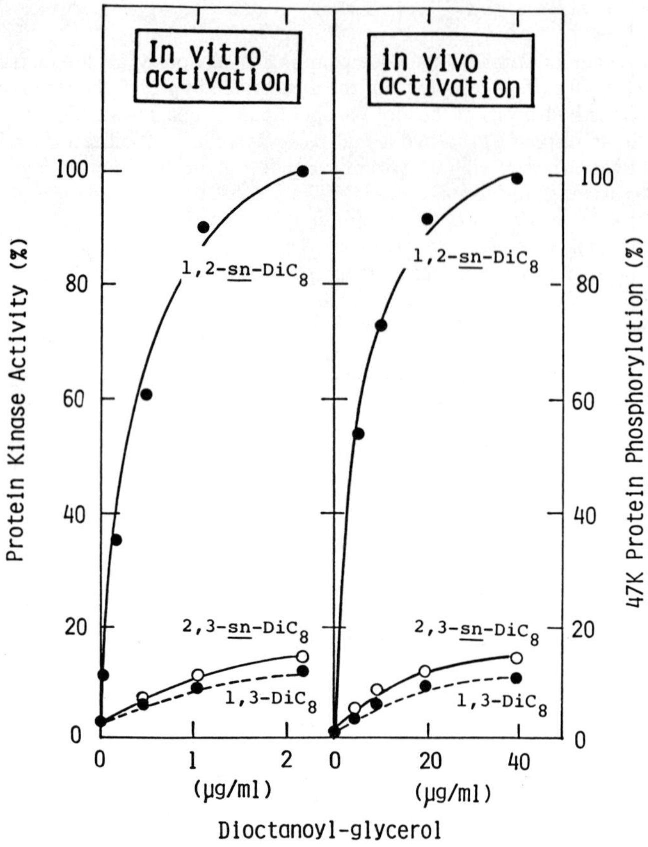

FIGURE 2. Stereospecificity of diacylglycerol in the activation of protein kinase C. In *in vitro* assays the enzyme activity was determined under the conditions similar to those given in FIGURE 1. In *in vivo* assays, human platelets were isolated and labeled with $^{32}P_i$ and then stimulated by diacylglycerol indicated. The phosphorylation of 47-kDa protein, a specific platelet protein kinase C substrate, was determined by measuring the incorporation of radioactive phosphate into that protein, which was separated by SDS-polyacrylamide gel electrophoresis and subjected to autoradiography, followed by densitometric tracing at 430 nm. Detailed conditions will be described elsewhere. *1,2-sn-DiC*$_8$, 1,2-*sn*-dioctanoylglycerol; *2,3-sn-DiC*$_8$, 2,3-*sn*-dioctanoylglycerol; and *1,3-DiC*$_8$, 1,3-dioctanoylglycerol.

and a heparin-releasable hepatic lipase has a 2,3-*sn*-configuration. On the other hand, many digestive lipases such as pancreatic and lysosomal lipases do not appear to produce any stereospecific diacylglycerol, and 2-monoacylglycerol is a major product that is formed rapidly by way of 1,2-*rac*-diacylglycerol. There must be special compartmentation or interaction of the active 1,2-*sn*-diacylglycerol with membrane phospholipid bilayer. The precise biochemical mechanism of this enzyme activation remains largely unexplored.

The first event of stimulus-response coupling is obviously an interaction of the agonist with its cell surface receptor, but the mechanism by which the receptor-agonist complex initiates hydrolysis of inositol phospholipids is not known. During the early phase of cellular responses, the two signal pathways, Ca^{2+} mobilization and protein kinase C activation, open rapidly within seconds but normally close very quickly, since the appearances of inositol-1,4,5-trisphosphate and 1,2-diacylglycerol are both transient. A major mechanism for terminating the signal flow via inositol-1,4,5-trisphosphate is thought to be removal of its 5-phosphate.[2,4,5] Although there must be more than one mechanism eventually leading to the required intracellular concentration of Ca^{2+}, the Ca^{2+} signal in most tissues is spike-like in nature. In experiments given in FIGURE 3, this Ca^{2+} spike has been successfully observed with the photoprotein, aequorin.[10]

Similarly, the diacylglycerol once produced in membranes disappears within a few seconds, or at most, several minutes of its formation. This rapid disappearance of

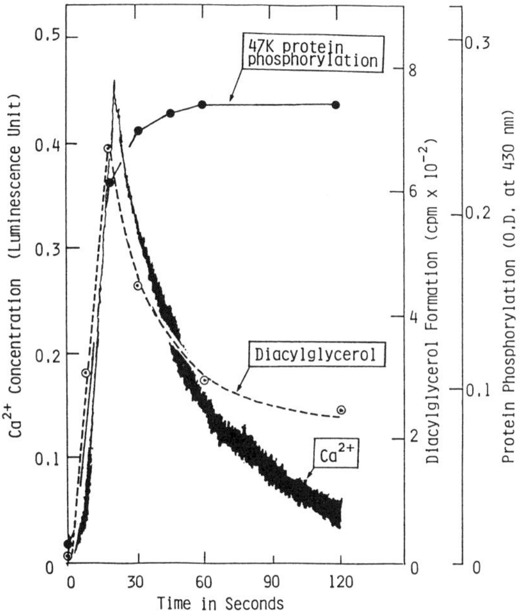

FIGURE 3. Thrombin-induced increase in cytosolic Ca^{2+}, diacylglycerol formation, and protein kinase C specific protein phosphorylation in human platelets. Platelets were isolated, labeled with either [^{3}H]arachidonate or $^{32}P_i$, and stimulated by thrombin. The radioactive diacylglycerol was extracted directly from the platelets, isolated by thin-layer chromatography, and determined. The phosphorylation of 47-kDa protein was determined as described in FIGURE 2. In a separate set of experiments platelets were loaded by aequorin, and stimulated by thrombin. The luminescence was traced using a Packard luminometer analyzer, Pico-Lite.

TABLE 1. Proposed Roles of Protein Kinase C in Secretory Responses

Tissues and Cells	Responses[a]
Endocrine Systems	
adrenal medulla	catecholamine secretion
adrenal cortex	aldosterone section
pancreatic islets	insulin release
insulinoma cells	insulin release
pituitary cells	pituitary hormone release
	growth hormone release
	luteinizing hormone release
	prolactin release
	thyrotropin release
parathyroid cells	parathyroid hormone release
thyroid C cells	calcitonin release
Exocrine Systems	
pancreas	amylase secretion
parotid gland	amylase and mucin secretion
submandibular gland	mucin secretion
gastric gland	pepsinogen secretion
	gastric acid secretion
alveolar cells	surfactant secretion
Nervous Systems	
ileal nerve endings	acetylcholine release
neuromuscular junction	transmitter release
caudate nucleus	acetylcholine release
PC12 cells	dopamine release
neurons	dopamine release
Inflammation and Immune Systems	
platelets	serotonin release
	lysosomal enzyme release
	arachidonate release
neutrophils	superoxide release
	lysosomal enzyme release
basophils	histamine release
mast cells	histamine release

[a] References are given in Kikkawa and Nishizuka.[11]

diacylglycerol is due both to its conversion back to inositol phospholipids through phosphatidic acid and to its further degradation to arachidonic acid, which in turn generates other messengers such as prostaglandins. Thus, protein kinase C reveals its activity only for a short time after the stimulation of receptors. The consequence of this enzyme activation may persist for a longer period, however, depending upon the biological stability of the phosphate that is covalently attached to the substrate protein molecule.

As mentioned above, the role of protein kinase C in cell surface signal transduction has been shown for the release reaction, secretion, and exocytosis of cellular constituents from various endocrine and exocrine tissues as well as for the activation of many other cellular functions. In TABLE 1, some examples are given that are reported in the literature.[11,12] Under appropriate conditions, the two signal pathways, Ca^{2+}

mobilization and protein kinase C activation, can be opened selectively and independently by the application of Ca^{2+} ionophore such as A23187 or ionomycin for the former, and permeable diacylglycerol or tumor-promoting phorbol ester (a substitute for diacylglycerol) for the latter. By using this procedure, it became possible to show that the two signal pathways are both essential and often act synergistically to elicit full cellular responses.[1] Such a potential role of protein kinase C in stimulus-response coupling has been extrapolated to neural functions, particularly transmitter release in both peripheral and central nervous systems as well as modulation of ion conductance in membranes.[12] In fact, the monoclonal antibodies have confirmed the presence of a protein kinase C positive immunoreactive material in many regions of the cytoplasm. The precise intracellular localization of protein kinase C appears to vary with cell types, and the enzyme is seemingly absent or poorly represented in the nucleus. FIGURE 4 shows a typical feature of rat Purkinje cells, and a large quantity of this protein kinase is associated with dendrites and axons.

Several lines of evidence available thus far suggest that protein kinase C modulates ion conductance by phosphorylating membrane proteins such as channels, pumps, and ion exchange proteins. It is proposed that protein kinase C plays a role in extrusion of Ca^{2+} immediately after its mobilization into the cytosol, and that Ca^{2+}-transport ATPase or its regulatory protein is a possible target of this protein kinase. As noted above, when receptors are stimulated the appearance of Ca^{2+} is very transient. In various cell types, the cytosolic Ca^{2+} concentration is frequently decreased by the addition of phorbol ester. In addition, it is postulated that protein kinase C may take part in the enhancement of Ca^{2+} entry. This proposal was based on the observation that microinjection of phorbol ester or protein kinase C itself into bag cell neurons from *Aplysia* enhances the voltage-sensitive Ca^{2+} current. Similarly, a possible role of protein kinase C in activating Na^{+}-transport ATPase in peripheral nerve has been suggested. The Na^{+}/H^{+} exchange protein appears to be another target that is activated by phorbol ester or by permeable diacylglycerol, and thus protein kinase C may function to increase cytoplasmic pH. An additional example is obtained with photoreceptor cells of *Hermissenda* and rat hippocampal pyramidal neurons. In these cells, treatment with phorbol ester or permeable diacylglycerol, as well as intracellular injection of protein kinase C, reduces Ca^{2+}-dependent K^{+} conductance. These possible roles of protein kinase C in the modulation of membrane functions are reviewed.[11,12] However, the biochemical target proteins of this enzyme in the modulation of these membrane functions remain largely unknown.

In biological systems, positive signals are normally followed by immediate negative feedback control to prevent overshoot, and to repeat responses to subsequent signals. Although much emphasis has been placed on the positive role of protein kinase C, major functions of this enzyme appear to be intimately related to such feedback control or down-regulation. The dual actions of protein kinase C are attributed entirely to the functional consequence of the phosphorylation of target molecule. Recent evidence obtained for several tissues including secretory cells such as platelets, neutrophils, and pituitary cells, strongly suggests that protein kinase C exerts negative feedback control on the receptors that are related to inositol phospholipid breakdown. The roles of this enzyme in such dual actions have been outlined.[12]

Signal-induced inositol phospholipid breakdown is usually associated with arachidonic acid release and cyclic GMP formation. Thus, protein kinase C activation, Ca^{2+} mobilization, arachidonic acid release, and cyclic GMP formation appear to be integrated in a single receptor cascade as schematically shown in FIGURE 5. Arachidonic acid may be derived from inositol phospholipids through two consecutive reactions catalyzed by phospholipase C followed by diacylglycerol lipase. However, this fatty acid may also be released from phosphatidylethanolamine as well as from phosphatidylcholine. It is likely that, when the receptor is stimulated, both phospho-

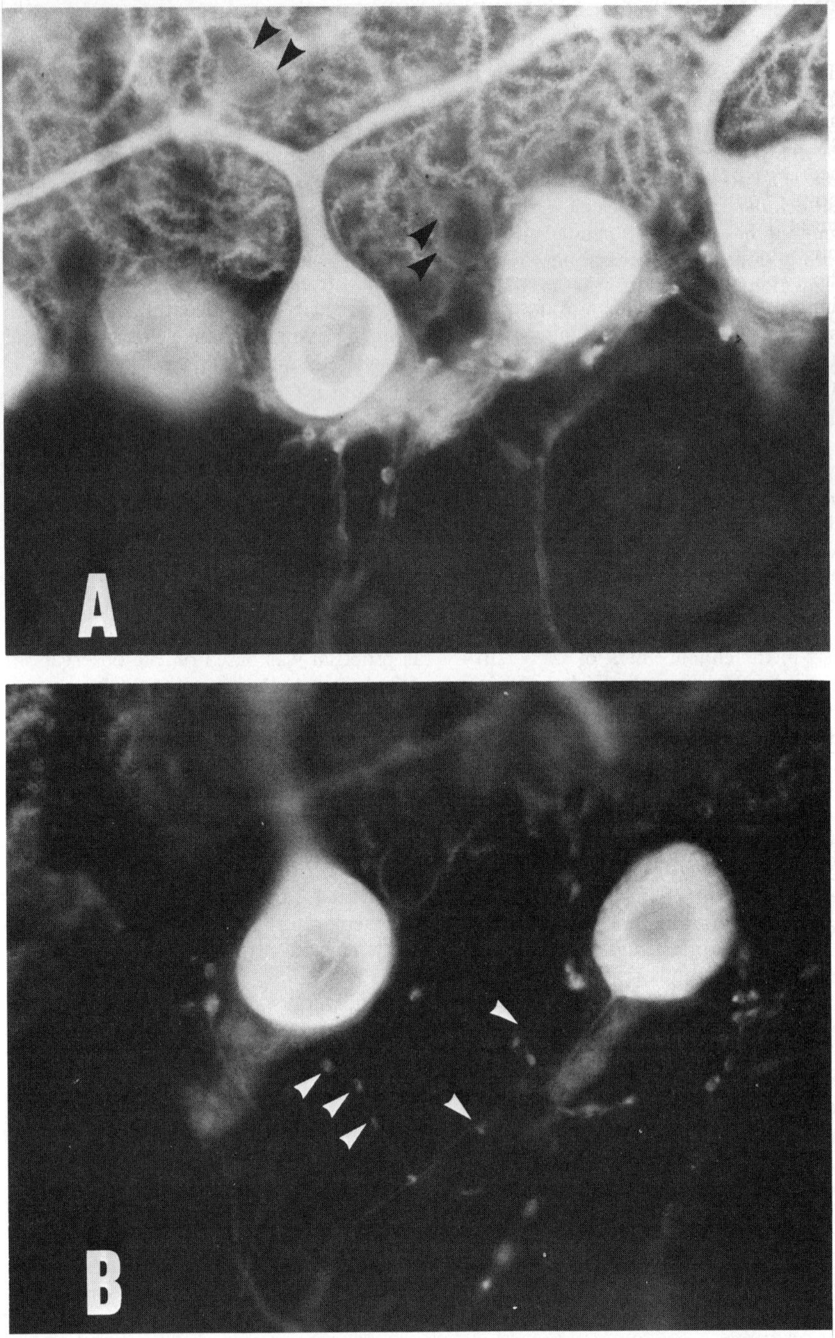

FIGURE 4. Immunofluorescence staining of rat Purkinje cells stained by monoclonal antibodies against protein kinase C. The detailed properties of the antibodies used and the conditions for staining will be described elsewhere. **(A)** Dendrites. Arrows show cell membrane of basket cells. **(B)** Cell bodies and axons. Arrows show axons.

lipases C and A_2 act in concert. Although Ca^{2+} causes direct activation of guanylate cyclase in some tissues, arachidonic acid peroxide and prostaglandin endoperoxide may also serve as activators for guanylate cyclase. It has been proposed that cyclic GMP may have a function to act as a "negative" rather than a "positive" intracellular mediator, providing again an intracellular feedback control that prevents over-response.[13] It is known that sodium nitroprusside, which induces a marked elevation of cyclic GMP levels, is a powerful inhibitor of smooth muscle contraction and platelet activation. As shown in FIGURE 6, 8-bromo cyclic GMP inhibits signal-induced inositol phospholipid breakdown, and thereby counteracts the activation of protein kinase C. Similar inhibitory actions are observed for several cyclic GMP-elevating agents such as sodium nitroprusside and nitroglycerol. Cyclic GMP may also take part in decrease in the intracellular Ca^{2+}, but crucial information on the role of this cyclic nucleotide in secretory tissues is still limited.

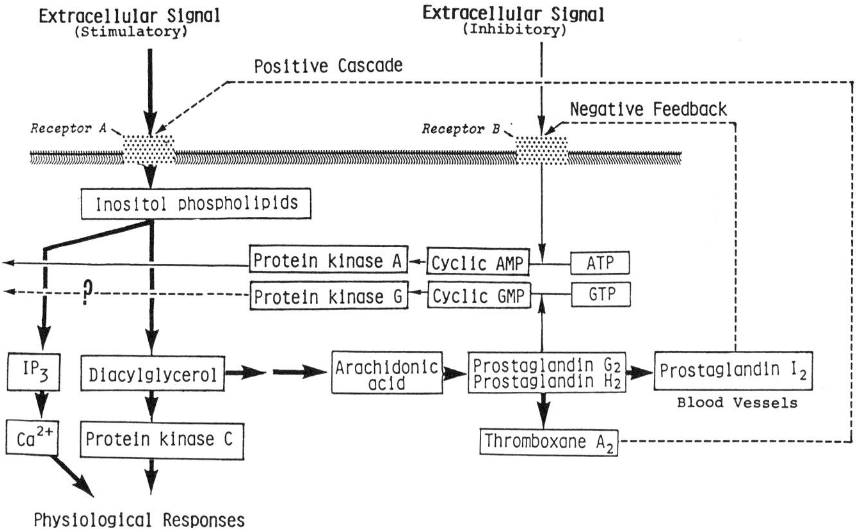

FIGURE 5. A possible metabolic cascade of inositol phospholipids and its feedback control by arachidonate metabolites and cyclic nucleotides. Platelets are used as a model system. IP_3, inositol-1,4,5-trisphosphate; protein kinase A and protein kinase G are cyclic AMP-dependent and cyclic GMP-dependent protein kinases, respectively.

FIGURE 7 illustrates the mode of cellular responses, which may be divided into several groups. In bidirectional control systems in many tissues, the signals that cause inositol phospholipid hydrolysis promote the activation of cellular functions, whereas the signals that produce cyclic AMP usually antagonize such activation. In platelets and neutrophils, for example, agonists that elevate cyclic AMP all block the signal-induced disappearance of inositol phospholipids, diacylglycerol formation, protein kinase C activation, and release reactions as shown in FIGURE 6. This inhibitory action of cyclic AMP extends to the mobilization of Ca^{2+}, presumably through the decreased formation of inositol-1,4,5-trisphosphate and through the activation of cyclic

AMP-dependent protein kinase (protein kinase A). Protein kinase A has the potential to decrease cytosolic Ca^{2+} concentration by phosphorylating the regulatory components of the Ca^{2+}-activated ATPase, thereby enhancing its catalytic activity. In other cell types that appear to be under an inverse form of bidirectional control, such as Leydig cells and ovarian granulosa cells, protein kinase C appears to inhibit and desensitize the adenylate cyclase system. On the other hand, in several other cell types including pinealocytes and pituitary cells, protein kinase C greatly potentiates cyclic AMP production. There is no obvious example as yet of a tissue in which cyclic AMP potentiates signal-induced turnover of inositol phospholipids. However, these two signal transduction pathways frequently act in concert in many endocrine cells such as pancreatic islets for the release of hormones. The evidence presented thus far is still incomplete, but it is reasonable to assume that various combinations of the two receptor systems may operate positively and sometimes be intensified in many secretory processes.[11,12] Additional major interactions of the signaling systems exist between calcium and cyclic AMP.[14,15] Further exploration of various interactions and networks of the signaling systems in individual cell types is of crucial importance for understanding the molecular basis of transmembrane control of secretory responses.

Protein kinase C has a broad substrate specificity *in vitro,* phosphorylating seryl and threonyl residues but not tyrosyl residue of endogenous proteins. Protein kinase C and protein kinase A relay information along different signal pathways within the cell as discussed above, but these two enzymes sometimes cause apparently similar cellular responses. These protein kinases often share the same phosphate acceptor proteins, even the same seryl and threonyl residues in a single protein molecule for

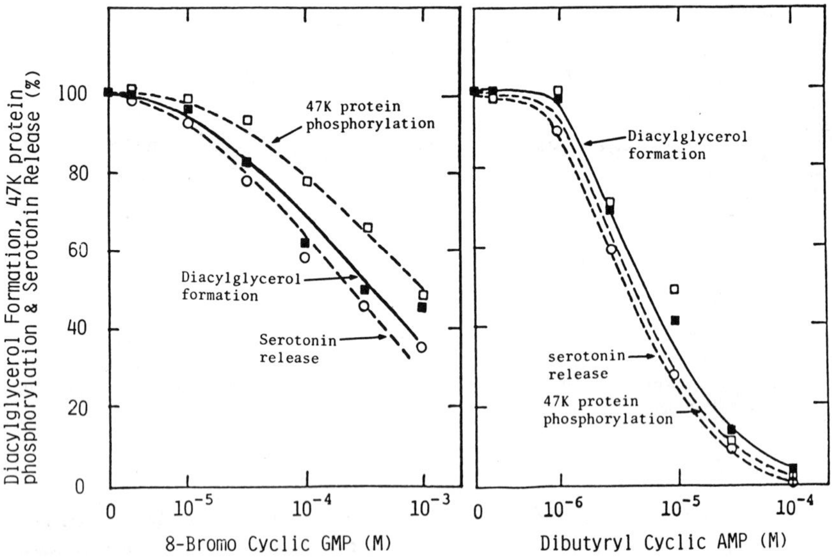

FIGURE 6. Inhibition of thrombin-induced activation of platelets by 8-bromo cyclic GMP and dibutyryl cyclic AMP. Diacylglycerol formation and 47-kDa protein phosphorylation were assayed under the conditions given in FIGURES 3 and 2, respectively. Serotonin release was assayed by measuring the radioactive serotonin that was taken up previously. Platelets were stimulated by thrombin in the presence of either 8-bromo cyclic GMP or dibutyryl cyclic AMP as indicated. The detailed experimental conditions have been described elsewhere.[7]

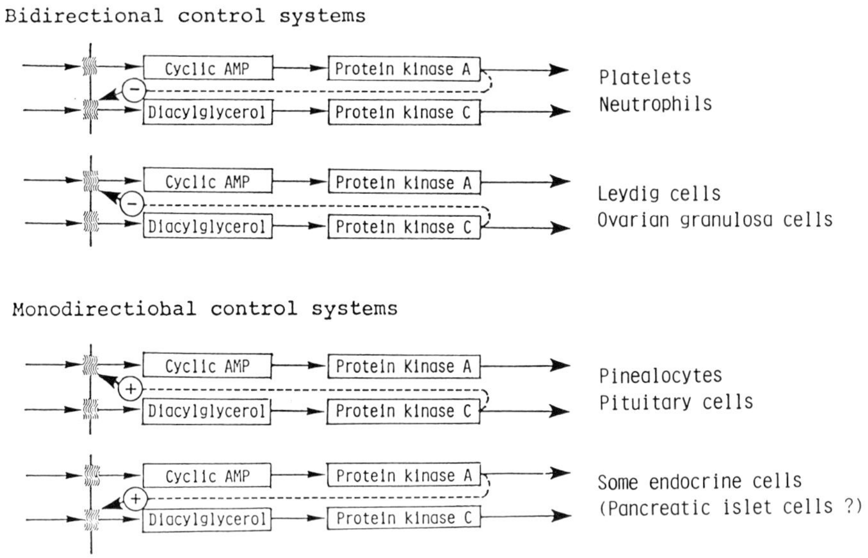

FIGURE 7. Mode of interaction of two major signal-transducing systems. Protein kinase A is cyclic AMP-dependent protein kinase.

phosphorylation.[16] The primary structure in the vicinity of the phosphorylation site is one of the determinant factors for substrate recognition.[17] The topographical arrangement or subcellular localization of the enzyme and its target proteins is another crucial factor for determining substrate specificity.

This article has briefly summarized some of our current knowledge on protein kinase C. The biochemical basis for the actions of protein kinase C outlined above is not fully substantiated, but the enzyme appears to play a crucial role in stimulus-response coupling in many tissues. Most likely, the signal-induced breakdown of inositol phospholipids initiates a cascade of events, starting with Ca^{2+} mobilization and protein kinase C activation. It seems premature to discuss the detailed relationship between the role of Ca^{2+} and that of protein kinase C. Each of the signal pathways plays diverse roles in controlling biochemical reactions. Nevertheless, it is reasonable to conclude that the protein phosphorylation catalyzed by this protein kinase may exert profound modulation of various Ca^{2+}-mediated processes, particularly release reaction, secretion, and exocytosis.

REFERENCES

1. Nishizuka, Y. 1984. The role of protein kinase C in cell surface signal transduction and tumour promotion. Nature **308:** 693-698.
2. Berridge, M. J. & R. F. Irvine. 1984. Inositol trisphosphate, a novel second messenger in cellular signal transduction. Nature **312:** 315-321.
3. Michell, R. H., C. J. Kirk, L. M. Jones, C. P. Downes & J. A. Creba. 1981. The stimulation of inositol lipid metabolism that accompanies calcium mobilization in stim-

ulated cells: Defined characteristics and unanswered questions. Phil. Trans. R. Soc. London. **B296:** 123-137.
4. Fisher, S. K., L. A. A. Van Rooijen & B. W. Agranoff. 1984. Renewed interest in the polyphosphoinositides. Trends Biochem. Sci. **9:** 53-56.
5. Majerus, P. W., D. B. Wilson, T. E. Connolly, T. E. Bross & E. J. Neufeld. 1985. Phosphoinositide turnover provides a link in stimulus-response coupling. Trends Biochem. Sci. **10:** 168-171.
6. Nishizuka, Y. 1983. Calcium, phospholipid turnover and transmembrane signalling. Phil. Trans. R. Soc. Lond. **B302:** 101-112.
7. Sano, K., Y. Takai, J. Yamanishi & Y. Nishizuka. 1983. A role of calcium-activated, phospholipid-dependent protein kinase in human platelet activation: Comparison of thrombin and collagen actions. J. Biol. Chem. **258:** 2010-2013.
8. Rando, R. R. & N. Young. 1984. The stereospecific activation of protein kinase C. Biochem. Biophys. Res. Commun. **122:** 818-823.
9. Nomura, H., H. Nakanishi, K. Ase, U. Kikkawa & Y. Nishizuka. 1986. Inositol phospholipid turnover in stimulus-response coupling. *In* Progress in Hemostasis and Thrombosis, Vol. 8. B. S. Coller, Ed.: 143-158. Grune & Stratton. Orlando, Fla.
10. Johnson, P. C., J. A. Ware, P. B. Cliveden, M. Smith, A. M. Dvorak & E. W. Salzman. 1985. Measurement of ionized calcium in blood platelets with the photoprotein aequorin. J. Biol. Chem. **260:** 2069-2076.
11. Kikkawa, U. & Y. Nishizuka. 1986. The role of protein kinase C in transmembrane signalling. Ann. Rev. Cell Biol. **2:** 149-178.
12. Nishizuka, Y. 1986. Studies and perspectives of protein kinase C. Science. **233:** 305-312.
13. Nishizuka, Y. 1983. Phospholipid degradation and signal translation for protein phosphorylation. Trends Biochem. Sci. **8:** 13-16.
14. Berridge, M. J. 1975. The interaction of cyclic nucleotides and calcium in the control of cellular activity. *In* Advances in Cyclic Nucleotide Research, Vol. 6. P. Greengard & G. A. Robison, Eds.: 1-98. Raven. New York.
15. Rassmussen, H. 1981. Calcium and cAMP as Synarchic Messengers. Wiley. New York.
16. Kishimoto, A., K. Nishiyama, H. Nakanishi, Y. Uratsuji, H. Nomura, Y. Takeyama & Y. Nishizuka. 1985. Studies on the phosphorylation of myelin basic protein by protein kinase C and adenosine 3′:5′-monophosphate-dependent protein kinase. J. Biol. Chem. **260:** 12492-12499.
17. Krebs, E. G. & J. A. Beavo. 1979. Phosphorylation-dephosphorylation of enzymes. Ann. Rev. Biochem. **48:** 923-959.

An Integrated Approach to Secretion

Phosphorylation and Ca^{2+}-Dependent Binding of Proteins Associated with Chromaffin Granules

MICHAEL J. GEISOW AND ROBERT D. BURGOYNE[a]

Laboratory of Protein Structure
National Institute for Medical Research
Mill Hill, London NW7 1AA, United Kingdom

[a]MRC Secretory Control Group
The Physiological Laboratory
University of Liverpool
Liverpool L69 3BX, United Kingdom

Despite the wealth of knowledge about the pharmacology and morphology of secretion, the molecular events that take place between stimulus of chromaffin cells and secretion of catecholamines remain unknown. Our approach to the definition of these events has been directed at the level of the chromaffin granule membrane. Specifically, we have explored its molecular interactions with cytoplasmic components.

We have examined in some detail interactions with cellular enzymes, particularly kinases and phosphatases, with cytoskeletal elements, especially actin- and clathrin-associated proteins, and with Ca^{2+}-regulated proteins.

Important outcomes of these studies have been the identification and characterization of a novel class of Ca^{2+}-binding proteins and the recognition of Ca^{2+}-dependent and Ca^{2+}-independent steps in acetylcholine-evoked secretion from chromaffin cells. These new proteins and processes are being further assessed for their precise role in secretion.

MATERIALS AND METHODS

Chromaffin cells were dissociated from bovine adrenal medullae and maintained in culture[1] or resuspended in Krebs-Ringer buffer for release experiments,[2] followed by fluorimetric analysis of catecholamines[3] or fixation and transmission electron microscopy[1] (TEM). Chromaffin granules were isolated on sucrose and Percoll gradients.[4] Cytoplasmic proteins from chromaffin and other tissues were obtained by homogenization and high-speed centrifugation and Ca^{2+}-dependent membrane binding proteins by EGTA chelation of Ca^{2+}.[5]

Coated vesicles were obtained from bovine adrenal medullae.[6] Calmodulin was obtained from bovine brain.[7]

Standard conditions for phosphorylation of membrane and cytosolic proteins were used,[4,6,7] Amersham γ-[^{32}P]ATP (tricine salt) being the phosphate donor. Antisera to Ca^{2+}-binding proteins were raised in rabbits and characterized with respect to antigen specificity.[5,8] Immunofluorescence, immunoprecipitation, and immunoblotting methods were as described previously.[5]

After purification of Ca^{2+}-dependent membrane binding proteins,[5,8,10] the molecules were reduced and *S*-carboxymethylated by standard chemistry. Proteins were fragmented by cyanogen bromide or TLCK-trypsin and peptides isolated by HPLC on Zorbax C_8 reversed-phase columns eluted with acetonitrile gradients in 0.1% v/v trifluoroacetic acid. Peptides were sequenced using an Applied Biosystems model 470A protein sequenator and PTH amino acids identified by HPLC on Waters Nova-Pak C_{18} reversed-phase columns. Limited proteolysis and peptide mapping was performed, as previously described.[5]

Ca^{2+}-dependent protein binding to chromaffin granule membranes[8] or liposomes[9] was followed by sedimentation and by chromatography on Sephacryl S-300.

RESULTS AND DISCUSSION

Morphology of Chromaffin Cells during Secretion

Isolated chromaffin cells stimulated with 0.5 mM carbachol produced morphological changes characteristic of exocytosis. In transmission electron microscopic (TEM) sections, infoldings and membrane blebs indicated an increase in surface membrane area corresponding to fusion of granule and plasma membranes. In resting cells there was a distinct 400-nm zone under the plasma membrane in which few granules were found. Following stimulation, there was a small but significant increase in the number of granules in this zone (FIG. 1A). The addition of trifluoperazine (TFP) to the incubation medium produced no discernable change in the morphology of resting cells. When carbachol was also present, there was no secretion but there was a highly significant increase in the number of granules present in the 400-nm zone described above (FIGS. 1A and 1B).

Effective chelation of external calcium ions by EGTA neither altered the appearance of the resting cells nor affected the migration of granules into the cortical zone in response to carbachol (FIG. 1). Within 30 s of carbachol stimulation in the absence of TFP, granule membrane antigens appeared on the cell surface, where they were accessible to extracellular antibodies (not shown). At the same time there was a sixfold increase in the number of surface-associated coated pits, followed by a slower increase in coated vesicles. The total amount of coated membrane in the cortical zone then decreased, with a half time of about 15 min (FIG. 2). Enhancing the contrast of coated structures in the cortical zone with tannic acid processing for TEM indicated that the majority of the coated structures remain surface-associated, with many apparent coated vesicles being in reality deep-coated pits.[1]

A series of recent studies have shown that nicotinic agonists produce small and transient increases in cell free Ca^{2+}, $[Ca^{2+}]_i$, under conditions of maximal secretion.[10,11] Secretion evoked in bovine chromaffin cells by the nicotinic receptor is absolutely dependent upon extracellular Ca^{2+}, and internal Ca^{2+} pools do not appear to be

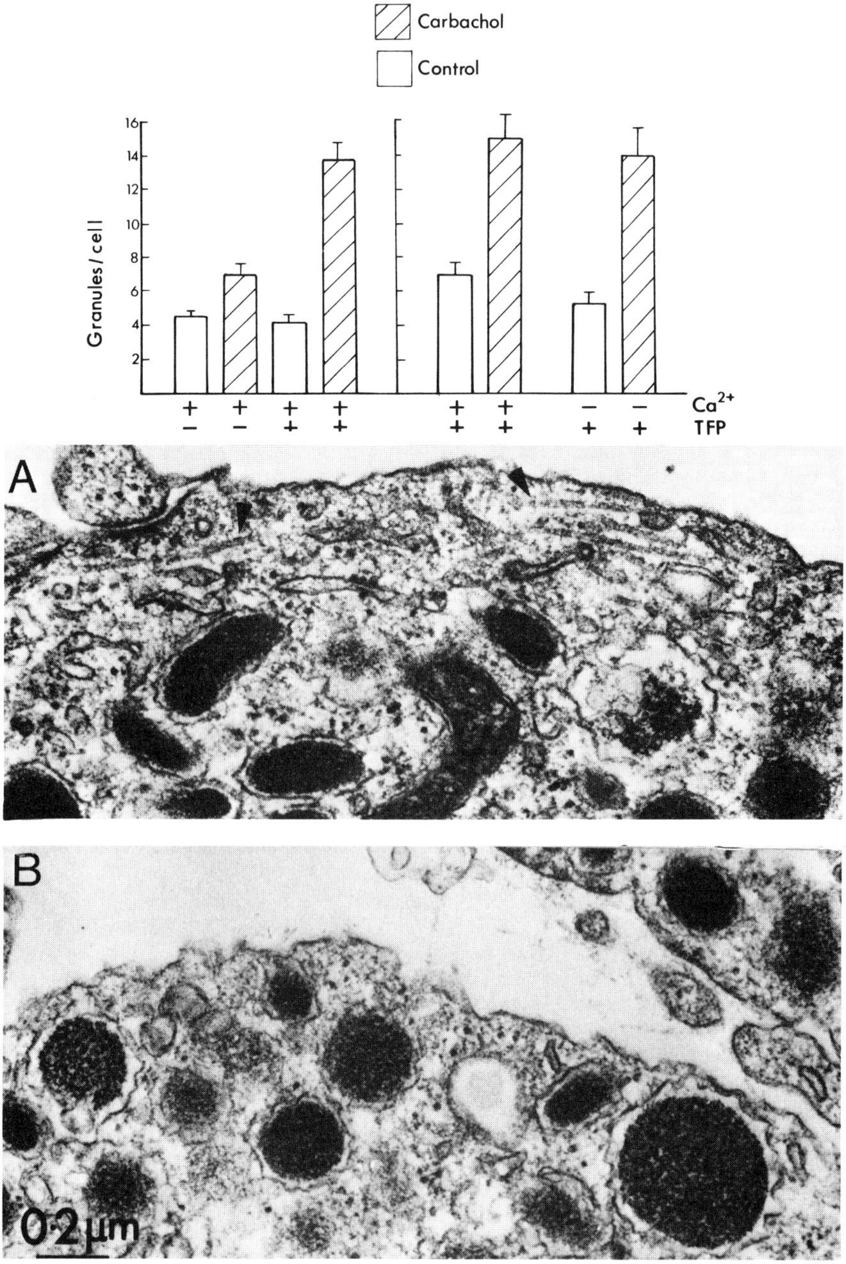

FIGURE 1. Migration of chromaffin granules into the cortical zone in response to carbachol. The histogram (top) shows the number of granules per cell section within 400 nm of surface membrane in the presence or absence of 0.1 mM carbachol. Extracellular Ca^{2+} was 2 mM (+) or 1 mM EGTA (−). Trifluoperazine (TFP) was present at 1 μM (+). Cell section **A** shows resting cell in presence of 25 μM TFP. Section **B** shows carbachol-stimulated cell in presence of 25 μM TFP.

released.[12] These observations bear upon the cortical zone changes in carbachol-stimulated chromaffin cells in the presence of TFP. They make it unlikely that the migration of granules into the cortical zone is an event that depends solely upon a change in Ca^{2+}, and support the suggestion that an unidentified signal is involved in exocytosis from this cell type.[10,13] Perhaps blockade of secretion by TFP reflects inhibition of an enzymatic event, since TFP is a potent inhibitor of protein kinases and phospholipases.

Assembly of coated pits may well be a Ca^{2+}-dependent event, since calmodulin is specifically associated with the protein coat in adrenal coated vesicles.[6] If this is indeed the case, then the extremely rapid assembly of coat proteins at the plasma membrane correlates well with the transient increase in $[Ca^{2+}]_i$. Although $[Ca^{2+}]_i$ rapidly decreases, once formed, coated vesicles do not depend upon Ca^{2+} for stability *in vitro.*

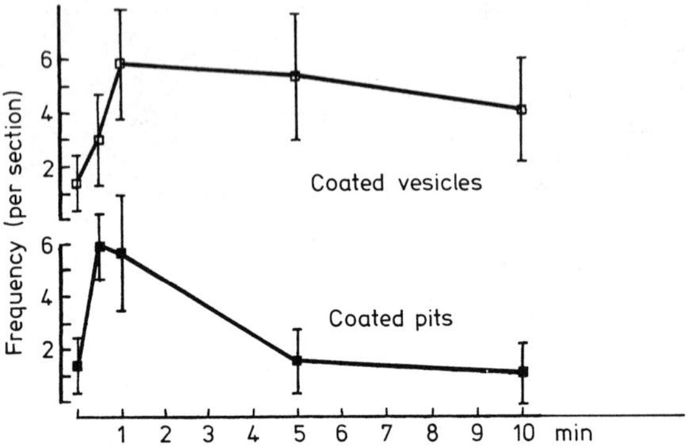

FIGURE 2. Coated membrane structures within 400 nm of the cell surface after carbachol stimulation for various times. The frequency per section represents an average over 25 cells. Coated vesicles had the appearance of a closed vesicle in section and were scored as such. Coated pits had obvious connections to the cell surface.

Phosphorylation of Proteins Associated with the Chromaffin Granule Membrane

Chromaffin granule membrane was prepared by repeated cycles of hypotonic lysis and sedimentation. In the absence of exogenous kinases, a number of proteins present in the membrane preparations incorporated $^{32}P_i$ from γ-[^{32}P]ATP (FIG. 3). The phosphorylations were time dependent; maximal after 2.5 min followed by a loss of $^{32}P_i$. This loss reflected the presence of endogenous phosphatase activity rather than loss of substrate, since phosphatase inhibitors like NaF prevented the decay of phosphoprotein present in the preparation. Phosphorylation of 27-kDa, 53-kDa, and 59-kDa acceptors was sensitive to Ca^{2+}. In this case, addition of bovine brain calmodulin increased the level of phosphorylation, and TFP decreased it. Although it has not been specifically identified, the membranes probably possess a Ca^{2+}- and calmodulin-dependent protein kinase. Acrylamide gel overlay techniques demonstrated the pres-

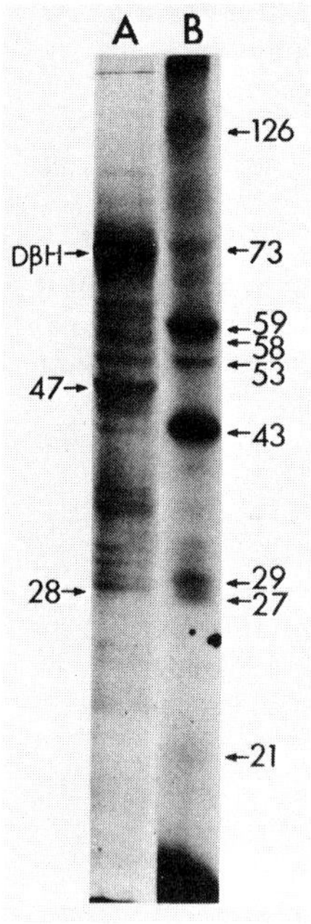

FIGURE 3. Phosphoproteins present in the chromaffin granule membrane. (**A**) Coomassie blue stain. (**B**) ^{32}P-autoradiograph; 5 μM cAMP was present in the phosphorylation buffer.

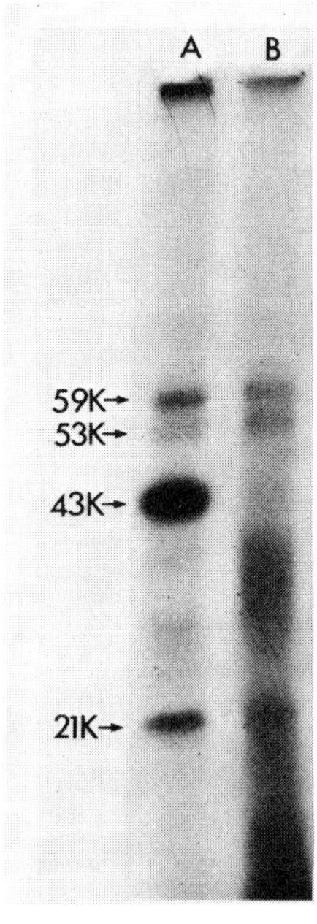

FIGURE 4. Comparison of ^{32}P-labeled polypeptides in granule membranes prepared by two different techniques. (**A**) Usual 1.7 M sucrose step gradient. (**B**) An extra Percoll gradient was used before pelleting the granules through sucrose. The 43-kDa phosphoprotein is selectively removed by the additional stage of purification.

ence of Ca^{2+}-and TFP-sensitive calmodulin binding proteins of 69 kDa and 50 kDa, while solution-binding studies indicated the presence of a high-affinity, low-capacity, and protease-sensitive site for calmodulin binding.[7,14] The phosphorylation of an 18-kDa acceptor was stimulated by Ca^{2+} in the presence of TFP.[15]

Cyclic AMP also regulated the level of phosphorylation of membrane-associated proteins. Phosphorylation of a 59-kDa acceptor was half-maximally stimulated by 0.5 μM cAMP (resting cell cAMP is typically 0.2-3.0 μM). At somewhat higher cAMP levels, 58-kDa and 126-kDa acceptors increased their incorporation of $^{32}P_i$. These effects were almost certainly due to endogenous cAMP-dependent kinase activity, since addition of the specific protein inhibitor of this kinase prevented the phosphorylations described.[6]

In many experiments, a cAMP-sensitive phosphate acceptor of 43 kDa was observed. This protein was shown to be the α-subunit of pyruvate dehydrogenase, since its phosphorylation was Mg^{2+}-independent (a characteristic of this mitochondrial protein). This was confirmed by phosphopeptide mapping using the corresponding band isolated from bovine adrenal mitochondria.[4] The 43-kDa phosphoprotein was largely depleted from granule membrane functions derived from granules purified on Percoll gradients (FIG. 4). In general, the phosphate acceptors associated with the granule membranes, mitochondria, and cytosol differed in relative amounts and identities. In the cytosol, only the 59-kDa and 58-kDa cAMP-sensitive phosphoproteins seemed to share common characteristics with counterparts in the granule membranes (FIG. 5).

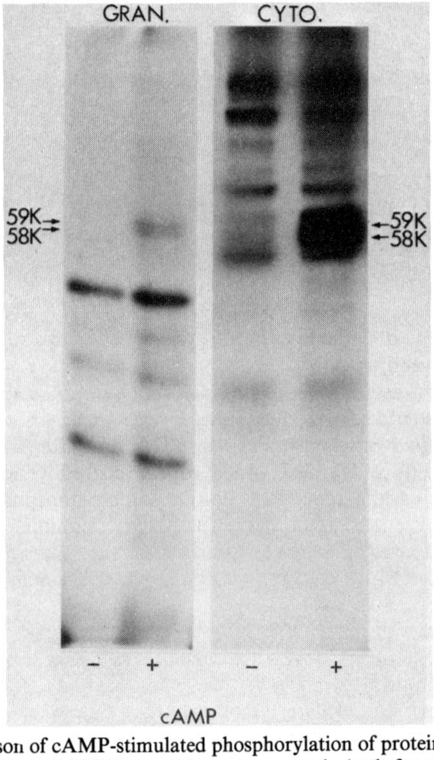

FIGURE 5. Comparison of cAMP-stimulated phosphorylation of proteins in granule membrane and cytosol. The 58-kDa and 59-kDa proteins are present in both fractions.

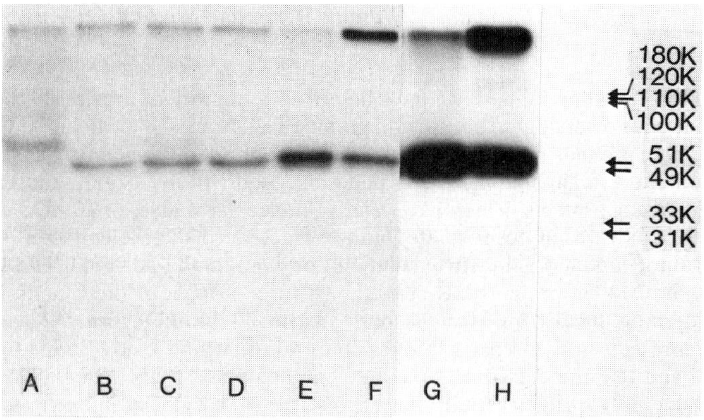

FIGURE 6. Autoradiograph showing phosphorylation of adrenal coated vesicle proteins. Adrenal coated vesicles (c.v., 20 μg protein) were incubated alone, with 2 μg of cAMP-dependent protein kinase or with 50 μg medullary cytosol. Lanes: A, cytosol + 5 μM cAMP; B, c.v. + cytosol + 1 mM Ca^{2+}; C, c.v. + cytosol + 5 mM cAMP; D, c.v. + cytosol + 1 mM EGTA; E, c.v. + protein kinase catalytic subunit; F, c.v. (no additions); G-H, 10 times longer autoradiograph exposure of lanes E and F. K = kDa.

The granule membrane phosphate acceptors were characterized as far as was possible by two-dimensional gel electrophoresis and by immunological methods. The 59-kDa protein appears to be tyrosine hydroxylase on the evidence of immunoprecipitation with a specific serum; similarly, a minor 73-kDa phosphoprotein appears to be dopamine β-hydroxylase. The 58-kDa phosphoprotein band may contain a type II regulatory subunit of cAMP-dependent kinase (which is phosphorylated by the catalytic subunit), on the basis of evidence presented elsewhere.[16] All the phosphate acceptor sites on granule membrane appear to be cytosolically oriented. No alteration in the pattern of phosphorylation was observed when the reaction was started immediately after lysis of membrane vesicles. Nor were there alterations in the pattern of phosphorylation in the presence of the protonophore FCCP, which blocks active ATP uptake by a granule membrane transport system.[4]

Coated vesicles from bovine adrenal medullae contain chromaffin granule membrane marker proteins[1] and therefore present granule membrane proteins in a different cellular context for the assessment of endogenous phosphate acceptors. No Ca^{2+}-calmodulin-dependent, cAMP-dependent, or cAMP-independent phosphoproteins could be detected in coated vesicle membrane proteins even after long autoradiographic exposure. Trace amounts of cytosol added to the coated vesicles resulted in the appearance of a major cAMP-dependent 59-kDa soluble phosphate acceptor, but no phosphorylation was observed when the pure catalytic subunit of cAMP-dependent protein kinase was added to the coated vesicles. The only phosphate acceptor in the coated vesicles was a 51-kDa coat protein, which may result from autophosphorylation of an intrinsic coat protein kinase (FIG. 6).

In summary, chromaffin granule membranes clearly contain associated independent, Ca^{2+}-, calmodulin-, and cAMP-dependent kinases as well as phosphatase activity. The only significant endogenous phosphate acceptors, however, represent proteins of catecholamine biogenesis, the protein kinases or their regulatory subunits, and major cell phosphoproteins that could represent contaminants or nonspecifically bound components of the granule membrane preparation.

Ca^{2+}-Dependent Chromaffin Granule Membrane-Binding Proteins

In addition to calmodulin,[7,14] we have described a number of previously unknown proteins that bind reversibly to chromaffin granule membranes in a calcium-dependent manner.[17] These proteins represent quantitatively important interactions in comparison with calmodulin. Similar findings were made independently by Creutz and his associates.[18] The major proteins, which have relative molecular masses of 70 kDa, 36 kDa, and 32.5 kDa, have each been isolated and characterized.[5,8] Like calmodulin, these proteins are not unique to the chromaffin cell or even to the adrenal gland. One of the two major 70-kDa proteins has been shown to be a form of the F-actin binding protein caldesmon, first isolated from smooth muscle in a 140-kDa form.[19] This protein probably represents the adrenal cytosolic component, which binds both to granule membranes and to immobilized calmodulin affinity columns.[14] The 70-kDa caldesmon is a heat-stable polypeptide (FIG. 7) and is present within cytoskeletal gels assembled

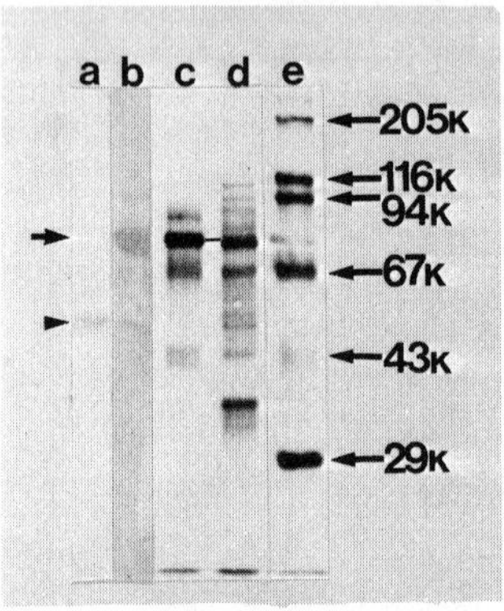

FIGURE 7. Separation of the granule-binding protein fraction into (**a,c**) heat-stable and (**b,d**) heat-labile fractions visualized by protein staining with ponceaus (**c,d**) or by immunoperoxidase staining with antiserum against the 70-kDa phospholipid-binding polypeptide (**a,b**) of nitrocellulose blots. The major 70-kDa polypeptide (*arrow*) in the heat-stable fraction is caldesmon, and the 70-kDa polypeptide, migrating with slightly higher mobility in the heat-labile fraction, is recognized by the 70-kDa antiserum. A minor cross-reacting polypeptide (*arrowhead*) is a proteolytic fragment generated during heat treatment.

from adrenal medulla.[19] Immunoperoxidase staining of cultured chromaffin cells indicated that this protein is concentrated within the cell cortex (FIG. 8). Since the cross-linking of F-actin by caldesmon is inhibited by Ca^{2+}-calmodulin, a reduction in actin cross-linking to allow granule movement could be one of the events preceding exocytosis. The movement of granules in the cortical zone in the presence of TFP and in the absence of external calcium (FIG. 1) suggests that the cytoskeleton may also be controlled by other second messengers.

The other 70-kDa protein does not cross-react with adrenal medulla caldesmon, migrates with slightly higher mobility on SDS-PAGE, and is heat-labile (FIG. 7). This molecule, together with the other major granule membrane binding components, are all recognized on Western blots by calelectrin antiserum (FIG. 9). Calelectrin is an analogous Ca^{2+}-dependent vesicle-binding protein from the electric tissue of the ray,

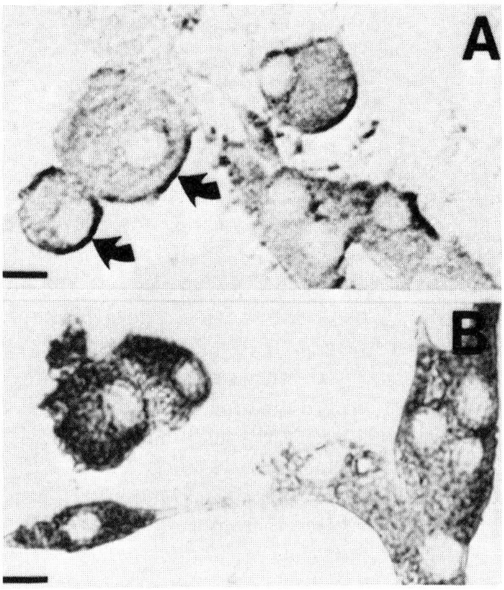

FIGURE 8. Immunoperoxidase localization of (**A**) caldesmon and (**B**) calmodulin in cultured chromaffin cells. Calmodulin is distributed throughout the cell, whereas caldesmon is concentrated at the cell periphery (*arrows*). *Scale bar* = 10 μM.

Torpedo marmorata. As in the adrenal medulla, this tissue is heavily invested with cholinergic neurons. The discovery of this immunological relationship, together with the observation that the 70-kDa, 36-kDa, and 32.5-kDa proteins and calelectrin potentiate aggregation of chromaffin granules in the presence of Ca^{2+}, has led us to refer to these proteins under the generic name of "annexins." Each annexin binds Ca^{2+} independently of membranes with affinities ($K_{Ca}{}^{2+}$) ranging from 2 μM to 200 μM. The interaction of annexins with chromaffin granule membranes is not affected by protease treatment of the membranes prior to binding. Annexins bind effectively to liposomes made from extracts of chromaffin granule lipids or to liposomes reconstituted from pure lecithin containing one or more types of acidic phospholipid (*e.g.*, phos-

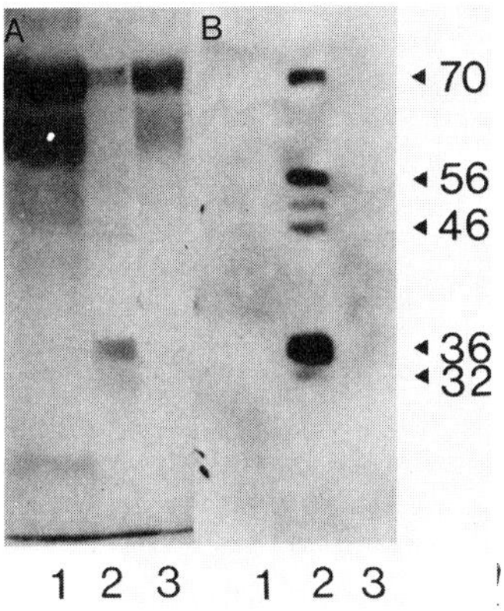

FIGURE 9. Cross-reaction of polyclonal antiserum to *Torpedo* calelectrin with Ca^{2+}-dependent chromaffin granule membrane binding proteins. Lanes 1 and 3, granule content proteins used as controls. Lane 2, the Ca^{2+}-binding proteins. Western blots were stained with amidoblack (**A**), or with anticalelectrin and then ^{125}I-protein A (**B**).

phatidyl serine, phosphatidyl inositol, or phosphatidic acid). There is functional analogy with the granule aggregating protein synexin, described by Creutz *et al.*,[20] but structural relationship remains to be proved.

Identity of p36 with the Substrate of Protein Tyrosine Kinases

The 36-kDa annexin (p36) has an isoelectric point of 7 to 7.4 on two-dimensional gels and a native molecular mass of 80-90 kDa by gel permeation chromatography. In the presence of Ca^{2+} and granule membranes or protein kinase C, it was phosphorylated. The identification of this protein was aided by the demonstration that a similar Ca^{2+}-regulated protein in pig intestinal epithelial cells was a substrate for Rous sarcoma transforming virus protein tyrosine kinase, $pp60^{v-src}$[21]. Adrenal p36 was immunoprecipitated and stained on Western blots by an antiserum specific for the Rous sarcoma virus tyrosine kinase substrate (FIG. 10), and this protein may be identical to the analogous Ca^{2+}-dependent substrate, CB9, reported by Creutz.[22]

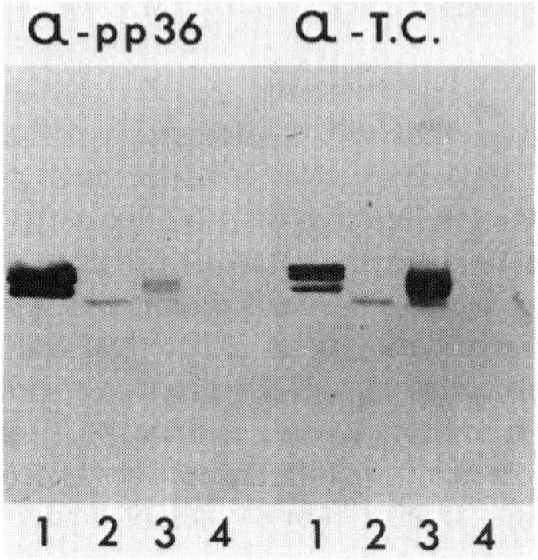

FIGURE 10. Cross-reaction of adrenal medulla p36 with an antiserum to the protein tyrosine kinase ($pp60^{src}$) substrate. Lane 1, medulla p36 and some p33 (a breakdown product of p36); lane 2, endonexin; lane 3, calelectrin; lane 4, granule content control. The antiserum to p36 only gives a weak cross-reaction with calelectrin, but anticalelectrin reacts strongly with p36.

Immunofluorescence, using specific antisera on cultured chromaffin cells, demonstrated that all annexins were present in chromaffin cells. In fibroblasts, where cytochemical localization is more easily determined, most of the annexins were present at the cytoplasmic face of the plasma membrane. The 32.5-kDa annexin (endonexin) gave a weak, vesicular pattern of immunostaining rather like the distribution of endoplasmic reticulum.

Structural Homologies of Annexins

Limited proteolysis of annexins under native or denaturing conditions produced fragments that could be stained on Western blots by anticalelectrin. Immunoreactive peptides were grouped with approximate correspondence to one half and one quarter the initial molecular mass, suggesting both repetition of the epitopes concerned and a potentially simple domain structure for annexins. The sequences of cyanogen bromide and tryptic peptides produced from p36, calelectrin, and p32.5 (endonexin) contained a closely homologous repeat.[9] A consensus sequence can be described from these, and four versions of the consensus sequence are present in lipocortin, a potent phospholipase A_2 inhibitor protein that is widely distributed in cells and tissues.[23] As yet no other known protein contains the annexin consensus sequence.

Alignment of the partial annexin sequences with the four homologous 70-amino-acid sequences of lipocortin (FIG. 11) shows that there is a second region of strong

```
                G               R                Y                  G
PSSDVA-ALHKAIMVKGVDEATIIDILTKRNNAQRQQIKAAYLQETGKPLDETLKKALTGHLEEVVLALLKT
PAQFDADELRAAMKGLGTDEDTLIEILASRTNKEIRDINRVYREELKRDLAKDITSDTSGDFRNALLSLAKG        1
ADSDARALYEAGERRKGTDVNVFNTILTTRSYPQLRRVFQKYTKYSKHDMNKVLDLELKGDIEKCLTAIVKCATSK
PAFF-AEKLHQAMKGVGTRHKALIRIMVSRSEIDMNDIKAFYQKMYGISLCQAILDETKGDYEKILVALCGGN
              GLGTDEDAIINVLAYR                         TSGSFEDALLAIVK
            MKGLGTDDDTLIRVMVSRAEI             MDDKLSEKSGNFEQS                   2
            MKGAGTDESTLIEILASRGPEEI
              GAGTDEGSLIEILASR
              GAGTSENVLIEILASR                                                  3
              GFGTDEDVILDLLTQR
            MLGLGTDEDRLIEIIL                                                    4
                      WISIMTERSVPHLQKVFSRYKS
```

FIGURE 11. Amino acid sequence alignments of Ca^{2+}-dependent phospholipid-binding proteins. Sequences: *1,* human lipocortin; *2,* endonexin; *3,* calelectrin; *4,* p36.

homology. This takes the form of a conserved hydrophobic sequence of precisely six amino acids. This feature is also present in phospholipid transport proteins and in phospholipase A_2 itself. There are no Ca^{2+}-binding sequences apparent in lipocortin or in any of the fragments of annexins, and it seems likely that Ca^{2+} is not bound as in calmodulin or parvalbumin, which possess Ca^{2+}-binding loops called "EF hands."

The partial sequences of annexins, together with the results of limited proteolysis, indicate that they too are constructed of at least one lipocortin-like 70-residue repeat. Two such repeats may constitute an active domain, since a 15-kDa fragment of lipocortin has full biological activity. In view of the many similarities of these molecules, we have also included lipocortin under our term "annexin," and the proteins so designated are included in TABLE 1. Synexins are included because these proteins also potentiate the aggregation of membranes by Ca^{2+}, but the sequences are not known.

Both lipocortin and p36 are substrates for protein kinase C and protein tyrosine kinases (*e.g.,* EGF-receptor kinase). This has been shown to be an important regulatory process in platelets, where lipocortin (a 40-kDa protein) normally complexes and inhibits membrane phospholipase A_2. Phosphorylation of lipocortin by protein kinase C reverses this inhibition and leads to the production of free fatty acid by phospholipase A_2, the prostaglandin cascade, and secretion of platelet granules. The role of p36 has not yet been determined but it potentially acts as a cross-linker of F-actin to the lipid bilayer, which is regulated by Ca^{2+} and by protein phosphorylation.

TABLE 1. Biochemical Characteristics of Ca^{2+}- and Phospholipid-Dependent Membrane Binding Proteins.[a]

Protein	Mol wt	$K_{Ca^{2+}}$, [M]	Lipids	Sequence	Refs.
p68, p70, protein III	67,000	1.5×10^{-6}	PI	+[b]	5,8
Synexin I	47,000	2×10^{-4}	PS	−[c]	20
Lipocortin, p35	40,000	nd[d]	nd	+	23
Protein I (p36 subunit)	36,000	1×10^{-4}	PS, PI	+	21
Calelectrin	34,000	10^{-5}-10^{-4}	PS	+	8
P32.5, endonexin, protein II	32,500	2.5×10^{-5}	PI, PA	+	5, 8, 9

[a] Alternative names are given in column 1. "Lipids" refers to phospholipids required for liposome binding.. "Sequence" refers to presence of consensus sequence shown in FIGURE 11.

[b] + = present.

[c] − = absent.

[d] nd = not determined.

At present the common domain shared by annexins appears to be involved in Ca^{2+}-dependent recruitment of the appropriate annexin to plasma or organelle membranes. This is an important function, since recruitment of a substrate to a membrane surface bearing kinase activities, like the chromaffin granule, constitutes an efficient regulatory process. Obviously the precise roles of the annexins described in the secretory process will have to wait for the determination of each of their functions. At this stage, however, these Ca^{2+}- and phospholipid-dependent molecules appear to be strongly implicated in the stimulus-secretion response.

SUMMARY

Pharmacological and morphological studies of secretion from bovine chromaffin cells indicate that the nicotinic receptor initiates intracellular signaling. An increase in $[Ca^{2+}]_i$ is a necessary but not sufficient element for secretion. Receptor-dependent, but intracellular-Ca^{2+}-independent alterations of the organization of cortical zone allows close approach of chromaffin granules and plasma membrane. Chromaffin granules possess both Ca^{2+}-dependent and Ca^{2+}-independent protein kinases, but the major substrates for these kinases appear to be associated with other organelles or to be soluble cytosolic proteins. Quantitatively the most important Ca^{2+}-dependent cytosolic components that interact with the chromaffin granule do not show strict specificity for this secretory organelle, but are widely distributed in different cell types and are localized at or close to the plasma membrane in intact chromaffin and other cells. These molecules are closely related in biochemical properties and sequence and include substrates for membrane-associated kinases. Two of these proteins (caldesmon and p36) have binding sites for F-actin; the others described have binding sites for acidic phospholipids.

REFERENCES

1. Geisow, M. J., J. Childs & R. D. Burgoyne. 1985. Cholinergic stimulation of chromaffin cells induces rapid coating of the plasma membrane. Eur. J. Cell Biol. **38:** 51-56.
2. Burgoyne, R. D., M. J. Geisow & J. Barron. 1982. Dissection of stages in exocytosis in the adrenal chromaffin cell with use of trifluoperazine. Proc. R. Soc. Lond. **B216:** 111-115.
3. Burgoyne, R. D. & K. M. Norman. 1984. Biochim. Biophys. Acta **805:** 37-43.
4. Burgoyne, R. D. & M. J. Geisow. 1982. Phosphoproteins of the adrenal chromaffin granule membrane. J. Neurochem. **39:** 1387-1396.
5. Geisow, M. J., J. Childs, B. Dash, A. Harris, G. Panayotou, T. C. Südhof & J. H. Walker. 1984. Cellular distribution of three mammalian Ca^{2+}-binding proteins related to *Torpedo calelectrin.* EMBO J. **3:** 2969-2974.
6. Geisow, M. J. & R. D. Burgoyne. 1984. Calmodulin binding and protein phosphorylation in adrenal medulla coated vesicles. FEBS Lett. **169:** 127-132.
7. Burgoyne, R. D. & M. J. Geisow. 1981. Specific binding of calmodulin to and protein phosphorylation in adrenal chromaffin granule membranes. FEBS Lett. **131:** 127-131.
8. Südhof, T. C., M. Ebbecke, J. H. Walker, U. Fritsche & C. Boustead. 1984. Isolation of mammalian calelectrins: A new class of ubiquitous Ca^{2+}-regulated proteins. Biochemistry **23:** 1103-1109.

9. GEISOW, M. J., U. FRITSCHE, J. M. HEXHAM, B. DASH & T. JOHNSON. 1986. A consensus amino acid sequence repeat in *Torpedo* and mammalian Ca^{2+}-dependent membrane-binding proteins. Nature **320:** 636-638.
10. BURGOYNE, R. D. 1984. The relationship between secretion and intracellular free calcium in bovine adrenal chromaffin cells. Bioscience Reports **4:** 605-611.
11. BURGOYNE, R. D. & T. R. CHEEK. 1985. Is the transient nature of the secretory response of chromaffin cells due to inactivation of calcium channels? FEBS Lett. **182:** 115-118.
12. CHEEK, T. R. & R. D. BURGOYNE. 1985. Effect of activation of muscarinic receptors on intracellular free calcium and secretion in bovine adrenal chromaffin cells. Biochim. Biophys. Acta **846:** 167-173.
13. BAKER, E. M., T. R. CHEEK & R. D. BURGOYNE. 1985. Cyclic AMP inhibits secretion from bovine adrenal chromaffin cells evoked by carbamylcholine but not by high K^+. Biochim. Biophys. Acta **846:** 388-393.
14. GEISOW, M. J. & R. D. BURGOYNE. 1983. Recruitment of cytosolic proteins to a secretory granule membrane depends on Ca^{2+}-calmodulin. Nature **301:** 432-435.
15. BURGOYNE, R. D. & M. J. GEISOW. 1982. Effect of Ca^{2+}, calmodulin and trifluoperazine on protein phosphorylation in adrenal chromaffin granule membranes. Trans. Biochem. Soc. **10:** 267-268.
16. TREIMAN, M., W. WEBER & M. GRATZL. 1983. 3′5′-cAMP and Ca^{2+}-calmodulin dependent endogenous protein phosphorylation activity in membranes of the bovine chromaffin secretory vesicles. J. Neurochem. **40:** 661-668.
17. GEISOW, M. J. & R. D. BURGOYNE. 1982. Calcium-dependent binding of cytosolic proteins by chromaffin granules from adrenal medulla. J. Neurochem. **38:** 1735-1741.
18. CREUTZ, C. E., L. G. DOWLING, J. J. SANDO, C. VILLAR PALASI, J. H. WHIPPLE & W. J. ZAKS. 1983. Characterization of the chromobindins. J. Biol. Chem. **258:** 14664-14674.
19. BURGOYNE, R. D., T. R. CHEEK & K.-M. NORMAN. 1986. Identification of a secretory granule-binding protein as caldesmon. Nature **319:** 68-70.
20. CREUTZ, C. E., C. J. PAZOLES & H. B. POLLARD. 1978. Identification and purification of an adrenal medullary protein (synexin) that causes calcium-dependent aggregation of isolated chromaffin granules. J. Biol. Chem. **253:** 2858-2866.
21. GERKE, V. & K. WEBER. 1984. Identity of p36K phosphorylated upon Rous sarcoma virus transformation with a protein purified from brush borders: Calcium-dependent binding to non-erythroid spectrin and F-actin. EMBO J. **3:** 227-233.
22. SUMMERS, T. A. & C. E. CREUTZ. 1985. Phosphorylation of a chromaffin granule binding protein by protein kinase. C. J. Biol. Chem. **260:** 2437-2443.
23. WALLNER, B. P., R. J. MATTALIANO, C. HESSION, R. L. CATE, R. TIZARD, L. K. SINCLAIR, C. FOELLER, E. P. CHOW, J. L. BROWNING, K. L. RAMACHANDRAN & R. B. PEPINSKY. 1986. Cloning and expression of human lipocortin, a phospholipase A_2 inhibitor with potential anti-inflammatory activity. Nature **320:** 77-81.

Differential Role of Calcium in Stimulus-Secretion-Synthesis Coupling in Lactotrophs and Corticotrophs of Rat Anterior Pituitary

JITENDRA R. DAVE,[a,b] LEE E. EIDEN,[c] DAVID LOZOVSKY,[b,d] JAMES A. WASCHEK,[c] AND ROBERT L. ESKAY[b]

[b]*Neurochemistry Section*
Laboratory of Clinical Studies
National Institute on Alcoholism and Alcohol Abuse
Bethesda, Maryland 20892

[c]*Laboratory of Cell Biology*
National Institute of Mental Health
Bethesda, Maryland 20892

[d]*Addiction Research Center*
NIDA
Baltimore, Maryland 21224

Hormone secretion and biosynthesis are continuous events occurring in neuroendocrine cells. Although both processes have been studied in great detail in pituitary cells and pituitary tumor cell lines in culture, little emphasis has been placed on the coupling of secretion and biosynthesis of peptide hormones in pituitary cells after exposure to secretogogues. Calcium and cAMP have been reported to play important regulatory roles in hormone secretion from the anterior pituitary gland.[1–4] Earlier findings indicated that extracellular calcium is essential for basal or stimulated prolactin release and for optimal ACTH release induced by hypothalamic extracts *in vitro.* The calcium channel blocker verapamil was reported to inhibit potassium- or ouabain-enhanced ACTH release; however, verapamil was unable to block crude-hypothalamic-extract-, vasopressin-, or dibutyryl-cAMP-induced ACTH release.[5] These observations prompted the conclusion that ouabain and potassium elicit release of ACTH by a calcium-dependent mechanism, whereas hypothalamic releasing factors and their second messengers elicit ACTH release in a completely calcium-independent manner. In contrast, Giguere *et al.*[6] have reported that CRF-stimulated ACTH release is totally calcium dependent. In order to resolve this apparent contradiction, we have examined basal and CRF-stimulated β-endorphin secretion in rat anterior pituitary cells in the

[a]Address correspondence to: Dr. J. R. Dave, LPPS, NIAAA, 12501 Washington Ave., Rockville, Md. 20852.

presence or absence of extracellular calcium. We have concomitantly measured prolactin secretion and compared the calcium requirements for secretion from corticotrophs and lactotrophs with the calcium requirements for biosynthesis of each of these pituitary hormones.

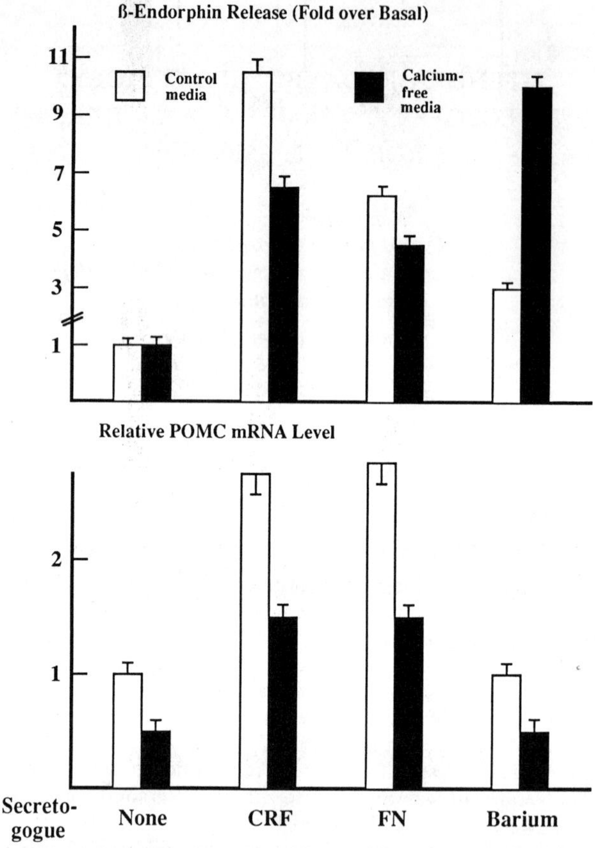

FIGURE 1. Effect of calcium on basal and secretogogue-mediated secretion of β-endorphin and pro-opiomelanocortin (POMC) mRNA levels in cultured rat anterior pituitary cells. Dispersed anterior-pituitary cells obtained from adult male rats were plated ($n = 3$) at 30,000-50,000 cells per well for secretion studies and 300,000-500,000 cells per well for mRNA studies and tested 4-5 days after plating. For secretion of peptides, cells were incubated with test agents for 1 h and for mRNA determinations for 24 h. β-Endorphin release was determined by radioimmunoassay and POMC mRNA levels by Northern and slot blots prepared from total nucleic acids hybridized with a ^{32}P-labeled 923-base-pair *Eco*RI-*Hind*III restriction fragment of mouse POMC cDNA, as described earlier.[7,8] Autoradiograms were scanned densitometrically, corrected for total RNA applied, and expressed as relative POMC mRNA levels.

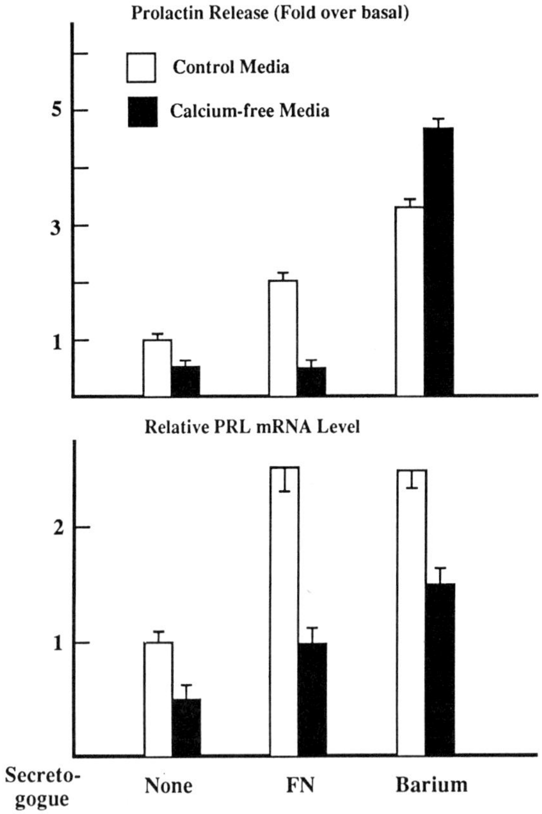

FIGURE 2. Effect of calcium on basal and secretogogue-mediated prolactin secretion and mRNA levels in cultured rat anterior pituitary cells. Prolactin release was determined by radioimmunoassay and prolactin mRNA levels were determined by Northern and slot blots prepared from total nucleic acids hybridized with a ^{32}P-labeled 900-base-pair *Pst*I restriction fragment of rat prolactin cDNA. For additional details see FIGURE 1.

RESULTS AND DISCUSSION

We have evaluated the role of calcium in basal and secretogogue-stimulated release of β-endorphin and prolactin and the synthesis of their respective messenger RNAs in primary cultured rat anterior pituitary cells. Treatment of anterior pituitary cells with calcium-free media for 1 hour partially blocked corticotropin-releasing factor (CRF, 10 nM) stimulated β-endorphin release, with no effect on basal release of β-endorphin (FIG. 1). The calcium agonist barium (1 mM) produced a twofold increase in β-endorphin release, which was enhanced in the absence of calcium. Omission of calcium from the culture media caused a 50% decrease in pro-opiomelanocortin (POMC) mRNA levels, without affecting the proportional increase in POMC mRNA elicited by CRF. Treatment of cells with barium had no effect on POMC mRNA levels (FIG. 1). In contrast to the observed effects of calcium-free media on corticotrophs, this test condition decreased basal secretion of prolactin from lactotrophs by 50-70% and completely blocked forskolin (FN, 10 μM) stimulated prolactin secretion (FIG. 2). The addition of barium to the culture media of pituitary cells produced a twofold increase in prolactin release. The stimulation of prolactin release by barium was not diminished by omission of calcium from the medium. Treatment of cells with calcium-free media decreased basal levels of prolactin mRNA by 25% or 50%, respectively, and partially decreased FN- or barium-stimulated prolactin mRNA levels (FIG. 2). These findings demonstrate a calcium-dependent as well as calcium-independent component of CRF-stimulated β-endorphin secretion and CRF-stimulated POMC mRNA elevation. Barium stimulation of β-endorphin release, but not biosynthesis, suggests that calcium plays a second messenger role in secretion, but is merely a permissive factor in POMC biosynthesis. In contrast, prolactin secretion and biosynthesis appear to be totally calcium-dependent processes, both mimicked by the calcium agonist barium.

REFERENCES

1. VALE, W., R. BURGUS & R. GUILLEMIN. 1967. Potassium induced stimulation of thyrotropin release *in vitro:* Requirement for presence of calcium and inhibition by thyroxin. Experientia **23:** 855.
2. WAKABAYASHI, K., A. K. IBRAHIM & S. M. MCCANN. 1969. *In vitro* response of the rat pituitary to gonadotropin releasing factors and to ions. Endocrinology **84:** 1046.
3. ZIMMERMAN, G. & N. FLEISCHER. 1970. Role of calcium ions in the release of ACTH from rat pituitary tissue *in vitro.* Endocrinology **87:** 426.
4. MCLEOD, R. M. & E. H. FONTHAM. 1970. Influence of ionic environment on the *in vitro* synthesis and release of pituitary hormones. Endocrinology **86:** 863.
5. ETO, S., J. M. WOOD, M. HUTCHINS & N. FLEISCHER. 1974. Pituitary $^{45}Ca^{++}$ uptake and release of ACTH, GH, and TSH: Effect of verapamil. Am. J. Physiol. **226:** 1315.
6. GIGUERE, V., G. LEFEVRE & F. LABRIE. 1982. Site of calcium requirement for stimulation of ACTH release in rat anterior pituitary cells in culture by synthetic ovine corticotropin-releasing factor. Life Sci. **31:** 3057.
7. DAVE, J. R., L. E. EIDEN, J. W. KARANIAN & R. L. ESKAY. 1986. Ethanol exposure decreases pituitary corticotropin-releasing factor binding, adenylate cyclase activity, pro-opiomelanocortin biosynthesis, and plasma β-endorphin levels in the rat. Endocrinology **118:** 280.
8. DAVE, J. R., L. E. EIDEN & R. L. ESKAY. 1986. Differential effects of isoproterenol, A23187 and barium on immunoreactive β-endorphin release and pro-opiomelanocortin synthesis in cultured anterior pituitary and AtT-20 cells. ICSU Reports **4:** 34.

Biochemical and Morphological Correlates of Thyroliberin Release in Hypothalamic Cell Cultures Grown in a Serum-Free Medium

C. LOUDES, A. FAIVRE-BAUMAN, A. BARRET, R. PICART, AND A. TIXIER-VIDAL

Groupe de Neuroendocrinologie Cellulaire et Moléculaire,
U. A. CNRS 1115, Collège de France
75231 Paris, France

The aim of this work was to devise a culture system to study biochemical and morphological mechanisms of thyroliberin (TRH) release. Mouse hypothalamic cells taken on the 16th fetal day develop in a serum-free medium and form synapses after 10-12 days *in vitro*. In order to obtain fully differentiated synapses, the culture medium must include three critical elements: T3, corticosterone, and polyunsaturated fatty acids.[1,2] In this medium, TRH cell content increases up to two weeks and TRH release in response to chemical depolarization occurs at the same time as synaptogenesis is achieved.[3] We therefore used this model to study (1) the response to depolarization and to several neurotransmitters tested alone or in combination with depolarization and (2) the response to kinase C activation, using phorbol esters.

K^+-Evoked Depolarization

Twelve-day-old cultures were incubated at 37°C, for 3-minute periods, in a buffer containing 3 mM (control), or 60 mM K^+, and TRH was radioimmunoassayed[4] in the medium and in the cells at the end of the incubation. Its biochemical identity was checked by HPLC. Cells release in 3 min 10-15% of their content, and are able to respond to repeated stimulations without any change in their TRH content. K^+-evoked TRH release is abolished by the absence of extracellular Ca^{2+}, or the presence of cobalt.[3] At the ultrastructural level, depolarization is accompanied by a depletion of small synaptic vesicles in the nerve terminals, and strikingly, return to the control medium induces in 3 min a massive restoration of vesicles, thus confirming the plasticity of the system.[2]

Several neurotransmitters or amino acids were tested, among them serotonin, GABA, histamine, glutamate, and acetylcholine. Dopamine was found efficient on TRH release in a dose-dependent manner, but only when used in combination with K^+ depolarization. This effect of dopamine is blocked by fluphenazine, but not by sulpuride, suggesting that Di receptors might be involved. Thus, in addition to voltage-dependent Ca^{2+} channels, other mechanisms are implicated in TRH release; we tested the possible involvement of protein kinase C activators (Loudes *et al.,* in preparation).

Kinase C Activation

Diacyl glycerol could trigger TRH release, but not as constantly as TPA, a phorbol ester, the response to which exhibits a time and a dose dependency. Its maximum effect is reached at 10^{-7} M and after 15-30 min. Inactive analogues of phorbol esters were inefficient. The Ca^{2+} dependency is not so strict as for depolarization: the absence of calcium reduces TPA response only to one third of its value. In contrast to K^+ depolarization, TPA does not induce a drastic depletion of synaptic vesicles, at least in all nerve endings.

The effect of TPA is not enhanced by dopamine, in contrast with what was observed with potassium. Combination of the two types of stimulation, TPA and K^+, were not additive either (Loudes *et al.,* in preparation).

In conclusion, the present data show that TRH release is mediated by at least two different mechanisms: activation of voltage-dependent Ca^{2+} channels and activation of protein kinase C. It is suggested that dopamine, which enhances K^+ response although being inactive by itself, could stimulate phosphatidyl inositol breakdown, which is presently under investigation. Lastly different vesicular mechanisms seem to be involved in the effect of K^+ depolarization and TPA, respectively.

REFERENCES

1. FAIVRE-BAUMAN, A., J. PUYMIRAT, C. LOUDES & A. TIXIER-VIDAL. 1984. Differentiated mouse fetal hypothalamic cells in serum free medium. *In* Cell Culture Methods for Molecular and Cell Biology, Vol. 4. D. W. Barnes, S. A. Sirbasku & G. H. Sato, Eds.: 37-56, Alan R. Liss. New York.
2. TIXIER-VIDAL, A., R. PICART, C. LOUDES & A. FAIVRE-BAUMAN. 1986. Effects of polyunsaturated fatty acids and hormones on synaptogenesis in serum-free medium cultures of mouse fetal hypothalamic cells. Neuroscience **17:** 115-132.
3. LOUDES, C., A. FAIVRE-BAUMAN, D. GROUSELLE, J. PUYMIRAT & A. TIXIER-VIDAL 1983. Release of immunoreactive TRH in serum-free cultures of mouse hypothalamic cells. Dev. Brain Res. **9:** 231-234.
4. GROUSELLE, D., A. TIXIER-VIDAL & P. PRADELLES. 1982. A new improvement of the sensitivity and specificity of radioimmunoassay for thyroliberin. Application to biological samples. Neuropeptides **3:** 29-44.

A Receptor for Extracellular Cations on Parathyroid Cells[a]

E. F. NEMETH

Department of Physiology and Biophysics
Case Western Reserve University
School of Medicine
Cleveland, Ohio 44106

It is now well established that extracellular Ca^{2+} acts as the principal physiological secretagogue regulating parathyroid hormone secretion, although it is far from clear how this extracellular signal acts at the cellular level to control secretion. It has been assumed that the initial action of Ca^{2+} involves interaction with a cell surface receptor,[1-3] yet there is no reason to suppose that the parathyroid cell detects changes in the concentration of extracellular Ca^{2+}, thereby altering secretion, by this means. Any hypothesis advancing the existence of such a plasma membrane "Ca^{2+} receptor" clearly hinges on the demonstration that extracellular Ca^{2+} can elicit intracellular changes by an action solely at the cell surface. Some data addressing this issue have recently been obtained by measuring the concentration of intracellular free Ca^{2+} ($[Ca^{2+}]_i$) in parathyroid cells using the fluorescent indicator fura-2.

In dissociated bovine parathyroid cells loaded with fura-2, increasing the concentration of extracellular Ca^{2+} evokes rapid, transient increases that are followed by lower, yet sustained, steady-state increases in $[Ca^{2+}]_i$. Rapid, transient increases in $[Ca^{2+}]_i$, which mimic those elicited by extracellular Ca^{2+}, are also seen when the concentration of extracellular Mg^{2+} is increased or when Sr^{2+} is added. Unlike Ca^{2+}, however, these other divalent cations do not cause large increases in steady-state $[Ca^{2+}]_i$. Cytosolic Ca^{2+} transients evoked by extracellular Ca^{2+} are suppressed in the presence of high concentrations of Mg^{2+} or Sr^{2+}. Conversely, high concentrations of extracellular Ca^{2+} block cytosolic Ca^{2+} transients elicited by Mg^{2+} and by Sr^{2+}. All these divalent cations therefore appear to act at a common cellular site to cause a rapid and transient increase in $[Ca^{2+}]_i$.[4]

Transient and steady-state increases in $[Ca^{2+}]_i$ induced by extracellular Ca^{2+} arise from two pharmacologically distinct mechanisms because they are differentially susceptible to inhibition by La^{3+}, which generally blocks Ca^{2+} influx in various cells and which depresses sustained but not transient increases in $[Ca^{2+}]_i$ in the parathyroid cell (FIG. 1). Cytosolic Ca^{2+} transients elicited by Mg^{2+} and by Sr^{2+} are likewise refractory to inhibition by La^{3+}. Steady-state levels of cytosolic Ca^{2+} are therefore dependent on extracellular Ca^{2+} and possibly reflect Ca^{2+} influx. In contrast, cytosolic Ca^{2+} transients can be evoked in the absence of any change in the concentration of extracellular Ca^{2+}, do not require cellular influx of Ca^{2+}, and likely arise from the mobilization of some cellular store of Ca^{2+}. Indeed, when the concentration of extracellular Ca^{2+} is reduced to submicromolar levels by the addition of EGTA, Mg^{2+}

[a]This work was supported by NIH grant no. AM-33928.

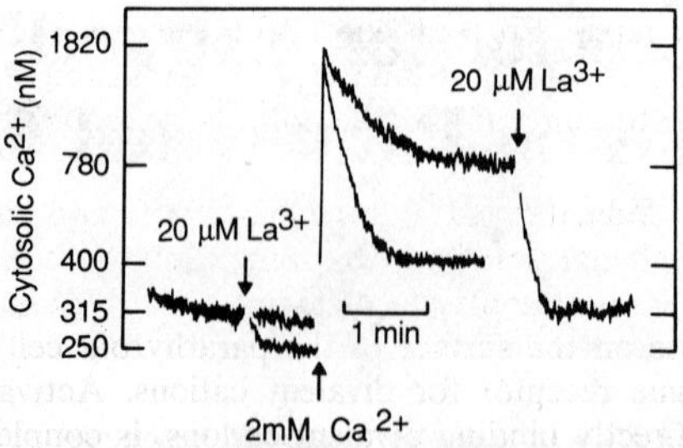

FIGURE 1. Preferential inhibitory effects of La^{3+} on steady-state $[Ca^{2+}]_i$ in fura-2-loaded parathyroid cells. Cells were initially equilibrated in phosphate-free buffer containing 0.5 mM $CaCl_2$. Two superimposed traces are shown. The addition of La^{3+} prior to increasing extracellular Ca^{2+} to 2.5 mM caused a small decrease in steady-state $[Ca^{2+}]_i$ and greatly depressed the steady-state $[Ca^{2+}]_i$ achieved at this higher level of extracellular Ca^{2+}. Cytosolic Ca^{2+} transients were unaffected by pretreatment with La^{3+}. The addition of La^{3+} after increasing the extracellular Ca^{2+} concentration caused a prompt fall in $[Ca^{2+}]_i$ to values approaching those obtained in the presence of low extracellular Ca^{2+} concentrations.

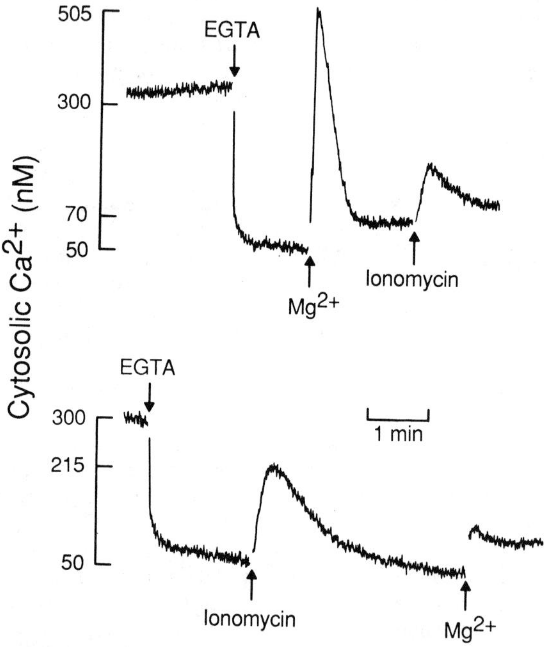

FIGURE 2. Cytosolic Ca^{2+} transients induced by Mg^{2+} persist in the absence of extracellular Ca^{2+}. Cells loaded with fura-2 were equilibrated in buffer containing 0.5 mM $CaCl_2$ before the addition of sufficient EGTA to reduce the extracellular Ca^{2+} concentration to 1 μM. The subsequent addition of 7 mM $MgCl_2$ (to achieve a final concentration of 8 mM) elicited a rapid and transient increase in $[Ca^{2+}]_i$ (*top trace*). The addition of ionomycin (1 μM) also evoked a transient rise in $[Ca^{2+}]_i$ and inhibited the response to Mg^{2+} (*bottom trace*).

(or Sr^{2+}) still evokes a rapid and transient 10-fold increase in $[Ca^{2+}]_i$. Pretreatment with ionomycin, to dissipate cellular stores of Ca^{2+}, inhibits the Mg^{2+}-induced cytosolic Ca^{2+} transient (FIG. 2). Ionomycin likewise inhibits cytosolic Ca^{2+} transients elicited by extracellular Ca^{2+}.

The results clearly indicate that extracellular Ca^{2+} and various other divalent cations can elicit rapid changes in $[Ca^{2+}]_i$ by some common mechanism that does not involve influx of Ca^{2+} across the plasma membrane. All these cations must therefore act initially at some site on the surface of the parathyroid cell, and this site can be regarded as a membrane receptor for divalent cations. Activation of this putative receptor, perhaps by directly binding divalent cations, is coupled to the rapid mobilization of cellular Ca^{2+}, although it is uncertain if this transient change in $[Ca^{2+}]_i$ is directly linked to the regulation of parathyroid hormone secretion.[5]

REFERENCES

1. LOPEZ-BARNEO, J. & C. M. ARMSTRONG. 1983. J. Gen. Physiol. **82:** 269-294.
2. WALLACE, J. & A. SCARPA. 1983. J. Biol. Chem. **258:** 6288-6292.
3. NEMETH, E. F., J. WALLACE & A. SCARPA. 1986. J. Biol. Chem. **261:** 2668-2674.
4. NEMETH, E. F. & A. SCARPA. 1986. *In* Calcium Ion: Membrane Transport and Cellular Regulation. F. Carpenedo, P. Debetto, M. Floreani & S. Luciani, Eds.: 27-33. University of Padua Press. Padua, Italy.
5. NEMETH, E. F. & A. SCARPA. 1987. Are changes in intracellular free calcium necessary for regulating secretion in parathyroid cells? This volume.

Concluding Remarks

Neuroendocrine Secretory Vesicles: No Longer a Black Box

ROBERT G. JOHNSON, JR.

Howard Hughes Medical Institute
Massachusetts General Hospital
Boston, Massachusetts 02114

INTRODUCTION

The papers in this volume have outlined the salient properties and functions of neuroendocrine secretory vesicles. The clinical messengers contained within the intravesicular space of these secretory vesicles play important roles within the body as neurotransmitters, hormones, or modulators. The previous articles constitute a splendid review of the integrative biology of the chemical messengers as observed through the study primarily of neuroendocrine secretory vesicles. Neuroendocrine secretory vesicles can, therefore, serve as paradigms for the elucidation of neuroendocrine homeostasis and physiology. Several common themes emerge from the presentations.

The structure, composition, and function of neuroendocrine vesicles from a wide variety of embryologically disparate tissues are remarkably alike. Cholinergic, adrenergic, peptidergic, and purinergic secretory vesicles all have similar fundamental properties, compositions, and cellular mechanisms. We have gained significant insights into the biogenesis, development, plasticity, bioenergetics, pharmacology, movement, and exocytosis of these vesicles. Neuroendocrine vesicles are certainly no longer a "black box." As is apparent from the reports and discussions contained in these proceedings, however, many fundamental questions remain unanswered.

INTRAVESICULAR PROTEINS AND PEPTIDES

One of more perplexing problems facing investigators is the physiological role of the chromogranins, the family of acidic proteins, ubiquitous in neuroendocrine secretory vesicles, that contain peptides and are present in enormous quantities within the intravesicular space. Explanations of their function as a buffer, binding site, or peptide precursor are all inadequate. Three classes of chromogranins exist (previously termed chromogranins A, B, and C). They are significantly different, as has been determined by molecular weight and existing sequence data. There is no known role of the chromogranins as hormones, neurotransmitters, or modulators. A physiologic role of degradation products of these proteins has not been established. Unfortunately, the recent cloning of the chromogranin A has not per se led to any significant insights. One advancement has been clarification of the confusing nomenclature (see TABLE 1).

The processing of peptide precursors within the secretory vesicles has been well documented. For some peptides, such as insulin, the processing is well understood. For others, such as those in the enkephalin family, various degrees of processing in the respective tissues and a multiplicity of possible enzymes have been found. Regulation of these enzymes and intravesicular control and regulation of peptide processing remain enigmatic, particularly in the adrenal medulla. The roles of other intravesicular constituents such as proteoglycans are also not well understood.

TABLE 1

Previous Nomenclature		Proposed Nomenclature
chromogranin A	secretory protein I	chromogranin A
chromogranin B	secretogranin I	chromogranin B
chromogranin C	secretogranin II	secretogranin A

REFERENCE: L. E. Eiden *et al.* 1986. *Neuroscience.* In press.

SECRETORY VESICLE PHENOTYPE

Secretory vesicles can contain various chemical messengers: classical neurotransmitters, hormones, peptides, purines, etc. Even in the same vesicle, multiple potential neurotransmitters, hormones, or modulators can exist. The need for multiple messengers within the same granule and the independent regulation of these messengers is only in the infancy of characterization. Moreover, in cells that demonstrate plasticity, the actual chemical messengers within the secretory vesicles can be altered. The universality of the various components of the secretory vesicle membrane (H^+-ATPase, p38, SV-2, etc.) is not well established, nor is the intracellular signal that regulates phenotypic changes.

MOVEMENT AND EXOCYTOSIS

Although an understanding of the movement of a vesicle down an axon has rapidly been achieved in recent years, there remains a myriad of proteins and effector mechanisms near the surface of the cell responsible for the positioning and fusion of the vesicle with the plasma membrane.

Elegant studies concerned with phosphorylation of secretory vesicle proteins have provided a lovely model for investigation of the spatial positioning of the vesicle. The molecular mechanisms, however, have not yet been elucidated. The mechanism of fusion of vesicle and plasma membrane is likewise still unknown.

CONCLUSION

I am optimistic that the material presented, discussed, and analyzed in this New York Academy of Sciences conference will lead to the design of even more ambitious experiments that will answer some of these important fundamental questions concerning the cellular and molecular biology of neurotransmitter- and hormone-containing secretory vesicles.

Index of Contributors